# Kinetic Theory of **Nucleation**

# Kinetic Theory of Nucleation

Eli Ruckenstein
Gersh Berim

CRC Press
Taylor & Francis Group
Boca Raton London New York

CRC Press is an imprint of the
Taylor & Francis Group, an **informa** business

CRC Press
Taylor & Francis Group
6000 Broken Sound Parkway NW, Suite 300
Boca Raton, FL 33487-2742



Printed on acid-free paper
Version Date: 20160426

International Standard Book Number-13: 978-1-4987-6283-0 (Hardback)



**Visit the Taylor & Francis Web site at**
**http://www.taylorandfrancis.com**

**and the CRC Press Web site at**
**http://www.crcpress.com**

# Contents

Preface xv

Authors xix

1 Kinetic Theory of Liquid-to-Solid Nucleation 1

Introduction 1
References 2
1.1 A Possible Nucleation Type of Mechanism for the Growth of Small Aerosol Particles 3
*Ganesan Narsimhan and Eli Ruckenstein*
1.1.1 Introduction 3
1.1.2 Comparison between Rates of Formation and Dissociation of a Doublet Consisting of Equal-Size Aerosol Particles 4
1.1.2.1 Rate of Formation of a Doublet 4
1.1.2.2 Rate of Dissociation of a Doublet 5
1.1.3 Two Alternate Mechanisms of Aerosol Growth 9
1.1.4 Conclusions 10
References 11
1.2 A New Approach for the Prediction of the Rate of Nucleation in Liquids 12
*Ganesan Narsimhan and Eli Ruckenstein*
1.2.1 Introduction 12
1.2.2 Structure of Molecular Clusters 13
1.2.2.1 Amorphous Spherical Cluster 13
1.2.2.2 Spherical fcc Crystalline Cluster 14
1.2.3 Interaction Potential of a Surface Monomer 14
1.2.4 First Passage Time Analysis for Estimation of Rate of Dissociation of a Surface Monomer 15
1.2.5 Solubility 19
1.2.6 Rate of Nucleation 20
1.2.7 Interfacial Tension of Macroscopic Clusters 22
1.2.8 Comparison of the Prediction with the Classical Nucleation Theory 24
1.2.9 Conclusions 28
References 28
1.3 A Kinetic Theory of Nucleation in Liquids 30
*Eli Ruckenstein and Bogdan Nowakowski*
1.3.1 Introduction 30
1.3.2 Interaction between a Molecule and a Cluster 30
1.3.3 Kinetics of Dissociation 32
1.3.4 Rate of Condensation 34

1.3.5 Critical Nucleus ... 35
1.3.6 Limit of Large Clusters and Thermodynamic Relations ... 35
1.3.7 Rate of Nucleation ... 36
1.3.8 Numerical Results and Comparison with the Classical Theory ... 37
1.3.9 Conclusions ... 39
Appendix 1A: Asymptotic Expansion of the Equation for Radius of Critical Cluster (Equation 1.125) ... 40
References ... 41
1.4 Rate of Nucleation in Liquids for Face-Centered Cubic and Icosahedral Clusters ... 42
*Bogdan Nowakowski and Eli Ruckenstein*
1.4.1 Introduction ... 42
1.4.2 Theory of Nucleation—Mean Passage Time Approach ... 42
1.4.3 Structure of Clusters ... 44
1.4.4 Interaction of the Surface Atom with the Cluster ... 45
1.4.5 Results and Discussion ... 46
1.4.6 Conclusions ... 49
References ... 49
1.5 Comparison among Various Approaches to the Calculation of the Nucleation Rate ... 51
*Bogdan Nowakowski and Eli Ruckenstein*
1.5.1 Introduction ... 51
1.5.2 Zeldovich Expansion ... 51
1.5.3 Goodrich and Modified Goodrich Expansions ... 52
1.5.4 Rate of Nucleation ... 53
1.5.5 Discussion ... 54
References ... 54
1.6 A Kinetic Approach to the Theory of Heterogeneous Nucleation on Soluble Particles during the Deliquescence Stage ... 55
*Yuri Djikaev and Eli Ruckenstein*
1.6.1 Introduction ... 55
1.6.2 Background of Deliquescence Theory ... 56
1.6.2.1 Thermodynamics of Deliquescence ... 57
1.6.2.2 Kinetic Equation of Deliquescence ... 58
1.6.3 A New Approach to Kinetics of Deliquescence ... 59
1.6.3.1 Kinetics of Fluctuational Deliquescence ... 60
1.6.3.2 Kinetics of Barrierless Deliquescence ... 64
1.6.4 Numerical Evaluations ... 65
1.6.5 Conclusions ... 69
Acknowledgments ... 69
Appendix 1B: Derivation of Equation 1.185 ... 69
Appendix 1C: Equilibrium Distribution of Composite Droplets ... 70
Appendix 1D: Solution of Equation 1.185 ... 72
Appendix 1E: Explicit Expressions for the Coefficients of Polynomial (Equation 1.208) ... 74
References ... 74
1.7 Effect of Solute–Solute and Solute–Solvent Interactions on Kinetics of Nucleation in Liquids ... 76
*Eli Ruckenstein and Gersh Berim*
1.7.1 Introduction ... 76
1.7.2 Background ... 76
1.7.2.1 Interaction Potentials ... 76
1.7.2.2 Nucleation Rate ... 78
1.7.2.3 Dissociation Rate ... 79
1.7.2.4 Condensation Rate ... 81
1.7.3 Numerical Estimation of Effect of Solute–Solute and Solute–Solvent Interactions on Kinetics of Nucleation ... 81
1.7.4 Conclusion ... 84
References ... 84

1.8 Thermodynamics of Heterogeneous Crystal Nucleation in Contact and Immersion Modes ..... 85
*Yuri Djikaev and Eli Ruckenstein*

1.8.1 Introduction ..... 85
1.8.2 Free Energy of Heterogeneous Formation of Crystal Nuclei in Contact and Immersion Modes ..... 87
1.8.2.1 Foreign Particle Completely Immersed in the Liquid ..... 88
1.8.2.2 Foreign Particle in Contact with a Liquid–Vapor Interface ..... 91
1.8.2.3 Comparison of Immersion and Contact Modes ..... 94
1.8.3 Numerical Evaluations and Discussions ..... 95
1.8.4 Concluding Remarks ..... 97
Acknowledgments ..... 98
Appendix 1F: Crystal Nuclei and Their Works of Formation ..... 98
References ..... 102
1.9 Correction to "Thermodynamics of Heterogeneous Crystal Nucleation in Contact and Immersion Modes" ..... 103
*Yuri Djikaev and Eli Ruckenstein*

1.10 Homogeneous Crystal Nucleation in Droplets as a Method for Determining Line Tension of a Crystal–Liquid–Vapor Contact ..... 104
*Yuri Djikaev and Eli Ruckenstein*

1.10.1 Introduction ..... 104
1.10.2 Background: The Thermodynamics and Kinetics of Homogeneous Crystal Nucleation in a Droplet ..... 105
1.10.2.1 Free Energy of Homogeneous Formation of Crystal Nuclei in Surface-Stimulated and Volume-Based Modes ..... 105
1.10.2.2 Effect of Surface-Stimulated Mode on Kinetics of Crystal Nucleation ..... 109
1.10.3 Determination of Line Tension of an Ice–Water–Vapor Contact ..... 110
1.10.3.1 Numerical Evaluations ..... 111
1.10.4 Concluding Remarks ..... 112
References ..... 112

2 Kinetic Theory of Vapor-to-Liquid Nucleation ..... 115

Reference ..... 115
2.1 A Kinetic Approach to the Theory of Nucleation in Gases ..... 116
*Bogdan Nowakowski and Eli Ruckenstein*

2.1.1 Introduction ..... 116
2.1.2 Evaporation Rate from a Cluster ..... 116
2.1.3 Rate of Growth of a Cluster ..... 119
2.1.4 Rate of Nucleation ..... 119
2.1.5 Results and Comparison with the Classical Theory and Experiment ..... 121
Appendix 2A ..... 124
References ..... 125
2.2 A Unidimensional Fokker–Planck Approximation in the Treatment of Nucleation in Gases ..... 127
*Eli Ruckenstein and Bogdan Nowakowski*

2.2.1 Introduction ..... 127
2.2.2 Kinetic Equation in Energy Space ..... 127
2.2.3 Rate of Evaporation of Molecules from a Cluster ..... 131
2.2.4 Rate of Growth of a Cluster ..... 131
2.2.5 Rate of Nucleation ..... 132
2.2.6 Results and Discussion ..... 133
2.2.7 Conclusion ..... 135
References ..... 135

2.3 Homogeneous Nucleation in Gases: A Three-Dimensional Fokker–Planck Equation for Evaporation from Clusters ..... 136
*Bogdan Nowakowski and Eli Ruckenstein*

2.3.1 Introduction ..... 136
2.3.2 Equation for Probability Density Distribution Function of Energy ..... 136
2.3.3 Kinetics of Growth (Evaporation) of a Single Cluster ..... 140
2.3.4 Rate of Nucleation ..... 142
2.3.5 Numerical Results and Discussion ..... 143
References ..... 145
2.4 Time-Dependent Cluster Distribution in Nucleation ..... 146
*Bogdan Nowakowski and Eli Ruckenstein*

2.4.1 Introduction ..... 146
2.4.2 Equations for the Population of Clusters ..... 146
2.4.3 Solution of Time-Dependent Equations ..... 149
2.4.4 Results of Numerical Calculations ..... 150
2.4.5 Conclusion ..... 154
References ..... 155
2.5 Kinetic Theory of Nucleation Based on a First Passage Time Analysis: Improvement by the Density-Functional Theory ..... 156
*Yuri Djikaev and Eli Ruckenstein*

2.5.1 Introduction ..... 156
2.5.2 Theoretical Basis ..... 156
2.5.2.1 RNN Approach to Kinetics of Nucleation ..... 156
2.5.2.2 DFT Methods in Nucleation Theory ..... 159
2.5.2.3 Improvement of RNN Kinetic Theory of Nucleation by DFT Methods ..... 160
2.5.3 Numerical Results ..... 161
2.5.4 Concluding Remarks ..... 163
Acknowledgments ..... 163
References ..... 163
2.6 Kinetic Theory of Binary Nucleation Based on a First Passage Time Analysis ..... 165
*Yuri Djikaev and Eli Ruckenstein*

2.6.1 Introduction ..... 165
2.6.2 Kinetics of Binary Nucleation Based on First Passage Time Analysis ..... 166
2.6.2.1 Thermodynamics of Binary Nucleation Based on Capillarity Approximation ..... 167
2.6.2.2 A New Approach to Kinetics of Binary Nucleation ..... 168
2.6.3 Model Binary System with a Surface Active Component ..... 174
2.6.3.1 Density Functional Formalism ..... 175
2.6.4 Numerical Evaluations ..... 177
2.6.5 Concluding Remarks ..... 179
Acknowledgments ..... 180
Appendix 2B: Derivation of Equation 2.183 ..... 180
Appendix 2C: Equilibrium Distribution of Binary Clusters ..... 181
Appendix 2D: Steady-State Solution of Equation 2.183 ..... 182
References ..... 184
2.7 A Kinetic Treatment of Heterogeneous Nucleation ..... 186
*Bogdan Nowakowski and Eli Ruckenstein*

2.7.1 Introduction ..... 186
2.7.2 Rate of Evaporation from the Cluster ..... 186
2.7.3 Growth of Cluster ..... 188
2.7.4 Rate of Nucleation ..... 189
2.7.5 Results of Calculations and Discussion ..... 191

Appendix 2E: Interaction Potential between a Molecule and a Spherical Cap Cluster ... 192
References ... 193
2.8 New Approach to the Kinetics of Heterogeneous Unary Nucleation on Liquid Aerosols of a Binary Solution ... 194
*Yuri Djikaev and Eli Ruckenstein*
2.8.1 Introduction ... 194
2.8.2 Kinetics of Nucleation ... 196
2.8.3 A New Approach to Kinetics of Heterogeneous Nucleation ... 198
2.8.4 Numerical Evaluations ... 200
2.8.5 Conclusions ... 203
Acknowledgments ... 203
Appendix 2F: Kinetic Equation of Heterogeneous Nucleation ... 203
Appendix 2G: Emission Rate via a First Passage Time Analysis ... 204
Appendix 2H: Equilibrium Distribution of Heterogeneously Formed Droplets ... 206
References ... 207
2.9 Kinetics of Heterogeneous Nucleation on a Rough Surface: Nucleation of a Liquid Phase in Nanocavities ... 208
*Eli Ruckenstein and Gersh Berim*
2.9.1 Introduction ... 208
2.9.2 The System and the Interaction Potentials ... 209
2.9.3 Basic Equations of the Kinetic Theory ... 210
2.9.4 Results ... 213
2.9.5 Discussion ... 216
Acknowledgments ... 217
References ... 217
2.10 Kinetic Theory of Heterogeneous Nucleation: Effect of Nonuniform Density in the Nuclei ... 218
*Gersh Berim and Eli Ruckenstein*
2.10.1 Introduction ... 218
2.10.2 The System, the Interaction Potentials, and the Definition of the Nucleus ... 218
2.10.3 Basic Equations of KNT ... 220
2.10.4 Results and Conclusion ... 221
Appendix 2I: Contributions to the Free Energy of the System and Derivation of Euler–Lagrange Equation for FDD ... 224
Appendix 2J: Basic Equations of KNT ... 225
References ... 228

3 Application of the Kinetic Nucleation Theory to Protein Folding 229

References ... 229
3.1 Model for the Nucleation Mechanism of Protein Folding ... 231
*Yuri Djikaev and Eli Ruckenstein*
3.1.1 Introduction ... 231
3.1.2 Heteropolymer Chain as a Protein Model and a CNT-Based Model for the Nucleation Mechanism of Protein Folding ... 232
3.1.2.1 Heteropolymer as a Protein Model ... 232
3.1.3 A Kinetic Approach to the Nucleation Mechanism of Protein Folding ... 234
3.1.3.1 Potential Well around a Cluster of Native Residues ... 234
3.1.3.2 Determination of Emission and Absorption Rates ... 238
3.1.3.3 Equilibrium Distribution and Steady-State Nucleation Rate ... 238
3.1.3.4 Evaluation of Protein Folding Time ... 241
3.1.4 Numerical Evaluations ... 241
3.1.5 Conclusions ... 243
Appendix 3A: Derivation of Equation 3.11 from Equation 3.9 ... 245

Appendix 3B: Derivation of Expressions for Emission and Absorption Rates by a Mean First Passage Time Analysis....247
Acknowledgment....249
References....249
3.2 A Ternary Nucleation Model for the Nucleation Pathway of Protein Folding....251
*Yuri Djikaev and Eli Ruckenstein*
3.2.1 Introduction....251
3.2.2 Previous Models for the Nucleation Mechanism of Protein Folding....252
3.2.2.1 Heteropolymer as a Protein Model....252
3.2.2.2 A CNT-Based Model for Nucleation Mechanism of Protein Folding....253
3.2.2.3 A Model for Nucleation Mechanism of Protein Folding Based on a First Passage Time Analysis....254
3.2.3 A Ternary Nucleation Mechanism of Protein Folding....255
3.2.3.1 Potential Well around a Cluster of Native Residues....255
3.2.3.2 Determination of Emission and Absorption Rates....258
3.2.3.3 Equilibrium Distribution and Steady-State Nucleation Rate....259
3.2.3.4 Evaluation of Protein Folding Time....261
3.2.4 Numerical Evaluations....262
3.2.5 Conclusions....267
Appendix 3C: Derivation of Equations 3.42 and 3.43 from Equations 3.40 and 3.41....268
Appendix 3D: Derivation of Expressions for Emission and Absorption Rates by a Mean First Passage Time Analysis....271
Appendix 3E: Equilibrium Distribution of Ternary Clusters....272
Appendix 3F: Kinetic Equation of Ternary Nucleation....274
References....275
3.3 Effect of Ionized Protein Residues on the Nucleation Pathway of Protein Folding....277
*Yuri Djikaev and Eli Ruckenstein*
3.3.1 Introduction....277
3.3.2 Ternary Nucleation Model of Protein Folding Based on a First Passage Time Analysis...278
3.3.2.1 Heteropolymer as a Protein Model....278
3.3.2.2 Potential Field for a Protein Residue around a Cluster (Folded Part of the Protein)....279
3.3.3 Taking Account of the Ionizability of Protein Residues....281
3.3.4 Determination of Emission and Absorption Rates....283
3.3.5 Evaluation of Protein Folding Time....285
3.3.6 Numerical Evaluations....286
3.3.7 Conclusions....289
References....290
3.4 Temperature Effects on the Nucleation Mechanism of Protein Folding and on the Barrierless Thermal Denaturation of a Native Protein....292
*Yuri Djikaev and Eli Ruckenstein*
3.4.1 Introduction....292
3.4.2 Temperature Expansions for Protein Folding Time....295
3.4.2.1 First Passage Time Analysis-Based Model for the Nucleation Mechanism of Protein Folding....295
3.4.2.2 Determination of $W^+ = W^+(\nu)$ and $W^- = W^-(\nu)$ via a First Passage Time Analysis....296
3.4.2.3 Optimal Temperature of Protein Folding....298
3.4.2.4 Taylor Series Expansions for Protein Folding Time....300
3.4.3 Time of Protein Denaturation via Spinodal Decomposition....301
3.4.3.1 Taylor Series Expansions for Emission and Absorption Rates....303
3.4.3.2 Threshold Temperature and Time of Thermal Denaturation....304

3.4.4 Numerical Evaluations ....304
3.4.4.1 Results for Protein Folding ....306
3.4.4.2 Results for Protein Unfolding ....308
3.4.5 Conclusions ....309
Appendix 3G: Three Constituents of the Potential Field around the Cluster for a Selected Bead ....310
Appendix 3H: Derivation of Expressions for Emission and Absorption Rates by a Mean First Passage Time Analysis ....316
Appendix 3I: Derivation of Equation 3.102 and Relevant Details ....317
Appendix 3J: Coefficients in Taylor Series Expansion (Equation 3.110) ....319
References ....322
3.5 First Passage Time Analysis of Protein Folding via Nucleation and of Barrierless Protein Denaturation ....324
*Yuri Djikaev and Eli Ruckenstein*
3.5.1 Introduction ....324
3.5.2 MFPTA in Protein Folding/Unfolding Problems ....326
3.5.2.1 Heteropolymer as a Protein Model ....326
3.5.2.2 Potential Well around a Cluster of Native Residues ....326
3.5.2.3 Rates of Emission and Absorption of Residues by a Cluster ....328
3.5.3 Protein Folding ....329
3.5.3.1 Evaluation of Protein Folding Time in the Framework of Ternary Nucleation Formalism ....329
3.5.3.2 Temperature Dependence of Protein Folding Time ....330
3.5.4 Barrierless Denaturation ....333
3.5.4.1 Time of Barrierless Protein Denaturation ....333
3.5.4.2 Threshold Temperatures $T_u^-$ and $T_u^+$ of Cold and Hot Denaturation and the Unfolding Times ....334
3.5.4.3 Taylor Series Expansions for Emission and Absorption Rates ....335
3.5.5 Numerical Evaluations ....335
3.5.5.1 Results for Protein Folding ....336
3.5.5.2 Results for Protein Unfolding ....340
3.5.6 Conclusions ....342
References ....344
3.6 Effect of Hydrogen Bond Networks on the Nucleation Mechanism of Protein Folding ....346
*Yuri Djikaev and Eli Ruckenstein*
3.6.1 Introduction ....346
3.6.2 Probabilistic Model for the Effect of Water–Water Hydrogen Bonding on the Interaction of Solute Particles ....347
3.6.2.1 Networks of Water–Water Hydrogen Bonds around Solutes ....350
3.6.2.2 Case of Protein Folding via Nucleation ....352
3.6.3 Modified Model for NMPF ....353
3.6.3.1 Ternary Heteropolymer as a Protein Model ....353
3.6.3.2 Hydrogen Bond Contribution to the Potential Field around the Folded Part of a Protein ....354
3.6.4 Numerical Evaluations ....356
3.6.5 Conclusions ....359
Appendix 3K: Explicit Expressions for $V_0(r)$, $V_m(r)$, and $V_s^{r\infty}(r)$ ....360
Appendix 3L: Evaluation of Protein Folding Time in the Framework of Ternary Nucleation Formalism ....362
References ....364
4 Kinetics of Cluster Growth on a Lattice 367
References ....368
4.1 Influence of Cluster Shape upon Its Growth in a Two-Dimensional Ising Model ....370
*Gersh Berim and Eli Ruckenstein*
4.1.1 Introduction ....370

4.1.2 The Model and the Basic Kinetic Equations ....... 370
4.1.3 Clusters with Simple Shape in the Ising Model with Nearest Neighbor Interactions ....... 374
4.1.3.1 Rectangular Cluster ....... 374
4.1.3.2 Triangular Isosceles Right Angle Cluster ....... 375
4.1.3.3 Zigzag Cluster ....... 376
4.1.3.4 Rough Boundary Cluster ....... 376
4.1.4 Small Clusters of Arbitrary Shape in the Two-Dimensional Ising Model ....... 377
4.1.4.1 Number of Configurations ....... 377
4.1.4.2 Five-Spin Cluster ....... 378
4.1.5 Discussion and Conclusions ....... 381
References ....... 382
4.2 Effect of Shape on the Critical Nucleus Size in a Three-Dimensional Ising Model: Energetic and Kinetic Approaches ....... 383
*Gersh Berim and Eli Ruckenstein*
4.2.1 Introduction ....... 383
4.2.2 Formulation of the Problem ....... 383
4.2.3 Clusters with Simple Topology ....... 385
4.2.3.1 Cubic Cluster ....... 386
4.2.3.2 Stairlike Cluster ....... 388
4.2.3.3 Pyramidal Cluster ....... 388
4.2.4 Small Clusters of Arbitrary Shape ....... 389
4.2.4.1 Number of Configurations ....... 390
4.2.4.2 Excitation Energy and Initial Rate ....... 391
4.2.5 Conclusion ....... 391
References ....... 392
4.3 Shape-Dependent Small Cluster Kinetics in the Two-Dimensional Ising Model beyond the Classical Approximations ....... 393
*Gersh Berim and Eli Ruckenstein*
4.3.1 Introduction ....... 393
4.3.2 Model 1 ....... 394
4.3.2.1 Kinetic Equations ....... 394
4.3.2.2 Quasi-Steady State ....... 398
4.3.2.3 Transient Regime in Small Cluster Kinetics ....... 399
4.3.3 Model 2 ....... 400
4.3.4 Conclusion ....... 403
References ....... 404
4.4 A Closed Reduced Description of the Kinetics of Phase Transformation in a Lattice System Based on Glauber's Master Equation ....... 405
*Gersh Berim and Eli Ruckenstein*
4.4.1 Introduction ....... 405
4.4.2 Kinetic Equations and Closing Procedure ....... 406
4.4.2.1 Kinetic Equations ....... 406
4.4.2.2 Closing Procedure ....... 407
4.4.2.3 Initial and Final Conditions ....... 408
4.4.3 Phase Transformation on the One-Dimensional Lattice ....... 408
4.4.3.1 Exact Solution for Growth Decay Regime ....... 410
4.4.3.2 Small Deviation from Equilibrium ....... 412
4.4.3.3 Reversal of External Field ....... 413
4.4.4 Size Distribution of the Solid Phase Clusters ....... 414
4.4.4.1 Size Distribution ....... 414
4.4.4.2 Average Cluster Size ....... 416
4.4.5 Conclusion ....... 418
Acknowledgment ....... 418

Appendix 4A: Operator $R_j(\mu)$ and Equilibrium Averages ... 418
Appendix 4B: Relations between Equilibrium Averages ... 420
Appendix 4C: Small Deviations from Equilibrium ... 421
References ... 421
4.5 Kinetics of Phase Transformation on a Bethe Lattice ... 423
*Gersh Berim and Eli Ruckenstein*

4.5.1 Introduction ... 423
4.5.2 Kinetic Equations for the IMBL ... 423
4.5.2.1 General Case ... 423
4.5.2.2 The Case $q = 3$ ... 425
4.5.3 Phase Transformation in the Absence of Supersaturation ... 426
4.5.3.1 Small Deviations from Equilibrium at $T > T_c$ ... 427
4.5.3.2 Small Deviations from Equilibrium at $T = T_c$ ... 427
4.5.4 Kinetics for $T < T_c$ in the General Case ... 428
4.5.4.1 Phase Transformation ... 428
4.5.4.2 Cluster Dynamics ... 429
4.5.5 Kinetics in Growth-Decay Regime ... 432
4.5.5.1 Phase Transformation ... 432
4.5.5.2 Cluster Dynamics ... 432
4.5.6 Kinetics in the Nonsplitting, Noncoagulation Regime ... 434
4.5.7 Discussion ... 434
Acknowledgment ... 435
Appendix 4D: Some Equilibrium Averages for the IMBL ... 435
Appendix 4E: Sketch of the Steps to Calculate Right-Hand Sides of the Kinetic Equations ... 436
Appendix 4F: Calculation of the Number of Clusters and of Average Cluster Size ... 437
References ... 438
4.6 Phase Transformation in a Lattice System in the Presence of Spin-Exchange Dynamics ... 439
*Gersh Berim and Eli Ruckenstein*

4.6.1 Introduction ... 439
4.6.2 Background ... 440
4.6.3 Kinetics after Field Reversal (Liquid–Solid Transformation) ... 442
4.6.4 Temperature Quench in the Absence of an External Field (Phase Separation) ... 443
4.6.5 Kinetics after Temperature Quenching and External Field Change ... 445
4.6.6 Conclusion ... 446
Acknowledgment ... 446
Appendix 4G: Coefficients of the Kinetic Equations ... 447
Appendix 4H: Average Cluster Size ... 447
References ... 448
4.7 Kinetics of Phase Transformation on a Bethe Lattice in the Presence of Spin Exchange ... 449
*Gersh Berim and Eli Ruckenstein*

4.7.1 Introduction ... 449
4.7.2 Kinetic Equations ... 450
4.7.3 Phase Transformation after Field Reversal for $T < T_c$ ... 454
4.7.4 Temperature Quench in the Absence of an External Field ... 456
4.7.4.1 The Case $T_f < T_c$, $B = 0$ ... 457
4.7.4.2 The Case $T_f > T_c$, $B = 0$ ... 458
4.7.5 Conclusion ... 459
Acknowledgment ... 460
Appendix 4I: Operators $R_{a,b}(v)$ and Their Properties ... 460
Appendix 4J: Some Details of the Closure Procedure of Kinetic Equations ... 461
References ... 462

Index 463

# Preface

The phenomenon of nucleation—emerging of nuclei into a new phase during a first-order transition—attracted attention over the past decades because of the key role it plays in various natural and technological processes, including phase transitions in the atmosphere and confined nanoscale systems. The classical nucleation theory (CNT) was developed by Völmer and Weber [1], Farkas [2], and Becker and Döring [3], and later refined by Kuhrt [4], Lothe and Pound [5], and Reiss [6], etc.

The CNT is based on the idea that a new phase appears first as small clusters of molecules which can grow or shrink depending on the state of the initial phase. The appearance and initial growth of the cluster is due exclusively to local density fluctuations because the initial increase of the cluster size is thermodynamically unfavorable. However, after the cluster attains a critical size, its growth becomes thermodynamically favorable, and the cluster will grow irreversibly, thus depleting the initial (metastable) phase. The free energy of formation of a critical cluster provides the height of the activation barrier to nucleation. A cluster of critical size (critical cluster) is often named a nucleus. The nucleation rate $J$ defined as the number of nuclei appearing per unit time and unit volume in the CNT is provided by the expression $J = K\exp[-W_\ast / kT]$, where $k$ is the Boltzmann constant, $T$ the absolute temperature, $K$ the kinetic prefactor, and $W_\ast$ the height of the nucleation barrier. The latter is defined as the free energy of formation of a critical cluster and in the framework of the CNT is calculated using the macroscopic concept of surface tension which, obviously, is not applicable to objects of atomic size. The use of the concept of surface tension constitutes one of the major shortcomings of the CNT.

The alternate approach, the kinetic nucleation theory of Ruckenstein–Narsimhan–Nowakowski (RNNT), does not involve the macroscopic concept of surface tension and determines the critical nucleus from the balance of evaporation and condensation rates on the surface of the nucleus. It was first developed using the assumption of uniform density in the cluster but was extended to nonuniform clusters both for homogeneous and heterogeneous nucleation.

RNNT was applied to various nucleation phenomena, starting with the more simple case of homogeneous nucleation in gases and liquids up to such complex cases as protein folding.

This book contains the papers written by Professor Ruckenstein and his coworkers on the kinetic theory of nucleation and cluster growth. Below, a brief outline of the new approach is presented. In addition, each chapter is preceded by a separate introduction.

## P.1 Outline of the New Approach

The main idea of RNNT is to compare the rates of evaporation ($\alpha_j$) of molecules from and condensation ($\beta_j$) on the surface of a cluster containing $j$ molecules. If the former rate is smaller (greater) than the latter, the cluster grows (shrink). When those rates become equal ($\alpha_j = \beta_j$), the cluster is the critical one. For uniform clusters, the rate of evaporation $\alpha$ was calculated either on the basis of the Fokker–Plank equation for the joint distribution function in the phase space (for liquid-to-solid phase transition) or using the master equation in energy space derived from the Fokker–Plank equation (for vapor-to-liquid phase transition). The condensation rate $\beta$ was obtained from the stationary solution of the diffusion equation in an external field (liquid-to-solid transformation) or using the kinetic theory of gases (vapor-to-liquid phase transition).

During the nucleation stage of a first-order phase transition, the new phase appears as a manifold of clusters of various sizes. Denoting by $f_i$ the number of clusters consisting of $i$ ($i = 1, 2, \ldots$) molecules, the flux $I_i$ of clusters passing from the population of clusters $f_{i-1}$ to the population $f_i$ can be defined as [7]

$$I_i = \beta_{i-1} f_{i-1} - \alpha_i f_i. \tag{0.1}$$

The change of $f_i$ as a function of time is described by the equation

$$\frac{df_i}{dt} = -(I_{i+1} - I_i) = \beta_{i-1} f_{i-1} - (\beta_i + \alpha_i) f_i + \alpha_{i+1} f_{i+1}. \tag{0.2}$$

Assuming a smooth dependence of $f_i$ on $i$, Equation 0.2 can be transformed into the following equation:

$$\frac{\partial f(g,t)}{\partial t} = -\frac{\partial I(g,t)}{\partial g} = \frac{1}{2}\frac{\partial^2}{\partial g^2}(\alpha(g)+\beta(g))f(g,t) - \frac{\partial}{\partial g}(\beta(g)-\alpha(g))f(g,t), \tag{0.3}$$

where the continuous variable $g$ has replaced the discrete variable $i$ (see Sections 1.1 and 1.3, Chapter 1).

The rate of nucleation, $I$, is provided by the stationary value of the flux $I(g,t)$:

$$I = \frac{\frac{1}{2}\beta(1)\rho_s}{\int_0^\infty \exp[-2w(g)]dg}, \tag{0.4}$$

where $\beta(1)$ is the condensation rate calculated for a "cluster" consisting of a single molecule of solute and

$$w(g) = \int_0^g \frac{\beta(g)-\alpha(g)}{\beta(g)+\alpha(g)} dg. \tag{0.5}$$

Equation 0.4 represents the stationary solution of Equation 0.3 for the boundary conditions $f(g) \to 0$ at $g \to \infty$ (large clusters) and $f(g) \to \rho_s$ for $g \to 1$ (smallest clusters, monomers). The first boundary condition means that clusters with very large sizes are absent in the system. The second boundary condition is based on the assumption that nucleation does not cause substantial depletion of small clusters and their number density $\rho_s$ remains the initial one (in calculations, the limit 1 can be replaced by zero).

Note that in Equation 0.5, the rates of dissociation, $\alpha(j)$, and condensation, $\beta(j)$, although functions of the discrete number of molecules in the cluster, are considered continuous functions of $g$, or, equivalently, of the cluster radius $R$. The function $w(g)$ exhibits a sharp minimum at the critical cluster size $g = g_\star$ and the integral in Equation 0.4 can be calculated by the steepest descent method.

The rate of nucleation acquires finally the form

$$I = \frac{1}{2}\beta(1)\rho_s \left(\frac{w''(g_\star)}{\pi}\right)^{1/2} \exp[2w(g_\star)]. \tag{0.6}$$

To determine the function $w(g)$ given by Equation 0.5, one should first find the rates $\alpha(j)$ and $\beta(j)$. The corresponding procedures developed by RNNT are presented in Chapters 1 and 2 for liquid-to-solid and vapor-to-liquid transitions, respectively. In Chapter 3, RNNT is applied to protein folding.

Along with developments of kinetic theory of nucleation, this volume contains papers in which the kinetic of the cluster growth is considered on the basis of several lattice models (Chapter 4). The molecules at the sites of the lattice can belong to two different phases (solid and liquid, or liquid and vapor) and kinetics of phase change is described by the appropriate master equation.

## References

1. Völmer, M., Weber, A., Keimbildung in übersüttigten Gebilden, *Z. Phys. Chem.* 119, 277–301 (1926).
2. Farkas, L., Keimbildungsgeschwindigkeit in übersüttigten Dämpfen, *Z Phys Chem.* 125, 236–242 (1927).
3. Becker, R., Döring, W., Kinetic treatment of grain-formation in super-saturated vapours, *Ann. Phys.* 24, 719–752 (1935).
4. Kuhrt, F., Das Tröpfchenmodell realer Gase, *Z. Phys.* 131, 185–205 (1952).
5. Lothe, J., Pound, G., Reconsiderations of nucleation theory, *J. Chem. Phys.* 36, 2080–2085 (1962).
6. Reiss, H., Replacement free-energy in nucleation theory, *Adv. Coll. Interface Sci.* 7, 1–66 (1977).
7. Abraham, F.F., *Homogeneous Nucleation Theory*, Academic, New York, 1974.

# Authors

**Eli Ruckenstein** is a State University of New York (SUNY) Distinguished Professor at the SUNY at Buffalo. He has published more than 1,000 papers in numerous areas of engineering science and has received a large number of awards from the American Chemical Society and the American Institute of Chemical Engineers. Ruckenstein has also received the Founders Gold Medal Award from the National Academy of Engineering and the National Medal of Science from President Clinton. He is a member of the National Academy of Engineering and of the American Academy of Art and Sciences.

**Gersh Berim** earned his PhD in Theoretical and Mathematical Physics from Kazan State University in Russia in 1978. He has authored or coauthored 70+ papers. His research interests include statics, kinetics, and dynamics of low-dimensional spin systems, kinetic theory of nucleation, and wetting at the nanoscale. He has been working Dr. Ruckenstein's group at SUNY at Buffalo since 2001. Previously, he was a visiting research scholar at the Institute of Physics "Gleb Wataghin" of the University of Campinas, Brazil (1995–1996), and at the Institute of Theoretical Physics I of the University of Erlangen, Germany (1998–2000).

# 1

# Kinetic Theory of Liquid-to-Solid Nucleation

## Introduction

In this chapter, the basic concepts of the kinetic nucleation theory are presented and applied to liquid-to-solid phase transformations (Sections 1.1 through 1.3). Both homogeneous (Sections 1.1 through 1.5) and heterogeneous (Sections 1.6 through 1.10) nucleations are considered. The dissociation of molecules from the surface of the cluster of the solid phase is considered to be caused by Brownian motion that is governed, in general, by the Fokker–Planck equation for the joint distribution function in the phase space [1,2]. For homogeneous nucleation, the latter equation can be reduced to the backward Smoluchowski equation, which provides the mean passage of time for a molecule to overcome the potential barrier generated by a spherical cluster and, hence, the rate of dissociation ($\alpha_j$) of the surface molecules.

The rate of condensation ($\beta_j$) of molecules on the surface of the cluster is considered to be governed by diffusion. The equality between the dissociation and condensation rates provides the size of the critical cluster. As a result, the function $w(g)$ that is needed to calculate the nucleation rate (see Preface, Equation 0.5) can be obtained.

In the classical nucleation theory, the clusters are assumed amorphous droplets. However, this assumption is not always accurate. Indeed, face-centered cubic (fcc) and icosahedral structures were detected experimentally and in dynamic simulations of the nucleation process [3–7]. The kinetic nucleation theory can avoid the assumption of uniformity of clusters and allows users to calculate the nucleation rate for nonuniform clusters (Section 1.4). An interesting feature of the obtained results is the irregular behavior of nucleation rate as a function of supersaturation for fcc clusters, caused by the irregular dependence of the mean interaction potential between a surface atom and cluster on the cluster size.

The kinetic nucleation theory also demonstrated its efficiency in the treatment of the deliquescence stage of heterogeneous nucleation on soluble particles that was previously examined on the basis of the classical theory [8]. As a result of kinetic considerations, the occurrence of nonsharp deliquescence was qualitatively explained for the first time (Section 1.6).

The kinetic nucleation theory provides the possibility to consider various interactions in the system and estimate their contributions to the nucleation rate. In particular, in Section 1.7, the contribution to the nucleation rate of the interaction of solute molecules with those outside the cluster is compared to the contribution of the interaction of solute molecules with those of the solvent.

In this chapter, the heterogeneous nucleation of crystals that are either completely immersed into the liquid droplet or have some contact with the droplet surface is also included (Section 1.8), even though its treatment is based on the classical nucleation theory. Its consideration on the basis of the kinetic nucleation theory is a challenge for future research. Another problem involves the application of heterogeneous nucleation as an experimental method for obtaining information about such thermodynamic quantities as surface tension. As an example, the line tension of a crystal–liquid–contact line was considered (Section 1.9).

The results presented in this chapter were obtained using the Goodrich approach [9] for the transformation of Equation 0.2 to the differential Equation 0.3 (see Preface). Shizgal and Barrett [10] noted that the Goodrich expansion does not recover the equilibrium solution of Equation 0.2. In contrast, the approach suggested by Zeldovich [11] is compatible with the equilibrium solution. (We use the Zeldovich approach in the next chapter to describe the vapor-to-liquid nucleation.) However, as shown in Section 1.10, the Zeldovich and Goodrich approaches provide almost the same results for a wide range of energy parameters characterizing intermolecular interactions.

## References


1. Chandrasekhar, S., Stochastic problems in physics and astronomy, *Rev. Mod. Phys.* 15, 1–89 (1943).
2. Gardiner, C.W., *Handbook of Stochastic Methods*, Springer, New York, 1983.
3. Kittel, C., *Introduction to Solid State Physics*, Wiley, New York, 1976.
4. Farges, J., De Feraudy, M.F., Raoult, B., Torchet, G., *Adv. Chem. Phys.* 70, 45–74 (1988).
5. Tanemura, M., Himatari, Y., Matsuda, H., Ogawa, T., Ogita, N., Ueda, A., Geometrical analysis of crystallization of the soft-core model, *Prog. Theor. Phys.* 58, 1079–1095 (1977).
6. Hsu, C.S., Rahman, A., Crystal nucleation and growth in liquid rubidium, *J. Chem. Phys.* 70, 5234–5240 (1979).
7. Hsu, C.S., Rahman, A., Interaction potentials and their effect on crystal nucleation and symmetry, *J. Chem. Phys.* 71, 4974–4986 (1979).
8. Shchekin, A.K., Shabaev, I., in: *Nucleation Theory and Applications*, Schmelzer, J.W., Röpke, G., Priezjev, V.B., Eds., JINR, Dubna, 2006.
9. Goodrich, F.C., Nucleation rates and the kinetics of particle growth: II. The birth and death process, *Proc. R Soc London* 277, 167–182 (1964).
10. Shizgal, B., Barret, J.C., Time dependent nucleation, *J. Chem. Phys.* 91, 6505–6518 (1989).
11. Zeldovich, J.B., Toward the theory of formation of a new phase. Cavitation, *J. Exp. Theor. Phys.* 12, 525–538 (1942).

## 1.1 A Possible Nucleation Type of Mechanism for the Growth of Small Aerosol Particles

*Ganesan Narsimhan and Eli Ruckenstein**

The collisions between small aerosol particles do not necessarily result in coagulation, since the dispersion forces responsible for coagulation are relatively weak, i.e., the interaction potential well due to the van der Waals attraction and Born repulsion is not deep. In addition, because of weak dispersion forces, the constituents of a doublet of small particles can derive sufficient energy from the collisions with the molecules of the suspending medium to overcome the potential well and dissociate. The rate of formation of a doublet, i.e., the rate of coagulation per particle, is calculated by accounting for the sticking probability in terms of the depth of the interaction potential well. The average dissociation time of a doublet has been determined as a first passage time in the energy of the relative motion of the two constituent particles. For sufficiently small particles, the rate of dissociation of a doublet is much larger than its rate of formation. This unstable doublet reaches, therefore, an equilibrium with the single particles over times much shorter than the time scale of coagulation. Because of the dramatic decrease in the rate of dissociation of a doublet with an increase in the size of its constituents, there exists a critical particle radius above which the doublet is stable, i.e., its rate of formation becomes greater than its rate of dissociation. When the doublet is unstable, aerosol growth occurs via a nucleation mechanism, i.e., via successive equilibria of unstable agglomerates leading to a critical size aggregate, which constitutes a nucleus for growth. On the other hand, when the doublet is stable, the singlets themselves act as nuclei for aerosol growth. (

### 1.1.1 Introduction

Aerosol particles of extremely small sizes (molecular clusters) can form via physical or chemical nucleation. The subsequent growth of these particles is brought about both by the condensation of monomers on the surface of the particles as well as by the Brownian coagulation of the particles themselves. Brownian coagulation of such small particles as a result of collisions has been shown to affect both aerosol formation and its subsequent growth [1]. The short-range dispersion forces are mainly responsible for the coagulation of aerosol particles. It is usually assumed that every collision between two aerosol particles leads to irreversible coagulation. The above assumption is justified when the particles are large, since the interparticle dispersion forces are strong in this case—i.e., the interaction potential well (due to the van der Waals attraction and Born repulsion) is very deep. However, not every collision between two sufficiently small particles results in coagulation, since the interaction potential well is not sufficiently deep, i.e., the interparticle forces are relatively weak. The collision efficiency, i.e., the sticking probability, has been calculated in terms of the interaction potential between the particles [2,3].

Moreover, the molecules of the suspending medium continually transfer energy through collisions to the constituents of the coagulated particle pairs. If the constituent particles are sufficiently small, the particle pair can derive sufficient energy from the molecules of the suspending medium to overcome the interaction potential well and, hence, dissociate. In other words, the coagulation of sufficiently small particles is reversible. As shown previously [4], the time scale of oscillations in the potential well of the constituents of the doublet is much shorter than the time scale of Brownian motion. As a result, the relative motion between the two particles of the doublet in the interaction potential well can be described by a one-dimensional Fokker–Planck equation in the particle energy space. The average lifetime of a doublet consisting of equal-size particles has been calculated as a first passage time in the energy of the relative motion of the two constituent particles [4].

Therefore, the doublets of sufficiently small particles are continually formed (due to coagulation) and destroyed (because of their dissociation). If the rate of formation of a doublet (i.e., the rate of coagulation per particle) is greater than its rate of dissociation (i.e., the rate of dissociation per doublet), the doublet is stable. As a result, the concentration of doublets increases continuously with time. As will be shown later, the time scale of evolution of these stable doublets is of the order of the time scale of coagulation. Of course, these stable doublets interact with the particles to form triplets that, in turn, form quadruplets, etc., resulting in the growth of the particle size.

On the other hand, if the rate of formation of a doublet is much smaller than its rate of dissociation, the doublet is unstable. It will be shown later that the unstable doublets reach a dynamic equilibrium with the singlets in an extremely short time (of the order of the time scale of dissociation). The equilibrium

* *J. Colloid Sci.* 116, 288 (1987). Republished with permission.

concentration of these unstable doublets is small and depends on the relative magnitudes of the rates of formation and dissociation. Since the dissociation rate of a doublet decreases rather dramatically with increasing particle size (because of the rapid increase in the depth of the interaction potential well with increasing particle size), there exists a critical particle size above which the coagulated particle pair is stable, i.e., its rate of formation is much greater than its rate of dissociation. This critical particle size is analogous to the critical cluster size for the growth of molecular clusters in nucleation. The unstable doublets undergo collisions with the particles to form triplets that, if unstable, attain equilibrium with singlets, again in a very short time, with their concentration being much lower than that of the doublets. The triplets, in turn, form quadruplets and so on, until a stable agglomerate is formed that constitutes a nucleus for further growth. Hence, the aerosol growth occurs either only by coagulation or through the formation of a stable nucleus, depending on whether or not the doublets are stable.

A comparison between the rates of formation and dissociation of doublets as well as the calculation of the critical particle size are presented in the next section. The subsequent section discusses the implications of the above two mechanisms of aerosol growth on the calculation of the Brownian coagulation coefficient from the evolution of particle number concentration.

### 1.1.2 Comparison between Rates of Formation and Dissociation of a Doublet Consisting of Equal-Size Aerosol Particles

#### *1.1.2.1 Rate of Formation of a Doublet*

Since the doublets are continually being formed by coagulation, their rate of formation depends on the rate of coagulation. The rate of formation of a doublet $r_f$ (or, equivalently, the rate of coagulation per particle) can be expressed in terms of the coagulation coefficient $\beta$ and the particle number concentration $n$ as

$$r_f = \beta n. \tag{1.1}$$

The initial particle number concentration $n_0$ is related to the volume fraction $\phi_p$ of the particles via the expression

$$n_0 = \frac{\phi_p}{(4/3)\pi R^3}, \tag{1.2}$$

where $R$ is the radius of the particles.

When the particles are sufficiently small, i.e., the Knudsen number ($= \lambda_g/R$, where $\lambda_g$ is the mean free path of the suspending medium) is large (free molecular limit), the particles can be considered gigantic molecules and the rates of collision $r_c$ between the particles can be obtained from the kinetic theory of gases as [5]

$$r_c = 4\pi R^2 \left( \frac{16kT}{\pi m_p} \right)^{1/2}, \tag{1.3}$$

where $k$ is the Boltzmann constant, $T$ is the absolute temperature of the suspending medium, and $m_p$ is the mass of the particle.

As pointed out earlier, not every collision between the small particles leads to coagulation because the interaction potential well (due to the van der Waals attraction and Born repulsion) is not deep. Consequently, the coagulation coefficient $\beta$ can be written as

$$\beta = \gamma r_c = \gamma 4\pi R^2 \left( \frac{16kT}{\pi m_p} \right)^{1/2}, \tag{1.4}$$

where $\gamma$ is the sticking probability, i.e., the probability of coagulation upon collision.

Examining the relative Brownian motion between two singlets and accounting for the energy exchange between the two colliding particles through a reflecting boundary condition, the expression for the sticking probability $\gamma$, in the free molecular limit, is given by [2]

$$\gamma = 1 - e^{-\Phi_0}(1+\Phi_0), \tag{1.5}$$

where $\Phi_0$ is the dimensionless depth of the interaction potential well (= $\phi_0/kT$, with $\phi_0$ being the depth of the potential well).

Combining Equations 1.4 and 1.5, the expression for the coagulation coefficient β becomes:

$$\beta = \{1 - e^{-\Phi_0}(1+\Phi_0)\}4\pi R^2\left(\frac{16kT}{\pi m_p}\right)^{1/2}. \tag{1.6}$$

### 1.1.2.2 Rate of Dissociation of a Doublet

As pointed out earlier, when the constituent particles of a doublet are small, they can derive sufficient energy from collisions with the molecules of the suspending medium to overcome the interaction potential well and thus dissociate. If the energy imparted by the collisions of the molecules of the suspending medium is smaller than the depth of the interaction potential well, the two particles of the doublet will exhibit an oscillatory motion within the potential well. For sufficiently small particles, the time scale of these oscillations is much smaller than the time scale of the Brownian motion of the constituent particles [4]. Because of this disparity between the two time scales, the relative Brownian motion between the two particles of a doublet can be described by a one-dimensional Fokker–Planck equation in the energy space of the relative motion of the two. The average dissociation time $\langle T\rangle$ can, therefore, be calculated as a first passage time (i.e., the time at which the energy of the relative motion becomes equal to the depth of the interaction potential well) in the energy of the relative motion. The expression for the average dissociation time $\langle T\rangle$, in terms of the dimensionless energy $E$ and the time scale of Brownian motion $\zeta^{-1}$, is given by [4]

$$\langle T\rangle = \frac{2\zeta^{-1}}{\sqrt{\pi}}\int_0^{\Phi_0} E^{1/2}e^{-E}\int_E^{\Phi_0} e^x \int_0^x e^{-y} \times \frac{\psi'(y)}{\psi(y)}\,\mathrm{d}y\mathrm{d}x\mathrm{d}E, \tag{1.7}$$

and the dimensionless energy $E$ is given by

$$E = \frac{1}{2}m_r\upsilon^2/kT + \Phi, \tag{1.8}$$

where $m_r(= m_p/2)$ is the reduced mass, $\upsilon$ is the relative velocity between the two constituent particles of the doublet, $\Phi$ (= $\phi/kT$) is the dimensionless interaction potential,

$$\psi(E) = \int_0^{\Phi^{-1}(E)} \sqrt{2(E-\Phi)}\mathrm{d}r, \tag{1.9}$$

and

$$\psi'(E) = \int_0^{\Phi^{-1}(E)} \frac{1}{\sqrt{2(E-\Phi)}}\mathrm{d}r. \tag{1.10}$$

The time scale of the Brownian motion $\zeta^{-1}$ is related to the diffusion coefficient $D$ of the particle via the expression

$$\zeta^{-1} = \frac{kT}{m_p D}, \tag{1.11}$$

where the diffusion coefficient expressed in terms of the Knudsen number Kn and viscosity of the suspending medium η is

$$D = \frac{kT}{6\pi R\eta}\left(\frac{5+4\text{Kn}+6\text{Kn}^2+18\text{Kn}^3}{5-\text{Kn}+(8+\pi)\text{Kn}^2}\right). \tag{1.12}$$

The rate of dissociation $r_d$ of a doublet is the reciprocal of the average dissociation time $\langle T \rangle$, i.e.,

$$r_d = \frac{1}{\langle T \rangle}. \tag{1.13}$$

The rates of formation and dissociation of a doublet consisting of equal-size aerosol particles, in air at 1 atm and 298 K, have been calculated for a Hamaker constant of $10^{-12}$ erg from Equations 1.6 through 1.13. The rate of formation of a doublet $r_f$ is plotted in Figure 1.1 as a function of $R$ for different values of $\phi_p$. The rate of formation of a doublet is found to decrease with increasing particle size at constant $\phi_p$ because of the predominant effect of the decrease in the particle number concentration (Equation 1.2). The rate of formation of a doublet does not show as strong a dependence on particle size as does the rate of dissociation (Figure 1.1). As expected, the rate of formation of a doublet is found to be higher at larger values of the particle volume fraction because of a greater number of collisions between the particles.

The rate of dissociation $r_d$ is also plotted in Figure 1.1 as a function of the radius $R$, and it is found to be very sensitive to particle size. Indeed, it decreases dramatically with increasing $R$ because of the rapid increase in the depth of the interaction potential well.

It is interesting to note that the rate of formation of a doublet is smaller than the rate of dissociation for sufficiently small particles. The difference between the rates of dissociation and formation decreases as the particle size increases, eventually becoming equal at a critical particle radius $R^*$. For particles smaller than $R^*$, the doublets are unstable and they reach a dynamic equilibrium with the single particles. For particles greater than $R^*$, however, the doublet is stable and, therefore, its concentration increases with time. The variation of the critical radius $R^*$ with the particle volume fraction $\phi_P$ is plotted in Figure 1.2. $R^*$ is found to be a weak function of $\phi_p$, decreasing from 44 to 18 Å as $\phi_p$ increases from $10^{-8}$ to $10^{-4}$.

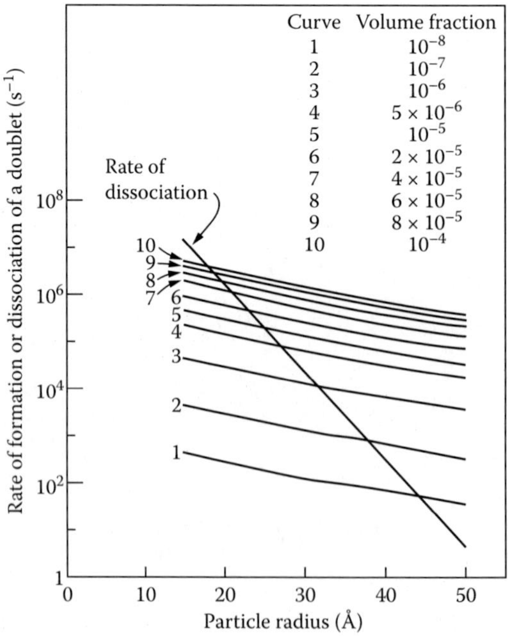


Figure 1.1

Comparison between the rates of formation and dissociation of a doublet consisting of equal-size aerosol particles of unit density, in air at 1 atm and 298 K, for a Hamaker constant of $10^{-12}$ erg and for different particle volume fractions.

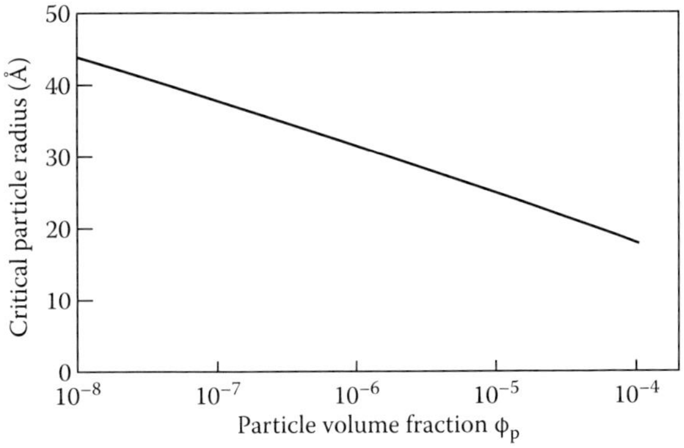


Figure 1.2

The variation of the critical particle radius with the particle volume fraction for particles of unit density, in air at 1 atm and 298 K, and for a Hamaker constant of $10^{-12}$ erg.

The time evolution of the number concentrations of particles and doublets is given by

$$\frac{dn}{dt} = -\beta n^2 - \frac{2}{\langle T \rangle} n_d, \tag{1.14}$$

and

$$\frac{dn_d}{dt} = \frac{1}{2}\beta n^2 - \frac{n_d}{\langle T \rangle}, \tag{1.15}$$

where $n$ and $n_d$ are the number concentrations of particles and doublets, respectively, at time $t$.

Introducing the dimensionless variables, $n^* = n/n_0$, $n_d^* = n_d/n_0$, $\tau = \beta n_0 t$, $\alpha = 1/\beta n_0 \langle T \rangle$, where $n_0$ is the initial number concentration of particles, $1/(\beta n_0)$ is the time scale of coagulation, and $\alpha$ is the ratio of the rates of dissociation and formation of a doublet.

Equations 1.14 and 1.15 can be rewritten as

$$\frac{dn^*}{d\tau} = -n^{*2} + 2\alpha n_d^* \tag{1.16}$$

and

$$\frac{dn_d^*}{d\tau} = \frac{1}{2} n^{*2} - \alpha n_d^*. \tag{1.17}$$

Equations 1.16 and 1.17 are solved for the initial conditions

$$n^* = 1 \quad n_d^* = 0 \quad \text{for} \quad \tau = 0 \tag{1.18}$$

to yield

$$n_d^* = \frac{[1 - \exp\{-\alpha^{1/2}(\alpha+2)^{1/2}\tau\}]}{\{(\alpha+2) + \alpha^{1/2}(\alpha+2)^{1/2}\} - [\{(\alpha+2) - \alpha^{1/2}(\alpha+2)^{1/2}\}\exp\{-\alpha^{1/2}(\alpha+2)^{1/2}\tau\}]}. \tag{1.19}$$

The value of the parameter $\alpha$ determines whether the doublets are stable or unstable. For $\alpha > 1$ ($<1$), the doublets are unstable (stable). For $\alpha \gg 1$, the dimensionless time scale of evolution of the number concentration of the doublets is of the order of $1/\alpha$, whereas for $\alpha \ll 1$, the dimensionless time scale of evolution is of the order of $1/\alpha^{1/2}$. In other words, the time scale of evolution of the highly unstable doublets is much shorter than the time scale of coagulation, whereas the highly stable doublets evolve over much longer times. Therefore, the unstable doublets attain equilibrium in an extremely short time (of the order of the time scale of dissociation, i.e., in $10^{-7}$ to $10^{-4}$ s).

Since the number concentration of the stable doublets increases continuously with time, the number of collisions between doublets and single particles would become significant at large times, resulting in the formation of triplets, quadruplets, etc. Equation 1.19 accounts only for the presence of single particles and doublets. It can, therefore, describe the evolution of the number concentration of stable doublets only for short times.

The expression for the dimensionless equilibrium concentration of the doublets $n^*_{d_{eq}}$ is given by

$$n^*_{d_{eq}} = \frac{1}{(\alpha+2)+\alpha^{1/2}(\alpha+2)^{1/2}} \quad \alpha > 1. \tag{1.20}$$

For $\alpha \gg 1$, $n^*_{d_{eq}} \sim \dfrac{1}{\alpha}$.

The parameter $\alpha$, being the ratio of the rates of dissociation and formation of the doublet, depends on the particle size as well as on the particle volume fraction. The equilibrium doublet concentration $n^*_{d_{eq}}$ is plotted in Figure 1.3 against the particle size for different particle volume fractions. The dotted portions of the curves in Figure 1.3 represent the hypothetical equilibrium concentrations of stable doublets. The equilibrium doublet concentration is found to be extremely small for small particles and low volume fractions, to increase dramatically with increasing particle size, and to exhibit a weak dependence on the particle volume fraction.

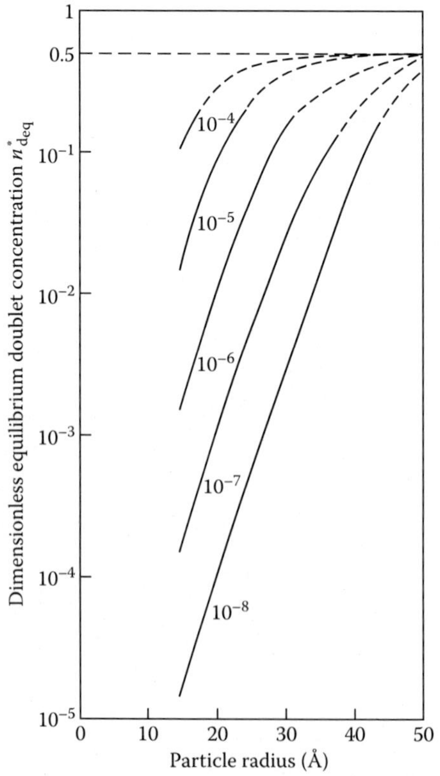


Figure 1.3

The variation of the dimensionless equilibrium doublet concentration with the radius of particles of unit density, in air at 1 atm and 298 K, for a Hamaker constant of $10^{-12}$ erg and for different initial particle volume fractions.

### 1.1.3 Two Alternate Mechanisms of Aerosol Growth

Let us consider a system of monodispersed aerosol particles. Because of the collisions between the particles, they will coagulate to form doublets. The rate of formation of a doublet depends, of course, on the particle radius, the particle density, the viscosity and temperature of the suspending medium, the Hamaker constant, and the particle number concentration. If the rate of formation of a doublet is much less than its rate of dissociation, the doublets are unstable. These unstable doublets attain equilibrium with the particles over a time scale that is much shorter than the time scale of coagulation. Their equilibrium concentrations are very small and depend on the relative magnitudes of their rates of formation and dissociation. The doublets continually undergo collisions with the particles to form triplets. The triplets, if unstable, quickly reach equilibrium with the particles, their concentrations being much smaller than those of the doublets. Over time scales of the order of the time scale of coagulation (i.e., much higher than the time scales of dissociation), the system will consist of extremely low concentrations of triplets, quadruplets, etc., in equilibrium with single particles. This continues until stable $i_c$-tuplets are formed, which grow in size. In other words, the formation and growth of stable $i_c$-tuplets occur via the formation of extremely low concentrations of $i$-tuplets ($i < i_c$). Therefore, the smallest stable $i_c$-tuplet acts as a nucleus for aerosol growth. The rate of dissociation of agglomerates exhibits a very strong dependence on their size compared to their rate of formation. Therefore, $\alpha$ is expected to be much larger than unity for all unstable agglomerates (up to $(i_c - 1)$-tuplet) and becomes much smaller than unity for $i_c$-tuplet.

The system can be represented schematically as

$$\begin{aligned}
&[1]+[1] \rightleftharpoons [2] \\
&[2]+[1] \rightleftharpoons [3] \\
&\vdots \\
&[i-1]+[1] \rightleftharpoons [i] \\
&\vdots \\
&[i_c-1]+[1] \rightarrow [i_c],
\end{aligned} \tag{1.21}$$

where $[i]$ denotes an $i$-tuplet. The dimensionless equilibrium concentration of an $i$-tuplet ($i < i_c$) can be expressed in terms of the dimensionless single particle concentration as

$$n_i^* = K_2 K_3 \ldots K_i n_1^{*i}, \tag{1.22}$$

with the equilibrium constants $\{K_i\}$, $i = 1, 2, \ldots$, in Equation 1.22 given by

$$K_i = \beta_{i-1,1}\, n_0 \langle T_i \rangle. \tag{1.23}$$

Here, $\beta_{i1}$ is the coagulation coefficient between an $i$-tuplet and a single particle, $n_0$ is the initial single particle concentration, and $\langle T_i \rangle$ is the average dissociation time for an $i$-tuplet to dissociate to an $(i \sim 1)$-tuplet.

The dynamic equations for the dimensionless concentrations of single particles and stable $i_c$-tuplets can be written as

$$\frac{dn_{i_c}^*}{d\tau} = -\left(\frac{\beta_{i_c 1}}{\beta_{11}}\right) n_{i_c}^* n_1^* + \left(\frac{\beta_{i_c-1,1}}{\beta_{11}}\right) K n_i^{*i_c} \tag{1.24}$$

and

$$\frac{dn_i^*}{d\tau} = -\sum_{j=2}^{i_c-1} L_j j^2 n_i^{*(j-1)} \frac{dn_i^*}{d\tau} - \left(\frac{\beta_{i_c 1}}{\beta_{11}}\right) n_{i_c}^* n_1^* - i_c \left(\frac{\beta_{i_c-1,1}}{\beta_{11}}\right) K n_1^{*i_c}. \tag{1.25}$$

Equation 1.25 is valid at short times since the growth of $(i_c + 1)$ and larger tuplets is neglected. In Equations 1.24 and 1.25, the constants $L_j$ and $K$ are given by

$$L_j = K_2 K_3 \ldots K_j,\ K = L_{(i_c-1)}.$$

If the singlet concentration is sufficiently large, one can neglect the change in the singlet concentration at short times to obtain

$$n_{i_c} = \left( \frac{\beta_{i_c-1,1}}{\beta_{i_c 1}} \right) K n_0 [1 - \exp(-\beta_{i_c 1} n_0 t)]\ . \tag{1.26}$$

For large $i_c$, $\beta_{i_c 1} \sim \beta_{i_c-1,1}$, and

$$n_{i_c} \simeq K n_0 [1 - \exp(-\beta_{i_c 1} n_0 t)]. \tag{1.27}$$

The time scale for nucleation is small compared to the time scale for coagulation. It is therefore given by

$$\frac{dn_{i_c}}{dt} \simeq \beta_{i_c 1} K n_0^2 \tag{1.28}$$

and the rate of change of singlet concentration is given by

$$\frac{dn_1}{dt} \simeq -i_c \frac{dn_{i_c}}{dt} = -i_c \beta_{i_c 1} K n_0^2. \tag{1.29}$$

On the other hand, if the rate of formation of a doublet is much greater than its rate of dissociation, the doublet will be stable. The doublet concentration increases continuously with time. The doublets will also interact with the particles resulting in the formation of triplets, quadruplets, etc. Therefore, the aerosol growth, in this case, does not require the formation of a nucleus. The single particles themselves act as nuclei for aerosol growth. The evolution of the concentration of the particles over a time of the order of the time scale of coagulation is given, in this case, by

$$\frac{dn}{dt} = -\beta_{11} n^2, \tag{1.30}$$

where $\beta_{11}$ is the coagulation coefficient of the single particles.

### 1.1.4 Conclusions

A comparison between the rates of formation and dissociation of a doublet consisting of sufficiently small equal-size aerosol particles of unit density, in air at 1 atm and 298 K, has been made for a Hamaker constant of $10^{-12}$ erg. The rate of formation of a doublet (the rate of coagulation per particle) has been calculated, accounting for the sticking probability in terms of the depth of the interaction potential well. The average dissociation time of a doublet has been obtained as a first passage time in the energy of the relative motion of the two constituent particles. The rate of dissociation of a doublet is found to decrease dramatically with increasing particle size because of the rapid increase in the depth of the interaction potential well. The rate of formation of a doublet, however, though a strong function of the particle volume fraction, does not exhibit as strong a dependence on particle size as does the rate of dissociation. For a doublet consisting of very small particles, the rate of dissociation is found to be greater than its rate of formation, thus rendering the doublet unstable. As the particle size increases, the difference between the rates of dissociation and formation of a doublet decreases. A critical particle size is found to exist above which the rate of formation of a doublet is greater than its rate of dissociation—i.e., the doublet is stable. The unstable doublets reach equilibrium with the singlets (particles) over times much shorter than the time scale of coagulation (i.e., in $10^{-7}$ to $10^{-4}$ s), their equilibrium concentration being dependent on the relative magnitudes of their rates of dissociation and formation. Because of successive collisions with particles, the doublets form agglomerates of different sizes. These agglomerates, if unstable, quickly reach equilibrium, their concentrations decreasing as the agglomerate size increases. The smallest stable agglomerate, formed via successive equilibria of smaller sized unstable agglomerates, constitutes a nucleus for the aerosol growth. When the doublets themselves are stable, however, the singlets (particles) themselves act as the nuclei for the aerosol growth. Therefore, the aerosol growth occurs by two different mechanisms depending on the stability of the doublet.

## References


1. Gelbard, F., Seinfeld, J.H., *J. Colloid Interface Sci.* 68, 363 (1979).
2. Narsimhan, G., Ruckenstein, E., *J. Colloid Interface Sci.* 104, 344 (1985).
3. Narsimhan, G., Ruckenstein, E., *J. Colloid Interface Sci.* 107, 174 (1985).
4. Narsimhan, G., Ruckenstein, E., *J. Colloid Interface Sci.* 116, 279 (1987).
5. Hidy, G.M., Brock, J.R., *The Dynamics of Aerocolloidal Systems*, Pergamon, Oxford, 1970.

## 1.2 A New Approach for the Prediction of the Rate of Nucleation in Liquids

*Ganesan Narsimhan and Eli Ruckenstein**

A formalism is developed for the prediction of the rate of nucleation in liquids which as opposed to the usual quasi-thermodynamic treatment does *not* use (1) the macroscopic concept of interfacial tension for small clusters and (2) the principle of detailed balance. The rate of dissociation of a surface monomer from a cluster is calculated by using a first passage time analysis. An expression for the critical cluster size is obtained in terms of the characteristics of the interaction potential between the monomers of a molecular cluster that, for illustrative purposes, is taken here as the square well potential, A generalized Kelvin equation is derived to obtain the dependence of the solubility on the cluster size. Finally, an expression for the rate of nucleation is obtained on the basis of the kinetics of dissociation. The predicted nucleation rates for spherical clusters with uniform distribution of molecules are found to be much larger than those predicted by the classical theory. In contrast to the classical theory, the nucleation rates exhibit a gradual increase at very low supersaturations, increase catastrophically at a critical supersaturation, and taper off gradually at high supersaturations. The predicted nucleation rates for approximately spherical clusters with face-centered cubic (fcc) packing are found to be much smaller than those for spherical amorphous clusters and exhibit behavior qualitatively different from that of the latter. The classical theory is shown to overpredict the interfacial tension and hence underpredict the nucleation rates. Since the spherical amorphous cluster and a cluster with a definite structure such as fcc are probably appropriate models for large and small clusters, respectively, the actual nucleation rates should lie between the two, coinciding with the former at low and the latter at high supersaturations. The classical theory for the rate of nucleation can be recovered from the present formalism when the critical cluster size is sufficiently large, i.e., larger than $10^6$ monomers. Such a condition, however, is satisfied only for extremely small supersaturations. (© *1989 Academic Press, Inc.*)

### 1.2.1 Introduction

It is well known that the clusters that form in a supersaturated solution grow only if their size exceeds a critical value. Volmer and Weber [1] assumed that the rate of nucleation is proportional to the probability of forming clusters of critical size and therefore the rate of nucleation is proportional to $\exp(-\Delta W^*/kT)$, with $\Delta W^*$ being the free energy of formation of a critical cluster. A more complete kinetic approach to nucleation was initiated by Farkas [2] that served as a basis for subsequent treatments of Becker and Döring [3], Zeldovich [4], and Frenkel [5]. In the "classical" theory of homogeneous nucleation, largely based on the treatment of Becker and Döring [3], the nucleation rate is calculated on the basis of a population balance for the size distribution of clusters generated by condensation and evaporation of atoms onto and from the clusters. Central to this theory are (1) the use of the principle of detailed balance and of a hypothetical equilibrium size distribution to relate the rates of dissociation of clusters to their rates of formation and (2) the use of the Kelvin equation to relate the vapor pressure of small liquid droplets to their radius of curvature and surface tension. Turnbull and Fisher [6] have extended the above theory to nucleation in liquids. Modifications and extensions of the classical theory have been presented by Kuhrt [7,8], Lothe and Pound [9], Reiss [10], and Katz and Donohue [11]. The use of the macroscopic property of interfacial tension for small clusters is the main drawback of the classical theory of nucleation. While it has been demonstrated from a thermodynamic point of view [12] as well as by statistical mechanics [13] that the interfacial tension of small bodies decreases with decreasing radius, the applicability of the latter result to very small clusters is still questionable.

In the present paper, a kinetic theory of nucleation in a supersaturated solution based on the molecular interactions and the packing geometry in the molecular clusters is developed. The proposed approach does not employ the macroscopic interfacial tension for small clusters. More importantly, the rate of dissociation is calculated directly from a first passage time analysis. In contrast, in the classical theory it is calculated by using the principle of detailed balance and a hypothetical equilibrium size distribution of clusters. A brief description of the structure of the clusters is presented in the next section. The rate of dissociation of a surface monomer from a cluster is calculated in the subsequent section by employing a first passage time analysis. Further, expressions for the solubility and the critical cluster size are given in terms of intermolecular interactions, modeled as a square well potential. Finally, an expression for the rate of nucleation is derived on

* *J. Colloid Interface Sci.* 128, 549 (1989). Republished with permission.

the basis of population balances that account for the formation and dissociation of clusters. A comparison of the predicted nucleation rates for two cluster geometries with the classical theory concludes the paper.

### 1.2.2 Structure of Molecular Clusters

A molecular cluster contains a group of solute molecules held together by intermolecular attractive forces and constitutes a precursor for the formation of a new phase. In order to investigate the effect of the cluster packing on the rate of nucleation, two different packings are examined: (1) a spherical cluster with uniformly distributed molecules (denoted further as amorphous spherical cluster) and (2) an approximately spherical cluster with fcc packing (denoted further as fcc crystalline spherical cluster). The intermolecular potential is assumed to be a square well potential and only the nearest neighbor interactions are taken into account. The results of the calculations for a more realistic Lennard–Jones (6–12) potential will be reported later.

#### *1.2.2.1 Amorphous Spherical Cluster*

Let us consider a spherical cluster of radius $R$ consisting of $i$ monomers and denote the interaction distance of the monomers by $\eta$. The $i$ monomers are assumed to be uniformly distributed within the spherical cluster with a coordination number $C$. Therefore, every interior monomer will interact with $C$ nearest neighbors. The interaction volume $V_{\text{int}}$ of a surface monomer is shown by the shaded region in Figure 1.4 and geometric considerations lead to

$$V_{\text{int}} = \frac{2}{3}\pi\eta^3\left[1 - \frac{1}{2}\left(\frac{\eta}{R}\right)\right]. \tag{1.31}$$

Since the interaction volume of a monomer is equal to $\frac{4}{3}\pi\eta^3$, the number of nearest neighbors $C_s$ of a surface monomer is given by

$$C_s = \frac{C}{2}\left[1 - \frac{1}{2}\left(\frac{\eta}{R}\right)\right]. \tag{1.32}$$

Since $R/\eta \sim i^{1/3}$, the above equation can be recast in terms of the cluster size $i$ as

$$C_s = \frac{C}{2}\left[1 - \frac{1}{2i^{1/3}}\right]. \tag{1.33}$$

The number of surface monomers $N_s$ for a spherical amorphous cluster consisting of $i$ monomers is given by

$$\begin{aligned} N_s &\simeq i - (i^{1/3} - 1)^3 \\ &= 1 + 3i^{2/3} - 3i^{1/3}. \end{aligned} \tag{1.34}$$

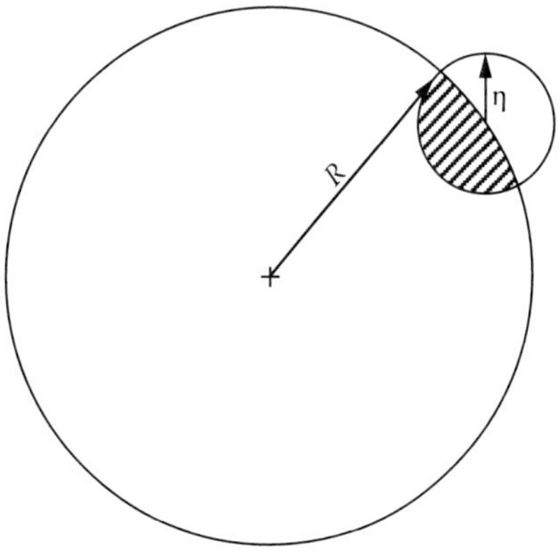


Figure 1.4

A surface monomer with an interaction range $\eta$ belonging to a spherical amorphous cluster of radius $R$. The shaded region represents the interaction volume between the surface monomer and the molecules of the cluster.

For sufficiently large clusters,

$$N_s \simeq 3i^{2/3}. \tag{1.35}$$

### 1.2.2.2 Spherical fcc Crystalline Cluster

Clusters with an fcc structure can be built by adding shells of monomers of increasing order around a central monomer. We follow the procedure of Dave and Abraham [14] in constructing such approximately spherical clusters. We define the dimensionless coordinates $x$, $y$, $z$ of a monomer in terms of the coordinates $x'$, $y'$, $z'$ and the interaction distance $\eta$ as $x = x'/\eta$, $y = y'/\eta$, $z = z'/\eta$, where $x$, $y$, $z$ are integers. From the geometry of this cluster, one can show that the dimensionless coordinates of a monomer belonging to the $n$th shell satisfy the two constraints

$$\text{(a) } x^2 + y^2 + z^2 = n, \; n = 1, 2, 3, \ldots$$

$$\text{(b) } x + y + z \text{ is even.} \tag{1.36}$$

The positions of the monomers in a given shell are given by all possible permutations of $x$, $y$, and $z$, which satisfy the above two constraints. The number of nearest neighbors of an interior monomer in an fcc packing is 12. Therefore, the number of surface monomers $N_s$ in a cluster consisting of $i$ monomers can be calculated by identifying the monomers with less than 12 nearest neighbors on the basis of the coordinates of the monomers. If $N_j$ is the number of surface monomers with $j$ (< 12) nearest neighbors, then the average number of nearest neighbors $C_s$ can be calculated from

$$N_s C_s = \sum_{j=1}^{11} j N_j. \tag{1.37}$$

## 1.2.3 Interaction Potential of a Surface Monomer

A monomer located in the interior of a cluster will interact with $C$ nearest monomers of the cluster. However, a surface monomer will interact with its nearest neighbors in the cluster as well as with the molecules of the surrounding medium. The liquid is characterized by a mean value of nearest neighbors that varies from 11 to about 4 as the temperature rises [15]. Let $N_1$ denote the average number of nearest neighbors of solvent molecules for a solute molecule in solution (i.e., not a member of any cluster). Since a surface monomer belonging to a cluster has $C_s$ nearest neighbor monomers in the cluster, it is reasonable to assume that the number of nearest neighbors of solvent molecules $C_1$ for a surface monomer is given by

$$C_1 = N_1\left(1 - \frac{C_s}{C}\right). \tag{1.38}$$

Therefore, the interaction potential $\psi$ of a surface monomer can be written as

$$\psi = -C_s\epsilon' - N_1\left(1 - \frac{C_s}{C}\right)\epsilon_s', \tag{1.39}$$

where $\epsilon'$ and $\epsilon_s'$ are the strengths of the monomer–monomer and monomer–solvent interactions, respectively. The interaction potential $\psi_d$ of a dissociated monomer is given by

$$\psi_d = -N_1\epsilon_s'. \tag{1.40}$$

Relative to the solvent, a surface monomer belonging to a cluster is captured in a potential well (Figure 1.5) of depth given by

$$\phi = \psi - \psi_d = -C_s\left(\epsilon' - \frac{N_1}{C}\epsilon_s'\right). \tag{1.41}$$

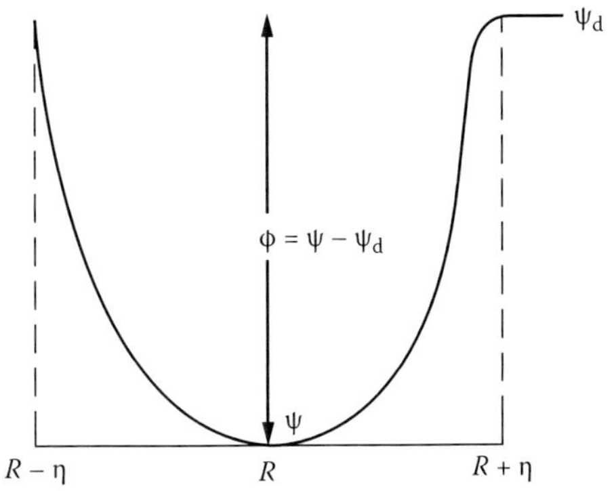


Figure 1.5

The interaction potential of a surface monomer belonging to a spherical cluster of radius *R*.

We now define the effective monomer–solvent interaction $\epsilon_s = (N_1/C)\epsilon_s'$. Equation 1.41 can be rewritten as

$$\begin{aligned} \phi &= -C_s(\epsilon' - \epsilon_s) \\ &= -C_s\epsilon, \end{aligned} \tag{1.42}$$

where $\epsilon = \epsilon' - \epsilon_s$ is the effective monomer–monomer interaction in the cluster. The effective interaction parameter $\epsilon$ accounts for (1) the difference between the monomer–monomer and the monomer–solvent interactions and (2) the number of nearest neighbors of the solvent medium. As will be shown later, the interfacial tension of a macroscopic cluster can be easily related to this effective monomer–monomer interaction.

The interaction potential of a surface monomer belonging to a spherical cluster of radius *R* is represented in Figure 1.5 by considering the monomers as points experiencing square well interaction potential. This interaction potential has a hard sphere repulsion at the distance $R - \eta$, as a result of Born repulsion, a minimum at *R*, the equilibrium position of the surface monomer, and becomes equal to the average interaction potential of the solute in solution $\psi_d$ at $R + \eta$. Even though the monomer–monomer and monomer–solvent interactions are square well potentials, the effective interaction potential of a surface monomer is not square well because of the change in the number of nearest neighbors of monomer and solvent molecules away from its equilibrium position. The actual variation depends on the structure of the cluster. If the depth of the interaction potential well is much greater than $kT$, it will be shown later that the variation of the interaction potential with position does not affect the dissociation rate of a surface monomer. It should be noted that the rate of nucleation changes insignificantly if the volume exclusion of the monomer is taken into account.

### 1.2.4 First Passage Time Analysis for Estimation of Rate of Dissociation of a Surface Monomer

Because of the collisions with the molecules of the surrounding medium, a surface monomer belonging to a cluster exhibits Brownian motion within the interaction potential well. So long as the surface monomer is within the interaction potential well, it would belong to the cluster. However, as soon as the monomer leaves (due to Brownian motion) the potential well or, equivalently, the distance between the surface monomer and the cluster exceeds the range of interactions, the surface monomer can be considered dissociated.

The Brownian motion of the surface monomer within the interaction potential well can in general be described by the Fokker–Plank equation [16]

$$\frac{\partial f}{\partial t} + \mathbf{V} \cdot \nabla_r f - \left(\frac{\nabla \phi}{m}\right) \cdot \nabla_v f = \zeta \nabla_v \cdot \left[\mathbf{V} f + \left(\frac{kT}{m}\right) \nabla_v f\right], \tag{1.43}$$

where $f(\mathbf{r}, \mathbf{V}, t)$ is the distribution function with respect to the monomer position $\mathbf{r}$ and velocity $\mathbf{V}$ at time $t$, $m$ is the mass of the solute molecule, $\zeta$ is the friction coefficient, $k$ is the Boltzmann constant, $T$ is the temperature, and $\nabla_r$ and $\nabla_v$ refer to the gradients with respect to position and velocity coordinates, respectively.

The relaxation of Brownian motion, i.e., the time over which the particle momentum is correlated, is of the order of $\zeta^{-1}$. For diffusion in liquids, this relaxation time is extremely small (~$10^{-21}$ s). Therefore, the time scales of interest are very much larger than the relaxation time. Consequently, the particle would undergo an extremely large number of collisions with the molecules of the medium over the time scales of interest. For this reason, the particle velocity distribution can be assumed to be Maxwellian. For a Maxwellian velocity distribution, the Fokker–Plank equation for the Brownian motion of a surface monomer in the interaction potential well reduces to the Smoluchowski equation

$$\frac{\partial n(\mathbf{r},t)}{\partial t} = \nabla_{\mathbf{r}} \cdot \left( D\nabla_{\mathbf{r}} n + \frac{D}{kT} n\nabla\phi \right), \tag{1.44}$$

with $n(\mathbf{r}, t)$ and $D$ being the number density and the diffusion coefficient of the surface monomer, respectively.

Collisions with the molecules of the medium would also result in the translational and rotational Brownian motion of the cluster. If the cluster is much larger than the molecules of the medium, its translational and rotational Brownian motion can be neglected—i.e., the molecular cluster can be assumed to be stationary. Such an assumption is equivalent to neglecting the replacement free energy in the calculation of the free energy of a molecular cluster [10].

The position of a surface monomer due to Brownian motion can be described by the conditional probability density $w(\mathbf{r}, t|\mathbf{r}_0, 0)$.

The quantity $w(\mathbf{r}, t|\mathbf{r}_0, 0)d\mathbf{r}$ represents the probability that at time $t$ the position of the particle is between $\mathbf{r}$ and $\mathbf{r} + d\mathbf{r}$ when its initial position was $\mathbf{r}_0$. As already noted, the Fokker–Plank equation for the evolution of the above conditional probability density acquires, in the present case, the Smoluchowski form [17]:

$$\frac{\partial}{\partial t} w(\mathbf{r},t\,|\,\mathbf{r}_0,0) = \nabla \cdot \left[ D\nabla w + \frac{D}{kT} w\nabla\phi \right]. \tag{1.45}$$

Assuming that the interaction potential is spherically symmetric, the range of interaction of a surface monomer with the cluster will depend only on the radial distance from the center of the cluster. Consequently, a surface monomer will dissociate whenever the radial distance exceeds the range of the interaction forces, and the lifetime of a surface monomer will depend only on the evolution of the radial distance. Thus, Equation 1.45 becomes

$$\frac{\partial}{\partial t} w(r,t\,|\,r_0,0) = \frac{D}{r^2}\frac{\partial}{\partial r^2}\left( r^2 \frac{\partial w}{\partial r} \right) + D\frac{\partial}{\partial r}\left( w\frac{\partial \Phi}{\partial r} \right), \tag{1.46}$$

where $\Phi = \phi/kT$ is the dimensionless interaction potential and $w(r,t|r_0, 0)\,dr$ is the conditional probability that the radial distance of the monomer at time $t$ is between $r$ and $r + dr$ given that its initial radial distance was $r_0$. In order to calculate the lifetime of a surface monomer (i.e., the time at which the radial distance of the monomer just exceeds the interaction distance), it is more convenient to deal with the backward evolution of the conditional probability density $w(r,t|r_0, t')$ (i.e., the evolution with respect to $t'$ for fixed $r$ and $t$), because the final condition at the moment of dissociation is known.

The corresponding backward Fokker–Plank equation in the Smoluchowski approximation is given by [17]

$$\frac{\partial}{\partial t'} w(r,t\,|\,r_0,t') = D\frac{d\Phi}{dr_0}\frac{\partial w}{\partial r_0} - \frac{D}{r_0^2}\frac{\partial}{\partial r_0}\left( r_0^2 \frac{\partial w}{\partial r_0} \right). \tag{1.47}$$

Since the Brownian motion of the surface monomer is time homogeneous, the above equation can be recast in the form

$$\frac{\partial}{\partial r}w(r,\tau|r_0,0)=-D\frac{\mathrm{d}\Phi}{\mathrm{d}r_0}\frac{\partial w}{\partial r_0}+\frac{D}{r_0^2}\frac{\partial}{\partial r_0}\left(r_0^2\frac{\partial w}{\partial r_0}\right), \tag{1.48}$$

where $\tau = t - t'$. A surface monomer belonging to a spherical cluster of radius $R$ can be considered to be associated with the cluster as long as its radial distance is less than $(R + \eta)$. Let $\tau$ be the time at which the surface monomer dissociates when its initial radial position was $r_0$ $(R - \eta \le r_0 \le R + \eta)$. Therefore, the probability for the time of dissociation $\tau \ge t$ is given by

$$\mathrm{Prob}(\tau\ge t|r_0,0)=G(r_0,t)=\int_{R-\eta}^{R+\eta} w(r,t|r_0,0)\mathrm{d}r \tag{1.49}$$

since the monomer is not dissociated, and its position at time $t$ can be located at any point between $R - \eta$ and $R + \eta$. Integrating Equation 1.48 with respect to $r$, one obtains

$$\frac{\partial}{\partial\tau}G(r_0,\tau)=-D\frac{\mathrm{d}\Phi}{\mathrm{d}r_0}\frac{\partial}{\partial r_0}G(r_0,\tau)+\frac{D}{r_0^2}\frac{\partial}{\partial r_0}\left(r_0^2\frac{\partial G}{\partial r_0}\right). \tag{1.50}$$

Since $w(r, 0|r_0, 0) = \delta(r - r_0)$,

$$G(r_0,0)=\begin{cases}1 & \text{for } R-\eta\le r_0\le R+\eta\\ 0 & \text{otherwise.}\end{cases} \tag{1.51}$$

If the initial radial position of a surface monomer were $(R + \eta)$, it would dissociate instantaneously, i.e.,

$$G[(R + \eta), \tau] = 0 \text{ for all } \tau. \tag{1.52}$$

The radial distance of a surface monomer cannot be less than $(R - \eta)$ because of the hard sphere repulsive part of the interaction potential. Therefore, the flux of the monomer at $(R - \eta)$ is zero, i.e.,

$$\frac{\partial}{\partial r}G(r,\tau)\Big|_{r=R-\eta}=0 \text{ for all } \tau. \tag{1.53}$$

The boundary conditions 1.52 and 1.53 represent absorbing and reflecting boundary conditions at $(R + \eta)$ and $(R - \eta)$, respectively.

Since $G(r_0, \tau)$ is the probability that the first passage time (time of dissociation) of a surface monomer whose initial position is $r_0 \ge \tau$, the probability that the dissociation time is between 0 and $\tau$ is $1 - G(r_0, \tau)$, and the mean first passage time $T(r_0)$ (not to be confused with the temperature $T$) is given by

$$\begin{aligned}T(r_0)&=\int_0^\infty t\frac{\partial}{\partial t}\{1-G(r_0,t)\}\mathrm{d}t\\ &=-\int_0^\infty t\frac{\partial}{\partial t}G(r_0,t)\mathrm{d}t\\ &=\int_0^\infty G(r_0,t)\mathrm{d}t.\end{aligned} \tag{1.54}$$

Integrating Equation 1.50 with respect to $\tau$ and noting that $G(r_0, \infty) = 0$, one obtains

$$-1 = -D\frac{d\Phi}{dr_0}\frac{dT}{dr_0} + \frac{D}{r_0^2}\frac{d}{dr_0}\left(r_0^2\frac{dT}{dr_0}\right) \quad (1.55)$$

with the boundary conditions

$$T(R + \eta) = 0 \quad (1.56)$$

and

$$\left.\frac{dT}{dr}\right|_{r=R-\eta} = 0. \quad (1.57)$$

Solving Equation 1.55 for the boundary conditions 1.56 and 1.57, one obtains

$$T(r_0) = \frac{1}{D}\int_{r_0}^{(R+\eta)} e^{\Phi(x)} \int_{R-\eta}^{x} y^2 e^{-\Phi(y)}\, dy\, dx. \quad (1.58)$$

It was pointed out earlier that even though the interaction potential between two atoms is assumed square well, the interaction potential of a surface monomer is not a square well potential because of the variation of the number of nearest neighbors of solute and solvent molecules away from the equilibrium position. However, one expects the variation to be gradual near the equilibrium position and sharp near $(R + \eta)$. If $\phi_0$ is the depth of the interaction potential well, the effective interaction potential will be close to $-\phi_0$ and will rise sharply to zero in the vicinity of $(R + \eta)$. If the depth of the potential well is much greater than $kT$, the exponential term in the inner integral of Equation 1.58 will be close to $e^{\Phi_0}$, with $\Phi_0$ being the dimensionless depth of the interaction potential well ($= \phi_0/kT$). Therefore, Equation 1.58 can be simplified to give

$$T(r_0) = \frac{e^{\Phi_0}R^2}{3D}\int_{r_{0/R}}^{(1+\eta/R)} \frac{e^{\Phi(x)}}{x^2} \times \left\{x^3 - \left(1 - \frac{\eta}{R}\right)^3\right\} dx. \quad (1.59)$$

The exponential term within the integral is very small except in the vicinity of $(R + \eta)$, where it is close to unity. The contribution to the above integral comes only from the region near $(R + \eta)$. Therefore,

$$T(r_0) = T^0 \simeq \frac{e^{\Phi_0}R^2}{3D} \times \frac{\left[(1+\eta/R)^3 - (1-\eta/R)^3)\right]}{(1+\eta/R)^2}, \quad (1.60)$$

which, since $\eta/R \sim i^{-1/3}$, becomes

$$T^0 \simeq \frac{e^{\Phi_0}\eta^2 i^{2/3}}{3D} \times \frac{\left[(1+i^{-1/3})^3 - (1-i^{-1/3})^3\right]}{(1+i^{-1/3})^2}. \quad (1.61)$$

As expected, the average lifetime of a surface monomer is independent of its initial position in the interaction potential well.

For sufficiently large clusters, i.e., $i \gg 1$, Equation 1.61 leads to

$$T^0 \simeq \frac{2e^{\Phi_0}}{D}\eta^2 i^{1/3}. \quad (1.62)$$

The rate of dissociation $\alpha_i$ of a monomer from a cluster consisting of $i$ monomers is therefore given by

$$\alpha_i = \frac{N_s}{T^0} = \frac{3N_s D}{R^2} \times \frac{e^{-\Phi_0}(1+\eta/R)^2}{\left[(1+\eta/R)^3 - (1-\eta/R)^3\right]}. \tag{1.63}$$

For a spherical amorphous cluster, Equations 1.33, 1.34, 1.42, and 1.63 can be combined to give

$$\alpha_i = \frac{3D\exp\left\{-(\epsilon C/2kT)(1-1/2i^{1/3})\right\} \times (1+i^{-1/3})^2 i^{1/3}\left\{1-(1-i^{-1/3})^3\right\}}{\eta^2\left\{(1+i^{-1/3})^3 - (1-i^{-1/3})^3\right\}}. \tag{1.64}$$

For sufficiently large clusters, i.e., $i \gg 1$, the above expression reduces to

$$\alpha_i^0 \simeq \frac{3Di^{1/3}}{2\eta^2}\exp\left\{-\frac{\epsilon C}{2kT}\left(1-\frac{1}{2i^{1/3}}\right)\right\}. \tag{1.65}$$

For spherical fcc crystalline clusters, the rate of dissociation $\alpha_i$, which is given by Equation 1.63, can be calculated using the values of $N_s$ and $\Phi_0$ as determined from the coordinates of the surface monomers. For sufficiently large clusters, the expression reduces to

$$\alpha_i^0 \simeq \frac{8RD}{\eta^3}\exp\left(-\frac{8.7\epsilon}{kT}\right). \tag{1.66}$$

The rate of addition of a monomer to a cluster containing $i$ monomers $\beta_i$ is given by

$$\beta_i = 4\pi RDn_0, \tag{1.67}$$

where $R$ is the radius of the cluster and $n_0$ is the number concentration of the solute in solution.

For a spherical amorphous cluster, the radius $R$ is given by

$$R = \eta i^{1/3}. \tag{1.68}$$

For a spherical fcc crystalline cluster, the radius $R$ is given by

$$R = \eta\sqrt{m}, \tag{1.69}$$

where $m$ is the order of the outermost shell.

### 1.2.5 Solubility

The rate of dissociation of a surface monomer from a cluster depends both on the intermolecular potential and on the structure of the cluster. We have derived equations for the rate of dissociation for a spherical amorphous cluster as well as for a spherical fcc crystalline cluster. It is, of course, possible to derive similar equations for other structures of clusters (namely, body-centered cubic [bcc], cubic packing, etc.). The rate of formation of a cluster is strongly influenced by the solute concentration. The rate of dissociation of small clusters is greater than their rate of formation. This renders these small clusters unstable. As the cluster size increases, the difference between their rates of dissociation and formation becomes smaller, eventually becoming zero at the critical cluster size. Clusters greater than the critical cluster continually grow since their rate of formation is greater than their rate of dissociation. The size of the critical cluster depends on the solute concentration and can be obtained by equating the rates of formation and dissociation. The solubility can be defined as the solute concentration for which the critical cluster size tends to infinity. Therefore, for spherical amorphous clusters, the solubility $n_s$ is given by

$$n_s = \frac{3}{8\pi\eta^3}\exp\left(-\frac{\epsilon C}{2kT}\right), \tag{1.70}$$

and for spherical fcc crystalline clusters,

$$n_s = \frac{8}{4\pi\eta^3}\exp\left(\frac{8.7\epsilon}{kT}\right). \tag{1.71}$$

The size of the critical cluster $i$ can be obtained by equating the rate of formation $\beta_{i^*-1}$, to the rate of dissociation $\alpha_{i^*}$:

$$4\pi R_{i^*-1} D n_0 = \frac{3N_s D}{R_{i^*}^2} \times \frac{e^{-\Phi_0}(1+\eta/R_{i^*})^2}{\left[(1+\eta/R_{i^*})^3 - (1-\eta/R_{i^*})^3\right]}. \tag{1.72}$$

For spherical amorphous clusters, the above equation reduces to

$$4\pi\eta D(i^*-1)^{1/3} n_0 = \frac{3D\exp\left\{-(\epsilon C/2kT)(1-1/2i^{*1/3})\right\}\times(1+i^{*-1/3})^2}{\eta^2\left\{(1+i^{*-1/3})^3-(1-i^{*-1/3})^3\right\}\times i^{*1/3}\{1-(1-i^{*-1/3})^3\}}. \tag{1.73}$$

Introducing the supersaturation ratio $s = n_0/n_s$ and combining Equations 1.70 and 1.73, one obtains

$$\frac{s((i^*-1)/i^*)^{1/3}\left\{(1+i^{*-1/3})^3-(1-i^{*-1/3})^3\right\}}{2\left\{1-(1-i^{*-1/3})^3\right\}(1+i^{*-1/3})^2} = \exp\left(\frac{\epsilon C}{4kTi^{*1/3}}\right). \tag{1.74}$$

When the critical cluster size is very large, i.e., $i^* \gg 1$, Equation 1.74 reduces to

$$R^* = \left(\frac{\epsilon C\eta}{4kT\ln s}\right), \tag{1.75}$$

$R^*$ being the radius of the critical cluster. It is of interest to compare the above equation with the following Kelvin equation, which contains the interfacial tension $\sigma$ and molecular volume $\upsilon$:

$$R^* = \left(\frac{2\sigma\upsilon}{kT\ln s}\right). \tag{1.76}$$

Equation 1.75 constitutes a modified Kelvin equation in terms of the effective monomer–monomer interaction and the coordination number.

### 1.2.6 Rate of Nucleation

In a supersaturated solution, molecular clusters of different sizes are continually formed and destroyed, leading eventually to the formation and growth of clusters greater in size than the critical cluster. The formation and growth of clusters can be viewed as a birth and death process in the phase space of cluster size [18]. The flux of clusters $I_i(t)$ from size $i - 1$ to size $i$ at time $t$ is given by

$$I_i(t) = \beta_{i-1} f_{i-1}(t) - \alpha_i f_i(t) \tag{1.77}$$

and the rate of change of the number concentration $f_i(t)$ of clusters of size $i$ at time $t$ by

$$\frac{d}{dt} f_i(t) = -(I_{i+1} - I_i). \tag{1.78}$$

Since the dependence of $\alpha_i$ and $\beta_i$ on the cluster size $i$ is sufficiently smooth, one can approximate them by the continuous functions $\alpha(g)$ and $\beta(g)$, where $g$ is a continuous variable. The functions $f_i(t)$ and $I_i(t)$ can also be replaced by the continuous functions $f(g, t)$ and $I(g, t)$. Expanding the right-hand side of Equation 1.78 about the midpoint $i+\frac{1}{2}$ gives

$$\frac{\partial}{\partial t} f(g,t) = -\left[\frac{\partial}{\partial g} I(g,t)\right]_{g+1/2}. \tag{1.79}$$

Replacing $i$ by $g+\frac{1}{2}$ in Equation 1.77 and expanding in Taylor series about $g$ yields

$$I\left(g+\frac{1}{2}, t\right) = -\frac{1}{2}\frac{\partial}{\partial g}(\beta+\alpha) f(g,t) + (\beta-\alpha) f(g,t). \tag{1.80}$$

Combining Equations 1.79 and 1.80, one obtains

$$\frac{\partial f(g,t)}{\partial t} = \frac{1}{2}\frac{\partial^2}{\partial g^2}(\beta+\alpha) f(g,t) - \frac{\partial}{\partial g}(\beta-\alpha) f(g,t). \tag{1.81}$$

The formation of critical clusters occurs via the growth of smaller clusters, which are unstable—i.e., their rates of dissociation are much greater than their rates of formation. The rate of nucleation is therefore dominated by the bottleneck in the neighborhood of the critical cluster size $g^*$, where the rates of formation and dissociation are comparable. Moreover, the time scales of formation and dissociation of smaller unstable clusters are much smaller than the time scale of formation and growth of the critical cluster. It is, therefore, reasonable to expect that a supersaturated solution would attain a quasi steady state after a short lapse of time. From Equation 1.80, the expression for the steady-state nucleation flux $I_s$ can be written in terms of the steady-state cluster size distribution $f_s(g)$ as

$$I_s = -\frac{1}{2}\frac{d}{dg}(\beta+\alpha) f_s(g) + (\beta-\alpha) f_s(g). \tag{1.82}$$

Since the change in the solute concentration due to the formation and growth of clusters is very small, especially at high supersaturation ratios, the solute concentration $n_0$ can be assumed constant. Moreover, the probability of formation of very large clusters (much larger than the critical cluster size) is extremely small. Therefore, the steady-state nucleation rate $I_s$ can be obtained by solving Equation 1.82 with the boundary conditions

$$f_s(G_0) = 0, \; G_0 \gg g^* \tag{1.83}$$

$$f_s(0) = n_0 \tag{1.84}$$

to yield

$$I_s = \frac{\frac{1}{2}\beta_0 n_0}{\int_0^{G_0} \exp\{-2w(g)\} dg}. \tag{1.85}$$

Here

$$w(g) = \int_0^g \frac{\beta - \alpha}{\beta + \alpha} \mathrm{d}g \tag{1.86}$$

and $\beta_0$ refers to the coagulation coefficient of solute molecules.

The Brownian coagulation coefficient of solute molecules $\beta_0$ can be calculated in terms of their diffusion coefficient $D$ by using the expression

$$\beta_0 = 16\,\pi\eta D n_0 \tag{1.87}$$

and the solute concentration $n_0$ is related to the solubility $n_s$ through

$$n_0 = s n_s. \tag{1.88}$$

Using Equations 1.87 and 1.88, the steady-state rate of nucleation $I_s$ can be expressed in terms of the supersaturation ratio $s$ as

$$I_s = \frac{8\pi\eta D s n_s}{\int_0^{G_0} e^{-2w(g)} \mathrm{d}g}. \tag{1.89}$$

Since $w(g)$ exhibits a rather sharp maximum at the critical cluster size $g^*$, the integral in Equation 1.86 can be evaluated by the method of steepest descent, i.e.,

$$w(g) \simeq w(g^*) + \frac{1}{12} \frac{\ln s}{g^*} (g - g^*)^2. \tag{1.90}$$

As a result,

$$I_s \simeq \frac{1}{2} \left( \frac{\ln s}{6\pi g^*} \right)^{1/2} \beta_0 s n_s e^{2w(g^*)}, \tag{1.91}$$

where

$$w(g^*) \approx \int_0^{g^*} \frac{\beta - \alpha}{\beta + \alpha} \mathrm{d}g. \tag{1.92}$$

### 1.2.7 Interfacial Tension of Macroscopic Clusters

Thus far, we have derived an expression for the steady-state rate of nucleation in terms of the effective monomer–monomer interaction and the structure of the clusters. In order to compare the rate of nucleation with the classical nucleation theory, it is necessary to relate the interfacial tension of a macroscopic cluster to the effective monomer–monomer interaction and the structure of the clusters. Let us consider a macroscopic spherical amorphous cluster consisting of $i (i \gg 1)$ monomers. The energy of formation $\Delta U_i$ of this cluster can be expressed as

$$\begin{aligned} \Delta U_i &= -\frac{C}{2}(i - N_s)\epsilon - \frac{C_s}{2} N_s \epsilon \\ &= -\frac{C}{2} i\epsilon + N_s \left[ \frac{C}{2} \left( 1 - \frac{C_s}{C} \right) \epsilon \right]. \end{aligned} \tag{1.93}$$

Combining Equations 1.32, 1.34, and 1.93, one obtains

$$\Delta U_i = -\frac{C}{2} i\epsilon + (1 + 3i^{2/3} - 3i^{1/3}) \times \left[\frac{C}{4}(1 + (1/2)i^{-1/3})\epsilon\right]. \tag{1.94}$$

The second term in the above equation refers to the surface energy $\Delta U_s$, which—neglecting any entropic contribution—can be related to the interfacial tension $\sigma$ by

$$\Delta U_s = 4\pi R^2 \sigma, \tag{1.95}$$

with $R$ being the radius of the cluster. Since $R = \eta i^{1/3}$, Equation 1.95 can be recast in the form

$$\Delta U_s = 4\pi \eta^2 i^{2/3} \sigma. \tag{1.96}$$

Comparing Equations 1.95 and 1.96, one obtains

$$\sigma = \frac{C\epsilon}{16\pi\eta^2}\{3 - 3i^{-1/3} + i^{-2/3}\} \times (1 + (1/2)i^{-1/3}), \tag{1.97}$$

which for large macroscopic clusters reduces to

$$\sigma = \frac{3C\epsilon}{16\pi\eta^2}. \tag{1.98}$$

According to the classical nucleation theory, the steady-state rate of nucleation $I_s$ is given by

$$I_s = \left(\frac{2D}{\eta^2 \upsilon}\right)\left(\frac{\ln s}{6\pi g^*}\right)^{1/2} \exp\left(-\frac{\Delta G^*}{kT}\right), \tag{1.99}$$

where

$$g^* = \frac{2\mu\sigma^3\upsilon^2}{(kT\ln s)^3}, \tag{1.100}$$

$$\Delta G^* = \frac{\mu\sigma^3\upsilon^2}{(kT\ln s)^2}, \tag{1.101}$$

$g^*$ is the critical cluster size, $\upsilon(= (4\pi/3)\eta^3)$ is the molecular volume, and $\mu$ (the geometric shape factor) is $16\pi/3$ for a spherical cluster.

It is to be noted that the interfacial tension becomes independent of cluster size and equal to the macroscopic value only for extremely large macroscopic clusters. In other words, Equation 1.94 reduces to Equation 1.95 only for clusters consisting of a number of monomers as large as $10^6$. One can show that Equation 1.91 reduces to the classical equation (Equation 1.99) only for extremely large values of $g^*$. However, $g^*$ is very large only for extremely low supersaturations. Even for moderately large clusters, Equation 1.95 overpredicts the interfacial tension [19]. Consequently, the classical theory overpredicts both $g^*$ and $\Delta G^*$. As a result, the classical theory underpredicts the rate of nucleation except for extremely small supersaturation ratios for which the critical cluster size is very large.

### 1.2.8 Comparison of the Prediction with the Classical Nucleation Theory

Steady-state nucleation rates have been calculated for spherical amorphous clusters as well as for spherical clusters with fcc packing, for different effective monomer–monomer interactions. All calculations have been performed for a coordination number of 12. The diffusion coefficient $D$ and the molecular volume $\upsilon$ of solute molecules are assumed to be $10^{-5}$ cm²/s and $4.3 \times 10^{-23}$ cm³, respectively. For a spherical cluster, the number of nearest neighbors $C_s$ for a surface monomer and the number of surface monomers $N_s$ were calculated from Equations 1.33 and 1.34, respectively. Spherical fcc clusters were constructed so as to satisfy the constraints provided by Equation 1.37. $N_s$ was calculated by identifying the monomers with less than 12 nearest neighbors from the positions of the monomers. $C_s$ was evaluated from Equation 1.37. The depth of the interaction potential well of a surface monomer was calculated by employing Equation 1.42. The rates of nucleation for different supersaturation ratios and effective monomer–monomer interactions were calculated from Equations 1.63, 1.67, 1.86, and 1.89 and compared with nucleation rates as calculated from Equations 1.98 through 1.101 derived on the basis of the classical theory. In Figure 1.6 the solubility is plotted against the dimensionless monomer–monomer interaction $\epsilon/kT$ for the two kinds of clusters. The solubility decreases exponentially with an increase in $\epsilon/kT$ and is much lower for the fcc clusters because of a more compact packing. Typical variations of the rates of formation and dissociation of clusters with size are presented in Figure 1.7. As expected, the rate of dissociation of a cluster is independent of the supersaturation ratio. It decreases rapidly with size for small clusters and more gradually for larger clusters. In contrast, the rate of formation of a cluster is a strong function of the supersaturation ratio. The rate of formation shows a stronger dependence on size as the supersaturation ratio increases. The critical size is provided by the intersection of the curves for the rates of formation and dissociation as indicated by the arrows in Figure 1.7. Typical variations of the critical size with supersaturation are presented in Figure 1.8. The critical size decreases with an increase in supersaturation, the change being less pronounced at higher supersaturations. The critical size is larger for stronger monomer–monomer interactions, or equivalently, for higher interfacial tensions. Figure 1.9 shows a plot of steady-state nucleation rate against supersaturation for spherical amorphous clusters. It shows that

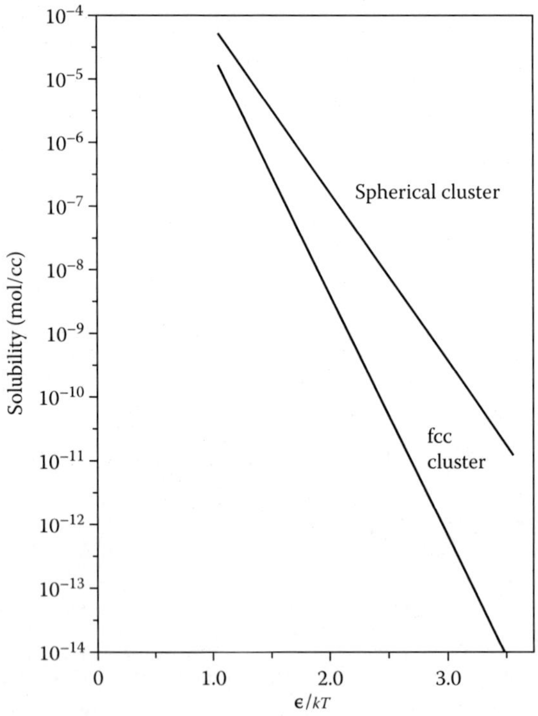


Figure 1.6

The variation of solubility with an effective monomer–monomer interaction energy for spherical amorphous and fcc clusters.

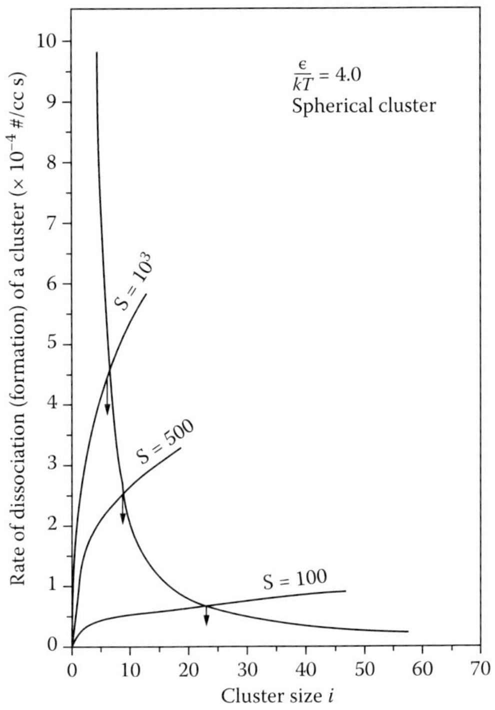


Figure 1.7

Variations of the rates of formation and dissociation of a spherical amorphous cluster with size for different supersaturation ratios.

the nucleation rate increases catastrophically at a critical supersaturation ratio but it tapers off gradually at high supersaturations. However, the catastrophic increase for weaker monomer–monomer interactions occurs only at significantly higher nucleation rates. For small interaction parameters, the rate of nucleation increases gradually with supersaturations at very low values of $s$ and increases catastrophically at a critical supersaturation. The region of initial gradual increase in the nucleation rate becomes narrower as the

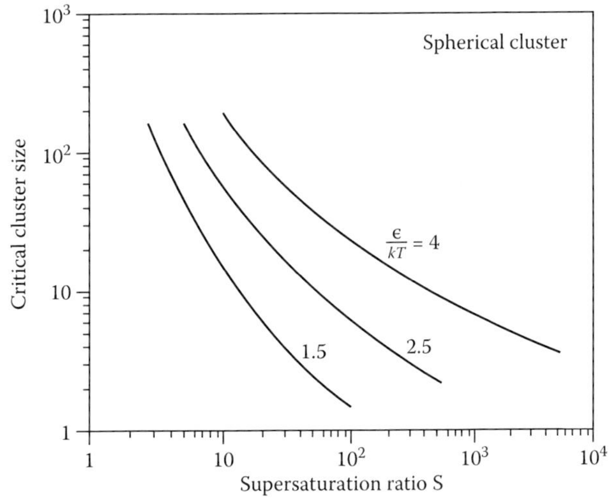


Figure 1.8

Variation of the critical cluster size of spherical amorphous clusters with supersaturation ratio for different dimensionless effective monomer-monomer interaction energies.

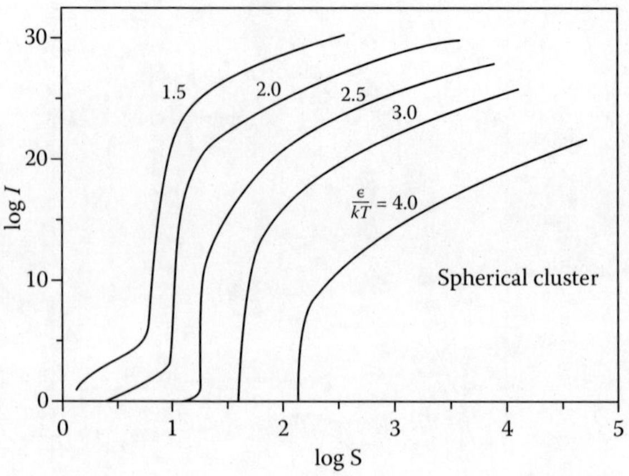


Figure 1.9

Plot of steady-state nucleation rate versus supersaturation ratio for spherical amorphous clusters for different dimensionless effective monomer–monomer interaction energies.

monomer–monomer interaction becomes stronger, eventually vanishing at very high interactions. In contrast, this region of initial gradual increase in nucleation rate is absent in the classical theory. A similar plot of nucleation rate versus supersaturation ratio for spherical fcc crystalline cluster is presented in Figure 1.10. The behavior of the nucleation rates is in this case qualitatively different from that for a spherical amorphous cluster, as can be seen by comparing Figures 1.9 and 1.10. For clusters with an fcc structure and for small interaction parameters, the nucleation rate increases catastrophically at a critical supersaturation ratio. The extent of catastrophic increase in the nucleation rate decreases as the monomer–monomer interaction becomes stronger, eventually vanishing at very high interactions. In other words, for very strong monomer interactions, the increase in nucleation rate with supersaturation is no longer catastrophic but gradual. The critical supersaturation ratio for an fcc cluster is an order-of-magnitude higher than that for a spherical cluster, as can be seen from Figures 1.9 and 1.10. The depth of the interaction potential well of a surface monomer depends on the structure of molecular clusters via the number of nearest neighbors of a surface monomer. Since the rate of nucleation exhibits an exponential dependence on the depth of the interaction potential well, it is not surprising that different cluster structures yield very different rates of nucleation. Comparisons of the rates of nucleation for spherical amorphous and spherical crystalline clusters with the classical nucleation theory are presented for two different interaction parameters in Figures 1.11 and 1.12. Even though the qualitative behavior of the curves for the spherical amorphous cluster and that for the classical theory are similar except at very small supersaturations, the predicted nucleation rates for a spherical cluster are about $10^{20}$ times larger than those obtained from the classical theory.

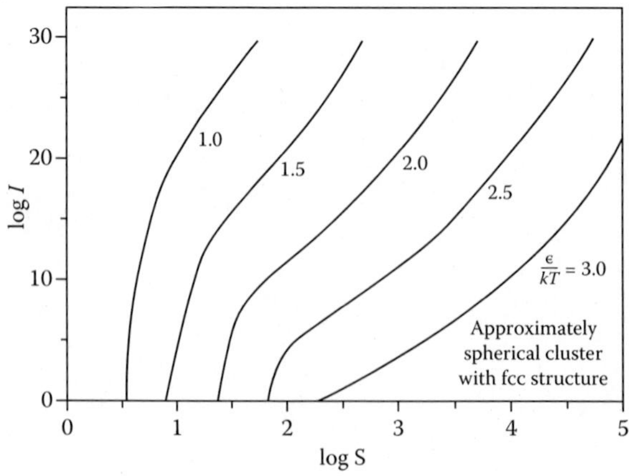


Figure 1.10

Plot of steady-state nucleation rate versus supersaturation ratio for approximately spherical clusters with fcc packing for different dimensionless effective monomer–monomer interaction energies.

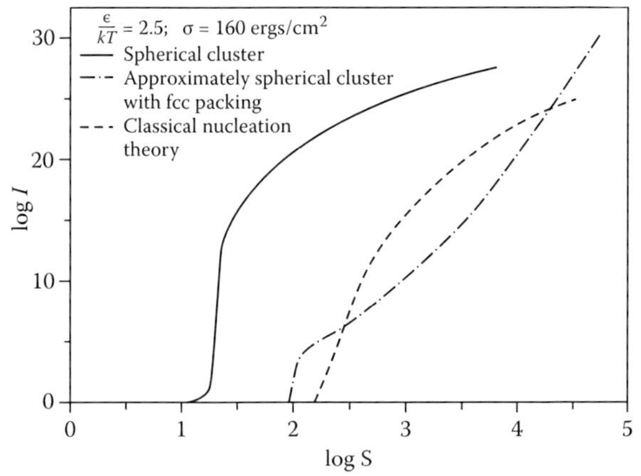


Figure 1.11

Comparison of the nucleation rates for spherical amorphous and fcc dusters with the classical nucleation theory for $\epsilon/kT = 2.5$.

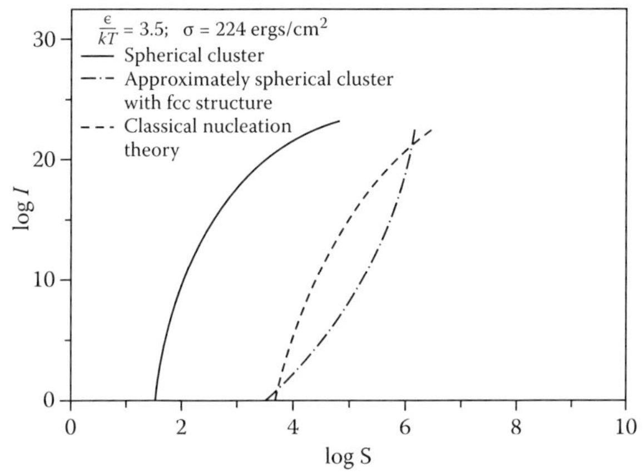


Figure 1.12

Comparison of the nucleation rates for spherical amorphous and fcc dusters with the classical nucleation theory for $\epsilon/kT = 3.5$.

The critical supersaturation ratio for the spherical amorphous cluster is an order of magnitude smaller than that predicted by the classical theory. As pointed out earlier, the difference between the nucleation rates for spherical amorphous clusters and classical theory can be attributed to the fact that the classical theory overpredicts the interfacial tension and the free energy barrier for nucleation and consequently grossly underpredicts the nucleation rates. On the other hand, the qualitative behavior of the curves for fcc clusters and that for the classical theory are very different even though the magnitudes of the rate of nucleation are comparable. Experimental observations regarding nucleation rates indicate [20] that (1) the critical supersaturation ratios are much smaller than the values predicted by the classical theory, (2) the variation of nucleation rates with $s$ is gradual at low supersaturation ratios, and (3) the nucleation rate becomes catastrophic only for supersaturation ratios greater than $10^3$. Such a behavior was attributed to a heterogeneous nucleation at low supersaturations that was considered dominant due to the presence of impurities. It is customary to infer the interfacial tension by comparing the experimental data with the classical theory at very high supersaturations. Such a comparison would yield inaccurate values of interfacial tensions because of the inapplicability of the classical theory for the reasons stated above. The spherical amorphous cluster might be an appropriate model for low supersaturations since the critical sizes for such conditions are large and therefore it is likely that the cluster structure is amorphous. On the other hand, at very high supersaturations, the critical size is small. It is therefore likely that clusters organize in a more definite structure. If the structure of very small molecular clusters is fcc, and that of very large ones is amorphous, the actual nucleation rates should lie between the predicted curves for spherical amorphous and spherical fcc clusters coinciding with the former at low supersaturations and with the latter at high supersaturations. Experimental data could not be compared with the nucleation rates for spherical and fcc clusters because information regarding the intermolecular interaction parameters is not available.

### 1.2.9 Conclusions

A kinetic theory of nucleation in liquids is proposed that, in contrast to the conventional quasi-thermodynamic theory, calculates directly the dissociation rates of clusters in terms of the interaction potential. In addition, the macroscopic concept of interfacial tension is not employed in the present formalism. Expressions for the interaction potential of a surface monomer belonging to a cluster have been derived in terms of intermolecular interactions assumed as square well potentials for (1) spherical clusters with uniform distributions of molecules (amorphous) and (2) approximately spherical clusters with fcc packing. The rate of dissociation of a surface monomer from a cluster is calculated employing a first passage time analysis. A modified Kelvin equation is derived to relate the dependence of the solubility on the cluster size. The rate of dissociation of clusters is found to decrease with cluster size, its dependence being more pronounced at smaller sizes. The critical cluster size is found to be smaller for larger supersaturation ratios as well as for weaker monomer–monomer interactions. An expression for the rate of nucleation is derived in terms of molecular interactions by considering the growth of the clusters as a birth and death process. The predicted nucleation rates for spherical clusters with uniform distributions of molecules exhibits a gradual increase at very low supersaturations, increases catastrophically at a critical supersaturation, and tapers off gradually at high supersaturations. The region of initial gradual increase becomes narrower as the monomer–monomer interaction becomes stronger, eventually vanishing at sufficiently high interactions. In contrast, such a region of gradual increase is absent from the classical theory. The nucleation rates for spherical amorphous clusters are found to be about $10^{20}$ times larger than those in the classical theory with the critical supersaturation for the former being an order of magnitude smaller than that for the latter. The macroscopic concept of interfacial tension employed by the classical theory is believed to be valid only for extremely large clusters consisting of at least $10^6$ monomers. The classical theory overpredicts the interfacial tension and the free energy barrier to nucleation and consequently underpredicts the rate of nucleation. The nucleation rates for the spherical clusters with fcc packing are much smaller than those for spherical amorphous clusters. They exhibit a catastrophic increase at a critical supersaturation, tapering off at higher supersaturations. The catastrophic change decreases as the monomer–monomer interaction becomes stronger, eventually vanishing at very large interactions. The behavior of nucleation rates for spherical amorphous and that for fcc clusters are found to be qualitatively different. This shows that the nucleation rates are fairly sensitive to the structure of molecular clusters. Since the spherical amorphous cluster and a cluster with a definite structure such as fcc are probably appropriate models for large and small clusters, respectively, it is envisaged that the actual nucleation rate may lie between the predicted curves for spherical amorphous and fcc clusters, coinciding with the former at low supersaturation ratios and with the latter at high supersaturation ratios. In addition, it is expected that the rate of nucleation would be sensitive to the form of the interaction potential and calculations for other interaction potentials are now being carried out.

## References


1. Volmer, M., Weber, A., Z. *Phys. Chem* (*Leipzig*) 119, 277 (1926).
2. Farkas, L., Z *Phys. Chem.* (*Leipzig*) *Abt. A* 125, 236 (1927).
3. Becker, R., Döiring, W., *Ann. Phys.* (*Leipzig*) 24, 719 (1935).
4. Zeldovich, J.B., *Acta Physicochim.* (*URSS*) 18, 1 (1943).
5. Frenkel, J., *Kinetic Theory of Liquids*, Dover, New York, 1946.
6. Turnbull, D., Fisher, J.C., *J. Chem. Phys.* 17, 71 (1949).
7. Kuhrt, F., Z. *Phys.* 131, 185 (1952).
8. Kuhrt, F., Z. *Phys.* 131, 205 (1952).
9. Lothe, J., Pound, G.M., *J. Chem. Phys.* 36, 2080 (1962).
10. Reiss, H., *Adv. Colloid Interface Sci.* 7, 1 (1977).
11. Katz, J.L., Donohue, M.D., *Adv. Chem. Phys.* 40, 137 (1979).
12. Tolman, R.C., *J. Chem. Phys.* 17, 333 (1949).
13. Kirkwood, J.G., Buff, F.P., *J. Chem. Phys.* 17, 338 (1949).

14. Dave, J.V., Abraham, F., *Surf. Sci.* 26, 557 (1971).
15. Walton, A.G., in *Nucleation*, A.C. Zettlemoyer, Ed., Dekker, New York, 1969.
16. Chandrasekhar, S., *Rev. Mod. Phys.* 15, 1 (1943).
17. Gardiner, C.W., *Handbook of Stochastic Methods*, Springer-Verlag, New York/Berlin, 1983.
18. Goodrich, F.C., *Proc. R. Soc. London Ser. A* 277, 167 (1964).
19. Abraham, F., *Homogeneous Nucleation Theory*, Academic Press, New York, 1974.
20. Nielson, A.E., *Kinetics of Precipitation*, Pergamon, Elmsford, NY, 1964.

## 1.3 A Kinetic Theory of Nucleation in Liquids

*Eli Ruckenstein and Bogdan Nowakowski* *

A kinetic theory of nucleation is developed in which the rate of dissociation of the nuclei is calculated on the basis of a mean first passage time analysis. In contrast to the classical theory of nucleation, the theory does not use the macroscopic concept of interfacial tension and the principle of detailed balance. It involves, instead, the interaction between a molecule and a cluster. An equation is derived for the critical radius, which in the limit of large clusters yields the Kelvin equation as well as an expression for the macroscopic surface tension. Numerical calculations assuming an interaction potential between two molecules that combines the dispersive attraction with the rigid core repulsion are performed. The predictions of the theory are consistent with the classical results in the limit of large critical clusters, i.e., for small supersaturations. For small critical clusters, the present theory provides much higher rates of nucleation than the classical one. This difference can be attributed to the fact that the classical theory uses the macroscopic interfacial tension, which overpredicts the surface energy of small clusters, and consequently provides lower values for the nucleation rate. The present paper extends the approach initiated in a previous paper (G. Narsimhan and E. Ruckenstein, *J. Colloid Interface Sci.* 128, 549, 1989) to a more realistic interaction potential. (© *1990 Academic Press, Inc.*)

### 1.3.1 Introduction

The kinetics of the formation and growth of small nuclei of a new phase constitutes one of the main problems in the understanding of phase transitions. The classical approach to nucleation theory was initiated by Farkas [1] and subsequently developed by Becker and Döring [2], Zeldovich [3], and Frenkel [4]. The theory relies substantially on (1) the use of the principle of detailed balance and of a hypothetical equilibrium size distribution of clusters to relate the rate constants of dissociation and condensation, and (2) the assumption that the macroscopic concept of surface tension can be satisfactorily applied to estimate the free energy of formation of small clusters. Modifications and extensions of the classical theory have been presented by Kuhrt [5,6], Lothe and Pound [7], Reiss [8], and Katz and Donohue [9].

The use of macroscopic interfacial tension for small clusters is the main drawback of the classical theory of nucleation. In fact, it has been argued on the grounds of thermodynamics [10] as well as statistical mechanics [11] that the macroscopic surface tension may not be quite adequate for small clusters.

Recently, a kinetic theory of nucleation in a supersaturated solution, which does not use the macroscopic concept of surface tension, but rather is based on molecular interactions, was developed [12]. In addition, the rate of dissociation was calculated kinetically (and not quasithermodynamically as in the classical theory) from the mean passage time of the escaping molecules. For the sake of simplicity, the square-well interaction potential was used in the calculations. Unfortunately, there is an error in the calculations (see erratum [13]). In the present paper, a more realistic interaction potential is used and, in addition, the computational error of Ref. [12] is corrected.

The problem studied is the precipitation of a solute from a supersaturated solution. The main assumptions regarding the molecular interactions as well as the calculation of the interaction potential between a molecule and a cluster are presented in the next section. In the subsequent section, a kinetic equation is established for the mean passage time, and on this basis the rate of dissociation is calculated. Further, the rate of nucleation is obtained by means of a master equation for the particle size distribution function. Finally, the predictions of the theory are presented and compared with the results of the classical nucleation theory.

### 1.3.2 Interaction between a Molecule and a Cluster

As a model of the cluster, we adopt the approximation of an amorphous, continuously uniform solid. As was shown in the previous paper [12], the structure of the cluster substantially affects the rate of nucleation. For this reason, calculations are also carried out for fcc crystalline clusters as well as icosahedral configurations, the latter being known to be energetically preferred by small clusters [14,15].

* *J. Colloid Interface Sci.* 137, 583 (1990). Republished with permission.

The following form for the interaction potential between two molecules is assumed:

$$\phi_{12}(r_{12}) = \begin{cases} \infty & \text{for} \quad r_{12} < \eta \\ -\epsilon\left(\dfrac{\eta}{r_{12}}\right)^6 & \text{for} \quad r_{12} \geqq \eta, \end{cases} \tag{1.102}$$

where $r_{12}$ denotes the distance between the centers of the molecules. Equation 1.102 combines the rigid core repulsion of range $\eta$ and the London–van der Waals attraction, the strength of which is determined by the constant $\epsilon$. The effect of the presence of the surrounding liquid medium is accounted for by the appropriate estimation of the effective interaction constant $\epsilon$. The resultant interaction potential exerted by a cluster on a single molecule can be obtained by the standard method of integration assuming pairwise additivity of the interactions:

$$\phi(r) = \int_{v'} dr' \phi_{12}(|\mathbf{r} - \mathbf{r}'|)\rho(\mathbf{r}'), \tag{1.103}$$

where the coordinates are taken with respect to the center of the cluster, and $\rho(\mathbf{r})$ denotes the density of molecules at the given location. Integration in Equation 1.103 extends over the whole volume of the cluster, excluding the region prohibited by the hard core repulsion. The continuum approximation is adopted and the density $\rho$ of the molecules in the cluster is assumed constant. Only the surface layer and the space outside the particle are of interest in the considered problem, and Equation 1.102 yields the following result for the above regions:

$$\Phi(r) = \begin{cases} \dfrac{\pi\mu\epsilon}{12}\left(\dfrac{R}{r}\right)\left(3\left(\dfrac{r+R}{R}\right)\left(\dfrac{r-R}{\eta}\right) - 8\dfrac{r}{R} + 6\dfrac{\eta}{R} - \left(\dfrac{r+3R}{R}\right)\left(\dfrac{\eta}{r+R}\right)^3\right) & \\ & \text{for} \quad R \leqq r \leqq R+\eta \quad (1.104) \\ -\dfrac{4\pi\mu\epsilon}{3}\left(\dfrac{\eta}{r-R}\right)^3\left(\dfrac{R}{r+R}\right)^3 & \text{for} \quad r > R+\eta, \quad (1.105) \end{cases}$$

where $\mu = \rho\eta^3$ and $R$ is the radius of the spherical cluster.

According to Equations 1.104 and 1.105, the interaction potential splits into two distinct branches. In the surface layer, $R \leqq r \leqq R + \eta$, the interaction potential exhibits the steepest variation; this region presents the main barrier to be overpassed by the dissociating molecule. In the outer space, $r > R + \eta$, Equation 1.105 corresponding to the dispersive attraction between a molecule and a particle of radius $R$ is valid.

The total number of molecules $i$ contained in a nucleus is related to its radius by

$$i = \frac{4\pi\mu}{3}\left(\frac{R}{\eta}\right)^3, \tag{1.106}$$

and the number of surface molecules can be calculated from

$$\begin{aligned} N_s &= \frac{4\pi\rho}{3}\left(R^3 - (R-\eta)^3\right) \\ &= 4\pi\rho R^2\eta\left(1 - \frac{\eta}{R} + \frac{1}{3}\left(\frac{\eta}{R}\right)^2\right). \end{aligned} \tag{1.107}$$

One may note that $N_s$ is proportional to the surface of the cluster only in the limit of large particles.

### 1.3.3 Kinetics of Dissociation

Due to collisions with other molecules, a surface molecule of the cluster performs Brownian motion. As long as it remains within the potential well that exists in the surface layer, it sticks to the other molecules of the cluster and therefore can be regarded as belonging to it. However, as soon as it overcomes the potential well and reaches the outer boundary of the surface layer, $r = R + \eta$, it can be considered as dissociated. The dissociation process can be analyzed by means of the kinetic theory presented below.

The Brownian motion of a molecule is governed in general by the Fokker–Planck equation for the joint distribution function in the phase space [16,17]. However, in liquids the relaxation time for the velocity distribution function is extremely short and negligible in comparison to the characteristic time scale of the dissociation process. Under these conditions, the Fokker–Planck equation reduces to the Smoluchowski equation, which describes diffusion in an external field [16,17]. In the case of spherical symmetry, it has the form

$$\frac{\partial n(r,t)}{\partial t} = Dr^{-2}\frac{\partial}{\partial r}\left(r^2 e^{-\Phi(r)}\frac{\partial}{\partial r}e^{\Phi(r)}n(r,t)\right). \tag{1.108}$$

Here, $n(r, t)$ is the number density of the diffusing molecules, $D$ denotes the diffusion coefficient, and $\Phi = \phi/kT$, where $k$ is the Boltzmann constant and $T$ denotes the temperature of the medium in $K$.

The solution of Equation 1.108 for the initial condition $n(r, 0) = \delta(r - r_0)$ (where $\delta$ denotes the delta function), i.e., the Green's function, is the transition probability $p(r, t|r_0)$ that a molecule initially at the distance $r_0$ will be located after time $t$ at the point $r$. The mean passage time depends on the initial position $r_0$ of the molecule. It is therefore convenient to use the backward Smoluchowski equation [17–19], which provides the dependence of $p(r, t|r_0)$ on the initial position $r_0$:

$$\frac{\partial p(r,t|r_0)}{\partial t} = Dr_0^{-2}e^{\Phi(r_0)} \times \frac{\partial}{\partial r_0}\left(r_0^2 e^{-\Phi(r_0)}\frac{\partial}{\partial r_0}p(r,t|r_0)\right). \tag{1.109}$$

The probability that a molecule initially at point $r_0$ within the surface layer will remain in this region after time $t$ is given by the so-called survival probability,

$$Q(t|r_0) = \int_R^{R+\eta} r^2 \rho(r,t|r_0)\mathrm{d}r. \tag{1.110}$$

Consequently, the probability that the dissociation time is between 0 and $t$ is $1 - Q(t|r_0)$, and the probability density for the dissociation time is given by $-\partial Q/\partial t$. The mean passage time is given by the first moment,

$$\begin{aligned}\tau(r_0) &= -\int_0^\infty t\frac{\partial Q(t|r_0)}{\partial t}\mathrm{d}t \\ &= \int_0^\infty Q(t|r_0)\mathrm{d}t.\end{aligned} \tag{1.111}$$

The equation for $\tau(r_0)$ is conventionally obtained by integrating Equation 1.109 over $r$ and $t$ over the entire range and using the boundary conditions $Q(0|r_0) = 1$ and $Q(t|r_0) \to 0$ as $t \to \infty$ for any $r_0$. This yields [17–21]

$$-Dr_0^{-2}e^{\Phi(r_0)}\frac{\partial}{\partial r_0}\left(r_0^2 e^{-\Phi(r_0)}\frac{\partial}{\partial r_0}\tau(r_0)\right) = 1. \tag{1.112}$$

The boundary conditions for $\tau(r_0)$ follow from the boundary conditions imposed on the function $p(r, t|r_0)$. Assuming a reflecting inner wall of the surface layer, one obtains

$$\left.\frac{d\tau}{dr}\right|_{r=R} = 0. \quad (1.113)$$

In contrast, the molecules reaching the outer boundary of the layer separate from the cluster and diffuse further in the outer space. The diffusive flux in the external field can be obtained from Equation 1.108 and is given by

$$j(r,t) = -De^{\Phi(r)}\frac{\partial}{\partial r}e^{\Phi(r)}n(r,t). \quad (1.114)$$

Assuming quasi steady state and integrating with the boundary condition $n(r) \to 0$ as $r \to \infty$, Equation 1.114 yields

$$j(r) = \frac{De^{\Phi(r)}}{\int_r^\infty x^{-2}e^{\Phi(x)}\,dx}r^{-2}n(r). \quad (1.115)$$

Equation 1.115 for $r = R + \eta$ becomes the radiation boundary condition

$$j(R + \eta) = k_0 n\,(R + \eta), \quad (1.116)$$

where

$$k_0 = \frac{De^{\Phi(R+\eta)}}{\int_{R+\eta}^\infty x^{-2}e^{\Phi(x)}\,dx}(R+\eta)^{-2}. \quad (1.117)$$

Equation 1.116 leads to the following boundary condition for the mean passage time:

$$\left.\frac{d\tau}{dr}\right|_{r=R+\eta} = -\frac{k_0}{D}\tau(R+\eta). \quad (1.118)$$

The boundary condition for perfect absorption (used in Ref. [12]) is recovered from Equation 1.118 in the limit of $k_0 \to \infty$.

The solution of Equation 1.112 satisfying boundary conditions 1.113 and 1.118 has the form

$$\tau(r_0) = \frac{1}{D}\int_{r_0}^{R+\eta}\frac{dy}{p_{eq}(y)}\int_R^y dx\,p_{eq}(x) + [k_0 p_{eq}(R+\eta)]^{-1}. \quad (1.119)$$

Here, the equilibrium probability distribution $p_{eq}$, is given by

$$p_{eq}(x) = x^2 e^{-\Phi(x)}/Z, \quad (1.120)$$

where $Z$ is the partition function,

$$Z=\int_{R}^{R+\eta} x^2 e^{-\Phi(x)}\,\mathrm{d}x. \tag{1.121}$$

Equation 1.119 provides the mean passage time for a specific initial position $r_0$. The average mean passage time is obtained from Equation 1.119 after integration over the initial distribution function. Since the relaxation time to achieve equilibrium is much shorter than the dissociation time, equilibrium distribution is assumed for $r_0$. The average mean passage time is therefore given by

$$\begin{aligned}\langle\tau\rangle &= \int_{R}^{R+\eta} \tau(r_0)p_{\mathrm{eq}}(r_0)\mathrm{d}r_0\\ &=\frac{1}{D}\int_{R}^{R+\eta}\mathrm{d}r_0 p_{\mathrm{eq}}(r_0)\int_{r_0}^{R+\eta}\frac{\mathrm{d}y}{p_{\mathrm{eq}}(y)}\times\int_{R}^{y}\mathrm{d}x p_{\mathrm{eq}}(x)+[k_0 p_{\mathrm{eq}}(R+\eta)]^{-1}.\end{aligned} \tag{1.122}$$

The rate of dissociation of the surface molecules is related to $\langle\tau\rangle$ via the kinetic law of the first-order reactions

$$\begin{aligned}\alpha &= \frac{N_s}{\langle\tau\rangle}\\ &= \frac{4\pi\rho R^2\eta\left(1-\frac{\eta}{R}+\frac{1}{3}\left(\frac{\eta}{R}\right)^2\right)}{\frac{1}{D}\int_{R}^{R+\eta}\mathrm{d}r_0 p_{\mathrm{eq}}(r_0)\int_{r_0}^{R+\eta}\frac{\mathrm{d}y}{p_{\mathrm{eq}}(y)}}\times\int_{R}^{y}\mathrm{d}x p_{\mathrm{eq}}(x)+[k_0 p_{\mathrm{eq}}(R+\eta)]^{-1}.\end{aligned} \tag{1.123}$$

Equation 1.123 constitutes the main result of the theory. The rate of dissociation depends on the form of the intermolecular potential and on the size of the cluster.

### 1.3.4 Rate of Condensation

The rate of condensation on the surface of the cluster is controlled by the diffusion process [22]. The expression for the condensation rate $\beta$ can be obtained by means of the stationary solution of the diffusion equation in an external field, Equation 1.108, and has the form

$$\beta=\frac{4\pi D n_0}{\int_{R+\eta}^{\infty} r^{-2}e^{\Phi(r)}\,\mathrm{d}r}=\gamma 4\pi D R n_0, \tag{1.124}$$

where $n_0$ denotes the number density of the condensing molecules in solution at an infinite distance from the cluster.

According to Equation 1.124, the rate of condensation is affected by the presence of the potential field $\Phi$. Table 1.1 lists some values of $\gamma$. In the limit $\eta/R\rightarrow 0$, the factor $\gamma$ tends to unity and Equation 1.124 leads to the well-known result for the condensation rate in the absence of external field [22].

Table 1.1 Values of Factor $\gamma$ in Equation 1.124, Calculated for Various Values of $\eta/R$, and for $\epsilon/kT = 4$ and $\mu = 1.2$ in Equation 1.105

| $\eta/R$ | $\gamma$ |
|---|---|
| 0.5 | 1.65 |
| 0.2 | 1.30 |
| 0.1 | 1.16 |
| 0.05 | 1.086 |
| 0.02 | 1.036 |
| 0.01 | 1.018 |
| 0.005 | 1.0096 |
| 0.002 | 1.0037 |
| 0.001 | 1.0018 |

### 1.3.5 Critical Nucleus

The clusters change their size as a result of two concurrent processes: dissociation and condensation. For a given concentration of the solute, the rate of condensation increases with the radius of the particle. In contrast, the dissociation rate is higher for smaller nuclei, because the well of the interaction potential is less deep. A critical cluster size is determined by the condition of (unstable) equilibrium, for which the rates of condensation and dissociation are balanced, $\alpha_{i^*} = \beta_{i^*-1}$. Using Equations 1.123 and 1.124, the following relation for the critical radius $R_*$ is obtained:

$$\frac{4\pi\rho R_\star^2\eta\left(1-\frac{\eta}{R_\star}+\frac{1}{3}\left(\frac{\eta}{R_\star}\right)^2\right)}{\frac{1}{D}\int_{R_*}^{R_*+\eta}\mathrm{d}r_0 p_{\mathrm{eq}}(r_0)\int_{r_0}^{R_*+\eta}\frac{\mathrm{d}y}{p_{\mathrm{eq}}(y)}}\times\int_{R_*}^{y}\mathrm{d}x p_{\mathrm{eq}}(x)+[k_0 p_{\mathrm{eq}}(R_\star+\eta)]=\gamma 4\pi D R_\star n_0. \tag{1.125}$$

As a result of the unstable thermodynamic conditions, clusters greater than the critical one grow, whereas those smaller diminish in size.

### 1.3.6 Limit of Large Clusters and Thermodynamic Relations

Equation 1.125 has a rather complicated form and the detailed calculations must be performed numerically. However, some analytical results can be obtained in the limit of large clusters, $\eta/R_* \to 0$. In this limit, one can recover some classical thermodynamic formulas and one can thus compare the present approach and the conventional theory of nucleation.

As the concentration of the solute in the liquid diminishes, the radius of the critical cluster increases. The solubility can be defined as the saturation number density of solute that coexists with particles of infinite radius or simply with a bulk condensed material. This asymptotic limit can be calculated from Equations 1.123 through 1.125. Expanding Equation 1.125 with respect to $\eta/R_*$ and retaining the first two terms, one obtains

$$\frac{4\pi\rho R_\star D\left(1-\frac{\eta}{R_\star}+\frac{1}{3}\left(\frac{\eta}{R_\star}\right)^2\right)}{e^{4A/3}(1-e^{-A})A^{-1}+\frac{\eta}{R_\star}\left(3(1-(1+A)e^{-A})A^{-2}-e^{4A/3}+e^{A/3}-\frac{1}{2}e^{-A}+\frac{2(\sinh A-A)}{(1-e^{-A})A^2}\right)}=4\pi n_0 R_\star D, \tag{1.126}$$

where

$$A=\mu\pi\epsilon/2kT. \tag{1.127}$$

The details of the derivation of Equation 1.126 are presented in Appendix 1A.

The equation for the saturation number density $n_s$ is obtained from Equation 1.126 in the limit $\eta/R \to 0$, i.e., for large clusters. This yields

$$n_s = \frac{\rho A}{(1-e^{-A})} e^{-4A/3}. \tag{1.128}$$

Using Equation 1.128, the following relation is obtained from Equation 1.126:

$$s = 1 + \left( A - 1 - \frac{3}{A} + \frac{3 + \frac{1}{2}A}{e^A - 1} - \frac{2(\sinh A - A)}{A(1-e^{-A})^2} e^{-4A/3} \right) \frac{\eta}{R_*}, \tag{1.129}$$

where $s = n_0/n_s$ is the supersaturation ratio.

Equation 1.129 corresponds to the classical Kelvin relation for large clusters,

$$s = \exp\left( \frac{2\sigma}{\rho kTR_*} \right) \approx 1 + \frac{2\sigma}{\rho kTR_*}. \tag{1.130}$$

The following formula for the macroscopic surface tension $\sigma$ can be derived by comparing Equations 1.129 and 1.130:

$$\sigma = \frac{1}{2} kT\rho\eta \left( A - 1 - \frac{3}{A} + \frac{3 + \frac{1}{2}A}{e^A - 1} - \frac{2(\sinh A - A)}{A(1-e^{-A})^2} e^{-4A/3} \right). \tag{1.131}$$

Equation 1.131 will be useful in the comparison between the present and the classical theory of nucleation.

### 1.3.7 Rate of Nucleation

The expressions derived for the rates of condensation and dissociation can be used to write equations for populations of clusters of various sizes [23,24]. The flux of clusters passing from the population of $(i-1)$-mers to $i$-mers along the coordinate representing the size of the clusters has the form

$$I_i = \beta_{i-1} f_{i-1} - \alpha_i f_i, \tag{1.132}$$

where $f_i$ denotes the population of clusters consisting of $i$ molecules. The rate of change of $f_i$ is given by the equation

$$\begin{aligned} \frac{df_i}{dt} &= -(I_{i+1} - I_i) \\ &= \beta_{i-1} f_{i-1} - (\alpha_i + \beta_i) f_i + \alpha_{i+1} f_{i+1}. \end{aligned} \tag{1.133}$$

Assuming a relatively smooth dependence of $f_i$ on $i$, the finite difference equation (Equation 1.133) can be easily rewritten into a continuum form. After expanding Equation 1.133 around the midpoint $i$, one obtains

$$\begin{aligned} \frac{\partial f(g,t)}{\partial t} &= -\frac{\partial I(g,t)}{\partial g} \\ &= \frac{1}{2} \frac{\partial^2}{\partial g^2} (\beta + \alpha) f(g,t) - \frac{\partial}{\partial g} (\beta - \alpha) f(g,t), \end{aligned} \tag{1.134}$$

where the continuum variable $g$ replaces the discrete variable $i$.

The rate of nucleation is given by the flux of clusters along the coordinate $g$, and has the form

$$I(g,t) = -\frac{1}{2}\frac{\partial}{\partial g}(\beta+\alpha)f(g,t) + (\beta-\alpha)f(g,t). \tag{1.135}$$

Before the nucleating clusters become stable, they must pass through the bottleneck around the critical cluster size $g_*$, where the rates of formation and dissociation are comparable. The time scales of formation and dissociation of the smaller clusters are much shorter than the characteristic times for formation and growth of the critical cluster. Therefore, one can expect that the process of nucleation would attain a quasi steady state after a short lapse of time. Using the boundary condition $f(g) \to 0$ as $g \to \infty$, the stationary solution of Equations 1.134 and 1.135 yields the following expression for the rate of nucleation:

$$I = \frac{\frac{1}{2}\beta_0 n_0}{\int_0^\infty \exp[-2w(g)]\mathrm{d}g}, \tag{1.136}$$

where

$$w(g) = \int_0^g \frac{\beta-\alpha}{\beta+\alpha}\mathrm{d}g. \tag{1.137}$$

Since the function $w(g)$ exhibits a rather sharp minimum at the critical cluster size $g = g_*$, the integral in Equation 1.136 can be evaluated by the method of steepest descent. One arrives then to the following final result for the rate of nucleation:

$$I = \frac{1}{2}\beta_0 n_0 \left(\frac{w''(g_*)}{\pi}\right)^{1/2} \exp[2w(g_*)]. \tag{1.138}$$

At this point, we write the result of the classical theory of nucleation [22,23]:

$$I_{\mathrm{cl}} = \frac{1}{2\pi}\beta_0 n_0 \left(\frac{kT}{2\sigma\eta^2}\right)^{1/2} \times \ln(s)\exp\left(-\frac{16\pi\sigma^3}{3(kT)^3\rho^2\ln^2 s}\right). \tag{1.139}$$

It is expected that Equations 1.138 and 1.139 will approach one another in the limit of large critical clusters, i.e., for small supersaturations.

### 1.3.8 Numerical Results and Comparison with the Classical Theory

Numerical calculations were performed to obtain the rates of dissociation and condensation from Equations 1.123 and 1.124, respectively, the radius of the critical cluster according to Equation 1.125, and the rate of nucleation as given by Equation 1.138.

In the calculations, the parameter $\mu = \rho\eta^3$ was taken equal to 1.2, corresponding to the random close packing of spheres [25]. Various values of the parameter $E = \epsilon/kT$ were chosen, so as to cover the range of the most typical values of the surface tension, according to Equation 1.131.

Figure 1.13 presents the dependencies of the dissociation and condensation rates on the relative radius of the cluster $R/\eta$, for $E = \epsilon/kT = 4$. The rate of dissociation initially decreases rapidly for small clusters, passes through a minimum, and then grows slowly. In contrast, the rate of condensation steadily increases as the size of the particle becomes larger. The two curves always intersect if the solute is supersaturated, $s > 1$. This provides the critical radius of the cluster, which decreases with increasing supersaturation.

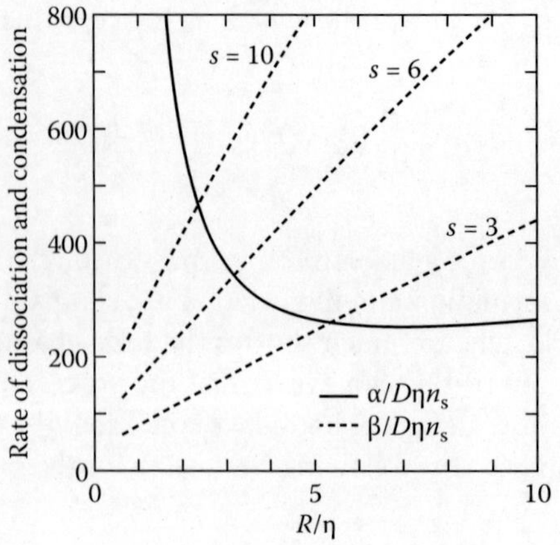


Figure 1.13

Rate of dissociation $\alpha/D\eta n_s$, and rate of condensation $\beta/D\eta n_s$ for various supersaturation ratios, $s$, against the radius $R/\eta$ of the cluster. The value $E=\epsilon/kT=4$ was chosen in the calculations.

The variation of the critical radius $R_*/\eta$ with the supersaturation ratio is presented in Figure 1.14, which shows that the critical radius is greater for stronger molecular interactions.

Figure 1.15 presents a plot of the nucleation rate against the supersaturation ratio, for various values of the parameter $E=\epsilon/kT$, which characterizes the strength of the interactions. For a given supersaturation, the rate of nucleation is smaller for stronger interactions, a result that is consistent with the predictions of the classical theory.

Figure 1.16 compares the nucleation rates calculated on the basis of the present and the classical theory of nucleation. The value $E=\epsilon/kT=4$ was chosen, since it provides the right order of magnitude for the interfacial tension. One may observe that both curve approach one another in the limit of small supersaturations, or equivalently, for large critical clusters. In contrast, for larger supersaturations the difference between the two theories becomes substantial, with the present theory providing much larger values for the rate of nucleation. For example, for $\ln s = 3$ ($s \cong 20$) calculations indicate that $I/I_{cl} \cong 10^{20}$. It is relevant to note that experiment shows [26] that appreciable nucleation occurs at supersaturation ratios smaller than those predicted by the classical theory. While this may be due to the involvement of heterogeneous nucleation, another explanation is also possible. Indeed, the above difference could also be attributed to the fact that the classical theory overpredicts the interfacial tension and the free energy of formation of the small clusters and, consequently, grossly underpredicts the rate of nucleation.

The present approach differs substantially from that employed in the classical theory of nucleation. However, a number of approximations are still involved, such as the assumption of an amorphous cluster

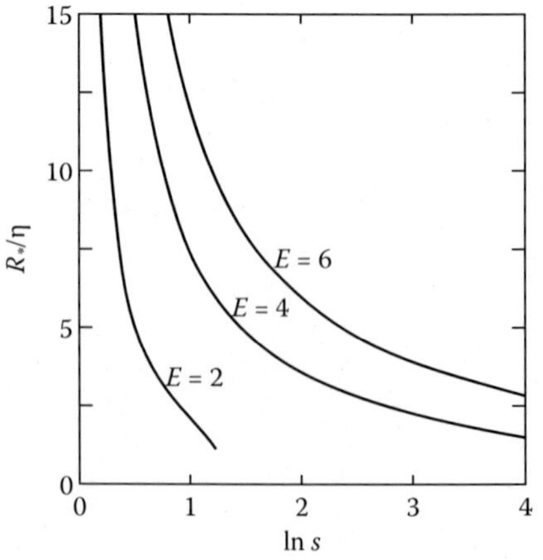


Figure 1.14

Radius of the critical cluster $R_*/\eta$ as a function of the supersaturation ratio for various values of $E=\epsilon/kT$.

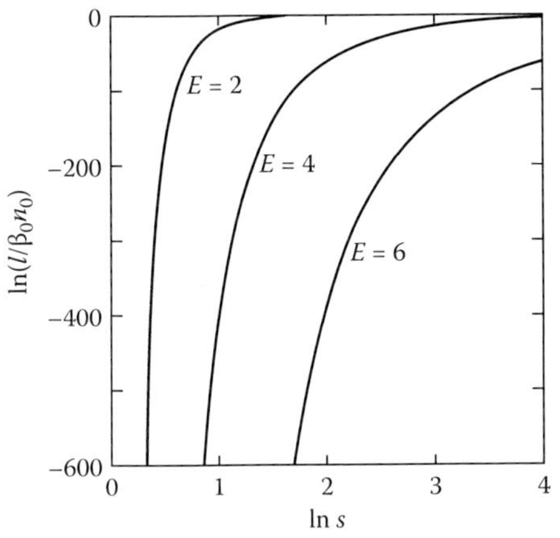


Figure 1.15

Rate of condensation versus the supersaturation ratio, calculated for various values of $E = \epsilon/kT$.

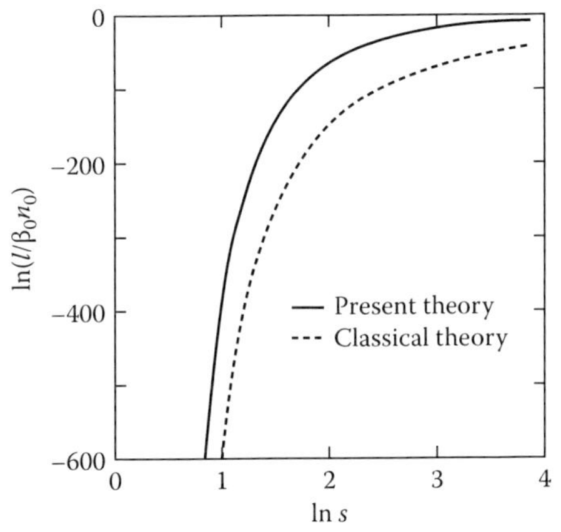


Figure 1.16

Rate of nucleation predicted by the present theory, Equation 1.138, and by the classical theory, Equation 1.139, plotted versus supersaturation ratio and for $E = \epsilon/kT = 4$.

and the approximation of the free energy of formation of the cluster by the internal energy. A detailed comparison with experiment is delayed until additional improvements of the theory have been made, the main goal of the present paper being to illustrate the basics of the novel kinetic theory.

### 1.3.9 Conclusions

A theory of nucleation in which the rate of dissociation is calculated on the basis of a mean passage time analysis is developed. In contrast to the classical, quasithermodynamic theory of nucleation, the concept of interfacial tension is not employed, but instead the interaction of a molecule with a cluster is taken into account. General equations have been derived for (1) the rate of dissociation of molecules from the cluster, (2) the radius of the critical cluster, and (3) the rate of nucleation.

Numerical calculations using an interaction potential between two molecules that combines the rigid core repulsion with the dispersive attraction between molecules have been performed. The classical Kelvin relation for the radius of the critical cluster is recovered in the limit of large clusters. Accordingly, the present and classical theories of nucleation agree in the limit of small super saturations, i.e., for large critical clusters. In contrast, the present theory predicts a much higher rate of nucleation for large supersaturations (by a factor of about $10^{20}$) than does the classical theory. This difference is caused by the use of the macroscopic value for the interfacial tension in the classical theory, which overpredicts the free energy of formation of small clusters. Consequently, the classical theory overpredicts the critical cluster size and leads to much smaller values for the rate of nucleation.

## Appendix 1A: Asymptotic Expansion of the Equation for Radius of Critical Cluster (Equation 1.125)

Equation 1.125 can be rearranged in the form

$$\frac{4\pi\rho R_*^2\eta\left(1-\frac{\eta}{R_*}+\frac{1}{3}\left(\frac{\eta}{R_*}\right)^2\right)}{X_1+X_2}=4\pi n_0 R_* D, \tag{1A.1}$$

where

$$X_1=\frac{\int_{R_*}^{R_*+\eta} dr_0 r_0^2 e^{-\Phi(r_0)}\int_{r_0}^{R_*+\eta} dy y^{-2} e^{\Phi(y)}\int_{R_*}^{y} dx x^2 e^{-\Phi(x)}}{D\gamma\int_{R_*}^{R_*+\eta} dx x^2 e^{-\Phi(x)}} \tag{1A.2}$$

and

$$X_2=\frac{\int_{R_*}^{R_*+\eta} dx x^2 e^{-\Phi(x)}}{DR_*}. \tag{1A.3}$$

Introducing the new variables $u$ and $v$ by means of the equations

$$x = R_* + \eta u$$

and

$$y = R_* + \eta u, \tag{1A.4}$$

Equations 1A.2 and 1A.3 become

$$X_1=\frac{\eta^2}{D}\frac{\int_0^1 dv\left(1+\frac{\eta}{R_*}\upsilon\right)^{-2} e^{\Phi(\upsilon)}\left(\int_0^{\upsilon} du\left(1+\frac{\eta}{R_*}u\right)^2 e^{-\Phi(u)}\right)^2}{\gamma\int_0^1 du\left(1+\frac{\eta}{R_*}u\right)^2 e^{-\Phi(u)}} \tag{1A.5}$$

and

$$X_2=\frac{R_*\eta}{D}\int_0^1 du\left(1+\frac{\eta}{R_*}u\right)^2 e^{-\Phi(u)}. \tag{1A.6}$$

In terms of the new variables, the interaction potential acquires the form

$$\Phi(u)=\frac{A}{6\left(1+\frac{\eta}{R_{\star}}u\right)}\left(3\left(2+\frac{\eta}{R_{\star}}u\right)u+6\frac{\eta}{R_{\star}}-8\left(1+\frac{\eta}{R_{\star}}u\right)-\left(\frac{\eta}{R_{\star}}\right)^{3}\frac{u+3}{(u+1)^{3}}\right), \tag{1A.7}$$

where $A=\mu\pi\epsilon/2kT$. Equations 1A.5 and 1A.6 are convenient to examine the limit $\eta/R_{\star}\to 0$. The interaction potential $\Phi(u)$ remains a well-behaved function for $\eta/R_{\star}=0$, and the integrals of Equations 1A.5 and 1A.6 converge in this limit to finite values. We confine ourselves to the first two terms in the expansion of Equation 1.125 with respect to $\eta/R_{\star}$. One may observe that $X_1$ is one order higher in $\eta/R_{\star}$ than $X_2$. Since $\gamma = 1$ for $\eta/R_{\star}\to 0$, one obtains in this limit

$$X_1=\frac{\eta^2}{D}\frac{(\sinh A-A)}{(1-e^{-A})A^2}. \tag{1A.8}$$

Expanding $X_2$ in series yields

$$X_2=\frac{R_{\star}\eta}{D}\left(e^{4A/3}(1-e^{-A})A^{-1}+\frac{\eta}{R_{\star}}\left(3(1-(A+1)e^{-A})A^{-2}-e^{4A/3}+e^{4/3}-\frac{1}{2}e^{-A}\right)\right). \tag{1A.9}$$

Introducing Equations 1A.8 and 1A.9 in the equation for the critical radius (Equation 1A.1), one arrives finally to Equation 1.126.

## References


1. Farkas, L., Z. *Phys. Chem. Leipzig Abt. A* 125, 236 (1927).
2. Becker, R., Döring, W., *Ann. Phys. Leipzig* 24, 719 (1935).
3. Zeldovich, J.B., *Acta Physicochim. URSS* 18, 1 (1943).
4. Frenkel, J., *Kinetic Theory of Liquids*, Dover, New York, 1946.
5. Kuhrt, F., Z. *Phys.* 131, 185 (1952).
6. Kuhrt, F., Z. *Phys.* 131, 205 (1952).
7. Lothe, J., Pound, G.M., *J. Chem. Phys.* 36, 2080 (1962).
8. Reiss, H., *Adv. Colloid Interface Sci.* 7, 1 (1977).
9. Katz, J.L., Donohue, M.D., *Adv. Chem. Phys.* 40, 137 (1979).
10. Tolman, R.C., *J. Chem. Phys.* 17, 333 (1949).
11. Kirkwood, J.G., Buff, F.P., *J. Chem. Phys.* 17, 338 (1949).
12. Narsimhan, G., Ruckenstein, E., *J. Colloid Interface Sci.* 128, 549 (1989). (Section 1.2 of this volume.)
13. Narsimhan, G., Ruckenstein, E., Erratum *J. Colloid Interface Sci.* 132, 289 (1989).
14. Hoare, M.R., *Adv. Chem. Phys.* 40, 49 (1979).
15. Burton, J.J., *Nature* (*London*) 229, 335 (1971).
16. Chandrasekhar, S., *Rev. Mod. Phys.* 15, 1 (1949).
17. Gardiner, C.W., *Handbook of Stochastic Methods*, Springer, New York, 1983.
18. Agmon, N., *J. Chem. Phys.* 81, 3644 (1984).
19. Weiss, G.H., *Adv. Chem. Phys.* 13, 1 (1967).
20. Weiss, G.H., *J. Stat. Phys.* 42, 1 (1986).
21. Deutch, J.M., *J. Chem. Phys.* 73, 4700 (1980).
22. Walton, A.G., in *Nucleation*, A.C. Zettlemoyer, Ed., Dekker, New York, 1969.
23. Abraham, F., *Homogeneous Nucleation Theory*, Academic Press, New York, 1974.
24. Goodrich, F.C., *Proc. R. Soc. London A* 277, 167 (1964).
25. Kittel, C., *Introduction to Solid State Physics*, Wiley, New York, 1976.
26. Nielson, A.E., *Kinetics of Precipitation*, Pergamon, Elmsford, NY, 1964.

## 1.4 Rate of Nucleation in Liquids for Face-Centered Cubic and Icosahedral Clusters

*Bogdan Nowakowski and Eli Ruckenstein**

A recent theory developed by the authors (Ruckenstein, E. and Nowakowski, B., *J. Colloid Interface Sci.*, 137, 583 [1990]) was used to predict the rate of nucleation for approximately spherical face-centered cubic (fcc) crystalline and icosahedral clusters. The critical radius and the nucleation rate were calculated as a function of the supersaturation ratio for clusters constructed by adding complete layers of atoms successively to a central atom. It was found that the results for the fcc clusters are not smooth, monotonic functions of supersaturation, but exhibit an irregular behavior. This feature is explained in terms of the irregular dependence of the mean interaction potential between a surface atom and a cluster on the cluster size. The corresponding plots for the icosahedral clusters are monotonic, since layer-by-layer addition preserves the elements of symmetry of the atoms on the surface of the clusters. Irregularities are expected, however, to occur for both structures when the clusters are formed by adding single atoms instead of complete layers. Experimental results indicate that the predicted nonmonotonic behavior is possible. (© *1990 Academic Press, Inc.*)

### 1.4.1 Introduction

The investigation of formation and growth of small clusters is an essential part of nucleation theory. The classical theory [1–4] treats this problem by (1) assuming the clusters to be amorphous droplets and using the macroscopic concept of surface tension to calculate the free energy of small clusters and (2) using the principle of detailed balance and a hypothetical equilibrium size distribution of particles to relate the rates of dissociation and condensation.

Lately, a kinetic approach to the theory of nucleation in liquids was developed [5,6] that avoids the macroscopic concept of surface tension and in which the rate of dissociation was calculated on the basis of the mean passage time of escaping molecules. An amorphous, continuous particle was previously employed as a model of the cluster. In the present paper, this assumption is relaxed by considering clusters with the fcc and the icosahedral configurations. The fcc structure is frequently encountered in bulk atomic crystals [7], but the icosahedral configurations are recognized to be energetically preferred by the small aggregates of atoms [8,9]. In fact, comprehensive numerical calculations [10] indicate that the icosahedral clusters containing as many as about $10^4$ atoms are still energetically more stable than the fcc structure. However, fcc (and even bcc) clusters of much smaller sizes were detected experimentally in beams of clusters [11] as well as by the numerical simulations of the nucleation process by molecular dynamics [12–15]. In the present paper, both structures are therefore considered; thus, the effect of the discrete nature of the nuclei is taken into account.

The following section summarizes the main results of the previously developed theory of nucleation [5,6]. In the subsequent section, details regarding the construction of the fcc and the icosahedral clusters are presented, and the interaction between a cluster and its surface atom is calculated. Finally, the rates of nucleation are calculated and compared with those obtained for the amorphous cluster as well as with the predictions of the classical theory.

### 1.4.2 Theory of Nucleation—Mean Passage Time Approach

We summarize here the main results of the previously presented theory [6], which is based on the calculation of the mean passage time for the atoms dissociating from the cluster.

We consider a spherical cluster immersed in a liquid medium. A surface atom of the cluster performs Brownian motion in the potential well generated in the surface layer by the interactions of the atom with the remainder of the cluster. An atom can be regarded as dissociated when it passes over the outer boundary of the well. The rate of dissociation is related to the mean passage time for this threshold distance. The Brownian motion of an atom is governed by the Fokker–Planck equation, which under the present conditions can be reduced to the Smoluchowski equation for diffusion in a potential field. The solution of the Smoluchowski equation for the delta functions as the initial condition is the probability $p(r,t|r_0)$ of transition from the initial distance $r_0$ to the final distance $r$. Since the mean passage time depends on the initial position $r_0$, it is convenient to employ the backward Smoluchowski equation [6,16], which expresses the dependence of the transition probability $p(r,t|r_0)$ on $r_0$:

* *J. Colloid Interface Sci.* 139, 500 (1990). Republished with permission.

$$\frac{\partial p(r,t|r_0)}{\partial t} = Dr_0^{-2}e^{\Phi(r_0)}\frac{\partial}{\partial r_0}\left(r_0^2 e^{-\Phi(r_0)}\frac{\partial}{\partial r_0}p(r,t|r_0)\right). \tag{1.140}$$

Here, $D$ is the diffusion coefficient; $\Phi = \phi/k_B T$, where $\phi$ is the potential field in which the atom moves; $k_B$ is the Boltzmann constant; and $T$ denotes the temperature in K. The mean passage time is related to the rate of change of the probability of finding the atom within the well via the expression

$$\tau(r_0) = -\int_0^t \mathrm{d}t\left(t\frac{\partial}{\partial t}\int_R^{R+\eta} r^2 p(r,t|r_0)\mathrm{d}r\right). \tag{1.141}$$

In Equation 1.141, $R$ is the radius of the cluster and $\eta$ is the width of the well taken equal to the diameter of an atom. The equation for the mean passage time, derived from the backward Smoluchowski equation, has the form [6,16]

$$-Dr_0^{-2}e^{\Phi(r_0)}\frac{\partial}{\partial r_0}\left(r_0^2 e^{-\Phi(r_0)}\frac{\partial}{\partial r_0}\tau(r_0)\right) = 1. \tag{1.142}$$

The solution to Equation 1.142 is obtained by assuming a reflecting inner boundary of the well ($r_0 = R$) and a radiation boundary condition at the outer boundary of the well ($r_0 = R + \eta$). One thus obtains for the mean passage time,

$$\tau(r_0) = \frac{1}{D}\int_{r_0}^{R+\eta}\frac{\mathrm{d}y}{p_{\mathrm{eq}}(y)}\int_R^y \mathrm{d}x p_{\mathrm{eq}}(x) + [kp_{\mathrm{eq}}(R+\eta)]^{-1}, \tag{1.143}$$

where $p_{\mathrm{eq}}$ is the equilibrium probability distribution within the well and $k$ denotes the coefficient

$$k = \frac{De^{\Phi(R+\eta)}}{\int_{R+\eta}^{\infty} x^{-2}e^{\Phi(x)}\mathrm{d}x}(R+\eta)^{-2}. \tag{1.144}$$

The rate of dissociation $\alpha$ can be expressed in terms of $\tau(r_0)$ as

$$\alpha = \frac{N_s}{\tau(r_0)} = \frac{DN_s}{\int_R^{R+\eta}\mathrm{d}y p_{\mathrm{eq}}^{-1}(y)\int_R^y \mathrm{d}x p_{\mathrm{eq}}(x) + D[kp_{\mathrm{eq}}(R+\eta)]^{-1}}, \tag{1.145}$$

where $N_s$ denotes the number of surface atoms. The dissociating atoms are assumed to be positioned initially on the surface of the cluster ($r_0 = R$).

The rate of condensation in liquids is diffusion controlled and is given by the equation

$$\beta = \frac{4\pi D n_0}{\int_{R+\eta}^{\infty} r^{-2}e^{\Phi(r)}\mathrm{d}r}, \tag{1.146}$$

where $n_0$ is the number density of atoms of condensing species at a large distance from the cluster.

For a given supersaturation, there exists a critical cluster for which the rates of dissociation and condensation are equal. From the relation thus obtained for the radius of the critical cluster $R_*$, the saturation number density of solute $n_s$ can be calculated in the limit of large clusters [6]:

$$n_s = \frac{N_s}{4\pi R_*^2 \int_{R_*}^{R_*+\eta} e^{-\Phi(r)}\,dr} (R_*/\eta \gg 1). \tag{1.147}$$

For $R_*/\eta \gg 1$, the result of this theory is expected to coincide with the classical result. In particular, the classical Kelvin relation between the supersaturation ratio and the critical radius is recovered and can be employed to calculate the macroscopic surface tension [6].

The rates of dissociation and condensation govern the evolution of the size distribution function of the clusters,

$$\frac{df_i}{dt} = \beta_{i-1} f_{i-1} - (\alpha_i + \beta_i) f_i + \alpha_{i+1} f_{i+1}, \tag{1.148}$$

where $f_i$ is the population of clusters containing $i$ atoms. The rate of nucleation follows from the stationary solution of Equation 1.148 and is given by

$$I = \frac{\frac{1}{2}\beta_0 n_0}{\int_0^\infty \exp[-2w(g)]dg} \cong \frac{1}{2}\beta_0 n_0 \left(\frac{w''(g_*)}{\pi}\right)^{1/2} \exp[2w(g_*)], \tag{1.149}$$

where the continuum variable $g$ corresponds to the discrete number $i$ of atoms in a cluster, and

$$w(g) = \int_0^g \frac{\beta - \alpha}{\beta + \alpha} dg. \tag{1.150}$$

The classical theory of nucleation provides the expression

$$I_{cl} = \frac{1}{2\pi}\beta_0 n_0 \left(\frac{k_B T}{2\sigma\eta^2}\right)^{1/2} \times \ln(s) \exp\left(-\frac{16\pi\sigma^3}{3(k_B T)^3 \rho^2 \ln^2 s}\right), \tag{1.151}$$

where $s$ is the supersaturation ratio, $\sigma$ is the macroscopic surface tension, and $\rho$ is the number density of atoms in the clusters. Both theories are expected to coincide in the limit of large clusters, i.e., for small supersaturations.

### 1.4.3 Structure of Clusters

Approximately spherical fcc clusters are constructed by adding successively complete layers of atoms around the central atom of a cluster. The atoms of diameter $\eta$ occupy discrete lattice sites, whose coordinates are expressed as integers multiplied by the distance $\xi = \eta/\sqrt{2}$, i.e., $x = i_x\xi$, $y = i_y\xi$ and $z = i_z\xi$. The $n$th layer consists of atoms whose coordinates satisfy the equation [17]

$$i_x^2 + i_y^2 + i_z^2 = 2n. \tag{1.152}$$

In the bulk, any atom in a fcc structure has 12 (maximum possible number) nearest neighbors; this number diminishes for the atoms of the layers located in the neighborhood of the surface of the cluster. The atoms that have up to eight bonds with their neighbors constitute the strict surface of the cluster. The atoms with more than eight neighbors are (partly) covered by atoms of outermost layers; hence, they cannot directly dissociate from the cluster without the removal of the covering atom(s). In the limit of large clusters, the distribution of surface atoms with respect to the number of neighbors is unique and has been calculated by Dave and Abraham [17].

Similar multilayer features are exhibited by the structure of the icosahedral clusters [18]. The first layer consists of 12 atoms located at the vertices of the icosahedron around the central atom. The atoms of the second layer are positioned on the icosahedron of exactly twice the size of the first, those of the third layer are positioned on the icosahedron three times the size of the first, and so on. The line passing through the center of the cluster and the vertices of the icosahedral layers is a fivefold symmetry axis, and the distance between the neighboring atoms on it has the minimum value—the diameter of an atom. As the vertex–center–vertex angle in an icosahedron is approximately 63°26′, the atoms on the same shell will not be close-packed, but will be spaced out at intervals of 1.05146 times their diameter. Generally, the $n$th layer consists of $10n^2 + 2$ atoms, 12 of which are located on the vertices of the icosahedron, $30(n - 1)$ on its edges, and $10(n - 1)(n - 2)$ on its faces. Although the local order of atoms in the icosahedral clusters exhibits almost crystalline symmetry, these structures cannot be classified as crystals because of the absence of translational symmetry. Extensive numerical calculations proved [10] that icosahedral clusters consisting of up to 13 layers are energetically preferred to the fcc clusters, the latter becoming more stable for greater sizes.

The clusters were built by adding successively complete layers around the central atom. This procedure leads to different characteristics of the surfaces in the two kinds of structures.

In the icosahedral cluster, the added, outermost layer covers all the atoms of the previous layers. By adding layer by layer, the icosahedral cluster is growing homogeneously, preserving its shape and its elements of symmetry. The number of bonds of the surface atoms located on the vertices, edges, and faces of the icosahedron remains the same at each of these positions, regardless of the size of the cluster. The number of bonds is six for the atoms on vertices, eight for those on edges, and nine for those on faces.

Contrary to this, the layer added in the fcc cluster does not provide a complete surface shell that covers all the atoms of the previous layers. The cluster is not expanding homogeneously, and adding a new layer can change the shape and the symmetry of the cluster. The cluster of order 1 (one layer around the central atom) is equivalent to a cubo-octahedron, that of order 2 is equivalent to a regular octahedron, and clusters consisting of 4, 5, and 10 layers acquire the shape of a cubo-octahedron [17]. It follows from Equation 1.152 that about $(2n^{1/2} - 1)$ layers contribute to the surface of the fcc cluster, which is irregular and "rough." The number of bonds of the surface atoms in the fcc cluster varies between 3 and 8, and the partition of the surface atoms with respect to the number of neighbors is an irregular function of size in the range of small clusters [17].

### 1.4.4 Interaction of the Surface Atom with the Cluster

We assume that the atoms forming the clusters are rigid spheres of diameter $\eta$, exhibiting London–van der Waals attraction. The pair interaction potential between atoms is given by

$$\phi_{12}(r_{12})\begin{cases} \infty & \text{for} \quad r_{12} < \eta \\ -\epsilon\left(\dfrac{\eta}{r_{12}}\right)^6 & \text{for} \quad r_{12} \geqq \eta, \end{cases} \tag{1.153}$$

where $\epsilon$ is the interaction constant and $r_{12}$ is the distance between the centers of the atoms. The effective interaction potential $\phi$ is calculated assuming pairwise additivity for a surface atom of a cluster consisting of complete layers. Because of the strong dependence of the interaction potential on the interatomic distance $r_{12}$, any atom is affected mostly by the interactions with its nearest neighbors. The overall interaction of a surface atom with the cluster can be separated into two terms. The prevailing contribution $\phi^{(n)}$ comes from the nearest neighbors of a given atom, while the remainder of the cluster provides the background contribution $\phi^{(b)}$:

$$\phi = \phi^{(n)} + \phi^{(b)}. \tag{1.154}$$

The term $\phi^{(n)}$ is calculated on the basis of the mean number of bonds of a surface atom

$$\phi^{(n)} = \sum_j N_j \phi_j / \sum_j N_j, \tag{1.155}$$

where $N_j$ is the number of surface atoms having $j$ neighbors, and $\phi_j$ is the interaction potential of a surface atom that interacts with its $j$ neighbors.

The remainder of the cluster is treated as a continuum medium, and its shape is approximated by a sphere whose mean radius $R$ can be calculated on the basis of the number of atoms in the cluster and their number density $\rho$. The corresponding contribution to the interaction potential is calculated by integrating the pair interaction potential $\phi_{12}$ over a volume $V'$ that excludes the nearest-neighbor shell:

$$\phi^{(b)}(r) = \int_{V'} d\mathbf{r}' \phi_{12}(\mathbf{r} - \mathbf{r}')\rho(\mathbf{r}') \tag{1.156}$$

Equation 1.156 yields the result

$$\begin{aligned}\phi^{(b)}(r) = \frac{\pi\mu\epsilon\eta^3}{2}\Bigg(&\frac{(2R - r_1)R}{\left(r_1^2 r + (r-R)^2 R\right)} \\ &\times\left(\frac{r_1}{R(r_1 + r - R)^2} - \frac{2 + r_1/R}{(R+r)^2}\right) \\ &-\frac{1}{3}\left(\frac{1}{(r_1 + r - R)^3} - \frac{1}{(R + r^3)}\right)\Bigg),\end{aligned} \tag{1.157}$$

where $r_1$ is the distance to the second nearest-neighbor atoms, and $\mu = \rho\eta^3$. For the fcc structure $\mu = \sqrt{2}$, and for the icosahedral clusters $\mu = 1.3132$.

### 1.4.5 Results and Discussion

The rate of nucleation, given by Equation 1.149, and the critical radius were calculated numerically for both the fcc and the icosahedral structures and for various values of the interaction constant $\epsilon$. The quantities $\alpha$ and $\beta$ have been calculated for clusters consisting of complete layers, and their intermediate values have been obtained by interpolation. The results thus obtained are compared with the predictions of the present theory for an amorphous, continuous cluster and with the classical theory.

Figure 1.17 presents the dependence of the critical radius, $R_*/\eta$, on the supersaturation ratio $s$ for $E = \epsilon/k_B T = 4$. The nonmonotonic dependence of $R_*$ on the supersaturation for the fcc crystalline cluster can be attributed to the dependence of the critical radius on the rate of dissociation. The rate of dissociation, Equation 1.145, being an exponential function of the interaction potential $\phi$, is very sensitive to even small variations in the interaction potential. The interaction of a surface atom is greatly affected by the number of nearest neighbors (bonds). It has been mentioned previously that for the fcc cluster the distribution of the surface atoms with respect to the number of bonds is not a monotonic, smooth function of cluster size, even when the clusters are constructed by adding complete layers of atoms. Consequently, the mean number of bonds of a surface atom does not increase monotonically, but exhibits some fluctuations, as one can see in Table 1.2. This effect is exponentially amplified and is particularly significant in the range of small clusters, where the discrete nature of clusters is more important.

In the classical theory, the relation between $R_*$ and supersaturation $s$ is given by the Kelvin equation, which is also recovered in the present theory in the limit of large clusters [6]. Hence, the dependence of $R^*$ on supersaturation, provided by the present theory, can be employed in the limit of large clusters (where the

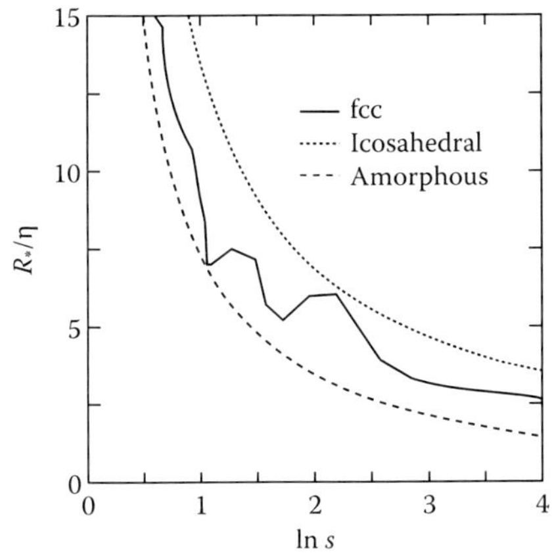


Figure 1.17

Radius of the critical cluster, $R_*/\eta$, as a function of the supersaturation ratio $s$ for the fcc, the icosahedral, and the amorphous clusters. Interaction constant $E = \epsilon/k_BT = 4$.

Table 1.2 Mean Number of Neighbors of a Surface Atom as a Function of the Number of Layers $n$ Contained in fcc and Icosahedral Clusters

| | Mean Number of Neighbors of a Surface Atom | |
|---|---|---|
| *n* | fcc | Icosahedral |
| 1 | 5.0 | 6.0 |
| 2 | 6.0 | 7.429 |
| 3 | 5.6 | 7.957 |
| 4 | 6.571 | 8.222 |
| 5 | 6.333 | 8.381 |
| 6 | 5.727 | 8.486 |
| 7 | 6.667 | 8.561 |
| 8 | 5.778 | 8.617 |
| 9 | 6.800 | 8.660 |
| 10 | 6.727 | 8.695 |
| 20 | 6.593 | 8.849 |
| 40 | 6.941 | 8.925 |
| 100 | 6.916 | 8.970 |
| 300 | 6.924 | 8.990 |

classical Kelvin relation is valid) to calculate the macroscopic surface tension. The values of σ thus obtained are listed in Table 1.3. The surface tensions for the fcc and the icosahedral structures are greater than that for the amorphous droplet. In the amorphous cluster, the number density of atoms is lower than those in the fcc and the icosahedral packings, and this results in a weaker energy of binding of atoms, hence in a lower value of σ.

Table 1.3 Surface Tension $\sigma\eta^2/K_BT$ for fcc, Icosahedral, and Amorphous Clusters for Various Values of the Interaction Constant $E = \epsilon/k_BT$

| | $\sigma\eta^2/K_BT$ | | |
|---|---|---|---|
| | $E = 2$ | $E = 4$ | $E = 6$ |
| fcc | 4.38 | 6.27 | 9.51 |
| Icosahedral | 6.13 | 8.85 | 13.49 |
| Amorphous | 1.86 | 3.69 | 6.03 |

Figure 1.18a through c presents the rate of nucleation against the supersaturation ratio for the fcc cluster, and Figure 1.19a through c presents that for the icosahedral cluster. For comparison, Figures 1.18 and 1.19 also contain the rates of nucleation for an amorphous cluster provided by the present theory and the predictions of the classical theory of nucleation, calculated for the values of the surface tension obtained for the fcc and the icosahedral clusters (Table 1.3). A nonmonotonic behavior corresponding to that observed for the critical radius is also present in the plot of the rate of nucleation for the fcc cluster (Figure 1.18). Both effects have a common physical origin, namely, the nonmonotonic changes in the average interaction potential of a surface atom with the cluster size. The construction of the fcc clusters carried out by adding full layers of atoms does not provide a monotonic increase in the mean number of bonds for a surface atom. In contrast, the average number of bonds for a surface atom in the icosahedral configuration varies monotonically with the size of the clusters, built by adding layer by layer. Consequently, the rate of nucleation for the icosahedral structure (Figure 1.19) is a smooth function of the supersaturation ratio. Thus, the results obtained for the two structures are different.

For clusters constructed by adding single atoms (instead of layers), nonmonotonic behaviors of the critical radius and the rate of nucleation are expected to occur for both structures. In fact, maxima have been observed experimentally in the size distribution function of clusters formed by rare gas atoms [19,20], alkali metals [21], and carbon [22]. This is a result of the formation of strongly bound atomic aggregates for specific numbers of atoms. The construction of clusters layer by layer, used in the present paper, substantially simplifies the problem, since by adding single atoms the number of possible configurations becomes exceedingly high.

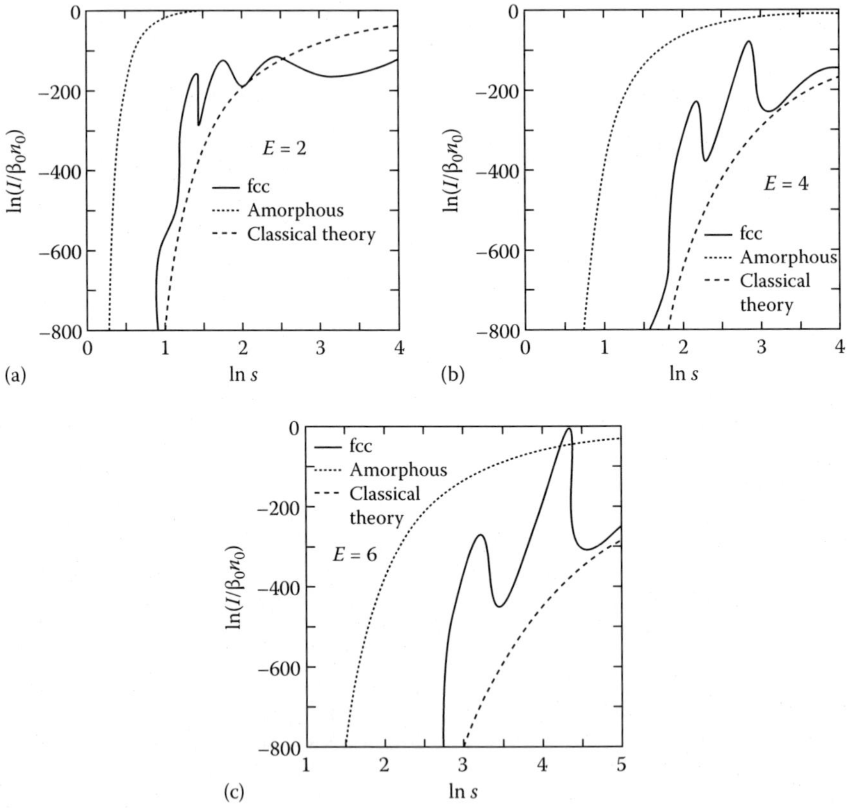


Figure 1.18

Rate of nucleation provided by the present theory, for the fcc and the amorphous clusters, and by the classical theory, calculated for various values of the interaction constant $E = \epsilon/k_BT$: (a) $E = 2$, (b) $E = 4$, (c) $E = 6$.

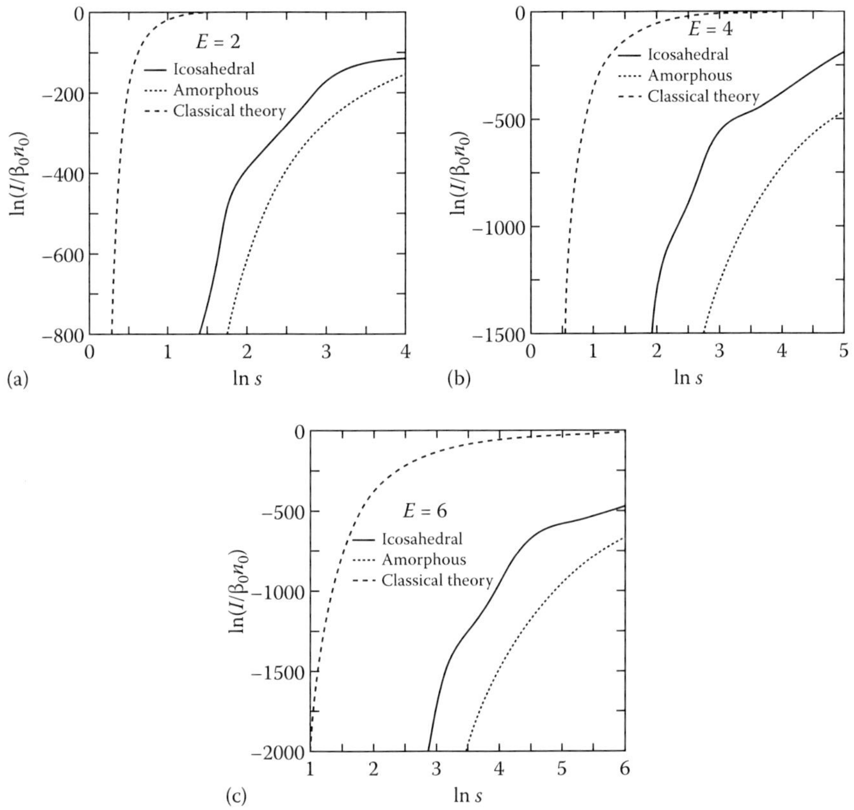


Figure 1.19

Rate of nucleation provided by the present theory, for the icosahedral and the amorphous clusters, and by the classical theory, calculated for various values of the interaction constant $E = \epsilon/k_B T$: (a) $E = 2$, (b) $E = 4$, (c) $E = 6$.

### 1.4.6 Conclusions

A previously developed kinetic theory of nucleation [6] was extended to investigate the rate of nucleation for clusters with fcc or icosahedral configurations. The clusters were constructed by adding successively complete layers of atoms around a central atom. The critical radius and the nucleation rate were calculated assuming a pair interaction potential that combines the rigid core repulsion with the dispersive attraction. It was found that the critical radius and the rate of nucleation for the fcc clusters are not smooth, monotonic functions of supersaturation, but exhibit an irregular behavior. This feature is attributed to the irregular dependence of the mean interaction potential between a surface atom and a cluster on the cluster size. In contrast, the corresponding plots for the icosahedral clusters are smooth, since the layer by layer addition preserves the elements of symmetry for the atoms on the surface of the cluster. Irregularities probably do occur for both structures when the clusters are formed by adding single atoms instead of full layers. Experimental results confirm the predicted nonmonotonic behavior.

## References

1. Farkas, L., *Z. Phys. Chem. (Leipzig) Abt. A* 125, 236 (1927).
2. Becker, R., Doring, W., *Ann. Phys. (Leipzig)* 24, 719 (1935).
3. Zeldovich, J.B., *Acta Physicochim. (USSR)* 18, 1 (1943).
4. Frenkel, J., *Kinetic Theory of Liquids*, Dover, New York, 1946.
5. Narsimhan, G., Ruckenstein, E., *J. Colloid Interface Sci.* 128, 549 (1989). (Section 1.2 of this volume.)

6. Ruckenstein, E., Nowakowski, B., *J. Colloid Interface Sci.* 137, 583 (1990). (Section 1.3 of this volume.)
7. Kittel, C., *Introduction to Solid State Physics*, Wiley, New York, 1976.
8. Hoare, M.R., *Adv. Chem. Phys.* 40, 49 (1979).
9. Burton, J.J., *Nature* (*London*) 229, 335 (1971).
10. Xie, J., Northby, J.A., *J. Chem. Phys.* 91, 612 (1989).
11. Forges, J., De Feraudy, M.F., Raoult, B., Torchet, G., *Adv. Chem. Phys.* 70, 45 (1988).
12. Tanemura, M., Himatari, Y., Matsuda, H., Ogawa,T., Ogita, N., Ueda, A., *Prog. Theor. Phys.* 58, 1079 (1977).
13. Tanemura, M., Himatari, Y., Matsuda, H., Ogawa,T., Ogita, N., Ueda, A., *Prog. Theor. Phys.* 59, 323 (1979).
14. Hsu, C.S., Rahman, A.J., *J. Chem. Phys.* 70, 5234 (1979).
15. Hsu, C.S., Rahman, A.J., *J. Chem. Phys.* 71, 4974 (1979).
16. Gardiner, C.W., "Handbook of Stochastic Methods, Springer-Verlag," New York, 1983.
17. Dave, J.V., Abraham, F.F., *Surf. Sci.* 26, 557 (1971).
18. Mackay, A.L., *Acta Crystallogr.* 15, 916 (1962).
19. Ding, A., Hesslich, J., *Chem. Phys. Lett.* 94, 54 (1983).
20. Stephens, P.W., King, J.G., *Phys. Rev. Lett.* 51, 1538 (1983).
21. Knight, W.D., de Heer, W.A., Saunders, W.A., Chou, M.Y., Cohen, M.L., *Phys. Rev. Lett.* 52, 2141 (1984).
22. Zhang, Q.L., O'Brien, S.C., Heath, J.R., Liu, Y., Curl, R.F., Kioto, H.W., Smalley, R.E., *J. Phys. Chem.* 90, 525 (1986).

## 1.5 Comparison among Various Approaches to the Calculation of the Nucleation Rate

*Bogdan Nowakowski and Eli Ruckenstein**

The discrete, finite difference equations for cluster distribution are transformed into partial differential equations using various expansions. The rate of nucleation is calculated by combining each of these expansions with the kinetic theory of nucleation in liquids developed recently by the authors (Ruckenstein, E., and Nowakowski, B., *J. Colloid Interface Sci* 137, 583 [1990]). Rates of nucleation calculated using expansions compatible with the discrete equilibrium distribution are compared with those obtained using the conventional expansion of finite difference equations to differential equation, and to the classical theory. The results obtained on the basis of the kinetic approach are very different from those based on the classical theory, regardless of the expansion used. The comparison between the different expressions demonstrates that the rate of nucleation can be very sensitive to the expansion used for the transformation of the discrete equations for cluster distribution into the continuous equation. Since the expansions compatible with the equilibrium distribution of clusters are expected to be closer to the finite difference equations for cluster distribution, they should be preferred to the other kind of expansions.

### 1.5.1 Introduction

In the nucleation theory, the rate of the process is calculated from the stationary solution of the equations governing the dynamics of the populations of clusters. Basically, these are the discrete birth and death equations for the concentrations $n_i(t)$ of the clusters consisting of $i$ molecules [1]:

$$\frac{dn_i}{dt} = \beta_{i-1}n_{i-1} - \alpha_i n_i - \beta_i n_i + \alpha_{i+1}n_{i+1} = I_{i-1} - I_i. \tag{1.158}$$

Here, $\alpha_i$ and $\beta_i$ denote the rate of dissociation from and condensation on a cluster containing $i$ molecules, respectively, and $I_i$, is the net flux of clusters passing from $i$ population to $(i + 1)$ population:

$$I_i = \beta_i n_i - \alpha_{i+1}n_{i+1}. \tag{1.159}$$

Assuming a smooth dependence on $i$, it is convenient to transform the discrete set of Equation 1.158 into a partial differential equation, which has the form of the widely studied Fokker–Planck equation [2]. This transformation can be carried out in various ways. Zeldovich [3] was among the first to perform such a transformation, and later Goodrich [4] presented an alternative approach. Both of these transformations have been carried out in the framework of the classical theory of nucleation, which expresses $\alpha_i$ in terms of $\beta_{i-1}$ with the use of the principle of detailed balance and of the macroscopic surface thermodynamics.

The authors recently developed a kinetic theory of nucleation in liquids [5], which provides an independent expression for $\alpha_i$ derived on the basis of a mean passage time technique, without using the macroscopic surface tension. The Goodrich expansion of Equation 1.158 was used in the calculation of the rate of nucleation. It has been noticed lately [6] that the Goodrich expansion does not recover the equilibrium solution of Equation 1.158, and therefore some modifications of his approach are necessary to satisfy this requirement. The scope of this note is to use expansions of Equation 1.158 that are compatible with the equilibrium solution, such as the Zeldovich and a modified Goodrich expansion, in the framework of our recent kinetic theory [5], to calculate the rate of nucleation.

### 1.5.2 Zeldovich Expansion

The equilibrium solution, $n_i^{eq}$, which follows from the condition $I_{eq} = 0$, satisfies the detailed balance equation

$$\beta_i n_i^{eq} = \alpha_{i+1}n_{i+1}^{eq}. \tag{1.160}$$

In the Zeldovich expansion, Equation 1.160 is used to recast Equation 1.158 in the form

$$\frac{dn_i}{dt} = \beta_i n_i^{eq}\left(\frac{n_{i+1}}{n_{i+1}^{eq}} - \frac{n_i}{n_i^{eq}}\right) - \beta_{i-1}n_{i-1}^{eq}\left(\frac{n_i}{n_i^{eq}} - \frac{n_{i-1}}{n_{i-1}^{eq}}\right). \tag{1.161}$$

* *J. Colloid Interface Sci.* 142, 599 (1991). Republished with permission.

Replacing the index $i$ by the continuous variable $g$, expanding Equation 1.161 around the midpoint $g = i$, and retaining only the two highest-order terms, one obtains the following equation for the function $P(g, t) = n(g, t)/n^{eq}(g)$:

$$\frac{\partial P(g,t)}{\partial t} = \frac{1}{n^{eq}(g)}\frac{\partial}{\partial g}\left(\beta(g)n^{eq}(g)\frac{\partial}{\partial g}P(g,t)\right). \tag{1.162}$$

Due to the procedure used, the Zeldovich expansion is compatible with the equilibrium solution. The Zeldovich classical theory of nucleation employs Equation 1.162, in which the hypothetical equilibrium distribution is calculated by using the macroscopic thermodynamics for the excess free energy of the clusters [1]. In the framework of our kinetic theory of nucleation, the equilibrium distribution can be determined from the rates of dissociation and condensation, which are obtained independently. Indeed, the equilibrium distribution of clusters, calculated by iteration from Equation 1.160, has the form

$$n_i^{eq} = n_i^{eq}\prod_{j=2}^{i}\frac{\beta_{j-1}}{\alpha_j} \equiv n_i^{eq}\exp(h(i)). \tag{1.163}$$

The function $h(i)$, introduced here to simplify the notation, is defined as

$$h(i) = \sum_{j=2}^{i}(\ln\beta_{j-1} - \ln\alpha_j), \tag{1.164}$$

where $\beta_{j-1}$ and $\alpha_j$ are given by Equations 1.123 and 1.124.

Equation 1.162 can be rewritten in the form of the Fokker–Planck equation

$$\frac{\partial n(g,t)}{\partial t} = \frac{\partial}{\partial g}(A(g)n(g,t)) + \frac{\partial^2}{\partial g^2}(B(g)n(g,t)) = -\frac{\partial}{\partial g}I(g,t). \tag{1.165}$$

The coefficients $A(g)$ and $B(g)$ in the Zeldovich expansion have the form

$$A_Z(g) = -\beta'(g) - \beta'(g)h'(g) \tag{1.166}$$

and

$$B_Z(g) = \beta(g), \tag{1.167}$$

where $h(g)$ is the continuous counterpart of the function defined by Equation 1.164.

### 1.5.3 Goodrich and Modified Goodrich Expansions

The Goodrich transformation involves the expansion of each of the terms in Equation 1.158 around the midpoint $i$; after retaining only the first two terms, this procedure yields Equation 1.165 with the following coefficients:

$$A_G(g) = \alpha(g) - \beta(g) \tag{1.168}$$

and

$$B_G(g) = (\alpha(g) + \beta(g))/2. \tag{1.169}$$

In the continuum formulation, the net flux of clusters is given by

$$I(g,t) = -A(g)n(g,t) - \frac{\partial}{\partial g}(B(g)n(g,t)). \quad (1.170)$$

Equation 1.170 provides the following expression for the hypothetical equilibrium distribution ($I = 0$):

$$n^{\text{eq}}(g) = n^{\text{eq}}(1)\frac{B(1)}{B(g)}\exp\left(-\int_1^g \mathrm{d}g' \frac{A(g')}{B(g')}\right). \quad (1.171)$$

For Equation 1.171 to acquire the form of Equation 1.163, the coefficients $A$ and $B$ must satisfy the condition

$$A(g) + B(g)h'(g) + B'(g) = 0 \quad (1.172)$$

$A_G(g)$ and $B_G(g)$, given by Equations 1.168 and 1.169, respectively, do not satisfy this condition, and therefore Equation 1.171 with $A_G(g)$ and $B_G(g)$ does not reproduce the equilibrium distribution (Equation 1.163) of Equation 1.158. This feature can be, however, corrected by appropriate modifications of the expressions *of* $A_G(g)$ and/or $B_G(g)$. One possible choice is to retain Equation 1.169 for $B_G(g)$ and calculate $A_G(g)$ from Equation 1.172. It has been shown [6] that in the framework of the classical nucleation theory, the results obtained using the continuous Equation 1.165 with these modified coefficients are in good agreement with the numerical solution of the original discrete equations (Equation 1.158).

### 1.5.4 Rate of Nucleation

The rate of nucleation, i.e., the steady-state net flux of clusters, as obtained from Equation 1.170, is given by

$$I = \frac{n_1 B(1)}{\int_1^\infty \mathrm{d}g \exp\left(\int_1^g \mathrm{d}g' \frac{A(g')}{B(g')}\right)} = \frac{n_1 B(1)}{\int_1^\infty \mathrm{d}g \exp\left(-h(g) - \ln\frac{B(g)}{B(1)}\right)}, \quad (1.173)$$

where $A$ and $B$ denote $A_Z$ and $B_Z$ (Equations 1.166 and 1.167) when the Zeldovich expansion is used, and $A_G$ and $B_G$ (Equations 1.169 and 1.172) when the modified Goodrich expansion is used. The function under the exponential in Equation 1.173

$$H(g) = -h(g) - \ln\frac{B(g)}{B(1)} \quad (1.174)$$

exhibits a rather sharp maximum for the critical cluster size $g = g^*$. Therefore, the integral in Equation 1.173 can be evaluated by the method of steepest descent. This yields the result

$$I \cong n_1 B(1)\left(-\frac{H''(g_*)}{2\pi}\right)^{1/2} \exp(-H(g_*)). \quad (1.175)$$

The macroscopic thermodynamical approach employed in the classical theory of nucleation provides the following result for the rate of nucleation in liquids [7]:

$$I_{\text{cl}} = \frac{1}{2}\beta_1 n_1 \left(\frac{kT}{2\sigma d^2}\right)\ln(s)\exp\left(-\frac{16\pi\sigma^3}{3(kT)^3\rho^2\ln^2 s}\right), \quad (1.176)$$

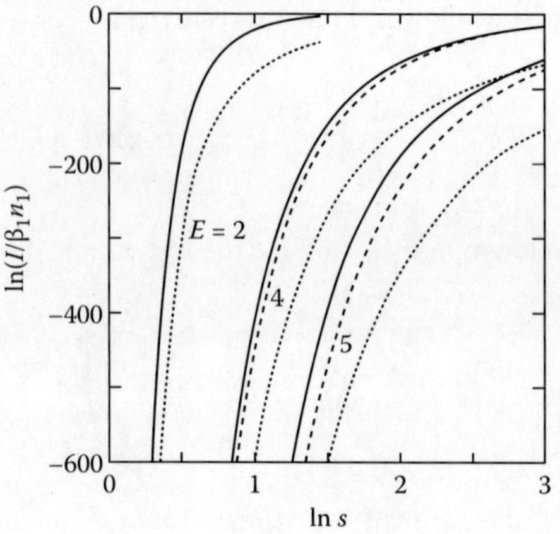


Figure 1.20

Rate of nucleation versus the supersaturation ratio for various values of the interaction parameter $E = \epsilon/kT$. Calculations were performed using the Goodrich expansion (—), the modified Goodrich expansion (---), and the classical theory of nucleation (...).

where $s$ is the supersaturation, $T$ denotes the temperature in degrees Kelvin, $k$ is the Boltzmann constant, $\sigma$ is the surface tension, $\rho$ is the density in the clusters, and $d$ is the diameter of molecules.

### 1.5.5 Discussion

Both the modified Goodrich and Zeldovich expansions were combined with our kinetic theory [5] for the rate of dissociation to calculate the rate of nucleation in liquids. The explicit forms of the expressions for the condensation and dissociation rates provided by the theory can be found elsewhere [5]. The calculations based on Equation 1.173 showed that the nucleation rates obtained with both the Zeldovich and modified Goodrich expansions practically coincide. Furthermore, it has been verified that the steepest descent formula (Equation 1.175) provides a fairly accurate approximation to the more exact Equation 1.173, The rate of nucleation is mainly determined by the exponential factor in Equation 1.175. Since at the critical size $\alpha(g_*) \simeq \beta(g_*)$, one can note using Equations 1.167, 1.169, and 1.174 that the function $H(g_*)$ in the exponential is practically the same for the Zeldovich and modified Goodrich expansions. This explains why the two expansions provide rates of nucleation that practically coincide. Representative results for the nucleation rate calculated on the basis of Equation 1.173 obtained by means of the modified Goodrich expansion are presented in Figure 1.20, where they are compared with those based on the original Goodrich expansion as well as with the predictions of the classical nucleation theory, given by Equation 1.176. The parameter $E$ is given by $E = \epsilon/kT$, where $\epsilon$ is the strength of the binary van der Waals attraction $V(r) = -\epsilon(d/r)^{-6}$. Regardless of the expansion used, the rates of nucleation based on our kinetic theory are much higher than those provided by the classical theory. The difference between the results obtained using the original and modified Goodrich expansions can be important, particularly for the larger values of parameter $E$. Although the results obtained using the modified Goodrich expansion, instead of the original one, are somewhat closer to the classical nucleation theory, particularly for the lower supersaturations, the difference between the results still remains large. The modified Goodrich expansion is to be preferred to the original Goodrich expansion, because it is compatible with the discrete equilibrium distribution, and therefore it is likely to constitute a better approximation of the discrete Equation 1.158. Interestingly, the Zeldovich and modified Goodrich expansions, both of which satisfy this requirement, lead to practically the same results for the nucleation rate.

## References

1. Abraham, F.F., *Homogeneous Nucleation Theory*, Academic Press, New York, 1974.
2. Risken, H., *The Fokker–Planck Equation*, Springer, New York, 1989.
3. Zeldovich, J.B., *J. Exp. Theor. Phys. (Russ.)* 12, 525 (1942).
4. Goodrich, F.C., *Proc. R. Soc. London Ser. A* 277, 167 (1964).
5. Ruckenstein, E., Nowakowski, B., *J. Colloid Interface Sci.* 137, 583 (1990). (Section 1.3 of this volume.)
6. Shizgal, B., Barrett, J.C., *J. Chem. Phys.* 91, 6505 (1989).
7. Walton, A.G., in *Nucleation*, A.C. Zettlemoyer, Ed., Dekker, New York, 1969.

# 1.6 A Kinetic Approach to the Theory of Heterogeneous Nucleation on Soluble Particles during the Deliquescence Stage

*Yuri Djikaev and Eli Ruckenstein**

Deliquescence is the dissolution of a solid nucleus in a liquid film formed on the nucleus due to vapor condensation. Previously, the kinetics of deliquescence was examined in the framework of the capillarity approximation, which involves the thermodynamic interfacial tensions for a thin film and the approximation of uniform density therein. In the present paper, we propose a kinetic approach to the theory of deliquescence that avoids the use of the above macroscopic quantities for thin films. The rates of emission of molecules from the liquid film into the vapor and from the solid core into the liquid film are determined through a first passage time analysis, whereas the respective rates of absorption are calculated through the gas kinetic theory. The first passage time is obtained by solving the single-molecule master equation for the probability distribution of a "surface" molecule moving in a potential field created by the cluster. Furthermore, the time evolution of the liquid film around the solid core is described by means of two mass balance equations that involve the rates of absorption and emission of molecules by the film at its two interfaces. When the deliquescence of an ensemble of solid particles occurs by means of large fluctuations, the time evolution of the distribution of composite droplets (liquid film + solid core) with respect to the independent variables of state is governed by a Fokker–Planck kinetic equation. When both the vapor and the solid soluble particles are single component, this equation has the form of the kinetic equation of binary nucleation. A steady-state solution for this equation is obtained by the method of separation of variables. The theory is illustrated with numerical calculation regarding the deliquescence of spherical particles in a water vapor with intermolecular interactions of the Lennard–Jones kind. The new approach allows one to qualitatively explain an important feature of experimental data on deliquescence, namely, the occurrence of nonsharp deliquescence, a feature that the previous deliquescence theory based on classical thermodynamics could not account for. (© *2006 American Institute of Physics.* doi: 10.1063/1.2202326.)

## 1.6.1 Introduction

During the initial stage of formation of a composite droplet on a solid soluble particle, the particle itself dissolves gradually while a liquid coating is formed around it. This process is usually called "deliquescence."

At present, there exists a well-developed theory of heterogeneous condensation (both unary and multicomponent) for almost all kinds of nucleating centers (for a short review, see, e.g., Refs. [1–3]), but the deliquescence aspect of the process was rarely considered. Experimentally, the deliquescence of large particles (micron size) has been studied extensively by electrodynamic levitation [4–6], whereas the deliquescence of nanometer particles has been examined using a tandem differential mobility analyzer (see, e.g., Refs. [7–12]). In recent years, some progress was also reported regarding the theory of deliquescence of single-component solid soluble particles located in a unary vapor [13–20]. The theory in Refs. [13–15] was developed by using a one-dimensional approach that assumes that "phase" equilibrium is rapidly attained between the solid core and the liquid film at every size of the composite droplet ("solid core + liquid film"). In Refs. [16–18], a two-dimensional theory of deliquescence was proposed without assuming a solid core–liquid film phase equilibrium. Both approaches have been recently thoroughly analyzed by Shchekin and Shabaev [19,20].

Both the one-dimensional and two-dimensional approaches to the thermodynamics and kinetics of deliquescence assume that the composite droplet can be treated using the classical thermodynamics in the framework of the capillarity approximation (CA) [21] and, in particular, that the concept of surface tension is applicable to nanosize droplets and to liquid films as thin as several molecular diameters.

As an alternative to the classical kinetic theory of nucleation, Ruckenstein and coworkers [22–27] proposed a new approach that involves molecular interactions and does not employ the traditional thermodynamics, thus avoiding the use of the concept of surface tension for tiny clusters. In the new kinetic theory [22–27] the rate of emission of molecules by a new-phase particle is determined using a mean first passage time analysis. This time is calculated by solving the single-molecule master equation for the probability distribution function of a surface layer molecule moving in a potential field created by the cluster. Ruckenstein's theory was developed for both liquid-to-solid and vapor-to-liquid phase transitions. Recently [28,29], a further development of that kinetic theory for vapor-to-liquid nucleation has been carried out by combining it with the density functional theory (DFT).

* *J. Chem. Phys.* 124, 194709 (2006). Republished with permission.

In the present paper, we extend Ruckenstein's approach to the kinetic theory of deliquescence. The new approach allows one to avoid the assumptions involved in the CA-based theory. In particular, the surface tensions "liquid–vapor" and "solid–liquid" for deliquescing nanosize particles are no longer used in the new theory.

The paper is structured as follows. Section 1.6.2 briefly outlines the existing thermodynamic theory of deliquescence as well as the kinetics of nucleational deliquescence that involves large fluctuations. Section 1.6.3 presents a new kinetic theory of deliquescence based on a first-passage time analysis. The theory is developed for the case of a single component soluble particle in a single component vapor. In Section 1.6.4, the new theory of deliquescence is illustrated with numerical calculations carried out for the deliquescence of a spherical solid soluble particle in a water vapor. Section 1.6.5 summarizes the conclusions.

## 1.6.2 Background of Deliquescence Theory

Let us consider a vapor of component 1. As usual, the vapor is assumed to behave as an ideal gas. The deliquescence process depends on the temperature $T$ and saturation ratio $\zeta_1$, defined as

$$\zeta_1 = P_1/P_{1\infty}, \qquad (1.177)$$

where $P_1$ is the pressure of molecules of component 1 in the vapor and $P_{1\infty}$ is the pressure of the vapor in equilibrium with its pure bulk liquid.

If the vapor contains foreign nuclei and its saturation ratio $\zeta_1$ is high enough (but still insufficient to trigger homogeneous nucleation), composite droplets can form as a result of condensation of molecules from the vapor onto nuclei. The condensation of atmospheric species onto the nucleus is accompanied by the dissolution of the nucleus itself. Gradually, a liquid coating forms around the nucleus that results in a composite particle (Figure 1.21a) consisting of a soluble crystalline core of radius $R$ surrounded by a spherical film of

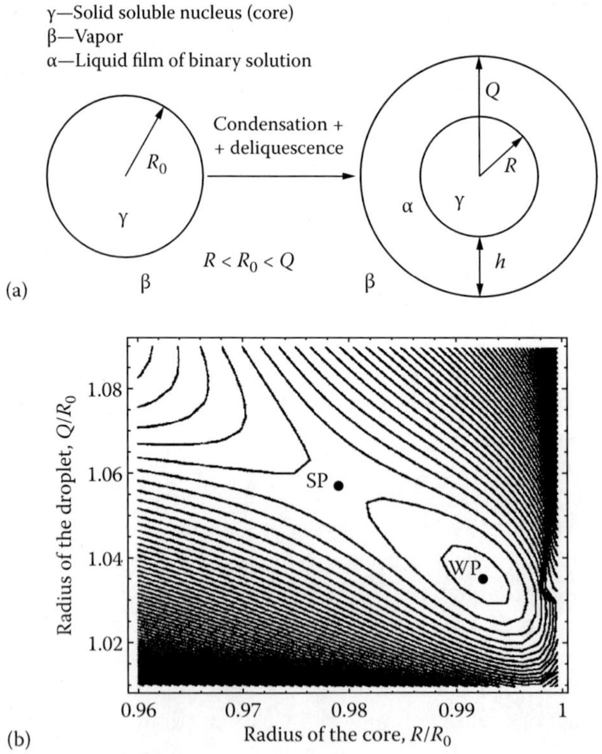


Figure 1.21

(a) Scheme of the droplet formation on a soluble particle of initial radius $R_0$. The composite droplet consists of a solid core of radius $R$ and a liquid film of thickness $h$, the external radius of the droplet being $Q$. (b) Contour plot of the free energy surface for the droplet formation on a model particle of initial radius $R_0$ = 10 nm in a water vapor at $T$ = 298.15 K and $\zeta$ = 0.63 when only the deliquescence of nuclei occurs but the formation of irreversibly growing droplets is impossible. The case corresponds to fluctuational deliquescence.

liquid solution of thickness $h$, with the external radius $Q = R + h$. The entire particle is surrounded by the vapor of component 1. The molecules of the crystal core (component 2) constitute the nonvolatile solute in the film.

Let us denote by $n_i$ ($i$ = 1, 2) the number of molecules of component $i$ that have passed from the vapor and the solid core, respectively, into the liquid film.

As already mentioned, there are one-dimensional and two-dimensional approaches to the thermodynamics and kinetics of deliquescence [13–20]. In the framework of the former, the soluble solid core and the liquid film are assumed to be in equilibrium at any moment. Consequently, the state of the composite droplet can be characterized by a single independent variable. In the latter approach, the theory does not assume equilibrium between the soluble core and the liquid film and two independent variables characterize the state of the composite droplet. Below, we will outline the two-dimensional theory of deliquescence because it is more general and hence more appropriate to be compared with the new approach proposed in Subsection 1.6.3.

### *1.6.2.1 Thermodynamics of Deliquescence*

Assuming that neither component in the film is surface active, the external and internal radii of the film, $Q$ and $R$, are related to the variables $n_1$ and $n_2$ by

$$R = [3(N_2 - n_2)\upsilon_2^{\gamma}/4\pi]^{1/3},$$
$$Q = \left[3(n_1\upsilon_1 + n_2\upsilon_2 + (N_2 - n_2)\upsilon_2^{\gamma})/4\pi\right]^{1/3}, \qquad (1.178)$$

where $\upsilon_2^{\gamma}$ is the volume per molecule in the dry solid nucleus containing $N_2$ molecules, $\upsilon_i$ ($i$ = 1, 2) is the volume per molecule of component $i$ in the film, and $\upsilon_2$ is the sum of the volumes of the ions into which one solute molecule dissociates.

The solute molecules were assumed to be either nondissociated or completely dissociated. The free energy $F$ of formation of a drop is a function of two independent variables, $F = F(R,Q)$. An explicit expression for $F(R,Q)$ was derived in the framework of the capillarity approximation whereby the interfaces are regarded as sharp, the composite droplet as incompressible, and the solute concentration in the film as uniform. The dry solid particle was characterized by the solid–vapor surface tension and the liquid film by the surface tensions of the liquid–vapor and liquid–solid interfaces. The soluble particle was assumed completely wettable by the solution (contact angle = 0), and the effect of the disjoining pressure $\Pi = \Pi(h,\chi)$ in the film of thickness $h$ and composition $\chi = n_1/(n_1 + n_2)$ ($\Pi > 0$ for stable films on lyophilic surfaces) was taken into account [1,13–20]. The function $F = F(R,Q)$ provides a three-dimensional free energy surface the global behavior of which was studied in Refs. [16–18]. This surface becomes a two-dimensional one of unary heterogeneous condensation as soon as the solid core has completely dissolved.

Most solid soluble nuclei in the atmosphere deliquesce at saturation ratios of water $\zeta_1$ well below 1. Under such conditions, the formation of irreversibly growing composite droplets (as in heterogeneous nucleation) is impossible. However, the free energy surface of deliquescence, determined by $F(R,Q)$, has the same form as the free energy surface of heterogeneous binary condensation [30,31]. It can have either one well point and one saddle point or none of them (this implies that there is no miscibility gap for the binary solution). In the former case, a solid particle deliquesces in a fluctuational way via nucleation, i.e., by overcoming the activation barrier by means of large fluctuations. In the latter case, it deliquesces in a barrierless way by small fluctuations (i.e., via the spinodal decomposition) [16].

Whether the deliquescence of a given dry particle has a nucleational or barrierless character depends on the saturation ratio of the vapor and on the physicochemical parameters of the particle. The minimum value $\zeta_d$ of the saturation ratio, at which deliquescence becomes barrierless, is usually called the "deliquescence point" and depends on the size of the particle. This threshold value of the saturation ratio represents the main characteristic of the barrierless deliquescence. However, if deliquescence is nucleational, the particle evolution occurs in a more complex way and its proper theoretical description involves both thermodynamics and kinetics.

Let us use subscripts "e" and "c" to mark the values of quantities at the well and saddle points, respectively, of the free energy surface of nucleational deliquescence (Figure 1.21b). In a two-dimensional plane of the variables $R$ and $Q$ the coordinates $R_e$, $Q_e$ of the well point correspond to a liquid film in metastable equilibrium. The coordinates $R_c$, $Q_c$ of the saddle point correspond to a liquid film in unstable equilibrium. These coordinates are such that $R_e > R_c$, $Q_e < Q_c$. The initial deliquescence of the nucleus and growth of the composite droplet are thermodynamically favorable, so that the droplet quickly finds itself at the bottom of the well. The further dissolution of the core and growth of the composite droplet are thermodynamically unfavorable since the incorporation of a new molecule from the core or vapor into the liquid film demands work. Owing to fluctuations, the growth of the droplet and dissolution of the core may still continue until the droplet attains a critical size corresponding to the saddle point, after which the growth and dissolution become thermodynamically favorable again and occur irreversibly. The difference $\Delta F = F_c - F_e$ between the values of the free energy in the saddle and well points ($F_c$ and $F_e$, respectively) represents the height of the activation barrier to deliquescence and plays a key role in the kinetics of the process. For deliquescence to occur in a nucleational mode, the condition $\Delta F/k_BT \gg 1$ ($k_B$ is the Boltzmann constant) must be fulfilled.

The free energy $F_e$ of formation of a droplet at the well point is negative and its absolute value is much greater than $k_BT$ below the threshold value $\zeta_d$ of the vapor saturation ratio. Thus, all particles quickly absorb so many molecules of vapor that they find themselves at the bottom of the well. As a result, an equilibrium distribution is established that has the form $n_e(R,Q) \propto e^{-[F(R,Q)-F_e]/k_BT}$.

The vicinity of the saddle point is most interesting for the kinetics of deliquescence, much like it is in the case of binary nucleation. The evolution of droplets in this vicinity determines the nucleation rate [15,18–20].

### *1.6.2.2 Kinetic Equation of Deliquescence*

The kinetics of deliquescence has been so far developed for the simplest case in which the evolution of composite droplets occurs via the following four elementary events of mass exchange between the liquid film and the vapor as well as between the liquid film and the solid core, absorption and emission of a molecule of component 1 of the vapor by the liquid film; absorption and emission of a molecule of component 2 of the solid core by the liquid film. For the sake of simplicity, the multimer absorption and emission were neglected (without affecting the kinetics of the process significantly, multimer events complicate the problem mathematically).

In the framework of the capillarity approximation, it was assumed that the characteristic time of the internal relaxation processes in a film is very short in comparison with the characteristic time between two successive elementary interaction of the film with the vapor and solid core media. This allowed one to assume that the film is in internal thermodynamic equilibrium at any time.

Let us denote by $g(n_1, n_2, t)$ the distribution of composite droplets with respect to the variables $n_1$ and $n_2$ at time $t$. The time evolution of the function $g(n_1, n_2, t)$ is governed by a two-dimensional discrete equation of balance. Denoting by $W_i^+(n_1,n_2)$ and $W_i^-(n_1,n_2)$ ($i = 1, 2$) the numbers of molecules of component $i$ that a liquid film absorbs and emits, respectively, per unit time, one can write the discrete equation of balance governing the time evolution of the distribution $g(n_1, n_2, t)$ in the form

$$\frac{\partial g(n_1,n_2,t)}{\partial t} = \Theta_1 + \Theta_2, \tag{1.179}$$

where

$$\begin{aligned}\Theta_1 = {} & W_1^+(n_1-1,n_2)g(n_1-1,n_2,t) - W_1^+(n_1,n_2)g(n_1,n_2,t) \\ & + W_1^-(n_1+1,n_2)g(n_1+1,n_2,t) \\ & - W_1^-(n_1,n_2)g(n_1,n_2,t),\end{aligned} \tag{1.180}$$

$$\begin{aligned}\Theta_2 = {} & W_2^+(n_1,n_2-1)g(n_1,n_2-1,t) - W_2^+(n_1,n_2)g(n_1,n_2,t) \\ & + W_2^-(n_1 n_2+1)g(n_1,n_2+1,t) \\ & - W_2^-(n_1,n_2)g(n_1,n_2,t).\end{aligned} \tag{1.181}$$

The terms $\Theta_1$ and $\Theta_2$ on the right-hand side of Equation 1.179 describe the evolution of the ensemble of composite droplets due to the mass transfer into and from the films by the molecules of components 1 and 2, respectively.

Assuming $n_1 \gg 1$, $n_2 \gg 1$, one can rewrite Equations 1.180 and 1.181 in a continuous form by using Taylor series expansions on their right-hand sides, retaining only the lowest order terms, and substituting the results into Equation 1.179. The resulting partial differential equation has the Fokker–Planck form

$$\frac{\partial g(n_1,n_2,t)}{\partial t} = -\frac{\partial}{\partial n_1} J_1(n_1,n_2,t) - \frac{\partial}{\partial n_2} J_2(n_1,n_2,t), \tag{1.182}$$

where

$$\begin{aligned} J_i(n_1,n_2,t) = &-A_i(n_1,n_2)g(n_1,n_2,t) \\ &-\frac{\partial}{\partial n_i}\left(B_i(n_1,n_2)g(n_1,n_2,t)\right). \end{aligned} \tag{1.183}$$

Here, the coefficients $A_i(n_1,n_2)$ and $B_i(n_1,n_2)$ are determined by the emission and absorption rates $W_i^-$ and $W_i^+$ of the liquid cluster. The steady-state solution of the kinetic equation 1.182 in the vicinity of the saddle point subject to the conventional boundary conditions

$$\begin{aligned} &\frac{g(n_1,n_2)}{g_e(n_1,n_2)} \to 1 \quad (n_1,n_2 \to 0), \\ &\frac{g(n_1,n_2)}{g_e(n_1,n_2)} \to 0 \quad (n_1,n_2 \to \infty), \end{aligned} \tag{1.184}$$

where $g_e(n_1,n_2)$ is the equilibrium distribution, provides the steady-state deliquescence rate [18–20].

Equation 1.182 has the well-known form of the kinetic equation of binary nucleation [32–35]. It is a two-dimensional Fokker–Planck equation and can be solved, e.g., by using the well-known method of complete separation of variables (see, e.g., Refs. [18,36–39]). However, there is a hidden difference between the kinetic equation of deliquescence and the kinetic equation of binary nucleation. Namely, in the kinetic equation of deliquescence the two components of the liquid films come from *two* different mother phases, while in the kinetic equation of binary nucleation both components come from one *common* mother phase.

To solve Equation 1.182 subject to boundary conditions (Equation 1.184), it is necessary to know its coefficients $A_i(n_1,n_2)$, $B_i(n_1,n_2)$ ($i$ = 1, 2), i.e., the emission and absorption rates $W_i^-(n_1,n_2)$ and $W_i^+(n_1,n_2)$. The rate of absorption $W_1^+(n_1,n_2)$ of condensate molecules by the film and the rate of emission $W_2^-(n_1,n_2)$ of solute molecules from the film are calculated by using the kinetic theory of gases, but for determining the emission rate $W_1^-(n_1,n_2)$ and the absorption rate $W_2^-(n_1,n_2)$ the existing deliquescence theory relies on the capillarity approximation that uses the concept of interfacial tension. The emission rate $W_1^-(n_1,n_2)$ and the absorption rate $W_2^+(n_1,n_2)$ are assumed to be independent of the state of the actual mother phase and equal to the rates that the film would have if it were in equilibrium with imaginary mother phases. Thus, the determination of $W_1^-(n_1,n_2)$ and $W_2^+(n_1,n_2)$ is based (via the Gibbs–Kelvin relation) on the interfacial tensions of the film. Because the concept of interfacial tension may not be adequate for too thin films with too large curvatures (such as those involved in deliquescence of nanosize particles) [40,41], not to mention the assumption (of the capillarity approximation) that they are equal to the interfacial tensions of planar interfaces, its use constitutes a drawback of the previous theory of deliquescence [13–20].

### 1.6.3 A New Approach to Kinetics of Deliquescence

In order to avoid the use of macroscopic thermodynamics in the binary nucleation theory, we have recently presented [29] a kinetic theory based on a first passage time analysis, which is a natural extension to binary systems of the approach originally developed by Ruckenstein and coworkers [22–27] for unary nucleation. The key idea of that approach consists of calculating the emission rates $W_i^-(n_1,n_2)$ of a binary cluster directly rather than indirectly, as usually done in the classical theory, by solving the Fokker–Planck or the Smoluchowski equation governing the motion of a single molecule in a potential well around the cluster.

### 1.6.3.1 Kinetics of Fluctuational Deliquescence

Before applying the new approach (based on the first passage time analysis) to the kinetics of nucleational deliquescence, we note that in the vicinity of the saddle point equation 1.182 can be rewritten as

$$\frac{\partial g(n_1,n_2,t)}{\partial t} = \left\{ W_{1c}^{+} \frac{\partial}{\partial n_1}\left[\frac{\partial}{\partial n_1} + \frac{\partial G}{\partial n_1}\right] + W_{2c}^{+} \frac{\partial}{\partial n_2}\left[\frac{\partial}{\partial n_2} + \frac{\partial G}{\partial n_2}\right]\right\} g(n_1,n_2,t) \tag{1.185}$$

(see Appendix 1B), where $G = -\ln g_e(n_1,n_2)$. The quantity $G$ in the present theory plays a role similar to the difference $F(n_1,n_2) - F_e$ for the free energy of droplet formation in the previous theory of deliquescence and can be expected to have a shape similar to the latter. The essence of the new method, as an alternative to the CA-based theory, consists of constructing the equilibrium distribution of clusters, $g_e(n_1,n_2)$ [and the function $G(n_1,n_2)$] without employing the classical thermodynamics for thin films of large curvature, hence *without* using macroscopic interfacial tensions thereof.

The above differential equation describes the time evolution of the ensemble of composite droplets that form on solid soluble nuclei and therefore it can be referred to as the kinetic equation of deliquescence. In order to proceed to its solution it is necessary to determine the condensate and solute absorption coefficients $W_{1c}^{+}$ and $W_{2c}^{+}$ for the saddle point and the equilibrium distribution $g_e(n_1,n_2)$, which also depend on all the emission and absorption coefficients $W_i^{+}, W_i^{-}$ $(i = 1,2)$.

One usually assumes that the exchange of molecules between the liquid droplet and the vapor occurs in a free-molecular regime (by no means it is always the case; this exchange may occur in a diffusion regime as well if the density of the vapor is high enough). In this regime the condensate absorption coefficient $W_1^{+}$ is given by the well-known expression [42]

$$W_1^{+} = \frac{1}{4}\alpha_1 \upsilon_{T1} n_1 S_Q, \tag{1.186}$$

where $\alpha_1$ is the sticking coefficient of component 1, $\upsilon_{T1}$ is the mean thermal velocity of the vapor molecules, and $S_Q = 4\pi Q^2$ is the surface area of the composite droplet (i.e., the area of the outer interface of the liquid film). If the exchange of molecules occurs in the diffusion regime, the coefficient $W_1^{+}$ can be easily calculated as well.

The problem of finding the solute emission coefficient $W_2^{-}$ from the film to the solid core is much more complex [43]. In the roughest approximation, one can assume that the solute molecules in the liquid film constitute an ideal gas in a solvent medium, and consider their integration into the crystal structure of the solid core as the absorption of ideal vapor molecules by a condensed phase cluster. In this approximation, the solute emission rate of the film $W_2^{-}$ is given by $W_2^{-} = \frac{1}{4}\alpha_2 \upsilon_{T2} n_1 S_R$, where $\alpha_2$ is the sticking coefficient of component 2, $\upsilon_{T2}$ is the mean thermal velocity of solute molecules, and $S_R = 4\pi R^2$ is the surface area of the solid core (i.e., the area of the inner interface of the liquid film). Unlike the CA-based kinetics of deliquescence, in the new approach the rates of condensate emission $W_1^{-}$ from the liquid film and the solute emission $W_2^{+}$ from the solid core will be determined via a first passage time analysis.

#### 1.6.3.1.1 First Passage Time Analysis

Consider a spherical cluster immersed in an initial phase. The cluster may be single component (solid core) or composite (solid core + a film of binary solution). A molecule located in the surface layer of the cluster performs thermal chaotic motion in a spherically symmetric potential well $\phi_i(r)$ created in the surface layer by its interactions with the cluster and surrounding medium. Neglecting the internal structure of molecules and assuming pairwise additivity of the intermolecular interactions, the potential $\phi_i(r)$ is given by

$$\phi_i(r) = \sum_j \int_V d\mathbf{r}' \rho_j(r') \phi_{ij}(|\mathbf{r}' - \mathbf{r}|). \tag{1.187}$$

Here, $\mathbf{r}$ is the coordinate of the surface molecule $i$, $\rho_j(r')$ ($j = 1, 2$) is the number density of molecules of component $j$ at point $\mathbf{r}'$ (spherical symmetry is assumed), and $\phi_{ij}(|\mathbf{r}' - \mathbf{r}|)$ is the interaction potential between two molecules of components $i$ and $j$ located at points $\mathbf{r}$ and $\mathbf{r}'$, respectively. The integration in Equation 1.187 is over the whole volume of the system, but the contribution of the medium in which the cluster is immersed can be assumed to be negligible or can be accounted for by an appropriate choice of the energy parameters in the intermolecular potential. The molecule is considered as belonging to the cluster as long as it remains in the potential well, and as dissociated from the cluster when it passes over the outer boundary of the well. The rate of emission, $W_i^-$, is determined by the mean first passage time necessary for this threshold point to be reached.

The mean first passage time of a molecule dissociating from the cluster is calculated on the basis of a kinetic equation governing the chaotic motion of the molecule escaping from the potential well generated around the cluster. The chaotic motion of a molecule is governed by the Fokker–Planck equation for the single-molecule distribution function with respect to its coordinates and momenta, i.e., in the phase space [44–46]. Prior to a dissociation event, the evolution of a molecule (belonging to the cluster) occurs in the surface layer of a condensed phase cluster, where the relaxation time for its velocity distribution is extremely short and negligible compared to the characteristic time scale of the dissociation process. Under these conditions, the Fokker–Planck equation reduces to the Smoluchowski equation, which involves diffusion in an external field [45,46]. In the case of spherical symmetry, it can be written in the form [22–24]

$$\frac{\partial p_i(r,t|r_0)}{\partial t} = D_i^{\mathrm{w}} r^{-2} \frac{\partial}{\partial r}\left( r^2 e^{-\Phi_i(r)} \frac{\partial}{\partial r} e^{-\Phi_i(r)} p_i(r,t|r_0)\right), \tag{1.188}$$

where $p_i(r,t|\, r_0)$ is the probability of observing a molecule of species $i$ between $r$ and $r + dr$ at time $t$ given that initially it was at a radial distance $r_0$, $D_i^{\mathrm{w}}$ is its diffusion coefficient in the well, and $\Phi_i(r) = \phi_i(r)/kT$.

The mean passage time depends on the initial position (distance from the center of the cluster) $r_0$ of the molecule. It is convenient to use the backward Smoluchowski equation [22–24], which expresses the dependence of the transition probability $p_i(r,t|r_0)$ on $r_0$,

$$\frac{\partial p_i(r,t|r_0)}{\partial t} = D_i^{\mathrm{w}} r_0^{-2} e^{\Phi_i(r_0)} \frac{\partial}{\partial r_0}\left( r_0^2 e^{-\Phi_i(r_0)} \frac{\partial}{\partial r_0} p_i(r,t|r_0)\right). \tag{1.189}$$

The probability that a molecule $i$, initially at a distance $r_0$ within the surface layer, will remain in this region after time $t$ is given by the so-called survival probability [25–27],

$$q_i(t|r_0) = \int_S^{S+\lambda_i^{\mathrm{w}}} \mathrm{d}r r^2 p_i(r,t|r_0) \quad \left(S < r_0 < S + \lambda_i^{\mathrm{w}}\right), \tag{1.190}$$

where $S$ is the radius of the cluster and $\lambda_i^{\mathrm{w}}$ is the width of the potential well defined such that a molecule $i$ can be assumed to have dissociated from the cluster at the distance $S + \lambda_i^{\mathrm{w}}$ from its center. The width $\lambda_i^{\mathrm{w}}$ should be chosen such that $\lambda_i^{\mathrm{w}}/S \ll 1$ and $\Phi_i\left(S + \lambda_i^{\mathrm{w}}\right)/\Phi_i^{\mathrm{w}} \ll 1$, where $\Phi_i^{\mathrm{w}}$ is the depth of the well. Having chosen $\lambda_i^{\mathrm{w}}$, the probability for the dissociation time to be between 0 and $t$ is equal to $1 - q_i(t\,|r_0)$, and the probability density for the dissociation time is given by $-\partial q_i/\partial t$. The first passage time is provided by [22–24]

$$\tau_i(r_0) = -\int_0^{\infty} t \frac{\partial q_i(t|r_0)}{\partial t} \mathrm{d}t = \int_0^{\infty} q_i(t|r_0) \mathrm{d}t. \tag{1.191}$$

The equation for the first passage time is obtained by integrating the backward the Smoluchowski equation 1.189 with respect to $r$ and $t$ over the entire range and using the boundary conditions $q_i(0|r_0) = 1$ and $q_i(t|\, r_0) \to 0$ as $t \to \infty$ for any $r_0$. This yields [25–27]

$$-D_i^{\mathrm{w}} r_0^{-2} e^{\Phi(r_0)} \frac{\partial}{\partial r_0}\left( r_0^2 e^{-\Phi_i(r_0)} \frac{\partial}{\partial r_0} \tau_i(r_0)\right) = 1. \tag{1.192}$$

One can solve Equation 1.192 by assuming a reflecting inner boundary of the well ($d\tau_i/dr_0 = 0$ at $r_0 = S$) and the radiation boundary condition at the outer boundary $\left[\tau_i(r_0)=0 \text{ at } r_0 = S+\lambda_i^{\text{w}}\right]$. One thus obtains for the first passage time,

$$\tau_i(r_0)=\frac{1}{D_i^{\text{w}}}\int_{r_0}^{S+\lambda_i^{\text{w}}} dy y^{-2} e^{\Phi_i(y)} \int_S^y dx x^2 e^{-\Phi_i(x)}. \tag{1.193}$$

The average dissociation time, $\bar{\tau}_i$, or the mean first passage time, is obtained by averaging $\tau_i(r_0)$ with the Boltzmann factor over all possible initial positions $r_0$,

$$\bar{\tau}_i=\frac{1}{Z}\int_S^{S+\lambda_i^{\text{w}}} dr_0 r_0^2 e^{-\Phi_i(r_0)}\tau_i(r_0) \tag{1.194}$$

with

$$Z=\int_S^{S+\lambda_i^{\text{w}}} dr_0 r_0^2 e^{-\Phi_i(r_0)}. \tag{1.195}$$

Clearly, $\bar{\tau}_1 \equiv \bar{\tau}_1(R,Q) \equiv \bar{\tau}_1(n_1,n_2)$ and $\bar{\tau}_2 \equiv \bar{\tau}_2(R)$.
Defining the quantity $\omega_i$ as

$$\omega_i \equiv 1/D_i^{\text{w}}\bar{\tau}_i, \tag{1.196}$$

$W_1^-$, the rate of emission of molecules of component 1 from the film, and $W_2^+$, the rate of emission of molecules of component 2 from the solid core, can be expressed as

$$W_1^- = \frac{N_1^{\text{w}}}{\bar{\tau}_1} = N_1^{\text{w}} D_1^{\text{w}}\omega_1,\ W_2^+ = \frac{N_2^{\text{w}}}{\bar{\tau}_2} = N_2^{\text{w}} D_2^{\text{w}}\omega_2, \tag{1.197}$$

where $N_i^{\text{w}}$ denotes the number of molecules in the well. Clearly, $\omega_1$ and $W_1^-$ are functions of $R$, $Q$, whereas $\omega_2$ and $W_2^+$ are functions of $R$.

For given metastabilities of the vapor with respect to the liquid film and the solution in the liquid film with respect to the solid core, the rates of emission and absorption are equal to each other for the critical film the composition of which will be denoted by $\chi_c$. The equality $W_{ic}^- = W_{ic}^+$ allows one to obtain the following expressions for the absorption rate $W_1^+(R,Q)$, emission rate $W_2^-(R,Q)$, and saturation ratio $\zeta_i$ ($i = 1, 2$):

$$W_1^+(R,Q)=N_1^{\text{w}} D_1^{\text{w}}\frac{Q^2}{Q_c^2}\omega_1(R_c,Q_c), \tag{1.198}$$

$$\begin{gathered} W_2^-(R,Q)=N_2^{\text{w}} D_2^{\text{w}} R^2/R_c^2\omega_2(R_c), \\ \zeta_1=\omega_1(R_c,Q_c)/\omega_{1\infty}(\chi_c),\ \zeta_2=\omega_2(R_c)/\omega_{2\infty}, \end{gathered} \tag{1.199}$$

where the subscript $\infty$ marks the quantities for a flat interface.

For $Q_c/\eta > R_c/\eta \gg 1$, the results of this theory are expected to coincide with those of the CA based theory. In particular, the classical Kelvin relations between the saturation ratios and the critical radii $R_c$,$Q_c$ are recovered and can be employed to evaluate the macroscopic surface tension [22–27].

*1.6.3.1.2 Equilibrium Distribution and Steady-State Nucleation Rate* The quantity $N_i^{\mathrm{w}}$ in expression 1.197 for $W_i^-$ depends on $\lambda_j^{\mathrm{w}}$. However, owing to Equation 1.198, neither $N_i^{\mathrm{w}}$ nor $D_i^{\mathrm{w}}$ enters into the final theoretical expressions since they depend on $W_i^-(i=1,2)$ only via the ratios $W_i^-/W_i^+(i=1,2)$ in the quantity $G = -\ln g_e(n_1, n_2)$ (see Equation 1.185). Actually, according to the principle of detailed balance at equilibrium both $\Theta_1$ and $\Theta_2$, given by Equations 1.180 and 1.181, are equal to zero. Thus, one can obtain the following expression for the equilibrium distribution $g_e(n_1,n_2)$ (see Appendix 1C):

$$g_e(n_1,n_2) = N\tilde{p}(n_1,n_2) \Big/ \sum_{n_1=0}^{\infty}\sum_{n_2=0}^{N_2} \tilde{p}(n,n_2)$$

$$\tilde{p}(n_1,n_2) = \frac{\omega_{1c}^{n_1-1}\omega_{2c}^{n_2}}{Q_{1c}^{2(n_1-1)}R_c^{n_2}} \frac{R_0}{Q^2(n_1,n_2)} \frac{Q^2(1,n_2)}{R^2(n_2)} \tag{1.200}$$

$$\times \prod_{i=1}^{n_1-1} \frac{Q^2(n_1-i+1,n_2)}{\omega_1(n_1-i+1,n_2)} \prod_{j=1}^{n_2} \frac{R^2(n_2-j+1)}{\omega_2(1,n_2-j+1)}.$$

Thus, knowing $\omega_1$, $\omega_2$, $Q$, and $R$ as functions of $n_1$ and $n_2$, one can find $G = -\ln g_e(n_1,n_2)$ from Equation 1.200. Clearly, $G$ in the present theory plays a role similar to the free energy of a cluster formation in the CA based theory. The crucial advantage of the present theory is that it does not employ the macroscopic interfacial tension for thin films with a large curvature.

The function $G(n_1,n_2)$ determines a three-dimensional surface that can be expected to have a shape similar to the free energy surface in the CA-based deliquescence theory, with well and saddle points at $n_{1e}$, $n_{2e}$ and $n_{1c}$, $n_{2c}$, respectively, and paths of steepest descent and steepest ascent [16–18]. In order to find the steady-state nucleation rate, Equation 1.185 has to be solved in the vicinity of the saddle point. In this vicinity, the function $G(n_1,n_2)$ (as the free energy of cluster formation in the CA-based theory) is accurately represented by its bilinear form, so that one can use the method of complete separation of variables to solve Equation 1.189 subject to the boundary conditions (Equation 1.184). In our case, the key idea of the method consists of finding such variables of state of the composite droplet that would simultaneously diagonalize the matrix of the second derivatives of the function $G(n_1,n_2)$ at the saddle point and the matrix of the diffusion coefficients (multiplying the second-order derivatives) in Equation 1.185. At the first step, the new variables $x$ and $y$ are introduced by the rotation transformation of the variables $n_1,n_2$, whereas at the second step the rotated variables $x,y$ are subjected to the Lorentz transformation that has the property to leave the difference $x^2 - y^2$ invariant. As a result (see Appendix 1D), the two-dimensional kinetic equation 1.185 is reduced to a one-dimensional one with a well-known steady-state solution. The steady-state rate of deliquescence is thus given by the expression

$$J_s = ACe^{-G_c}, \tag{1.201}$$

where

$$A = (a/2\varepsilon^2)\{\varepsilon^2 - 1 - p + [(\varepsilon^2 + 1 + p)^2 - 4\varepsilon^2]^{1/2}\}, \tag{1.202}$$

$$a = W_1^+c_{11}^2 + W_2^+c_{12}^2,$$

$$\varepsilon = -\left(W_1^+c_{11}^2 + W_2^+c_{12}^2\right)/\left(W_1^+c_{11}c_{21} + W_2^+c_{12}c_{22}^2\right), \tag{1.203}$$

$$p = W_1^+W_2^+(c_{11}c_{22} - c_{12}c_{21})^2/\left(W_1^+c_{11}c_{21} + W_2^+c_{12}c_{22}^2\right)^2,$$

$$c_{11} = \sqrt{\frac{1}{2}\left|G''_{xx}\right|}\cos\beta, \quad c_{12} = \sqrt{\frac{1}{2}\left|G''_{xx}\right|}\sin\beta,$$

$$c_{21} = -1\sqrt{\frac{1}{2}\left|G''_{yy}\right|}\sin\beta, \; c_{22} = \sqrt{\frac{1}{2}\left|G''_{yy}\right|}\cos\beta, \tag{1.204}$$

$$|G''_{xx}| = -\left(G''_{11}\cos^2\beta + G''_{12}\sin 2\beta + G''_{22}\sin^2\beta\right), \tag{1.205}$$

$$|G''_{yy}| = G''_{11}\sin^2\beta - G''_{12}\sin 2\beta + G''_{22}\cos^2\beta), \tag{1.206}$$

$$G''_{ij} = \partial^2 G/\partial n_i \partial n_j \quad (i,j = 1,2), \quad C = \left|c_{11}c_{22} - c_{12}c_{21}\right|^{-1}, \tag{1.207}$$

with the angle $0 \le \beta \le \pi/2$ between the path of the steepest descent and the $n_1$ axis (constant in the vicinity of the saddle point) provided by the solution of the equation $\tan 2\beta = 2G''_{12}/(G''_{11} - G'_{22})$.

*1.6.3.1.3 Profile of Solute Concentration within the Film* In the CA-based theory of deliquescence, the solute concentration within the film was assumed to be uniform. This assumption is clearly very convenient from the modeling standpoint but too rough to be adequate in most physical situations. An approximate treatment of this nonuniformity is provided below.

Let us assume that the solute density profile within the film can be relatively well fit by the parabolic polynomial

$$\rho_2(r) = Ar^2 + Br + C, \tag{1.208}$$

where the coefficients $A$, $B$, and $C$ are functions of the variables of state of the composite droplet, i.e., $A = A(R,Q)$, $B = B(R,Q)$, $C = C(R,Q)$ and are determined by the following two boundary and one normalization conditions:

$$D_2 \left.\frac{d\rho_2}{dr}\right|_R = j_2(R,Q), \tag{1.209}$$

$$d\rho_2/dr|_Q = 0, \tag{1.210}$$

$$\int_R^Q d\mathbf{r}\rho_2(r) = \rho_2^\gamma \frac{4\pi}{3}\left(R_0^3 - R^3\right), \tag{1.211}$$

where $D_2$ is the diffusion coefficient of the solute molecules in the film (assumed constant), $j_2(R,Q) \equiv (1/4\pi R^2)\left[W_2^-(R,Q) - W_2^+(R,Q)\right]$ is the flux of solute molecules onto the solid core at its surface, and $\rho_2^\gamma$ is the density of the solid core (assumed uniform). The explicit forms of functions $A(R,Q)$, $B = B(R,Q)$, and $C = C(R,Q)$ following from Equations 1.209 through 1.211 are given in Appendix 1E.

### 1.6.3.2 Kinetics of Barrierless Deliquescence

The formalism presented in Section 1.6.3.1 is applicable to deliquescence when the function $G(R,Q)$ has well and saddle points and the difference $G_c - G_e$ is of the order of at least several $k_BT$. Only in this case one can subject the kinetic equation 1.185 to the first of the boundary conditions in Equation 1.184 and obtain Equations 1.201 through 1.207. If $G_c - G_e$ is less than or of the order of $k_BT$, an ensemble of initially dry particles deliquesces practically without having to overcome an activation barrier. In the formalism of Section 1.6.3.1, the value $\zeta_d$ of the saturation ratio $\zeta_1$ (i.e., the deliquescence point) is determined by the condition $G_c - G_e = 0$ (in the CA-based theory, the deliquescence point is determined by the condition $F_c - F_e = 0$). Thus, if $\zeta_1 \simeq \zeta_d$, the ensemble of deliquescing particles remains monodisperse during the whole process if it was monodisperse initially. Therefore, it is sufficient to know the evolution of a single deliquescing particle to know the evolution of the whole ensemble. The previous theory of deliquescence was not able to address this problem. In this subsection, we propose an approximate method for examining the temporal evolution of a single deliquescing particle.

The evolution of the film on the solid core occurs via the absorption and emission of solute and condensate molecules. In addition to the flux $j_2$ of solute molecules from the liquid solution in the film towards the solid core, let us introduce an analogous quantity $j_1(R,Q) \equiv (1/4\pi Q^2)\left[W_1^+(R,Q) - W_1^-(R,Q)\right]$, representing the flux of condensate molecules from the vapor towards the liquid film. Note that both quantities $j_1$ and $j_2$ are fluxes per unit area (spherical symmetry is assumed).

The temporal evolution of the total volume of solute molecules in the film is governed by the ordinary differential equation

$$\frac{d}{dt}\left[\frac{4\pi}{3}(Q^3-R^3)-\upsilon_1\int_R^Q dr\rho_1(r)\right]=-4\pi R^2 j_2(R,Q)\upsilon_2, \tag{1.212}$$

while the evolution of the total volume of condensate molecules in the film is governed by the equation

$$\frac{d}{dt}\left[\frac{4\pi}{3}(Q^3-R^3)-\upsilon_2\int_R^Q dr\rho_2(r)\right]=-4\pi Q^2 j_1(R,Q)\upsilon_1. \tag{1.213}$$

The solution of this set of differential equations subject to some initial conditions provides the dependence of the internal and external radii of the film on time, $R = R(t)$ and $Q = Q(t)$. The natural initial conditions for these functions are those corresponding to an initial dry solid particle, that is, $R(t = 0) = R_0$ and $Q(t = 0) = R_0$. Consequently,

$$R(t)=\left[Q^3(t)-\frac{3}{4\pi}\upsilon_1 n_1(R(t),Q(t))-3\upsilon_2\int_0^t dt' R^2(t')j_2(R(t'),Q(t'))\right]^{1/3}, \tag{1.214}$$

$$Q(t)=\left[R^3(t)+\frac{3}{4\pi}\upsilon_2 n_2(R(t),Q(t))+3\upsilon_1\int_0^t dt' Q^2(t')j_1(R(t'),Q(t'))\right]^{1/3}, \tag{1.215}$$

which are more convenient than the differential Equations 1.212 and 1.213 for solving by iterations. It should be noted that the couple of Equations 1.212 and 1.213 (or Equations 1.214 and 1.215) are valid only when an increase of $Q$ occurs for $j_1 > 0$ and a decrease of $R$ occurs for $j_2 < 0$. However, these equations can be generalized by including stochastic terms representing the fluctuational increase in $Q$ and fluctuational decrease in $R$ when $j_1 < 0$ and $j_2 > 0$. Such a generalization leads to more complicated equations (both conceptually and mathematically) and is not considered in the present paper.

### 1.6.4 Numerical Evaluations

In this section, we will present some numerical results of the application of our theory to the deliquescence of spherical nuclei consisting of molecules that do not dissociate in the solution, placed in a water vapor at $T = 298.15$ K. Of course, it would be more interesting to examine the deliquescence of soluble nuclei of real substances [such as NaCl or $(NH_4)_2SO_4$], which are of great interest for atmospheric phenomena. However, such systems would require a more sophisticated theory because the dissociation of such molecules is expected to play a role in their deliquescence behavior.

The interactions between a pair of molecules of components $i$ and $j$ were modeled via the Lennard–Jones (LJ) potentials

$$\phi_{ij}=4\epsilon_{ij}\left[\left(\frac{\eta_{ij}}{r}\right)^{12}-\left(\frac{\eta_{ij}}{r}\right)^{6}\right]\quad(i,j=1,2), \tag{1.216}$$

where $\epsilon_{ij}$ and $\eta_{ij}$ are energy and length parameters, respectively, satisfying the simple mixing rules

$$\epsilon_{ij}=\sqrt{\epsilon_{ii}\epsilon_{jj}},\ \eta_{ij}=\frac{1}{2}(\eta_i+\eta_j), \tag{1.217}$$

where $\eta_i$ $(i = 1,2)$ is the length parameter of the interaction potential between two molecules of component $i$.

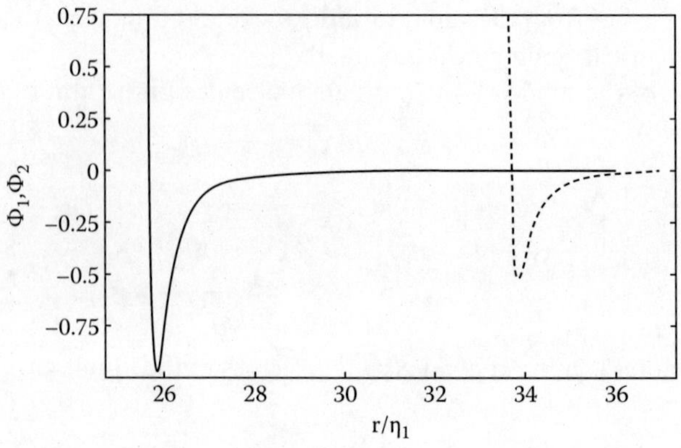


Figure 1.22

Effective potentials $\Phi_1(r)$ (dashed curve) for a condensate molecule around the liquid film and $\Phi_2(r)$ (solid curve) for a solute molecule around the solid core as functions of the distance from the center of the composite droplet with $R/\eta_1 = 25$ and $Q/\eta_1 = 33$.

Calculating the effective potential field (see Equation 1.187) in the vicinity of the composite droplet (Figure 1.22), the interactions of the condensate molecule with the vapor molecules were neglected because the number density of the latter is negligible compared to that of the droplet. The effective potential field in the vicinity of the solid core was calculated (Figure 1.22) by taking into account the interactions of a solute molecule with the solid core. The interactions of the solute molecule with those in the film cannot be neglected in this case but their contribution is assumed to be included in the potential field by an appropriate choice of $\epsilon_{22}$, the solute–solute energy parameter in Equation 1.216.

All numerical calculations were carried out for the following values of the system parameters:

$$\eta_1 = 3.5\times10^{-8}\ \text{cm},\ \eta_2/\eta_1 = 1.44,$$

$$\epsilon_{11}/k_BT = 0.26,\ \epsilon_{22}/k_BT = 1,$$

$$R_0/\eta_1 = 30,\ \upsilon_1/\eta_1^3 = 0.96,\ \upsilon_2^\gamma/\eta_1^3 = 2.12,$$

$$\upsilon_2/\eta_1^3 = 2.14,$$

$$D_1^w = 1\times10^{-5}\ \text{cm}^2/\text{s},\ D_2 = 1.5\times10^{-5}\ \text{cm}^2/\text{s},$$

$$D_2^w = 5\times10^{-5}\ \text{cm}^2/\text{s}.$$

This set of parameters was chosen as representative for water (component 1) and some large nondissociating molecules (component 2). The results of the numerical calculations carried out for a single particle are presented in Figures 1.22 through 1.27.

Figure 1.23 presents the rate of emission of solute molecules from the solid core, i.e., the rate of absorption of solute molecules by the liquid film, per unit area, $w_2(R,Q) = w_2^+(R,Q)/4\pi R^2$, calculated as outlined in Section 1.6.3.1. The dashed curve in Figure 1.23 represents the emission rate for the values of parameters presented above. In order to simulate the experimentally observed fact that the solubility of particles, composed of dissociating molecules, decreases with decreasing size [43], an $R$-dependent interaction parameter for the solute–solute interactions $\epsilon_{22}^R = \epsilon_{22}[1+18.4(R/\eta_1)^{-3}]$ was used to obtain the solid curve in Figure 1.23.

Figure 1.24 shows the flux of solute molecules, $j_2(R,Q)$, as a function of the solid core radius $R$ for five different values of the external radius $Q$ of the film. The solid curve corresponds to $Q/\eta_1 = 33.19$, the long-dashed curve to $Q/\eta_1 = 34.61$, the short-dashed curve to $Q/\eta_1 = 36.12$, the dot-dashed curve to $Q/\eta_1 = 39.72$, and the dotted curve to $Q/\eta_1 = 46.70$. The initial dissolution rate of the core is very large for an external radius corresponding to a monolayer of water on the original particle, $Q \simeq R_0 + \eta_1$. Strictly speaking, the

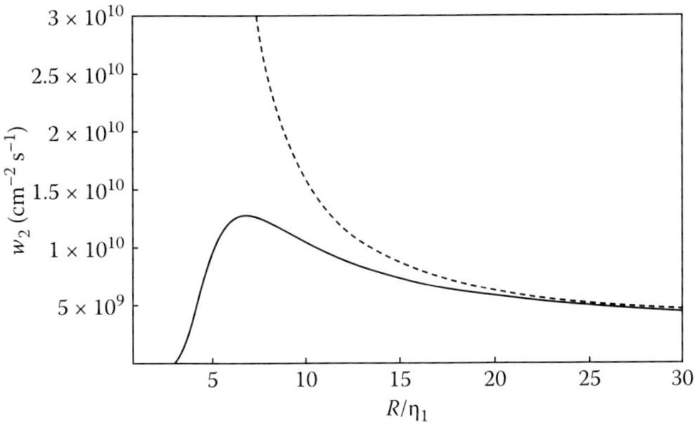


Figure 1.23

Rate of emission of solute molecules from the solid core (i.e., the rate of absorption of solute molecules by the liquid film) per unit area, $w_2(R,q) = w_2^+(R,Q)/4\pi R^2$, calculated as outlined in Section 1.6.3.1, as a function of the core radius $R$.

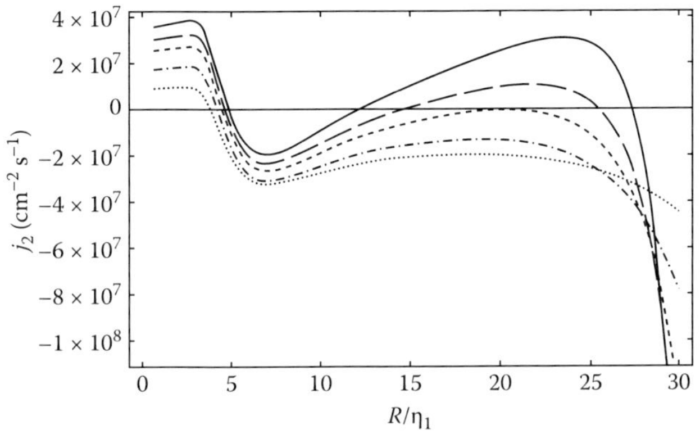


Figure 1.24

Flux of solute molecules $j_2(R,Q)$ onto the solid core as a function of the core radius $R$ for five different values of the external radius $Q$ of the film. The solid curve corresponds to $Q/\eta_1 = 33.19$, the long-dashed curve to $Q/\eta_1 = 34.61$, the short-dashed curve to $Q/\eta_1 = 36.12$, the dot-dashed curve to $Q/\eta_1 = 39.72$, and the dotted curve to $/\eta_1 = 46.70$.

present theory is not applicable to such thin films. The complete dissolution of the core can take place only if $Q$ is larger than a threshold value $Q_{th}/\eta_1 \simeq 36.12$. (Note that the core can dissolve even when $j_2(R,Q) > 0$, by means of fluctuations. In this case, the time evolution of the core radius $R$ would occur by a modified version of Equation 1.212 or Equation 1.214, which would include a stochastic term accounting for the possibility of a fluctuational decrease in $R$ [as opposed to a deterministic one for $j_2(R,Q) < 0$].)

The curve of zero flux of condensate molecules, $j_1(R,Q)$, is presented in Figure 1.25 for $\zeta_1 = 0.67$. In the "negative region" $j_1(R,Q) < 0$, i.e., the external radius cannot increase. In the "positive region" $j_1(R,Q) > 0$, i.e., the external radius can increase. For any given $R$ the external radius $Q$ of the droplet cannot grow larger than a maximum value dependent on $R$ and provided by the boundary between the positive and negative regions. The largest value $Q_{max}$ of the external radius for the selected $\zeta_1$ corresponds to $R = 0$. (Note again that the droplet can grow by means of fluctuations even when $j_1(R,Q) < 0$. In this case, the time evolution of the droplet radius $Q$ would be governed by a modified version of Equation 1.213 or Equation 1.215, which would include a stochastic term accounting for the possibility of a fluctuational increase in $Q$ [as opposed to a deterministic one for $j_1(R,Q) > 0$].)

The maximum external radius $Q_{max}$ of the droplet as a function of the vapor saturation ratio is plotted in Figure 1.26. The qualitative behavior of the function $Q_{max} = Q_{max}(\zeta_1)$ is very similar to the experimental data on deliquescence of sodium chloride and potassium chloride particles [7–12] in an undersaturated

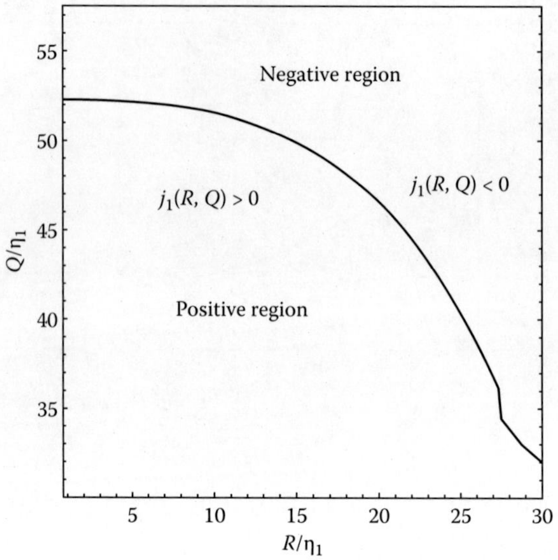


Figure 1.25

Contour plot of the flux of condensate molecules $j_1(R, Q)$ onto the droplet as a function of $R$ and $Q$ at fixed $\zeta_1 = 0.67$ (with $\zeta_d \simeq 0.708$). In the "negative region" $j_1(R, Q) < 0$ and the external radius tends to decrease, while in the "positive region" $j_1(R, Q) > 0$ and the external radius tends to increase.

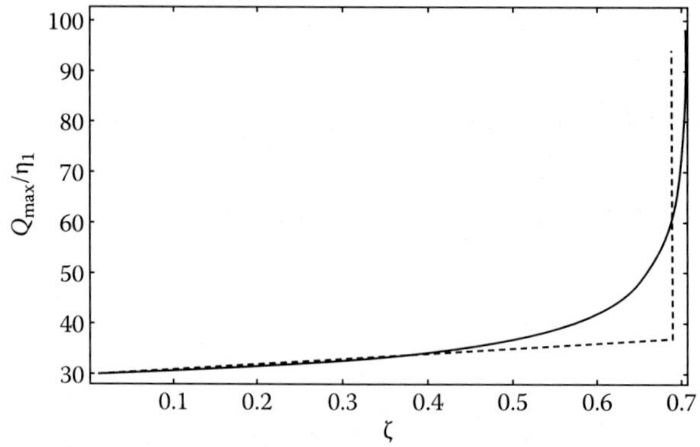


Figure 1.26

Maximum external radius $Q_{max}$ of the droplet as a function of the vapor saturation ratio $\zeta_1$ (with $\zeta_d \simeq 0.708$).

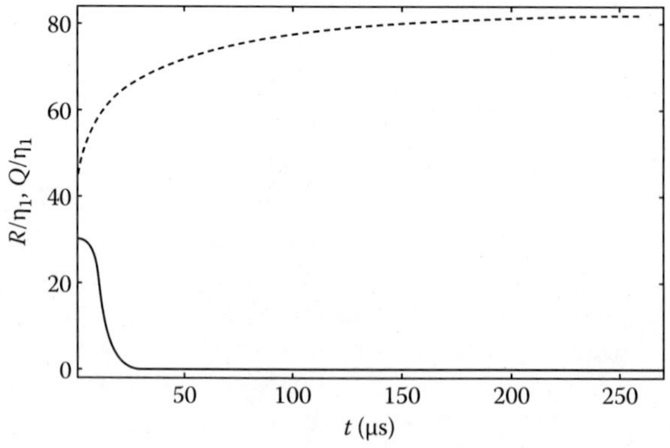


Figure 1.27

Time evolution of $R$ and $Q$, the internal and external radius of the film, for the case where deliquescence occurs in a barrierless way at the vapor saturation ratio $\zeta_1 = 0.73$ (with $\zeta_d \simeq 0.708$). The solid curve is for $R(t)$ and the dashed curve is for $Q(t)$.

water vapor. In those experiments, the growth factor (i.e., the ratio $Q/R_0$) of the deliquescing particle was monitored as a function of the relative humidity, and in many cases a nonsharp deliquescence transition was observed, like the one predicted by our theory in Figure 1.26. Starting with a dry particle and gradually increasing the relative humidity, the growth factor slightly increased (by about 10%). When the relative humidity approached a threshold value, a steeper (but nonsharp) increase in the growth factor was observed over a narrow range of relative humidities. The previous theory of deliquescence [13–20] based on the capillarity approximation, could not explain this steep nonsharp character of the experimental results. Instead, it predicted a sharp increase of $Q/R_0$ at the deliquescence point $\zeta_d$ (dashed curve).

Figure 1.27 presents the time evolution of the external and internal radii of the film when deliquescence occurs in a barrierless way. The functions $R = R(t)$ and $Q = Q(t)$ (as obtained by iterations from Equations 1.214 and 1.215) have very different characteristic time scales. For a given saturation ratio $\zeta_1 = 0.73$, the dissolution of the solid core occurs on time scales of the order of 10 μs, whereas the increase in the external radius of the droplet requires time of the order of 100 μs.

### 1.6.5 Conclusions

Deliquescence is the very first stage of heterogeneous condensation on a solid soluble nucleus. At this stage the solid nucleus dissolves in a liquid film formed by the vapor molecules condensing on the nucleus. Considering the deliquescence of single-component nuclei in a single component vapor (the presence of an inert gas does not affect the process), we presented a new approach to the kinetics of deliquescence that does not use the classical thermodynamics and, instead, is based on a first passage time analysis.

The nuclei were assumed to be spherical, completely wettable, and soluble in the condensing liquid. In this model, a droplet is a composite particle consisting of a soluble crystalline core surrounded by a spherical film of liquid binary solution. The entire droplet is surrounded by the vapor.

Previously, the kinetics of nucleational deliquescence was examined in the framework of the capillarity approximation that involves the interfacial tensions of films even at the early stages of deliquescence when the films are very thin. Besides, in the previous theory of deliquescence the distribution of solute molecules within the film was assumed to be uniform.

In the present paper, we propose a new approach to the kinetic theory of deliquescence that does not use the classical thermodynamics and avoids the above approximation. The rates of emission of molecules from a liquid film into the vapor and from the solid core into the film are determined via the mean first passage time analysis. This time is calculated by solving the single-molecule master equation for the probability distribution of a "surface" molecule moving in a potential field created by the cluster. The temporal evolution of a liquid film around the solid core is described by a couple of first order ordinary differential equations involving the rates of absorption and emission of molecules by a liquid film. This approach can be applied to deliquescence in either mode (nucleational or barrierless). In the nucleational mode, the temporal evolution of the distribution of composite droplets with respect to independent variables of state is governed by the Fokker–Planck kinetic equation. In the simplest case where both the vapor and solid soluble particles are single component, this equation has the form of the kinetic equation of binary nucleation. A steady-state solution for this equation is obtained by the separation of variables.

The theoretical results have been illustrated by numerical calculation for the deliquescence of spherical particles in a water vapor with the intermolecular interactions of the LJ kind. The new approach allows one to qualitatively explain an important feature of experimental data on deliquescence, namely, the occurrence of nonsharp deliquescence over a range (albeit narrow) of vapor saturation ratios. The previous deliquescence theory (based on the classical thermodynamics) was unable to explain this kind of deliquescence often observed in experiments on sodium chloride and potassium chloride particles.

## Acknowledgments

This work was supported by the National Science Foundation through the Grant No. CTS-0000548.

## Appendix 1B: Derivation of Equation 1.185

To derive Equation 1.185 from Equations 1.182 and 1.183, let us first transform $\Theta_i$ ($i$ = 1, 2) (Equations 1.180 and 1.181) by using Taylor series expansions. One obtains $B_i(n_1,n_2)=\frac{1}{2}\left(W_i^+(n_1,n_2)+W_i^-(n_1,n_2)\right)$. On the other hand, according to the principle of detailed balance (see Equations 1C.1 and 1C.2 in Appendix 1C),

$$
\begin{aligned}
&W_1^-(n_1+1,n_2)/W_1^+(n_1,n_2)\\
&\quad=\exp[-G(n_1,n_2)+G(n_1+1,n_2)]\\
&\quad\simeq\exp[\partial G(n_1,n_2)/\partial n_1],\\
&W_2^-(n_1,n_2+1)/W_2^+(n_1,n_2)\\
&\quad=\exp[-G(n_1,n_2)+G(n_1,n_2+1)]\\
&\quad=\exp[\partial G(n_1,n_2)/\partial n_2].
\end{aligned}
$$

The first derivatives $\partial G(n_1,n_2)/\partial n_i$ ($i$ = 1, 2) are equal to zero at the saddle point and are close to zero in its vicinity where, in addition, $W_1^+(n_1+1,n_2)\simeq W_1^+(n_1,n_2)$ and $W_2^+(n_1,n_2+1)\simeq W_2^+(n_1,n_2)$. Therefore, one can write

$$W_i^-(n_1,n_2)/W_i^+(n_1,n_2)\simeq 1+o(\partial G(n_1,n_2)/\partial n_i)\,, \tag{1B.1}$$

where $o(x)$ denotes a quantity of the same order of magnitude as $x$. Neglecting the terms of the order $o[\partial G(n_1, n_2)/\partial n_i]$, one can assume that the sum $W_i^+(n_1,n_2)+W_i^-(n_1,n_2)$ in the vicinity of the saddle point (where the kinetic Equation 1.185 has to be solved) is equal to $2W_i^+(n_1,n_2)$. Furthermore, in the vicinity of the saddle point $W_i^+(n_1,n_2)\simeq W_j^+(n_{1c},n_{2c})\equiv W_{ic}^+$, hence

$$B_i\simeq W_{ic}^+ \quad (i=1,2). \tag{1B.2}$$

Because at equilibrium $J_1 = 0$ and $J_2 = 0$, Equation 1.183 leads to

$$A_i(n_1,n_2)g_e(n_1,n_2)+B_i(n_1,n_2)\frac{\partial g_e(n_1,n_2)}{\partial n_i}=0\ (i=1,2)\,, \tag{1B.3}$$

whence one obtains

$$
\begin{aligned}
A_i(n_1,n_2)&=-B_i(n_1,n_2)\frac{1}{g(n_1,n_2)}\frac{\partial g_e(n_1,n_2)}{\partial n_i}\\
&\simeq W_{ic}^+\frac{\partial G(n_1,n_2)}{\partial n_i}\,,
\end{aligned}
\tag{1B.4}
$$

where $G(n_1,n_2) = -\ln g_e(n_1,n_2)$. The coefficients $A_i(n_1,n_2)$ and $B_i(n_1,n_2)$ being the same for equilibrium and nonequilibrium states, $J_i$ ($i$ = 1, 2) in Equation 1.183 acquires the form

$$J_i(n_1,n_2,t)=-W_{ic}^+\left(\frac{\partial}{\partial n_i}+\frac{\partial G(n_1,n_2)}{\partial n_i}\right)g(n_1,n_2,t). \tag{1B.5}$$

The substitution of this equation into Equation 1.182 leads to Equation 1.185.

## Appendix 1C: Equilibrium Distribution of Composite Droplets

Let us derive Equation 1.200. According to the principle of detailed balance,

$$W_1^+(n_1-1,n_2)g_e(n_1-1,n_2)=W_1^-(n_1,n_2)g_e(n_1,n_2), \tag{1C.1}$$

$$W_2^+(n_1,n_2-1)g_e(n_1,n_2-1)=W_2^-(n_1,n_2)g_e(n_1,n_2). \tag{1C.2}$$

These equalities can be rewritten as

$$\frac{g_e(n_1,n_2)}{g_e(n_1-1,n_2)}=\frac{W_1^+(n_1-1,n_2)}{W_1^-(n_1,n_2)} \tag{1C.3}$$

and

$$\frac{g_e(n_1,n_2)}{g_e(n_1,n_2-1)}=\frac{W_2^+(n_1,n_2-1)}{W_2^-(n_1,n_2)}. \tag{1C.4}$$

By applying Equation 1C.3 to the films $(n_1 - i, n_2)$ with $i = 2, 3, \ldots, n_1$, multiplying the right- and left-hand sides of all equations, one obtains

$$\frac{g_e(n_1,n_2)}{g_e(1,n_2)}=\prod_{i=1}^{n_1-1}\frac{W_1^+(n_1-i,n_2)}{W_1^-(n_1-i+1,n_2)}. \tag{1C.5}$$

By applying Equation 1C.4 to the films $(1, n_2 - j)$ with $j = 1, 2, \ldots, n_2$ and multiplying the right- and left-hand sides of all equations, one obtains

$$\frac{g_e(1,n_2)}{g_e(1,0)}=\prod_{j=1}^{n_2}\frac{W_2^+(1,n_2-j)}{W_2^-(1,n_2-j+1)}. \tag{1C.6}$$

Using the normalization condition $\Sigma_{n_1=0}^{\infty}\Sigma_{n_2=0}^{N_2}g_e(n_1,n_2)=N$, where $N$ is the total number of initial soluble particles in unit volume of the system, one obtains

$$g_e(1,0)=N\Big/\sum_{n_1=0}^{\infty}\sum_{n_2=0}^{N_2}\prod_{i=1}^{n_1-1}\frac{W_1^+(n_1-i,n_2)}{W_1^-(n_1-i+1,n_2)}\times\prod_{j=1}^{n_2}\frac{W_2(1,n_2-j)}{W_2^-(1,n_2-j+1)}, \tag{1C.7}$$

so that the equilibrium distribution of clusters acquires the form

$$g_e(n_1,n_2)=N\tilde{p}(n_1,n_2)\Big/\sum_{n_1'=0}^{\infty}\sum_{n_2'=0}^{N_2}\tilde{p}(n_1',n_2'), \tag{1C.8}$$

where

$$\tilde{p}(n_1,n_2)=\prod_{i=1}^{n_1-1}\frac{W_1^+(n_1-i,n_2)}{W_1^-(n_1-i+1,n_2)}\prod_{j=1}^{n_2}\frac{W_2^+(1,n_2-j)}{W_2^-(1,n_2-j+1)}. \tag{1C.9}$$

Equation 1C.9 can be rewritten as

$$\tilde{p}(n_1,n_2)=\frac{W_1^+(1,n_2)}{W_1^+(n_1,n_2)}\frac{W_2^+(1,0)}{W_2^+(1,n_2)}\times\prod_{i=1}^{n_1-1}\frac{W_1^+(n_1-i+1,n_2)}{W_1^-(n_1-i+1,n_2)}\prod_{j=1}^{n_2}\frac{W_2^+(1,n_2-j+1)}{W_2^-(1,n_2-j+1)}. \tag{1C.10}$$

Finally, substituting Equations 1.197 and 1.198 into the right-hand side of Equation 1C.10, and then substituting the result into Equation 1C.8, one obtains Equation 1.200.

## Appendix 1D: Solution of Equation 1.185

Equation 1.185 subject to boundary conditions (Equation 1.184) can be solved by using the separation of variables that was previously used for various problems in the kinetics of mulicomponent and nonisothermal nucleation [18,36–39]. In the context of the kinetics of deliquescence, the key idea of the method consists of finding such variables of state of the binary cluster that would simultaneously diagonalize the matrix of the diffusion coefficients in Equation 1.185 and the matrix of the second derivatives of the function $G(n_1,n_2) = -\ln g_e(n_1,n_2)$ at the saddle point (recall that Equation 1.185 has to be solved in the vicinity of the saddle point).

First, let us introduce $x$ and $y$ as the new variables of state of a cluster defined as

$$x = c_{11}(n_1 - n_{1c}) + c_{12}(n_2 - n_{2c}),$$

$$y = c_{21}(n_1 - n_{1c}) + c_{22}(n_2 - n_{2c}), \tag{1D.1}$$

with $c_{11}$, $c_{12}$, $c_{21}$, and $c_{22}$ given by Equation 1.204. The bilinear form

$$G(n_1,n_2) = G_c + \frac{1}{2}\frac{\partial^2 G}{\partial n_1^2}\bigg|_c (n_1 - n_{1c})^2 + \frac{\partial^2 G}{\partial n_1 \partial n_2}\bigg|_c \times (n_1 - n_{1c})(n_2 - n_{2c}) + \frac{1}{2}\frac{\partial^2 G}{\partial n_2^2}\bigg|_c (n_2 - n_{2c})^2$$

of the function $G(n_1,n_2)$ in the vicinity of the saddle point acquires the following diagonal form in variables $x$ and $y$,

$$G = G_c - x^2 + y^2, \tag{1D.2}$$

so that they become separated in the equilibrium distribution,

$$n_e(x,y) = Ce^{x^2}e^{-y^2}, \tag{1D.3}$$

where $C$ is the Jacobian of the variables transformation (Equation 1D.1). This allows one to identify $x$ as an unstable variable and $y$ as stable variable.

The kinetic equation 1.185 in variables $x,y$ acquires the canonical form of the Fokker–Planck equation

$$\frac{\partial n(x,y,t)}{\partial t} = a\left\{\frac{\partial}{\partial x}\left(\frac{\partial}{\partial x} - 2x\right) - \varepsilon^{-1}\left[\frac{\partial}{\partial x}\left(\frac{\partial}{\partial y} + 2y\right) + \frac{\partial}{\partial y}\left(\frac{\partial}{\partial x} - 2x\right)\right] + \varepsilon^{-2}(1+p)\frac{\partial}{\partial y}\left(\frac{\partial}{\partial y} + 2y\right)\right\} n(x,y,t), \tag{1D.4}$$

where $n(x,y,t)$ is the distribution of clusters with respect to $x$ and $y$ at time $t$, and the positive parameters $a$, $\varepsilon$, and $p$ are given by Equation 1.203.

Next, it is necessary to find such a transformation of variables $x,y$ that would diagonalize the matrix of diffusion coefficients of Equation 1D.4 while leaving the form 1D.2 invariant. This goal is achieved by using the Lorentz transformation

$$u = (1 - \alpha^2)^{-1/2}(x + \alpha y),\ z = (1 - \alpha^2)^{-1/2}(y + \alpha x), \tag{1D.5}$$

where $u$ and $z$ are the new variables and $\alpha$ is a transformation parameter ($-1 < \alpha < 1$).

The Jacobian of transformation 1D.5 is equal to unity; hence, $n(x,y,t) = n(u,z,t)$,

$$G = G_c - u^2 + z^2, \tag{1D.6}$$

so that the variables $x$ and $y$ are separated in the equilibrium distribution,

$$n_e(u,z) = Ce^{u^2}e^{-z^2}. \tag{1D.7}$$

Choosing the transformation parameter

$$\alpha = (1/2\varepsilon)\{\varepsilon^2 + 1 + p - [(\varepsilon^2 + 1)^2 - 4\varepsilon^2]^{1/2}\}, \tag{1D.8}$$

the kinetic equation 1D.4 becomes

$$\frac{\partial n(u,z,t)}{\partial t} = A\left[\frac{\partial}{\partial u}\left(\frac{\partial}{\partial u} - 2u\right) + \frac{(1-\varepsilon\alpha)^2}{\alpha^2 p}\frac{\partial}{\partial z}\left(\frac{\partial}{\partial z} + 2z\right)\right]n(u,z,t), \tag{1D.9}$$

with $A$ given by Equation 1.202. Equation 1D.9 is subjected to the boundary conditions for any $z$:

$$\frac{n(u,z,t)}{n_e(u,z)} = \begin{cases} 1 & (u \to -\infty) \\ 0 & (u \to \infty) \end{cases} \tag{1D.10}$$

One can show that the solution of Equation 1D.9 subject to boundary conditions (Equation 1D.10) has the form

$$n(u,z,t) = \sqrt{\pi}q(u,t)e^{-z^2}, \tag{1D.11}$$

where the function $q(u,t)$ is governed by the one-dimensional kinetic equation

$$\partial_t q(u,t) = A\frac{\partial}{\partial u}\left(\frac{\partial}{\partial u} - 2u\right)q(u,t) \tag{1D.12}$$

and satisfies the boundary conditions

$$\frac{q(u,t)}{q_e(u)} = \begin{cases} 1 & (u \to -\infty) \\ 0 & (u \to \infty). \end{cases} \tag{1D.13}$$

From Equation 1D.11, one obtains the expression

$$q(u,t) = \int_{-\infty}^{\infty} dz\, n(u,z,t), \tag{1D.14}$$

which clearly shows that $q(u, t)$ is the distribution of clusters with respect to the unstable variable $u$. Equation 1D.11 can be rewritten as

$$\partial_t q(u,t) = -\frac{\partial}{\partial u} J_u, \; J_u = -A\left(\frac{\partial}{\partial u} - 2u\right) q(u,t), \tag{1D.15}$$

where $J_u = J_u(u,t)$ is the one-dimensional flux density of clusters along the $u$ axis integrated with respect to the stable variable $z$.

The steady-state solution of the one-dimensional Equation 1D.15 subject to boundary conditions (Equation 1D.13) leads to Equation 1.201 for the steady-state nucleation rate $J_s$.

## Appendix 1E: Explicit Expressions for the Coefficients of Polynomial (Equation 1.208)

Solving Equations 1.209 through 1.211 with respect to $A$, $B$, and $C$, one obtains

$$A \equiv A(R,Q) = -\frac{\frac{1}{4}\upsilon_{T2} C(R,Q) - W_2^-(R,Q)}{2D_2(Q-R) - \frac{1}{4}R(2Q-R)}, \tag{1E.1}$$

$$B \equiv B(R,Q) = -2A(R, Q)Q, \tag{1E.2}$$

$$C \equiv C(R,Q) = \frac{80\left(R_0^3 - R^3\right)s_1(R,Q)/\upsilon_2^\gamma + W_2^-(R,Q)s_2(R,Q)}{80(Q^3 - R^3)s_1(R,Q) + s_2(R,Q)}, \tag{1E.3}$$

where $s_1(R,Q) = 2D(Q-R) - \frac{1}{4}\upsilon_{T2}R(2Q-R)$, $s^2(R, Q) = 30(Q^4 - R^4)Q - 12(Q^5 - R^5)$.

## References


1. Djikaev, Y.S., Donaldson, D.J., *J. Geophys. Res. D* 104, 14283 (1999).
2. Djikaev, Y.S., Donaldson, D.J., *J. Chem. Phys.* 113, 6822 (2000).
3. Djikaev, Y.S., Donaldson, D.J., *J. Geophys. Res. D* 106, 14447 (2001).
4. Tang, I.N., Munkelwitz, H.R., Wang, N., *J. Colloid Interface Sci.* 114, 409 (1986).
5. Tang, I.N., Munkelwitz, H.R., *Atmos. Environ., Part A* 27A, 467 (1993).
6. Tang, I.N., Munkelwitz, H.R., *J. Geophys. Res.* 99, 18 (1994).
7. McMurry, P., Stolzenburg, M.R., *Atmos. Environ.* 23, 497 (1989).
8. Svenningsson, I.B., Doctoral thesis, Department of Nuclear Physics, University of Lund, 1997.
9. Richardson, C.B., Spann, J.F., *J. Aerosol Sci.* 15, 563 (1984).
10. Hämeri, K., Väkevä, M., Hanssen, H.-C., Laaksonen, A., *J. Geophys. Res.* 105, 22231 (2000).
11. Hämeri, K., Laaksonen, A., Väkevä, M., Suni, T., *J. Geophys. Res.* 106, 20749 (2001).
12. Romakkaniemi, S., Hämeri, K., Väkevä, M., Hanssen, H.-C., Laaksonen, A., *J. Phys. Chem. A* 105, 8183 (2001).
13. Kuni, F.M., Shchekin, A.K., Rusanov, A.I., *Colloid J. Russ. Acad. Sci.* (Engl. Transl.) 55, 686 (1993).
14. Shchekin, A.K., Rusanov, A.I., Kuni, F.M., *Colloid J. Russ. Acad. Sci.* (Engl. Transl.) 55, 776 (1993).
15. Shchekin, A.K., Shabaev, I., *Colloid J. Russ. Acad. Sci.* (Engl. Transl.) 60, 111 (1998).
16. Djikaev, Y.S., Bowles, R., Reiss, H., Hämeri, K., Laaksonen, A., Väkevä, M., *J. Phys. Chem. B* 105, 7708 (2001).
17. Djikaev, Y.S., Bowles, R., Reiss, H., *Physica A* 298, 155 (2001).
18. Djikaev, Y.S., *J. Chem. Phys.* 116, 9865 (2002).
19. Shchekin, A.K., Shabaev, I., in *Nucleation Theory and Applications*, Schmelzer, J.W.P., Röpke, G., Priezjev, V.B., Eds., p. 177, Dubna, JINR, 2006.
20. Shchekin, A.K., Shabaev, I., *Colloid J. Russ. Acad. Sci.* (Engl. Transl.) 60, 111 (1998).
21. Lothe, J., Pound, G.M.J., in *Nucleation*, Zettlemoyer, A.C., Ed., Marcel-Dekker, New York, 1969.

22. Narsimhan, G., Ruckenstein, E., *J. Colloid Interface Sci.* 128, 549 (1989). (Section 1.2 of this volume.)
23. Ruckenstein, E., Nowakowski, B., *J. Colloid Interface Sci.* 137, 583 (1990). (Section 1.3 of this volume.)
24. Nowakowski, B., Ruckenstein, E., *J. Colloid Interface Sci.* 139, 500 (1990). (Section 1.4 of this volume.)
25. Nowakowski, B., Ruckenstein, E., *J. Chem. Phys.* 94, 1397 (1991). (Section 2.1 of this volume.)
26. Ruckenstein, E., Nowakowski, B., *Langmuir* 7, 1537 (1991). (Section 2.2 of this volume.)
27. Nowakowski, B., Ruckenstein, E., *J. Chem. Phys.* 94, 8487 (1991). (Section 2.3 of this volume.)
28. Djikaev, Y.S., Ruckenstein, E., *J. Chem. Phys.* 123, 214503 (2005). (Section 2.5 of this volume.)
29. Djikaev, Y.S., Ruckenstein, E., *J. Chem. Phys.* 124, 124521 (2006). (Section 2.6 of this volume.)
30. Konopka, P., *J. Aerosol Sci.* 28, 1411 (1997).
31. Gorbunov, B., *J. Chem. Phys.* 110, 10035 (1999).
32. Reiss, H., *J. Chem. Phys.* 18, 840 (1950).
33. Stauffer, D., *J. Aerosol Sci.* 7, 319 (1976).
34. Melikhov, A.A., Kurasov, V.B., Dzhikaev, Y.S., Kuni, F.M., *Khim Fiz.* 9, 1713 (1990).
35. Melikhov, A.A., Kurasov, V.B., Dzhikaev, Y.S., Kuni, F.M., *Zh. Tekh. Fiz.* 61, 27 (1991).
36. Kuni, F.M., Melikhov, A.A., *Theor. Math. Phys.* 81, 1182 (1989).
37. Kuni, F.M., Melikhov, A.A., Novozhilova, T.Yu., Terent'ev, I.A., *Theor. Math. Phys.* 83, 530 (1990).
38. Djikaev, Y.S., Kuni, F.M., Grinin, A.P., *J. Aerosol Sci.* 30, 265 (1999).
39. Djikaev, Y.S., Teichmann, J., Grmela, M. *Physica A* 267, 322 (1999).
40. Tolman, R.C., *J. Chem. Phys.* 17, 333 (1949).
41. Kirkwood, J.G., Buff, F.P., *J. Chem. Phys.* 17, 338 (1949).
42. Abraham, F.F., *Homogeneous Nucleation Theory*, Academic, New York, 1974.
43. Mullin, J.W., *Crystallysation*, Butterworth, London, 1972.
44. Chandrasekhar, S., *Rev. Mod. Phys.* 15, 1 (1949).
45. Gardiner, C.W., *Handbook of Stochastic Methods*, Springer, New York, 1983.
46. Agmon, N., *J. Chem. Phys.* 81, 3644 (1984).

## 1.7 Effect of Solute–Solute and Solute–Solvent Interactions on Kinetics of Nucleation in Liquids

*Eli Ruckenstein and Gersh Berim**

One of the assumptions of the theory of nucleation developed by Ruckenstein et al. is that the main contribution to the nucleation rate of a solid phase from a solution comes from the interaction of a solute molecule with those in a cluster (nucleus) of the solid phase. This assumption is avoided in this paper by including the interactions of the solute molecule with those outside the cluster and with the molecules of the solvent. For each of the above interactions, the rate of nucleation changes when compared to the original theory by several orders of magnitudes when calculated at a fixed density of the solute, but changes less than 1 order of magnitude when calculated as a function of supersaturation. Such changes are usually small compared with the absolute magnitude of the nucleation rate.

### 1.7.1 Introduction

In a couple of papers [1,2], a kinetic theory of nucleation in liquids was developed in which the rate of dissociation of nuclei was calculated on the basis of a mean first passage time analysis. The theory did not employ the macroscopic concept of surface tension and was based on molecular interactions only. To simplify the problem, only the interactions between the molecules of solute and those of a cluster of the new solid phase were taken into account. The interactions between the solute molecules themselves as well as the interactions of these molecules with those of the solvent were taken into account only indirectly by modifying the energy parameter of the solute–cluster potential. In the present paper the theory is improved by including those interactions.

### 1.7.2 Background

#### *1.7.2.1 Interaction Potentials*

The considered system consists of a cluster of the new phase that is assumed to be an amorphous, uniform, spherical solid of density $\rho_c$ and radius $R$ surrounded by a liquid that consists of molecules of solvent of density $\rho_w$ and molecules of solute of density $\rho_s$. Note that the latter densities are connected by the relation

$$\rho_w \upsilon_2 + \rho_s \upsilon_s = 1, \tag{1.218}$$

where $\upsilon_2$ and $\upsilon_s$ are the volumes of a molecule of solvent and a molecule of solute, respectively. The dissociation of molecules from the cluster is supposed to occur only from the surface layer of thickness $\eta$, whereas the molecules beneath this layer do not dissolve. The interaction potential between two molecules of solute has the form

$$\phi_1(r_{12}) = \begin{cases} -\epsilon_1 \left(\dfrac{\eta}{r_{12}}\right)^6, & r_{12} \geq \eta, \\ \infty, & r_{12} < \eta \end{cases} \tag{1.219}$$

where $r_{12} = |\mathbf{r} - \mathbf{r}'|$ denotes the distance between the centers of the interacting molecules, $\eta$ being the diameter of the hard core repulsion, and $\epsilon_1$ is the energy parameter of the London–van der Waals attraction. In Ref. [2], the effect of the presence of the surrounding liquid medium was accounted for by the appropriate estimation of the effective interaction constant $\epsilon_1$. In the present paper, we introduce explicitly a new potential, $\phi_2(r_{12})$, which accounts for the interactions between a molecule of solute and the molecules of solvent, which will be taken to be of the same form as the potential $\phi_1(r_{12})$, but with another energy parameter $\epsilon_1 \neq \epsilon_2$ instead of $\epsilon_1$ (see Equation 1.219). Even though the real solute–solvent interaction potential might have a different form, this approximation allows one to simplify the calculations. The interaction between a molecule of solute with the solute molecules located outside the cluster are also taken into account. The net potential exerted on a single molecule of solute is given by the expression

* *J. Colloid Interface Sci.* 342, 528 (2010). Republished with permission.

$$\phi(r)=\int_{v_c} d\mathbf{r}'\phi_1(|\mathbf{r}-\mathbf{r}'|)\rho_c(\mathbf{r}')+\int_{v'} d\mathbf{r}'\phi_2(|\mathbf{r}-\mathbf{r}'|)\rho_w(\mathbf{r}')+\int_{v'} d\mathbf{r}'\phi_1(|\mathbf{r}-\mathbf{r}'|)\rho_s(\mathbf{r}') \tag{1.220}$$

where $\mathbf{r}$ provides the location of the solute molecule, $v_c$ and $v'$ are the volumes of the cluster and of the remaining of the system, respectively, and the coordinates are taken with respect to the center of the cluster. The three integrals in Equation 1.220 represent the interaction potentials of a solute molecule with the cluster, with the molecules of the solvent outside the cluster, and with the solute molecules located outside the cluster, respectively. (It is assumed that the molecules of the solvent do not penetrate into the cluster.) Only the first of the three potentials in Equation 1.220 was taken into account in Refs. [1,2]. To evaluate the second integral of Equation 1.220, it is convenient to write it in the form

$$\int_{v'} d\mathbf{r}'\phi_2(|\mathbf{r}-\mathbf{r}'|)\rho_w(\mathbf{r}')=\int_{v} d\mathbf{r}'\phi_2(|\mathbf{r}-\mathbf{r}'|)\rho_w(\mathbf{r}')-\int_{v_c} d\mathbf{r}'\phi_2(|\mathbf{r}-\mathbf{r}'|)\rho_w(\mathbf{r}'), \tag{1.221}$$

where $v$ is the volume of the entire system, which is considered to be infinite. For uniform densities of the cluster, solute, and solvent ($\rho_c(\mathbf{r}) = \rho_c$, $\rho_s(\mathbf{r}) = \rho_s$, $\rho_w(\mathbf{r}) = \rho_w$) the case considered below, the first right-hand side integral of Equation 1.221 is a constant that can be omitted. Consequently,

$$\int_{v'} d\mathbf{r}'\phi_2(|\mathbf{r}-\mathbf{r}'|)\rho_w(\mathbf{r}')=-\rho_w\int_{v_c} d\mathbf{r}'\phi_2(|\mathbf{r}-\mathbf{r}'|) \tag{1.222}$$

i.e., the potential of interaction of a solute molecule with the entire volume of the solvent is given by that for the interaction of this molecule with the solvent that occupies the volume $v_c$ of the cluster, taken with opposite sign. The third integral in Equation 1.220 can be rewritten in a similar way as

$$\int_{v'} d\mathbf{r}'\phi_1(|\mathbf{r}-\mathbf{r}'|)\rho_s(\mathbf{r}')=-\rho_s\int_{v_c} d\mathbf{r}'\phi_1(|\mathbf{r}-\mathbf{r}'|). \tag{1.223}$$

The explicit analytical expression for the net potential $\phi(r)$ depends on whether the center of the selected solute molecule is located inside the surface layer ($R \le r \le R + \eta$) of the cluster or outside of it ($r \ge R + \eta$). Introducing the dimensionless potential $\Phi(r) = \phi(r)/k_B T$, where $k_B$ is the Boltzmann constant, one obtains

$$\Phi(r)=\begin{cases}\dfrac{\pi}{12}\mu\varepsilon_1\left[-8+3\dfrac{r}{\eta}\left(1-\dfrac{R^2}{r^2}\right)+6\dfrac{\eta}{r}-\dfrac{r+3r}{r(r+R)^3}\eta^3\right], & R\le r\le R+\eta\\ -\dfrac{4\pi}{3}\mu\varepsilon_1\left(\dfrac{\eta}{r-R}\right)^3\left(\dfrac{R}{r+R}\right)^3, & r>R+\eta\end{cases} \tag{1.224}$$

where $\varepsilon_1 = \epsilon_1/k_B T$,

$$\mu=\mu_c+\delta_\epsilon\eta^3/\upsilon_2-\mu_s\left(1-\delta_\epsilon\frac{\upsilon_2}{\upsilon_1}\right), \tag{1.225}$$

$\mu_c = \rho_c\eta^3$, $\delta_\epsilon = \epsilon_2/\epsilon_1$, $\mu_s = \rho_s\eta^3$. The general form of the net interaction potential, Equation 1.224, coincides with that of the potential provided by Equation 3 of Ref. [2], which was calculated by taking into account the interaction of a solute molecule with the cluster only. In the latter case, the parameter $\mu$ in Equation 1.224 is equal with the density of the molecules in the cluster, whereas in the present case $\mu$ is given by Equation 1.225.

### 1.7.2.2 Nucleation Rate

To calculate the nucleation rate, let us consider the populations $f_i$ of clusters of various sizes consisting of $i$ molecules ($i = 1, 2, \ldots$) [3,4]. The flux of clusters passing from the populations $f_{i-1}$ to $f_i$ is given by the equation

$$I_i = \beta_{i-1} f_{i-1} - \alpha_i f_i, \tag{1.226}$$

where $f_i$ is the number of clusters consisting of $i$ molecules, $\alpha_i$ and $\beta_i$ are the rates of dissociation and condensation of molecules from and on the surface of a cluster from population $f_i$, respectively, both being dependent on the cluster size. The change of $f_i$ as function of time is described by the equation

$$\frac{df_i}{dt} = (I_{i+1} - I_i) = \beta_{i-1} f_{i-1} - (\alpha_i + \beta_i) f_i + \alpha_{i+1} f_{i+1}. \tag{1.227}$$

Assuming a smooth dependence of $f_i$ on $i$, Equation 1.227 can be transformed into the following form [2]

$$\frac{\partial f(g,t)}{\partial t} = -\frac{\partial I(g,t)}{\partial t} = \frac{1}{2}\frac{\partial^2}{\partial g^2}(\beta+\alpha) f(g,t) - \frac{\partial}{\partial g}(\beta-\alpha) f(g,t), \tag{1.228}$$

where the continuum variable $g$ replaces the discrete variable $i$. The rate of nucleation, $I$, is provided by the stationary value of the flux

$$I(g,t) = -\frac{1}{2}\frac{\partial}{\partial g}[(\beta+\alpha) f(g,t)] + (\beta-\alpha) f(g,t) \tag{1.229}$$

of clusters along the coordinate $g$ [2,5]. To find the stationary solution of Equation 1.229, hence the solution for $I(g, t) \equiv I = \text{const}$ and $f(g, t) \equiv f(g)$, boundary conditions for $f(g)$ at $g \to \infty$ (large clusters) and $g \to 1$ (smallest clusters, monomers) should be satisfied. Regarding the first boundary condition, it is assumed that clusters with very large sizes are absent in the system. Consequently, $f(g) \to 0$ as $g \to \infty$. Regarding the small clusters (monomers), it is assumed that nucleation does not cause their substantial depletion, hence, that their number density remains the initial one. Consequently, $f(g) \to \rho_s$ as $g \to 1$ (in calculations, the limit 1 can be replaced by zero). Straightforward calculations provide the following expression for the rate of nucleation

$$I = \frac{\frac{1}{2}\beta_1 \rho_s}{\int_0^{\infty} \exp[-2w(g)]\,dg}, \tag{1.230}$$

where $\beta_1$ is the condensation rate calculated for a cluster consisting of one molecule of the solute (see Section 1.7.2.4 for details) and

$$w(g) = \int_0^g \frac{\beta-\alpha}{\beta+\alpha} dg. \tag{1.231}$$

In Equation 1.231, the rates of dissociation ($\alpha$) and condensation ($\beta$), although functions of the discrete number $i$ of the molecules in the cluster, are considered continuous functions of $i$, or, equivalently, of the cluster radius $R$. Equations 1.230 and 1.231 will be used below to calculate the nucleation rate.

### 1.7.2.3 Dissociation Rate

The similarity between the interaction potential, Equation 1.224, and that used in Ref. [2] provides the possibility to use the results of Ref. [2] for calculating the rate of dissociation $\alpha$ of the molecules from the external layer of the cluster. Dissociation is assumed to be due to Brownian motion of the surface molecules that is the result of collisions of those molecules with others. In Ref. [2], it was suggested to describe the Brownian motion using the Smoluchowski equation under an external field [6,7], which in the case of spherical symmetry has the form

$$\frac{\partial n(r,t)}{\partial t} = Dr^{-2}\frac{\partial}{\partial r}\left(r^2 e^{-\Phi(r)} n(r,t)\right), \tag{1.232}$$

where $n(r,t)$ is the number density of diffusing molecules and $D$ is the diffusion coefficient. The transition probability $p(r, t|r_0)$ that a molecule initially at distance $r_0$ will be located after time $t$ at point $r$ satisfies the backward Smoluchowski equation [7–9]:

$$\frac{\partial p(r,t\,|\,r_0)}{\partial t} = Dr_0^{-2}e^{\Phi(r_0)} \times \frac{\partial}{\partial r_0}\left(r_0^2 e^{-\Phi(r_0)}\frac{\partial}{\partial r_0}p(r,t\,|\,r_0)\right). \tag{1.233}$$

The probability that a molecule initially at point $r_0$ will leave the surface layer of the cluster (dissociate) during a time between 0 and $t$ is $1 - Q(t|r_0)$, where $Q(t|r_0)$ is the so-called survival probability

$$Q(t\,|\,r_0) = \int_R^{R+\eta} r^2 p(r,t\,|\,r_0)\mathrm{d}r. \tag{1.234}$$

The probability density for the dissociation time is given by $-\partial Q/\partial t$ and the mean passage time can be calculated with the formula [2]

$$\tau(r_0) = -\int_0^\infty t\frac{\partial Q(t\,|\,r_0)}{\partial t}\mathrm{d}t = \int_0^\infty Q(t\,|\,r_0)\mathrm{d}t. \tag{1.235}$$

Integrating Equation 1.233 over $r$ and $t$ over their entire ranges and using the boundary conditions $Q(0|r_0) = 1$ and $Q(t|r_0) \to 0$ as $t \to \infty$ for any $r_0$, one arrives at the following equation for $\tau(r_0)$ [2,7–10]:

$$-Dr_0^{-2}e^{\Phi(r_0)}\frac{\partial}{\partial r_0}\left(r_0^2 e^{-\Phi(r_0)}\frac{\partial}{\partial r_0}\tau(r_0)\right) = 1. \tag{1.236}$$

The boundary conditions for $\tau(r_0)$ follow from the boundary conditions imposed on the function $p(r, t|r_0)$. Considering the inner wall of the surface layer as a reflecting boundary, one obtains

$$\left.\frac{\mathrm{d}\tau}{\mathrm{d}r}\right|_{r=R} = 0. \tag{1.237}$$

The molecules reaching the outer boundary of the surface layer separate from the cluster and diffuse further in the outer space. The diffusive flux under an external field can be obtained from Equation 1.232 and is given by the equation

$$j(r,t) = De^{-\Phi(r)}\frac{\partial}{\partial r}e^{-\Phi(r)}n(r,t). \tag{1.238}$$

Assuming a quasi steady state and integrating with the boundary condition $n(r) \to 0$ as $r \to \infty$, the following expression for the diffusive flux is obtained:

$$j(r) = \frac{De^{\Phi(r)}}{\int_r^{\infty} x^{-2} e^{\Phi(x)} \mathrm{d}x} r^{-2} n(r). \tag{1.239}$$

The latter equation at $r = R + \eta$ becomes the radiation boundary condition

$$j(R+\eta) = k_0 n(R+\eta), \tag{1.240}$$

where

$$k_0 = \frac{De^{\Phi(R+\eta)}}{(R+\eta)^2 \int_{R+\eta}^{\infty} x^{-2} e^{\Phi(x)} \mathrm{d}x}. \tag{1.241}$$

As a consequence of Equation 1.240, the following boundary condition for the mean passage time can be written:

$$\left.\frac{\mathrm{d}\tau}{\mathrm{d}r}\right|_{r=R+\eta} = -k_0 D^{-1} \tau(R+\eta). \tag{1.242}$$

Using the boundary conditions Equations 1.237 and 1.242, the following solution of Equation 1.236 is obtained:

$$\tau(r_0) = \frac{1}{D} \int_{r_0}^{R+\eta} \frac{\mathrm{d}y}{p_{\mathrm{eq}}(y)} y \int_R^{y} \mathrm{d}x p_{\mathrm{eq}}(x) + [k_0 p_{\mathrm{eq}}(R+\eta)]^{-1}, \tag{1.243}$$

where $p_{\mathrm{eq}}$ is the equilibrium probability distribution given by

$$p_{\mathrm{eq}}(x) = x^2 e^{-\Phi(x)} / Z$$
$$\text{and } Z = \int_R^{R+\eta} x^2 e^{-\Phi(x)} / \mathrm{d}x \tag{1.244}$$

Let us assume that the relaxation time to achieve equilibrium is much shorter than the dissociation time. Then the average mean passage time $\langle\tau\rangle$ can be obtained from Equation 1.243 after integration over the initial distribution of $r_0$, which is assumed to be the equilibrium one, i.e., $\langle\tau\rangle = \int_R^{R+\eta} \tau(r_0) p_{\mathrm{eq}}(r_0) \mathrm{d}r_0$. As a result, the average mean passage time is given by

$$\langle\tau\rangle = \frac{1}{D} \int_R^{R+\eta} \mathrm{d}r_0 p_{\mathrm{eq}}(r_0) \int_{r_0}^{R+\eta} \frac{\mathrm{d}y}{p_{\mathrm{eq}}(y)} \int_R^{y} \mathrm{d}x p_{\mathrm{eq}}(x) + \left[k_0 p_{\mathrm{eq}}(R+\eta)\right]^{-1}. \tag{1.245}$$

The rate of dissociation of the surface molecules is provided by the equation [2]

$$\alpha = \frac{N_s}{\langle \tau \rangle}, \tag{1.246}$$

where $N_s$ is the number of surface molecules that is given by the expression

$$N_s = 4\pi\mu_c(R/\eta)^2(1 + \eta/R + \eta^2/3R^2). \tag{1.247}$$

The rate of dissociation depends on the form of the intermolecular potential and on the size of the cluster.

#### 1.7.2.4 Condensation Rate

The expression for the condensation rate β was obtained from the stationary solution of the diffusion equation in an external field, Equation 1.232, and has the form

$$\beta = \gamma 4\pi D R \rho_s, \tag{1.248}$$

where $\gamma^{-1} = R\int_{R+\eta}^{\infty} r^{-2} e^{\Phi(r)}\, dr$. In the limit $R \to \infty$, the factor γ tends to unity.

The size $R_c$ of the critical nucleus can be calculated from the balance between dissociation and condensation rates ($\alpha = \beta$) provided by Equations 1.246 and 1.248. In the limit of large clusters, $R_c$ is provided by the equation [2]

$$\frac{4\pi\rho_c R_c D\left(1+\eta/R_c+\frac{1}{3}(\eta/R_c)^2\right)}{e^{A/3}(e^A-1)/A+\frac{\eta}{R_c}\left(3(1-(1+A)e^{-A})/A^2-e^{4A/3}+\frac{1}{2}e^{A/3}-\frac{1}{2}e^{-A}+\frac{2(\sinh A-A)}{(1-e^{-A})A^2}\right)} = 4\pi\rho_s R_c D, \tag{1.249}$$

where

$$A = \mu\pi\epsilon/2kT. \tag{1.250}$$

The important consequence from this equation is the expression for the saturation number density $n_s$ of the solute, which can be obtained from Equation 1.249 in the limit of large clusters ($\eta/R \to 0$):

$$n_s = \frac{\rho A}{1-e^{-A}} e^{-4A/3}. \tag{1.251}$$

When solving Equation 1.251, the density of the solute, $\rho_s$, which is present in the parameter μ (see Equation 1.225), should be replaced by $n_s$.

### 1.7.3 Numerical Estimation of Effect of Solute–Solute and Solute–Solvent Interactions on Kinetics of Nucleation

To clarify the importance of the interactions of a solute molecule with other solute molecules located outside the cluster and with the molecules of solvent, the nucleation rate $I$ was calculated separately for three different cases. The first one (i) coincides with the case considered in Ref. [2], where only the interaction of the solute molecule with those in the cluster was taken into account. In the second case (ii), in addition to that interaction the interaction of the solute molecule with the solute molecules outside the cluster is included. In the third case (iii), the interaction of the solute molecule with the molecules of the solvent is added to the two interactions mentioned above. To distinguish between these three cases, the subscripts 0, 1, and 2, respectively, will be used for the supersaturation $s$, critical radius $R_c$, nucleation rate $I$, and saturated density of the solute $n_s$. By comparing the results for cases (ii) and (iii) with those for case (i), one can estimate the contribution of the additional interactions to the nucleation rate.

In all calculations, the parameter $\mu_c$ was taken as 1.2, as was selected in Ref. [2]. The energy parameter $\epsilon_1$ of the solute–solute interaction was selected to be $\varepsilon_1 = 2$ and $\varepsilon_1 = 4$, where $\varepsilon_1 = \epsilon_1/k_BT$ and the parameters $v_2$ and $v_s$ were selected equal to $\eta^3$.

Because the results for case (i) can be obtained using the corresponding equations of Ref. [2], let us analyze first the role of the interaction of a solute molecule with other solute molecules located outside the cluster (case (ii)) by neglecting the interaction of the molecule with those of the solvent. To analyze this case, one should take $\delta_\epsilon = 0$ in the expression of $\mu$, the latter parameter thus becoming $\mu = \mu_c - \mu_s$. As an immediate consequence of the change in the value of the parameter $\mu$, one can note the change in the value of $n_s$ of the saturation density calculated with Equation 1.251. For example, at $\varepsilon_1 = 2$; $v_{s0} = 0.03038$ ($v_s = \eta_s\eta^3$) for case (i) and $v_{s1} = 0.03415$ if the solute–solute interaction is taken into account (case (ii)). This affects the supersaturations $s \equiv \rho_s/n_s = \mu_s/v_s$ which are different for the same solute density $\rho_s$. The same observations can be made regarding the critical cluster size $R_c$ and the rate of nucleation $I$.

In Table 1.4, the nucleation rates $I_0$ and $I_1$ are listed for several values of $\mu_s$ along with the supersaturations $s_0$ and $s_1$ and the critical cluster radii $R_{c0}$ and $R_{c1}$. This table shows that for a given $\mu_s$ the solute–solute interactions decrease the nucleation rate, and that the ratio $I_1/I_0$ decreases with decreasing $\mu_s$. For $\mu_s \leq v_{s1} = 0.03415$, this ratio becomes zero. The latter result occurs because for $\mu_s < v_{s1}$ the solution is undersaturated ($s_1 < 1$) in case (ii) but oversaturated ($s_0 > 1$) in case (i). One should note that the changes in the nucleation rate due to the solute–solute interactions are small (less than 1 order of magnitude, which for homogeneous nucleation constitutes a small number) for almost all values of $\mu_s$.

At the lower temperature ($\varepsilon_1 = 4$); $v_{s0} \simeq v_{s1}$($v_{s0} = 3.897 \times 10^{-4}$, $v_{s1} = 3.909 \times 10^{-4}$) and the difference between $s_0$ and $s_1$ for a given $\mu_s$ is negligible. In this case, Table 1.4 shows that $I_0 \simeq I_1$.

The small contribution of solute–solute interactions to the nucleation rate can be explained by the relatively small density of the solute molecules outside the cluster and, consequently, the small change in the total potential energy of a solute molecule due to this interaction.

Similar to the solute–solute interactions, the solute–solvent ones also affect (at a selected $\mu_s$) the calculated values of the saturation density $n_s$ and, as a consequence, the supersaturation $s$, critical cluster size $R$, and the rate of nucleation $I$. The dependence of $v_{s2}$ on $\delta_\epsilon = \epsilon_2/\epsilon_1$ is shown in Figure 1.28 for $\varepsilon_1 = 2$ and $0 \leq \delta_\epsilon \leq 0.5$.

The quantity $v_s$ increases with increasing $\delta_\epsilon$ and rapidly acquires the solid-like magnitude ~0.2. For this reason, our considerations are restricted to the range of small values of $\delta_\epsilon (\delta_\epsilon \leq 0.1)$.

In Table 1.5, the results for case (iii) are presented along with those for case (i), marked with subscript 0. The data in this table are calculated for $\varepsilon_1 = 2$ and $\delta_\epsilon = 0.03$ and $\delta_\epsilon = 0.1$. In both cases, the concentration $\mu_s$ is taken larger than $v_{s2}$. Note that, for comparison between cases (i) and (iii) at $\varepsilon_1 = 3$ and various solute number densities $\mu_s$, and comparison between cases (i) and (iii) at $\varepsilon_1 = 4$ and various solute number densities $\mu_s \leq v_{s2}$, the nucleation rate $I_2$ is equal to zero because $s_2$ is smaller than one, and $R_{c2} = \infty$. In all cases, $I_2 \ll I_0$ and the ratio $I_2/I_0$ decreases with decreasing $\mu_s$ as in case (ii), varying from $I_2/I_0 \sim 10^{-1}$ for larger $\mu_2$ to $10^{-5}$ for $\delta_\epsilon = 0.03$, or to $10^{-6}$ for $\delta_\epsilon = 0.1$ for $\mu_s$ close to $v_{s2}$.

In Table 1.6, similar results are presented for $\varepsilon_1 = 4$. In this case, $I_2/I_0$ varies from $I_2/I_0 \sim 10^{-1}$ for large $\mu_s$ to $10^{-8}$ for $\delta_\epsilon = 0.03$, or to $10^{-11}$ for $\delta_\epsilon = 0.1$.

Table 1.4 Comparison between Cases (i) and (ii) at $\varepsilon_1 = 2$ and $\varepsilon_1 = 4$ and Various Solute Number Densities $\mu_s$

| $\mu_s$ | $s_0$ | $s_1$ | $2I_0/\rho_s\beta_1$ | $2I_1/\rho_s\beta_1$ | $I_1/I_0$ | $R_{c0}/\eta$ | $R_{c1}/\eta$ |
|---|---|---|---|---|---|---|---|
| $\varepsilon_1 = 2$, $v_{s1} = 0.03415$, $v_{s0} = 0.03038$ | | | | | | | |
| $3.5 \times 10^{-2}$ | 1.15 | 1.025 | $4.30 \times 10^{-3}$ | $3.9 \times 10^{-4}$ | $9.1 \times 10^{-2}$ | 12.5 | 86 |
| $4.0 \times 10^{-2}$ | 1.32 | 1.170 | $2.52 \times 10^{-2}$ | $5.8 \times 10^{-3}$ | $2.3 \times 10^{-1}$ | 6.1 | 12.1 |
| $5.0 \times 10^{-2}$ | 1.64 | 1.464 | $1.05 \times 10^{-1}$ | $5.3 \times 10^{-2}$ | $5.0 \times 10^{-1}$ | 2.5 | 3.9 |
| $\varepsilon_1 = 4$, $v_{s1} = 3.909 \times 10^{-4}$, $v_{s0} = 3.897 \times 10^{-4}$ | | | | | | | |
| $3.99 \times 10^{-4}$ | 1.0238 | 1.021 | $1.3 \times 10^{-13}$ | $6.1 \times 10^{-14}$ | $4.7 \times 10^{-1}$ | 290 | 310 |
| $4.50 \times 10^{-4}$ | 1.155 | 1.151 | $5.3 \times 10^{-8}$ | $4.6 \times 10^{-8}$ | $8.7 \times 10^{-1}$ | 44 | 45 |
| $5.00 \times 10^{-4}$ | 1.282 | 1.279 | $1.8 \times 10^{-6}$ | $1.6 \times 10^{-6}$ | $8.9 \times 10^{-1}$ | 23.5 | 23.8 |
| $1.00 \times 10^{-4}$ | 2.570 | 2.560 | $3.1 \times 10^{-3}$ | $3.0 \times 10^{-3}$ | $9.7 \times 10^{-1}$ | 5.1 | 5.2 |
| $5.00 \times 10^{-4}$ | 12.80 | 12.80 | $2.10 \times 10^{-1}$ | $2.05 \times 10^{-1}$ | $9.8 \times 10^{-1}$ | 1.25 | 1.28 |

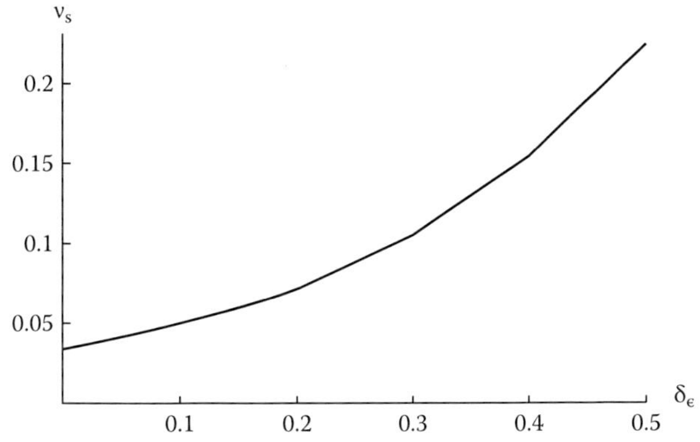


Figure 1.28

Dimensionless saturation density $\mu_s = n_s\eta^3$ as function of the parameter $\delta_\epsilon = \epsilon_2/\epsilon_1$ at $\epsilon_1/k_BT = 2$.

Table 1.5 Comparison between Cases (i) and (iii) at $\varepsilon_1 = 2$ and Various Solute Number Densities $\mu_s$

| $\mu_s$ | $s_0$ | $s_2$ | $2I_0/\rho_s\beta_1$ | $2I_2/\rho_s\beta_1$ | $I_2/I_0$ | $R_{c0}/\eta$ | $R_{c1}/\eta$ |
|---|---|---|---|---|---|---|---|
| $\varepsilon_2/\varepsilon_1 = 0.03$, $v_{s2} = 0.03820$, $v_{s0} = 0.03038$ | | | | | | | |
| $3.85 \times 10^{-2}$ | 1.267 | 1.0078 | $1.94 \times 10^{-2}$ | $1.3 \times 10^{-6}$ | $6.7 \times 10^{-5}$ | 7.0 | 255 |
| $3.90 \times 10^{-2}$ | 1.283 | 1.021 | $2.18 \times 10^{-2}$ | $4.4 \times 10^{-5}$ | $2.0 \times 10^{-3}$ | 6.9 | 99 |
| $4.00 \times 10^{-2}$ | 1.32 | 1.047 | $2.52 \times 10^{-2}$ | $2.96 \times 10^{-4}$ | $1.2 \times 10^{-2}$ | 6.1 | 43 |
| $4.50 \times 10^{-2}$ | 1.481 | 1.178 | $6.13 \times 10^{-2}$ | $8.4 \times 10^{-3}$ | $1.4 \times 10^{-1}$ | 3.5 | 11.5 |
| $5.00 \times 10^{-2}$ | 1.64 | 1.309 | $1.05 \times 10^{-1}$ | $2.8 \times 10^{-2}$ | $2.7 \times 10^{-1}$ | 2.5 | 5.5 |
| $6.00 \times 10^{-2}$ | 1.97 | 1.57 | $2.06 \times 10^{-1}$ | $9.2 \times 10^{-2}$ | $4.5 \times 10^{-1}$ | 1.0 | 2.6 |
| $\varepsilon_2/\varepsilon_1 = 0.1$, $v_{s2} = 0.04966$, $v_{s0} = 0.03038$ | | | | | | | |
| $4.98 \times 10^{-2}$ | 1.639 | 1.0028 | $1.04 \times 10^{-1}$ | $8.8 \times 10^{-7}$ | $8.5 \times 10^{-6}$ | 2.51 | 820 |
| $5.00 \times 10^{-2}$ | 1.645 | 1.0069 | $1.05 \times 10^{-1}$ | $1.02 \times 10^{-5}$ | $9.7 \times 10^{-5}$ | 2.50 | 262 |
| $5.50 \times 10^{-2}$ | 1.810 | 1.076 | $1.54 \times 10^{-1}$ | $5.23 \times 10^{-3}$ | $3.4 \times 10^{-2}$ | 1.8 | 15.5 |
| $6.60 \times 10^{-2}$ | 1.970 | 1.208 | $2.06 \times 10^{-1}$ | $2.14 \times 10^{-2}$ | $1.0 \times 10^{-1}$ | 1.0 | 7.6 |

Table 1.6 Comparison between Cases (i) and (iii) at $\varepsilon_1 = 4$ and Various Solute Number Densities $\mu_s$

| $\mu_s$ | $s_0$ | $s_2$ | $2I_0/\rho_s\beta_1$ | $2I_2/\rho_s\beta_1$ | $I_2/I_0$ | $R_{c0}/\eta$ | $R_{c1}/\eta$ |
|---|---|---|---|---|---|---|---|
| $\varepsilon_2/\varepsilon_1 = 0.03$, $v_{s2} = 4.9 \times 10^{-4}$, $v_{s0} = 3.897 \times 10^{-4}$ | | | | | | | |
| $5.0 \times 10^{-4}$ | 1.282 | 1.020 | $1.77 \times 10^{-6}$ | $1.07 \times 10^{-13}$ | $6.0 \times 10^{-8}$ | 24 | 350 |
| $5.5 \times 10^{-4}$ | 1.411 | 1.122 | $2.28 \times 10^{-5}$ | $2.10 \times 10^{-8}$ | $9.2 \times 10^{-4}$ | 18 | 65 |
| $7.0 \times 10^{-4}$ | 1.790 | 1.427 | $2.66 \times 10^{-4}$ | $2.25 \times 10^{-5}$ | $8.5 \times 10^{-2}$ | 9.9 | 17 |
| $8.0 \times 10^{-4}$ | 2.05 | 1.613 | $8.02 \times 10^{-4}$ | $1.36 \times 10^{-4}$ | $1.7 \times 10^{-1}$ | 5.7 | 11 |
| $2.0 \times 10^{-3}$ | 5.13 | 4.080 | $3.79 \times 10^{-2}$ | $2.40 \times 10^{-2}$ | $6.3 \times 10^{-1}$ | 2.6 | 3.5 |
| $6.0 \times 10^{-3}$ | 15.4 | 12.20 | $2.68 \times 10^{-1}$ | $2.2 \times 10^{-1}$ | $8.2 \times 10^{-1}$ | 1.1 | 1.5 |
| $\varepsilon_2/\varepsilon_1 = 0.1$, $v_{s2} = 8.3 \times 10^{-4}$, $v_{s0} = 3.897 \times 10^{-4}$ | | | | | | | |
| $8.4 \times 10^{-4}$ | 2.16 | 1.011 | $1.04 \times 10^{-3}$ | $7.80 \times 10^{-14}$ | $7.5 \times 10^{-11}$ | 7.1 | 450 |
| $8.6 \times 10^{-4}$ | 2.20 | 1.035 | $1.31 \times 10^{-3}$ | $2.97 \times 10^{-11}$ | $2.3 \times 10^{-8}$ | 6.8 | 160 |
| $9.0 \times 10^{-4}$ | 2.31 | 1.083 | $1.73 \times 10^{-3}$ | $5.10 \times 10^{-9}$ | $2.9 \times 10^{-6}$ | 6.0 | 65 |
| $1.5 \times 10^{-3}$ | 3.85 | 1.805 | $1.65 \times 10^{-2}$ | $7.80 \times 10^{-4}$ | $4.7 \times 10^{-2}$ | 3.5 | 8.0 |
| $3.0 \times 10^{-3}$ | 7.70 | 3.610 | $9.18 \times 10^{-2}$ | $2.64 \times 10^{-2}$ | $2.9 \times 10^{-1}$ | 1.8 | 3.2 |
| $5.0 \times 10^{-3}$ | 12.8 | 6.020 | $2.10 \times 10^{-1}$ | $9.72 \times 10^{-2}$ | $4.6 \times 10^{-1}$ | 1.3 | 2.1 |
| $7.0 \times 10^{-3}$ | 18.0 | 8.430 | $3.24 \times 10^{-1}$ | $1.79 \times 10^{-1}$ | $5.5 \times 10^{-1}$ | 1.0 | 1.6 |

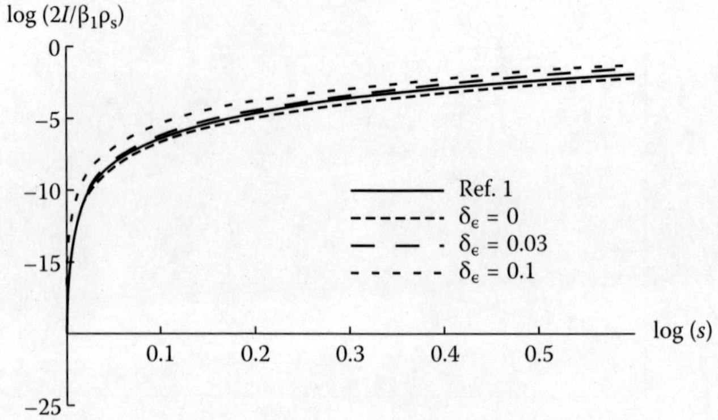


Figure 1.29

Nucleation rate $I$ calculated at $\epsilon_1/k_BT = 2$ as function of the supersaturation $s$ for cases (i)–(iii). Case (i) considered in Ref. [2] is presented by the solid line. Case (ii) is represented by the curve for $\delta_\epsilon = 0$. Case (iii) is represented by curves computed for $\delta_\epsilon = 0.03$ and $\delta_\epsilon = 0.1$.

In Figure 1.29, the nucleation rate is presented at $\varepsilon_1 = 2$ as a function of the supersaturation $s$ for the cases: (i), solid line; (ii), line corresponding to $\delta_\epsilon = 0$; and (iii), two lines for $\delta_\epsilon = 0.03$ and $\delta_\epsilon = 0.1$.

The largest difference between the various nucleation rates at the same supersaturation does not exceed 1 order of magnitude for all considered supersaturations.

### 1.7.4 Conclusion

In the present paper, a modification of the Ruckenstein, Narasimhan, and Nowakowski [1,2] nucleation theory is developed that, in addition to the solute–cluster interactions, accounts for the solute–solute and solute–solvent interactions. It is shown that at the same supersaturation these additional interactions do not affect significantly the rate of nucleation and hence, the approximation made in Refs. [1,2] by neglecting them is valid in most cases.

## References


1. Narsimhan, G., Ruckenstein, E., *J. Colloid Interface Sci.* 128, 549 (1989). (Section 1.2 of this volume.)
2. Ruckenstein, E., Nowakowski, B., *J. Colloid Interface Sci.* 137, 583 (1990). (Section 1.3 of this volume.)
3. Abraham, F., *Homogeneous Nucleation Theory*, Academic Press, New York, 1974.
4. Goodrich, F.C., *Proc. R. Soc. London A* 277, 167 (1964).
5. Ruckenstein, E., Djikaev, Y.S., *Adv. Colloid Interface Sci.* 118, 51 (2005).
6. Chandrasekhar, S., *Rev. Mod. Phys.* 15, 1 (1949).
7. Gardiner, C.W., *Handbook of Stochastic Methods*, Springer, New York, 1983.
8. Agmon, N., *J. Chem. Phys.* 81, 3644 (1984).
9. Weiss, G.H., *J. Stat. Phys.* 42, 1 (1986).
10. Deutch, J.M., *J. Chem. Phys.* 73, 4700 (1980).

# 1.8 Thermodynamics of Heterogeneous Crystal Nucleation in Contact and Immersion Modes

*Yuri Djikaev and Eli Ruckenstein**

One of the most intriguing problems of heterogeneous crystal nucleation in droplets is its strong enhancement in the contact mode (when the foreign particle is presumably in some kind of contact with the droplet surface) compared to the immersion mode (particle immersed in the droplet). Heterogeneous centers can have different nucleation thresholds when they act in contact or immersion modes. The underlying physical reasons for this enhancement have remained largely unclear. In this paper, we present a model for the thermodynamic enhancement of heterogeneous crystal nucleation in the contact mode compared to the immersion one. To determine if and how the surface of a liquid droplet can thermodynamically stimulate its heterogeneous crystallization, we examine crystal nucleation in the immersion and contact modes by deriving and comparing with each other the reversible works of formation of crystal nuclei in these cases. The line tension of a three-phase contact gives rise to additional terms in the formation free energy of a crystal cluster and affects its Wulff (equilibrium) shape. As an illustration, the proposed model is applied to the heterogeneous nucleation of hexagonal ice crystals on generic macroscopic foreign particles in water droplets at $T = 253$ K. Our results show that the droplet surface does thermodynamically favor the contact mode over the immersion one. Surprisingly, the numerical evaluations suggest that the line tension contribution (from the contact of three water phases [vapor–liquid–crystal]) to this enhancement may be of the same order of magnitude as or even larger than the surface tension contribution.

## 1.8.1 Introduction

The size, composition, and phases of aerosol and cloud particles affect the radiative and chemical properties [1,2] of clouds and hence have a great impact on Earth's climate as a whole. On the other hand, the composition, size, and phases of atmospheric particles are determined by the rate at and mode in which these particles form and evolve [2–4].

Water constitutes an overwhelmingly dominant chemical species that participates in atmospheric processes. Consequently, great importance is attributed to studying aqueous aerosols and cloud droplets as well as their phase transformations. In a number of important cases, atmospheric particles appear to freeze homogeneously [4–6]. For example, the conversion of supercooled water droplets into ice at temperatures below about −30°C is known to occur homogeneously, mainly because the concentrations of the observed ice particles in the clouds often exceed the number densities of preexisting particles capable of nucleating ice [4,5]. Also, it has been suggested that aqueous nitric acid-containing cloud droplets in the polar stratosphere freeze into nitric acid hydrates via homogeneous nucleation [6]. Understanding how nitric acid clouds form and grow in the stratosphere is a topic of current interest because such clouds participate in the heterogeneous chemistry that leads to springtime ozone depletion over the polar regions [2].

However, most phase transformations in aqueous cloud droplets occur as a result of heterogeneous nucleation on preexisting macroscopic particles, macromolecules, or even ions [3]. Heterogeneous nucleation of ice on a microscopic foreign particle can be considered a result of the adsorption of water molecules on a substrate that serves as a template. If the substrate stimulates the adsorption of water molecules in a configuration close enough to the crystalline structure of ice, then the energy barrier between phases is substantially reduced [7]. Recent work on the heterogeneous nucleation of ice in the atmosphere is motivated by the evidence, primarily from modeling studies, that heterogeneous freezing may significantly impact the radiative properties, in both the visible and infrared regions, of cirrus clouds. The leading candidates for heterogeneous nucleating centers are the mineral dusts (fly ash and metallic particles) and emissions from aircraft, primarily soot [8–10]. Interest in cirrus clouds motivated several laboratory studies as well [11,12]. They showed that the presence of various foreign inclusions shifts the apparent freezing temperature of droplets upward by as much as 10°C.

Most investigators targeted particulates as the primary heterogeneous ice nucleating centers in the atmosphere. Recently, however, increasing attention has focused on the role of films of high-molecular-weight organic compounds located on droplets. Such compounds are emitted into the atmosphere, especially in regions that are influenced by biomass burning [13]. It was reported, for example, that the films of long-chain alcohols and some other organic species can catalyze ice nucleation in droplets at a supercooling of only 1°C [14,15].

* *J. Phys. Chem. A* 112, 11677 (2008). Republished with permission.

So far, the physical mechanism underlying heterogeneous crystal nucleation in droplets remains rather obscure [2]. As an additional mystery, many heterogeneous centers have different nucleation thresholds when they act in different modes, contact or immersion, indicating that the mechanisms may be actually different for the different modes. In the contact mode, the ice nucleating particle contacts the water droplet, i.e., touches or intersects its surface, whereas in the immersion mode the particle is immersed in the water droplet (Figure 1.30) [2,16]. The same particle can trigger the freezing of a supercooled water droplet at a higher temperature in the contact mode than in the immersion one [2,6,17].

The cause of this enhancement is unknown, but it provides a hint that the water surface could be of special interest in ice nucleation. Several investigators have put forward conjectures on the mechanism of contact nucleation, all of which depend on the contact of a particle impinging upon the droplet surface from air [2]. One of these hypotheses is based on the partial solubility of small solid particles whereby the active sites at the surface of a particle are subject to erosion after it becomes immersed in water [18,19]. Another hypothesis suggests [20] that only those particles enhance nucleation in the contact mode that exhibit a strong affinity for water. During the initial contact with the droplet (before the equilibrium adsorption is achieved), such particles might strongly lower the free energy barrier to ice nucleation at its surface. Another interesting explanation [21] suggests that the contact mode enhancement of crystal nucleation is due to the mechanically forced rapid spreading of water along a hydrophobic solid surface that forces its local wetting and thereby temporarily creates local high interface-energy zones increasing the probability of crystal nucleation. While acceptable for some particular cases, all those explanations have some inconsistencies and limitations, and so far no rigorous (and general enough) theoretical model of this phenomenon has been proposed.

As a related problem, recently a thermodynamic theory was developed [22,23] that determines the condition under which the surface of a droplet can stimulate homogeneous crystal nucleation therein so that the homogeneous formation of a crystal nucleus with one of its facets at the droplet surface (surface-stimulated mode) is thermodynamically favored over its formation with all the facets within the liquid phase (volume-based mode). For both unary and multicomponent droplets, that condition has the form of an inequality

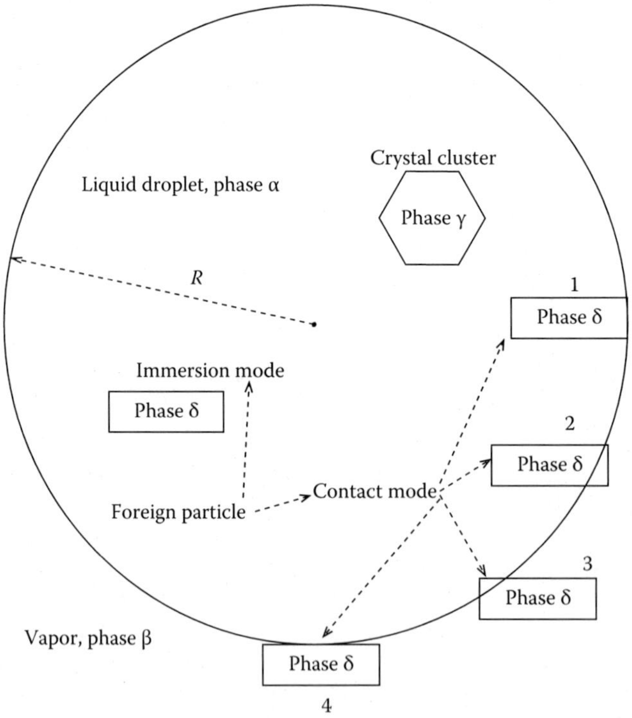


Figure 1.30

Heterogeneous crystal nucleation in a liquid droplet surrounded by vapor: "immersion" mode, the crystal cluster forms with one of its facets on a foreign particle, completely immersed in a liquid droplet, i.e., all other crystal facets interface the liquid; "contact" mode, the foreign particle is in contact with the droplet surface; the cluster forms with one of the crystal facets on the particle, another facet at the liquid–vapor interface, and all other facets making the "crystal–liquid" interface. Cases 1 through 4 represent a few of possible variations of the "foreign particle–droplet surface" contact.

that coincides with the condition of partial wettability of at least one of the facets of the crystal nucleus by its own melt [24]. This effect was experimentally observed for several systems [25–28], including water–ice [29] at temperatures at or below 0°C.

Clearly, the mode of crystal nucleation is most likely determined by both thermodynamic and kinetic factors. However, the partial wettability of a solid by its melt and the confining constraint may help to explain why, in molecular dynamics simulations of various kinds of supercooled liquid droplets [30,31], crystal nuclei appear preferentially close to the surface. Since smaller droplets have a higher surface/volume ratio, the per-droplet nucleation rate in small droplets tends to be higher than those in the bulk. Hence, it is experimentally easier to observe the crystallization of aerosols, having a large collective surface area, than those having a large volume. Recent experiments [17] on the heterogeneous freezing of water droplets in both immersion and contact modes have also provided evidence that the rate of crystal nucleation in the contact mode is much higher because the droplet surface may stimulate heterogeneous crystal nucleation in a way similar to the enhancement of the homogeneous process.

In this paper, we extend the approach, previously developed in Refs. [22,23], to heterogeneous crystal nucleation on a solid particle (in both immersion and contact modes) and present a thermodynamic model thereof in the framework of the classical nucleation theory (CNT) using the capillarity approximation. Our main thrust is to demonstrate that line tensions associated with three-phase contact lines involved in the process may play as important a role as the surface tensions. Taking into account the line tension effects, a set of modified Wulff's relations (which determine the equilibrium shape of a crystal cluster) is obtained. For particular cases where a cluster of hexagonal ice (Ih) forms with its basal facet on a foreign particle in the immersion and contact modes, we present the works of formation of a cluster as explicit functions of its single variable of state (e.g., its radius or number of molecules therein). Our thermodynamic analysis suggests that, indeed, the droplet surface can thermodynamically enhance crystal nucleation in the contact mode compared to the immersion mode. Whether this occurs or not for a particular foreign particle is determined, however, by the interplay between the surface tensions and line tensions involved in this process.

The paper is structured as follows. In Section 1.8.2, we derive and compare with each other the expressions for the free energy of formation of a crystal nucleus on a solid (say, dust) particle in the immersion and contact modes. For the sake of simplicity, in this work we consider only unary systems, i.e., pure water droplets, but the generalization to multicomponent droplets can be carried out as well. Only one kind of foreign nucleating centers is considered, namely, those completely wettable by water. In Section 1.8.3, the model is applied to the formation of crystal clusters of hexagonal ice (Ih) on a foreign solid particle. In the immersion mode, one of the basal facets of an Ih cluster is formed on the foreign particle. In the contact mode, the same basal facet is formed on the foreign particle and, in addition, a prismal facet (assumed to be only partially wettable by water) at the droplet surface. In these cases, explicit expressions for the works of formation of a crystal cluster are presented as functions of just one variable of state of the cluster. Numerical evaluations are discussed in Section 1.8.4, and the results and conclusions are summarized in Section 1.8.5.

## 1.8.2 Free Energy of Heterogeneous Formation of Crystal Nuclei in Contact and Immersion Modes

To determine if and how the surface of a liquid droplet can thermodynamically stimulate its heterogeneous crystallization, it is necessary to consider the formation of a crystal cluster in the two modes (Figure 1.30). In the "immersion" mode, the crystal cluster is formed with one of its facets on a foreign particle that is completely immersed in a liquid droplet; all other crystal facets interface the liquid. In the "contact" mode, the foreign particle touches (i.e., is in contact with) the droplet surface and the cluster forms with one of the crystal facets on the particle (as in the immersion mode), another facet at the liquid–vapor interface, and all other facets making the "crystal–liquid" interface. In these two cases the reversible works of formation of a crystal nucleus (critical cluster) should be derived and compared with each other. This can be carried out in the framework of CNT for both unary and multicomponent droplets. In this paper, we consider the crystallization of unary droplets.

The droplet surface can incur some deformation if its crystallization is initiated at its surface. Thus, the thermodynamic analysis of this case is considerably more complicated when compared to the case where the crystal forms at the surface of a bulk liquid. However, for large enough droplets, one can assume the droplet surface to be flat and avoid the complexity of taking into account the droplet deformation upon crystallization. This assumption is reasonable under conditions relevant to the freezing of atmospheric droplets, because crystal nuclei are usually of subnanometer or nanometer size, whereas the droplets themselves are in submicrometer to micrometer size range. Besides, heterogeneous particles serving as nucleating centers can be considered as macroscopic particles of linear sizes much greater than the crystal nuclei; hence, the part of its surface on which the crystal nucleus forms can be considered to be flat as well.

Let us consider a single-component bulk liquid. A macroscopic foreign particle is either completely immersed in the liquid or in contact with the liquid–vapor interface. Crystallization will take place in this liquid if it is in a metastable (supercooled) state. The reversible work of crystal formation, $W$, can be found as the difference between $X_{\rm fin}$, the appropriate thermodynamic potential of the system in its final state (liquid + crystal), and $X_{\rm in}$, the same potential in its initial state (liquid): $W = X_{\rm fin} - X_{\rm in}$. Since the density of the liquid may be different from that of the solid, the volume of the liquid may change upon crystallization if the process is not constrained to be conducted at constant volume. In such a case, strictly speaking, one cannot calculate $W$ as the difference in the Helmholtz free energies. As an approximation, the use of the Helmholtz free energy is still acceptable since, in the thermodynamic limit (i.e., the system volume and number of molecules tend to infinity but the number density of molecules remains finite), the change in the total volume of the system is usually negligible. A better choice for the thermodynamic potential is the Gibbs free energy if the system is in contact with a pressure reservoir. However, in the thermodynamic limit [32], the use of either the Gibbs or Helmholtz free energy or grand thermodynamic potential is acceptable for the evaluation of $W$.

Neglecting the density difference between liquid and solid phases and assuming the crystallization process to be isothermal, one can say that the temperature, total volume, and number of molecules in the system are constant. Thus, the reversible work of formation of a crystal cluster can be evaluated as the difference between $F_{\rm fin}$, the Helmholtz free energy of the system in its final state (liquid + crystal + foreign particle), and $F_{\rm in}$, in its initial state (liquid + foreign particle):

$$W = F_{\rm fin} - F_{\rm in}. \tag{1.252}$$

#### 1.8.2.1 Foreign Particle Completely Immersed in the Liquid

Consider a bulk liquid (in a container) whose upper surface is in contact with its vapor phase of constant pressure and temperature. A macroscopic foreign particle is completely immersed in this liquid. Clearly, for this system to be in mechanical and thermodynamic equilibrium, the particle must be completely wettable by the liquid. Upon sufficient supercooling, a crystal nucleus may form heterogeneously with one of its facets on the foreign particle. The crystal is considered to be of arbitrary shape with $\lambda$ facets (Figure 1.31). We will assign the subscript "$\lambda$" to the facet that is in contact with the foreign particle (Figure 1.32).

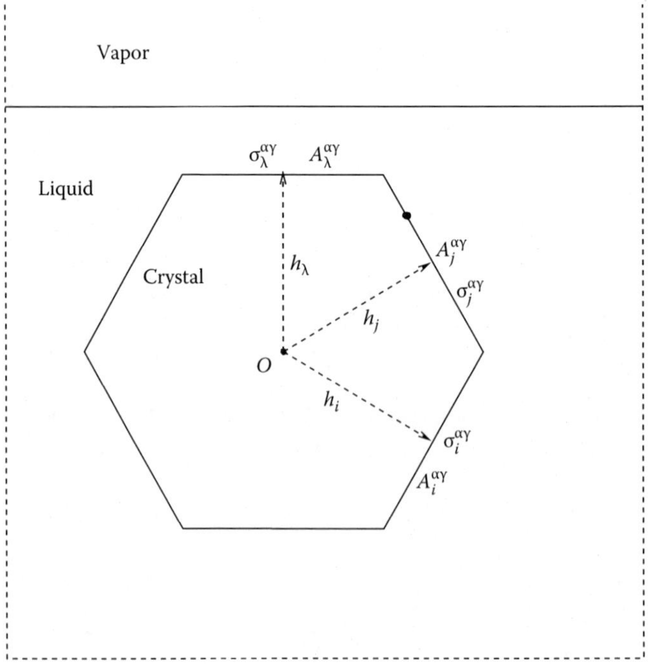


Figure 1.31

Illustration of Wulff's relations. The surface area and surface tension of facet $i$ are denoted by $A_i$ and $\sigma_i$, respectively; $h$ is the distance from facet $i$ to reference point $O$.

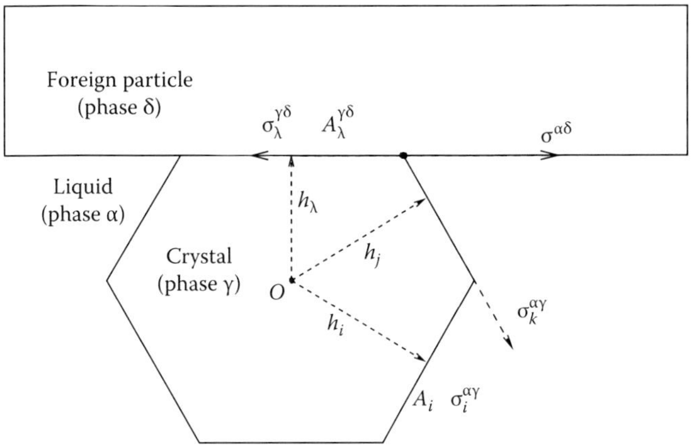


Figure 1.32

Heterogeneous formation of a crystal nucleus on a foreign particle completely immersed in the liquid.

Let us introduce the superscripts α, β, γ, and δ to denote quantities in the liquid, vapor, crystal nucleus, and foreign particle, respectively. Double superscripts will denote quantities at the corresponding interfaces, and triple superscripts at the corresponding three-phase contact lines. The surface area and surface tension of facet $i$ ($i$ = 1, ..., λ) will be denoted by $A_i$ and $\sigma_i$, respectively. (Anisotropic interfacial free energies are believed to be particularly important in determining the character of the nucleation process.) Hereafter, we adopt the definition of the surface tension of a solid, $\sigma^{\text{solid}}$, as given in Chapter 17 of Ref. [24]—namely, $\sigma^{\text{solid}} = f' + \sum_i \Gamma_i \mu_i'$, where $f'$, Γ, and μ′ are respectively the surface free energy per unit area, adsorption, and chemical potential of component $i$, all attributed to the dividing surface between solid and fluid. In the following, we will neglect the adsorption at the solid–fluid interfaces. Thus, by definition, the surface tension of the solid will be equal to the surface free energy per unit area.

Let us denote the number of molecules in the crystal cluster by $v$. Neglecting the density change upon freezing, the reversible work of heterogeneous formation of the crystal (with its facet λ on the foreign particle) is given by the expression

$$W^{\text{imm}} = v[\mu^\gamma(P^\gamma,T) - \mu^\alpha(P^\alpha,T)] - V^\gamma(P^\gamma - P^\alpha)$$
$$+\sum_{i=1}^{\lambda-1} \sigma_i^{\alpha\gamma} A_i^{\alpha\gamma} + \sigma_\lambda^{\gamma\delta} A_\lambda^{\gamma\delta} - \sigma^{\alpha\delta} A_\lambda^{\gamma\delta} + \tau^{\alpha\gamma\delta} L^{\alpha\gamma\delta}, \tag{1.253}$$

where μ, $P$, $V$, and $T$ are the chemical potential, pressure, volume, and temperature, respectively, and τ is the line tension associated with a three-phase contact line [33,34] of length $L$. Note that hereinafter the "edge effects" (due to the edges between the adjacent "homogeneously" formed facets of the crystal) are not considered.

The necessary and sufficient conditions for the equilibrium shape (known as the Wulff form) of the crystal are represented by a series of equalities, referred to as Wulff's relations (see, e.g., Ref. [24]), which can be regarded as a series of equilibrium conditions on the crystal "edges" formed by adjacent facets. For example, on the edge between homogeneously formed facets $i$ and $i$ + 1 the equilibrium condition is

$$\frac{\sigma_i^{\alpha\gamma}}{h_i} = \frac{\sigma_{i+1}^{\alpha\gamma}}{h_{i+1}} \quad (i = 1,\ldots,\lambda),$$

where $h_i$ is the distance from facet $i$ to a point $O$ within the crystal (see Figure 1.31) resulting from the Wulff construction [24].

When one of the facets (facet λ) constitutes the crystal–foreign particle interface while all the others lie within the liquid phase (see Figure 1.32), the shape of the crystal will differ from that in which all facets are in contact with the liquid. Modified Wulff's relations, taking into account the effect of the line tension on the equilibrium shape of the crystal, can be obtained by using a generalized version of the Gibbs–Curie theorem [35,36] (see also, e.g., Chapter 17 of Ref. [24]): the equilibrium shape of a crystal of given volume $V^\gamma$ should minimize the function

$$\phi = \sum_{i=1}^{\lambda-1} \sigma_i A_i + (\sigma_\lambda - \sigma^{\alpha\beta}) A_\lambda + \sum_{k\in\{\kappa\}} \tau_{\lambda k} L_{\lambda k}. \tag{1.254}$$

On the right-hand side of Equation 1.254, the summation goes over the facets of the crystal in the first term and over the pieces of a three-phase contact line in the third term, where $\{\kappa\}$ is the set of facets adjacent to facet $\lambda$, and $\tau_{\lambda k}$ and $L_{\lambda k}$ are the line tension and length of the edge formed by adjacent facets $\lambda$ and $k$ (as already mentioned, the foreign particle is assumed to be much larger than the crystal clusters). The necessary and sufficient condition for the function to have an extremum subject to the constraint $V^\gamma$ = constant is

$$\frac{\partial(\phi - mV^\gamma)}{\partial h_i} = 0, \tag{1.255}$$

where $m$ is a Lagrange multiplier. Thus, one can obtain the following modified Wulff relations:

$$\begin{aligned} &\frac{1}{h_1}\left(\sigma_1^{\alpha\gamma} - \sum_{k\in\{\kappa\}} \tau_{\lambda k}\frac{\partial L_{\lambda k}}{\partial A_1}\right) = \frac{1}{h_2}\left(\sigma_2^{\alpha\gamma} - \sum_{k\in\{\kappa\}} \tau_{\lambda k}\frac{\partial L_{\lambda k}}{\partial A_2}\right) \\ &= \ldots = \frac{1}{h_\lambda}\left(\sigma_\lambda^{\gamma\delta} - \sigma^{\alpha\delta} - \sum_{k\in\{\kappa\}} \tau_{\lambda k}\frac{\partial L_{\lambda k}}{\partial A_\lambda}\right) = \frac{m}{2}. \end{aligned} \tag{1.256}$$

Equation 1.256 can be derived from Equation 1.257 by taking into account that $\partial V^\gamma/\partial h_i = A_i$, $\partial A_i/\partial h_j = \partial A_j/\partial h_i$, and $A_i = ½\Sigma_j h_j \partial A_i/\partial h_j = ½\Sigma_j h_j \partial A_j/\partial h_i$.

In the above considerations, it is assumed that the mechanical effects within the crystal (e.g., stresses) reduce to an isotropic pressure $P$. In this case [24],

$$P^\gamma - P^\alpha = m. \tag{1.257}$$

The relations implied in Equation 1.257 for a crystal are equivalent to Laplace's equation applied to a liquid. Thus, just as for a droplet, one can expect to find a high pressure within a small crystal. It is this pressure that increases the chemical potential within the crystal.

Note that the line tension contributions to the free energy of crystal formation were omitted in the model for homogeneous crystal nucleation in the surface-stimulated mode [22,23] because they were assumed to be negligible compared to the volume and surface contributions. However, this assumption may no longer be valid for heterogeneous crystal nucleation because the nucleus is now much smaller (compared to the homogeneously formed one) and hence the contributions of three-phase contact lines can be more important [37–39].

Equation 1.253 can be rewritten in the form

$$W^{\text{imm}} = \nu[\mu^\gamma(P^\alpha,T) - \mu^\alpha(P^\alpha,T)] + \sum_{i=1}^{\lambda-1} \sigma_i^{\alpha\gamma} A_i^{\alpha\gamma} + \sigma_\lambda^{\gamma\delta} A_\lambda^{\gamma\delta} - \sigma^{\alpha\delta} A_\lambda^{\gamma\delta} + \tau^{\alpha\gamma\delta} L^{\alpha\gamma\delta}. \tag{1.258}$$

In this equation, the first term represents the excess Gibbs free energy of the molecules in the crystal compared to their Gibbs free energy in the liquid state. This term is related to the enthalpy of fusion $\Delta q$ by (see, e.g., Ref. [24])

$$\mu^\gamma(P^\alpha,T) - \mu^\alpha(P^\alpha,T) = -\int_{T_0}^{T} \Delta q \frac{dT'}{T'}, \tag{1.259}$$

where $T_0$ is the melting temperature of the bulk solid ($T < T_0$) and $\Delta q < 0$.

If the supercooling $T - T_0$ is not too large or, alternatively, if in the temperature range between $T$ and $T_0$ the enthalpy of fusion does not change significantly, Equation 1.259 takes the form

$$\mu^{\gamma}(P^{\alpha},T) - \mu^{\alpha}(P^{\alpha},T) = -\Delta q \ \ln\Theta \tag{1.260}$$

with $\Theta = T/T_0$. Thus, one can rewrite Equation 1.258 in the following form:

$$W^{\text{imm}} = -v\Delta q \ln\Theta + \sum_{i=1}^{\lambda-1} \sigma_i^{\alpha\gamma} A_i^{\alpha\gamma} + \sigma_{\lambda}^{\gamma\delta} A_{\lambda}^{\gamma\delta} - \sigma^{\alpha\delta} A_{\lambda}^{\gamma\delta} + \tau^{\alpha\gamma\delta} L^{\alpha\gamma\delta}. \tag{1.261}$$

By definition, the critical crystal (i.e., nucleus) is in unstable equilibrium with the surrounding melt. For such a crystal, the first term in Equation 1.253 vanishes. On the other hand, for a crystal with one of its facets constituting a crystal–foreign particle interface, and the others interfaced with the liquid, one can show that

$$\begin{aligned} V^{\gamma}(P^{\gamma} - P^{\alpha}) = & \frac{2}{3}\left( \sum_{i=1}^{\lambda-1} \sigma_i^{\alpha\gamma} A_i^{\alpha\gamma} + \sigma_{\lambda}^{\gamma\delta} A_{\lambda}^{\gamma\delta} - \sigma_{\lambda}^{\alpha\delta} A_{\lambda}^{\gamma\delta} \right) \\ & - \frac{2}{3} \sum_{k\in\{\kappa\}} \tau_{\lambda k} \frac{\partial L_{\lambda k}}{\partial A_k} A_k. \end{aligned} \tag{1.262}$$

This equality can be derived by representing $V^{\gamma}$ as the sum $1/3\,\Sigma_{i=1}^{\lambda} h_i A_i$ of the volumes of $\lambda$ pyramids with their bases at the crystal facets and their apexes at point $O$. The difference $P^{\gamma} - P^{\alpha}$ for every term in this sum is replaced by the right-hand side of the corresponding equality in Equation 1.257. Substituting Equation 1.262 into Equation 1.253, one obtains the following expression for the reversible work $W_{\star}^{\text{imm}}$ of formation of a critical crystal:

$$W_{\star}^{\text{imm}} = \frac{1}{3}\left( \sum_{i=1}^{\lambda-1} \sigma_i^{\alpha\gamma} A_i^{\alpha\gamma} + \sigma_{\lambda}^{\gamma\delta} A_{\lambda}^{\gamma\delta} - \sigma^{\alpha\delta} A_{\lambda}^{\gamma\delta} \right) + \tau^{\alpha\gamma\delta} L^{\alpha\gamma\delta} + \frac{2}{3} \sum_{k\in\{\kappa\}} \tau_{\lambda k} \frac{\partial L_{\lambda k}}{\partial A_k} A_k \tag{1.263}$$

or, alternatively,

$$W_{\star}^{\text{imm}} = \frac{1}{2} V_{\star}^{\gamma}(P_{\star}^{\gamma} - P^{\alpha}) + \tau^{\alpha\gamma\delta} L^{\alpha\gamma\delta} + \sum_{k\in\{\kappa\}} \tau_{\lambda k} \frac{\partial L_{\lambda k}}{\partial A_k} A_k. \tag{1.264}$$

(Hereafter, the asterisk subscript indicates the quantities for the nucleus; it is omitted on the right-hand side of expressions for $W_{\star}^{\text{imm}}$ to avoid the overcrowding of indices.)

Clearly, in the atmosphere the crystal cluster forms not in the bulk liquid but within a liquid droplet (see Figure 1.30) that is itself surrounded by a vapor phase. The reasoning here is almost identical to the preceding if, again, we neglect the density difference between the liquid and crystal phases. One can easily show that all above equations, starting with Equation 1.253 and including Equations 1.263 and 1.264 for the reversible work $W_{\star}^{\text{imm}}$ of formation of the critical crystal, remain valid except that $P^{\alpha}$ and $P^{\beta}$, pressures in the liquid and vapor phases, are related by the Laplace equation $P^{\alpha} = P^{\beta} + 2\sigma^{\alpha\beta}/R$, with $R$ being the radius of the droplet (assumed to remain constant during freezing).

### *1.8.2.2 Foreign Particle in Contact with a Liquid–Vapor Interface*

Now let us consider a foreign particle that is not immersed in a bulk liquid but is in some kind of contact with the liquid–vapor interface (Figures 1.30 and 1.33), with the vapor phase being at constant pressure and temperature. (Clearly, the particle must be completely wettable by the liquid in order for the same particle to

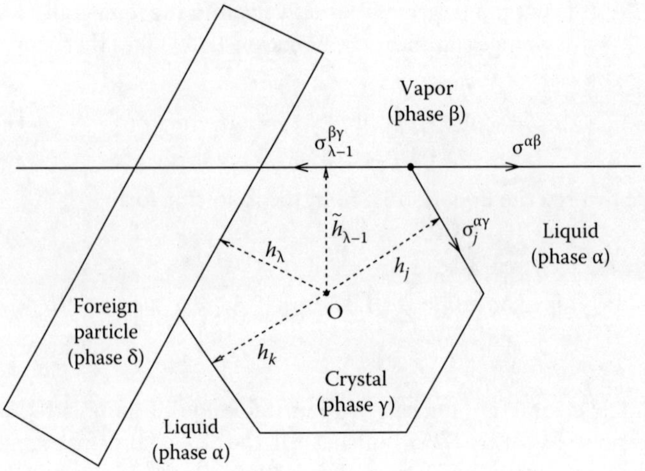


Figure 1.33

Heterogeneous formation of a crystal nucleus on a foreign particle in contact with the liquid–vapor interface.

be able to be in mechanical and thermodynamic equilibrium in both immersion and contact modes.) Upon sufficient supercooling, a crystal nucleus may form heterogeneously with one of its facets (marked with the subscript "λ") on the foreign particle and another one at the vapor–liquid interface. The latter facet will be marked with the subscript $\lambda - 1$. All the other $\lambda - 2$ facets lie within the liquid phase.

Again, neglecting the density change upon freezing, the reversible work of heterogeneous formation of the crystal with its facet λ on the foreign particle, the facet $\lambda - 1$ interfacing vapor (i.e., in the contact mode), and all the others within the liquid phase will be given by the expression

$$\begin{aligned} W^{\text{con}} &= \nu\left[\mu^{\gamma}(P^{\gamma},T)-\mu^{\alpha}(P^{\alpha},T)\right]-V'^{\gamma}(P^{\gamma}-P^{\alpha}) \\ &+\sum_{i=1}^{\lambda-2}\sigma_i^{\alpha\gamma}A_i^{\alpha\gamma}+\sigma_{\lambda-1}^{\beta\gamma}A_{\lambda-1}^{\beta\gamma}-\sigma^{\alpha\beta}A_{\lambda-1}^{\beta\gamma}+\sigma_{\lambda}^{\gamma\delta}A_{\lambda}^{\gamma\delta} \\ &-\sigma^{\alpha\delta}A_{\lambda}^{\gamma\delta}+\tau^{\alpha\beta\gamma}L^{\alpha\beta\gamma}+(\tau^{\beta\gamma\delta}-\tau^{\alpha\beta\gamma})L^{\beta\gamma\delta}+\tau^{\alpha\gamma\delta}L'^{\alpha\lambda\delta}. \end{aligned} \tag{1.265}$$

Considering the contact mode, the prime will indicate quantities whereof the values may differ from those in the immersion mode. The equilibrium shape of the crystal (the Wulff form) will differ from that formed heterogeneously in the immersion mode (i.e., when all facets, except for facet λ, are in contact with the liquid). Corresponding Wulff's relations, taking into account the effect of line tensions on the equilibrium shape of the crystal, can be again derived by using an extended version of the Gibbs–Curie theorem [24,35,36], leading to

$$\begin{aligned} &\frac{1}{h_1'}\left(\sigma_1^{\alpha\gamma}-\sum_{j\in\{\omega\}}\tau_{\lambda-1,j}\frac{\partial L_{\lambda-1,j}}{\partial A_1}-\sum_{k\in\{\kappa\}}\tau_{\lambda k}\frac{\partial L_{\lambda k}}{\partial A_1}\right)=\frac{1}{h_2'} \\ &\left(\sigma_2^{\alpha\gamma}-\sum_{j\in\{\omega\}}\tau_{\lambda-1,j}\frac{\partial L_{\lambda-1,j}}{\partial A_2}-\sum_{k\in\{\kappa\}}\tau_{\lambda k}\frac{\partial L_{\lambda k}}{\partial A_2}\right)=\ldots=\frac{1}{h_{\lambda-1}'} \\ &\left(\sigma_{\lambda-1}^{\gamma\beta}-\sigma^{\alpha\beta}-\sum_{j\in\{\omega\}}\tau_{\lambda-1,j}\frac{\partial L_{\lambda-1,j}}{\partial A_{\lambda-1}}-\sum_{k\in\{\kappa\}}\tau_{\lambda k}\frac{\partial L_{\lambda k}}{\partial A_{\lambda-1}}\right)=\frac{1}{h_\lambda'} \\ &\left(\sigma_{\lambda}^{\gamma\delta}-\sigma^{\alpha\delta}-\sum_{j\in\{\omega\}}\tau_{\lambda-1,j}\frac{\partial L_{\lambda-1,j}}{\partial A_{\lambda}}-\sum_{k\in\{\kappa\}}\tau_{\lambda k}\frac{\partial L_{\lambda k}}{\partial A_{\lambda}}\right)=\frac{m'}{2}, \end{aligned} \tag{1.266}$$

where $\{\omega\}$ is the set of facets adjacent to facet $\lambda - 1$, $\tau_{\lambda-1,j}$ and $L_{\lambda-1,j}$ are the line tension and length of the edge formed by adjacent facets $\lambda - 1$ and $j$. Note that $m'$ in Equation 1.266 is the Lagrange multiplier analogous (but not identical) to $m$, both arising from the minimization of the free energy of formation of a crystal cluster subject to the constraint $V^\gamma$ = constant. Consequently, Equation 1.257 (the equivalent of Laplace's equation applied to crystals) becomes

$$P^\gamma - P^\alpha = m'. \tag{1.267}$$

Equation 1.265 can be rewritten as

$$\begin{aligned} W^{\text{con}} &= v[\mu^\gamma(P^\alpha,T) - \mu^\alpha(P^\alpha,T)] \\ &+ \sum_{i=1}^{\lambda-2} \sigma_i^{\alpha\gamma} A_i^{\alpha\gamma} + \sigma_\lambda^{\beta\gamma} A_{\lambda-1}^{\beta\gamma} - \sigma^{\alpha\beta} A_{\lambda-1}^{\beta\gamma} + \sigma_\lambda^{\gamma\delta} A_\lambda^{\gamma\delta} \\ &- \sigma^{\alpha\delta} A_\lambda^{\gamma\delta} + \tau^{\alpha\beta\gamma} L^{\alpha\beta\gamma} + (\tau^{\beta\gamma\delta} - \tau^{\alpha\beta\delta}) L^{\beta\gamma\delta} + \tau^{\alpha\gamma\delta} L'^{\alpha\gamma\delta}. \end{aligned} \tag{1.268}$$

Furthermore, using Equation 1.260, one can rewrite Equation 1.268 in the following form:

$$\begin{aligned} W^{\text{con}} &= -v\Delta q \ln\Theta + \sum_{i=1}^{\lambda-2} \sigma_i^{\alpha\gamma} A_i^{\alpha\gamma} + \sigma_{\lambda-1}^{\beta\gamma} A_{\lambda-1}^{\beta\gamma} - \sigma^{\alpha\beta} A_{\lambda-1}^{\beta\gamma} + \sigma_\lambda^{\gamma\delta} A_\lambda^{\gamma\delta} \\ &- \sigma^{\alpha\delta} A_\lambda^{\gamma\delta} + \tau^{\alpha\beta\gamma} L^{\alpha\beta\gamma} + (\tau^{\beta\gamma\delta} - \tau^{\alpha\beta\delta}) L^{\beta\gamma\delta} + \tau^{\alpha\gamma\delta} L'^{\alpha\gamma\delta}. \end{aligned} \tag{1.269}$$

For a crystal with one of its facets constituting a solid–vapor interface, and the others interfaced with the liquid, one can obtain

$$\begin{aligned} V'^\gamma(P^\gamma - P^\alpha) &= \frac{2}{3}\left(\sum_{i=1}^{\lambda-2} \sigma_i^{\alpha\gamma} A_i^{\alpha\gamma} + \sigma_{\lambda-1}^{\beta\gamma} A_\lambda^{\beta\gamma} - \sigma^{\alpha\beta} A_{\lambda-1}^{\beta\gamma} + \sigma_\lambda^{\gamma\delta} A_\lambda^{\gamma\delta} - \sigma^{\alpha\delta} A_\lambda^{\gamma\delta}\right) \\ &- \frac{2}{3}\sum_{j\in\{\omega\}} \tau_{\lambda-1,j} \frac{\partial L_{\lambda-1,j}}{\partial A_j} A_j - \frac{2}{3}\sum_{k\in\{\kappa\}} \tau_{\lambda k} \frac{\partial L_{\lambda k}}{\partial A_k} A_k. \end{aligned} \tag{1.270}$$

This equality makes it possible to represent the reversible work $W$ of formation of a critical crystal by the expression

$$\begin{aligned} W_*^{\text{con}} &= \frac{1}{3}\left(\sum_{i=1}^{\lambda-2} \sigma_i^{\alpha\gamma} A_i^{\alpha\gamma} + \sigma_{\lambda-1}^{\beta\gamma} A_{\lambda-1}^{\beta\gamma} - \sigma^{\alpha\beta} A_{\lambda-1}^{\beta\gamma} + \sigma_\lambda^{\gamma\delta} A_\lambda^{\gamma\delta} - \sigma^{\alpha\delta} A_\lambda^{\gamma\delta}\right) \\ &+ \tau^{\alpha\beta\gamma} L^{\alpha\beta\gamma} + (\tau^{\beta\gamma\delta} - \tau^{\alpha\beta\delta}) L^{\beta\gamma\delta} + \tau^{\alpha\gamma\delta} L'^{\alpha\gamma\delta} \\ &+ \frac{2}{3}\sum_{j\in\{\omega\}} \tau_{\lambda-1,j} \frac{\partial L_{\lambda-1,j}}{\partial A_j} A_j + \frac{2}{3}\sum_{k\in\{\kappa\}} \tau_{\lambda k} \frac{\partial L_{\lambda k}}{\partial A_k} A_k \end{aligned} \tag{1.271}$$

or, alternatively, as

$$\begin{aligned} W_*^{\text{con}} &= \frac{1}{2} V_*'^\gamma (P_*^\gamma - P^\alpha) + \tau^{\alpha\beta\gamma} L^{\alpha\beta\gamma} + (\tau^{\beta\gamma\delta} - \tau^{\alpha\beta\delta}) L^{\beta\gamma\delta} + \tau^{\alpha\gamma\delta} L'^{\alpha\gamma\delta} \\ &+ \sum_{j\in\{\omega\}} \tau_{\lambda-1,j} \frac{\partial L_{\lambda-1,j}}{\partial A_j} A_j + \sum_{k\in\{\kappa\}} \tau_{\lambda k} \frac{\partial L_{\lambda k}}{\partial A_k} A_k. \end{aligned} \tag{1.272}$$

Equations 1.271 and 1.272 are similar to Equations 1.263 and 1.264, which apply to heterogeneous crystal nucleation in the immersion mode. Along with Equation 1.264, Equation 1.272 will be most useful in later discussions.

### 1.8.2.3 Comparison of Immersion and Contact Modes

Clearly, to calculate $W_*^{\text{imm}}$ and $W_*^{\text{con}}$ it is necessary to know not only the physicochemical characteristics of the forming crystals (such as $\Delta q$, $\sigma$'s, and $\tau$'s) but also the shape and size of the crystal nuclei. The latter, however, can be accurately determined analytically if the former are known (see Appendix 1F).

The reversible works of heterogeneous formation of crystal nuclei in the immersion and contact modes can be also compared by using the above results. The difference between the internal pressure of the nucleus and the external pressure does not depend on whether the nucleus forms in the immersed mode (let us denote it by $(P_*^{\gamma} - P^{\alpha})^{\text{imm}}$) or in the contact mode (denoted by $(P_*^{\gamma} - P^{\alpha})^{\text{con}}$). Indeed, by using Equation 1.260 and the equilibrium condition for the nucleus

$$\mu^{\gamma}(P^{\gamma},T) - \mu^{\alpha}(P^{\alpha},T) = 0 \tag{1.273}$$

and assuming the crystal to be incompressible, one can show that in both cases the difference $P_*^{\gamma} - P^{\alpha}$ for the nucleus is determined by the supercooling of the liquid, so that

$$(P_*^{\gamma} - P^{\alpha})^{\text{con}} = (P_*^{\gamma} - P^{\alpha})^{\text{imm}} = \frac{\Delta q}{\upsilon^{\gamma}} \ln \Theta, \tag{1.274}$$

where $\upsilon^{\gamma}$ is the volume per molecule in the crystal phase. According to Equations 1.264 and 1.272,

$$\begin{aligned} W_*^{\text{con}} - W_*^{\text{imm}} &= \frac{1}{2}(V_*'^{\gamma} - V_*^{\gamma})(P_*^{\gamma} - P^{\alpha}) + \tau^{\alpha\gamma\delta}(L_{\lambda}'^{\alpha\gamma\delta} - L_{\lambda}^{\alpha\gamma\delta}) \\ &+ (\tau^{\beta\gamma\delta} - \tau^{\alpha\beta\delta})L'^{\beta\gamma\delta} + \tau^{\alpha\beta\gamma}L_{\lambda-1}'^{\alpha\gamma\delta} \\ &+ \sum_{j\in\{\omega\}} \tau_{\lambda-1,j}' \frac{\partial L_{\lambda-1,j}'}{\partial A_j'} A_j' + \sum_{k\in\{\kappa\}} \left( \tau_{\lambda k}' \frac{\partial L_{\lambda k}'}{\partial A_k'} A_k' - \tau_{\lambda k} \frac{\partial L_{\lambda k}}{\partial A_k} A_k \right) \end{aligned} \tag{1.275}$$

(recall that primes are used to distinguish between the same quantities in the contact and immersion modes).

For homogeneous crystal nucleation, the difference $W_*^{\text{com}} - W_*^{\text{imm}}$ (i.e., the right-hand side of Equation 1.275) was obtained and analyzed without taking the line tension effects into account [22,23]. In such an approximation, the first equality in Equation 1.274 is equivalent to

$$h_{\lambda-1}' = \frac{\sigma_{\lambda-1}^{\beta\gamma} - \sigma^{\sigma\beta}}{\sigma_{\lambda-1}^{\alpha\gamma}} h_{\lambda-1}. \tag{1.276}$$

On the other hand, by virtue of Equations 1.256, 1.257, 1.266, 1.267, and 1.274, $h_i = h_i'$ for $i = 1, \ldots, \lambda - 2, \lambda$. This would mean that the Wulff shape of the crystal in the contact mode is obtained by simply changing the height of the $(\lambda - 1)$th pyramid of the Wulff crystal in the immersion mode. It would be hence clear that if $\sigma_{\lambda-1}^{\beta\gamma} - \sigma^{\alpha\beta} < \sigma_{\lambda-1}^{\beta\gamma}$, then

$$h_{\lambda-1}' < h_{\lambda-1} \Rightarrow V_*'^{\gamma} < V_*^{\gamma}. \tag{1.277}$$

Because of Equations 1.276 and 1.277, if

$$\sigma_{\lambda-1}^{\beta\gamma} - \sigma^{\alpha\beta} < \sigma_{\lambda-1}^{\alpha\gamma}, \tag{1.278}$$

then the first term on the right-hand side of Equation 1.275 would be negative. Thus, one could conclude that if the condition in Equation 1.278 is fulfilled, it is thermodynamically more favorable for the crystal nucleus to form with its facet $\lambda - 1$ at the surface rather than within the liquid. Inequality 1.278 coincides with the condition of partial wettability of the $(\lambda - 1)$th facet of the crystal by its own liquid phase [24].

However, the presence of the line tension contribution on the right-hand side of Equation 1.278 makes it impossible to draw unambiguous conclusions concerning the difference $W_{\star}^{\text{con}} - W_{\star}^{\text{imm}}$ for heterogeneous crystal nucleation even when inequality 1.278 is fulfilled. Although the first term on the right-hand side of Equation 1.275 is negative and gives rise to the thermodynamic propensity of the crystal nucleus to form with facet $\lambda - 1$ at the droplet surface, the line tension contributions can be both negative and positive because any of the line tensions involved can be either negative or positive [37–39]. Moreover, depending on the temperature at which the crystallization occurs, the sign of the line tension may change.

As clear from Equations 1.256 and 1.261 and from Equations 1.266 and 1.269, taking into account the effect of the line tension of three-phase contacts on both $W^{\text{imm}}$ and $W^{\text{con}}$ has a twofold result. On one hand, there appear explicit contributions to the free energy of formation of a crystal cluster due to the line tension. The corresponding terms on the right-hand sides of Equations 1.261 and 1.269 have the form "line tension × length of the line." On the other hand, the line tension has a more subtle effect on $W^{\text{imm}}$ and $W^{\text{con}}$ because, as clear from Equations 1.256 and 1.266, it affects the equilibrium shape of the crystal cluster. These two effects are jointly represented on the right-hand side of Equations 1.264 and 1.272: the former (hereafter referred to as "the primary" effect of line tension) gives rise to terms proportional to the lengths of three-phase contact lines, while the latter (hereafter referred to as "the secondary" effect of the line tension) results in terms proportional to the first derivatives of the contact line lengths with respect to the areas of facets.

On the right-hand side of Equation 1.275, the second, third, and fourth terms represent the primary effect of the line tension on the difference $W_{\star}^{\text{con}} - W_{\star}^{\text{imm}}$, whereas the last two terms (the fifth and sixth ones) represent the secondary effect of line tension. To roughly evaluate the fifth term, consider facet $\lambda - 1$ such that $\tau^{\beta\gamma\delta} = \tau_{\lambda-1}$ for $j = \lambda$ and $\tau^{\alpha\beta\gamma} = \tau_{\lambda-1,j}$ for all other $j$ values from the set $\{\omega\}$. Roughly assuming that $L_{\lambda-1,j}(j\epsilon\{\omega\})$ is proportional to $A_j^{1/2}$, we have $A_j\, \partial L_{\lambda-1,j}/\partial A_j \simeq ½ L_{\lambda-1,j}$. Therefore, the fifth term (representing the secondary effect of line tension) on the right-hand side of Equation 1.275 is smaller than the sum $\tau^{\beta\gamma\delta}L'^{\beta\gamma\delta} + \tau^{\alpha\beta\gamma}L_{\lambda-1}^{\beta\gamma\delta}$ (representing the primary effect of line tension) therein by a factor of 1/2. Thus, the former can be neglected compared to the latter in rough evaluations involving the uncertainty in the order of magnitude of $\tau$ values. Likewise, one can show that the sixth term (due to the secondary effect of line tension) on the right-hand side of Equation 1.275 is negligible compared to the corresponding contributions in the second and third terms (due to the primary line tension effects) therein. Because all the following numerical evaluations will involve line tensions with large uncertainties in their orders of magnitude, the secondary effects of the line tension (e.g., the fifth and sixth terms on the right-hand side of Equation 1.275) will be neglected hereafter.

### 1.8.3 Numerical Evaluations and Discussions

Whereas a detailed treatment is presented in Appendix 1F, the lack of information regarding the physical parameters required us to use an approximate evaluation in what follows.

To illustrate the above theory with numerical calculations, let us consider the formation of Ih crystals on foreign particles in water droplets in both immersion and contact modes at some temperature $T$ K (specified below). By evaluating the difference $W_{\star}^{\text{con}} - W_{\star}^{\text{imm}}$ as a whole and the relative importance of the volume, surface, and contact line contributions thereto, one can shed some light on physical mechanisms underlying the enhancement of crystal nucleation in the contact mode compared to the immersion one.

To estimate the difference between $W_{\star}^{\text{com}}$ and $W_{\star}^{\text{imm}}$, it is necessary to know the shape and the sizes of nuclei in both modes as well as the density of Ih, its enthalpy of melting, and all the surface and line tensions involved. This set of physicochemical characteristics also determines the shape and the size of the nuclei. Thus, information on $\rho$, $\Delta q$, $\sigma^{\alpha\beta}$, $\sigma^{\alpha\delta}$, $\sigma_{\text{b}}^{\alpha\gamma}$, $\sigma_{\text{p}}^{\alpha\gamma}$, $\sigma_{\text{b}}^{\gamma\delta}$, $\sigma_{\text{p}}^{\beta\gamma}$, $\tau^{\alpha\beta\gamma}$, $\tau^{\beta\gamma\delta}$, $\tau^{\alpha\gamma\delta}$, and $\tau^{\beta\gamma\delta}$ is needed to evaluate $W_{\star}^{\text{con}} - W_{\star}^{\text{imm}}$ and the relative importance of various terms on the right-hand side of Equation 1.275. The subscripts b and p mark quantities for the basal and prismal facets, respectively.

Experimental data on $\rho^{\gamma} \equiv 1/\upsilon^{\gamma}$, $\Delta q$, and $\sigma^{\alpha\beta}$ are readily available (even as functions of temperature). For our evaluations, they were taken to be $\rho^{\gamma} = 0.92\, N_{\text{A}}/18)$ cm$^{-3}$, $\Delta q \simeq (333.55 \times 10^{7}\, N_{\text{A}}/18)$ erg (where $N_{\text{A}}$ is the Avogadro number), and $\sigma^{\alpha\beta} = 83$ dyn/cm. Some data on the crystal–liquid and crystal–vapor surface tension have been also reported (see Ref. [2] for a short review). These data suffice to evaluate the size of the critical cluster for homogeneous crystal nucleation. Such estimates can serve as a reference point to evaluate the relative importance of the line tension contributions on the right-hand side of Equation 1.275 for $W_{\star}^{\text{con}} - W_{\star}^{\text{imm}}$.

Indeed, consider the homogeneous freezing of water droplets at $T_{\text{hm}} = 233$ K and assume that in the surface-stimulated mode one of the basal facets of the Ih crystal constitutes a part of the droplet surface. The height of the basal pyramid of the crystal cluster will be denoted by $\tilde{h}_{\text{b}}$ when the basal facet is at the droplet surface and by $h_b$ when the entire crystal is immersed in the droplet. According to Pruppacher and Klett [2], at this temperature $\sigma_b^{\alpha\gamma} \simeq 19.2$ dyn/cm, $\sigma_p^{\alpha\gamma} \simeq 20.5$ dyn/cm, and $\sigma_b^{\beta\gamma} \simeq 102$ dyn/cm, $\sigma_p^{\beta\gamma} \simeq 111$ dyn/cm. For the sake of rough evaluations neglecting the line tension contribution to $W_*^{ss}$ (free energy of formation of a crystal nucleus in the surface-stimulated mode), by virtue of Equation 1F.2 (Appendix 1F; and equations thereafter) one can obtain for the crystal nuclei in the volume-based and surface-stimulated modes: $h_* \equiv 2h_b \simeq 23 \times 10^{-8}$ cm, $\tilde{h}_* \equiv \tilde{h}_b + h_b \simeq 20 \times 10^{-8}$ cm, $W_*^{vb} \simeq 47kT_{hm}$, $W_*^{ss} \simeq 40.5kT_{hm}$, $\Delta W_* \equiv W_*^{ss} - W_*^{vb} \approx -6.5kT_{hm}$ ($W_*^{\Delta b}$ is the free energy of formation of a crystal nucleus in the volume-based mode).

These evaluations are for homogeneous crystal nucleation in the volume-based versus surface-stimulated modes that correspond to the immersion and contact modes, respectively, of the heterogeneous crystal nucleation. At any particular temperature, the critical crystal of heterogeneous nucleation is much smaller than for homogeneous nucleation. On the other hand, the size of the nucleus increases with increasing temperature (i.e., decreasing supercooling). Thus, one can expect that for any foreign particle there exists a temperature 233 K < $T_{ht}$ < 273 K such that the linear size of the crystal nucleus (hence the number of molecules therein, $\nu_c$) for heterogeneous nucleation is comparable to that estimated above for homogeneous nucleation at $T_{hm}$ = 233 K.

Let us assume that for a selected foreign particle the temperature $T_{ht} \simeq 253$ K with the thermal energy unit $k_BT_{ht} = 3.5 \times 10^{-14}$ erg. At this temperature, the first term on the right-hand side of Equation 1.275 (hereafter referred to as the "surface-stimulation term") can be roughly assumed to be equal to $\Delta W_*$ because both quantities represent the surface contribution to the difference between the free energy of nucleus formation in the surface-stimulated and volume-based modes (for heterogeneous and homogeneous nucleation, respectively). Thus, according to the above estimates,

$$\frac{(1/2)(V_*'^{\gamma} - V_*^{\gamma})(P_*^{\gamma} - P^{\alpha})}{k_BT_{ht}} \approx \frac{\Delta W_*}{k_BT_{ht}} \approx -6.5\frac{T_{hm}}{T_{ht}} \approx -6.0. \tag{1.279}$$

As mentioned above, this contribution is negative if facet $\lambda - 1$ (formed at the liquid–vapor interface) is only partially wettable by its melt (i.e., water), which is the case with the basal facet of the crystals of hexagonal ice. Consequently, the droplet surface always makes the contact mode of heterogeneous crystal nucleation in water droplets thermodynamically more favorable than the immersion mode, regardless of the nature of the foreign particle.

It is almost impossible to provide general unambiguous estimates for the line tension contributions to $W_*^{con} - W_*^{imm}$ in Equation 1.275. Indeed, the line tension is notorious not only for the lack of reliable experimental data but (mostly) for its ability of being both negative and positive and take values in the range from $10^{-1}$ to $10^{-5}$ dyn (see, e.g., Refs. [37–39]; according to Widom [37], e.g., a positive line tension prevents a liquid droplet from spreading at the solid–vapor interface, while a negative line tension facilitates such spreading). Nevertheless, some estimates can provide useful insight into the problem of "contact mode versus immersion mode" of heterogeneous crystal nucleation.

In the second term on the right-hand side of Equation 1.275, $L'^{\alpha\gamma\delta} - L^{\alpha\gamma\delta}$ represents the difference between the lengths of the "liquid–crystal–foreign particle" contact line in the contact and immersion modes. It is negative because the length in the contact mode is smaller than the length in the immersion mode. Considering, as above, that it is the basal facet of the hexagonal ice crystal that forms at the liquid–vapor interface (with one of the six prismal facets formed on the foreign particle), one can conclude that $L'^{\alpha\gamma\delta} - L^{\alpha\gamma\delta} \approx -a = -12.2 \times 10^{-8}$ cm. As for the line tension $\tau^{\alpha\gamma\delta}$, its sign can be expected to be positive [35–38], but we are not aware of any experimental or theoretical data reported for "foreign particle–crystal–vapor" three-phase contact regions. Assuming that $\tau^{\alpha\gamma\delta}$ can be anywhere in the range from $10^{-1}$ to $10^{-5}$ dyn, it is still most likely to be closer to $10^{-5}$ than to $10^{-1}$ dyn because two out of three phases in contact are solid phases involving little inhomogeneities of density profiles in the contact region. Thus, one can cautiously suggest that (1) this three-phase contact line impedes the heterogeneous crystal nucleation in the immersion mode versus contact mode and (2) possible values of the term $\tau^{\alpha\gamma\delta}(L'^{\alpha\gamma\delta} - L^{\alpha\gamma\delta})/k_BT$ may be somewhere in the range from −100 to −10.

The third term on the right-hand side of Equation 1.275 is due to the three-phase contact line "liquid–vapor–foreign particle. The length of this line, $L^{\beta\gamma\delta}$, is equal to $-(L'^{\alpha\gamma\delta} - L^{\alpha\gamma\delta})$, evaluated in the above paragraph, i.e., $L^{\beta\gamma\delta} \approx 12.2 \times 10^{-8}$ cm. Furthermore, the density inhomogeneities in the "foreign particle–crystal–vapor" contact region can be expected to be negligible compared to those in the "foreign particle–crystal–liquid,"

"liquid–vapor–foreign particle," or "liquid–vapor–liquid." Therefore, one can consider that the line tension $\tau^{\beta\gamma\delta}$ is negligible when compared to $\tau^{\beta\gamma\delta}$, so that the third term becomes $(\tau^{\beta\gamma\delta} - \tau^{\alpha\gamma\delta})L^{\beta\gamma\delta} \simeq -\tau^{\alpha\beta\delta}L^{\beta\gamma\delta}$. Depending on the wettability of the foreign particle by liquid water in the water vapor, $\tau^{\alpha\beta\delta}$ can be positive as well as negative. It was noted, however, that for the same foreign particle to be able to serve as an equilibrium nucleating center in both immersion and contact modes it has to be completely wettable by water. Because of this, one can suggest [35–38] that the line tension $\tau^{\alpha\beta\delta} < 0$ with its absolute value closer to $10^{-5}$ dyn than to $10^{-1}$ dyn. Besides, the third term on the right-hand side of Equation 1.275 can be expected to provide a contribution to $W_{\star}^{\text{con}} - W_{\star}^{\text{imm}}$ that is close to the contribution from the second term in absolute value and has the opposite sign. The approximate compensation of the second and third terms can thus be expected (if such compensation does not occur, the effect of the line tension on $W_{\star}^{\text{con}} - W_{\star}^{\text{imm}}$ will be even bigger).

The fifth term on the right-hand side of Equation 1.275 is due to the three-phase contact line "liquid–vapor–crystal." Considering again the basal facet of the hexagonal ice crystal forms at the liquid–vapor interface (with one of the six prismal facets on the foreign particle), the length of this contact line is approximately $L^{\alpha\beta\gamma} \approx 5a$, that is, $L^{\alpha\beta\gamma} \approx 63.0 \times 10^{-8}$ cm. As the basal facet of an Ih crystal is partially wettable by liquid water with the contact angle (measured inside the liquid phase) less than $\pi/2$, one can consider $\tau^{\alpha\beta\gamma}$ to be negative [33,34]. Even assuming for the value of $\tau^{\alpha\beta\gamma}$ the smallest experimentally reported order of magnitude, $10^{-5}$ erg, one can conclude that (1) this contact line significantly enhances the contact mode of heterogeneous crystal nucleation compared to the immersion mode and (2) the absolute value of the line tension contribution to $W_{\star}^{\text{con}} - W_{\star}^{\text{imm}}$ from the "vapor–liquid–ice" contact line can be greater than that of the surface-stimulation term (first term on the right-hand side of Equation 1.275) by 1 order of magnitude. Thus, this line tension contribution to $W_{\star}^{\text{con}} - W_{\star}^{\text{imm}}$ can dominate the surface-stimulation term.

Evaluations for the case where the nucleus of an Ih crystal is formed (1) in the immersion mode with the basal facet on the foreign particle and (2) in the contact mode with the basal facet on the particle and one of its prismal facet at the droplet–vapor interface can be carried out in a similar fashion. Besides, one can consider the case where in the contact mode one prismal facet forms on the foreign particle and another at the droplet surface. Curiously, in this situation the foreign particle does not even have to be in contact with the droplet surface. Moreover, the crystal nucleus may form with one of its basal facets on the foreign particle and the other at the droplet surface, and in this case, the foreign particle cannot be in contact with the droplet surface at all (unless it has a very irregular, noncompact shape). In the latter case, the term "contact mode" is not even appropriate. Two common features of all these "contact mode" situations are that (1) one of the crystal facets always forms at the droplet surface and (2) there always exists a contact "vapor–liquid–crystal" of three water phases. Both of these factors (the latter even significantly stronger than the former) thermodynamically favor the formation of a crystal nucleus in the contact mode compared to the immersion one. The correctness of the term "contact mode" becomes, however, questionable, at least from a thermodynamic standpoint. One trivial exception from the above consideration is the case where the surface of the foreign particle touches the liquid–vapor interface from outside in parallel orientation (see Figure 1.30, case 4). In this situation the same facet of the crystal nucleus forms at the droplet surface and on the foreign particle, and there is no thermodynamic advantage for this mode compared to the immersion mode (when the crystal nucleus forms with the same facet on the same surface of the foreign particle).

As outlined above and in Appendix 1F, one can carry out accurate numerical evaluations (of $a_{\star}$ and $h_{\star}$ as well as $W_{\star}^{\text{con}} - W_{\star}^{\text{imm}}$) if information on $\rho$, $\Delta q$, $\sigma^{\alpha\beta}$, $\sigma^{\alpha\delta}$, $\sigma_{\text{b}}^{\alpha\gamma}$, $\sigma_{\text{p}}^{\alpha\gamma}$, $\sigma_{\text{b}}^{\gamma\delta}$, $\sigma_{\text{p}}^{\beta\gamma}$, $\tau^{\alpha\beta\gamma}$, $\tau^{\alpha\beta\delta}$, $\tau^{\alpha\gamma\delta}$, and $\tau^{\beta\gamma\delta}$ were available. However, virtually no reliable data are currently available for any solid–ice and solid–water interfacial tensions and line tensions in "solid substrate–ice–liquid water–water vapor" systems. One can choose them somewhat arbitrarily, the main criterion being a reasonable agreement of the estimates extracted from Equation 1F.1 in Appendix 1F with those obtained in this section on the basis of $a_{\star}$, $h_{\star}$, and $\Delta W_{\star}$ for homogeneous ice nucleation. Considering the nucleation of ice crystals on a foreign particle such that at a given temperature $\sigma^{\alpha\delta} = 40$ dyn/cm, $\sigma_{\text{b}}^{\alpha\gamma} = 23$ dyn/cm, $\sigma_{\text{p}}^{\alpha\gamma} = 24$ dyn/cm, $\sigma_{\text{b}}^{\gamma\delta} = 50$ dyn/cm, $\sigma_{\text{p}}^{\beta\gamma} = 102$ dyn/cm, $\tau^{\alpha\beta\gamma} = -10^{-4}$ dyn, $\tau^{\alpha\beta\delta} = 7 \times 10^{-5}$ dyn, $\tau^{\alpha\gamma\delta} = 10^{-5}$ dyn, and $\tau^{\beta\gamma\delta} = 5 \times 10^{-6}$ dyn (reasonable choice according to scarce data available in literature), equations for $a_{\star}$ and $h_{\star}$ would provide $a_{\star} \simeq 29 \times 10^{-8}$ cm, $h_{\star} \simeq 34.5 \times 10^{-8}$ cm, and $W_{\star}^{\text{con}} - W_{\star}^{\text{imm}} \simeq -12k_BT_{\text{ht}}$, which are consistent with expectations.

### 1.8.4 Concluding Remarks

Previously, in the framework of CNT a criterion was found for when the surface of a droplet can stimulate crystal nucleation therein so that the formation of a crystal nucleus with one of its facets at the droplet surface is thermodynamically favored (i.e., occurs in a surface stimulated mode) over its formation with all the facets within the liquid phase (i.e., in a volume-based mode). For both unary [22] and multicomponent [23] droplets, this criterion coincides with the condition of partial wettability of at least one of the crystal facets by the melt (the contact angle, measured inside the liquid phase, must be greater than zero).

A theory of crystal nucleation in droplets is very complex even for the homogeneous case. Not surprisingly, the presence of foreign particles, serving as nucleating centers, makes the crystal nucleation phenomenon (and hence its theory) significantly more involved. Numerous aspects of heterogeneous crystal nucleation still remain obscure. One of most intriguing problems in this field remains the strong enhancement of heterogeneous crystallization in the contact mode compared to the immersion one. It has been observed that the same nucleating center initiates the crystallization of a supercooled droplet at a higher temperature in the contact mode (with the foreign particle just *in contact* with the droplet surface) compared to the immersion mode (particle *immersed* in the droplet) [2]. Many heterogeneous centers have different nucleation thresholds when they act in contact or immersion modes, indicating that the mechanisms may be actually different for the different modes. Underlying physical reasons for this enhancement have remained largely unclear, but the phenomenon of surface-stimulated (homogeneous) crystal nucleation had strongly suggested that the droplet surface could enhance heterogeneous nucleation in a way similar to the enhancement of the homogeneous process.

In this paper we have treated heterogeneous crystal nucleation on a solid particle (in both immersion and contact modes) and have presented a thermodynamic model shedding some light on the mechanism of the enhancement of this process in the contact mode. We have shown that the line tension has a twofold effect on the free energy of formation of a crystal cluster. A set of modified Wulff's relations determining the equilibrium shape of a crystal cluster is derived by using a generalized Gibbs–Curie theorem [24,35,36] taking into account the effect of line tensions involved. Our thermodynamic analysis suggests that the droplet surface can indeed thermodynamically enhance crystal nucleation in the contact mode compared to the immersion mode. Whether this occurs or not for a particular foreign particle is determined, however, by the interplay between various surface tensions and four line tensions involved in this process. For example, the effect can be expected to strongly depend on the degree of wettability of the foreign particle by water (the quantitative measure of wettability (or hydrophobicity or hydrophilicity) being the contact angle [2,3,33,34]). As clear from our model, the droplet surface may stimulate the heterogeneous crystal nucleation even in the case where the foreign particle is actually completely immersed therein but is situated closely enough to the surface. This suggests that the term "contact mode enhancement" is probably not very appropriate for this phenomenon.

As a numerical illustration of the proposed model, we have considered heterogeneous nucleation of Ih crystals on generic macroscopic foreign particles in water droplets at $T = 253$ K. Our results suggest that while the droplet surface always stimulates crystal nucleation on foreign particles in the "contact mode," the line tension contribution to this phenomenon (due to the contact of three water phases, "vapor–liquid–crystal") may be as important as the surface tension contribution.

Clearly, there is a great variety of factors affecting real atmospheric phenomena. The model presented above is an attempt to elucidate the effect of some of these factors on heterogeneous crystal nucleation in droplets. The more details that are included in the model, the more complex the model becomes. However, there is probably no alternative way to develop an increasingly adequate model of freezing of atmospheric droplets.

## Acknowledgments

The authors thank F.M. Kuni and R.S. Kabisov for helpful discussions.

## Appendix 1F: Crystal Nuclei and Their Works of Formation

The shape of the crystal nucleus is determined by Wulff's relations Equations 1.256 and 1.266. For example, since the shape of an ice crystal cluster is known (assumed to be a hexagonal prism), its state is completely determined by two geometric variables (provided that its density and temperature are given), e.g., the height of the prism and the length of a side of a (regular) hexagon (the base of the prism). However, owing to Wulff's relations, Equations 1.256 and 1.266, only one of these two variables is independent. Therefore, both works $W^{\text{imm}}$ and $W^{\text{con}}$ are functions of only one independent variable, say, variable $a$, the length of a side of the hexagon. The concrete form of the functions $W^{\text{imm}} = W^{\text{imm}}(a)$ and $W^{\text{con}} = W^{\text{con}}(a)$ depends on the mutual orientation and location of the crystal cluster and foreign particle (and droplet surface in the case of $W^{\text{con}}$).

For instance, consider a crystal cluster formed with one of its basal facets on a foreign particle in the immersion mode. For the contact mode, let us consider the same basal facet on the foreign particle and a prismal facet (assumed to be only partially wettable by water) at the droplet surface. Mark the basal facets with subscripts 1 and 8 and the prismal facets with subscripts 2, …, 7 (Figure 1.34).

As agreed upon above, facet 8 forms on the foreign particle. Clearly, in the immersion mode $\sigma_p^{\alpha\gamma} \equiv \sigma_2^{\alpha\gamma} = \cdots = \sigma_7^{\alpha\gamma}$, $A_p^{\alpha\gamma} \equiv A_2^{\alpha\gamma} = \cdots A_7^{\alpha\gamma}$, $A_b^{\alpha\gamma} \equiv A_1^{\alpha\gamma} = A_8^{\alpha\gamma}$. In the contact mode, the prismal facet 7 (assumed to be only partially wettable by liquid water) represents the crystal–vapor interface; hence, $\sigma_p^{\alpha\gamma} \equiv \sigma_2^{\alpha\gamma} = \cdots = \sigma_6^{\alpha\gamma}$, $\sigma_p^{\beta\gamma} = \sigma_7^{\beta\gamma}$. Unlike the crystal cluster in the immersion mode, the basal facet in the contact mode is not a regular hexagon, $A_7'^{\alpha\gamma} \equiv A_1'^{\alpha\gamma} = A_8'^{\alpha\gamma} =$ and $A_b'^{\alpha\gamma} < A_b^{\alpha\gamma}$, i.e., the surface area of the basal facets in the contact mode is smaller than that in immersion mode, according to Equation 1.276. Let us mark two prismal facets adjacent to facet 7 by subscripts 2 and 6. Clearly, $A_p^{\alpha\gamma} = A_3^{\alpha\gamma} = A_4^{\alpha\gamma} = A_5^{\alpha\gamma}$, $A_p'^{\alpha\gamma} \equiv A_2^{\alpha\gamma} = A_6^{\alpha\gamma} < A_p^{\alpha\gamma}$, $A_p^{\beta\gamma} \equiv A_7^{\beta\gamma} > A_7^{\alpha\gamma}$ (both inequalities are again due to Equation 1.276).

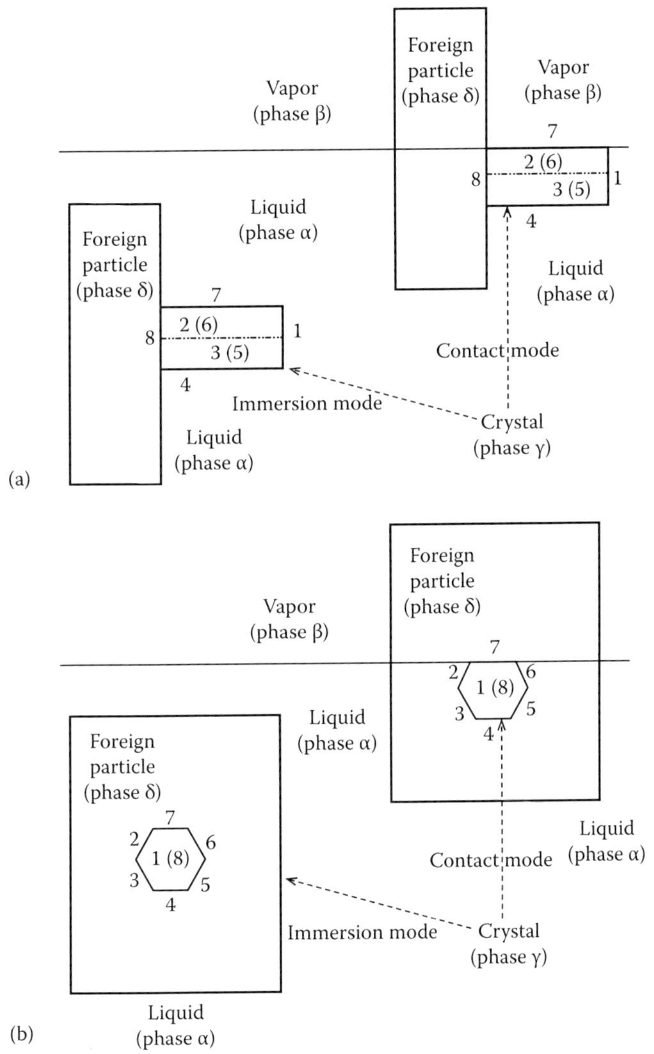


Figure 1.34

Heterogeneous formation of an Ih cluster on a foreign particle in the immersion and contact modes. The crystal cluster has a shape of a hexagonal prism. One of the basal facets (facet 8) is formed on the foreign particle, the other (facet 1) interfaces the liquid. Two of the prismal facets (with numbers 4 and 7) lie in the plane perpendicular to the figure. (a) A section along the axis of the prism and perpendicular to prismal facets 4 and 7. Prismal facets 5 and 6 cannot be seen by the reader, so their numbers are shown in the parentheses. (b) A view of basal facet 1 perpendicular thereto (basal facet 8, facing the foreign particle, cannot be seen hence its number is shown in the parentheses).

Let us use $a_i$ and $a_i'$ ($i = 2, ..., 7$) to denote the length of the edge formed by the basal facet with prismal facet $i$ in the immersion and contact modes, respectively. In the immersion mode, the base is a regular hexagon, i.e., $a \equiv a_2 = ... = a_7$. As can be seen clearly in Equation 1.276 and Figure 1.34b, in the contact mode $a_2' = a_6' < a$, $a_7' > a$, whereas $a_3' = a_4' = a_5' = a$ with

$$a_2' = a_6' = C_p a, \quad a_7' = (2 - C_p)a$$

and positive $C_p = \left(\sigma_p^{\beta\gamma} - \sigma^{\alpha\beta}\right)/\sigma_p^{\alpha\gamma} < 1$ (the vapor–crystal interfacial tension is greater than the liquid–crystal one).

In the first term on the right-hand sides of Equations 1.261 and 1.269, the number of molecules in the crystal cluster can be represented as $\nu = \rho^\gamma V^\gamma$ or $\nu = \rho^\gamma V'^\gamma$, respectively, where $\rho^\gamma$ is the number density of molecules in phase $\gamma$ (ice). The volume of an Ih crystal (shaped as a hexagonal prism) is equal to the product "height of the prism" × "surface area of the base." In both the immersion and contact modes, the surface area of the base (regular hexagon in the former and irregular in the latter) is proportional to $a^2$, although the coefficients of proportionality are different. In both cases, the height of the prism, $h$, is linearly related to $a$ according to Wulff's relations 1.256 and 1.266, respectively:

$$h = a\frac{\sqrt{3}}{2}\frac{\sigma_b^{\alpha\gamma} + \sigma_b^{\gamma\delta} - \sigma^{\alpha\delta}}{\sigma_p^{\alpha\gamma}}.$$

Thus, in both Equations 1.261 and 1.269, $\upsilon \propto \rho^\gamma a^3$. Likewise, one can show that all the surface tension and line tension terms on the right-hand sides of Equations 1.261 and 1.270 are proportional to $a^2$ and $a$, respectively. Therefore, the reversible works of formation of a crystal cluster in these modes can be written (tedious but simple algebra is omitted) as

$$\begin{aligned} W^{\text{imm}}(a) &= -I_3 a^3 + I_2 a^2 + I_1 a \\ W^{\text{con}}(a) &= -C_3 a^3 + C_2 a^2 + C_1 a, \end{aligned} \tag{1F.1}$$

where $I_3$, $I_2$, $I_1$, and $C_3$, $C_2$, $C_1$ are positive coefficients,

$$\begin{aligned} I_3 &= \frac{9}{4}\rho^\gamma \Delta q \ln(\Theta)\left(\sigma_b^{\alpha\gamma} + \sigma_b^{\gamma\delta} - \sigma^{\alpha\delta}\right)/\sigma_p^{\alpha\gamma} \\ I_2 &= \frac{3\sqrt{3}}{2}\left[2\sigma_b^{\alpha\lambda} + 3\left(\sigma_b^{\gamma\delta} - \sigma^{\alpha\delta}\right)\right] \\ I_1 &= 6\tau^{\alpha\gamma\delta}, \end{aligned}$$

and

$$\begin{aligned} C_3 &= \rho^\gamma \Delta q \ln(\Theta)\frac{\sqrt{3}}{2}2\left(3 - \frac{2}{\sqrt{3}}C_h + \frac{2C_h^2}{3}\right)C_h \\ C_2 &= \left(3 - \frac{2}{\sqrt{3}}C_h + \frac{2C_h^2}{3}\right)C_h + [\sigma_p^{\alpha\gamma}(3 + 2C_p) + (\sigma_p^{\beta\gamma} - \sigma^{\alpha\beta})(2 - C_p)C_h \\ C_1 &= 2(\tau^{\beta\gamma\delta} - \tau^{\alpha\beta\delta}) + 3\tau^{\alpha\gamma\delta} + 2\tau^{\alpha\beta\gamma} + (-(\tau^{\beta\gamma\delta} - \tau^{\alpha\beta\delta}) + 2\tau^{\alpha\gamma\delta} - \tau^{\alpha\beta\gamma})C_p + \tau^{\alpha\beta\gamma}2C_h, \end{aligned}$$

With $C_h = (3^{1/2}/2)(\sigma_b^{\alpha\gamma} + \sigma_b^{\gamma\delta} - \sigma^{\alpha\beta})/\sigma_p^{\alpha\gamma}$. Note that, although the linear (in $a$) contribution to $W^{con}(a)$ is due to the line tensions, the coefficient, $C_1$, depends not only on the line tensions but also on the surface tensions of both prismal and basal facets of the crystal cluster because the lengths of the three-phase contact lines, as functions of $a$, depend on those surface tensions (see Figures 1.33 and 1.34).

Using Equation 1F.1, one can find the length $a_*$ of a side of the hexagonal base of the crystal nucleus. According to CNT (whereof the framework has been adopted for our treatment), the free energy of formation of a cluster as a function of a single variable of state of the cluster attains a maximum for the critical cluster, i.e., nucleus. Therefore, $a_*$ can be found as the positive solution of the equation $dW^{imm}(a)/da\big|_{a_*} = -3I_3a_*^2 + 2I_2a_* + I_1 = 0$, or alternatively, $dW^{con}(a)/da\big|_{a_*} = -3C_3a_*^2 + 2C_2a_* + C_1 = 0$, which leads to $a_* = (2I_2 + (4I_2^2 + 12I_1I_3)^{1/2})6I_3$ or $a_* = (2C_2 + (4C_2^2 + 12C_1C_3)^{1/2})6C_3$ (see two paragraphs above Equation 1F.1). The height of the crystal nucleus (shaped as a hexagonal prism) is the same in both immersion and contact modes, $h_* = a_*(3^{1/2}/2)(\sigma_b^{\alpha\gamma} + \sigma_b^{\gamma\delta} - \sigma^{\alpha\delta})/\sigma_p^{\alpha\gamma}$.

The immersion and contact modes of the heterogeneous crystal nucleation are heterogeneous versions of the volume-based and surface-stimulated modes [22,23], respectively, of homogeneous crystal nucleation. The difference is that no foreign particles are involved in the homogeneous process. Let us denote the corresponding works of formation as $W^{vb} \equiv W^{vb}(a)$ (for the volume-based mode) and $W^{ss} \equiv W^{ss}(a)$ (for the surface-stimulated mode).

Clearly, homogeneous crystal nucleation in the surface-stimulated mode differs from the heterogeneous crystal nucleation in the immersion mode only in that one of the crystal facets (a basal one in our particular example) forms at the droplet surface (in the former case) rather than on an immersed foreign particle (in the latter case). Therefore, the function $W^{ss}(a)$ can be formally obtained from $W^{imm}(a)$ by replacing all the superscripts with the superscript β. On the other hand, in the volume-based mode of the homogeneous process all the crystal facets represent the liquid–crystal interface and the height ($h$) of the crystal cluster (shaped as a right prism with a regular hexagon at its base) is related to the radius of a regular hexagon ($a$) as $h = a(3^{1/2})(\sigma_b^{\alpha\gamma}/\sigma_p^{\alpha\gamma})$.

Thus, the functions $W^{ss}(a)$ and $W^{vb}(a)$ can be written in the form

$$W^{ss}(a) = -S_3a^3 + S_2a^2 + S_1a$$
$$W^{vb}(a) = -B_3a^3 + B_2a^2 + B_1a \qquad (1F.2)$$

with

$$S_3 = \frac{9}{4}\rho^\gamma \Delta q \ln(\Theta)(\sigma_b^{\alpha\gamma} + \sigma_b^{\alpha\beta} - \sigma^{\alpha\beta})/\sigma_p^{\alpha\gamma}$$
$$S_2 = \frac{3\sqrt{3}}{2}\left[2\sigma_b^{\alpha\gamma} + 3(\sigma_b^{\gamma\beta} - \sigma^{\alpha\beta})\right]$$
$$S_1 = 6\tau^{\alpha\gamma\beta}$$

and

$$B_3 = \frac{9}{2}\rho^\gamma \Delta q \ln(\Theta)\left(\sigma_b^{\alpha\gamma}/\sigma_p^{\alpha\gamma}\right)$$
$$S_2 = 9\sqrt{3}\sigma_b^{\alpha\gamma}$$
$$B_1 = 0$$

(there is no three-phase contact line in the volume-based mode of homogeneous crystal nucleation; hence, $B_1 = 0$). Again, one can find the radius $a_*$ of the hexagonal base of the crystal nucleus as the positive solution of the equation $dW^{ss}(a)/da\big|_{a_*} = -3S_3a_*^2 + 2S_2a_* + S_1 = 0$, or alternatively, $dW^{vb}(a)/da\big|_{a_*} = -3B_3a_*^2 + 2B_2a_* + B_1 = 0$, which lead to $a_\star = (2S_2 + (4S_2^2 + 12S_1S_3)^{1/2})/6S_3$ or $a_\star = (2B_2 + (4B_2^2 + 12B_1B_3)^{1/2})/6B_3$. The heights of the crystal nucleus (a hexagonal prism) are now different in the surface-stimulated and volume-based modes, namely, $h_\star^{ss} = a_\star(3^{1/2}/2)(\sigma_b^{\alpha\gamma} + \sigma_b^{\alpha\beta} - \sigma^{\alpha\beta})/\sigma_p^{\alpha\gamma}$ and $h_\star^{vb} = a_\star(3^{1/2})(\sigma_b^{\alpha\gamma}/\sigma_p^{\alpha\gamma})$, respectively.

## References


1. *IPCC, Climate Change 2001. The Scientific Bases*, Intergovernmental Panel on Climate Change; Cambridge University Press, Cambridge, UK, 2001.
2. Pruppazrefscher, H.R., Klett, J.D., *Microphysics of Clouds and Precipitation*, Kluwer Academic Publishers, Dordrecht, 1997.
3. Fletcher, N.H., *The Physics of Rainclouds*, Cambridge University Press, Cambridge, 1962.
4. Cox, S.K. *J. Atmos. Sci.* 28, 1513 (1971).
5. Jensen, E.J., Toon, O.B., Tabazadeh, A., Sachse, G.W., Andersen, B.E., Chan, K.R., Twohy, C.W., Gandrud, B., Aulenbach, S.M., Heymsfield, A., Hallett, J., Gary, B., *Geophys. Res. Lett.* 25, 1363 (1998).
6. Heymsfield, A.J., Miloshevich, L.M., *J. Atmos. Sci.* 50, 2335 (1993).
7. Ruckenstein, E., *J. Colloid Interface Sci.* 45, 115 (1973).
8. DeMott, P.J., Cziczo, D., Prenni, A., Murphy, D., Kreidenweis, S., Thomson, D., Borys, R., Rogers, D., *Proc. Natl. Acad. Sci. U. S. A.* 100, 14655–14660 (2003).
9. DeMott, P.J., Sassen, K., Poellot, M., Baumgardner, D., Rogers, D., Brooks, S., Prenni, A., Kreidenweis, S., *Geophys. Res. Lett.* 30, 1732 (2003b). doi:10.1029/2003GL017410.
10. Sassen, K., DeMott, P.J., Prospero, J., Poellot, M., *Geophys. Res. Lett.* 30, 1633 (2003). doi:10.1029/2003 GL017371.
11. Zuberi, B., Bertram, A., Koop, T., Molina, L., Molina, M., *J. Phys. Chem. A* 105, 6458–6464 (2001).
12. Hung, H.-M., Malinkowski, A., Martin, S., *J. Phys. Chem. A* 107, 1296–1306 (2003).
13. Elias, V., Simoneit, B., Pereira, A., Cabral, J., Cardoso, J., *Environ. Sci. Technol.* 33, 2369–2376 (1999).
14. Popovitz-Biro, R., Wang, J., Majewski, J., Shavit, E., Leiserowitz, L., Lahav, M., *J. Am. Chem. Soc.* 116, 1179–1191 (1994).
15. Fukuta, N., Experimental studies of organic ice nuclei, *J. Atmos. Sci.* 23, 191–196 (1966).
16. Vali, G., in *Nucleation and Atmospheric Aerosols*, Kulmala, M., Wagner, P., Eds., Pergamon: New York, 1996.
17. Shaw, R.A., Durant, A.J., Mi, Y., *J. Phys. Chem. B* 109, 9865 (2005).
18. Fletcher, N.H., *J. Atmos. Sci.* 27, 1098 (1970).
19. Guenadiev, N., *J. Rech. Atmos.* 4, 81 (1970).
20. Evans, L.F., *Preprints Conference on Cloud Physics (1970, Fort Collins, CO)*, American Meteorological Society, Boston, MA, 1970, p. 14.
21. Fukuta, N., *J. Atmos. Sci.* 32, 1597–2371 (1975).
22. Djikaev, Y.S., Tabazadeh, A., Hamill, P., Reiss, H., *J. Phys. Chem. A* 106, 10247 (2002).
23. Djikaev, Y.S., Tabazadeh, A., Reiss, H., *J. Chem. Phys.* 118, 6572–6581 (2003).
24. Defay, R., Prigogine, I., Bellemans, A., Everett, D.H., *Surface Tension and Adsorption*, John Wiley, New York, 1966.
25. Zell, J., Mutaftshiev, B., *Surf. Sci.* 12, 317 (1968).
26. Grange, G., Mutaftshiev, B., *Surf. Sci.* 47, 723 (1975).
27. Grange, G., Landers, R., Mutaftshiev, B., *Surf. Sci.* 54, 445 (1976).
28. Chatain, D., Wynblatt, P., in *Dynamics of Crystal Surfaces and Interfaces*, Duxbury, P.M., Pence, T.J., Eds., Springer, New York, 2002, pp. 53–58.
29. Elbaum, M., Lipson, S.G., Dash, J.G., *J. Cryst. Growth* 129, 491 (1993).
30. Chushak, Y.G., Bartell, L.S., *J. Phys. Chem. B* 103, 11196 (1999).
31. Zasetsky, A.Y., Remorov, R., Svishchev, I.M., *Chem. Phys. Lett.* 435, 50–53 (2007).
32. Rusanov, A.I., *Phasengleichgewichte und Grenzflachenerscheinungen*, Akademie Verlag, Berlin, 1978.
33. Rowlinson, J.S., Widom, B., *Molecular Theory of Capillarity*, Clarendon Press, Oxford, 1982.
34. Boruvka, L., Neumann, A.W., *J. Phys. Chem.* 66, 5464 (1977).
35. Gibbs, J.W., *The Collected Works of J. Willard Gibbs*, Longmans, Green and Co., New York, 1928.
36. Curie, P., *Bull. Soc. Miner. Fr.* 8, 195 (1885).
37. Widom, B., *J. Phys. Chem.* 99, 2803 (1995).
38. Aveyard, R., Clint, J.H., *J. Chem. Soc., Faraday Trans.* 92, 85–89 (1996).
39. Amirfazli, A., Neumann, A.W., *Adv. Colloid Interface Sci.* 110, 121–141 (2004).

## 1.9 Correction to "Thermodynamics of Heterogeneous Crystal Nucleation in Contact and Immersion Modes"

*Yuri Djikaev and Eli Ruckenstein**

The correct sign of the terms with the partial derivatives of the contact line lengths in Equation 1.256 must be "+" (plus) instead of "−" (minus). Consequently, the signs of the corresponding terms (involving the partial derivatives of the lengths of the contact lines) in Equations 1.262 through 1.264, 1.266, 1.270 through 1.272, and 1.275 must be replaced by the opposite ones. It should be noted that these terms (involving the partial derivatives of the lengths of the contact lines) represent the secondary effect of the line tension on crystal nucleation, which is much weaker than the primary effect (see the original paper). They have been neglected in Section 1.8.3 ("Numerical Evaluations and Discussions") of the original paper. (*J. Phys. Chem. A* 2008, 112, 46, 11677–11687. doi: 10.1021/jp803155f.)

* *J. Phys.Chem. A* 116, 1119 (2012). Republished with permission.

## 1.10 Homogeneous Crystal Nucleation in Droplets as a Method for Determining Line Tension of a Crystal–Liquid–Vapor Contact

*Yuri Djikaev and Eli Ruckenstein**

The line tension of a three-phase contact is believed to play an important role in phase transition and phase equilibria in multiphase nanoscale systems, hence the need in developing various methods for its experimental evaluation. In this paper, we suggest an indirect experimental method for determining the line tension of a solid–liquid–vapor contact based on experiments on homogeneous crystallization of droplets. The underlying idea explores our recent finding that the line tension can give rise to an important contribution to the free energy of formation of a crystal nucleus in a surface-stimulated mode when one of its facets forms at the droplet surface and thus represents a "crystal–vapor" interface. The proposed method requires experimental data on the rate of homogeneous crystal nucleation as a function of droplet size. However, it can provide a rough estimate of line tension even if the rate is known only for one droplet size. Using the method to examine experimental data on homogeneous crystal nucleation in droplets of 19, 49, and 60 μm radii at $T = 237$ K, we evaluated the line tension of ice–(liquid) water–(water) vapor contact to be positive and of the order of $10^{-11}$ N consistent with the current expectations.

### 1.10.1 Introduction

Wetting phenomena in many systems (e.g., adhesives, lubricants, melts in contact with solids, cell adhesion to biological membrane [1,2]) are often studied by means of contact angle-based methods. The interpretation of the observed contact angle in real systems is complicated by many factors such as heterogeneity of the system, presence of surfactants, smoothness of the surface, and line tension, which all can affect the observed contact angle. All these factors need to be studied to better understand the contact angle and wetting phenomena in general. The line tension factor is becoming particularly important nowadays due to the fast development of nanotechnologies because line tension can play a major role in the behavior of nanosize systems. It is thus important to clearly establish the magnitude and sign of line tension to determine its relevance to (nano)technological applications.

Gibbs [3] was the first to introduce the concept of line tension of a three-phase contact in the classical thermodynamics. However, "full-scale" studies of this quantity have started only relatively recently [4–6]. (Note that although the term "line tension" is mostly used in association with the contacts of three three-dimensional phases, it also often referred to the tension associated with the borderline of two two-dimensional phases; in what follows, only the former contact regions will be of interest, i.e., those of three three-dimensional phases.)

Thermodynamically, the line tension can be defined as follows. If three phases are in equilibrium, there may exist a three-phase contact region in addition to three two-phase interfaces. The distortion of the order parameter(s) spatial distribution(s) in the three-phase contact region gives rise to excess contributions to the extensive thermodynamic characteristics of the system. These are usually attributed to a three-phase contact line. Excess quantities per unit length of the contact line are called linear adsorptions. For three bulk phases in equilibrium, the interfaces between them can be treated as planar and the line of their intersection, that is, the contact line, as a straight line. The location of the contact line is arbitrary and among all linear adsorptions only one does not depend on its choice. This is the linear adsorption of the grand canonical potential also called "line tension" [4,5]. All other adsorptions depend on the location of the contact line and hence are functions of its two coordinates. There are large discrepancies among theoretical, simulated, and experimental data concerning the magnitude and sign of line tension [6].

The majority of theoretical [5,7–15]. and simulated [14–16] predictions for line tension (except for conditions close to a wetting transition) range from $10^{-12}$ to $10^{-10}$ N, but values as high as $10^{-6}$ N are also reported. Unlike the surface tension (which is always positive), the sign of the line tension can theoretically be either positive or negative [11,17–20].

There are various experimental techniques for determining the line tension in both liquid–liquid–fluid and solid–liquid–vapor systems (see Amirfazli and Neumann [6] for a recent review). This diversity is partly due to the small magnitude of the line tension that requires considerable ingenuity in the design of experiments. Most experimental methodologies fall under one of the following categories: thin film arrangements, drop size dependence of contact angles, critical particle/liquid lens size, and heterogeneous nucleation. In all of the above techniques, the line tension is found indirectly by measuring such parameters as the contact

* *Chem. Eng. Sci.* 64, 4498 (2009). Republished with permission.

angle or particle size, and inferring the line tension through appropriate thermodynamic or other theoretically developed correlations (e.g., the modified Young equation).

Most experiments in liquid–liquid–vapor and solid–liquid–vapor systems provided [8–10,21,22] positive values for the line tension ranging from $10^{-9}$ to $10^{-6}$ N. However, studying the line tension dependence on the contact angle [23–26], a change in sign was observed near the first-order wetting transition. For example, in Pompe's study [26], the line tension of hexaethylene glycol on a silicon wafer changes from $+2.5 \times 10^{-11}$ N for a contact angle below 6° to $-2 \times 10^{-10}$ for a contact angle above 6°. In this paper, we present a theoretical basis for determining the line tension of a "solid–liquid–vapor" contact region by means of experiments on homogeneous crystallization of droplets. The underlying idea is a natural result of recent advances in the thermodynamics [27] and kinetics [28] of surface-stimulated mode of homogeneous crystal nucleation in droplets.

Neglecting the effect of line tension associated with a crystal–liquid–vapor contact, it was previously found [29,30] that, if at least one of the facets of a crystal nucleus is only partially wettable by its own melt [31], then the droplet surface can stimulate *homogeneous* crystal nucleation so that the formation of a nucleus with that facet at the droplet surface (surface-stimulated mode) is thermodynamically favored over its formation with all the facets *within* the droplet (volume-based mode). However, as recently shown by Djikaev and Ruckenstein [27], the line tension can give rise to an important contribution to the free energy of formation of a "surface-stimulated" crystal nucleus. Moreover, as will be shown in the present paper, this contribution (equal to the product of the tension and length of the three-phase contact line) can be evaluated with the help of experimental data on the rate of freezing of droplets.

### 1.10.2 Background: The Thermodynamics and Kinetics of Homogeneous Crystal Nucleation in a Droplet

For the sake of concreteness, we will hereafter discuss the formation of ice crystals in liquid water droplets surrounded by water vapor. Water constitutes an overwhelmingly dominant chemical species in atmospheric processes. Consequently, great importance is attributed to studying aqueous aerosols and cloud droplets as well as their phase transformations. The size, composition, and phases of aerosol and cloud particles affect the radiative and chemical properties [32,33] of clouds and hence have a great impact on Earth's climate as a whole. On the other hand, the composition, size, and phases of atmospheric particles are determined by the rate at and mode in which these particles form and evolve [32–34]. Although most phase transformations in aqueous cloud droplets occur as a result of heterogeneous nucleation on preexisting macroscopic particles, macromolecules, or even ions [33], in a number of important cases atmospheric particles appear to freeze homogeneously [34,35].

#### *1.10.2.1 Free Energy of Homogeneous Formation of Crystal Nuclei in Surface-Stimulated and Volume-Based Modes*

To determine if and how the surface of a liquid droplet of radius $R$ can thermodynamically stimulate its homogeneous crystallization, it is necessary to compare the reversible works of formation of a crystal nucleus (critical crystal cluster) in two modes (Figure 1.35). In the volume-based mode, the crystal cluster is formed with all the facets interfacing the liquid. In the surface-stimulated mode, the crystal cluster forms with one of its facets at the liquid–vapor interface, whereas all the others form the "crystal–liquid" interface. For simplicity, the droplet surface can be considered to be planar.

Consider a bulk liquid (in a container) whose upper surface is in contact with its vapor phase of constant pressure and temperature. Upon sufficient supercooling, a crystal nucleus may form homogeneously with one of its facets on the foreign particle. The crystal is considered to be of arbitrary shape with $\lambda$ facets (Figure 1.36).

Let us introduce the superscripts $\alpha$, $\beta$, and $\gamma$ to denote quantities in the liquid, vapor, and crystal nucleus, respectively. Double superscripts will denote quantities at the corresponding interfaces, and triple superscripts at the corresponding three-phase contact lines.

The surface area and surface tension of facet $i$ ($i = 1, \ldots, \lambda$) will be denoted by $A_i$ and $\sigma_i$, respectively. (Anisotropic interfacial free energies are believed to be particularly important in determining the character of the nucleation process.)

The reversible work of homogeneous formation of a crystal in the volume-based mode (with all its facets within the liquid phase) is given by the expression [27,29]

$$W^{vb} = -\nu \Delta q \ln \Theta + \sum_{i=1}^{\lambda} \sigma_i^{\alpha\gamma} A_i^{\alpha\gamma}, \qquad (1.280)$$

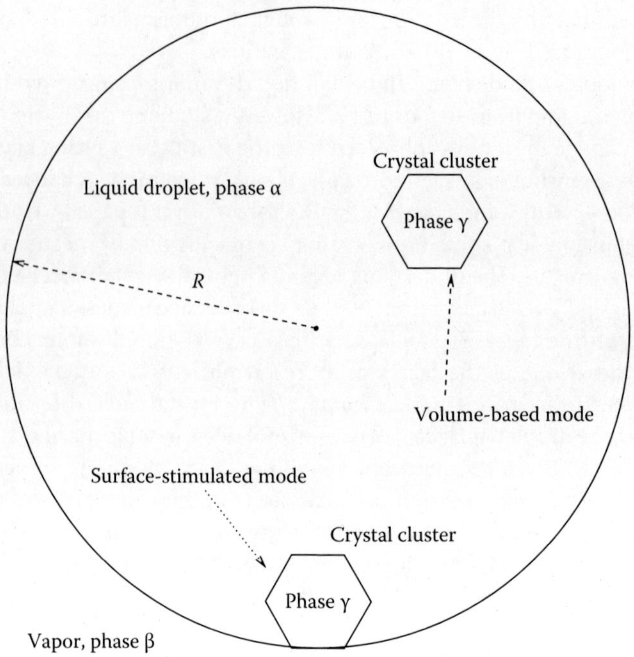


Figure 1.35

Homogeneous crystal nucleation in a liquid droplet surrounded by vapor. In the volume-based mode, the crystal cluster forms with all its facets completely immersed in a liquid droplet. In the surface-stimulated mode the cluster forms with one of its facets at the liquid–vapor interface, and all other facets representing the "crystal–liquid" interface.

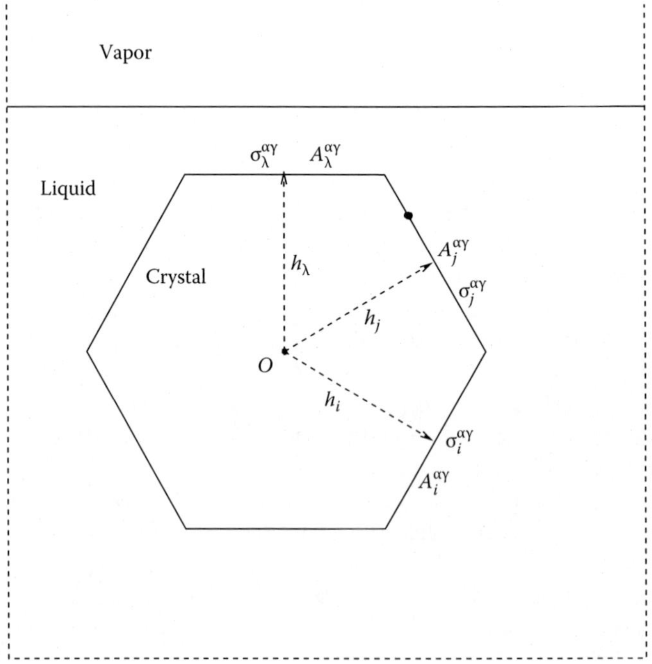


Figure 1.36

Illustration to Wulff's relations. The surface area and surface tension of facet *i* are denoted by $A_i$ and $\sigma_i$, respectively; $h_i$ is the distance from facet *i* to reference point *O*.

where $v$ is the number of molecules in the crystal cluster, $\Delta q < 0$ is the enthalpy of melting per molecule (see, e.g., Ref. [31]), $\Theta = T/T_0$, $T$ is the temperature of the system, $T_0$ is the melting temperature of the bulk ice ($T < T_0$). Equation 1.280 implies that the supercooling $T - T_0$ is not too large or, alternatively, in the temperature range between $T$ and $T_0$ the enthalpy of fusion does not change significantly.

The necessary and sufficient conditions for the equilibrium shape (known as the Wulff form) of the crystal are represented by a series of equalities, referred to as Wulff's relations (see, e.g., Ref. [31]), which can be regarded as a series of equilibrium conditions on the crystal "edges" formed by adjacent facets. For example, on the edge between homogeneously formed facets $i$ and $i + 1$ the equilibrium condition is

$$\frac{\sigma_i^{\alpha\gamma}}{h_i} = \frac{\sigma_{i+1}^{\alpha\gamma}}{h_{i+1}} \equiv \frac{m}{2} \quad (i = 1, \ldots, \lambda), \tag{1.281}$$

where $h_i$ is the distance from facet $i$ to a point $O$ within the crystal (see Figure 1.36) resulting from the Wulff construction.

In the above equations (as well as in those following), it is assumed that the mechanical effects within the crystal (e.g., stresses) reduce to an isotropic pressure $P^\gamma$, so that [31]

$$P^\gamma - P^\alpha = m. \tag{1.282}$$

The relations implied in Equation 1.282 for a crystal are equivalent to Laplace's equation applied to a liquid. Thus, just as for a droplet, one can expect to find a high pressure within a small crystal. It is this pressure that increases the chemical potential within the crystal.

In the surface-stimulated mode, when one of the facets (say, facet $\lambda$) of the crystal cluster constitutes the crystal–vapor interface while all the others lie within the liquid phase (see Figure 1.37), the reversible work of homogeneous formation of such a crystal can be written in the form [27]

$$W^{ss} = -\nu \Delta q \ln \Theta + \sum_{i=1}^{\lambda-1} \sigma_i^{\alpha\gamma} A_i^{\alpha\gamma} + \sigma_\lambda^{\beta\gamma} A_\lambda^{\beta\gamma} - \sigma^{\alpha\beta} A_\lambda^{\beta\gamma} + \tau^{\alpha\beta\gamma} L^{\alpha\beta\gamma}, \tag{1.283}$$

where $\tau$ is the line tension associated with a three-phase contact line [4,5] of length $L$. Again, in this equation, it is assumed that the supercooling $T - T_0$ is not too large so that the enthalpy of fusion is almost constant in the temperature range between $T$ and $T_0$.

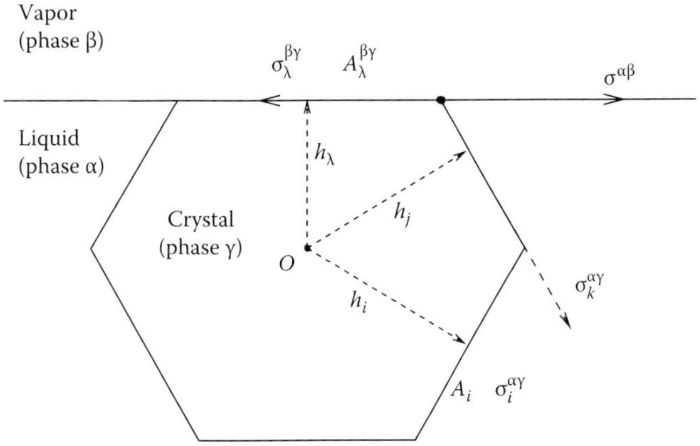


Figure 1.37

Illustration to modified Wulff's relation when the crystal forms with one of its facets at the liquid–vapor surface.

The shape of the crystal will now differ from that in which all facets are in contact with the liquid. As shown by Djikaev and Ruckenstein [27], the modified Wulff relations that take into account the effect of the line tension on the equilibrium shape of the crystal have the form

$$\frac{1}{h_1'}\left(\sigma_1^{\alpha\gamma}-\sum_{k\in\{k\}}\tau_{\lambda k}\frac{\partial L_{\lambda k}}{\partial A_1}\right)=\frac{1}{h_2'}\left(\sigma_2^{\alpha\gamma}-\sum_{k\in\{k\}}\tau_{\lambda k}\frac{\partial L_{\lambda k}}{\partial A_2}\right)$$
$$=\cdots=\frac{1}{h_\lambda'}\left(\sigma_\lambda^{\gamma\delta}-\sigma^{\alpha\delta}-\sum_{k\in\{k\}}\tau_{\lambda k}\frac{\partial L_{\lambda k}}{\partial A_\lambda}\right)=\frac{m'}{2}, \tag{1.284}$$

where the prime marks quantities for the surface-stimulated mode, $\{k\}$ is the set of facets adjacent to facet $\lambda$, $\tau_{\lambda k}$, and $L_{\lambda k}$ are the line tension and length of the edge formed by adjacent facets $\lambda$ and $k$ (as already mentioned, the droplet is assumed to be much larger than crystal clusters). Note that $m'$ in Equation 1.284 is a Lagrange multiplier analogous (but not identical) to $m$, both arising from the minimization [31,36,37] of the free energy of formation of a crystal cluster subject to the constraint $V^\gamma$ = const (see Djikaev and Ruckenstein [27] for more details) Consequently, Equation 1.282 (the equivalent of Laplace's equation for crystals) becomes

$$P^\gamma - P^\alpha = m'. \tag{1.285}$$

Let us assume that the shape of an ice crystal cluster is known, and it is a hexagonal prism. Hence, its state is completely determined by two geometric variables (provided that its density and temperature are given), e.g., the height of the prism and the length of a side of a (regular) hexagon (the base of the prism). However, owing to Wulff's relations, Equations 1.281 and 1.284, only one of these two variables is independent. Therefore, both works $W^{\mathrm{vb}}$ and $W^{\mathrm{s}}$ are functions of only one independent variable, say, variable $a$, the length of a side of the hexagon. The concrete form of the function $W^{\mathrm{ss}} = W^{\mathrm{ss}}(a)$ depends on the mutual orientation of the crystal cluster and droplet surface.

For instance, consider an ice cluster formed with one of its basal facets (assumed to be only partially wettable by water) at the droplet surface. Mark the basal facets with subscripts 1 and 2 and the prismal facets with subscripts 3, …, 8 (as agreed above, facet 2, which is partially wettable by water, forms at the droplet surface and represents the crystal–vapor interface). Clearly, $\sigma_{\mathrm{p}}^{\alpha\gamma}\equiv\sigma_3^{\alpha\gamma}=\cdots=\sigma_8^{\alpha\gamma}$, $A_{\mathrm{p}}^{\alpha\gamma}\equiv A_3^{\alpha\gamma}=\cdots=A_8^{\alpha\gamma}$, $A_{\mathrm{b}}^{\alpha\gamma}\equiv A_1^{\alpha\gamma}=A_2^{\alpha\gamma}$ (subscripts "b" and "p" mark quantities for the basal and prismal facets, respectively). In both the volume-based and surface-stimulated modes, the base is an identical regular hexagon with the side length $a \equiv a_3 = \cdots = a_8$, where $a_i$ ($i$ = 3, …, 8) is the length of the edge formed by the basal facet with prismal facet $i$.

Let us denote the volume of the crystal cluster in the volume-based and surface-stimulated modes by $V^\gamma$ and $V'^\gamma$, respectively. In the first term on the right-hand sides of Equations 1.280 and 1.283, the number $\nu$ of molecules in the crystal cluster can be represented as either $\rho^\gamma V^\gamma$ or $\rho^\gamma V'^\gamma$, respectively, where $\rho^\gamma$ is the number density of molecules in phase $\gamma$ (ice). The volume of an Ih crystal (shaped as a hexagonal prism) is equal to the product "height of the prism" × "surface area of the base." The surface area of the base (regular hexagon in the former and irregular in the latter) is proportional to $a^2$. In both cases, the height of the prism is linearly related to $a$ according to Wulff's relations 1.281 and 1.284. Indeed, it can be found as the sum of the heights of two basal pyramids (i.e., pyramids with the basal facets as their bases). In the volume-based mode, the heights of both basal pyramids, according to Wulff's relations 1.281, are equal to $a(\sqrt{3}/2)(\sigma_{\mathrm{b}}^{\alpha\gamma}/\sigma_{\mathrm{p}}^{\alpha\gamma})$. This is also the height of that pyramid in the surface-stimulated mode whereof the base is *within* the droplet. On the other hand, the height of the pyramid built on the basal facet at the droplet surface is equal to $a(\sqrt{3}/2)(\sigma_{\mathrm{b}}^{\beta\gamma}-\sigma^{\alpha\beta})/\sigma_{\mathrm{p}}^{\alpha\gamma}$. Therefore, in the surface-stimulated mode,

$$h' = a\frac{\sqrt{3}}{2}\frac{\sigma_{\mathrm{b}}^{\alpha\gamma}+\sigma_{\mathrm{b}}^{\beta\gamma}-\sigma^{\alpha\beta}}{\sigma_{\mathrm{p}}^{\alpha\gamma}} \tag{1.286}$$

while in the volume-based mode,

$$h = a\sqrt{3}\left(\sigma_{\mathrm{b}}^{\alpha\gamma}/\sigma_{\mathrm{p}}^{\alpha\gamma}\right) \tag{1.287}$$

(the effect of line tension on the shape of an equilibrium crystal is neglected as discussed by Djikaev and Ruckenstein [27]). Thus, in both Equations 1.280 and 1.283, $v \propto \rho^{\gamma} a^3$. Likewise, one can show that all the surface tension and line tension terms on the right-hand sides of Equations 1.280 and 1.283 are proportional to $a^2$ and $a$, respectively. Therefore, the functions $W^{ss}(a)$ and $W^{vb}(a)$, reversible works of formation of a crystal cluster in the surface-stimulated and volume-based modes, can be written (tedious but simple algebra is omitted) as [27]

$$W^{ss}(a) = -S_3 a^3 + S_2 a^2 + S_1 a, \; W^{vb}(a) = -B_3 a^3 + B_2 a^2 + B_1 a \tag{1.288}$$

with

$$S_3 = \frac{9}{4}\rho^{\gamma}\Delta q \ln(\Theta)\left(\sigma_b^{\alpha\gamma} + \sigma_b^{\gamma\beta} - \sigma^{\alpha\beta}\right)/\sigma_p^{\alpha\gamma},$$

$$S_2 = \frac{3\sqrt{3}}{2}\left[2\sigma_b^{\alpha\gamma} + 3\left(\sigma_b^{\gamma\beta} - \sigma^{\alpha\beta}\right)\right], \; S_1 = 6\tau^{\alpha\gamma\beta}$$

and

$$B_3 = \frac{9}{2}\rho^{\gamma}\Delta q \ln(\Theta)\left(\sigma_b^{\alpha\gamma}/\sigma_p^{\alpha\gamma}\right), \; B_2 = 9\sqrt{3}\,\sigma_b^{\alpha\gamma}, \; B_1 = 0$$

(there is no three-phase contact line in the volume-based mode of homogeneous crystal nucleation; hence, $B_1 = 0$). One can find the radius $a_*$ of the hexagonal base of the crystal nucleus as the positive solution to the equation $dW^{ss}(a)/da|_{a_*} = -3S_3 a_*^2 + 2S_2 a_* + S_1 = 0$, or alternatively, $dW^{vb}(a)/da|_{a_*} = -3B_3 a_*^2 + 2B_2 a_* + B_1 = 0$, which lead to $a_* = \left(2S_2 + \sqrt{4S_2^2 + 12S_1S_3}\right)/6S_3$ or $a_* = 2B_2/3B_3$. The heights of the crystal nucleus (a hexagonal prism) are different in the surface-stimulated and volume-based modes, according to Equations 1.286 and 1.287.

#### 1.10.2.2 Effect of Surface-Stimulated Mode on Kinetics of Crystal Nucleation

The classical expression for the rate of crystal nucleation conventionally used in atmospheric models as well as for treating experimental data, is derived by assuming that crystal nuclei form *within* the liquid [31–33]. However, under certain conditions [27–30], the formation of a crystal nucleus with one of its facets at the droplet surface is thermodynamically favored over its formation with all the facets *within* the droplet. This effect can become important when the crystallizing liquid is in a dispersed state, which is the case with the freezing of atmospheric droplets [32–35] and many experiments [38–40].

Assuming a monodisperse (or narrow enough Gaussian-like) distribution of liquid droplets, the average crystallization time of the ensemble equals that of a single droplet (this is a reasonable assumption in experiments because the droplet sizes are usually well controlled). Let us denote that time by $t_1$. For typical sizes of atmospherically relevant droplets the formation of a single crystal nucleus in a droplet immediately leads to the crystallization of the latter, i.e., the time of growth of a crystal nucleus to the size of the whole droplet is negligible compared to the time necessary for the first nucleation event in the droplet to occur (in experiments, this can be achieved by using special techniques [38–40]). Consequently,

$$t_1 = 1/I, \tag{1.289}$$

where $I$ is the per-droplet (pd) nucleation rate, i.e., the total number of crystal nuclei appearing in the whole volume of the liquid droplet per unit time.

Even when the droplet surface stimulates crystal nucleation therein, the crystal nucleus has to start its initial evolution *homogeneously* as a subcritical cluster in a spherical layer adjacent to the droplet surface (subsurface nucleation [SSN] layer [28]). When this crystal becomes large enough (due to fluctuational growth usual for the nucleation stage), one of its facets hits the droplet surface and at this moment or shortly thereafter it becomes a nucleus owing to a drastic change in its thermodynamic state.

Any crystalline cluster that starts its evolution with its center in the SSN layer can become a nucleus once one of its facets that stimulates nucleation (say, facet $\lambda$), meets the droplet surface. However, to become a surface-stimulated nucleus, the subcritical cluster must evolve in such a way that its facet $\lambda$ is parallel to the droplet surface at the time they meet. The orientation adjustment cannot be mechanical because this would require relatively long time scales, but may, or may not, occur by means of appropriate spatial distribution of density and structure fluctuations around the cluster.

In the framework of the SSN layer model, the pd rate of crystal nucleation can be written as

$$I = I^{vb} + I^{ss}, \quad (1.290)$$

where $I^{vb}$ is the pd rate of volume-based crystal nucleation (i.e., number of crystal nuclei that can form in the volume-based mode per unit time in the whole droplet) and $I^{ss}$ is the pd rate of surface-stimulated crystal nucleation (i.e., number of crystal nuclei forming in the surface-stimulated mode per unit time in the SSN layer). The rates $I^{vb}$ and $I^{ss}$ are simple functions of $R$,

$$I^{vb} = J_v \frac{4\pi}{3} R^3, \; I^{ss} = J_v 4\pi R^2 \delta e^{-\Delta w_*/kt}, \quad (1.291)$$

where $J_v$ is the rate of ice nucleation in the volume-based mode [41,42], $\delta$ is the effective thickness of the SSN layer, and $\Delta W_* \equiv W_*^{ss} - W_*^{vb}$, with $W_*^{vb}$ and $W_*^{ss}$ being the free energies of formation of nuclei in the volume-based and surface-stimulated (with some representative facet at the droplet surface) modes, respectively. Thus, one can represent Equation 1.290 in the form

$$\frac{I}{I^{vb}} = 1 + \kappa \frac{1}{R}, \quad (1.292)$$

where

$$\kappa = 3\delta e^{-\Delta w_*/kT}. \quad (1.293)$$

As will be seen below, $\kappa$ depends on the line tension of the ice–water–vapor contact, $\tau$. Note that on the right-hand side of Equation 1.292 terms of the order of $(\delta/R)^2$ and $(\delta/R)^3$ (arising because volume-based nucleation takes place in a sphere of radius $R - \delta$ while surface-stimulated nucleation occurs in a spherical shell of thickness $\delta$ around that sphere) are omitted because $\delta$ is of the order of nucleus size that is usually much smaller than $R$.

### 1.10.3 Determination of Line Tension of an Ice-Water-Vapor Contact

As clear from Equations 1.280, 1.281, 1.290 through 1.293, the pd rate of crystal nucleation, $I$, depends on the value of the tension $\tau \equiv \tau^{\alpha\beta\gamma}$ at the ice–water–vapor contact line. Thus, it appears that one can determine $\tau$ from experimental data on $I$, much like the liquid–vapor surface tension is estimated as a "kinetic parameter" adjusted to the experimentally obtained rates of ice nucleation in bulk supercooled water [32]. A theoretical basis for the procedure of determining $\tau$ is as follows.

For a given supercooling (i.e., given $\Theta$), the rate $J_v$ can be assumed to be known provided that the ice–liquid surface tension is known (see, e.g., Pruppacher and Klett [32]; Chapter 7.2 and Pruppacher [43]). The numerator on the left-hand side of Equation 1.292 is the overall per-particle rate of crystal nucleation that is effectively measured in experiments on freezing of droplets [39,40]. Thus, one can find a quasi-experimental dependence of the ratio $I/I^{vb}$ on $1/R$. Using then the linear LMS fit of the experimental data ($I/I^{vb}$ versus $1/R$), one can determine the coefficient $\kappa$ in Equation 1.293.

For simplicity, let us consider the case where only two basal facets of an ice crystal contribute to the surface-stimulated mode of nucleation, so that the effective thickness $\delta \approx h'_b$. If the both the basal and prismal facets contribute to the surface-stimulated nucleation, one could estimate $\delta$ as a simple arithmetic mean, $\delta \approx (h'_b + h'_p)/2$, but this will have virtually no effect on the estimate for $\tau$ (recall that $h'_b$ and $h'_p$ are the heights of the pyramids whereof the bases are the basal or prismal facets, respectively, forming a crystal–vapor interface; they differ from each other by less than 25%, i.e., $h_p'/h_b' \approx 0.76$). According to Equations 1.288 and 1.293, the coefficient κ depends on τ because so does $\Delta W_*$. The dependence of $\Delta W_*$ on τ has a simple linear form

$$\Delta W_* = Q_0 + Q_1\tau, \tag{1.294}$$

where

$$Q_0 = -(S_3 - B_3)a_\star^3 + (S_2 - B_2)a_\star^2,\ Q_1 = 6a_\star \tag{1.295}$$

and $a^\star = 2B_2/3B_3$. Thus, having found κ from experiments on crystallization of droplets, one can determine the line tension τ from

$$\tau = -\frac{1}{Q_1}\left[Q_0 + k_B T \ln\left(\frac{\kappa}{3h_b'}\right)\right] \tag{1.296}$$

with $h_b'$ and $a_\star$ related by $h_b' = a_\star(\sqrt{3}/2)\left(\sigma_b^{\beta\gamma} - \sigma^{\alpha\beta}\right)/\sigma_p^{\alpha\gamma}$.

#### 1.10.3.1 Numerical Evaluations

The main problem in using the above procedure is the lack of systematic experiments on the crystallization of droplets. To accurately determine τ, experiments must be carried out under the same external conditions (temperature, pressure, composition of the ambient medium, etc.) on droplets of various sizes that should not be too large (in order for the surface-stimulation effect to be significant). At present, however, such systematic experiments are rather rare. Nevertheless, one can use the above procedure to roughly evaluate τ even from experimental results at one droplet size because the free term in the linear dependence of $I/I^{vb}$ is known (equal to 1).

As an illustration, we used the experimental data reported by Krämer et al. [39] and Duft and Leisner [40] to evaluate the line tension of ice–(liquid) water–(water) vapor contact. In Krämer et al.'s study [39], homogeneous crystal nucleation rates were measured in single levitated water droplets of radius $R = 60 \pm 2$ μm in the range of temperatures between 236 and 237 K, where the observed nucleation rates varied by 2 orders of magnitude. Homogeneous crystal nucleation rates in supercooled water droplets of radii 19 ± 0.25 and 49 ±0.5. μm at $T = 237$ K were determined by Duft and Leisner [40] using the same technique. We considered the results reported for $T = 237$ K because this temperature is close to the temperature for which the data on the ice–water, ice–vapor, and water–vapor interfacial tensions are available [32]. The latter are needed to calculate $I^{vb}$ in the denominator on the left-hand side of Equation 1.292, as well as the coefficients $S_1$, …, $B_3$ in the expressions for $W^{ss}$ and $W^{vb}$ in Equation 1.288. The interfacial tensions were thus taken [32] to be $\sigma_b^{\alpha\beta} \simeq 88$ dyn/m, $\sigma_b^{\alpha\gamma} \simeq 23$ dyn/cm, $\sigma_p^{\alpha\gamma} \simeq 24$ dyn/cm, $\sigma_b^{\beta\gamma} \simeq 102$ dyn/cm, and $\sigma_p^{\beta\gamma} \simeq 111$ dyn/cm. The number densities of water and ice were taken as $\rho^\alpha = (1\ N_A/18)$ cm$^{-3}$ and $\rho^\gamma = (0.92\ N_A/18)$ cm$^{-3}$ (where $N_A$ is the Avogadro number), and the enthalpy of melting per molecule $\Delta q \simeq (333.55 \times 10^7\ N_A/18)$ erg (the latter is needed for $S_3$ and $B_3$). The activation barrier $\Delta F_{act}$ that a water molecule overcomes [32,41,42] upon its transition across the ice–liquid interface (i.e., from its equilibrium position in the liquid state to its equilibrium position in ice) was evaluated according to Pruppacher and Klett [32] and Pruppacher [43] ($\Delta F_{act}$ is needed to calculate $I_{vb}$).

Using these numerical values and experimental data for $I$ provided by Krämer et al. [39] and Duft and Leisner [40], we evaluated the line tension τ of ice–water–vapor contact region to be positive: $\tau \sim 1.16 \times 10^{-11}$ N (from Krämer et al. [39]) and $\tau \sim 1.46 \times 10^{-11}$ (from Duft and Leisner [40]). Both values (obtained by using Equation 1.296) fall well in the middle of the so far experimentally observed and theoretically predicted values for line tension in various systems. In calculating the former, κ was evaluated from Equation 1.292 because droplets of only one radius were studied by Krämer et al. [39], whereas the latter was calculated with κ obtained by fitting the experimental dependence $I$ versus $R$ (provided by Duft and Leisner [40]) with the polynomial of the form $A_2R^2 + A_3R^3$, whereof the coefficients $A_2$ and $A_3$, as clear from Equations 1.291 and 1.292, are expected to be $A_2 = (4\pi/3)\ J_v\kappa$ and $A_3 = (4\pi/3)J_v$. The fitting procedure also provided $J_v \simeq 2.7 \times 10^6$ cm$^{-3}$ s$^{-1}$, which is in excellent agreement with expectations [40].

The sign of τ is not affected by the uncertainty in experimental data on $I$: when the value of the latter is changed by 50%, the value of τ changes by less than 10%. The uncertainly in measuring the droplet radii in experiments has even a smaller effect on $\tau$. The most significant influence on $\tau$ thus extracted from experiments on droplet freezing is exerted by the value of $\sigma_b^{\alpha\gamma}$ and $\sigma_b^{\beta\gamma}$ used in calculating $S_3$ and $B_3$. For example, using a not unlikely value $\sigma_b^{\beta\gamma} \simeq 120$ dyn/cm (see Pruppacher and Klett [32], Section 5.7.1) leads to a negative line tension, $\tau \sim -4 \times 10^{-12}$ N. Thus, very accurate interfacial tensions are needed for the proposed method to provide accurate results for the ice–water–vapor line tension.

### 1.10.4 Concluding Remarks

The line tension can play a major role in phase transition and phase equilibria in multiphase nanosize systems. Due to the ever-increasing need in nanotechnologies, it is thus important to develop theoretical and experimental methods allowing one to clearly establish the magnitude and sign of line tension relevant to various (nano)technological applications.

At present, there are large discrepancies in both sign and magnitude of line tension data available from various theories, simulations, and experiments. This may reflect the diversity of systems studied. However, inconsistencies in experimental data provided by similar approaches and for similar systems can be caused by difficulties in sample preparation, inadequate experimental techniques, etc.

In this paper, we have presented a theoretical basis for determining the line tension of a solid–liquid–vapor contact region by means of experiments on homogeneous crystallization of droplets. The underlying idea is based on our recent finding [27] that the line tension effect can give rise to an important contribution to the free energy of formation of a crystal nucleus in a surface-stimulated mode when one of its facets forms at the droplet surface and thus represents a "crystal–vapor" interface. It should be noted that no assumptions are required as for whether the surface-stimulated mode predominates the volume-based one or not, although the accuracy of the results for the line tension does increase with increasing predominance (of the former over the latter).

An accurate application of the proposed method requires systematic experimental data on the per-droplet rate of homogeneous crystal nucleation as a function of droplet size under otherwise identical external conditions (temperature, pressure, etc.). However, it can provide a rough estimate for the line tension even if the rate is known only for one droplet size. If experimental (per-droplet) rates for a series of droplet sizes are available, the proposed method allows one to independently extract not only the line tension but also the rate of volume-based crystal nucleation under given conditions. As an illustration the method has been used to treat experimental data on homogeneous crystal nucleation in droplets of 19, 49, and 60 μm radii at $T = 237$ K. The line tension of ice–(liquid) water–(water) vapor contact thus obtained is predicted to be positive of the order of $10^{-11}$ N, which is consistent with the previously known data. The accuracy of the results for the line tension strongly depends on the accuracy of input data on the interfacial (solid–liquid and solid–vapor) tensions. By no means, however, should this sensitivity be considered a flaw of the method developed for evaluating the line tension (the results extracted for the volume-based crystal nucleation rate are independent of the input for interfacial tensions).

## References


1. Adamson, A.W., *Physical Chemistry of Surfaces*, 5th ed., Wiley, New York, 1990.
2. Duncan-Hewitt, W., Policova, Z., Cheng, P., Vargha-Butler, E.I., Neumann, A.W., *Colloids Surf.* 42, 391 (1989).
3. Gibbs, J.W., 1961. *The Scientific Papers of JW Gibbs*, vol. 1. Dover, New York, p. 288.
4. Boruvka, L., Neumann, A.W., *J. Chem. Phys.* 66, 5464 (1977).
5. Rowlinson, J.S., Widom, B., *Molecular Theory of Capillarity*, Clarendon Press, Oxford, 1982.
6. Amirfazli, A., Neumann, A.W., *Adv. Colloid Interface Sci* 110, 121–141 (2004).
7. Churaev, N.V., Starov, V.M., Derjaguin, B.V., *J. Colloid Interface Sci.* 89, 16 (1982).
8. Toshev, B.V., Platikanov, D., Scheludko, A., *Langmuir* 4, 4388 (1988).
9. Getta, T., Dietrich, S., *Phys. Rev. E* 57, 655 (1998).
10. Bauer, C., Dietrich, S., *Eur. Phys. J. B* 10, 767–779 (1999).
11. de Feijter, J.A., Vrij, A., *J. Electroanal. Chem.* 37, 9 (1972).
12. Harkins, W.D., *J. Chem. Phys.* 5, 135 (1937).
13. Widom, B., *J. Phys. Chem.* 99, 2803 (1995).
14. Bresme, F., Quirke, N., *Phys. Rev. Lett.* 80, 3791 (1998).
15. Bresme, F., Quirke, N., *J. Chem. Phys.* 110, 3536 (1999).
16. Djikaev, Y., *J. Chem. Phys.* 123, 184704 (2005).

17. Indekeu, J.O., *Physica (Amsterdam)* 183A, 439 (1992).
18. Varea, C., Robledo, A., *Phys. Rev. A* 45, 2645 (1992).
19. Szleifer, I., Widom, B., *Mol. Phys.* 75, 925, 1992.
20. Dobbs, H., *Langmuir* 15, 2586 (1999).
21. Aveyard, R., Clint, J.H., *J. Chem. Soc. Faraday Trans.* 92, 85–89 (1996).
22. Drelich, J., *Pol. J. Chem.* 71, 525 (1997).
23. Wang, J.Y., Betelu, S., Law, B.M., *Phys. Rev. Lett.* 83, 3677 (1999).
24. Wang, J.Y., Betelu, S., Law, B.M., *Phys. Rev. E* 63, 031601 (2001).
25. Pompe, T., Herminghaus, S., *Phys. Rev. Lett.* 85, 1930 (2000).
26. Pompe, T., *Phys. Rev. Lett.* 89, 076102 (2002).
27. Djikaev, Y.S., Ruckenstein, E., *J. Phys. Chem. A* 112, 11677–11687 (2008). (Section 1.8 of this volume.)
28. Djikaev, Y.S., *J. Phys. Chem. A* 112, 6592–6600 (2008).
29. Djikaev, Y.S., Tabazadeh, A., Hamill, P., Reiss, H., *J. Phys. Chem. A* 106, 10247 (2002).
30. Djikaev, Y.S., Tabazadeh, A., Reiss, H., *J. Chem. Phys.* 118, 6572–6581 (2003).
31. Defay, R., Prigogine, I., Bellemans, A., Everett, D.H., *Surface Tension and Adsorption*, Wiley, New York, 1966.
32. Pruppacher, H.R., Klett, J.D., *Microphysics of Clouds and Precipitation*, Kluwer Academic Publishers, Norwell, 1997.
33. Fletcher, N.H., *The Physics of Rainclouds*, Cambridge University Press, Cambridge, 1962.
34. Cox, S.K., *J. Atmos. Sci.* 28, 1513 (1971).
35. Heymsfield, A.J., Miloshevich, L.M., *J. Atmos. Sci.* 50, 2335 (1993).
36. Gibbs, J.W., *Collected Works*, vol. 2, Yale University Press, New Haven, CT, 1928.
37. Curie, P., *Bull. Soc. Fr. Miner.* 8, 195 (1885).
38. Bertram, A.K., Sloan, J.J., *J. Geophys. Res.* 103, 3553 (1998).
39. Krämer, B., Hübner, O., Vortisch, H., Wöste, L, Leisner, T., Schwell, M., Rühl, E., Bämgartel, H., *J. Chem. Phys.* 111, 6521 (1999).
40. Duft, D., Leisner, T., *Atmos. Chem. Phys.* 4, 1997–2000 (2004).
41. Turnbull, D., *J. Appl. Phys.* 21, 1022 (1950).
42. Turnbull, D., Fisher, J.C., *J. Chem. Phys.* 17, 71 (1949).
43. Pruppacher, H.R., *J. Atmos. Sci.* 52, 1924 (1995).

# 2

# Kinetic Theory of Vapor-to-Liquid Nucleation

In contrast to liquids, in gases the molecule–medium interaction is relatively weak. The molecular collisions are relatively rare, and between collisions the molecule moves at constant energy. In this case, the escape of a molecule from a cluster must be treated by using the Fokker–Planck equation subjected to the constraint of a slowly changing energy. The latter constraint allows one to reduce the Fokker–Planck equation to a kinetic equation in the energy space and calculate the evaporation rate from the cluster (Sections 2.1 through 2.3). The condensation rate can be calculated using the kinetic theory of gases [1].

Along with the nucleation rate, Ruckenstein–Narsimhan–Nowakowski theory (RNNT) provides the description of the time-dependent size distribution function of clusters, as well as the relaxation and lag times (Section 2.4).

The initial version of the kinetic nucleation theory (RNNT) was developed assuming that the nucleus (cluster) of the new phase is uniform. To avoid this assumption, the kinetic approach was combined with the density functional theory (Section 2.5) that allows one to find the density profile in the cluster, thus providing more accurate results for the potential field of the cluster, the emission rate, and nucleation rate.

Another extension of the kinetic nucleation theory is presented in Section 2.6, where the theory of binary nucleation is treated. In contrast to the classical nucleation theory for binary systems, which uses the macroscopic concept of surface tension for nanosize clusters, the new approach is based on molecular interactions and Fokker–Planck equation. The new theory is free from the inconsistencies of the classical theory.

The kinetic theory for heterogeneous vapor-to-liquid nucleation is developed in Sections 2.7 through 2.10. Initially (Section 2.7), a uniform cluster was considered, the motion of an evaporating molecule in the potential well being generated by its interaction with the cluster and substrate. The theory was applied (Section 2.8) to nucleation on the liquid aerosols of a binary solution and on rough planar surfaces (Section 2.9). Combining the kinetic theory with the density functional one, the effect of nonuniform density in the nuclei on the rate of heterogeneous nucleation was treated (Section 2.10).

## Reference


1. Abraham, F.F., *Homogeneous Nucleation Theory*, Academic, New York, 1974.

## 2.1 A Kinetic Approach to the Theory of Nucleation in Gases

*Bogdan Nowakowski and Eli Ruckenstein**

A theory of nucleation in gases is developed in which the rate of evaporation of the molecules from the cluster is calculated on the basis of a diffusion equation in the energy space. In contrast to the classical theory of nucleation, the theory does not use macroscopic thermodynamics. Numerical calculations were performed by assuming an intermolecular potential that combines the dispersive attraction with the rigid core repulsion. The predictions of the theory are consistent with the classical theory in the limit of small supersaturations, i.e., for large critical clusters. For small critical clusters, the present theory provides much higher rates of nucleation than the classical one. This difference can be attributed to the use of the macroscopic surface tension in the classical theory, which overpredicts the surface energy of small clusters, and consequently provides lower values for the nucleation rate. The theoretical results are compared with experimental data for nucleation of supersaturated vapors of lower alcohols carried in argon.

### 2.1.1 Introduction

The rates of growth and decay of small clusters constitute the basic problems of the nucleation theory. The classical theory of nucleation was initiated by Volmer and Weber [1] and Farkas [2], and formulated more completely by Becker and Döring [3]. To calculate the rate of evaporation of molecules from a cluster, the theory relies on the principle of detailed balance and on a hypothetical equilibrium distribution of clusters. In addition, macroscopic thermodynamics—particularly the macroscopic surface tension—is employed to estimate the free energy of formation of small clusters. Certain features of the classical theory have been modified and extended in the papers of Kuhrt [4], Lothe and Pound [5], and Reiss [6]. The assumption that macroscopic thermodynamics can be satisfactorily applied to small clusters constitutes the main drawback of the classical nucleation theory.

Recently, a kinetic approach to the theory of nucleation in liquids was proposed [7,8], which avoided the use of the macroscopic surface tension. In that approach, the rate of evaporation of molecules from a cluster was obtained directly by calculating the mean passage time of molecules that escape from the potential well generated in the neighborhood of the surface of the cluster. The Brownian motion of a molecule within the well is governed in general by the Fokker–Planck equation for the probability distribution function in phase space. In liquids, the strong interactions between a molecule and the medium lead to a fast relaxation to the equilibrium velocity distribution in each point of the system. The Fokker–Planck equation reduces in this case to the Smoluchowski equation [9] for the position probability of a molecule. On the other hand, in gases a molecule experiences a relatively weak interaction with the medium. The energy of the molecule changes only during the relatively rare molecular collisions; between successive collisions its motion in the potential field is at constant energy, with relatively rapid variations of position and momentum. Consequently, one can differentiate between the slowly changing energy and the fast changing variables—position and momentum. In contrast to nucleation in liquids, there is no longer a fast relaxation to the equilibrium velocity distribution and, therefore, the escape of a molecule from a cluster must be treated with the complete Fokker–Planck equation under the constraint of a slowly changing energy. However, since the energy is a slowly changing variable in comparison with position and momentum, one can employ a more direct formalism [10] based on a kinetic equation in the energy space.

In this study, the kinetic approach [7,8] suggested for liquids is extended to nucleation in gases. In the next section, the rate of evaporation of molecules from a cluster is calculated on the basis of a kinetic equation for the energy of molecules within the well of the interaction potential near the surface of the cluster in the gas phase. Furthermore, the rate of nucleation is calculated and the results obtained are compared with the predictions of the classical nucleation theory, as well as with experimental data.

### 2.1.2 Evaporation Rate from a Cluster

In the gas layer around a cluster assumed spherical, a potential well is generated by the attractive intermolecular forces due to the molecules of the cluster. The motion of a molecule with a total energy $0 > E > \Phi_0$

* *J. Chem. Phys.*, 94, 1397 (1991). Republished with permission.

(where $\Phi_0$ is the energy level at the bottom of the well) is confined within the potential well, unless it acquires sufficient energy from the collisions with the molecules of the gas to overcome the energy of the potential well. Since the molecules contained in the well can be regarded as bound to the cluster, the process of evaporation from a cluster can be related to the rate at which the molecules leave the well.

In gases, the interaction of a molecule with the medium is not very strong, and the time scale for the variation of the total energy of a molecule is much larger than the relaxation times for the distributions of rapidly changing positions and momenta. Therefore, the partial distribution functions for the fast changing variables remain approximately at equilibrium, whereas the energy changes slowly. The probability distribution of energy $P(E,t)$ can be described by the Markovian equation [10]

$$\frac{\partial}{\partial t}P(E,t)=\int_{\Phi_0}^{0}\mathrm{d}E'[K(E,E')P(E',t)-K(E',E)P(E,t)], \tag{2.1}$$

where $K(E',E)$ is the transition probability from energy $E$ to $E'$. The equilibrium distribution $P_{eq}(E)$ satisfies the detailed balance equation

$$K(E,E')P_{eq}(E')=K(E',E)P_{eq}(E). \tag{2.2}$$

The equilibrium distribution at temperature $T$ is given by the statistical mechanics formula

$$P_{eq}(E)=\Omega(E)\exp(-E/kT)/Z, \tag{2.3}$$

where $k$ is the Boltzmann constant, $\Omega(E)$ is the number of states of energy $E$, and $Z$ is the partition function

$$Z=\int_{\Phi_0}^{0}\mathrm{d}E\Omega(E)\exp(-E/kT). \tag{2.4}$$

For a single molecule with the Hamiltonian $H(p,x)$, $\Omega(E)$ is given by

$$\Omega(E)=\int_{\Gamma}\mathrm{d}p\mathrm{d}x\,\delta[E-H(p,x)], \tag{2.5}$$

where the integral is carried out over the entire phase space $\Gamma$.

In the limit of weak interactions with the medium, the rate of change of the energy is low, and the transition probability $K(E',E)$ is narrowly centered around $E'=E$. Consequently, Equation 2.1 can be expanded around $E'=E$, and using the detailed balance equation (Equation 2.2) one obtains, in the second-order approximation, the following diffusion equation in the energy space [10]:

$$\frac{\partial}{\partial t}P(E,t)=\frac{\partial}{\partial E}D(E)P_{eq}(E)\frac{\partial}{\partial E}\frac{P(E,t)}{P_{eq}(E)} \tag{2.6}$$

$$\equiv-\frac{\partial}{\partial E}j_E(E), \tag{2.7}$$

where $j_E(E)$ is the flux in the energy space, and the diffusion coefficient $D(E)$ is given by

$$D(E)=\frac{1}{2}\int\mathrm{d}E'K(E',E)(E-E')^2. \tag{2.8}$$

Assuming that each molecule that reaches the upper energy of the well ($E = 0$) separates from the cluster, one can write the absorption boundary condition

$$P(E)|_{E=0} = 0. \tag{2.9}$$

The stationary solution of Equation 2.6 that satisfies the boundary condition (Equation 2.9) has the form

$$P(E) = j_E P_{eq}(E) \int_E^0 \frac{dE'}{D(E')P_{eq}(E')}. \tag{2.10}$$

The rate $\alpha^1 = j_E$ at which a molecule leaves the well is obtained from Equation 2.10 by using the normalization condition $\int P(E)dE = 1$. This yields

$$\alpha^1 = \left[ \int_{\Phi_0}^0 dE P_{eq}(E) \int_E^0 \frac{dE'}{D(E')P_{eq}(E')} \right]^{-1}. \tag{2.11}$$

The overall evaporation rate per cluster $\alpha$ is related to the total number $N_w$ of vapor molecules in the well around a cluster. Denoting by $n_w$ the average density of vapor molecules in the well, the total number of molecules in a well of thickness $\lambda$, around the spherical cluster of radius $R$ is $N_w = 4\pi R^2 \lambda n_w$. Consequently, the overall dissociation rate $\alpha = \alpha^1 N_w$ is given by

$$\alpha = \frac{4\pi R^2 \lambda n_w}{\int_{\Phi_0}^0 dE P_{eq}(E) \int_E^0 [dE'/D(E')P_{eq}(E')]}. \tag{2.12}$$

The quantities $\lambda$ and $n_w$ are not known; however, as shown later, they are not present in the final equations.

The energy diffusion coefficient depends on the dynamics governing the motion of a molecule. The Brownian motion of a molecule in the well is described by the Fokker–Planck equation, which can provide the transition probability $K(p',x',p,x)$ for the phase variables $p,x$ [11]. The transition probability for the energy is then obtained from [12]

$$K(E',E) = \left\langle \int dp' dx' \delta[E' - H(p',x')] K(p',x',p,x) \right\rangle_E, \tag{2.13}$$

where $\langle \ \rangle_E$ denotes the average in the microcanonical ensemble of energy $E$. For Brownian molecules, the diffusion coefficient $D(E)$ calculated from Equation 2.8 has the form [12]

$$D(E) = \frac{kT\xi d}{m\Omega(E)} \int_{\Phi_0}^E dE'\Omega(E'), \tag{2.14}$$

where $m$ is the mass of the molecule, $\xi$ is the friction coefficient, and $d$ is the dimensionality of the geometrical space in which the Brownian motion takes place. For a three-dimensional space and a molecule moving in the spherically symmetric potential field $\Phi(r)$ of a cluster, $\Omega(E)$, defined by Equation 2.5, has the form

$$\Omega(E) = 8\pi^2 (2m)^{1/2} \int_R^{R_0} r^2 [E - \Phi(r)]^{1/2} dr, \tag{2.15}$$

where $R_0$ is determined from the condition $E - \Phi(R_0) = 0$.

### 2.1.3 Rate of Growth of a Cluster

The change of the size of a particular cluster is a result of the difference between the rates of evaporation $\alpha$ and condensation $\beta$ per cluster. The latter quantity for small spherical clusters in gases is given by the free molecular formula of the kinetic theory of gases [13]

$$\beta = \pi R^2 \bar{\upsilon} n_1, \tag{2.16}$$

where $\bar{\upsilon}$ is the mean thermal velocity of the vapor molecules, and $n_1$ is the number density of the vapor molecules (monomers) at a large distance from the cluster. The critical cluster size $R_*$ is determined by the condition of (unstable) equilibrium between evaporation and condensation, $\alpha_* = \beta_*$, and depends on the concentration of the condensing component in the gaseous phase. Clusters larger than the critical one grow, whereas those smaller decay. Using Equations 2.11 and 2.16, one can calculate a concentration of vapor $n_1$, in unstable equilibrium with a critical cluster of radius $R_*$

$$n_1 = n_w \frac{4\lambda}{\bar{\upsilon}} \alpha^1(R_*), \tag{2.17}$$

where $\alpha^1$ $(R_*)$ denotes the value of $\alpha^1$ given by Equation 2.11 for a cluster of radius $R_*$. The quantity $\alpha^1$ decreases with the size of the cluster, because the potential well is deeper for large clusters. Consequently, higher concentrations of vapor are needed to ensure unstable equilibrium with smaller critical clusters. The saturation number density $n_s$ is denned as the vapor concentration in equilibrium with a cluster of infinite size, i.e., with the bulk condensed phase. The supersaturation $s = n_1/n_s$ can be expressed using Equation 2.17 as

$$s = \frac{\alpha^1(R_*)}{\alpha^1(\infty)}. \tag{2.18}$$

One may note that the supersaturation does not depend on $\lambda n_w$. The supersaturation and the critical radius are related in the classical theory of nucleation via the Kelvin equation

$$s = \exp\left(\frac{2\sigma}{kT\rho R_*}\right), \tag{2.19}$$

where $\rho$ is the density of the condensed phase and $\sigma$ is the surface tension. The classical relation 2.19 is expected to be recovered in the present theory in the limit of large clusters for which the macroscopic concepts are valid. Consequently, Equations 2.18 and 2.19 can be used to calculate the surface tension, which constitutes the main parameter in the classical theory of nucleation.

### 2.1.4 Rate of Nucleation

The evolution of the population of clusters consisting of $i$ molecules, $n_i$, is provided in the equation

$$\begin{aligned} \frac{dn_i}{dt} &= [\beta_{i-1} n_{i-1}(t) - \alpha_i n_i(t)] - [\beta_i n_i(t) - \alpha_{i+1} n_{i+1}(t)] \\ &= I_{i-1}(t) - I_i(t). \end{aligned} \tag{2.20}$$

Here $I_i$ is the net flux of clusters passing from the $i$ population to $(i + 1)$ population:

$$I_i = \beta_i n_i - \alpha_{i+1} n_{i+1}. \tag{2.21}$$

$I_i$ vanishes at equilibrium leading to the detailed balance equation. Conventionally, the discrete variable $i$ is replaced by a continuum one, $g$, and the finite difference equation (Equation 2.20) is transformed into a differential equation after expanding α, β, and $n$ around the midpoint $i$. This familiar forward Kramers–Moyal expansion [9], if limited to the second-order terms only, yields the following equation for the continuum distribution function $n(g,t)$:

$$\frac{\partial n}{\partial t}=\frac{\partial}{\partial g}[A(g)n(g,t)]+\frac{\partial^2}{\partial g^2}[B(g)n(g,t)] \\ =-\frac{\partial}{\partial g}I(g,t). \tag{2.22}$$

The net flux of clusters $I$ expressed in terms of the continuum variable $g$ has the form

$$I(g,t)=-\frac{\partial}{\partial g}B(g)n(g,t)-A(g)n(g,t). \tag{2.23}$$

However, Equation 2.22 with the coefficients $A(g)$ and $B(g)$ obtained by the standard method of expansion does not recover the equilibrium solution of Equation 2.20, and certain modifications that remove this inconsistency have been proposed [14]. Details are presented in Appendix 2A.

The rate of nucleation is given by the steady-state flux of Equation 2.23:

$$I=\frac{\frac{1}{2}n_1\beta_1}{\int_1^\infty \mathrm{d}g\exp[H(g)]}, \tag{2.24}$$

where $H(g)$ is the smooth function constructed from its values in the discrete points:

$$H(i)=\sum_{j=2}^{i-1}\ln\left(\frac{\alpha_j}{\beta_j}\right)+\ln[\alpha_i/(\beta_i+\alpha_i)]. \tag{2.25}$$

One may note that because this expression contains only ratios, $H(i)$ is independent of $\lambda n_w$. The function $H(g)$ exhibits a rather sharp maximum at the critical cluster size $g_*$. This allows us to evaluate the integral in Equation 2.24 by the method of steepest descent, yielding the following result for the nucleation rate:

$$I=\frac{1}{2}n_1\beta_1\left(-\frac{H''(g_*)}{2\pi}\right)^{1/2}\exp[-H(g_*)]. \tag{2.26}$$

The classical theory provides the following expression for the rate of nucleation in the steepest descent approximation:

$$I_{\mathrm{cl}}=\frac{n_1\beta_1}{4\pi\rho\eta^3}\left(\frac{\sigma\eta^2}{kT}\right)^{1/2}\exp\left(-\frac{16\pi\sigma^3}{3(kT)^3\rho^2\ln^2 s}\right), \tag{2.27}$$

where η is the diameter of a vapor molecule.

The present theory is expected to approach the classical result for small supersaturations, i.e., large critical clusters, for which the macroscopic concept of surface tension can be applied.

## 2.1.5 Results and Comparison with the Classical Theory and Experiment

The binary interaction potential $\phi$ employed in the calculations combines the rigid core repulsion with the nonretarded dispersive attraction, and is given by

$$\phi(r_{12}) = \begin{cases} \infty & \text{for } r_{12} < \eta \\ -\epsilon\left(\dfrac{\eta}{r_{12}}\right)^6 & \text{for } r_{12} \geq \eta. \end{cases} \quad (2.28)$$

Here $r_{12}$ is the distance between the centers of two spherical molecules of diameter $\eta$, and the parameter $\epsilon$ determines the strength of the interactions. The effective potential $\Phi$ of a molecule in a well was calculated by summing the binary interactions among a molecule in the well and those in the cluster. For a compact, spherical, amorphous cluster of radius $R$ this yields

$$\Phi(r) = -\frac{4\pi\epsilon\rho\eta^3}{3}\left(\frac{\eta}{r-R}\right)^3\left(\frac{R}{r+R}\right)^3 \quad r > R+\eta. \quad (2.29)$$

The calculations were performed for $\epsilon/kT = 2, 4, 6, 10, 15$, and for a density in the cluster $\rho\eta^3 = 1.2$, corresponding to a random packing of rigid spheres [15]. The radius of a critical cluster was calculated using Equation 2.18 and the rates of nucleation with Equation 2.24 as well as with its steepest descent approximation, Equation 2.26. The rates of nucleation predicted by the classical theory were obtained from Equation 2.27.

Figure 2.1 shows the dependence of the critical radius $R_*/\eta$ on the supersaturation ratio. The Kelvin relation, Equation 2.19, was assumed to be valid for clusters containing more than 5000 molecules, for which $R_*/\eta > 10$. The values of the surface tension $\sigma\eta^2/kT$ were calculated from the best fit of the slope of the Kelvin equation (Equation 2.19) to the obtained numerical results, and are listed in Table 2.1.

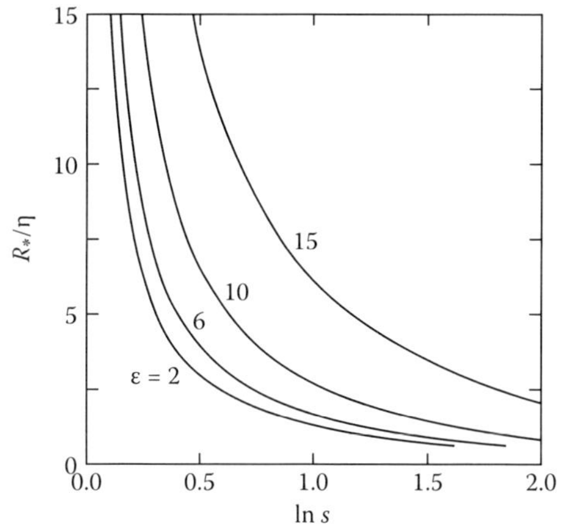


Figure 2.1

Radius of the critical cluster $R_*/\eta$ as a function of the supersaturation ratio for various values of $\varepsilon = \epsilon/kT$.

Table 2.1 Surface Tension $\sigma\eta^2/kT$ for Various Values of $\varepsilon = \epsilon/kT$

| $\epsilon/kT$ | $\sigma\eta^2/kT$ |
|---|---|
| 2 | 0.92975 ± 60 |
| 4 | 1.0444 ± 10 |
| 6 | 1.2310 ± 18 |
| 10 | 1.9317 ± 50 |
| 15 | 3.637 ± 12 |

Figure 2.2 presents the values of the rate of nucleation versus supersaturation calculated from Equation 2.24 for $\varepsilon = \epsilon/kT = 2, 6, 10$, and 15. It has been verified that the rates of nucleation calculated with Equation 2.26 differ by no more than 1% (on the logarithmic scale) from those presented in Figure 2.2.

Figure 2.3 compares the rates of nucleation provided by the present theory, Equation 2.24, and by the classical theory of nucleation for various values of $\epsilon/kT$. As expected, the results of both theories coincide for small supersaturations, i.e., for large critical clusters, since in such cases the surface tension of a cluster is well approximated by its bulk value. For large supersaturations, the present theory predicts much larger nucleation rates than the classical theory does. This can be attributed to the use of the macroscopic surface tension in the classical theory, which overpredicts the surface free energy of the small clusters, and thus provides lower values for the nucleation rates.

The available experimental results for the rates of nucleation in gases involve nonspherical molecules, such as the *n*-alkanes (hexane, …, nonane) [16–18] and *n*-alcohols (methanol, …, hexanol) [19,20]. In contrast, the results of the present theory are based on the interaction potential (Equation 2.28) valid for spherical molecules. For this reason, we confine the comparison between theoretical and experimental results to the lower alcohols, because in this case the deviation from sphericity is smaller. Experimental data for propanol, ethanol, and methanol in argon are available [20] as temperature versus supersaturated vapor pressure for a nucleation rate of $10^8$ $cm^{-3}$ $s^{-1}$. Figure 2.4 presents the theoretical results and the experimental data for the above conditions. The parameter $\epsilon$ in the interaction potential of Equation 2.28 was determined so as to provide the correct values of the surface tension given by Peters and Peikert [20] (see Tables 2.2 and 2.3), and the diameter $\eta$ of the hypothetical spherical

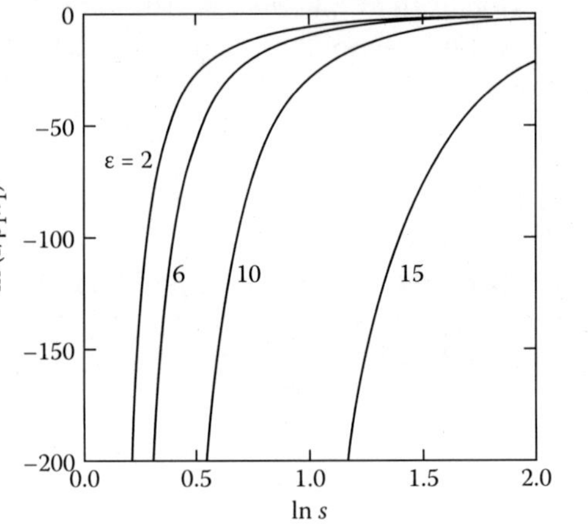


Figure 2.2

Rate of nucleation versus the supersaturation ratio, calculated for various values of $\varepsilon = \epsilon/kT$.

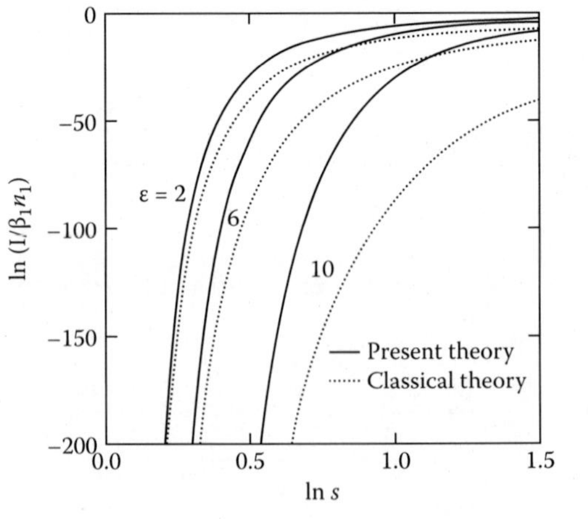


Figure 2.3

Rate of nucleation predicted by the present theory, Equation 2.24, and by the classical theory, plotted versus supersaturation ratio for various values of $\varepsilon = \epsilon/kT$.

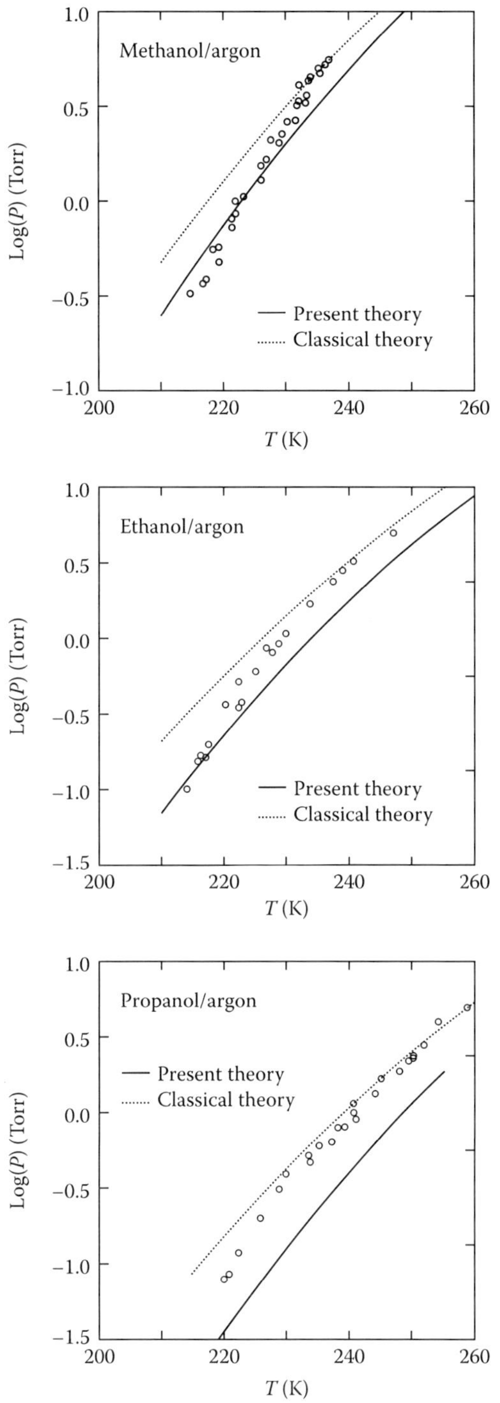


Figure 2.4

Supersaturated vapor pressure $P$ of propanol, ethanol, and methanol in argon as a function of temperature $T$ for rate of nucleation of $10^8$ $cm^{-3}$ $s^{-1}$. The symbols (○) represent the experimental results [20].

molecules was calculated from the density of the liquids. It is worth noting that although the interactions among the molecules of the considered compounds are much more complex than those modeled by Equation 2.28, the smaller, more spherical molecules of ethanol and methanol are expected to be better approximated by the simple London–van der Waals attraction. Therefore, even though the theory based on the interaction potential (Equation 2.28) cannot be strictly applied to predict the rates of nucleation of alcohol vapors, the increasing consistency of the experimental results for the lower alcohols with the present theory, and their simultaneous deviations from the classical predictions, might be interpreted as an indirect support for the present theory.

Table 2.2 Equations Used to Calculate Surface Tension, Equilibrium Vapor Pressure and Density of Methanol, Ethanol, and Propanol in Range of Temperatures of Interest

| | $\sigma$ ($\times 10^{-3}$ N/m) | ln $p_s$ (Torr) | $\rho$ (kg/m$^3$) |
|---|---|---|---|
| Methanol | 24.23–0.092 54 $T$ | 87.4776<br>–7543.0896/$T$<br>–10.06 ln $T$ | 801–0.9253 $T$<br>–4.1 × $10^{-4}$ $T^2$ |
| Ethanol | 23.88–0.088 07 $T$ | 82.0249<br>–7794.4744/$T$<br>–9.09 ln $T$ | 806.25–0.8450 $T$<br>+ 2.9 × $1.0^{-4}$ $T^2$ |
| Propanol | 25.28′–0.083 94 $T$ | 150.2480<br>–11286.4904/$T$<br>–19.19 ln $T$ | 799.5–0.82 ($T$ – 25) |

*Source:* Peters, F., Peikert, B., *J. Chem. Phys.*, 91, 5672, 1989. With permission.
*Note:* Temperature $T$ is in K for the vapor pressure and in °C for the other two quantities.

Table 2.3 Values of Interaction Constant $\epsilon$, Obtained on the Basis of Surface Tension for Methanol, Ethanol, and Propanol

| | Methanol | | Ethanol | | Propanol | |
|---|---|---|---|---|---|---|
| Temperature | $\sigma\eta^2/kT$ | $\epsilon/kT$ | $\sigma\eta^2/kT$ | $\epsilon/kT$ | $\sigma\eta^2/kT$ | $\epsilon/kT$ |
| 210 | 3.06 | 8.88 | 3.81 | 10.51 | 4.68 | 12.02 |
| 220 | 2.85 | 8.32 | 3.55 | 9.99 | 4.37 | 11.52 |
| 230 | 2.66 | 7.76 | 3.31 | 9.48 | 4.09 | 11.03 |
| 240 | 2.48 | 7.18 | 3.10 | 8.97 | 3.83 | 10.55 |
| 250 | 2.32 | 6.58 | 2.89 | 8.44 | 3.59 | 10.08 |
| 260 | 2.17 | 5.96 | 2.71 | 7.91 | 3.37 | 9.61 |
| 270 | 2.03 | 5.31 | 2.53 | 7.35 | 3.16 | 9.12 |

## Appendix 2A

At equilibrium the rates of evaporation and condensation are related via the detailed balance equation,

$$\beta_{i-1}n_{i-1}^{\text{eq}} = \alpha_i n_i^{\text{eq}}. \tag{2A.1}$$

The equilibrium distribution can be obtained by recurrence from Equation 2A.1, and has the form

$$n_i^{\text{eq}} = n_1^{\text{eq}} \prod_{j=2}^{i} \frac{\beta_{j-1}}{\alpha_j} = n_1^{\text{eq}} \exp[h(i)]. \tag{2A.2}$$

The conventional expansion of Equation 2.20 provides the following equation for the continuum distribution function of the clusters:

$$\begin{aligned} \frac{\partial n(g,t)}{\partial t} &= \frac{\partial}{\partial g}[A(g)n(g,t)] + \frac{\partial^2}{\partial g^2}[B(g)n(g,t)] \\ &= -\frac{\partial}{\partial g} I(g,t), \end{aligned} \tag{2A.3}$$

where the coefficients $A(g)$ and $B(g)$, expressed in terms of the evaporation and condensation rates, have the form

$$\begin{aligned} A(g) &= \alpha_g - \beta_g, \\ B(g) &= (\alpha_g + \beta_g)/2. \end{aligned} \tag{2A.4}$$

However, it has been noticed [14] that the truncated equation (Equation 2A.3) with the coefficients given by Equation 2A.4 does not lead to the equilibrium distribution function (Equation 2A.2). This inconsistency can be eliminated by modifying the coefficients in Equation 2A.3. To ensure the proper form for the equilibrium solution, the coefficients $A(g)$, $B(g)$ must satisfy the following equation resulting from the condition $I_{eq} = 0$:

$$A(g) + B(g)h'(g) + B'(g) = 0. \tag{2A.5}$$

An additional physical constraint is that the diffusion coefficient $B(g)$ must be positive. Shizgal and Barrett [14] compared the numerical solution of the exact set of Equation 2.20 and of the Fokker–Planck equation (Equation 2A.3) using a few chosen forms for $A(g)$ and $B(g)$. Good agreement was obtained using Equation 2A.4 for $B(g)$ and then calculating $A(g)$ from Equation 2A.5. After the coefficients are determined, the steady-state flux $I$ of Equation 2A.3 is given by

$$I = \frac{n_1 B(1)}{\int_1^{\infty} dg \exp\left\{\int_1^{g} dg' A(g')/B(g')]\right\}}. \tag{2A.6}$$

From Equation 2A.4, one obtains $B(1) = \beta_1/2$. It follows from Equation 2A.5 that the function under the exponential in Equation 2A.6 is given by

$$\int_1^{g} dg' \frac{A(g')}{B(g')} = -h(g) + h(1) - \ln\frac{B(g)}{B(1)}. \tag{2A.7}$$

Using Equation 2A.2, the right-hand side of the above equation acquires the following form for integer values $g = i$:

$$\int_1^{i} dg' \frac{A(g')}{B(g')} = \sum_{j=2}^{i-1} \ln\left(\frac{\alpha_j}{\beta_j}\right) + \ln\left(\frac{\alpha_i}{\alpha_i + \beta_i}\right). \tag{2A.8}$$

Denoting by $H(g)$ the smooth continuation constructed from the discrete expression (Equation 2A.8), the rate of nucleation can be finally written in the form

$$I = \frac{\frac{1}{2} n_i \beta_1}{\int_1^{\infty} dg' \exp[H(g')]}. \tag{2A.9}$$

## References


1. Volmer, M., Weber, A., *Z. Phys. Chem. (Leipzig)* 119, 227 (1926).
2. Farkas, L., *Z. Phys. Chem. (Leipzig)* 125, 236 (1927).
3. Becker, R., Döring, W., *Ann. Phys. (Leipzig)* 24, 719 (1935).
4. Kuhrt, F., *Z. Phys.* 131, 185, 205 (1952).
5. Lothe, J., Pound, G.M., *J. Chem. Phys.* 36, 2080 (1962).
6. Reiss, H., *Adv. Colloid Interface Sci.* 7, 1 (1977).
7. Narsimhan, G., Ruckenstein, E., *J. Colloid Interface Sci.* 128, 549 (1989). (Section 1.2 of this volume.)
8. Ruckenstein, E., Nowakowski, B., *J. Colloid Interface Sci.* 137, 583 (1990). (Section 1.3 of this volume.)
9. Gardiner, C.W., *Handbook of Stochastic Methods*, Springer, New York, 1983.
10. Keck, J., Carrier, G., *J. Chem. Phys.* 43, 2284 (1965).

11. Skinner, J.L., Wolynes, P.G., *J. Chem. Phys.* 72, 4913 (1980).
12. Borkovec, M., Berne, B.J., *J. Chem. Phys.* 82, 797 (1985).
13. Abraham, F.F., *Homogeneous Nucleation Theory*, Academic, New York, 1974.
14. Shizgal, B., Barrett, J.C., *J. Chem. Phys.* 91, 6505 (1989).
15. Kittel, C., *Introduction to Solid State Physics*, Wiley, New York, 1976.
16. Schmitt, J.L., Zablasky, R.A., Adams, G.W., *J. Chem. Phys.* 79, 4496 (1983).
17. Adams, G.W., Schmitt, J.L., Zalabsky, R.A., *J. Chem. Phys.* 81, 5074 (1984).
18. Wagner, P.E., Strey, R., *J. Chem. Phys.* 80, 5266 (1984).
19. Strey, R., Wagner, P.E., Schmeling, T., *J. Chem. Phys.* 84, 2325 (1986).
20. Peters, F., Peikert, B., *J. Chem. Phys.* 91, 5672 (1989).

## 2.2 A Unidimensional Fokker–Planck Approximation in the Treatment of Nucleation in Gases

*Eli Ruckenstein and Bogdan Nowakowski**

A unidimensional Fokker–Planck equation is used to calculate the rate of evaporation of molecules from clusters. The obtained result is further used to derive an expression for the rate of nucleation. This is compared with a previously derived expression based on a three-dimensional treatment as well as with the classical theory. The unidimensional and three-dimensional approaches lead to results that are very close to one another. For higher supersaturations, they are much greater than those provided by the classical theory.

### 2.2.1 Introduction

The kinetics of evaporation of molecules from the small clusters of an emerging new phase is one of the main problems in the nucleation theory. The first attempt to address this question was in the framework of the so-called classical theory of nucleation, which was initiated by Volmer and Weber [1] and Farkas [2] and formulated in a more comprehensive form by Becker and Döring [3]. In order to calculate the rate of evaporation of molecules, the classical approach relies on the principle of detailed balance and on an hypothetical equilibrium distribution of clusters. The appearance of molecular clusters is explained in terms of fluctuations whose probabilities are determined by the excess free energy of the created nuclei. The surface free energy of these small clusters is calculated by using the surface tension of a bulk phase. The classical theory of nucleation has been improved by Kuhrt [4], Lothe and Pound [5], and Reiss [6]. It is commonly acknowledged that the use of macroscopic thermodynamics for small clusters constitutes the main drawback of the classical nucleation theory.

Recently, a kinetic approach to the theory of nucleation was proposed [7–9] that avoided the use of macroscopic thermodynamics. In that approach, the rate of evaporation of molecules from a cluster was calculated on the basis of a kinetic equation governing the motion of the molecules escaping from the potential well generated around the cluster. The form of this kinetic equation depends on the strength of the interactions of a molecule with the medium and is therefore different in a liquid and a gas. The high dissipation of energy in liquids results in a fast relaxation to the equilibrium velocity distribution in each point of the system. As a result, the probability density for a molecule to have a certain position in the potential well can be calculated by using the Smoluchowski approximation to the Fokker–Planck equation [10]. In gases, because of the relatively infrequent collisions, one can differentiate between fast changing phase variables, position and momentum, and the slowly changing energy. Consequently, by disregarding the fast relaxation processes, the evolution of the system can be described by a master equation in the energy space [9].

The most detailed description of the random motion of a molecule is provided by the Fokker–Planck equation (FPE). As already noted, the Smoluchowski equation is obtained from the FPE in the high damping limit [10]. In this paper, we examine the low damping limit of the FPE, which corresponds to the physical conditions encountered in gases. A unidimensional FPE in position coordinate and velocity constitutes the starting point in the derivation of an expression for the rate of evaporation from clusters. The FPE for the phase space variables is transformed into a kinetic equation in the energy space that is valid in the long time scale approximation. In the subsequent sections, an expression for the nucleation rate is derived on the basis of the unidimensional treatment. Finally, the results obtained are compared with those obtained on the basis of our previous three-dimensional treatment [9] and the classical theory.

### 2.2.2 Kinetic Equation in Energy Space

Let us consider a molecule in a potential well that is generated in the gaseous layer around a cluster by the attractive interactions with the molecules of the cluster. If the energy of a molecule is $0 > E > \Phi_0$ (where $\Phi_0$ is the energy level at the bottom of the well), its motion is confined within the well, and the molecule can be regarded as bound to the cluster. The molecule can leave the well only if it acquires a sufficient amount of energy from the collisions with the molecules of the gaseous medium to overcome the potential energy of the well. Hence, the process of evaporation from a cluster can be related to the rate by which the molecules pass over the upper energetic boundary of the potential well, $E = 0$.

The random motion of a molecule is governed by the FPE (also often called the Kramers equation when it includes an external field) for the probability density function $p(x,v,t)$ with respect to the position $x$ and

* *Langmuir*, 7, 1537 (1991). Republished with permission.

velocity $v$ of a molecule, at time $t$, or by the corresponding equivalent dynamical Langevin equation [10,11]. Let us consider a spherically symmetric potential $\Phi(r)$ in the well around a cluster assumed spherical, where $r$ is the distance to the center of the cluster. If the thickness of the well, $\lambda_w$, is small in comparison to the radius of the cluster, $R$, the molecules move effectively in the one-dimensional potential, since the curvature effects, which are expected to be of the order of $\lambda_w/R$, are negligible. One can therefore reduce the FPE to the one-dimensional form

$$\frac{\partial p}{\partial t} = -\frac{\partial}{\partial r}\upsilon p + \frac{\partial}{\partial \upsilon}\left(\gamma\upsilon + \frac{1}{m}\frac{d\Phi}{dr}\right)p + \frac{\gamma kT}{m}\frac{\partial^2 p}{\partial \upsilon^2}, \tag{2.30}$$

where $p$ is the probability density for the molecule to be located at $r$ with a velocity $\upsilon$ at time $t$, $\gamma$ is the friction coefficient, $m$ is the mass of the molecule, $k$ is the Boltzmann constant, and $T$ denotes the temperature in degrees Kelvin.

Some qualitative considerations regarding the motion of a molecule in the potential well are useful in the development of approximate procedures for solving Equation 2.30. These molecules perform a kind of anharmonic oscillation, with a characteristic period estimated as $t_w = \lambda_w(m/kT)^{1/2}$. On the other hand, the molecules exchange energy with the medium and the strength of these interactions can be characterized by the correlation time $t_c = \gamma^{-1}$. In gases, the interactions with the medium are rather weak, due to the relatively infrequent molecular collisions. Consequently, the correlation time for Brownian motion, $t_c$, is long compared to the period of oscillations in the well, $t_w$, and a molecule performs many oscillations in the well before its energy changes significantly. As a result, one can differentiate between the fast changing phase variables, position and velocity, and the slowly changing energy. This distinction becomes even more clear after the introduction in the FPE (Equation 2.30) of the dimensionless variables:

$$x = r/\lambda_w \tag{2.31}$$

$$u = \upsilon\left(\frac{m}{kT}\right)^{1/2} \tag{2.32}$$

$$t = t/t_0 \tag{2.33}$$

and

$$\Psi = \Phi/kT, \tag{2.34}$$

where

$$t_0 = \lambda_w\left(\frac{m}{kT}\right)^{1/2}. \tag{2.35}$$

In terms of the new variables, Equation 2.30 acquires the form

$$\frac{\partial p}{\partial t} = -\frac{\partial}{\partial x}(up) + \frac{\partial}{\partial u}\frac{d\Psi}{dx}p + \Lambda\frac{\partial}{\partial u}\left(u + \frac{\partial}{\partial u}\right)p, \tag{2.36}$$

where the parameter

$$\Lambda = t_w/t_c \tag{2.37}$$

compares two time scales. Furthermore, Equation 2.36 is transformed from the variables $(x,u)$ to $(x,E)$, where $E$ is the dimensionless energy, scaled with $kT$, defined by

$$E = \tfrac{1}{2}u^2 + \Psi. \tag{2.38}$$

In terms of the new variables, the probability density $p(x,E,t)$ is given by [11]

$$p(x,E,t) = p(x,u,t)\frac{\partial(x,u)}{\partial(x,E)} = \frac{p(x,u,t)}{u} \tag{2.39}$$

from which, using Equation 2.38, one finally obtains

$$p(x,E,t) = \frac{p(x,u,t)}{(2(E-\Psi))^{1/2}}. \tag{2.40}$$

Converting from the velocity $u$ to the energy $E$ in Equation 2.36 and using Equation 2.40, one arrives at the following FPE for the probability density $p(x,E,t)$:

$$\begin{aligned}\frac{\partial p(x,E,t)}{\partial t} &= -2\frac{\partial}{\partial x}((2(E-\Psi))^{1/2}p(x,E,t)) \\ &+ 2\Lambda\left\{\frac{\partial}{\partial E}\left(E-\Psi-\frac{1}{2}\right)p(x,E,t) + \frac{\partial^2}{\partial E^2}(E-\Psi)p(x,E,t)\right\}.\end{aligned} \tag{2.41}$$

Taking advantage of the very different time scales for the changes in position and energy, one can assume that the probability density for the position $x$ relaxes rapidly in the fast time scale to its stationary solution $p_x(x|E)$, which contains as a parameter the slowly changing energy $E$, assumed approximately constant in the fast time scale. Therefore, the joint probability density function can be decomposed as

$$p(x, E, t) = p_x(x|E)P(E,t). \tag{2.42}$$

The conditional probability density $p_x(x|E)$ is obtained as a stationary solution of Equation 2.41,

$$\frac{\partial}{\partial x}[(2(E-\Psi))^{1/2}p_x(x|E)] = 0, \tag{2.43}$$

from which one immediately obtains

$$p_x(x|E) = \frac{C(E)}{(2(E-\Psi))^{1/2}}, \tag{2.44}$$

where $C(E)$ is a constant that can be determined from the normalization condition.

Combining Equations 2.42 and 2.44, the joint probability density can be expressed as

$$p(x,E,t) = \frac{P(E,t)}{\phi'(E)(2(E-\Psi))^{1/2}}, \tag{2.45}$$

where

$$\phi(E) = \int_{R_1}^{R_2}(2(E-\Psi))^{1/2}\,\mathrm{d}r \tag{2.46}$$

with $R_1$ and $R_2$ being the roots of the equation $E = \Psi(R)$. Introducing Equation 2.45 into the FPE (Equation 2.41), one obtains

$$\frac{1}{\phi'(E)(2(E-\Psi))^{1/2}}\frac{\partial}{\partial t}P(E,t)=\Lambda\frac{\partial}{\partial E}\left(\frac{[(2(E-\Psi))-1]P(E,t)}{\phi'(E)(2(E-\Psi))^{1/2}}\right) + \Lambda\frac{\partial^2}{\partial E^2}\left(\frac{(2(E-\Psi))^{1/2}P(E,t)}{\phi'(E)}\right) \tag{2.47}$$

A closed equation for the probability density with respect to energy $P(E,t)$ can be obtained after integrating Equation 2.47 with respect to the position of a molecule between $R_1$ and $R_2$. This yields

$$\frac{\partial}{\partial t}P(E,t)=\Lambda\frac{\partial}{\partial E}\left\{\left(\frac{\phi(E)}{\phi'(E)}-1\right)P(E,t)\right\}+\Lambda\frac{\partial^2}{\partial E^2}\left\{\frac{\phi(E)}{\phi'(E)}P(E,t)\right\} \equiv -\frac{\partial}{\partial E}j(E,t), \tag{2.48}$$

where $j$ is the flux in the energy space. A similar equation for $P(E,t)$ can be established from a master equation in the energy space. The latter procedure leads to the following diffusion-type equation [12]:

$$\frac{\partial}{\partial t}P(E,t)=\Lambda\frac{\partial}{\partial E}\left(D(E)P^{eq}(E)\frac{\partial}{\partial E}\frac{P(E,t)}{P^{eq}(E)}\right). \tag{2.49}$$

Here, $P^{eq}(E)$ is the equilibrium distribution given by

$$P^{eq}(E) = \Omega(E)\exp(-E)/Z \tag{2.50}$$

and $D(E)$ is the diffusion coefficient in the energy space [13]

$$D(E)=\frac{d\int_{\Psi_0}^{E}\Omega(E')\mathrm{d}E'}{\Omega(E)} \tag{2.51}$$

with $d$ the dimensionality of the space. In Equations 2.49 and 2.50, $\Omega(E)$ is the number of states of energy $E$ for a single molecule with the Hamiltonian $H(x,u)$

$$\Omega(E)=\int_{\Gamma}\mathrm{d}x\,\mathrm{d}u\,\delta(E-H(x,u)), \tag{2.52}$$

and $Z$ is the canonical partition function

$$z=\int_{\Psi_0}^{0}\mathrm{d}E\Omega(E)e^{-E}. \tag{2.53}$$

Equation 2.52 yields the following specific result for a molecule with the one-dimensional Hamiltonian $H(x,u) = 1/2u^2 + \Psi(x)$:

$$\Omega(E) = \int_{\Gamma} \frac{\mathrm{d}x}{u} = \int_{R_1}^{R_2} \frac{\mathrm{d}x}{(2(E-\Psi))^{1/2}}. \tag{2.54}$$

Comparing Equations 2.46 and 2.54, one finds that $\Omega(E)$ coincides with $\phi'(E)$. Consequently, the kinetic equation for energy, Equation 2.48, derived from the FPE coincides (in the one-dimensional case) with the diffusion equation for energy, Equation 2.49, with the diffusion coefficient $D(E)$ given by Equation 2.51.

### 2.2.3 Rate of Evaporation of Molecules from a Cluster

The rate of evaporation is obtained from the stationary solution of Equation 2.48. Assuming that each molecule reaching the upper boundary of the well ($E = 0$) separates from the cluster, the following absorption boundary condition can be written:

$$P(E)|_{E=0} = 0. \tag{2.55}$$

The steady-state solution of Equation 2.48 that satisfies condition 2.55 has the form

$$P(E) = j_E \phi'(E) e^{-E} \int_{E}^{0} \frac{\mathrm{d}E'}{\phi(E')e^{-E'}}, \tag{2.56}$$

where $j_E$ is the steady-state flux in the energy space. The rate of evaporation per molecule, $\alpha^1 = j_E$, is obtained from Equation 2.56 by using the normalization condition $\int P(E)\mathrm{d}E = 1$. This yields

$$\alpha^1 = \left\{ \int_{\Psi_0}^{0} \mathrm{d}E \phi'(E) e^{-E} \int_{E}^{0} \frac{\mathrm{d}E'}{\phi(E')e^{-E}} \right\}^{-1}. \tag{2.57}$$

Returning to the unscaled variables, $\alpha^1$ is replaced by $\gamma\alpha^1$. The overall evaporation rate per cluster $\alpha$ is given by $\alpha = N_w \gamma \alpha^1$, where $N_w$ is the number of vapor molecules in the well of a cluster. If $n_w$ is the average number density of vapor molecules in the well, the total number of molecules in the well is $N_w = n_w 4\pi R^2 \lambda_w$. Consequently, the dissociation rate per cluster is given by the expression

$$\alpha = \frac{\gamma 4\pi R^2 \lambda_w n_w}{\int_{\Psi_0}^{0} \mathrm{d}E \phi'(E) e^{-E} \int_{E}^{0} \frac{\mathrm{d}E'}{\phi(E')e^{-E'}}}. \tag{2.58}$$

Although the quantities $\lambda_w$ and $n_w$ are not known, they do not appear in the equations used to calculate the final results.

### 2.2.4 Rate of Growth of a Cluster

The rate of change of the cluster size is the difference between the rates of evaporation $\alpha$ and condensation $\beta$ per cluster. In the gaseous systems, the rate of impingement of molecules on a spherical cluster is given by the formula of the kinetic theory of gases [14],

$$\beta = \pi R^2 \bar{\upsilon} n_1, \tag{2.59}$$

where $\bar{\upsilon}$ is the mean thermal velocity of the vapor molecules and $n_1$ is the number density of vapor molecules at a large distance from the cluster. The critical cluster size $R_*$ is defined by the condition of (unstable) equilibrium between evaporation and condensation, $\alpha_* = \beta_*$, and is a function of the concentration of the condensing component in the gaseous phase. From this relation, one can calculate the concentration of vapor that coexists in unstable equilibrium with a critical cluster of radius $R_*$.

Using Equations 2.58 and 2.59, one obtains

$$n_1 = n_w \frac{4\lambda_w \gamma}{\bar{\upsilon}} \alpha^1(R_*). \tag{2.60}$$

The rate $\alpha^1$ decreases with increasing cluster size because the potential well is deeper for large clusters. Consequently, higher concentrations are needed to ensure the unstable equilibrium with smaller critical clusters. Clusters decay if their size is smaller than the critical one and grow when they are greater. The saturation number density $n_s$ is the vapor concentration that coexists with a cluster of infinite size, hence with the bulk condensed phase. The supersaturation $s$ is given by

$$s = \alpha^1(R_*)/\alpha^1(\infty). \tag{2.61}$$

In the classical theory of nucleation, the supersaturation and the critical radius are related by the Kelvin equation (Equation 2.47)

$$s = \exp\left(\frac{2\sigma}{kT\rho R_*}\right), \tag{2.62}$$

where $\rho$ is the number density of the condensed phase and $\sigma$ is the surface tension, which constitutes the basic thermodynamic parameter of the classical theory. Macroscopic thermodynamics is expected to be valid for large clusters. It is therefore expected that the present theory will coincide with the classical result in the limit of large clusters. Consequently, Equations 2.61 and 2.62 can be used to calculate the surface tension. One may note that its value does not depend on the quantities $\lambda_w$ and $n_w$.

### 2.2.5 Rate of Nucleation

The rates of evaporation $\alpha$ and condensation $\beta$ are involved in the equations governing the concentrations $n_i(t)$ of clusters consisting of $i$ molecules (Equation 2.47):

$$dn_i/dt = \beta_{i-1}n_{i-1} - \alpha_i n_i - \beta_i n_i + \alpha_{i+1}n_{i+1} = I_{i-1} - I_i, \tag{2.63}$$

where $I_i$ is the net flux of clusters passing from $i$ population to $(i + 1)$ population

$$I_i = \beta_i n_i - \alpha_{i+1}n_{i+1}. \tag{2.64}$$

The rate of nucleation is given by the steady-state flux $I$, given by the stationary solution of Equation 2.63. Traditionally, the set of discrete equations (Equation 2.63) are transformed into a partial differential equation for the continuous distribution function $n(g,t)$. Although various methods of such a conversion are possible [15], in this paper we use the Zeldovich approach [16] because in contrast to other expansions, it satisfies automatically the detailed balance principle for the equilibrium distribution $n_i^{eq}$ (that follows from the condition $I \equiv 0$):

$$\beta_i n_i^{eq} = \alpha_{i+1} n_{i+1}^{eq}. \tag{2.65}$$

Since the present theory provides independently the values of $\alpha$ and $\beta$, Equation 2.65 is used to calculate by recurrence the equilibrium distribution $n_i^{eq}$. This yields

$$n_i^{eq} = n_1^{eq} \prod_{j=2}^{i} \frac{\beta_{j-1}}{\alpha_j} \equiv n_1^{eq} \exp(h(i)). \tag{2.66}$$

In the Zeldovich expansion, Equation 2.65 is used to rewrite Equation 2.63 in the form

$$\frac{dn_i}{dt} = \beta_i n_i^{eq}\left(\frac{n_{i+1}}{n_{i+1}^{eq}} - \frac{n_i}{n_i^{eq}}\right) - \beta_{i-1} n_{i-1}^{eq}\left(\frac{n_i}{n_i^{eq}} - \frac{n_{i-1}}{n_{i-1}^{eq}}\right). \tag{2.67}$$

Replacing the discrete variable $i$ by the continuous variable $g$, and expanding Equation 2.67 around the midpoint $g = i$, one obtains, in the second-order approximation, the following equation for the distribution function $n(g,t)$:

$$\frac{\partial n(g,t)}{\partial t} = \frac{\partial}{\partial g}\left[\beta(g) n^{eq}(g)\frac{\partial}{\partial g}\left(\frac{n(g,t)}{n^{eq}(g)}\right)\right] = -\frac{\partial}{\partial g} I(g,t), \tag{2.68}$$

where the flux of clusters $I(g,t)$ in the continuum formulation has the form

$$I(g,t) = -\beta(g) n^{eq}(g)\frac{\partial}{\partial g}\left(\frac{n(g,t)}{n^{eq}(g)}\right). \tag{2.69}$$

The stationary solution of Equations 2.68 and 2.69 must satisfy the boundary conditions

$$n(1)/n^{eq}(1) = 1 \tag{2.70}$$

and

$$n(g)/n^{eq}(g) \to 0 \quad \text{for} \quad g \to \infty. \tag{2.71}$$

Equation 2.71 is valid if the very large clusters are removed from the system. The stationary solution of Equation 2.69, which satisfies the boundary conditions (Equations 2.70 and 2.71), has the form

$$I = \frac{1}{\int_1^\infty \frac{dg}{\beta(g) n^{eq}(g)}} = \frac{n_1\beta_1}{\int_1^\infty dg\left[\exp\left(-h(g) - \text{In}\frac{\beta(g)}{\beta(1)}\right)\right]}, \tag{2.72}$$

where the function $h(g)$ is defined by Equation 2.66.

Equation 2.72 provides the steady-state rate of nucleation (the rate of dissociation is involved in Equation 2.72 via the function $h(g)$). A similar result is obtained by using the modified Goodrich approach [15]. The results of the present theory will be compared with the classical theory of nucleation, which yields the following equation for the nucleation rate:

$$I_{cl} = \frac{n_1\beta_1}{4\pi\rho\eta^3}\left(\frac{\sigma\eta^2}{kT}\right)^{1/2} \exp\left(-\frac{16\pi\sigma^3}{3(kT)^3\rho^2\text{In}^2 s}\right), \tag{2.73}$$

where $\eta$ is the diameter of a molecule of the condensing species.

### 2.2.6 Results and Discussion

The calculations were performed for a binary interaction potential $\phi_{12}$ that combines the rigid core repulsion with the nonretarded dispersion attraction:

$$\phi_{12}(r_{12}) = \begin{cases} \infty & \text{for } r_{12} < \eta \\ -\epsilon\left(\frac{\eta}{r_{12}}\right)^6 & \text{for } r_{12} \geq \eta. \end{cases} \tag{2.74}$$

Here, $r_{12}$ is the distance between the centers of two interacting molecules and $\epsilon$ is a measure of the strength of these interactions. The potential $\Phi$ in a well around the cluster is calculated by summing the binary interaction (Equation 2.74) over the whole cluster. For a spherical amorphous cluster of radius $R$, with a uniform distribution of molecules within it, the integration yields

$$\Phi(r) = -\frac{4\pi\epsilon\rho\eta^3}{3}\left(\frac{\eta}{r-R}\right)^3\left(\frac{R}{r+R}\right)^3 \text{ for } r > R+\eta. \tag{2.75}$$

For amorphous bodies composed of randomly packed spheres, the number density is given by $\rho\eta^3 = 1.2$ [17].

The rate of nucleation was calculated by using Equation 2.72, with the rate of dissociation given by Equation 2.58. Figure 2.5 presents the nucleation rate plotted versus the supersaturation ratio $s$ for several values of the interaction parameter $\epsilon/kT$.

It is of interest to compare the results of the present theory with those provided by other approaches to the nucleation theory. Previously, the authors have developed a kinetic theory of nucleation in gases [9], based on a diffusion-like equation (Equation 2.49) in the energy space for the rate of evaporation. In that treatment, a three-dimensional motion of the molecules in the well was considered. Since the radial direction is the only one along which the potential energy in a well varies, one may expect the present one-dimensional

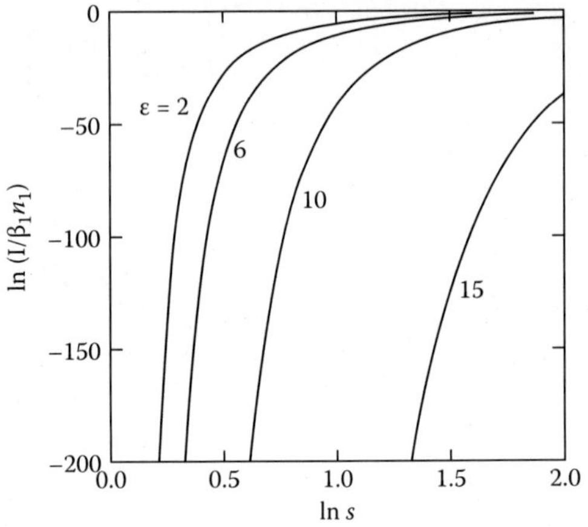


Figure 2.5

Nucleation rate versus supersaturation for various values of $\varepsilon = \epsilon/kT$.

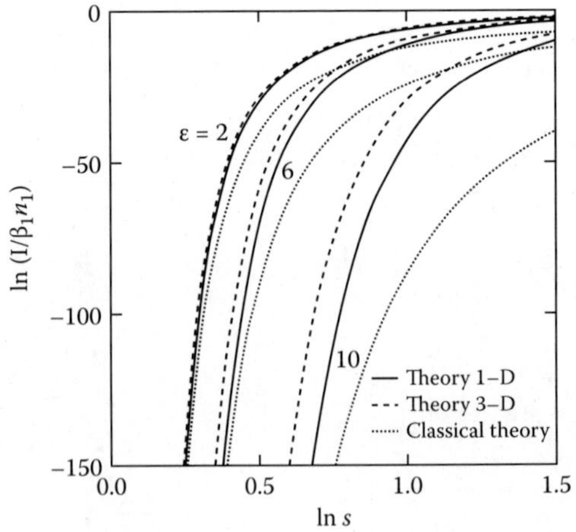


Figure 2.6

Comparison among the nucleation rates predicted by various approaches.

treatment for the rate of evaporation to constitute a fairly accurate approximation of the more exact three-dimensional solution. The unidimensional approximation used here is particularly justified, when the thickness of the well is small compared to the radius of the cluster; in such a system a molecule moves virtually in the one-dimensional potential well along the radial direction, and its motion along the lateral and longitudinal coordinates is independent and separable.

Figure 2.6 compares the nucleation rates provided by the present one-dimensional approach for the rate of evaporation and by the three-dimensional treatment [9]. The results obtained by using the classical theory of nucleation (Equation 2.73) are also presented. As already mentioned, the surface tension $\sigma$ involved in the classical theory was calculated from the best fit of the slope of the Kelvin equation (Equation 2.62) to the numerical results of the present theory for large clusters (clusters for which $R_*/\eta > 10$). One can notice that the one-dimensional and three-dimensional approaches provide very close results. For small supersaturations, the one-dimensional approximation tapers more rapidly, but all three approaches converge to common values for large critical clusters, because macroscopic thermodynamics is expected to be valid in such cases.

### 2.2.7 Conclusion

A kinetic approach to the theory of nucleation in gases is presented, in which the rate of evaporation of molecules from clusters is calculated independently, i.e., without using the detailed balance principle and macroscopic thermodynamics as the classical theory does. The motion of the evaporating molecules within the potential well around a cluster is described by using an FPE. Assuming that the effective thickness of the well is much smaller than the radius of the cluster, the motion of a molecule in the well is treated in a one-dimensional approximation, in which only the radial coordinate and velocity are relevant. The effective one-dimensional FPE leads, after averaging over a short time scale, to a kinetic equation in the energy space, which is shown to coincide with the diffusion equation in the energy space obtained by another formalism. The rate of nucleation has been found to provide values close to those predicted on the basis of a three-dimensional description of the motion of the evaporating molecules.

## References


1. Volmer, M., Weber, A., *Z. Phys. Chem. (Leipzig)* 119, 227 (1926).
2. Farkas, L., *Z. Phys. Chem. (Leipzig)* 125, 236 (1927).
3. Becker, R., Döring, W., *Ann. Phys. (Leipzig)* 24, 719 (1935).
4. Kuhrt, F., *Z. Phys.* 131, 185, 205 (1952).
5. Lothe, J., Pound, G.M., *J. Chem. Phys.* 36, 2080 (1962).
6. Reiss, H., *Adv. Colloid Interface Sci.* 7, 1 (1977).
7. Narsimhan, G., Ruckenstein, E., *J. Colloid Interface Sci.* 128, 549 (1989). (Section 1.2 of this volume.)
8. Ruckenstein, E., Nowakowski, B., *J. Colloid Interface Sci.* 137, 583 (1990). (Section 1.3 of this volume.)
9. Nowakowski, B., Ruckenstein, E., *J. Chem. Phys.* 94, 1397 (1991). (Section 2.1 of this volume.)
10. Gardiner, C., *Handbook of Stochastic Methods*, Springer, New York, 1983.
11. Risken, H., *The Fokker–Planck Equation*, Springer, New York, 1989.
12. Keck, J., Carrier, G., *J. Chem. Phys.* 43, 2284 (1965).
13. Borkovec, M., Berne, B.J., *J. Chem. Phys.* 82, 797 (1985).
14. Abraham, F.F., *Homogeneous Nucleation Theory*, Academic Press, New York, 1974.
15. Nowakowski, B., Ruckenstein, E., *J. Colloid Interface Sci.* 142, 599 (1991). (Section 1.5 of this volume.)
16. Zeldovich, J.B., *J. Exp. Theor. Phys. (Russian)* 12, 525 (1942).
17. Kittel, C., *Introduction to Solid State Physics*, Wiley, New York, 1976.

## 2.3 Homogeneous Nucleation in Gases: A Three-Dimensional Fokker–Planck Equation for Evaporation from Clusters

*Bogdan Nowakowski and Eli Ruckenstein**

A theory of homogeneous nucleation in gases is developed, based on independently obtained expressions for the rates of evaporation from and condensation on the clusters. The rate of evaporation was calculated using a diffusion equation in the energy space, derived from the three-dimensional Fokker–Planck equation averaged with respect to the distribution of the angular variables. The obtained rates of nucleation are compared with those obtained previously by the authors on the basis of a diffusion equation in the energy space and a unidimensional Fokker–Planck equation, as well as with the predictions of the classical theory. The present approach predicts rates of nucleation that are closer to the classical theory than those provided by the former approaches. Nevertheless, the difference between the present approach and the classical result is significant for small clusters, i.e., for high supersaturations.

### 2.3.1 Introduction

The rate of evaporation from small clusters of a new phase remains one of the central problems of the nucleation theory. The classical nucleation theory [1–3] addresses this question, relying on the detailed balance principle and on a hypothetical equilibrium distribution of clusters. The relative abundance of equilibrium clusters of different sizes is obtained in the classical theory in the framework of the fluctuation theory using the macroscopic surface tension to calculate the surface free energy of clusters.

A kinetic approach to the nucleation theory was developed recently [4–6], in which the rate of evaporation of molecules from a cluster was calculated, without employing macroscopic thermodynamics, on the basis of a kinetic equation governing the motion of a molecule in the potential well around the cluster. The Fokker–Planck equation [7] constitutes the kinetic equation that describes the motion of a molecule in the neighborhood of the cluster. This equation can be reduced to simpler forms by taking into account the characteristics of the interactions between the molecule and the medium. In gases, the interactions occur only during the relatively infrequent molecular collisions, whereas between successive collisions the motion takes place with rapid changes of position and momentum, but at constant energy. Consequently, the time scale for the variations of position and momentum is much shorter than the time scale for the variation of energy. The evolution of the motion of the molecule in the long time scale can be therefore described by a kinetic equation for the probability distribution function of the energy, because the probability distribution function of position and momentum relaxes rapidly to its stationary form for the given energy that remains approximately constant in the short time scale.

The authors recently presented results obtained for the nucleation rate in gases on the basis of a kinetic equation in the energy space [5]. In the one-dimensional case, the above kinetic equation can be derived from the Fokker–Planck equation [6,8]. However, the unidimensional approximation is valid essentially for large clusters, for which the thickness of the potential well is small compared to the radius of the cluster. In the present paper, we derive a diffusion equation in the energy space, starting from the three-dimensional Fokker–Planck equation, which is reduced to a unidimensional equation by assuming that the angular momenta achieve equilibrium in a very short time. Section 2.3.2 presents the details of the derivation of the kinetic equation in the energy space. In the subsequent section, its stationary solution is used to calculate the rate of evaporation of molecules from the cluster. Then, by combining this result with the rate of condensation, the critical cluster size for a given supersaturation is calculated, as well as the saturation density of the vapor. Finally, the dynamic equations for the size distribution function of clusters are solved for the stationary case to obtain the rate of nucleation using the independently calculated rates of evaporation and condensation.

### 2.3.2 Equation for Probability Density Distribution Function of Energy

In the gaseous layer around a cluster, a potential well is generated by the attractive interactions with the molecules of the cluster. The motion of a molecule in this potential well is described by the three-dimensional Fokker–Planck equation. If the vapor molecule has an energy $E$ lower than the upper boundary of the well, $E_b$, it moves within the well, and therefore can be regarded as bound to the cluster. It can leave the well and

* *J. Chem. Phys.*, 94, 8487 (1991). Republished with permission.

hence evaporate from the cluster only if it acquires enough energy from the medium to overcome the potential well. Consequently, the rate of evaporation is determined by the rate with which molecules pass over the upper energy of the well.

The random motion of the molecule in the well is governed by the Fokker–Planck equation for the probability density function $P(\mathbf{x},\mathbf{p},t)$ for the position $\mathbf{x}$ and momentum $\mathbf{p}$ of the molecule,

$$\frac{\partial P(\mathbf{x},\mathbf{p},t)}{\partial t}=-\frac{p_i}{m}\frac{\partial}{\partial x_i}P+\frac{\mathrm{d}\Phi}{\partial x_i}\frac{\partial}{\partial p_i}P+\gamma\frac{\partial}{\partial p_i}(p_iP)+m\gamma kT\frac{\partial^2 P}{\partial p_i\partial p_i}, \tag{2.76}$$

where $\gamma$ is the friction coefficient, $m$ is the mass of the molecule, $k$ is the Boltzmann constant, and $T$ is the temperature in K. Summation with respect to a repeated subscript is implied in Equation 2.76. The potential $\mathbf{\Phi}$ in the well is spherically symmetric, depending only on the radial distance $r$ from the center of the cluster. It is convenient to transform the Fokker–Planck equation (Equation 2.76) from the rectangular coordinates $\{\mathbf{x},\mathbf{p}\}$ to the spherical coordinates $\{r,\theta,\phi,p_r,J_\theta,J_\phi\}$. The angular momenta $J_\theta,J_\phi$ are generalized impulses conjugated to the angles $\theta$ and $\phi$ via the Hamilton equations

$$\frac{\mathrm{d}J_\theta}{\mathrm{d}t}=-\frac{\partial H}{\partial\theta}\ \text{and}\ \frac{\mathrm{d}J_\phi}{\mathrm{d}t}=-\frac{\partial H}{\partial\phi}, \tag{2.77}$$

where $H$ is the one-particle Hamiltonian. In spherical coordinates, $H$ has the form

$$H=\frac{1}{2m}\left(p_r^2+\frac{J_\theta^2}{r^2}+\frac{J_\phi^2}{r^2\sin^2\theta}\right)+\Phi(r). \tag{2.78}$$

In terms of the generalized coordinates $\{q_i,p_i,\}$ of the Hamilton equations, the Fokker–Planck equation can be written as [7–9]

$$\frac{\partial P}{\partial t}=\frac{\partial H}{\partial q_i}\frac{\partial P}{\partial p_i}-\frac{\partial H}{\partial p_i}\frac{\partial P}{\partial q_i}+\gamma\frac{\partial}{\partial p_i}(p_iP)+m\gamma kTg_{ij}\frac{\partial^2 P}{\partial p_i p_j}, \tag{2.79}$$

where $g_{ij}$ are the covariant components of the metric tensor in the generalized coordinate system. In spherical coordinates, the Fokker–Planck equation (Equation 2.79) has the form

$$\begin{aligned}\frac{\partial P(r,\theta,\phi,\upsilon,J_\theta,J_\phi,t)}{\partial t}=&-\upsilon\frac{\partial}{\partial r}P+\frac{1}{m}\left[\frac{\mathrm{d}\Phi}{\mathrm{d}r}-\frac{1}{mr^3}\left(J_\theta^2+\frac{J_\phi^2}{\sin^2\theta}\right)\right]\frac{\partial P}{\partial\upsilon}\\&-\mathrm{ctg}\theta\frac{2J_\phi^2}{r^2\sin^2\theta}\frac{\partial P}{\partial J_\theta}+\gamma\frac{\partial}{\partial\upsilon}(\upsilon P)+\gamma\frac{\partial}{\partial J_\theta}(J_\theta P)\\&+\gamma\frac{\partial}{\partial J_\phi}(J_\phi P)+\frac{\gamma kT}{m}\frac{\partial^2 P}{\partial\upsilon^2}\\&+m\gamma kT\left(r^2\frac{\partial^2 P}{\partial J_\theta^2}+r^2\sin^2\theta\frac{\partial^2 P}{\partial J_\phi^2}\right),\end{aligned} \tag{2.80}$$

where

$$\upsilon=p_r/m \tag{2.81}$$

is the radial velocity of the molecule.

In most physical situations, one can assume that the angular momenta achieve equilibrium distribution after a short lapse of time that is much shorter than the time scale of evolution in the radial direction. Under these conditions, one can decompose the distribution function as follows:

$$P(r,\theta,\phi,\upsilon,J_\theta,J_\phi,t)=\frac{1}{2\pi mkT}\bar{P}\times(r,\upsilon,t)\times\exp\left[-\frac{1}{2mr^2kT}\left(J_\theta^2+\frac{J_\phi^2}{\sin^2\theta}\right)\right]. \tag{2.82}$$

Introducing Equation 2.82 into Equation 2.80 and integrating with respect to the angular coordinates and momenta, one obtains the following equation for the radial distribution function $\bar{P}(r,\upsilon,t)$ [10]:

$$\frac{\partial\bar{P}}{\partial t}=-\upsilon\frac{\partial}{\partial r}\bar{P}+\frac{1}{m}\left(\frac{d\Phi}{dr}-\frac{2kT}{r}\right)\frac{\partial\bar{P}}{\partial\upsilon}+\gamma\frac{\partial}{\partial\upsilon}(\upsilon\bar{P})+\frac{\gamma kT}{m}\frac{\partial^2\bar{P}}{\partial\upsilon^2}. \tag{2.83}$$

Equation 2.83 has the form of the Fokker–Planck equation for a unidimensional motion in the effective potential $\Phi_{eff}(r)$ given by

$$\Phi_{eff}(r)=\Phi(r)-2kT\ln r. \tag{2.84}$$

Apart from the ordinary potential $\Phi$, the effective interaction potential (Equation 2.84) contains an additional term that arises as a result of the averaging of Equation 2.83 with respect to the angular variables. It represents a kind of repulsive force accounting for the centrifugal force that acts upon the molecule. This "centrifugal" repulsive potential exhibits a weak, logarithmic singularity at infinity, which does not, however, affect the calculations for reasons noted below.

The attractive potential $\Phi$ decays relatively rapidly with the distance $r$, for example, the most typical nonretarded van der Waals interactions fall off as $r^{-n}$, where $n$ ranges from 3 for the interactions of a molecule with a flat surface to 6 for its interactions with a very small cluster. For small values of $r$, the attractive interaction dominates the logarithmically increasing repulsive term, but for very large values of $r$ the repulsive force dominates. The effective potential is an increasing function for small values of $r$, passes through a maximum at some value $r_b$, and decreases for $r > r_b$. A typical dependence of the effective potential $\Phi_{eff}$ on the distance from the cluster surface is presented in Figure 2.7. A potential barrier is generated at $r_b$, which can be calculated from the condition $d\Phi_{eff}/dr = 0$ of vanishing of the force. The effective potential energy at this point constitutes the upper energetic limit of the well, $E_b = \Phi_{eff}(r_b)$. The barrier bounds the potential well that stretches between the surface of the cluster ($r = R$) and the top of the barrier ($r = r_b$). The molecules are confined within the well if their energy is insufficient to overcome the potential barrier $E_b$ at $r_b$.

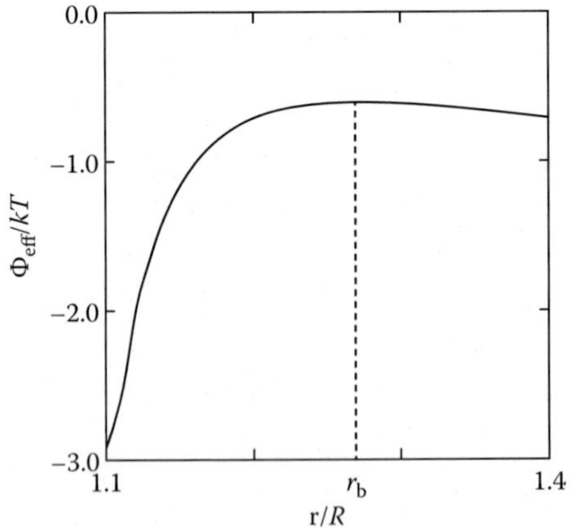


Figure 2.7

Typical effective interaction potential $\Phi_{eff}/kT$ versus the distance $r$ from the center of the spherical cluster of radius $R$. The attractive potential $\Phi$ was calculated using Equation 2.121 for an interaction parameter $\epsilon/kT = 6$, $\eta/R = 0.1$, and a density $\rho$ of the molecules in the cluster given by $\rho\eta^3 = 1.2$.

The motion of the molecule in the well involves two distinct time scales. The molecule performs a kind of anharmonic oscillation in the well, with a characteristic time estimated as $t_w = \lambda_w(m/kT)^{1/2}$, where $\lambda_w$ is the thickness of the well. The energy of these oscillations changes as a result of the interactions of the molecule with the medium. The rate of this change can be characterized by the period $t_c = \gamma^{-1}$. In gases, the interactions are not very strong, and, because the friction coefficient $\gamma$ is small, the time scale $t_c$ for the variation of the energy is much longer than the period of oscillations in the well, $t_w$. As a result, one can discriminate between the fast changing position and momentum and the slowly changing energy. The distinction between fast and slow variables becomes more apparent after the following scaled variables are introduced into Equation 2.83:

$$z = r/\lambda_w, \tag{2.85}$$

$$u = v(m/kT)^{1/2}, \tag{2.86}$$

$$t = t/t_0, \tag{2.87}$$

$$\Psi_{\text{eff}} = \Phi_{\text{eff}}/kT, \tag{2.88}$$

where

$$t_0 = \lambda_w(m/kT)^{1/2}. \tag{2.89}$$

Equation 2.83 can be transformed by using the scaled energy $E$, defined by

$$E = \frac{1}{2}u^2 + \psi_{\text{eff}}, \tag{2.90}$$

instead of the variable $u$ defined by Equation 2.86.

The probability distribution function becomes

$$\bar{P}(z,E,t) = \bar{P}(z,u,t)\frac{\partial u}{\partial E} = \frac{\bar{P}(z,u,t)}{[2(E-\psi_{\text{eff}})]^{1/2}}. \tag{2.91}$$

Using variables (Equations 2.85 through 2.89) and introducing the energy $E$ (Equation 2.90), one obtains from Equation 2.83 the following Fokker–Planck equation for the probability density $\bar{P}(z,E,t)$:

$$\frac{\partial \bar{P}(z,E,t)}{\partial t} = -2\frac{\partial}{\partial z}\left[(2(E-\psi_{\text{eff}}))^{1/2}\bar{P}\right] + 2\Lambda\left[\frac{\partial}{\partial E}\left(E-\psi_{\text{eff}}-\frac{1}{2}\right)\bar{P} + \frac{\partial^2}{\partial E^2}(E-\psi_{\text{eff}})\bar{P}\right], \tag{2.92}$$

where $\Lambda = t_w/t_c$ is the ratio of the time scales for the fast and slow variables.

Because the changes of position and momenta are rapid whereas those of the energy are slow, the probability density function can be decomposed as

$$\bar{P}(z,E,t) = W(z|E)\cdot Q(E,t), \tag{2.93}$$

where $W(z|E)$ is the probability density for the position z at the constant energy $E$ and $Q(E,t)$ is the probability density for the energy $E$ at time $t$. Since the scale of time for positions is short compared to the time scale for energy, the probability density $W(z|E)$ can be approximated by the stationary solution of the equation

$$\frac{\partial}{\partial z}[(2(E-\psi_{\text{eff}}))^{1/2}/W(z|E)] = 0. \tag{2.94}$$

The solution of Equation 2.94 has the form

$$W(z|E)=\frac{C(E)}{(2(E-\psi_{\text{eff}}))^{1/2}}, \tag{2.95}$$

where $C(E)$ is a normalization coefficient, which is a function of the energy $E$,

$$C(E)=\left\{\int_{z_1}^{z_2}\frac{dz}{(2(E-\psi_{\text{eff}}))^{1/2}}\right\}^{-1}, \tag{2.96}$$

and $z_1$, $z_2$ are the boundaries of the well for the energy $E$ given by the condition $E = \psi_{\text{eff}}(z)$.

Using Equation 2.95, the joint probability density (Equation 2.93) becomes

$$\bar{P}(z,E,t)=\frac{Q(E,t)}{G'(E)[2(E-\psi_{\text{eff}})]^{1/2}}, \tag{2.97}$$

where

$$G(E)=\int_{z_1}^{z_2}(2(E-\psi_{\text{eff}}))^{1/2}\,dz. \tag{2.98}$$

Introducing Equation 2.97 into the Fokker–Planck equation (Equation 2.92), one obtains

$$\begin{aligned}\frac{1}{G'(E)(2(E-\psi_{\text{eff}}))^{1/2}}\frac{\partial}{\partial t}Q(E,t)&=\Lambda\frac{\partial}{\partial E}\left(\frac{(2(E-\psi_{\text{eff}})-1)Q(E,t)}{G'(E)(2(E-\psi_{\text{eff}}))^{1/2}}\right)\\&+\Lambda\frac{\partial^2}{\partial E^2}\left(\frac{(2(E-\psi_{\text{eff}}))^{1/2}Q(E,t)}{G'(E)}\right).\end{aligned} \tag{2.99}$$

Since the time scales for the variation of $z$ and $E$ are very different, a closed equation for the probability density with respect to energy $Q(E,t)$ is obtained by integrating Equation 2.99 with respect to the position of the molecule between $z_1$ and $z_2$. The final equation is

$$\begin{aligned}\frac{\partial Q(E,t)}{\partial t}&=\Lambda\frac{\partial}{\partial E}\left(\left(\frac{G(E)}{G'(E)}-1\right)Q(E,t)\right)+\Lambda\frac{\partial^2}{\partial E^2}\left(\frac{G(E)}{G'(E)}Q(E,t)\right),\\&\equiv-\frac{\partial}{\partial E}j(E,t),\end{aligned} \tag{2.100}$$

where $j$ is the flux in the energy space.

### 2.3.3 Kinetics of Growth (Evaporation) of a Single Cluster

The rate of change of the size of the cluster is given by the difference between the rates of evaporation $\alpha$ and condensation $\beta$. The evaporation rate is obtained from the stationary solution of Equation 2.100. Assuming that every molecule reaching the upper energy boundary of the well, $E = E_b$, separates from the cluster, the following absorption boundary condition can be written

$$Q(E)|_{E=E_b}=0. \tag{2.101}$$

The stationary solution of Equation 2.100 that satisfies Equation 2.101 has the form

$$Q(E) = j_E G'(E)e^{-E} \int_{E}^{E_b} \frac{dE'}{G(E')e^{-E'}}, \tag{2.102}$$

where $j_E$ is the stationary flux of molecules along the energy coordinate. The dimensionless rate of evaporation per molecule $\alpha^1$ is equal to $j_E$, which can be determined using the normalization condition $\int Q(E)dE = 1$ for the one-particle probability distribution function. This provides the equation

$$\alpha^1 = \left[ \int_{\psi_0}^{E_b} dEG'(E)e^{-E} \int_{E}^{E_b} \frac{dE'}{G(E')e^{-E'}} \right]^{-1}, \tag{2.103}$$

where $\psi_0$ is the value of $\psi_{eff}$ at the bottom of the energy well. The overall evaporation rate per cluster in the real time scale is given by $\alpha = \gamma N_w \alpha^1$, where $N_w$ is the number of vapor molecules in the well around the cluster. Denoting by $n_w$ the average density of vapor molecules in the well, the total number of molecules in the well is $N_w = 4\pi R^2 \lambda_w n_w$. The overall evaporation rate per cluster (molecules/s) is therefore given by the expression

$$\alpha = \gamma 4\pi R^2 \lambda_w n_w \Bigg/ \int_{\omega_0}^{E_b} dEG'(E)e^{-E} \int_{E}^{E_b} \frac{dE'}{G(E')e^{-E'}}. \tag{2.104}$$

The actual values of $n_w$ and $\lambda_w$ are not known, but as shown later they are not needed to calculate the rate of nucleation.

The rate of condensation of molecules on the cluster depends on the mechanism that controls the transfer process of the vapor molecules through the medium. In gaseous systems, it is given by the formula derived from the kinetic theory of gases for the number of impingements on a spherical cluster [11]:

$$\beta = n_1 \pi R^2 \bar{\upsilon}. \tag{2.105}$$

Here, $\bar{\upsilon}$ is the mean thermal velocity of the vapor molecules, and $n_1$ is the (undisturbed) number density of the vapor at a large distance from the cluster.

In the thermodynamics of clusters, an important role is played by the critical cluster size $R_*$. For clusters with a radius $R > R_*$ condensation prevails over evaporation, because the potential well is deeper for larger clusters, and consequently it is more difficult for the evaporating molecules to overcome the energetic barrier. In contrast, evaporation prevails over condensation for clusters smaller than the critical one. The point of balance between evaporation and condensation provides the critical radius $R_*$. Using Equations 2.104 and 2.105, one obtains

$$n_1 = n_w (4\lambda_w \gamma / \bar{\upsilon}) \alpha^1 (R_*). \tag{2.106}$$

Because $\alpha^1$ increases with decreasing $R_*$, higher concentrations of vapor $n_1$ are needed to ensure the unstable equilibrium for smaller critical clusters. The saturation number density $n_s$ is the vapor concentration that coexists with a cluster of infinite size, i.e., with the bulk condensed phase. The supersaturation $s$ of the system is the ratio $n_1/n_s$,

$$s = \alpha^1(R_*)/\alpha^1(\infty). \tag{2.107}$$

In the classical theory of nucleation, the relation between supersaturation and the critical radius is provided by the Kelvin equation

$$s = \exp(2\sigma/kT\rho R_*), \tag{2.108}$$

where $\rho$ is the density of molecules in the clusters and $\sigma$ is the surface tension, which constitutes the basic thermodynamic parameter of the classical theory. The present theory is expected to recover the classical expression in the limit of large clusters, for which macroscopic thermodynamics is expected to become valid. Consequently, an expression for the surface tension can be obtained by comparing Equation 2.107 of the present theory with the Kelvin equation (Equation 2.108) for large radii.

### 2.3.4 Rate of Nucleation

The rates of evaporation $\alpha$ and condensation $\beta$ govern the overall evolution of a system containing clusters of various sizes. The dynamics of the concentrations $n_i(t)$ of clusters consisting of $i$ molecules is given by the equation [11]

$$\frac{dn_i}{dt} = \beta_{i-1}n_{i-1} - \alpha_i n_i - \beta_i n_i + \alpha_{i+1}n_{i+1} = I_{i-1} - I_i, \tag{2.109}$$

where $I_i$ is the net flux of particles passing from $i$ population to $(i + 1)$ population

$$I_i = \beta_i n_i - \alpha_{i+1}n_{i+1}. \tag{2.110}$$

The system relaxes rapidly to the steady state, and the rate of nucleation can be obtained from the stationary solution of Equation 2.109. For convenience, the discrete set of Equation 2.109 are usually transformed into a partial differential equation for the continuous distribution function $n(g,t)$, with the variable $g$ replacing the discrete variable $i$. Various methods for this transformation have been suggested [12,13]; in this paper, the Zeldovich [14] approach is used because it is compatible with the detailed balance principle for the equilibrium distribution $n_i^{\text{eq}}$,

$$\beta_i n_i^{\text{eq}} = \alpha_{i+1}n_{i+1}^{\text{eq}}, \tag{2.111}$$

and therefore is expected to provide better agreement with the discrete Equation 2.109 [12,13]. The detailed balance principle is a consequence of the vanishing of the flux I at equilibrium. In the classical theory of nucleation, Equation 2.111 serves as a basis to calculate the evaporation rate $\alpha$ in terms of $\beta$ and the equilibrium concentration $n_i^{\text{eq}}$ of clusters. The latter concentrations are calculated in the framework of the thermodynamic fluctuation theory. Since the present theory provides independently the values of $\alpha$ and $\beta$, the equilibrium distribution of clusters can be calculated by recurrence from Equation 2.111. This yields

$$n_i^{\text{eq}} = n_1^{\text{eq}} \prod_{j=2}^{i} \frac{\beta_{j-1}}{\alpha_j} \equiv n_1^{\text{eq}} \exp(h(i)). \tag{2.112}$$

In the Zeldovich expansion, the detailed balance Equation 2.111 is used to rewrite Equation 2.109 in the form

$$\frac{dn_i}{dt} = \beta_i n_i^{\text{eq}} \left( \frac{n_{i+1}}{n_{i+1}^{\text{eq}}} - \frac{n_i}{n_i^{\text{eq}}} \right) - \beta_{i-1} n_{i-1}^{\text{eq}} \left( \frac{n_i}{n_i^{\text{eq}}} - \frac{n_{i-1}}{n_{i-1}^{\text{eq}}} \right). \tag{2.113}$$

Converting Equation 2.113 to the continuous variable $g$, and expanding up to the second-order terms around the midpoint $g = i$, the following equation for the distribution function $n(g,t)$ is obtained:

$$\frac{\partial n(g,t)}{\partial t} = \frac{\partial}{\partial g}\left[ \beta(g) n^{\text{eq}}(g) \frac{\partial}{\partial g}\left( \frac{n(g,t)}{n^{\text{eq}}(g)} \right) \right] = -\frac{\partial}{\partial g} I(g,t). \tag{2.114}$$

The flux of clusters $I(g,t)$, expressed in the continuum representation has the form

$$I(g,t) = -\beta(g) n^{\mathrm{eq}}(g) \frac{\partial}{\partial g}\left(\frac{n(g,t)}{n^{\mathrm{eq}}(g)}\right). \tag{2.115}$$

Assuming that nucleation does not cause a substantial depletion of monomers, one can write

$$\frac{n(1)}{n^{\mathrm{eq}}(1)} = 1. \tag{2.116}$$

The second boundary condition accounts for the removal of very large clusters from the system

$$\frac{n(g)}{n^{\mathrm{eq}}(g)} \to 0, \text{ for } g \to \infty. \tag{2.117}$$

The stationary solution of Equation 2.115, which satisfies the boundary conditions 2.116 and 2.117, has the form

$$I = \left[\int_1^\infty \frac{dg}{\beta(g) n^{\mathrm{eq}}(g)}\right]^{-1} = n_1\beta_1 \Bigg/ \int_1^\infty dg \exp\left[h(g) - \ln\frac{\beta(g)}{\beta(1)}\right]. \tag{2.118}$$

Equation 2.118 provides the steady-state rate of nucleation. In contrast to the classical theory, the evaporation rate is involved directly in expression 2.118 (via the function $h(g)$ defined by Equation 2.112). The classical theory, which will be compared with the present theory, provides for the nucleation rate the expression [11]

$$I_{\mathrm{cl}} = \frac{n_1\beta_1}{4\pi\rho\eta^3}\left(\frac{\sigma\eta^2}{kT}\right)^{1/2} \exp\left(-\frac{16\pi\sigma^3}{3(kT)^3\rho^2 \ln^2 s}\right), \tag{2.119}$$

where $\eta$ is the diameter of the molecules of the condensing species.

### 2.3.5 Numerical Results and Discussion

The numerical calculations were performed using a binary interaction potential $\phi_{12}$ that combines the rigid core repulsion with the nonretarded dispersive attraction

$$\phi_{12}(r_{12}) = \begin{cases} \infty & \text{for } r_{12} < \eta \\ -\epsilon(\eta/r_{12})^6 & \text{for } r_{12} \geq \eta, \end{cases} \tag{2.120}$$

where $r_{12}$ is the distance between the centers of two interacting molecules, and $\epsilon$ measures the strength of these interactions. The interaction potential between a molecule and the cluster is obtained by summing the interactions (Equation 2.120) over all the molecules of the cluster. For a spherical cluster with a uniform distribution of molecules within it, the integration over the volume of the cluster yields

$$\Phi(r) = -\frac{4\pi\epsilon\rho\eta^3}{3}\left(\frac{\eta}{r-R}\right)^3\left(\frac{R}{r+R}\right)^3, r > R + \eta. \tag{2.121}$$

The density of randomly packed spherical molecules in a cluster is given by $\rho\eta^3 = 1.2$ [15].

The rate ratios $\beta_{i-1}/\alpha_i$ were calculated for given values of supersaturation, and the function $h(g)$ of Equation 2.112 thus obtained was used to calculate the rate of nucleation. Figure 2.8 presents the nucleation rate as a function of the supersaturation $s$ for several values of the interaction parameter $\varepsilon = \epsilon/kT$.

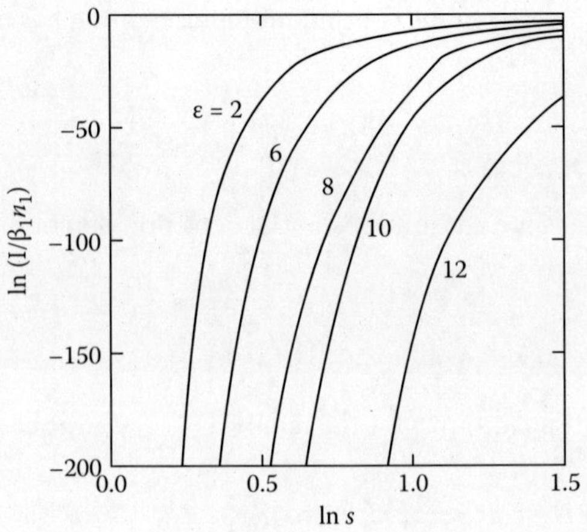


Figure 2.8

Nucleation rate, $\log(I/\beta_1 n_1)$, calculated using Equation 2.118, against the supersaturation $s$ for several values of the interaction parameter $\varepsilon = \epsilon/kT$.

Previously, the authors presented an approach to the theory of nucleation, in which the rate of evaporation was obtained from a kinetic equation in the energy space, which was based on different equations than those employed here. For the three-dimensional motion of a molecule, a diffusion-like equation in the energy space was obtained starting from a master equation involving transition probabilities calculated on the basis of the multidimensional Fokker–Planck equation [5,16,17]. If the thickness of the potential well is small compared to the radius of the cluster, the curvature effects in the three-dimensional space are expected to be negligible. For such systems a unidimensional approximation was developed, based on the one-dimensional Fokker–Planck equation [6].

Figure 2.9 presents nucleation rates calculated on the basis of the three-dimensional diffusion equation in the energy space [5], of the unidimensional Fokker–Planck model [6], and on the predictions of the present approach. For comparison, the results provided by the classical nucleation theory, Equation 2.119, are also included. The surface tension, which is the basic parameter of the classical theory, was calculated from the best fit of the Kelvin equation (Equation 2.108) to the results of the present theory in the range of large clusters, for which $R/\eta > 10$. In the calculation of the surface tension, the "centrifugal" force, accounting for the angular degrees of freedom, has not been taken into account. One can observe that the rates provided by the present approach are somewhat closer to the classical theory than those obtained previously by the authors [5,6]. All the curves converge for small supersaturation to a common limit, since for large critical clusters

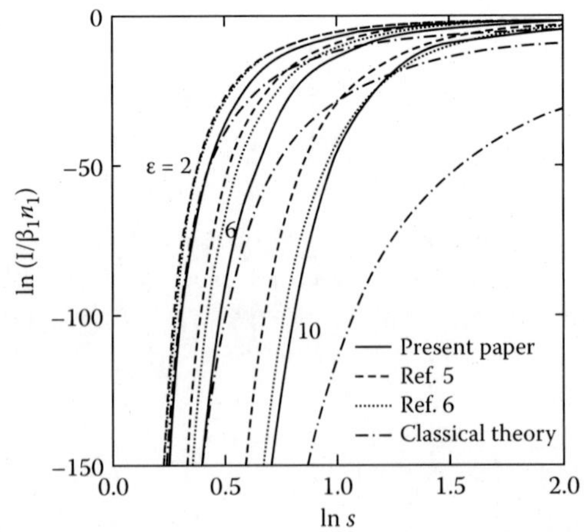


Figure 2.9

Comparison among the rates of nucleation against supersaturation $s$ calculated using Equation 2.118 of the present paper, and equations established previously on the basis of a diffusion equation in the energy space [5] and of a unidimensional Fokker–Planck model [6]. The predictions of the classical theory are also included.

macroscopic thermodynamics is expected to be valid, and hence the results of the classical nucleation theory should be recovered.

In the present approach, the averaging over the angular momentum introduces an effective potential $\Phi_{eff}$, Equation 2.84, containing an additional term, which corresponds to a kind of repulsive, "centrifugal" force (the second component in Equation 2.84). As a result, the rate of evaporation from a cluster of a given radius $R$ is enhanced by the centrifugal force, and hence the rate of nucleation is decreased by this effect when compared to that predicted by our previous tridimensional kinetic equation in the energy space.

## References


1. Volmer, M., Weber, A., *Z. Phys. Chem. (Leipzig)* 119, 227 (1926).
2. Farkas, L., *Z. Phys. Chem. (Leipzig)* 125, 236 (1927).
3. Becker, R., Döring, W., *Ann. Physik (Leipzig)* 24, 719 (1935).
4. Ruckenstein, E., Nowakowski, B., *J. Colloid Interface Sci.* 137, 583 (1990). (Section 1.3 of this volume.)
5. Nowakowski, B., Ruckenstein, E., *J. Chem. Phys.* 94, 1397 (1991). (Section 2.1 of this volume.)
6. Ruckenstein, E., Nowakowski, B., *Langmuir* 7, 1537 (1991). (Section 2.2 of this volume.)
7. Risken, H., *The Fokker–Planck Equation*, Springer, New York, 1989.
8. Stratonovich, R.L., *Topics in the Theory of Random Noise*, Gordon and Breach, New York, 1963.
9. Wang, M.C., Uhlenbeck, G.E., *Rev. Mod. Phys.* 17, 323 (1945).
10. Rodger, P.M., Sceats, M.G., *J. Chem. Phys.* 83, 3358 (1985).
11. Abraham, F.F., *Homogeneous Nucleation Theory*, Academic, New York, 1974.
12. Shizgal, B., Barrett, J.C., *J. Chem. Phys.* 91, 6505 (1989).
13. Nowakowski, B., Ruckenstein, E., *J. Colloid Interface Sci.* 142, 599 (1991). (Section 1.5 of this volume.)
14. Zeldovich, J.B., *J. Exp. Theor. Phys. (Russian)* 12, 525 (1942).
15. Kittel, C., *Introduction to Solid State Physics*, Wiley, New York, 1976.
16. Borkovec, M., Berne, B.J., *J. Chem. Phys.* 82, 797 (1985).
17. Borkovec, M., Berne, B.J., *J. Chem. Phys.* 84, 4327 (1986).

# 2.4 Time-Dependent Cluster Distribution in Nucleation

*Bogdan Nowakowski and Eli Ruckenstein**

The time-dependent size distribution function of clusters, as well as the relaxation and lag times, are calculated numerically in the framework of the recently developed kinetic theories of nucleation in liquids (E. Ruckenstein and B. Nowakowski, *J. Colloid Interface Sci.* 137, 583 [1990] [Section 1.3 of this volume]) and gases (B. Nowakowski and E. Ruckenstein, *J. Chem. Phys.* 94, 1397 [1991] [Section 2.1 of this volume]). The basic approach used in the calculation is the same as that used for the classical theory. Differences in the treatment occur as a result of the fact that, in contrast to the classical theory, the rates of condensation and evaporation are calculated independently. The size distribution function of clusters approaches monotonically the steady-state distribution. The nucleation rate relaxes monotonically to the steady-state value for clusters greater than the critical cluster $x^*$; for $x < x^*$, the nucleation rate exhibits an initial overshoot that exceeds the stationary value and then tapers to the steady-state rate from above. The relaxation times calculated for typical systems are of the order of 1 μs both in liquids and gases. The time lags were found to be of the order of, but larger than, the relaxation time and were found to increase with the size of the cluster. In contrast to the steady-state nucleation rates, which differ by orders of magnitude, the relaxation and lag times calculated in the framework of the kinetic theories are comparable to those obtained on the basis of the classical theory. (© *1991 Academic Press, Inc.*)

## 2.4.1 Introduction

In unstable, supersaturated systems, the emergence of a new phase passes through the stage of formation of small nuclei consisting of agglomerated molecules. In the classical theory of nucleation, due to Farkas [1], Becker and Döring [2], and developed further by Zeldovich [3] and Frenkel [4], the appearance of molecular clusters is explained in terms of fluctuations, whose probabilities are determined by the excess free energies of the created nuclei. The surface free energy of these small clusters is assumed to be equal to the surface tension of a bulk phase. In addition, the rate of evaporation of molecules from clusters is obtained on the basis of the principle of detailed balance using a hypothetical equilibrium distribution of clusters. The main objective of the nucleation theory is to predict the rate of formation of the new phase. Usually, this is calculated in the steady-state approximation, which is assumed to be attained within a very short lapse of time. The stationary approximation has been justified in the classical theory by solving numerically the equations that describe the time-dependent distribution of clusters [5–8].

Recently, a kinetic approach to nucleation theory has been developed [9,10] that avoids the use of the macroscopic surface tension, and in which the evaporation rate is calculated from the mean passage time of molecules evaporating from clusters. The steady-state rates of nucleation predicted by the kinetic theories differ by orders of magnitude from those obtained in the classical theory. Does this imply that the relaxation and lag times are also very different? In this paper we present numerical calculations for the time-dependent cluster size distribution and for the relaxation and lag times in the framework of the recent kinetic theories of nucleation in liquids [9] and gases [10].

The next section presents the basic equations governing the population of clusters. They have the form of a system of finite difference equations, which can be approximated by a single partial differential equation of the Fokker–Planck form for the distribution function of cluster sizes. The subsequent section presents the solution of the resulting equation obtained by the method of discrete ordinates [11,12]. Finally, the numerical results obtained for the time-dependent size distribution function are presented, and the relaxation time as well as the time lag to the steady nucleation rate are calculated. The basic approach used in the calculations is very similar to that already employed in the classical theory [7,8]. In contrast to the classical theory, the rates of condensation and evaporation on and from the cluster are calculated independently. This generates some differences in the treatment.

## 2.4.2 Equations for the Population of Clusters

It is usually assumed that the clusters change their size as a result of addition or loss of single molecules. The condensation of single molecules on a cluster depends on the mechanism of transport of molecules in the medium.

* *J. Colloid Interface Sci.*, 145, 1826 (1991). Republished with permission.

In liquids, the transport is diffusion-controlled, and the expression for the condensation rate $\beta_i$ is obtained from the solution of the diffusion equation. The result can be written in the form [9]

$$\beta_i = \gamma 4\pi R_i D n_1, \tag{2.122}$$

where $D$ is the diffusion coefficient, $n_1$ is the concentration of condensing molecules, and $R_i$ is the radius of a cluster that consists of $i$ molecules. Equation 2.122 also contains the correction factor $\gamma$, usually of the order of unity [9], which accounts for the interaction potential $\phi(r)$ between the molecule and the cluster:

$$\gamma = \left[ R_i \int_{R_i}^{\infty} r^{-2} \exp(-\phi(r)/kT)\,\mathrm{d}r \right]^{-1}. \tag{2.123}$$

Here, $k$ is the Boltzmann constant and $T$ denotes the temperature in degrees Kelvin.

In gases, the transport is under kinetic rather than diffusion control, and the rate of condensation is given by the free molecular formula of the kinetic theory of gases [13],

$$\beta_i = \pi R_i^2 \bar{\upsilon} n_1, \tag{2.124}$$

where $\bar{\upsilon}$ denotes the mean thermal velocity of molecules in the gas.

Another rate constant, which enters into the equations governing the evolution of cluster size distribution, is the rate of evaporation $\alpha_i$. In the classical theory of nucleation, the rate of evaporation $\alpha_i$ is expressed in terms of the rate of condensation $\beta_{i-1}$ using the detailed balance principle and a hypothetical equilibrium distribution of clusters. The hypothetical equilibrium distribution, in turn, is calculated using macroscopic thermodynamics for the excess free energy of the clusters.

In the kinetic theories of nucleation developed by the authors [9,10], the rate of evaporation is calculated from the mean passage time of a molecule crossing the outer boundary of a potential well surrounding the cluster. In contrast to the classical theory, in this theory the rate of evaporation is obtained independently and not via the rate of condensation. The binary interaction potential, which was used in the calculations, was assumed to combine the London dispersive attraction with the rigid core repulsion:

$$V(r) = \begin{cases} \infty & r < d \\ -\epsilon \left( \dfrac{d}{r} \right)^6 & r \geq d, \end{cases} \tag{2.125}$$

where $\epsilon$ is an interaction constant, and $d$ denotes the diameter of a molecule. Explicit expressions for the rate of evaporation in liquids are given by Equation 1.123 and by Equation 2.12 in gases, respectively. The calculations in this paper are performed with these expressions for an interaction constant $\epsilon/kT = 4$, which provides typical values for the surface tension [9,10].

The rate of change of the population of clusters consisting of $i$ molecules, $n_i$, is governed by the equation

$$\frac{\mathrm{d}n_i}{\mathrm{d}t} = \beta_{i-1} n_{i-1} - \alpha_i n_i - \beta_i n_i + \alpha_{i+1} n_{i+1} = I_{i-1} - I_i, \tag{2.126}$$

where $I_i$ denotes the net flux of clusters passing from $i$-population to $(i + 1)$-population

$$I_i = \beta_i n_i - \alpha_{i+1} n_{i+1}. \tag{2.127}$$

The flux (Equation 2.127) vanishes at equilibrium; this results in the detailed balance equation for the equilibrium distribution $n_i^{\mathrm{eq}}$

$$\beta_{i-1} n_{i-1}^{\mathrm{eq}} = \alpha_i n_i^{\mathrm{eq}}. \tag{2.128}$$

The hypothetical equilibrium distribution is obtained from Equation 2.128 by recurrence

$$n_i^{eq} = n_1^{eq} \prod_{j=2}^{i} \frac{\beta_{j-1}}{\alpha_j} \equiv n_1^{eq} \exp(h(i)). \tag{2.129}$$

In the classical approach, Equation 2.128 is used to express $\alpha_i$ in terms of $\beta_{i-1}$ and the equilibrium concentrations $n_{i-1}^{eq}$ and $n_i^{eq}$. The latter concentrations are calculated on the basis of the free energy of formation of the clusters. In contrast, in the present approach the quantities $\alpha$ and $\beta$ are calculated independently and, as a result, $n_i^{eq}$ is calculated directly using Equation 2.129.

For small clusters, the evaporation exceeds condensation, but the ratio $\beta_{j-1}/\alpha_j$ increases with the cluster size. The cluster of size $i_*$ for which $\beta_{i*-1} = \alpha_{i*}$ is the critical cluster size for which the equilibrium concentration reaches its minimum value. Above the critical cluster condensation prevails.

It is convenient to transform the system of finite difference equations (Equation 2.126) into the continuous differential equation

$$\frac{\partial n(x,t)}{\partial t} = \frac{\partial}{\partial x}(A(x)n(x,t)) + \frac{\partial^2}{\partial x^2}(B(x)n(x,t)) = -\frac{\partial}{\partial x} I(x,t), \tag{2.130}$$

where the continuous variable $x$ replaces the discrete variable $i$. In this formulation, the net flux of clusters is given by

$$I(x,t) = -A(x)n(x,t) - \frac{\partial}{\partial x}(B(x)n(x,t)). \tag{2.131}$$

Expanding each of the terms in Equation 2.126 around the midpoint $i$ and retaining only the first two terms, one obtains Equation 2.130 with the following coefficients:

$$A(x) = \alpha(x) - \beta(x) \tag{2.132}$$

and

$$B(x) = (\alpha(x) + \beta(x))/2. \tag{2.133}$$

However, the above expansion does not recover [8] the equilibrium distribution (Equation 2.129) of the original discrete equations (Equation 2.126). The equilibrium solution of Equation 2.130 can be found from Equation 2.131 using the condition $I^{eq} = 0$:

$$n^{eq}(x) = n^{eq}(1) \exp\left( -\int_1^x dx' \frac{A(x')}{B(x')} + \ln \frac{B(1)}{B(x)} \right). \tag{2.134}$$

Comparing Equations 2.129 and 2.134, one finds that in order to recover the proper equilibrium solution, the following relation must be satisfied:

$$\frac{A(x)}{B(x)} + \frac{B'(x)}{B(x)} = h'(x). \tag{2.135}$$

The coefficients $A(x)$ and $B(x)$ given by Equations 2.132 and 2.133, respectively, do not satisfy this condition. One possible modification of the above coefficients that satisfies Equation 2.135 consists in retaining Equation 2.133 for $B(x)$ and calculating $A(x)$ using Equation 2.135 [8]. The differential equation (Equation 2.130) can then be recast in the form

$$\frac{\partial P(x,t)}{\partial t} = \frac{1}{n^{eq}(x)} \frac{\partial}{\partial x}\left( B(x) n^{eq}(x) \frac{\partial}{\partial x} P(x,t) \right) \\ = LP(x,t), \tag{2.136}$$

where $P(x, t) = n(x, t)/n^{eq}(x)$ and $L$ is a Fokker–Planck operator. Although Equation 2.136 is the same as that employed in the framework of the classical theory, $n^{eq}(x)$ is calculated here via Equation 2.129. Equation 2.136 must be complemented with appropriate boundary conditions. First, it is assumed that the process of nucleation does not cause noticeable depletion of monomers, i.e., $n(1, t) = n^{eq}(1)$, or

$$P(1, t) = 1. \tag{2.137}$$

Second, all clusters whose size is greater than $x_f$ are assumed to be removed from the system:

$$P(x_f + 1, t) = 0. \tag{2.138}$$

A typical value assumed for $x_f$ is $2x_*$, where $x_*$ is the critical size in the continuous representation. The stationary solution of Equation 2.136 that satisfies the boundary conditions 2.137 and 2.138 is

$$P^{st}(x) = I^{st} \int_x^{x_f+1} \frac{dx'}{B(x') n^{eq}(x')}. \tag{2.139}$$

The stationary flux $I^{st}$, which provides the stationary nucleation rate, is obtained from Equation 2.139 by taking $x = 1$:

$$(I^{st})^{-1} = \int_1^{x_f+1} \frac{dx'}{B(x') n^{eq}(x')}. \tag{2.140}$$

### 2.4.3 Solution of Time-Dependent Equations

The solution of the time-dependent equations follows the approach used by Shugard and Reiss [7] and Shizgal and Barrett [8] in the framework of the classical theory. The difference consists in the manner in which $n^{eq}(x)$ is calculated. It is convenient to introduce the following time-dependent excess distribution function $P_e(x, t)$ with respect to the stationary solution:

$$P_e(x,t) = P(x,t) - P^{st}(x). \tag{2.141}$$

The function $P_e(x,t)$ satisfies Equation 2.136, but the boundary conditions 2.136 and 2.137 are replaced by the symmetrical boundary conditions

$$P_e(1,t) = P_e(x_f + 1, t) = 0. \tag{2.142}$$

The Fokker–Planck operator $L$ on the right hand side of Equation 2.136 is not self-adjoint, but can be transformed into a self-adjoint operator by introducing the function [7,14]

$$\hat{P}(x,t) = (n^{eq}(x))^{1/2} P_e(x,t). \tag{2.143}$$

With the new function $\hat{P}(x,t)$, Equation 2.136 acquires the form

$$\frac{\partial \hat{P}(x,t)}{\partial t} = (n^{eq}(x))^{1/2} L (n^{eq}(x))^{-1/2} \hat{P}(x,t), \tag{2.144}$$

where, because $n^{eq}(x)$ is the equilibrium solution of the Fokker–Planck operator $L$, the differential operator

$$\hat{L} = (n^{\text{eq}}(x))^{1/2} L (n^{\text{eq}}(x))^{-1/2} \tag{2.145}$$

is self-adjoint [7,14]. The solution of Equation 2.144 can be obtained in terms of the solutions of the equation

$$\hat{L}\phi_k(x) = \lambda_k \phi_k(x), \tag{2.146}$$

where $\lambda_k$ are the eigenvalues.

The general solution of Equation 2.144 can be written in the form

$$\hat{P}(x,t) = \sum_k c_k e^{\lambda_k t} \phi_k(x), \tag{2.147}$$

where the coefficients $c_k$ can be expressed in terms of the initial function $\hat{P}(x,0)$ as

$$c_k = \int_1^{x_f+1} \mathrm{d}x \hat{P}(x,0)\phi_k(x). \tag{2.148}$$

Using Equations 2.141, 2.142, and 2.147, the corresponding general solution $P(x, t)$ is obtained in the form

$$\frac{n(x,t)}{n^{\text{eq}}(x)} \equiv P(x,t) = P^{\text{st}}(x) + (n^{\text{eq}}(x))^{-1/2} \sum_k c_k e^{\lambda_k t} \phi_k(x), \tag{2.149}$$

where the coefficients $c_k$ are given by

$$c_k = \int_1^{x_f+1} \mathrm{d}x (P(x,0) - P^{\text{st}}(x))(n^{\text{eq}}(x))^{1/2} \phi_k(x). \tag{2.150}$$

In the calculation, it will be assumed that the initial distribution has the form

$$P(x,0) = \begin{cases} 1 & \text{for } x \leq x_0 \\ 0 & \text{for } x > x_0, \end{cases} \tag{2.151}$$

where $x_0$ is some arbitrary size, smaller than the critical size $x_*$. The eigenvalue equation (Equation 2.146) has been solved numerically by the discrete ordinate method [11,12], which is essentially equivalent to the expansion of the eigenvectors $\phi_k$ in the finite basis of some appropriate orthogonal functions. For a finite interval, such a basis is provided in a natural way by the Legendre polynomials. The self-adjoint operator $\hat{L}$ of Equation 2.146 can be approximated by a real symmetric matrix, whose order $N$ is given by the number of grid points employed in the integration of Equation 2.146 by the discrete ordinate method. In this paper, the solution was obtained for $N = 60$. In the context of the classical nucleation theory, it has been shown that the discrete ordinate method provides the solution of the eigenvalue problem (Equation 2.146), which is rapidly convergent with respect to the number $N$ of grid points [8]. Standard numerical routines can be used to find the eigenvalues $\lambda_k$ of the matrix $\hat{L}$ and the corresponding $N$-dimensional eigenvectors $\phi_k$.

### 2.4.4 Results of Numerical Calculations

The interaction parameter was chosen to be $\epsilon/kT = 4$; the number density of molecules in the clusters $\rho$ was taken as given by the equation $\rho d^3 = 1.2$, which corresponds to a random packing of spheres of diameter $d$ in the condensed phase [9]. These values of the parameters are typical for liquids and gases, since they provide reasonable values for the macroscopic surface tension calculated as shown previously [9,10].

It was assumed that the greatest clusters in the system consist of 100 molecules, $x_f = 100$, and the supersaturation $s$ was adjusted to provide a critical cluster size of $x_* = 50$. This condition results in a supersaturation $s = 11.7$ in liquids, and a supersaturation $s = 2.07$ in gases, respectively. The largest clusters present in the initial distribution were assumed to consist of $x_0 = 10$ atoms.

Figure 2.10a and b presents the relative populations of clusters, $n(x)/n(1)$, for a hypothetical equilibrium distribution and for the steady state in liquids and gases, respectively. For the equilibrium distribution $n^{eq}(x)$, a sharp minimum can be observed for the critical cluster size $x_* = 50$. In contrast, the stationary distribution $n^{st}(x)$ diminishes monotonically with increasing cluster size. A dramatic change in the ratio $n^{st}(x)/n^{eq}(x) = P^{st}(x)$ occurs in a narrow interval $\Delta x$ around the critical cluster size. With the estimate $\Delta x \cong 10$, one can observe (Figure 2.11), the steady-state distribution) that $n^{st}(x)/n^{eq}(x) \cong 1$ for $x_- = x_* - \Delta x$, and $n^{st}(x_+)/n^{eq}(x_+) \cong 0$ for $x_+ = x_* + \Delta x$; at the critical size $n^{st}(x_*)/n^{eq}(x_*) \cong \frac{1}{2}$.

To obtain the time-dependent solution, the eigenvalue problem (Equation 2.146) has been solved on the interval (1, 101) for the boundary conditions $\phi_k(1) = \phi_k(101) = 0$ for all $k$. The time $t$ was converted to the dimensionless scale $t_d$ by the substitution $t_d = tB(x_*)$, where $B(x_*)$ is the value of $B$ defined by Equation 2.133 for the critical size $x_* = 50$. The solution of Equation 2.146 by the discrete ordinates method provides eigenvalues with the corresponding eigenvectors satisfying the above-mentioned boundary conditions. All eigenvalues of the Fokker–Planck operator (Equation 2.145) are negative [10]. This feature ensures the decay of the transient component $P_e(x, t)$ of the time-dependent solution and the convergence to the steady state. A number of eigenvalues (for the time scale $t_d$) are listed in Table 2.4. The absolute magnitude of the lowest eigenvalues increases approximately proportional to $k$. This feature can be easily understood in terms of the

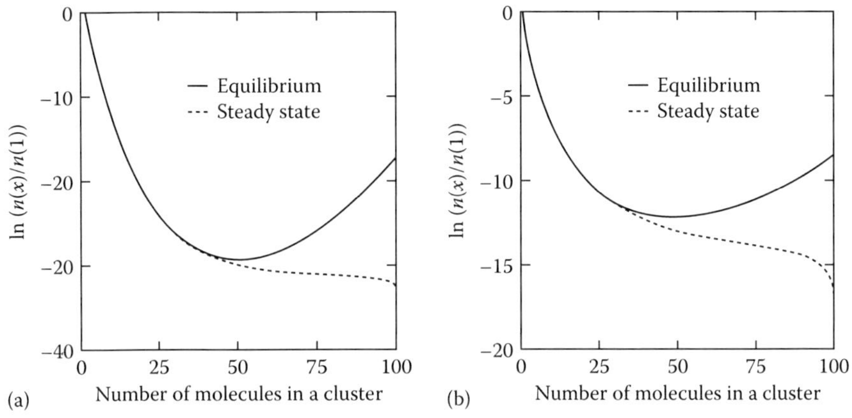


Figure 2.10

Relative equilibrium and steady-state distributions $n(x)/n(1)$ of clusters: (a) in liquid, (b) in gas.

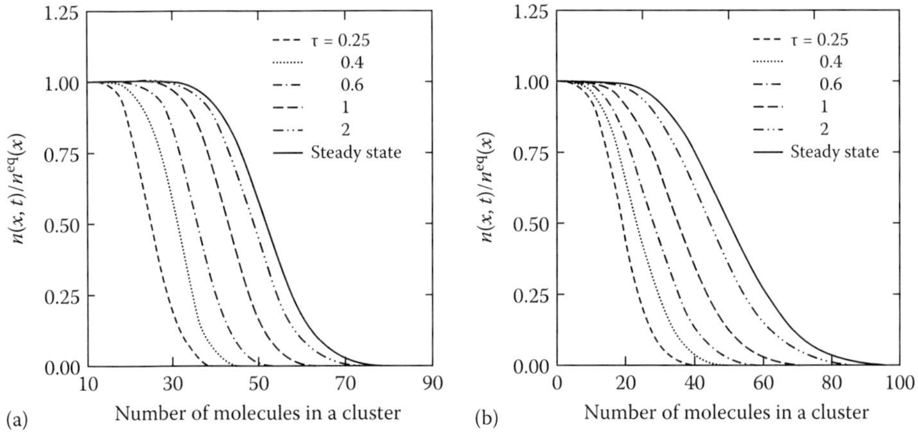


Figure 2.11

Distribution function $n(x, t)/n^{eq}(x)$ from the solution of Equation 2.136 for several values of the reduced time $\tau = t/\tau_1$: (a) for a liquid, (b) for a gas.

Table 2.4 Eigenvalues $\lambda_k$ of the Solutions of Equation 2.146

| | Eigenvalues $\lambda_k$ | |
|---|---|---|
| $k$ | Gas | Liquid |
| 1 | -0.042160 | -0.130014 |
| 2 | -0.090049 | -0.255128 |
| 3 | -0.147019 | -0.368998 |
| 4 | -0.217286 | -0.504490 |
| 5 | -0.303768 | -0.667368 |
| 7 | -0.514568 | -1.04943 |
| 10 | -0.935088 | -1.70085 |
| 15 | -1.94585 | -3.28639 |
| 20 | -3.31875 | -5.28994 |
| 25 | -5.17812 | -7.92770 |
| 30 | -7.41584 | -11.0123 |

corresponding Schrödinger-type equation, into which the Fokker–Planck equation (Equation 2.146) can be transformed [14]. The interaction potential term in this Schrodinger-type equation can be approximated by the parabolic shape of the potential energy of a harmonic oscillator; it is well known that the eigenvalues of a quantum harmonic oscillator are a linear function of their quantum number.

Bounds for the first eigenvalue $\lambda_1$ can be evaluated without using the numerical solution of Equation 2.136 by employing a variational formulation of the Fokker–Planck equation. The result is [15]

$$\frac{3.0}{\pi} < \frac{|\lambda_1|}{h''(x_*)} < \frac{3.4}{\pi}. \tag{2.152}$$

With the derivative $h''$ calculated from Equation 2.129, one obtains $h''(x_*) = 0.013199$ for a liquid and $h''(x_*) = 0.0041759$ for a gas. One can verify that the calculated eigenvalues $\lambda_1$ satisfy inequalities (Equation 2.152).

The main relaxation time $\tau_1$ of the nucleation process can be calculated from the first eigenvalue of the solution (Equation 2.149)

$$\tau_1 = (|\lambda_1|B(x_*))^{-1}.$$

In order to have an idea about the orders of magnitude of the relaxation times, the calculations have been carried out using the approximation $B(x_*) \cong \beta(x_*)$ for systems with the characteristics listed in Table 2.5. The results are

$\tau_1 = 1.2$ μs for a liquid
$\tau_1 = 1.7$ μs for a gas

The effectiveness of the transport mechanism and the concentration of the condensing species are the two main factors that affect the value of the relaxation time. The transport is more rapid in a gas than in a liquid; however, the concentration of the condensing molecules is higher in the liquid (additionally amplified by the higher supersaturation in the liquid in the cases considered here). As a result, in the cases considered here the relaxation times $\tau_1$ have comparable values in both media.

Table 2.5 Characteristics of the Sample Systems Used for Estimation of the Relaxation Time $\tau_1$ in Liquids and Gases

| Liquid | Gas |
|---|---|
| $D = 10^{-5}$ cm²/s | $d = 3.10^{-8}$ cm |
| $d = 10^{-7}$ cm | $n_s = 4 \times 10^{16}$ mol/cm³ |
| $n_1/\rho = 3.9 \times 10^{-4}$ (Equation 2.148; Ref. [8]) | $\bar{\upsilon} = 5 \times 10^4$ cm/s |
| $s = 11.7$ | $s = 2.06$ |

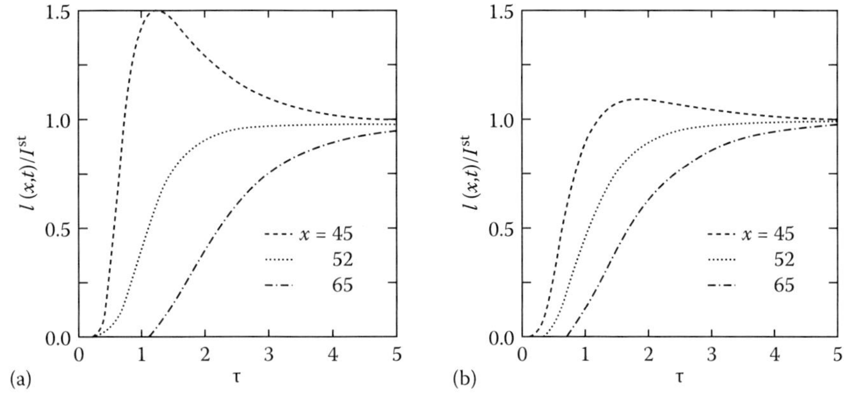


Figure 2.12

Relative flux $I(x,t)/I^{st}$ as a function of the reduced time variable $\tau = t/\tau_1$ for several cluster sizes $x$: (a) in a liquid, (b) in a gas.

Figure 2.11a and b presents the distribution function $n(x, t)/n^{eq}(x)$ for a liquid and a gas, respectively, for several values of the dimensionless time $\tau = t/\tau_1$. One can see that the distribution function approaches monotonically the steady state, and becomes close to it for $t > \tau_1$.

Figure 2.12a and b presents the time dependence (in terms of the dimensionless time $\tau = t/\tau_1$) of the flux $I(x, t)/I^{st}$ in a liquid and a gas, respectively, for three values of $x$. For clusters greater than the critical size $x \geqq x_*$, the flux is a steadily increasing function of time, approaching the stationary value for $t > \tau_1$. The relaxation is most rapid for the critical cluster. For clusters smaller than the critical, the flux exhibits an initial overshot that exceeds the steady-state value, and then tapers and approaches the stationary flux from above. This feature was also noted [6] for the time-dependent solution of the discrete equation (Equation 2.126). Let us demonstrate that the flux $I(x, t)$ can exhibit a maximum as a function of time. From Equations 2.130 and 2.136, one obtains

$$\begin{aligned}\frac{\partial I(x,t)}{\partial t} &= -\frac{\partial}{\partial t}\left( B(x)n^{eq}(x)\frac{\partial}{\partial x}P(x,t)\right) \\ &= B(x)n^{eq}(x)\frac{\partial}{\partial x}\left(\frac{1}{n^{eq}(x)}\frac{\partial}{\partial x}I(x,t)\right).\end{aligned} \tag{2.153}$$

$(-\partial/\partial x)I(x, t)$ is a monotonically decreasing function of $x$ (as follows from the diffusion-type Equation 2.136 and the initial condition [2.151]). The inverse of the equilibrium solution, $[n^{eq}(x)]^{-1}$, increases with $x$ for $x < x_*$, and decreases for $x > x_*$. Consequently, it follows from Equation 2.153 that the time derivative of $I(x, t)$ cannot vanish for $x > x_*$; it can, however, become zero at some moment for $x < x_*$. Thus, $I(x, t)$ approaches monotonically $I^{st}$ for $x > x_*$, but can exhibit a maximum for $x < x_*$.

Another time characteristic of the process of nucleation is provided by the time lag $T(x)$, which is defined by [13]

$$\lim_{t\to\infty}\int_0^t I(x,t')\mathrm{d}t' = I^{st}(t - T(x)). \tag{2.154}$$

Using the explicit form of the solution (Equation 2.149), one obtains the following expression for the time lag

$$T(x) = \sum_{k=1}^{N}\frac{c_k I_k(x)}{\lambda_k I^{st}}, \tag{2.155}$$

where

$$I_k(x) = -B(x)n^{eq}(x)\frac{\mathrm{d}}{\mathrm{d}x}\frac{\phi_k(x)}{(n^{eq}(x))^{1/2}}. \tag{2.156}$$

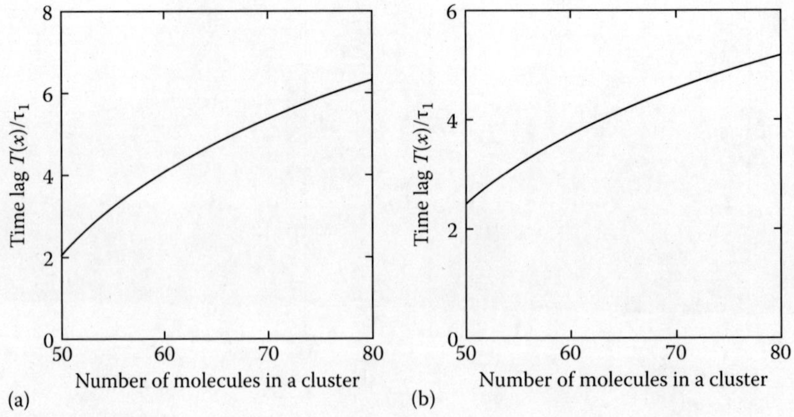


Figure 2.13
Reduced time lag $T(x)/\tau_1$ versus the size of the dusters (a) in a liquid and (b) in a gas.

The time lags were calculated using the numerical solutions (Equation 2.149), and the results are presented in Figure 2.13a and b for cluster sizes in the range $50 \leqq x \leqq 80$, for a liquid and a gas, respectively. The time lags are of the same order of magnitude as the principal relaxation times $\tau_1$, but are greater and increase with the size $x$ of the cluster. Because of the dependence of $T(x)$ on the cluster size, it can be regarded as the relaxation time for a cluster of a given size. However, for the overall, global relaxation process, the size-independent principal relaxation time $\tau_1$ appears to be a more appropriate characteristic than the time lag $T(x)$.

The relaxation times $\tau_1$, provided for the same systems by the classical theory of nucleation, can be estimated using inequalities (Equation 2.152). In the classical approach, the equilibrium distribution is given by

$$n_i^{\text{eq}} = n_i^{\text{eq}} \exp\left(-\left(\frac{36\pi}{\rho^2}\right)^{1/3} \frac{\sigma(i^{2/3}-1)}{kT} + (i-1)\ln s\right) \equiv n_1^{\text{eq}} \exp(h_{\text{cl}}(i)), \tag{2.157}$$

where $\sigma$ is the surface tension of the clusters considered equal to the macroscopic surface tension. This surface tension $\sigma$ is calculated in the kinetic theories of nucleation [9,10] in terms of $\epsilon/kT$ on the basis of the dependence of the critical radius $R_*$ on the supersaturation $s$ for large clusters, for which the classical Kelvin equation is expected to be valid. For the considered systems, one obtains

$$\frac{\sigma d^2}{kT} = 3.687$$

for the liquid (Equation 1.131) and

$$\frac{\sigma d^2}{kT} = 1.044$$

for the gas (Table 2.1).

Employing the classical theory, one obtains $h''_{\text{cl}}(x_*) = 0.0035231$ for the gas and 0.0104747 for the liquid, values that are very close to those obtained in the framework of the kinetic theory. As a result, the relaxation and lag times are comparable to the values obtained on the basis of the kinetic theories.

### 2.4.5 Conclusion

The time-dependent size distribution of clusters is calculated numerically by solving the partial differential equation describing the population of clusters. The kinetic theory of nucleation developed by the authors [9,10] is used in the calculations. The distribution of cluster size approaches monotonically the stationary state. The relaxation time calculated for typical physicochemical conditions are of the order of 1 μs. The nucleation rate approaches the steady state within the same time scale. It is a monotonically increasing function of time for clusters greater than the critical cluster; for clusters smaller than the critical, the nucleation rate exhibits an overshoot that exceeds the stationary value, to which it approaches from above after a few microseconds. These features are similar to those obtained in the framework of the classical theory. It is also of interest to note that although the steady-state rates of nucleation predicted by the kinetic theory are very different from those provided by the classical theory, the relaxation and lag times are comparable.

## References


1. Farkas, L., Z. *Phys. Chem. Abt. A (Leipzig)* 125, 236 J (1927).
2. Becker, R., Döring, W., *Ann. Phys. (Leipzig)* 24, 719 (1935).
3. Zeldovich, J.B., *Acta Physicochim. URSS* 18, 1 (1943).
4. Frenkel, J., *Kinetic Theory of Liquids*, Dover, New York, 1946.
5. Andres, R.P., Boudart, M., *J. Chem. Phys.* 42, 2057 (1965).
6. Abraham, F.F., *J. Chem. Phys.* 51, 1632 (1969).
7. Shugard, W.J., Reiss, H., *J. Chem. Phys.* 65, 2827 (1976).
8. Shizgal, B., Barrett, J.C., *J. Chem. Phys.* 91, 6505 (1989).
9. Ruckenstein, E., Nowakowski, B., *J. Colloid Interface Sci.* 137, 583 (1990). (Section 1.3 of this volume.)
10. Nowakowski, B., Ruckenstein, E., *J. Chem. Phys.* 94, 1397 (1991). (Section 2.1 of this volume.)
11. Shizgal, B., Blackmore, R., *J. Comp. Phys.* 55, 313 (1984).
12. Blackmore, R., Shizgal, B., *Phys. Rev. A* 31, 1855 (1985).
13. Abraham, F.F., *Homogeneous Nucleation Theory*, Academic Press, New York, 1974.
14. Risken, H., *The Fokker–Planck Equation*, Springer-Verlag, Berlin, 1989.
15. Binder, K., Stauffer, D., *Adv. Phys.* 25, 343 (1976).

# 2.5 Kinetic Theory of Nucleation Based on a First Passage Time Analysis: Improvement by the Density-Functional Theory

*Yuri Djikaev and Eli Ruckenstein**

A recent kinetic theory of nucleation (see, e.g., E. Ruckenstein and B. Nowakowski, *J. Colloid Interface Sci.* 137, 583 [1990][Section 1.3 of this volume]) is based on molecular interactions and avoids the traditional thermodynamics. The rate of emission of molecules from a cluster is found via a first passage time analysis. This time is calculated by solving the single-molecule master equation for the probability distribution function of a surface molecule located in the potential field created by the cluster. The liquid cluster was assumed to have sharp boundaries and uniform density. In the present paper, this assumption is removed by using the density-functional theory to find the density profiles. Thus, more accurate calculations of the potential field created by the cluster, its emission rate, and nucleation rate are obtained. The modified theory is illustrated by numerical calculations for a molecular pair interaction potential combining the dispersive attraction with the hard-sphere repulsion. (© *2005 American Institute of Physics*. doi: 10.1063/1.2135777.)

## 2.5.1 Introduction

Nucleation is the initial stage of any first-order phase transition. The thermodynamics and kinetics of nucleation have been studied since 1930s when the basis of the classical nucleation theory (CNT) was created [1–3]. Besides CNT, other theoretical and computational methods have been also applied to nucleation, such as the density-functional theory (DFT) [4,5] and computer simulations [6,7].

In the CNT, the kinetics of nucleation is based on the Gibbsian thermodynamics and on the capillarity approximation [8] and involves a particle of a nascent (stable) phase assumed to have sharp boundaries and uniform bulk properties. An alternative to the classical kinetic theory of nucleation was relatively recently proposed [9–14] that does not employ the traditional thermodynamics, thus avoiding the use of the surface tension for tiny clusters. In the Ruckenstein–Nowakowski–Narsimhan (RNN) [9–14] kinetic theory the rate of emission of molecules by a new-phase particle is determined by a mean first passage time analysis. This time is calculated by solving a single-molecule master equation for the probability distribution function of a surface molecule moving in a potential field generated by the cluster. The theory of Ruckenstein and coworkers was developed for both liquid-to-solid [9–11] and vapor-to-liquid [12–14] phase transitions. Recently, their approach was also applied by Shen and Debenedetti [15] to the kinetics of homogeneous bubble nucleation.

The present paper further develops the RNN kinetic theory of vapor-to-liquid nucleation. In the original version of the RNN theory the cluster (a spherical particle of a nascent liquid phase) was assumed to have sharp boundaries and (bulk) uniform density. This assumption simplifies the problem of calculating the effective potential field created by the cluster. In fact, it allows one to find an analytical expression for the effective potential as a function of the distance from the cluster surface. However, it may be inadequate when the thickness of the transition (interfacial) layer of the cluster becomes comparable with the cluster size or if the cluster is a multicomponent one. In this work (concerning only the single-component nucleation), this assumption is removed by using DFT methods. The more realistic density profile thus obtained allows for a more accurate calculation of the potential field created by the cluster, its rate of emission of molecules, and the nucleation rate.

The paper is structured as follows. In Section 2.5.2, we briefly describe the new approach to the kinetics of vapor-to-liquid nucleation, outline the application of DFT methods to nucleation, and propose the combination of the new kinetic theory of nucleation with the DFT methods. Numerical results obtained on the basis of such a combination are presented in Section 2.5.3, while Section 2.5.4 summarizes the results and conclusions.

## 2.5.2 Theoretical Basis

### 2.5.2.1 RNN Approach to Kinetics of Nucleation

Let us denote the number of molecules in a cluster by $\nu$ and the cluster distribution with respect to $\nu$ at time $t$ by $g(\nu,t)$. The goal of the kinetic theory of nucleation is to determine the temporal evolution of $g(\nu,t)$. The nucleation rate is found by solving the kinetic equation of nucleation for a steady-state regime. In the simplest case the kinetic equation of nucleation has the form

* *J. Chem. Phys.*, 123, 214503 (2005). Republished with permission.

$$\partial g(v, t)/\partial t = -\partial J(v, t)/\partial v,$$

$$J(v, t) = -A(v)g(v, t) - \partial(B(v)g(v, t))/\partial v, \tag{2.158}$$

where $J$ is the flux of the new-phase particles in the space of variable $v$. The coefficients $A(v)$ and $B(v)$ in the Fokker–Planck equation (Equation 2.158) are determined by the emission and absorption rates $W^-(v)$ and $W^+(v)$ of the cluster. Conventionally, the following boundary conditions are used:

$$\begin{aligned} g(v,t)/g_{eq}(v) &\to 1 \text{ for } v \to 0, \\ g(v,t)/g_{eq}(v) &\to 0 \text{ for } v \to \infty, \end{aligned} \tag{2.159}$$

where $g_{eq}$ is the equilibrium distribution.

The coefficients $A(v)$ and $B(v)$ in Equation 2.159 are determined by $W^-(v)$ and $W^+(v)$. The rate $W^+(v)$ is calculated by using the kinetic theory of gases, but in determining the emission rate $W^-(v)$ CNT relies on the capillarity approximation and makes use of the surface tension of the cluster: $W^-(v)$ is assumed to be independent of the supersaturation and equal to the absorption rate $W^+(v)$ that the cluster would have if it were in equilibrium with a vapor of particular supersaturation. Thus, the determination of $W^-(v)$ relies (via the Gibbs–Kelvin relation) on the surface tension of the cluster.

To avoid the use of macroscopic thermodynamics, Ruckenstein and coworkers proposed to calculate $W^-(v)$ independently of $W^+(v)$ by solving the (Fokker–Planck or Smoluchowski) equation governing the motion of a single molecule in a potential well around the cluster. This is achieved by calculating the mean passage time for a molecule dissociating from the cluster. A surface molecule of the cluster performs thermal chaotic motion in a spherically symmetric potential well $\Phi(r)$ created by the cluster. Neglecting the internal structure of molecules and assuming pairwise additivity of the molecular interactions, the potential $\Phi(r)$ is given by

$$\Phi(r) = \int_V d\mathbf{r}'\rho(r')\phi(|\mathbf{r}' - \mathbf{r}|), \tag{2.160}$$

where $\mathbf{r}'$ is the coordinate of the surface molecule, $\rho(r)$ is the number density of molecules at point $\mathbf{r}$ (spherical symmetry is assumed), and $\phi(|\mathbf{r}'-\mathbf{r}|)$ is the interaction potential between two molecules located at the points $\mathbf{r}'$ and $\mathbf{r}$. The integration in Equation 2.160 extends over the whole volume of the system, but the vapor-phase contribution is negligible. The molecule is considered as belonging to the cluster if it is in the potential well, and as dissociated from the cluster when it passes over the outer boundary of the well. The rate $W^-$ is determined by the mean passage time for this threshold distance (all distances are measured from the center of the cluster, unless otherwise specified).

The mean passage time of a molecule is calculated by using a kinetic equation governing the chaotic motion of the molecule escaping from the potential well generated around the cluster. The form of this equation depends on the strength of molecular interactions and is different in liquids and gases. In gases, the evolution of the molecule can be described by a master equation in the energy space [12]. This kinetic equation can be derived either from the one-dimensional Fokker–Planck equation [13] that is valid for large clusters for which the width of the potential well is small compared with their radii, or from the three-dimensional Fokker–Planck equation [14] that can be reduced to a unidimensional equation because the relaxation of the distribution of molecules with respect to the angular variables is very fast.

In the gaseous layer around a cluster, a potential well is generated by the attractive interactions with the cluster molecules. If the molecule has an energy $E$ lower than the upper boundary of the well, $E_b$, it moves within the well and can be regarded as bound to the cluster. It can leave the well (i.e., evaporate from the cluster) if it acquires enough energy from the medium to overcome the potential barrier. Thus, the rate of evaporation is determined by the rate with which the molecules pass over the upper energy of the well.

The random motion of a molecule in the well is characterized by its probability density distribution with respect to the position and momentum. The temporal evolution of this function is governed by a Fokker–Planck equation. Using the spherical coordinates $r$, $\Theta$, $\phi$, let us denote this function by $P(r, \Theta, \phi, p_r, J_\Theta, J_\phi, t)$, where $p_r$, $J_\Theta$, $J_\phi$, are the conjugate to $r$, $\Theta$, $\phi$. Let us also define the distribution function with respect to $r$ and $p_r$ as

$$\bar{P}(r,v,t)=\int \mathrm{d}\Theta \mathrm{d}\phi \mathrm{d}J_{\Theta}\,\mathrm{d}J_{\phi}P(r,\Theta,\phi,\upsilon,J_{\Theta},J_{\phi},t), \tag{2.161}$$

where $v = p_r/m$, $m$ being the mass of the molecule. Assuming that the relaxation with respect to the angular variables is much faster than the evolution with respect to $r$ and $p_r$, one can obtain [12–14,16] the following equation for $\bar{P}(r,v,t)$:

$$\begin{aligned}\partial\bar{P}/\partial t &= -\upsilon\partial\bar{P}/\partial r + m^{-1}(\mathrm{d}\Phi/\mathrm{d}r - 2kT/r)\partial\bar{P}/\partial\upsilon \\ &+ \gamma\partial(\upsilon\bar{P})/\partial\upsilon + (\gamma kT/m)\partial^2\bar{P}/\partial\upsilon^2,\end{aligned} \tag{2.162}$$

where $\gamma$ is the friction coefficient, $k$ is the Boltzmann constant, and $T$ is the temperature.

Equation 2.162 has the form of a Fokker–Planck equation for a unidimensional motion in the effective potential $\Phi_{\mathrm{eff}}(r) = \Phi(r) - 2kT \ln r + \text{const}$. The potential $\Phi_{\mathrm{eff}}(r)$ increases with increasing $r$ for $r < r_b$, attains a maximum at $r = r_b$, determined by the condition $\mathrm{d}\Phi_{\mathrm{eff}}/\mathrm{d}r = 0$, and decreases with increasing $r$ for $r > r_b$. The height of the potential barrier is $E_b = \Phi_{\mathrm{eff}}(r_b)$. Thus, the potential well stretches between the cluster surface ($r = R$) and the top of the barrier ($r = r_b$). The molecules are confined within the well if their energy is insufficient to overcome the potential barrier $E_b$ at $r_b$.

The motion of the molecule in the well involves two distinct time scales. The molecule performs a kind of anharmonic oscillation in the well, with a characteristic time estimated as $t_w = \lambda_w(m/kT)^{1/2}$, where $\lambda_w$ is the thickness of the well. The energy of these oscillations changes as a result of the interactions of the molecule with the medium. The rate of this change can be characterized by the period $t_c = \gamma^{-1}$. In gases, the interactions are not very strong, and, because the friction coefficient $\gamma$ is small, the time scale $t_c$ for the variation of the energy is much longer than the period of oscillations in the well, $t_w$. As a result, one can discriminate between the fast-changing position and momentum and the slowly changing energy.

The rate of change of the cluster size is determined by the difference between $W^-$ and $W^+$. Introducing the new scaled variables $z$, $E$, and $t$ as $z = r/\lambda_w$, $E = \frac{1}{2}mv^2/kT + \Psi_{\mathrm{eff}}$, and $t = t/t_0$ (with $\Psi_{\mathrm{eff}} = \Phi/kT$ and $t_0 = \lambda_w(m/kT)^{1/2}$) transforming the variables in Equation 2.162 to $z$, $E$, and $t$, and using the hierarchy of time scales in the evolution of $\bar{P}$ with respect to $z$ and $E$, one can find [12–14] the stationary flux per molecule in the energy space. This flux determines $W^-$ (by convention, a molecule is said to have left the cluster upon reaching the upper boundary of the well $E = E_b$),

$$\begin{aligned}W^- &= \gamma 4\pi R^2\lambda_w n_w W_1^-, \\ W_1^- &= \left[\int_{\Psi_0}^{E_b} \mathrm{d}E G'(E)e^{-E}\int_{E}^{E_b} \mathrm{d}E' e^{E'}/G(E')\right]^{-1},\end{aligned} \tag{2.163}$$

where $n_w$ is the average density of molecules in the well,

$$G(E) = \int_{z_1}^{z_2} (2(E - \Psi_{\mathrm{eff}}))1/2\mathrm{d}z,$$

$$G'(E) = \int_{z_1}^{z_2} \frac{\mathrm{d}z}{(2(E - \Psi_{\mathrm{eff}}))^{1/2}},$$

and $z_1$ and $z_2$ are the boundaries of the well for the energy level $E$, which are given by the condition $E = \Psi(z)$. The actual values of $n_w$ and $\lambda_w$ are not known, but they are not needed to calculate the nucleation rate [12–14].

The rate $W^+$ depends on the mechanism of the transfer process of vapor molecules to the cluster. In gaseous systems, it is given [17] by the expression $W^+ = n_1\pi R^2\bar{\upsilon}$, where $\bar{\upsilon}$ is the mean thermal velocity of vapor molecules, $n_1$ is the (undisturbed) number density of vapor molecules, and the mass accommodation coefficient is assumed to be unity.

The adsorption and emission rates are equal for a critical cluster with $R = R_{*}$, i.e., $W^{-}(R_{*}) = W^{+}(R_{*})$, which provides the equation $n_1 = n_w(4\lambda_w\gamma/\upsilon)W_1^{-}(R_{*})$ for $R_{*}$.

Let us denote the equilibrium number density of vapor molecules by $n_e$. This vapor coexists with a cluster of infinite size, i.e., with the bulk condensed phase. Since the vapor supersaturation is defined as $\zeta = n_1/n_e$, we have $\zeta = W_1^{-}(R_{*})/W_1^{-}(\infty)$. In the CNT, the relation between $\zeta$ and $R_{*}$ is provided by the Kelvin equation $\zeta = \exp(2\sigma/kT\rho R_{*})$, where $\rho$ is the density of molecules in the cluster and $\sigma$ is the surface tension of a flat interface. The RNN theory is expected to recover the classical expression in the limit of large clusters. Thus, an expression for the surface tension can be obtained by comparing the above two expressions for $\zeta$ for large radii.

Knowing $W^{-}$ and $W^{+}$, one can find the steady-state nucleation rate $J_s$ by solving Equations 2.158 and 2.159 (see, e.g., Refs. [12–14]),

$$J_s = \left[\int_0^{\infty} \mathrm{d}v / W^{+}(v) g_{eq}(v)\right]^{-1}$$
$$= n_1 W^{+}(1) \Big/ \int_0^{\infty} \exp[h(v) - \ln(W^{+}(v)/W^{+}(1))]\mathrm{d}v \tag{2.164}$$

$$h(v) = \ln \prod_{j=2}^{v} W^{+}(j-1)/W^{-}(j). \tag{2.165}$$

### 2.5.2.2 DFT Methods in Nucleation Theory

In the DFT-based approach to the thermodynamics of nucleation [18–22], the key assumption of the CNT—the capillarity approximation—is removed. The interface between the cluster and vapor has a finite width and is characterized by a density profile $\rho(\mathbf{r})$.

Let us consider a system where molecules interact via a pairwise potential $\phi(\mathbf{r})$ and suppose that $\phi(\mathbf{r})$ can be divided into a steeply repulsive (reference) part $\phi^{(1)}(\mathbf{r})$ and a small attractive (perturbative) part $\phi^{(2)}(\mathbf{r})$, $\phi(\mathbf{r}). = \phi^{(1)}(\mathbf{r}) + \phi^{(2)}(\mathbf{r})$.

In the grand canonical ensemble (constant chemical potential $\mu$, volume $V$, and temperature $T$), the equilibrium density profile $\rho(\mathbf{r})$ is determined by minimizing the grand canonical potential $\Omega[\rho(\mathbf{r})]$ (a functional of $\rho(\mathbf{r})$) with respect to $\rho(\mathbf{r})$. In this case, using the local-density approximation (LDA) and random-phase approximation (RPA) for the contributions to $\Omega[\rho(\mathbf{r})]$ from $\phi^{(1)}(\mathbf{r})$ and $\phi^{(2)}(\mathbf{r})$, respectively [19–25], the Euler–Lagrange equation leads to

$$\mu = \mu^{(1)}(\rho(\mathbf{r})) + \int \mathrm{d}\mathbf{r}' \phi^{(2)}(|\mathbf{r} - \mathbf{r}'|)\rho(\mathbf{r}'), \tag{2.166}$$

where $\mu^{(1)}$ is the local chemical potential of the reference system (usually hard-spheres fluid). Both LDA and RPA are sufficiently accurate for weakly inhomogeneous systems not too close to critical conditions. In the canonical ensemble (constant $V$, $T$, and number of molecules $N$) the equilibrium density profile $\rho(\mathbf{r})$ is determined by minimizing the Helmholtz free energy $F[\rho(\mathbf{r})]$ as a functional of $\rho(\mathbf{r})$. In this case the Euler–Lagrange equation leads to a more complicated equation [19,26].

The density profile of a planar gas–liquid interface can be found by solving Equation 2.166 by iterations [24,27,28] for a planar geometry $\rho(\mathbf{r}) = \rho(z)$. Substituting the solution of this equation into the functional $\Omega[\rho(\mathbf{r})]$, one can calculate the gas–liquid surface tension as $\sigma = (\Omega + pV)/A$, where $P$ and $V$ are the pressure and volume of the system and $A$ is the area of the interface. In an inhomogeneous fluid with density fluctuations in the form of spherical clusters, the density profile can be obtained by solving Equation 2.166 [21,22] (or its analog for an $NVT$ ensemble [19,26]) for a spherically symmetric distribution $\rho(\mathbf{r}) = \rho(r)$. Knowing the density profile, one can obtain the surface tension of the cluster if one identifies the radius of the cluster, $R$, as that of the surface of tension. For this surface (of tension) the Laplace equation holds, $\Delta P = 2\sigma_R/R$, for the difference in pressure within the cluster interior and far away from it. The surface tension and radius of the cluster are then given by

$$\sigma_R = \left( \frac{3\Delta\Omega(\Delta P)^2}{16\pi} \right)^{1/3}, \quad R = \left( \frac{3(\Delta\Omega)}{2\pi\Delta P} \right)^{1/3}, \tag{2.167}$$

where $\Delta\Omega$ is the work of formation of the (critical) nucleus. All the quantities on the right-hand sides of these equalities are calculated on the basis of density profiles.

#### 2.5.2.3 Improvement of RNN Kinetic Theory of Nucleation by DFT Methods

In the original version of the RNN theory, the spherical liquid cluster was assumed to have sharp boundaries wherein the number density of molecules is uniform and equal to that of a bulk liquid. This assumption simplifies the problem of calculating the effective potential field that is created by the cluster and wherein a surface-located molecule performs a chaotic motion. In fact, it allows one to find an analytical expression for the effective potential as a function of the distance from the cluster surface. For example, consider an intermolecular interaction potential $\phi(|\mathbf{r}'-\mathbf{r}|)$ (between two molecules at points $\mathbf{r}$ and $\mathbf{r}'$) that combines the rigid core repulsion with the nonretarded dispersive attraction:

$$\begin{aligned} &\phi(r_{12}) = \infty (r_{12} < \eta), \\ &\phi(r_{12}) = -4\epsilon(\eta/r_{12})^6 (r_{12} \geq \eta), \end{aligned} \tag{2.168}$$

where $\epsilon$ is an interaction constant and $\eta$ is the molecular diameter. For a spherical cluster with a uniform distribution of molecules the integration over the cluster volume in Equation 2.160 yields

$$\begin{aligned} &\Phi(r) = -(4\pi/3)4\epsilon\rho\eta^3[\eta/(r-R)]^3[R/(r+R)]^3 \\ &(r \geq R+\eta), \end{aligned} \tag{2.169}$$

where $\rho$ is the number density of molecules in the cluster.

The assumption $\rho(r)$ = const may be inadequate when the thickness of the transition (interfacial) layer between the cluster and vapor becomes comparable with the cluster size. However, this (density uniformity) assumption can be easily removed. It is simply necessary to calculate the potential field $\Phi(r)$ in Equation 2.160 by using the more realistic density distribution $\rho(r)$ provided by DFT. No analytical expression for the effective potential can be obtained in this case because $\rho(r)$ is determined by numerically solving Equation 2.166 (or its analog for an *NVT* ensemble [19,26]) by iterations. While a more accurate function $\Phi(r)$ is thus obtained, the formalism of the RNN theory remains unchanged.

The main numerical problem related with using the DFT methods is the calculation of the nucleation rate that requires the determination of the function $h(\nu)$ defined by Equation 2.165. To calculate this function for a given supersaturation, it is necessary to know the emission rates of clusters of all sizes, i.e., the function $W^-(j)$ for $2 < j < \nu$. However, for a given supersaturation (or chemical potential of the vapor phase), the DFT methods allow one to determine the density profile in a "cluster-vapor" system only for a unique cluster size, namely, for the critical cluster that is in unstable equilibrium with the vapor. Thus, for a given supersaturation $\zeta$, the DFT methods can provide the emission rate $W^-(j_*|\zeta)$ only for the nucleus (critical cluster). To construct the function $h(\nu)$, the following procedure can be used.

Let us denote by $W_1^-(j|\zeta_j)$ the emission rate per molecule for a nucleus of size $j$ corresponding to an arbitrary supersaturation $\zeta_j \neq \zeta$. This quantity can be determined by combining the DFT with the RNN theory, as outlined above. Since $\zeta_j = W_1^-(j|\zeta_j)/W_1^-(\infty)$, we have

$$\partial W_1^-(j|\zeta_j)/\partial\zeta_j = W_1^-(\infty), \tag{2.170}$$

so that $\partial^k W_1^-(j|\zeta_j)/\partial\zeta_j^k = 0 (k \geq 2)$. Therefore, the Taylor expansion $W_1^-(j/\zeta_j) = W_1^-(j|\zeta) + W_1^-(\infty)(\zeta_j - \zeta)$ is exact even if $\zeta_j - \zeta$ is not very small. Thus, we have

$$W_1^-(j|\zeta) = W_1^-(j|\zeta_j) + W_1^-(\infty)(\zeta - \zeta_j). \tag{2.171}$$

Carrying out DFT calculations in combination with the RNN theory for a series of supersaturations, one can find the values of $W_1^-(j|\zeta_j)$ for a set of supersaturations $\{\zeta_j\}$ corresponding to a set of cluster radii $\{R_j\}$. Then, using Equation 2.171, one can determine $W_1^-(R_j|\zeta)$ for the set $\{R_j\}$, which allows one to construct the whole function $W_1^-(R|\zeta)$ for a selected $\zeta$. In the function $h(v)$, the unknown parameters $\lambda_w$ and $n_w$ cancel in the numerator and denominator owing to $n_1 = n_w(4\lambda_w\gamma/\bar{\upsilon})W_1^-(R_*)$. Thus, $h(v)$ becomes completely determined and can be further used in Equation 2.164 for calculating the nucleation rate.

### 2.5.3 Numerical Results

To illustrate the effect of DFT methods on the RNN kinetic theory, we have carried out numerical calculations using the molecular interaction potential given by Equation 2.168. The results are presented in Figures 2.14 through 2.17.

Figure 2.14 shows the dependence of the potential field $\Phi(r)$ on the distance from the center of the cluster. The dashed curve corresponds to the combined RNN/DFT theory, whereas the solid curve represents the original RNN theory (based on the density uniformity assumption). In the combined RNN/DFT theory, the cluster radius was associated with the radius of the surface of tension $R_{st}$. For the intermolecular potential considered, the Tolman length $\delta = R_{eq} - R_{st}$ (where $R_{eq}$ is the radius of the equimolar dividing surface) changes from positive values $\delta \sim \eta$ for small clusters of size $(R_{st} + R_{eq})/2 \simeq 3\eta$ to negative values $\delta \sim -\eta$ for large clusters of $R_{st} \geq 100\eta$. As shown in Figure 2.14, the potential well $\Phi(r)$ created by the cluster is always markedly deeper for a uniform density than for a more realistic

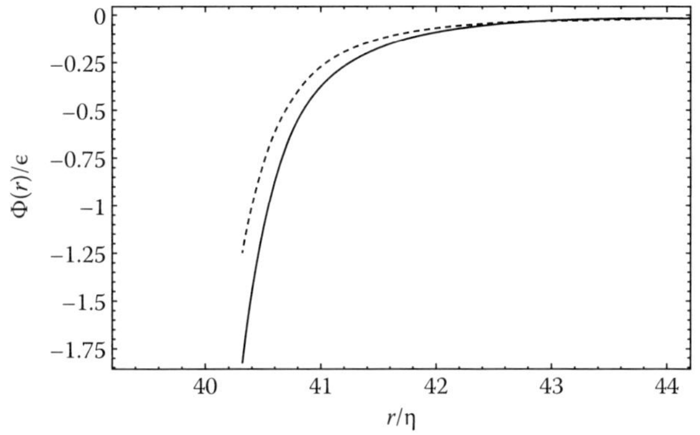


Figure 2.14

The cluster potential $\Phi(r)$ as a function of the distance from the center of the cluster for $R/\eta = 39.32$. The dashed and solid curves correspond to $\Phi(r)$ in the combined RNN/DFT theory and original RNN theory, respectively.

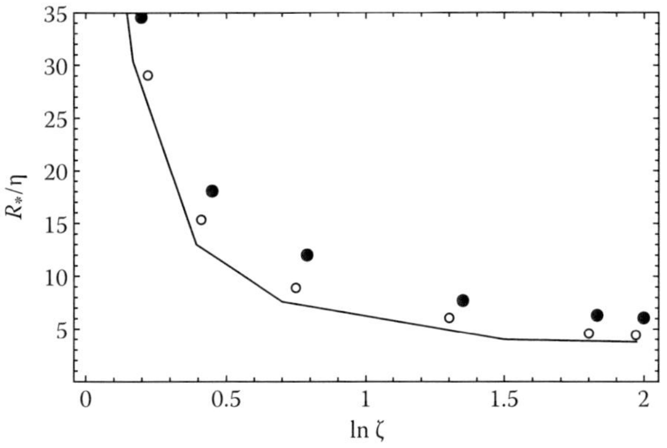


Figure 2.15

The radius of the critical cluster $R_*$ as a function of the supersaturation ratio $\zeta$ for various values of $\epsilon/kT$. The solid curve is for $\epsilon/kT = 1.408\ 45$, the circles for $\epsilon/kT = 3$, and the disks for $\epsilon/kT = 5$.

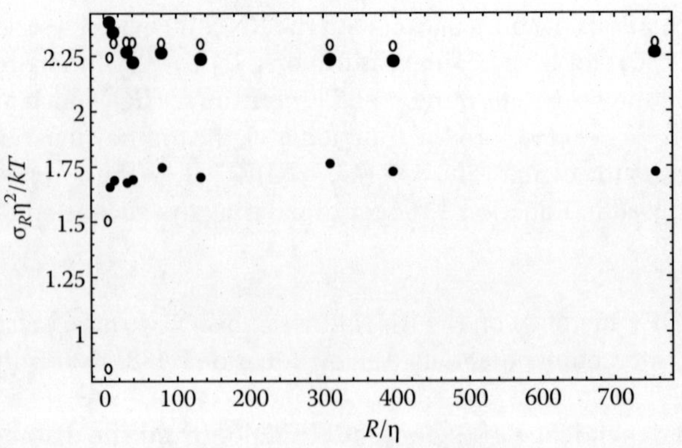


Figure 2.16

The dimensionless surface tension $\sigma_R\eta^2/kT$ of the cluster as a function of $R/\eta$. The circles represent the DFT data. The data obtained by using the Kelvin relation are shown as large dots (for the supersaturation determined by $\zeta = n_1/n_e$ and as small dots (for the supersaturation determined as $\zeta = W^-(R)/W^-(\infty)$).

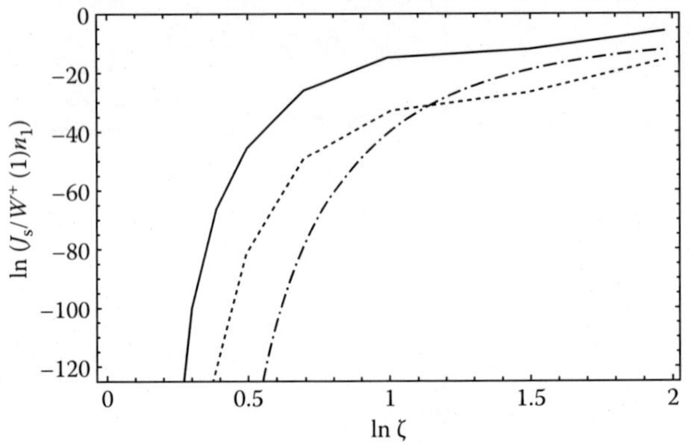


Figure 2.17

The nucleation rate (shown as $\ln(J_s/W^+(1)n_1)$) versus supersaturation $\zeta$ for the interaction parameter $\epsilon/kT = 1.408\ 45$. The solid curve represents the original version of the RNN theory, the dashed curve is for the DFT-based RNN theory, and the dot-dashed curve is for the CNT.

profile provided by DFT. Since a surface molecule can easily escape from a shallower well, the nucleation rate predicted by the combined RNN/DFT theory is expected to be lower than that predicted by the original RNN theory.

Figure 2.15 shows the dependence of the critical radius $R_*$ (associated with the radius of the surface of tension) on the supersaturation $\zeta$. The solid line is for $\epsilon/kT = 1.408\ 45$, the circles for $\epsilon/kT = 2$, and the disks for $\epsilon/kT = 3$.

Figure 2.16 shows the dependence of the surface tension on the cluster radius $R$ (all the data are for $\epsilon/kT = 1.408\ 45$). The circles show the data provided by DFT (this section), whereas the large and small dots are the data obtained by using the Kelvin relation that was assumed to be valid for clusters containing more than 5000 molecules (i.e., $R_*/\eta \geq 10$). The large dots represent the data for the case where supersaturation $\zeta$ is determined as the ratio $\zeta = n_1/n_e$ (with $n_1$ and $n_e$ provided by DFT), whereas the small dots are for the case where the supersaturation is calculated as $\zeta = W_1^-(R_*)/W_1^-(\infty)$. In this case, the "centrifugal" contribution $-2kT \ln r$ to $\Phi_{\text{eff}}$, which accounts for the angular degrees of freedom, was not taken into account. The two ways of determining $\zeta$ lead to rather different values of the surface tension (differing by about 25%), which can be probably explained by the inaccuracy in calculating $W^-(R_*)$ and $W^-(\infty)$ numerically.

Figure 2.17 compares the nucleation rates calculated by using the original version of the RNN theory (solid line), the DFT version of the RNN theory (dashed), and the CNT (dot-dashed line) (all the data are for $\epsilon/kT = 1.408\ 45$). One can see that the rates provided by the DFT version of the RNN theory are closer to those predicted by the CNT, although the discrepancy is relatively large for small supersaturations. On the other hand, the rates provided by the original RNN theory (based on a uniform density) are higher than those of the DFT-based RNN theory. This is because the potential well $\Phi(r)$ is significantly deeper in the former case, as shown in Figure 2.14. Clearly, it is harder for surface molecules to escape from a deeper potential well; hence, the disaggregation of such clusters occurs more slowly, which leads to higher nucleation rates.

### 2.5.4 Concluding Remarks

We have presented a development of the kinetic theory of Ruckenstein and coworkers for the case of vapor-to-liquid nucleation. The RNN theory is based on molecular interactions and does not employ the traditional thermodynamics, thus avoiding the use of the surface tension for tiny clusters. In that theory the rate of emission of molecules from a new-phase particle is determined by using the mean first passage time analysis. This time is calculated by solving a single-molecule master equation for the probability distribution function of a surface layer molecule moving in a potential field created by the rest of the cluster. The theory of Ruckenstein and coworkers was developed for both liquid-to-solid and vapor-to-liquid phase transitions.

In the original version of the RNN theory for vapor-to-liquid nucleation the cluster (a spherical particle of a nascent liquid phase) was assumed to have sharp boundaries and uniform density equal to that of the bulk liquid. This assumption significantly simplifies the problem of calculating the effective potential field created by the cluster and to which a surface-located molecule is subjected. It allows one to find an analytical expression for the effective potential as a function of the distance from the cluster surface. However, such an assumption may be inadequate when the thickness of the transition (interfacial) layer of the cluster becomes comparable with the cluster size.

We have shown that the (density uniformity) assumption can be removed from the theory by using the methods of density-functional theory for determining more realistic density profiles. Using these profiles, one can more accurately calculate the potential field created by the cluster and thus more accurately determine the rate of emission of molecules by the cluster. The modified theory is illustrated by numerical calculations for a molecular pair interaction potential combining the dispersive attraction with the hard-sphere repulsion. The numerical calculations indicate that the predictions of the modified RNN theory for the nucleation rate are lower than those of the uniform density version.

Finally, we note that extending the RNN theory to multicomponent systems, it is important to calculate the cluster potential field by using DFT-based density profiles, particularly when surface-active components are involved in nucleation. The multicomponent CNT contains some intrinsic inconsistencies related to the composition dependence of the surface tension, surface adsorption effects, etc., which are reflected in large discrepancies between theoretical predictions and experimental data for the nucleation rate. These can be expected to be avoided in a combined RNN/DFT kinetic theory.

## Acknowledgments

This work was supported by the National Science Foundation through the Grant No. CTS-0000548.

## References

1. Volmer, M., Weber, A., *Z. Phys. Chem. (Leipzig)* 119, 227 (1926).
2. Farkas, L., *Z. Phys. Chem. (Leipzig)* 125, 236 (1927).
3. Becker, R., Döring, W., *Ann. Phys.* 24, 719 (1935).
4. Evans, R., *Adv. Phys.* 28, 143 (1979).
5. Oxtoby, D., in: *Fundamentals of Inhomogeneous Fluids*, Henderson, D., Ed., Marcel Dekker, New York, 1992.
6. Allen, M.P., Tildesley, D.J., *Computer Simulation of Liquids*, Clarendon, Oxford, 1989.
7. Frenkel, D., Smit, B., *Understanding Molecular Simulation*, Academic, Boston, 1996.
8. Lothe, J., Pound, G.M.J., in: *Nucleation*, Zettlemoyer, A.C., Ed., Marcel Dekker, New York, 1969.
9. Narsimhan, G., Ruckenstein, E., *J. Colloid Interface Sci.* 128, 549 (1989). (Section 1.2 of this volume.)
10. Ruckenstein, E., Nowakowski, B., *J. Colloid Interface Sci.* 137, 583 (1990). (Section 1.3 of this volume.)
11. Nowakowski, B., Ruckenstein, E., *J. Colloid Interface Sci.* 139, 500 (1990). (Section 1.4 of this volume.)
12. Nowakowski, B., Ruckenstein, E., *J. Chem. Phys.* 94, 1397 (1991). (Section 2.1 of this volume.)

13. Ruckenstein, E., Nowakowski, B., *Langmuir* 7, 1537 (1991). (Section 2.2 of this volume.)
14. Nowakowski, B., Ruckenstein, E., *J. Chem. Phys.* 94, 8487 (1991). (Section 2.3 of this volume.)
15. Shen, V.K., Debenedetti, P.G., *J. Chem. Phys.* 118, 768 (2003).
16. Rodger, P.M., Sceats, M.G., *J. Chem. Phys.* 83, 3358 (1985).
17. Abraham, F.F., *Homogeneous Nucleation Theory*, Academic, New York, 1974.
18. Ramakrishnan, T.V., Yussouff, M., *Phys. Rev. B* 19, 2775 (1979).
19. Lee, D.J., Telo da Gama, M.M., Gubbins, K.E., *J. Chem. Phys.* 85, 490 (1986).
20. Haymet, A.D.J., Oxtoby, D.W., *J. Chem. Phys.* 74, 2559 (1981).
21. Oxtoby, D.W., Evans, R., *J. Chem. Phys.* 89, 7521 (1988).
22. Zeng, X.C., Oxtoby, D.W., *J. Chem. Phys.* 94, 4472 (1991).
23. Barker, J.A., Henderson, D., *J. Chem. Phys.* 47, 4714 (1967).
24. Telo da Gama, M.M., Evans, R., *Mol. Phys.* 43, 229 (1983).
25. Sullivan, D.E., *Phys. Rev. B* 20, 3991 (1979).
26. Talanquer, V., Oxtoby, D.W., *J. Chem. Phys.* 100, 5190 (1994).
27. Tarazona, P., Evans, R., *Mol. Phys.* 52, 847 (1984).
28. Tarazona, P., *Phys. Rev. A* 31, 2672 (1985).

# 2.6 Kinetic Theory of Binary Nucleation Based on a First Passage Time Analysis

*Yuri Djikaev and Eli Ruckenstein**

The binary classical nucleation theory (BCNT) is based on the Gibbsian thermodynamics and applies the macroscopic concept of surface tension to nanosize clusters. This leads to severe inconsistencies and large discrepancies between theoretical predictions and experimental results regarding the nucleation rate. We present an alternative approach to the kinetics of binary nucleation that avoids the use of classical thermodynamics for clusters. The new approach is an extension to binary mixtures of the kinetic theory previously developed by Narsimhan and Ruckenstein and Ruckenstein and Nowakowski (*J. Colloid Interface Sci.* 128, 549 [1989]; 137, 583 [1990] [Sections 1.2 and 1.3 of this volume]) for unary nucleation that is based on molecular interactions and in which the rate of emission of molecules from a cluster is determined via a mean first passage time analysis. This time is calculated by solving the single-molecule master equation for the probability distribution of a "surface" molecule moving in a potential field created by the cluster. The starting master equation is a Fokker–Planck equation for the probability distribution of a surface molecule with respect to its phase coordinates. Owing to the hierarchy of characteristic time scales in the evolution of the molecule, this equation can be reduced to the Smoluchowski equation for the distribution function involving only the spatial coordinates. The new theory is combined with density functional theory methods to determine the density profiles. This is essential for nucleation in binary systems particularly when one of the components is surface active. Knowing these profiles, one can determine the potential fields created by the cluster, its rate of emission of molecules, and the nucleation rate more accurately than by using the uniform density approximation. The new theory is illustrated by numerical calculations for a model binary mixture of Lennard–Jones monomers and rigidly bonded dimers of Lennard–Jones atoms. The amphiphilic character of the dimer component (i.e., its surface activity) is induced by the asymmetry in the interaction between a monomer and the two different sites of a dimer. The inconsistencies of the BCNT are avoided in the new theory. (© *2006 American Institute of Physics.* doi: 10.1063/1.2178317.)

## 2.6.1 Introduction

Nanoscale systems are of increasing interest both scientifically and technologically. Nanoscience demands a progressive reduction in the size of systems and the fabrication of the new materials often requires an accurate control over phase transitions therein. The initial and crucial step in many of these phase transitions is the formation of an embryo or nucleus of the new phase in the metastable mother phase, often multicomponent. This stage of the phase transition, nucleation, plays a very important role in many fields.

Recent progress in simulation and experiments has made possible the exploration of this process at a molecular level [1–5]. It is possible to observe the formation of tiny clusters and crystals containing as few as tens of molecules or less. The results of these experiments are challenging our understanding of the process. Nevertheless, the development of an adequate macroscopic nucleation theory remains of great importance because it provides a simple and virtually universal formalism applicable to first-order phase transitions in most complex systems [6,7].

At present, the theory of multicomponent nucleation is in a particularly unsatisfactory state. Consider, for instance, the simplest multicomponent system. Experimental data on the binary nucleation rate can differ by many orders of magnitude [8–12] from the predictions of binary classical nucleation theory (BCNT) based on the capillarity approximation [7]. As a possible explanation, it was suggested [8] that the large discrepancies between the BCNT and experiment are due to adsorption effects. These cannot be taken into account within the framework of the capillarity approximation in which the cluster is considered as a particle with a sharp boundary and with uniform intensive properties of the bulk liquid phase [7]. This approximation is intrinsically unable to take into account adsorption effects which, roughly speaking, can arise only from density nonuniformities in the transition region between the two phases. Besides, this approximation makes use of the macroscopic concept of surface tension in the thermodynamic treatment of small new-phase particles. Thus, a more realistic alternative to the capillarity approximation is needed to develop a binary nucleation theory that would allow one to more or less adequately take into account adsorption effects. This necessity led to the development of a revised BCNT [13] and of a new model for a binary cluster [14–18] that is based on the Gibbs concept of dividing surface and often referred to as the Gibbsian model.

* *J. Chem. Phys.*, 124, 124521 (2006). Republished with permission.

Although the Gibbsian model has necessary features to account for adsorption effects, it still treats the small cluster by means of the classical thermodynamics and uses the concept of surface tension for them. Moreover, a kinetic theory of binary nucleation based on the Gibbsian model is nonexistent.

As an alternative to the kinetic theory of nucleation in single component systems, a new approach to the kinetics of nucleation was proposed [19–26] that involves molecular interactions and does not employ the traditional thermodynamics, thus avoiding the use of the controversial concept of surface tension for tiny clusters involved in nucleation. In the new kinetic theory [19–26] (hereafter referred to as the Ruckenstein–Nowakowski–Narsimhan [RNN] theory) the rate of emission of molecules by a new-phase particle is determined via a mean first passage time analysis. This time is calculated by solving a single-molecule master equation for the probability distribution function of a surface layer molecule moving in a potential field created by the rest of the cluster. The theory of Ruckenstein and coworkers was developed for both liquid-to-solid and vapor-to-liquid phase transitions. In the former case [19–21], the single-molecule master equation is a Fokker–Planck equation in the phase space that can be reduced to the Smoluchowski equation owing to the hierarchy of characteristic time scales in the evolution of the single-molecule distribution function with respect to coordinates and momenta. In the latter case [22–24], the starting master equation was a Fokker–Planck equation for the probability distribution function of a surface layer molecule with respect to both its energy and phase coordinates. Unlike the case of liquid-to-solid nucleation, this Fokker–Planck equation cannot be reduced to the Smoluchowski equation, but the hierarchy of time scales does allow one to reduce it to a Fokker–Planck equation in the energy space.

Recently, we have presented a further development of the RNN kinetic theory of vapor-to-liquid nucleation. In the original version of the RNN theory the cluster (a spherical particle of a nascent liquid phase) was assumed to have sharp boundaries wherein the number density of molecules is uniform and equal to that of the bulk liquid. This assumption simplifies the problem of calculating the effective potential field created by the cluster. It even allows one to obtain an analytical expression for this potential as a function of the distance from the cluster surface. However, such an assumption may be inadequate if, e.g., the thickness of the transition (interfacial) layer of the cluster becomes comparable with the cluster size or if the cluster is a multicomponent one and contains surface active components. In Ref. [27] concerning the single-component nucleation, this assumption has been avoided by using the methods of density functional theory (DFT) to determine the density profiles in the system. Knowing these profiles, one can more accurately determine the potential field created by the cluster and thus more accurately calculate its rate of emission of molecules as well as the nucleation rate.

In the present paper, we use the RNN approach combined with DFT methods and present a new kinetic theory of binary nucleation based on a first passage time analysis. In the new theory, the application of thermodynamics to tiny nucleating clusters is avoided [19–26]. Since the surface tension concept is not used by the new theory, the major inconsistencies of the BCNT are eliminated (see Section 2.6.2 for more details). The classical thermodynamics is used only within the framework of DFT to find the density distributions in the whole macroscopic system (in which nucleation takes place). By using DFT instead of relying on the assumption of uniform density within a cluster, the new kinetic theory provides a tool to examine nucleation in systems in which the adsorption of one of the components at the cluster surface is significant and must be taken into account. Thus, this shortcoming of the BCNT is eliminated. Note that this particular goal can be also achieved if the traditional kinetic theory of nucleation is developed by using the DFT-based thermodynamics of nucleation (instead of the capillarity approximation) [28–30]. But such a hybrid kinetic theory still relies on the use of the controversial notion of surface tension for a tiny cluster, unlike the purely kinetic approach of the combined RNN/DFT theory.

The paper is structured as follows. In Section 2.6.2, after a brief outline of the BCNT ideas, we present a new kinetic theory of binary nucleation based on a first passage time analysis. Section 2.6.3 describes a particular model of a two-component system (monomers + dimers) where one of the components (the dimer) is surface active (the DFT procedure for finding the density profiles is also presented). Section 2.6.4 contains some results of numerical calculations based on the new theory. Section 2.6.5 discusses the results and summarizes the conclusions.

### 2.6.2 Kinetics of Binary Nucleation Based on First Passage Time Analysis

Consider a liquid cluster of a binary solution surrounded by the corresponding vapor mixture. The total volume of the system is $v$ and its temperature $T$ is assumed to be uniform and constant. The quantities associated with component $i$ ($i = 1, 2$) will be denoted with the subscript $i$.

The vapor mixture is assumed to be ideal and can be characterized by the partial saturation ratios $a_1$ and $a_2$

$$a_i = P_i/P_{ie} \ (i = 1, 2),$$

where $P_i$ and $P_{ie}$ are the partial and equilibrium vapor pressures of component $i$ of the vapor mixture. A pure vapor of component $i$ of pressure $P_{ie}$ would be in equilibrium with the plane surface of its liquid phase at a given temperature $T$.

In such a system (vapor + liquid cluster), there is a transition layer between the liquid density in the center of the cluster and the vapor density far away from the cluster. We will define a "cluster" as the part of the system where the density is significantly different from that of the bulk vapor.

### 2.6.2.1 Thermodynamics of Binary Nucleation Based on Capillarity Approximation

According to the BCNT, which is based on the capillarity approximation, the system is regarded to consist of two distinct subsystems: (1) cluster–liquid sphere of radius $R$ and of uniform densities of both components within and (2) vapor mixture of uniform densities of both components. The boundary between the cluster and the vapor mixture is assumed to be sharp. Therefore, $R$ represents the radius of the cluster of volume $\upsilon_R = 4\pi R^3/3$ and surface area $A = 4\pi R^2$.

Let us use the superscripts "l" and "υ" to indicate the liquid and vapor phases, respectively. The state of the cluster can be characterized by the numbers of molecules $v_1$ and $v_2$ it contains. The radius of the cluster is related to the numbers of molecules by $R = [3(v_1\upsilon_1 + v_2\upsilon_2)/4\pi]^{1/3}$, where υ's are the partial volumes per molecule in a bulk solution of mole fraction $\chi = v_1/(v_1 + v_2)$.

Let us denote by $W$ the reversible work of formation of a cluster. By definition, this work is the difference in the appropriate thermodynamic potential between the final and initial states of the system (i.e., the vapor mixture with and without a cluster). For the thermodynamic limit [31,32] one can write [13–16]

$$W \equiv W(\nu_1,\nu_2) = \sum_i \nu_i\left(\mu_i^{\mathrm{l}} - \mu_i^{\upsilon}\right) + \sigma A, \tag{2.172}$$

where $\mu^\upsilon = \mu^\upsilon(P, T)$ and $\mu^{\mathrm{l}} = \mu^{\mathrm{l}}(P, \chi, T)$ are the chemical potentials corresponding to the equilibrium between a vapor mixture of total pressure $P = P_1 + P_2$ and a bulk binary solution of uniform mole fraction $\chi$. In Equation 2.172, the cluster is assumed incompressible and in mechanical (but not chemical) equilibrium with the vapor mixture, and $\sigma$ and $A$ are the surface area and surface tension of the cluster, respectively.

The reversible work $W$ of formation of a new-phase cluster is a function of two independent variables. This function determines the free energy surface in a three-dimensional space. If the initial phase is metastable, the free energy surface has at least one saddle point [33–39] through which a path of steepest descent goes from the region of precritical clusters (where they tend to disaggregate) into the region of supercritical ones (where they tend to grow regularly and irreversibly). The coordinates of the saddle point of the nucleation barrier correspond to the variables of state of the critical new-phase cluster, often referred to as a nucleus, which is in unstable equilibrium with the surrounding initial phase. The free energy of nucleus formation, $W_c$ (i.e., the value of $W$ at the saddle point), determines the height of the nucleation barrier, which plays a crucial role in the whole BCNT.

Equation 2.172 leads to the following explicit expression for the free energy of formation $W$ as a function of $v_1$ and $v_2$:

$$W(v_1,v_2) = -v_1\Delta\mu_1 - v_2\Delta\mu_2 + b(\chi)(v_1 + v_2)^{2/3}, \tag{2.173}$$

where

$$\Delta\mu_1 = \frac{1}{kT}\ln\frac{a_1}{\chi f_1(\chi)}, \quad \Delta\mu_2 = \frac{1}{kT}\ln\frac{a_2}{(1-\chi)f_2(\chi)},$$

$$b(\chi) = \frac{4\pi\sigma}{kT}\left(\frac{3\upsilon(\chi)}{4\pi}\right)^{2/3},$$

$\upsilon(\chi) = \chi\upsilon_1 + (1 - \chi)\upsilon_2$ is the volume per molecule in a liquid solution of mole fraction $\chi$, with $\upsilon_i (i = 1, 2)$ the volume per molecule of component $i$; $f_i(\chi)$ is the activity coefficient of component $i$ in a solution of composition $\chi$; and $k$ is the Boltzmann constant.

Expression 2.173 for $W(\nu_1,\nu_2)$ involves some inconsistencies arising because of adsorption effects and composition dependence of the cluster surface tension that could not be avoided within the framework of the capillarity approximation [13,17,18,30]. One of the difficulties can be eliminated by replacing the approximation of uniform densities by a DFT method to calculate the height of the nucleation barrier [30] (see Section 2.6.4).

The second difficulty is related to the determination of the characteristics of the critical cluster (nucleus). The numbers of molecules in the nucleus, $\nu_{1c},\nu_{2c}$, are provided by the coordinates of the saddle point, that is, by the conditions [15–17]

$$\partial W/\partial \nu_i|_c = 0 \quad (i = 1, 2),\ \det W''|_c < 0, \tag{2.174}$$

where $W''$ is a $2 \times 2$ matrix of the second derivatives of $W$ with respect to $\nu_1,\nu_2$.

In the original BCNT, the Gibbs adsorption equation cannot be invoked, consequently, the extrema conditions $\partial W/\partial \nu_i = 0$ along with the Gibbs–Duhem relation lead to

$$\mu_i^l - \mu_i^\upsilon = -2\nu_i\sigma/R - A\partial\sigma/\partial\nu_i. \tag{2.175}$$

These equations should provide the characteristics of the binary nucleus, which should be equal to the coordinates of the saddle point of the free energy barrier $W$ (Equation 2.173). As illustrated in Ref. [18], this is indeed the case for nucleation in a water–methanol vapor mixture. However, according to the equilibrium requirement, if the cluster and the vapor mixture are in equilibrium (stable, metastable, or unstable), then $\mu_i^l(P^l) - \mu_i^\upsilon = 0\ (i = 1,2)$, or, equivalently,

$$\mu_i^l - \mu_i^\upsilon = -2\upsilon_i\sigma/R \quad (i = 1,2). \tag{2.176}$$

These equations, which determine the characteristics of the binary nucleus, should provide also the coordinates of the saddle point. As shown in Ref. [18] for nucleation in a water–methanol vapor mixture, these characteristics are not equal to the coordinates of the saddle point of the free energy surface, determined by Equation 2.173. Only if the composition dependence of the surface tension in the expression for $W$ is neglected, Equation 2.175 reduces to Equation 2.176. But no reasonable theoretical justification can be provided for neglecting $\partial\sigma/\partial\nu_i$ in Equation 2.175, although some nucleation experiments on water–alcohol mixtures [40] do indicate a flawed character of Equation 2.175.

### 2.6.2.2 A New Approach to Kinetics of Binary Nucleation

*2.6.2.2.1 Kinetic Equation* The goal of the kinetics of binary nucleation is to determine the temporal evolution of $g(\nu_1,\nu_2,t)$, the distribution function of binary clusters, formed during nucleation, with respect to their independent variables of state, $\nu_1,\nu_2$, at time $t$. The nucleation process quickly (compared to the duration of the whole stage) attains a steady state, so that only at the very beginning of nucleation the cluster distribution depends on time. This stage, known as transient nucleation [25,41], does not affect the nucleation rate, which is found by solving the kinetic equation of nucleation for a steady-state regime.

The kinetic equation of binary nucleation can be derived from the discrete conservation equation that governs the temporal evolution of the distribution function $g(\nu_1,\nu_2,t)$. In the simplest case, the metastability of the initial phase is assumed to be created instantaneously and to be constant during the whole nucleation process, the temperature of new-phase clusters is assumed to be constant and equal to that of the initial phase, and the exchange of matter between the new-phase clusters and the initial phase is assumed to occur via absorption and emission of monomers.

Considering only this simplest case (although kinetic theories for more complicated conditions are also available [42–44]), one can write the discrete conservation equation of binary nucleation as

$$\frac{\partial g(\nu_1,\nu_2,t)}{\partial t} = \Theta_1 + \Theta_2, \tag{2.177}$$

where

$$\Theta_1 = W_1^+(\nu_1 - 1, \nu_2)g(\nu_1 - 1, \nu_2, t) - W_1^+(\nu_1, \nu_2)g(\nu_1, \nu_2, t)$$
$$+ W_1^-(\nu_1 + 1, \nu_2)g(\nu_1 + 1, \nu_2, t) \tag{2.178}$$
$$- W_1^-(\nu_1, \nu_2)g(\nu_1, \nu_2, t),$$

$$\Theta_2 = W_2^+(\nu_1, \nu_2 - 1)g(\nu_1, \nu_2 - 1, t) - W_2^+(\nu_1, \nu_2)g(\nu_1, \nu_2, t)$$
$$+ W_2^-(\nu_1, \nu_2 + 1)g(\nu_1, \nu_2 + 1, t) \tag{2.179}$$
$$- W_2^-(\nu_1, \nu_2)g(\nu_1, \nu_2, t),$$

where $W_i^-$ $(\nu_1,\nu_2)$ and $W_i^+$ $(\nu_1,\nu_2)$ $(i = 1, 2)$ are the numbers of molecules of component $i$ that a cluster $\nu_1,\nu_2$ emits and absorbs, respectively, per unit time.

Assuming $\nu_1 \geq 1$, $\nu_2 \geq 1$, one can rewrite Equations 2.178 and 2.179 in a continuous form by using Taylor series expansions on the right-hand side thereof and retaining only the lowest order terms. The resulting partial differential equation has the Fokker–Planck form

$$\frac{\partial g(\nu_1, \nu_2, t)}{\partial t} = -\frac{\partial}{\partial \nu_1} J_1(\nu_1, \nu_2, t) - \frac{\partial}{\partial \nu_2} J_2(\nu_1, \nu_2, t) \tag{2.180}$$

with

$$J_i(\nu_1, \nu_2, t) = -A_i(\nu_1, \nu_2)g(\nu_1, \nu_2, t)$$
$$- \frac{\partial}{\partial \nu_i}(B_i(\nu_1, \nu_2)g(\nu_1, \nu_2, t)). \tag{2.181}$$

Here, the coefficients $A_i(\nu_1,\nu_2)$ and $B_i(\nu_1,\nu_2)$ are determined by the emission and absorption rates $W_i^-$ and $W_i^+$ of the liquid cluster.

In the vicinity of the saddle point the kinetic equation (Equation 2.180) can be further simplified by expansions in Taylor series and its steady-state solution subject to the conventional boundary conditions

$$\frac{g(\nu_1, \nu_2)}{g_e(\nu_1, \nu_2)} \to 1 \quad (\nu_1, \nu_2 \to 0),$$
$$\frac{g(\nu_1, \nu_2)}{g_e(\nu_1, \nu_2)} \to 0 \quad (\nu_1, \nu_2 \to \infty), \tag{2.182}$$

where $g_e(\nu_1,\nu_2)$ is the equilibrium distribution, provides the steady-state nucleation rate.

To solve Equation 2.180 subject to boundary conditions (Equation 2.182), it is necessary to know its coefficients $A_i(\nu_1,\nu_2)$,$B_i(\nu_1,\nu_2)$ $(i = 1, 2)$, i.e., the emission and absorption rates $W_i^-(\nu_1,\nu_2)$ and $W_i^+(\nu_1,\nu_2)$. The absorption rate $W_i^+(\nu_1,\nu_2)$ is calculated by using the kinetic theory of gases, but for determining the emission rate $W_i^-$ $(\nu_1,\nu_2)$, BCNT relies on the capillarity approximation and makes use of the concept of surface tension for the new-phase cluster: the emission rate $W_i^-$ $(\nu_1,\nu_2)$ is assumed to be independent of the state of the actual mother phase and be equal to the absorption rate that the cluster would have if it were in equilibrium with an imaginary mother phase of special metastability. Thus, the determination of $W_i^-$ $(\nu_1,\nu_2)$ is based (via the Gibbs–Kelvin relations) on the surface tension of the cluster. Because the concept of surface tension may not be adequate for too small clusters (such as those of interest in nucleation) [45,46], not to mention the assumption (of the capillarity approximation) that it is equal to the surface tension of a planar interface, its use constitutes a major drawback of the kinetics of nucleation in the BCNT.

In order to avoid the use of macroscopic thermodynamics and of the concept of surface tension of a cluster, an alternative approach was developed by Ruckenstein and coworkers to construct a kinetic theory of nucleation [19–26]. The key idea of that approach consists of calculating the emission and absorption rates $W_i^-(\nu_1,\nu_2)$ and $W_i^+(\nu_1,\nu_2)$ independently of each other by solving the Fokker–Planck or Smoluchowski equation governing the motion of a single molecule in a potential well around the cluster to obtain the emission rate.

Before presenting the extension of the method of Ruckenstein and coworkers to binary nucleation, we note that in the vicinity of the saddle point Equation 2.180 can be rewritten as

$$\frac{\partial g(\nu_1,\nu_2,t)}{\partial t}=\left\{W_{1c}^+\frac{\partial}{\partial \nu_1}\left[\frac{\partial}{\partial \nu_1}+\frac{\partial G}{\partial \nu_1}\right]+W_{2c}^+\frac{\partial}{\partial \nu_2}\left[\frac{\partial}{\partial \nu_2}+\frac{\partial G}{\partial \nu_2}\right]\right\}, \times g(\nu_1,\nu_2,t). \tag{2.183}$$

(see Appendix 2B), where $G = -\ln g_e(\nu_1,\nu_2)$. The essence of the method of Ruckenstein and coworkers as an alternative to the BCNT consists of constructing the equilibrium distribution of clusters, $g_e(\nu_1,\nu_2)$, without applying the classical thermodynamics to a single tiny cluster and (unlike the BCNT) *without* using the concept of surface tension.

*2.6.2.2.2 First Passage Time Analysis* Consider a spherical binary cluster immersed in a binary vapor mixture. A molecule of component $i$ ($i$ = 1, 2) located in the surface layer of the cluster performs thermal chaotic motion in a spherically symmetric potential well $\phi_i(r)$ resulting from the interactions of the molecule with the rest of the cluster. Assuming pairwise additivity of the intermolecular interactions, the potential $\phi_i(r)$ is given by

$$\phi_i(r)=\sum_j\int_V d\mathbf{r}'\rho_j(r')\phi_{ij}\left(|\mathbf{r}'-\mathbf{r}|\right). \tag{2.184}$$

Here, $\mathbf{r}$ is the coordinate of the surface molecule $i$, $\rho_j(r')$ ($j$ = 1, 2) is the number density of molecules of component $j$ at point $\mathbf{r}'$ (spherical symmetry is assumed), and $\phi_{ij}(|\mathbf{r}'-\mathbf{r}|)$ is the interaction potential between two molecules of components $i$ and $j$ at points $\mathbf{r}$ and $\mathbf{r}'$, respectively. The integration in Equation 2.184 is over the whole volume of the system, but the vapor phase contribution can be assumed to be negligible. The molecule is considered as belonging to the cluster as long as it remains in the potential well and as dissociated from the cluster when it passes over the outer boundary of the well (reasonably defined). The rate of emission, $W_i^-$, is determined by the mean passage time for this threshold distance of the outer boundary (henceforth all distances are measured from the center of the cluster, unless otherwise specified). In this way the kinetic theory of nucleation of Ruckenstein and coworkers avoids the use of macroscopic thermodynamics.

The mean first passage time of a molecule dissociating from the cluster is calculated on the basis of a kinetic equation governing the chaotic motion of the molecule escaping from the potential well generated around the cluster. The chaotic motion of a molecule is governed by the Fokker–Planck equation for the single-molecule distribution function with respect to its coordinates and momenta, i.e., in the phase space [47–49]. Prior to a dissociation event, the evolution of a molecule (belonging to the cluster) occurs in a liquidlike cluster surface layer, where the relaxation time for its velocity distribution function is extremely short and negligible compared to the characteristic time scale of the dissociation process. Under these conditions, the Fokker–Planck equation reduces to the Smoluchowski equation, which involves diffusion in an external field [48,49]. In the case of spherical symmetry it can be written in the form [19–21]

$$\frac{\partial p_i(r,t|r_0)}{\partial t}=D_i r^{-2}\frac{\partial}{\partial r}\left(r^2 e^{-\Phi_i(r)}\frac{\partial}{\partial r}e^{\Phi_i(r)}p_i(r,t|r_0)\right), \tag{2.185}$$

where $p_i(r,t|r_0)$ is the probability of observing a molecule between $r$ and $r + dr$ at time $t$ given that initially it was at a radial distance $r_0$, $D_i$ is the diffusion coefficient, and $\Phi_i(r) = \varphi_i(r)/kT$.

The mean passage time depends on the initial position (distance from the center of the cluster) $r_0$ of the molecule. It is convenient to use the backward Smoluchowski equation [19–21], which expresses the dependence of the transition probability $p(r,t|r_0)$ on $r_0$,

$$\frac{\partial p_i(r,t|r_0)}{\partial t} = D_i r_0^{-2} e^{\Phi_i(r_0)} \frac{\partial}{\partial r_0}\left( r_0^2 e^{-\Phi_i(r_0)} \frac{\partial}{\partial r_0} p_i(r,t|r_0) \right). \tag{2.186}$$

The probability that a molecule $i$, initially at a distance $r_0$ within the surface layer, will remain in this region after time $t$ is given by the so-called survival probability [19–21],

$$Q_i(t|r_0) = \int_R^{R+\lambda_i^w} dr r^2 p_i(r,t|r_0) \quad \left(R < r_0 < R+\lambda_i^w\right), \tag{2.187}$$

where $R$ is the radius of the cluster and $\lambda_i^w$ is the width of the potential well defined such that a molecule $i$ can be assumed to have dissociated from the cluster at the distance $R+\lambda_i^w$ from its center. The choice of $\lambda_i^w$ is somewhat arbitrary, depending on the definition of a cluster, but for a reasonable choice of $\lambda_i^w$ the theory is insensitive thereto. Having chosen $\lambda_i^w$, the probability for the dissociation time to be between 0 and $t$ is equal to $1-Q_i(t|r_0)$, and the probability density for the dissociation time to be between 0 and $t$ is given by $-\partial Q_i/\partial t$. The first passage time is provided by [19–21]

$$\tau_i(r_0) = -\int_0^\infty t \frac{\partial Q_i(t|r_0)}{\partial t} dt = \int_0^\infty Q_i(t|r_0) dt. \tag{2.188}$$

The equation for the first passage time is obtained by integrating the backward Smoluchowski equation (Equation 2.186) with respect to $r$ and $t$ over the entire range and using the boundary conditions $Q_i(0|r_0) = 1$ and $Q_i(t|r_0) \to 0$ as $t \to \infty$ for any $r_0$. This yields [19–21]

$$-D_i r_0^{-2} e^{\Phi(r_0)} \frac{\partial}{\partial r_0}\left( r_0^2 e^{-\Phi_i(r_0)} \frac{\partial}{\partial r_0} \tau_i(r_0) \right) = 1. \tag{2.189}$$

One can solve Equation 2.189 by assuming a reflecting boundary condition $d\tau_i/dr_0 = 0$ at $r_0 = R$ and a complete absorption boundary condition $\tau_i(r_0) = 0$ at $r_0 = R + \lambda_i^w$. One thus obtains for the first passage time,

$$\tau_i(r_0) = \frac{1}{D_i} \int_{r_0}^{R+\lambda_i^w} dy y^{-2} e^{\Phi_i(y)} \int_R^y dx x^2 e^{-\Phi_i(x)}. \tag{2.190}$$

The average dissociation time, $\bar{\tau}_i$, or the mean first passage time, is obtained by averaging $\tau_i(r_0)$ with the Boltzmann factor over all possible initial positions $r_0$,

$$\bar{\tau}_i = \frac{1}{Z} \int_R^{R+\lambda_i^w} dr_0 r_0^2 e^{-\Phi_i(r_0)} \tau_i(r_0), \tag{2.191}$$

$$Z = \int_R^{R+\lambda_i^w} dr_0 r_0^2 e^{-\Phi_i(r_0)}. \tag{2.192}$$

Clearly, $\bar{\tau}_i$ is a function of the cluster size and composition: $\bar{\tau}_j \equiv \bar{\tau}_i(R,\chi) \equiv \bar{\tau}_i(\nu_1,\nu_2)$. Defining the quantity $\omega_i \equiv \omega_i(R,\chi) \equiv \omega_i(\nu_1,\nu_2)$ as

$$\omega_i = 1/D_i\bar{\tau}_i, \tag{2.193}$$

the rate of dissociation $W_i^-$ can be expressed as

$$W_i^- = \frac{N_i^{\mathrm{w}}}{\bar{\tau}_i} = N_i^{\mathrm{w}} D_i \omega_i, \tag{2.194}$$

where $N_i^{\mathrm{w}}$ denotes the number of molecules in the well. The quantity $\omega_i$ has a simple physical meaning, representing the reciprocal of the square of the distance that a molecule of species $i$ covers in the time $\omega_i$. Clearly, $W_i^-$ is also a function of the cluster size and composition, $R$ and $\chi$ (or, equivalently, $\nu_1$ and $\nu_2$).

The absorption rate $W_i^+$ is calculated by using the kinetic theory of gases (see, e.g., Ref. [50]),

$$W_i^+ = \pi R^2 \rho_i^{\upsilon} \bar{\upsilon}_i, \tag{2.195}$$

where $\rho_i^{\upsilon}$ is the number density of molecules of component $i$ in the vapor mixture, $\bar{\upsilon}_i$ is their mean thermal velocity, and the sticking coefficient is assumed to be unity.

At a given metastability of the vapor mixture, for either component the rates of emission and absorption are equal to each other for the critical cluster, the size and composition of which will be denoted by $R_{\mathrm{c}}$ and $\chi_{\mathrm{c}}$. The equality $W_i^-(R_{\mathrm{c}},\chi_{\mathrm{c}}) = W_i^+(R_{\mathrm{c}},\chi_{\mathrm{c}})$ combined with Equations 2.194 and 2.195 allows one to obtain the following expressions for the absorption rate $W_i^+(R,\chi)$ and for the saturation ratio $a_i$:

$$W_i^+(R,\chi) = N_i^{\mathrm{w}} D_i \frac{R^2}{R_{\mathrm{c}}^2}\omega_i(R_{\mathrm{c}},\chi_{\mathrm{c}}) \quad (i=1,2), \tag{2.196}$$

$$\alpha_i = \omega_i(R_{\mathrm{c}}, \chi_{\mathrm{c}})/\omega_i(\infty, \chi_{\mathrm{c}}) \quad (i = 1, 2). \tag{2.197}$$

Equation 2.197 was obtained by assuming that the ratio $N_i^{\mathrm{w}}/4\pi R^2$ (i.e., the number of molecules in the well per unit interfacial area) does not depend on the cluster radius. In principle, this approximation can be avoided because DFT allows one to find the dependence of $N_i^{\mathrm{w}}/4\pi R^2$ on $R$.

For $R_{\mathrm{c}}/\eta \geq 1$, where $\eta$ is the largest molecular diameter, the results of this theory are expected to coincide with those of the BCNT. In particular, the classical Kelvin relations between the saturation ratios and the critical radius are recovered and can be employed to calculate the macroscopic surface tension [19–24].

The quantity $N_i^{\mathrm{w}}$ in expression 2.193 for $W_i^-$ depends on the cluster definition, whereas the quantity $D_i$ is not known at all. However, owing to Equation 2.196, both $N_i^{\mathrm{w}}$ and $D_i$ are absent in the final expression for the nucleation rate, which depends only on the ratios $W_i^-/W_i^+$ $(i = 1, 2)$ via the quantity $G = -\ln g_{\mathrm{e}}(\nu_1,\nu_2)$ (see Equation 2.183). According to the principle of detailed balance, one can obtain the following equivalent expressions (see Appendix 2C) for the equilibrium distribution $g_{\mathrm{e}}(\nu_1,\nu_2)$:

$$g_{\mathrm{e}}(\nu_1,\nu_2) = \rho_1^{\upsilon} \frac{\omega_{1c}^{\nu_1-1}\omega_{2c}^{\nu_2}}{R_{\mathrm{c}}^2(\nu_1+\nu_2-1)} \frac{R_1^2}{R^2(\nu_1,\nu_2)} \times \prod_{i=1}^{\nu_1-1} \frac{R^2(\nu_1-i+1,\nu_2)}{\omega_1(\nu_1-i+1,\nu_2)} \times \prod_{j=1}^{\nu_2} \frac{R^2(1,\nu_2-j+1)}{\omega_2(1,\nu_2-j+1)}, \tag{2.198}$$

$$g_e(\nu_1,\nu_2)=\rho_2^{\upsilon}\frac{\omega_{1c}^{\nu_1}\omega_{2c}^{\nu_2-1}}{R_c^2(\nu_1+\nu_2-1)}\frac{R_2^2}{R^2(\nu_1,\nu_2)}$$

$$\times\prod_{j=1}^{\nu_2-1}\frac{R^2(\nu_1,\nu_2-j+1)}{\omega_2(\nu_1,\nu_2-j+1)} \tag{2.199}$$

$$\times\prod_{i=1}^{\nu_1}\frac{R^2(\nu_1-i+1,1)}{\omega_1(\nu_1-i+1,1)},$$

where $R(\nu_1,\nu_2)$ is the radius of the cluster $\nu_1,\nu_2$, and $R_1 = R(1,0)$ and $R_2 = R(0,1)$. Thus, knowing $\omega_1$, $\omega_2$, and $R$ as functions of $\nu_1$ and $\nu_2$, one can find $G \equiv G(\nu_1,\nu_2) = -\ln g_e(\nu_1,\nu_2)$ by using either Equation 2.198 or Equation 2.199. In the present theory, the quantity $G$ plays a role similar to the free energy of cluster formation in the BCNT. The crucial advantage of the present theory is that it does not employ the concept of surface tension for tiny clusters.

The function $G(\nu_1,\nu_2)$ represents a three-dimensional surface that can be expected to have a shape similar to the free energy surface in the BCNT, with a saddle point at $\nu_{1c}$, $\nu_{2c}$ and paths of steepest descent and steepest ascent [33–39]. In order to find the steady-state nucleation rate, Equation 2.183 has to be solved in the vicinity of the saddle point. In this vicinity the function $G(\nu_1,\nu_2)$ (as the free energy of cluster formation in the BCNT) is accurately represented by its bilinear form, so that one can use the method of complete separation of variables [33,34,39,44] to solve Equation 2.183 subject to the boundary conditions (Equation 2.182). In our case, the key idea of the method consists of finding such variables of state of the binary cluster that would simultaneously diagonalize the matrix of the second derivatives of the function $G(\nu_1,\nu_2)$ at the saddle point and the matrix of the diffusion coefficients (multiplying the second-order derivatives) in Equation 2.183. At the first step, the new variables $x$ (not to be confused with the mole fraction) and $y$ are introduced by the rotation transformation of the variables $\nu_1,\nu_2$, whereas at the second step the rotated variables $x,y$ are subjected to the Lorentz transformation that has the property to leave the difference $x^2 - y^2$ invariant. As a result (see Appendix 2D), the two-dimensional kinetic equation (Equation 2.183) is reduced to a one-dimensional one with a well-known steady-state solution. The steady-state rate of binary nucleation is given by the expression

$$J_s = ACe^{-G_c}, \tag{2.200}$$

where

$$A = (a/2\varepsilon^2)\{\varepsilon^2 - 1 - p + [(\varepsilon^2 + 1 + p)^2 - 4\varepsilon^2]^{1/2}\}, \tag{2.201}$$

$$a = W_1^+c_{11}^2 + W_2^+c_{12}^2, \tag{2.202}$$

$$\varepsilon = -\left(W_1^+c_{11}^2 + W_2^+c_{12}^2\right)/\left(W_1^+c_{11}c_{21} + W_2^+c_{12}c_{22}\right),$$

$$p = W_1^+W_2^+(c_{11}c_{22} - c_{12}c_{21})^2/\left(W_1^+c_{11}c_{21} + W_2^+c_{12}c_{22}\right)^2, \tag{2.203}$$

$$c_{11} = \sqrt{\frac{1}{2}|G''_{xx}|}\cos\beta,\ c_{12} = \sqrt{\frac{1}{2}|G''_{xx}|}\sin\beta,$$

$$c_{21} = -\sqrt{\frac{1}{2}|G''_{yy}|}\sin\beta,\ c_{22} = \sqrt{\frac{1}{2}|G''_{yy}|}\cos\beta, \tag{2.204}$$

$$|G''_{xx}| = -\left(G''_{11}\cos^2\beta + G''_{12}\sin 2\beta + G''_{22}\sin^2\beta\right), \tag{2.205}$$

$$|G''_{yy}| = G''_{11}\sin^2\beta - G''_{12}\sin 2\beta + G''_{22}\cos^2\beta, \tag{2.206}$$

$$G''_{ij} = \partial^2 G/\partial v_i \partial v_j \ (i,j=1,2),\ C = |c_{11}c_{22} - c_{12}c_{21}|^{-1}, \tag{2.207}$$

with the angle $0 \le \beta \le \pi/2$ between the path of the steepest descent and the $v_1$ axis that is constant in the vicinity of the saddle point and is provided by the solution of the equation $\tan\, 2\beta = 2G''_{12}/\left(G''_{11} - G''_{22}\right)$. Note that the expression for $J$, Equation 2.200, has a well-known form for the binary nucleation rate [37,51], $J = KZg_{ec}$, where $g_{ec}$ is the equilibrium distribution at the saddle point, $Z$ is the "Zeldovich factor," and the factor $K$ is either 1 (i.e., included in $Z$ as in Ref. [51]) or represents an "average" growth rate [37] (a combination of $W_1^+$ and $W_2^+$).

*2.6.2.2.3 Determination of the Function $\omega_i(v_1,v_2)$ by DFT Methods* As clear from the foregoing, the knowledge of the functions $R(v_1,v_2)$ and $\omega_i(v_1,v_2)$ for a *given* metastability of the vapor mixture is essential in the new (based on the first passage time analysis) kinetic theory of binary nucleation. If the densities of both components are assumed to be uniform in the cluster, these functions can be easily found. In this case, $R(v_1,v_2) = [3(v_1\upsilon_1 + v_2\upsilon_2)/4\pi]^{1/3}$, whereas for any $v_1,v_2$ (and fixed $a_1,a_2$) $\omega_i(v_1,v_2)$ is obtained by using Equations 2.184 and 2.190 through 2.193.

As already mentioned, the uniform density approximation may not be adequate for binary nucleation involving surface active components. In this case, the density profiles in the cluster can be obtained by using DFT methods. This allows one to calculate the potential wells $\Phi_1(r)$ and $\Phi_2(r)$ more accurately. However, there is a difficulty in using DFT methods in the new kinetic theory. Indeed, to calculate the function $G(v_1,v_2)$ for given supersaturations $a_1$ and $a_2$, it is necessary to know the functions $\omega_1(v_1,v_2)$ and $\omega_2(v_1,v_2)$ for all $v_1$ and $v_2$ for the same $a_1$ and $a_2$. These functions cannot be obtained directly by DFT methods, because for given $a_1$ and $a_2$ (or chemical potentials of both components) DFT allows one to determine the density profiles in the system only for a unique cluster size and composition (i.e., for the critical cluster that is in unstable equilibrium with the vapor mixture). Thus, for given $a_1$ and $a_2$, the DFT methods can provide the mean first passage time $\bar{\tau}_i$ ($i$ = 1, 2) only for the nucleus (critical cluster), and hence, only one value, $\omega_{ic} = \omega_i(v_{1c},v_{2c})$ for the function $\omega_i(v_1,v_2)$ ($i$ = 1, 2) can be obtained. To construct the whole function $\omega_i(v_1,v_2)$, the following procedure can be used [27].

Let us denote by $R'$ and $\chi'$ the radius and composition of the critical cluster (nucleus) corresponding to the arbitrary supersaturations $a'_1,a'_2\left(a'_i \neq a_i\right)$($i$ = 1, 2) as outlined above, $\omega_i\left(R',\chi' \mid a'_1,a'_2\right)$ can be determined by using DFT in the first passage time analysis based theory (for clarity, $a'_1,a'_2$ are also listed as arguments of $\omega_i$). Since $a'_i = \omega_i\left(R',\chi' \mid a'_1,a'_2\right)/\omega_i(\infty,\chi')$ (Equation 2.197) and we have

$$\partial\omega_i\left(R',\chi'|a'_1,a'_2\right)/\partial a'_i = \omega_i(\infty,\chi'), \tag{2.208}$$

so that $\partial^k\omega_i\left(R',\chi'|a'_1,a'_2\right)/\partial a_i'^k = 0\ (k \ge 2)$ (note that Equation 2.208 is approximate because it is derived from the approximate Equation 2.197). Therefore, the Taylor expansion $\omega_i\left(R',\chi'|a'_1,a'_2\right) = \omega_i(R',\chi'|a_1,a_2) + \omega_i(\infty,\chi')\left(a'_i - a_i\right)$ is exact even when $a'_i - a_i$ is not small. Thus, we have

$$\omega_i(R',\chi'|a_1,a_2) = \omega_i\left(R',\chi'|a'_1,a'_2\right)\left(2a'_i - a_i\right)/a'_i. \tag{2.209}$$

Carrying out DFT calculations and using Equations 2.184 and 2.190 through 2.193 for a series of pairs $a'_1,a'_2$, one can find the values of $\omega_i\left(R',\chi' \mid a'_1,a'_2\right)$ ($i$ = 1, 2) for a set of pairs $\left\{a'_1,a'_2\right\}$ corresponding to a set of cluster radii and compositions $\{R', \chi'\}$. Then, using Equation 2.209, one can determine $\omega_i(R', \chi'|a_1, a_2)$ for the same set $\{R', \chi'\}$, which allows one to construct the whole function $\{R', \chi'\}$ for the selected $a_1$, $a_2$. Thus, $G(v_1,v_2)$ becomes completely determined and can be further used in Equations 2.200 through 2.207 for calculating the nucleation rate.

### 2.6.3 Model Binary System with a Surface Active Component

As an illustration, we will apply the above theory to binary nucleation in a model binary mixture containing an amphiphilic component. Both the phase behavior of and binary nucleation in such model systems were extensively studied by applying the density functional theory and the interaction site model

[28–30,52–54]. In the latter approach, pair potentials are written for each atom or atomic group separately, which makes it possible to treat the amphiphilic interactions without having to consider orientational distributions explicitly. The elongated shape of the real amphiphiles is thus incorporated in the system in a natural manner.

### 2.6.3.1 Density Functional Formalism

Consider a binary mixture of single-hard-sphere molecules and molecules of two bonded hard spheres. Let us denote the density distributions of monomers and the sites of diatomic chains by $\rho_0(\mathbf{r})$ and $\rho_j(\mathbf{r})$ ($j = 1, 2$), respectively. The total density of the hard-sphere components is $\rho(\mathbf{r}) = \Sigma_{i=0}^{2}\rho_i(\mathbf{r})$, and the packing fraction $\xi(\mathbf{r}) = \Sigma_{i=0}^{2}\upsilon_i\rho_i(\mathbf{r})$, where $\upsilon_i = \pi\eta_i^2/6$ is the molecular volume of the hard-sphere component $i$ with the diameter $\eta_i$. The Helmholtz free energy of this system as a functional of $\rho_0(\mathbf{r})$, $\rho_1(\mathbf{r})$, and $\rho_2(\mathbf{r})$ can be written as [28–30,52–54]

$$
\begin{aligned}
F[\rho_0(\mathbf{r}),\{\rho_j(\mathbf{r})\}] = {} & kT\int d\mathbf{r}\rho_0(\mathbf{r})(\ln\rho_0(\mathbf{r})-1) \\
& +kT\sum_{j=1}^{2}\int d\mathbf{r}\rho_j(\mathbf{r})\ln\gamma_j(\mathbf{r}) \\
& +kT\int d\mathbf{r}\psi(\xi(\mathbf{r}))\rho(\mathbf{r}) \\
& -kT\int\int d\mathbf{r}\,d\mathbf{r}'\gamma_1(\mathbf{r})\gamma_2(\mathbf{r}')s^{(2)}(\mathbf{r}'-\mathbf{r}) \\
& +\frac{1}{2}\sum_{i,j=0}^{2}\int\int d\mathbf{r}\,d\mathbf{r}'\phi_{ij}^{a}(|\mathbf{r}-\mathbf{r}'|)\rho_i(\mathbf{r})\rho_j(\mathbf{r}'),
\end{aligned}
\tag{2.210}
$$

where $\beta = 1/kT$ and $\gamma_j$ ($j = 1, 2$) are auxiliary functions that are determined by minimizing the free energy functional.

The first two terms on the right-hand side of Equation 2.210 are the ideal gas contributions to the free energy, whereas the third term represents the short-range repulsive interactions between particles. For the latter contribution, the local density approximation is used, which is reflected in the argument of $\Psi(\xi)$, the excess (compared to an ideal system) free energy of the hard-sphere system with a packing fraction $\xi$. The bonding between atom sites (causing a decrease in entropy) is taken into account by the fourth term, where $s^{(2)}(\mathbf{r}' - \mathbf{r})$ is the intramolecular correlation function. The last term represents the contribution to the free energy due to the attractive part $\phi_{ij}^{a}(|\mathbf{r}-\mathbf{r}'|)$ of the interaction potentials in the mean-field approximation.

For a system with constant chemical potentials, temperature, and volume (a grand canonical ensemble), the equilibrium properties of the fluid are obtained by minimizing the grand potential

$$
\Omega[\rho_0(\mathbf{r}),\{\rho_j(\mathbf{r})\}] = F[\rho_0(\mathbf{r}),\{\rho_j(\mathbf{r})\}] - \sum_{i=0}^{2}\mu_i\int d\mathbf{r}\rho_i(\mathbf{r})
\tag{2.211}
$$

(where $\mu_i$ is the chemical potential) with respect to the densities $\rho_i(\mathbf{r})$ ($i = 0, 1, 2$) and the auxiliary functions $\gamma_j(\mathbf{r})$ ($j = 1, 2$). The Euler–Lagrange equations following from the extrema conditions

$$
\frac{\partial\Omega[\rho_0(\mathbf{r}),\{\rho_j(\mathbf{r})\}]}{\partial\rho_i(\mathbf{r})} = 0 \quad (i = 0,1,2)
\tag{2.212}
$$

lead to

$$
\rho_0(\mathbf{r}) = e^{[\mu_0-U_0(\mathbf{r})]/kT}, \quad \gamma_j(\mathbf{r}) = e^{[\mu_j-U_j(\mathbf{r})]/kT} \quad (j=1,2),
\tag{2.213}
$$

where

$$U_i(r) = kT\Psi(\xi(\mathbf{r})) + kT\Psi'(\xi(\mathbf{r}))\upsilon_i\rho(r) + \sum_{j=0}^{2}\int d\mathbf{r}'\phi_{ij}^{a}(|\mathbf{r}-\mathbf{r}'|)\rho_j(\mathbf{r}')\,(i=0,1,2), \tag{2.214}$$

whereas the extrema conditions

$$\frac{\delta\Omega[\rho_0(\mathbf{r}),\{\rho_j(\mathbf{r})\}]}{\delta\gamma_j(\mathbf{r})} = 0 \quad (j=1,2) \tag{2.215}$$

result in

$$\rho_j(\mathbf{r}) = \gamma_j(\mathbf{r})\int d\mathbf{r}'\gamma_{j'}(\mathbf{r}')s^{(2)}(\mathbf{r}'-\mathbf{r})\,(j' \neq j = 1,2). \tag{2.216}$$

The equilibrium density distributions in an inhomogeneous fluid can be obtained by numerically solving Equations 2.213, 2.214, and 2.216 in the geometry of interest.

For homogeneous bulk phases Equations 2.213, 2.214, and 2.216 can be considerably simplified, because the monomer densities and the site densities are constant, $\rho_0(\mathbf{r}) = \rho_0$, $\rho_1 = \rho_2 = \rho_d$, so that Equation 2.216 reduces to $\gamma_1\gamma_2 = \rho_d$. Taking into account the condition of chemical equilibrium [28–30,52–54] $\mu_d = \mu_1 + \mu_2$, one obtains from Equation 2.213

$$\mu_0 = kT\ln\rho_0 + U_0,\ \mu_d = kT\ln\rho_d + U_1 + U_2. \tag{2.217}$$

The densities of equilibrium coexisting gas and liquid phases are found by equating the pressures $P = -\Omega[\rho_0,\rho_d]/V$ and chemical potentials of these phases at constant temperature. The phase behavior of this system was studied by Napari et al. [28] and Talanquer and Oxtoby [29].

For a system with constant numbers of molecules, temperature, and volume (a canonical ensemble), the equilibrium properties of the fluid can be obtained by minimizing the Helmholtz free energy $F[\rho_0(\mathbf{r}),\{\rho_j(\mathbf{r})\}]$ given by Equation 2.210. The minimization with respect to the densities must be carried out subject to conditions of constant numbers of monomers and dimer sites,

$$N_i = \int \rho_i(\mathbf{r})\,d\mathbf{r}(i=0,1,2). \tag{2.218}$$

In this case, the Euler–Lagrange equations that follow from the extrema conditions (Equation 2.212) and (Equation 2.215) lead to

$$\gamma_j(\mathbf{r}) = \frac{1}{\lambda_j}e^{-U_j(\mathbf{r})/kT}\ (j=1,2), \tag{2.219}$$

$$\rho_j(\mathbf{r}) = \gamma_j(\mathbf{r})\int d\mathbf{r}'\gamma_{j'}(\mathbf{r}')s^{(2)}(\mathbf{r}'-\mathbf{r})\quad (j' \neq j = 1,2), \tag{2.220}$$

$$\ln\rho_0(\mathbf{r}) + \Psi(\xi(\mathbf{r})) + \Psi'(\xi(\mathbf{r}))\upsilon_i\rho(\mathbf{r}) = \ln\left(\frac{1}{N_0}\int d\mathbf{r}e^{-U_0(\mathbf{r})/kT}\right)\left(-\frac{1}{kT}\right)\sum_{j=0}^{2}\int d\mathbf{r}'\phi_{0j}^{a}(|\mathbf{r}-\mathbf{r}'|)\rho_j(\mathbf{r}'), \tag{2.221}$$

where

$$\lambda_1 = \sqrt{I_1 I_2 / N_2 N_1}, \; \lambda_2 = \sqrt{I_1 N_2 / I_2 N_1}, \tag{2.222}$$

$$\begin{aligned} I_1 &= \int d\mathbf{r} e^{-U_1(\mathbf{r})/kT} \int d\mathbf{r}' e^{-U_2(\mathbf{r}')/kT} s^{(2)}(\mathbf{r}' - \mathbf{r}), \\ I_2 &= \int d\mathbf{r} \gamma_2(\mathbf{r}) \int d\mathbf{r}' \gamma_2(\mathbf{r}') e^{-[U_1(\mathbf{r}') - U_2(\mathbf{r}')]/kT} s^{(2)}(\mathbf{r}' - \mathbf{r}). \end{aligned} \tag{2.223}$$

If the nucleation of spherical clusters occurs in an open system (grand canonical ensemble) the set of Equations 2.213, 2.214, and 2.216 must be solved in spherical geometry, whereas for nucleation in a closed system (canonical ensemble) it has to be the set of Equations 2.219 through 2.221. In both cases, the solution is carried out numerically by using an iteration procedure. Solving Equations 2.213, 2.214, and 2.216 provides the density profiles in the system with a critical cluster (nucleus) in *unstable* equilibrium with the vapor mixture. In this case, the iteration procedure is also unstable and hence complicated [55]. To avoid this difficulty, it is more convenient to study nucleation in a closed system and find the density profiles by solving Equations 2.219 through 2.221. The solution would then correspond to a truly equilibrium cluster surrounded by a metastable vapor mixture. Considering this system as an open system (with chemical potentials corresponding thereto) and solving Equations 2.213, 2.214, and 2.216 exactly the same density profiles would be found except for the fact that they would be unstable.

If the spheres in the model are assumed to have the same diameter, $\eta = \eta_0 = \eta_1 = \eta_2$, the intramolecular correlation function for dimers can be expressed as [52]

$$s^{(2)}(r) = \frac{1}{4\pi\eta^2} \delta(r - \eta), \tag{2.224}$$

where $\delta(x)$ is the Dirac delta function. For the excess free energy per particle of hard-sphere system one can use the approximation of Mansoori et al. [56], which for equal-sized hard spheres reduces to

$$\Psi(\xi) = \frac{\xi(4 - 3\xi)}{(1 - \xi)^2}. \tag{2.225}$$

The interactions between a pair of particles of components $i$ and $j$ can be described, e.g., by the Lennard–Jones (LJ) potentials

$$\phi_{ij}(r) = 4\epsilon_{ij} \left[ \left( \frac{\eta_{ij}}{r} \right)^{12} - \left( \frac{\eta_{ij}}{r} \right)^{6} \right] \quad (i, j = 0, 1, 2), \tag{2.226}$$

where $\epsilon_{ij}$ and $\eta_{ij}$ are the energy and length parameters, respectively, with the mixing rules

$$\epsilon_{ij} = (1 - k_{ij}) \sqrt{\epsilon_{ii} \epsilon_{jj}}, \; \eta_{ij} = \frac{1}{2}(\eta_i + \eta_j). \tag{2.227}$$

The negative values of the parameter $k_{ij}$ mean positive deviations from the geometric mean rule of the energy parameters, whereas the positive values decrease the pair interactions between the molecular species $i$ and $j$. The Weeks–Chandler–Andersen (WCA) [57] separation will be used to divide the intermolecular potential into a reference (repulsive) and perturbation (attractive) parts.

### 2.6.4 Numerical Evaluations

As a numerical illustration of the above theory, we considered binary nucleation in a monomer(0)–dimer($d$) fluid system with $\epsilon_{00}/kT = 0.7$, $\epsilon_{11} = \epsilon_{22} = 0.6\epsilon$ ($\epsilon = \epsilon_{00}$). Nucleation in such systems was extensively studied with

DFT methods by Napari et al. [28] and Laaksonen and Napari [30]. Here, we considered only the case $k_{12} = 0$, $k_{01} = -0.4$, and $k_{02} = 0.5$, which corresponds to an enhanced surface activity of the dimer component (modeling an amphiphilic molecule). The dimer site 1 ("monophilic" site) is attracted to monomers much stronger than the dimer site 2 ("monophobic" site), thus inducing a preferential orientation of the dimers at the gas–liquid interface. In this respect such a model system simulates, for example, a typical binary mixture of water and alcohol molecules. The results of numerical calculations are presented in Figures 2.18 through 2.20.

Figure 2.18 shows the density profiles of monomers (solid curve) and dimer sites (dashed curve for monophilic sites and dot-dashed curve for monophobic) for the activities $a_0 \simeq 1.715$ and $a_d \simeq 0.173$, with $a_0$ and $a_d$ defined as $a_i = e^{\mu_i/kT}$ ($i = 0, d$). The vertical dashed line indicates the (arbitrary) position of the "cluster radius" (taken equal to the "equivolume" radius of the $K$-dividing surface [17,18,28,58]. The center of the cluster is depleted of dimers—most of them are concentrated at the cluster surface layer with the monophilic sites oriented towards the cluster center whereas the monophobic sites with lower interaction energy form the top surface layer and this reduces the surface energy of the cluster.

Figure 2.19 presents the dependence of the potential field $\Phi_i(r)$ ($i = 0, d$) on the distance from the cluster center. The solid curves correspond to the monomers [i.e., $\Phi_0(r)$], whereas the dashed curves to the dimers [i.e., $\Phi_d(r)$]. The two upper wells were calculated by using the DFT-based density profiles in the cluster (shown in Figure 2.18). The two lower potential fields were calculated on the basis of the uniform density

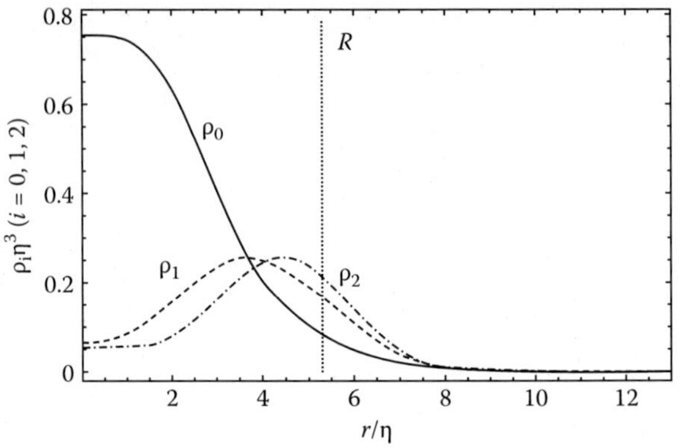


Figure 2.18

The density profiles of monomers (solid curve) and dimer sites (dashed curve for monophilic sites and dot-dashed curve for monophobic). Monomer and dimer activities are $a_0 \simeq 1.715$ and $a_d \simeq 0.173$, respectively. The vertical dashed line indicates the position of the "cluster radius" associated with the radius of the $K$-dividing surface [18,28,58].

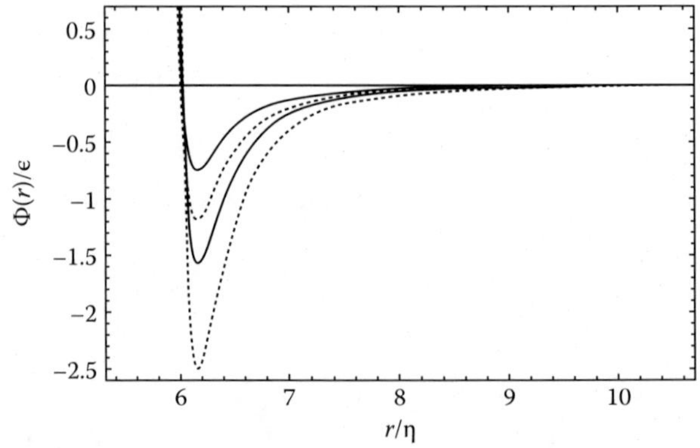


Figure 2.19

The dependence of the potential wells for monomers ($\Phi_0(r)$, solid curves) and dimers ($\Phi_d(r)$, dashed curves) on the distance from the cluster center. The two upper curves correspond to the DFT-based density profiles in the cluster (shown in Figure 2.18), whereas the two lower curves correspond to the uniform density approximation (assuming that all the excess numbers of monomers and dimers are uniformly distributed within a sphere of radius $R$).

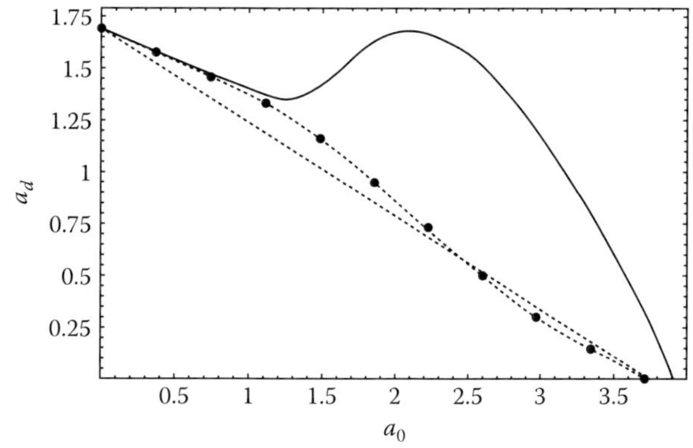


Figure 2.20

The activity plots for a constant ratio $J_s/AC \simeq 50 \pm 1$ (see Equation 2.200). The solid curve (with a hump) represents the BCNT, whereas the predictions of the theory presented here are shown as dots with an eye guiding dashed line ($a_0$ and $a_d$ are the activities of monomers and dimers, respectively).

approximation, by assuming that all the excess (with respect to the vapor mixture) numbers of monomers and dimers are uniformly distributed within a sphere of radius $R$. Note that the effective potential becomes repulsive and goes to infinity as the surface molecule approaches the core of the cluster. This is due to the repulsive part of the LJ potential (recall that the interaction of the molecule with the vapor medium is neglected). The potential wells for both monomers and dimers, $\Phi_0(r)$ and $\Phi_d(r)$, created by the cluster are markedly deeper in the uniform density approximation than when more realistic (DFT-based) profiles are used. Since a surface molecule can escape easier from a shallower well, the first passage time analysis based kinetic theory would significantly underestimate the nucleation rate in a water–alcohol-like binary system if the uniform density approximation were used for the cluster. The use of DFT-based density profiles is thus essential for the accurate calculation of the potential wells $\Phi_0(r)$, $\Phi_d(r)$ in the binary mixtures containing surface active components.

In Figure 2.20, another inconsistency of the BCNT (besides the one mentioned in Section 2.6.2) is examined in the framework of the model considered here. The activities of monomers and dimers, $a_0$ and $a_d$, are plotted for the constant ratio $J_s/AC \simeq 50 \pm 1$ (see Equation 2.200). As expected, the theoretical curve (dots with an eye guiding dashed line) behaves monotonically and has a negative slope in the whole range of activities. On the other hand, the BCNT-based curve has a region of positive slope that is unphysical because (1) it would require that for a given $a_d = \text{const}$ and increasing $a_1$ the nucleation rate to first increase, then decrease, and finally increase again and (2) it would correspond to a negative excess number of molecules in the critical (saddle point) cluster (nucleus). This inconsistency of the BCNT was thoroughly studied by Laaksonen et al. [17] and Laaksonen and Napari [30], who showed that it can be removed by using DFT methods (instead of the capillarity approximation). The use of DFT in our kinetic theory of binary nucleation (based on the first passage time analysis) also ensured that this inconsistency is avoided.

### 2.6.5 Concluding Remarks

The binary classical nucleation theory involves severe inconsistencies that are also reflected in large discrepancies between theoretical predictions and experimental results for the nucleation rate. This is because the BCNT is based on the Gibbsian thermodynamics and applies the macroscopic concept of surface tension to nanosize clusters. In this paper, we have presented an alternative approach to the kinetics of binary nucleation that avoids treating clusters by means of classical thermodynamics. Since the surface tension concept is not used by the new theory, the major inconsistencies of the BCNT are eliminated.

The new approach is a binary extension of the kinetic theory previously developed [19–26] for unary nucleation. In this approach based on molecular interactions the rate of emission of molecules from a cluster is determined via a mean first passage time analysis. This time is calculated by solving the single-molecule master equation for the probability distribution of a "surface" molecule moving in a potential field created by the cluster. The starting master equation is a Fokker–Planck equation for the probability distribution of a surface molecule with respect to its phase coordinates. Owing to the hierarchy of characteristic time scales in the evolution of the molecule, this equation reduces to the Smoluchowski equation for the distribution function in the coordinate space.

The new theory is combined with density functional theory methods for determining the density profiles in the system. These methods are essential for nucleation in binary systems particularly when one of the components is surface active. By using DFT instead of relying on the assumption of uniform density within a cluster, the new kinetic theory provides a tool necessary for studying nucleation in systems where the adsorption of one of the components on the cluster surface is significant. Knowing the real density profiles, one can determine the potential fields created by the cluster, its rate of emission of molecules, and the nucleation rate more accurately than on the basis of the uniform densities approximation for the cluster.

The new theory is illustrated by numerical calculations for a model binary mixture of LJ monomers and rigidly bonded dimers of LJ atoms. The amphiphilic character of the dimer component (i.e., its surface activity) is induced by an asymmetry in the interaction between a monomer and two different sites of a dimer. The inconsistencies of the BCNT are avoided in the new theory.

Clearly, the ultimate goal (and test) of the new kinetic theory of binary nucleation is its application to fluid–fluid phase transitions in real systems containing components that behave like surfactants in one of the phases. By avoiding the concept of surface tension for the new-phase cluster, one can expect to significantly improve the agreement between experimental data and theoretical predictions for the nucleation rate (at present, the discrepancy between the BCNT and experiment for some systems is by dozens of orders of magnitude [15–17]).

## Acknowledgments

This work was supported by the National Science Foundation through Grant No. CTS-0000548.

## Appendix 2B: Derivation of Equation 2.183

The goal of this appendix is to derive Equation 2.183 from Equations 2.180 and 2.181. Transforming $\theta_i$ ($i$ = 1, 2) (Equations 2.178 and 2.179) by using Taylor series expansions, one obtains $B_i(\nu_1,\nu_2)=-\frac{1}{2}\left(W_i^+(\nu_1,\nu_2)+W_i^-(\nu_1,\nu_2)\right)$. On the other hand, according to the principle of detailed balance (see Equations 2C.1 and 2C.2 in Appendix 2C),

$$\begin{aligned}W_1^-(\nu_1+1,\nu_2)/W_1^+(\nu_1,\nu_2)&=\exp[-G(\nu_1,\nu_2)+G(\nu_1+1,\nu_2)]\\&\simeq\exp[\partial G(\nu_1,\nu_2)/\partial\nu_1]\end{aligned}$$

and

$$\begin{aligned}W_2^-(\nu_1,\nu_2+1)/W_2^+(\nu_1,\nu_2)&=\exp[-G(\nu_1,\nu_2)+G(\nu_1,\nu_2+1)]\\&\simeq\exp[\partial G(\nu_1,\nu_2)/\partial\nu_2].\end{aligned}$$

The first derivatives $\partial G(\nu_1,\nu_2)/\partial\nu_i$ ($i$ = 1, 2) are equal to zero at the saddle point and are close to zero in its vicinity where, in addition, $W_1^+(\nu_1+1,\nu_2)\simeq W_1^+(\nu_1,\nu_2)$ and $W_2^+(\nu_1,\nu_2+1)\simeq W_2^+(\nu_1,\nu_2)$. Therefore, one can write

$$W_i^-(\nu_1,\nu_2)/W_i^+(\nu_1,\nu_2)\simeq 1+o(\partial G(\nu_1,\nu_2)/\partial\nu_i), \tag{2B.1}$$

where $o(x)$ denotes a quantity of the same order of magnitude as $x$. Neglecting the terms of the order $o(G(\nu_1,\nu_2)/\partial\,\nu_i)$, one can thus assume that the sum $W_i^+(\nu_1,\nu_2)+W_i^-(\nu_1,\nu_2)$ in the vicinity of the saddle point (where the kinetic equation [Equation 2.180] has to be solved) is equal to $2W_i^+(\nu_1,\nu_2)$. Furthermore, in the vicinity of the saddle point $W_i^+(\nu_1,\nu_2)\simeq W_i^+(\nu_{1c},\nu_{2c})\equiv W_{ic}^+$, hence

$$B_i\simeq -W_{ic}^+\ (i=1,2). \tag{2B.2}$$

Because at equilibrium $J_1 = 0$ and $J_2 = 0$, Equation 2.181 leads to

$$A_i(v_1,v_2)g_e(v_1,v_2)+B_i(v_1,v_2)\frac{\partial g_e(v_1,v_2)}{\partial v_i}=0 \ (i=1,2), \tag{2B.3}$$

whence one obtains

$$\begin{aligned} A_i(v_i,v_2) &= -B_i(v_1,v_2)\frac{1}{g_e(v_1,v_2)}\frac{\partial g_e(v_1,v_2)}{\partial v_i} \\ &\simeq W_{ic}^{+}\frac{\partial G(v_1,v_2)}{\partial v_i}, \end{aligned} \tag{2B.4}$$

where $G(v_1,v_2) = -\ln g_e(v_1,v_2)$. The coefficients $A_i(v_1,v_2)$ and $B_i(v_1,v_2)$ being the same for equilibrium and non-equilibrium states, $J_i$ ($i = 1, 2$) in Equation 2.181 acquires the form

$$J_i(v_i,v_2 t)=-W_{ic}^{+}\left(\frac{\partial}{\partial v_i}+\frac{\partial G(v_i,v_2)}{\partial v_i}\right). \tag{2B.5}$$

The substitution of this equation into Equation 2.180 leads to Equation 2.183.

## Appendix 2C: Equilibrium Distribution of Binary Clusters

Equations 2.198 and 2.199 can be derived by applying the principle of detailed balance to binary nucleation (as done, e.g., in Ref. [59]). According to this principle,

$$W_1^{+}(v_1-1,v_2)g_e(v_1-1,v_2)=W_1^{-}(v_1,v_2)g_e(v_1,v_2), \tag{2C.1}$$

$$W_2^{+}(v_1,v_2-1)g_e(v_1,v_2-1)=W_2^{-}(v_i,v_2)g_e(v_i,v_2). \tag{2C.2}$$

These equalities can be rewritten as

$$\frac{g_e(v_1,v_2)}{g_e(v_1-1,v_2)}=\frac{W_1^{+}(v_1-1,v_2)}{W_1^{-}(v_1,v_2)} \tag{2C.3}$$

and

$$\frac{g_e(v_1,v_2)}{g_e(v_1,v_2-1)}=\frac{W_2^{+}(v_1,v_2-1)}{W_2^{-}(v_1,v_2)}. \tag{2C.4}$$

By applying Equation 2C.3 to the clusters $(v_1 - i, v_2)$ with $i = 2, 3, \ldots, v_1$, multiplying the right- and left-hand sides of all equations, one obtains

$$\frac{g_e(v_1,v_2)}{g_e(1,v_2)}=\prod_{i=1}^{v_1-1}\frac{W_1^{+}(v_1-i,v_2)}{W_1^{-}(v_1-i+1,v_2)}. \tag{2C.5}$$

By applying Equation 2C.4 to the clusters $(1, \nu_2 - j)$ with $j = 1, 2, \ldots, \nu_2$ and multiplying the right- and left-hand side of all equations, one obtains

$$\frac{g_e(1,\nu_2)}{g_e(1,0)} = \prod_{j=1}^{\nu_2} \frac{W_2^+(1,\nu_2 - j)}{W_2^-(1,\nu_2 - j+1)}. \tag{2C.6}$$

The equilibrium distribution of clusters (1,0) is just the number density of molecules of component 1, i.e., $g_e(1,0) = \rho_1^{\upsilon}$. Therefore, the combination of Equations 2C.5 and 2C.6 provides

$$g_e(\nu_1,\nu_2) = \rho_1^{\upsilon} \prod_{i=1}^{\nu_1 - 1} \frac{W_1^+(\nu_1 - i,\nu_2)}{W_1^-(\nu_1 - i+1,\nu_2)} \prod_{j=1}^{\nu_2} \frac{W_2^+(1,\nu_2 - j)}{W_2^-(1,\nu_2 - j+1)}, \tag{2C.7}$$

which can be rewritten in the form

$$\begin{aligned} g_e(\nu_1,\nu_2) &= \rho_1^{\upsilon} \frac{W_1^+(1,\nu_2) W_2^+(1,0)}{W_1^+(\nu_1,\nu_2) W_2^+(1,\nu_2)} \\ &\times \prod_{i=1}^{\nu_1 - 1} \frac{W_1^+(\nu_1 - i+1,\nu_2)}{W_1^-(\nu_1 - i+1,\nu_2)} \\ &\times \prod_{j=1}^{\nu_2} \frac{W_2^+(1,\nu_2 - j+1)}{W_2^-(1,\nu_1 - j+1)}. \end{aligned} \tag{2C.8}$$

(Note that Equations 2C.7 and 2C.8 are very similar to those obtained by Wilemski and Wyslouzil [59].) Finally, substituting Equations 2.194 and 2.196 into the right-hand side of Equation 2C.8 and denoting $R_1 = R_1(1,0)$ (the radius of the cluster (1,0)), one obtains Equation 2.198. The derivation of Equation 2.199 can be carried out in a similar manner.

## Appendix 2D: Steady-State Solution of Equation 2.183

Equation 2.183 subject to boundary conditions (Equation 2.182) can be solved by using the separation of variables, which was previously used for various problems in the kinetics of multicomponent and nonisothermal nucleation [33,34,39,44]. In the context of the kinetics of binary nucleation, the key idea of the method consists of finding such variables of state of the binary cluster that would simultaneously diagonalize the matrix of the diffusion coefficients in Equation 2.183 and the matrix of the second derivatives of the function $G(\nu_1,\nu_2) = -\ln g_e(\nu_1,\nu_2)$ at the saddle point (recall that Equation 2.183 has to be solved in the vicinity of the saddle point).

First, let us introduce $x$ and $y$ as the new variables of state of a cluster defined as

$$\begin{aligned} x &= c_{11}(\nu_1 - \nu_{1c}) + c_{12}(\nu_2 - \nu_{2c}), \\ y &= c_{21}(\nu_1 - \nu_{1c}) + c_{22}(\nu_2 - \nu_{2c}), \end{aligned} \tag{2D.1}$$

with $c_{11}$, $c_{12}$, $c_{21}$, and $c_{22}$ given by Equation 2.204. The bilinear form

$$\begin{aligned} G(\nu_1,\nu_2) &= G_c + \frac{1}{2}\frac{\partial^2 G}{\partial \nu_1^2}\bigg|_c (\nu_1 - \nu_{1c})^2 \\ &+ \frac{\partial^2 G}{\partial \nu_1 \partial \nu_2}\bigg|_c (\nu_1 - \nu_{1c})(\nu_2 - \nu_{2c}) + \frac{1}{2}\frac{\partial^2 G}{\partial \nu_2^2}\bigg|_c (\nu_2 - \nu_{2c})^2 \end{aligned}$$

of the function $G(v_1,v_2)$ in the vicinity of the saddle point acquires the following diagonal form in variables $x$ and $y$:

$$G = G_c - x^2 + y^2, \tag{2D.2}$$

so that they become separated in the equilibrium distribution,

$$n_e(x,y) = Ce^{x^2}e^{-y^2}, \tag{2D.3}$$

where $C$ is the Jacobian of the variables transformation (Equation 2D.1). This allows one to identify $x$ as an unstable variable and $y$ as a stable variable.

The kinetic equation (Equation 2.183) in variables $x$ and $y$ acquires the canonical form of the Fokker–Planck equation

$$\begin{aligned}\frac{\partial n(x,y,t)}{\partial t} &= a\left\{\frac{\partial}{\partial x}\left(\frac{\partial}{\partial x} - 2x\right) - \varepsilon^{-1}\left[\frac{\partial}{\partial x}\left(\frac{\partial}{\partial y} + 2y\right) + \frac{\partial}{\partial y}\left(\frac{\partial}{\partial x} - 2x\right)\right]\right.\\ &\left.+\varepsilon^{-2}(1+p)\frac{\partial}{\partial y}\left(\frac{\partial}{\partial y} + 2y\right)\right\} n(x,y,t),\end{aligned} \tag{2D.4}$$

where $n(x,y,t)$ is the distribution of clusters with respect to $x$ and $y$ at time $t$, and the positive parameters $\alpha$, $\varepsilon$, and $p$ are given by Equation 2.201.

Next, it is necessary to find such a transformation of variables $x$ and $y$ that would diagonalize the matrix of diffusion coefficients of Equation 2D.4 while leaving the form (Equation 2D.2) invariant. This goal is achieved by using the Lorentz transformation

$$u = (1 - \alpha^2)^{-1/2}(x + \alpha y),\ z = (1 - \alpha^2)^{-1/2}(y + \alpha x), \tag{2D.5}$$

where $u$ and $z$ are the new variables and $\alpha$ is a transformation parameter ($-1 < \alpha < 1$).

The Jacobian of transformation (Equation 2D.5) is equal to unity; hence $n(x,y,t) = n(u,z,t)$ and

$$G = G_c - u^2 + z^2, \tag{2D.6}$$

so that the variables $x$ and $y$ are separated in the equilibrium distribution,

$$n_e(u,z) = Ce^{u^2}e^{-z^2}. \tag{2D.7}$$

Choosing the transformation parameter

$$\alpha = (1/2\varepsilon)\{\varepsilon^2 + 1 + p - [(\varepsilon^2 + 1)^2 - 4\varepsilon^2]^{1/2}\}, \tag{2D.8}$$

the kinetic equation (Equation 2D.4) becomes

$$\frac{\partial n(u,z,t)}{\partial t} = A\left[\frac{\partial}{\partial u}\left(\frac{\partial}{\partial u} - 2u\right) + \frac{(1-\varepsilon\alpha)^2}{\alpha^2 p}\frac{\partial}{\partial z}\left(\frac{\partial}{\partial z} + 2z\right)\right] n(u,z,t), \tag{2D.9}$$

with $A$ given by Equation 2.200. Equation 2D.9 is subjected to the boundary conditions

$$\frac{n(u,z,t)}{n_c(u,z)} = \begin{cases} 1 & (u \to -\infty) \\ 0 & (u \to \infty) \end{cases} \tag{2D.10}$$

for any $z$.

One can show that the solution of Equation 2D.9 subject to boundary conditions (Equation 2D.10) has the form

$$n(u,z,t)=\sqrt{\pi}q(u,t)e^{-z^2}, \tag{2D.11}$$

where the function $q(u,t)$ is governed by the one-dimensional kinetic equation

$$\partial_t q(u,t)=A\frac{\partial}{\partial u}\left(\frac{\partial}{\partial u}-2u\right)q(u,t) \tag{2D.12}$$

and satisfies the boundary conditions

$$\frac{q(u,t)}{q_e(u)}=\begin{cases}1 & (u\to-\infty)\\ 0 & (u\to\infty).\end{cases} \tag{2D.13}$$

From Equation 2D.11, one obtains the expression

$$q(u,t)=\int_{-\infty}^{\infty}\mathrm{d}z\,n(u,z,t), \tag{2D.14}$$

which clearly shows that $q(u, t)$ is the distribution of clusters with respect to the unstable variable $u$. Equation 2D.11 can be rewritten as

$$\partial_t q(u,t)=-\frac{\partial}{\partial u}J_u,\; J_u=-A\left(\frac{\partial}{\partial u}-2u\right)q(u,t), \tag{2D.15}$$

where $J_u = J_u(u, t)$ is the one-dimensional flux density of clusters along the $u$ axis integrated with respect to the stable variable $z$.

The steady-state solution of the one-dimensional Equation 2D.15 subject to boundary conditions (Equation 2D.13) leads to Equation 2.200 for the steady-state nucleation rate $J_s$.

## References


1. Allen, M.P., Tildesley, D.J., *Computer Simulation of Liquids*, Clarendon, Oxford, 1989.
2. Frenkel, D., Smit, B., *Understanding Molecular Simulation*, Academic, Boston, 1996.
3. Auer, S., Frenkel, D., *Nature (London)* 409, 1020 (2001).
4. Gasser, U., Weeks, E.R., Schofield, A., Pusey, P.N., Weitz, D.A., *Science* 292, 258 (2001).
5. Yau, S.T., Vekilov, P.G., *Nature (London)* 406, 494 (2000).
6. Schmelzer, J., Röpke, G., Priezzhev, V.B., Eds., *Nucleation Theory and Applications*, JINR, Dubna, 1999.
7. Lothe, J., Pound, G.M.J., in: *Nucleation*, Zettlemoyer, A.C., Ed., Marcel Dekker, New York, 1969.
8. Flageollet, C., Dihn Cao, M., Mirabel, P., *J. Chem. Phys.* 72, 544 (1980).
9. Wilemski, G., *J. Phys. Chem.* 91, 2492 (1987).
10. Wyslouzil, B.E., Seinfeld, J.H., Flagan, R.C., Okuyama, K., *J. Chem. Phys.* 94, 6827 (1991).
11. Laaksonen, A., *J. Chem. Phys.* 97, 1983 (1992).
12. Strey, R., Viisanen, Y., *J. Chem. Phys.* 99, 4693 (1993).
13. Wilemski, G., *J. Chem. Phys.* 80, 1370 (1984).
14. Nishioka, K., Kusaka, I., *J. Chem. Phys.* 96, 5370 (1992).
15. Debenedetti, P.G., *Metastable Liquids: Concepts and Principles*, Princeton University Press, Princeton, NJ, 1996.
16. Oxtoby, D.W., Kashchiev, D., *J. Chem. Phys.* 100, 7665 (1994).
17. Laaksonen, A., McGraw, R., Vehkamäki, H., *J. Chem. Phys.* 111, 2019 (1999).
18. Djikaev, Y.S., Napari, I., Laaksonen, A., *J. Chem. Phys.* 120, 9752 (2004).

19. Narsimhan, G., Ruckenstein, E., *J. Colloid Interface Sci.* 128, 549 (1989). (Section 1.2 of this volume.)
20. Ruckenstein, E., Nowakowski, B., *J. Colloid Interface Sci.* 137, 583 (1990). (Section 1.3 of this volume.)
21. Nowakowski, B., Ruckenstein, E., *J. Colloid Interface Sci.* 139, 500 (1990). (Section 1.4 of this volume.)
22. Nowakowski, B., Ruckenstein, E., *J. Chem. Phys.* 94, 1397 (1991). (Section 2.1 of this volume.)
23. Ruckenstein, E., Nowakowski, B., *Langmuir* 7, 1537 (1991). (Section 2.2 of this volume.)
24. Nowakowski, B., Ruckenstein, E., *J. Chem. Phys.* 94, 8487 (1991). (Section 2.3 of this volume.)
25. Nowakowski, B., Ruckenstein, E., *J. Colloid Interface Sci.* 145, 182 (1991). (Section 2.4 of this volume.)
26. Nowakowski, B., Ruckenstein, E., *J. Phys. Chem.* 96, 2313 (1992). (Section 2.7 of this volume.)
27. Djikaev, Y.S., Ruckenstein, E., *J. Chem. Phys.* 123, 214503 (2005). (Section 2.5 of this volume.)
28. Napari, I., Laaksonen, A., Strey, R., *J. Chem. Phys.* 113, 4476 (2000).
29. Talanquer, V., Oxtoby, D.W., *J. Chem. Phys.* 113, 7013 (2000).
30. Laaksonen, A., Napari, I., *J. Phys. Chem. B* 105, 11678 (2001).
31. Lee, D.J., Telo da Gama, M.M., Gubbins, K.E., *J. Chem. Phys.* 85, 490 (1986).
32. Rusanov, A.I., *Phasengleichgewichte und Grenzflachenerscheinungen*, Academie, Berlin, 1978.
33. Kuni, F.M., Melikhov, A.A., *Theor. Math. Phys.* 81, 1182 (1989).
34. Kuni, M., Melikhov, A.A., Novozhilova, T.Y., Terent'ev, I.A., *Theor. Math. Phys.* 83, 530 (1990).
35. Reiss, H., *J. Chem. Phys.* 18, 840 (1950).
36. Hirschfelder, J.O., *J. Chem. Phys.* 61, 2690 (1974).
37. Stauffer, D., *J. Aerosol Sci.* 7, 319 (1976).
38. Ray, A.K., Chalam, M., Peters, L.K., *J. Chem. Phys.* 85, 2161 (1986).
39. Melikhov, A.A., Kurasov, V.B., Dzhikaev, Y.S., Kuni, F.M., *Sov. Phys. Tech. Phys.* 36, 14 (1991).
40. Flageollet, C., Garnier, G.P., Mirabel, P., *J. Chem. Phys.* 78, 2600 (1983).
41. Wyslouzil, B.E., Wilemski, G., *J. Phys. Chem.* 105, 1090 (1996).
42. Kuni, F.M., Grinin, A.P., Djikaev, Y.S., *J. Aerosol Sci.* 29, 1 (1998).
43. Djikaev, Y.S., Kuni, F.M., Grinin, A.P., *J. Aerosol Sci.* 30, 265 (1999).
44. Djikaev, Y.S., Teichmann, J., Grmela, M., *Physica A* 267, 322 (1999).
45. Tolman, R.C., *J. Chem. Phys.* 17, 333 (1949).
46. Kirkwood, J.G., Buff, F.P., *J. Chem. Phys.* 17, 338 (1949).
47. Chandrasekhar, S., *Rev. Mod. Phys.* 15, 1 (1949).
48. Gardiner, C.W., *Handbook of Stochastic Methods*, Springer, New York, 1983.
49. Agmon, N., *J. Chem. Phys.* 81, 3644 (1984).
50. Abraham, F.F., *Homogeneous Nucleation Theory*, Academic, New York, 1974.
51. Trinkaus, H., *Phys. Rev. B* 27, 7372 (1983).
52. Cherepanova, T.A., Stekolnikov, A.V., *Mol. Phys.* 82, 125 (1994).
53. Chandler, D., McCoy, J.D., Singer, S.J., *J. Chem. Phys.* 85, 5971 (1986).
54. Chandler, D., McCoy, J.D., Singer, S.J., *J. Chem. Phys.* 85, 5977 (1986).
55. Oxtoby, D.W., Evans, R., *J. Chem. Phys.* 89, 7521 (1988).
56. Mansoori, G.A., Carnahan, N.F., Starling, K.E., Leland, T.W., *J. Chem. Phys.* 54, 1523 (1971).
57. Weeks, J.D., Chandler, D., Andersen, H.C., *J. Chem. Phys.* 54, 5237 (1971).
58. Ono, S., Kondo, S., in: *Encyclopedia of Physics*, Flugge, S., Ed., Springer, Berlin, 1960, Vol. 10, p. 134.
59. Wilemski, G., Wyslowzil, B.E., *J. Chem. Phys.* 103, 1127 (1995).

## 2.7 A Kinetic Treatment of Heterogeneous Nucleation

*Bogdan Nowakowski and Eli Ruckenstein**

An expression is derived for the rate of evaporation of a molecule interacting with a spherical cap cluster and the solid on which the cluster is located. The motion of the evaporating molecule in the potential well generated by its interaction with the cluster and substrate is described by a diffusion equation in the energy space. The rate of heterogeneous nucleation, calculated in the steady-state approximation, is strongly dependent on the interactions with the solid substrate. A comparison between the kinetic and classical theories shows that the former provides much larger values than the latter for small critical clusters (large supersaturations) but tends to the latter for large critical clusters.

### 2.7.1 Introduction

Nucleation in supersaturated vapors frequently occurs on the surface of the container, foreign particles, or other surface heterogeneities. The number of nucleation sites as well as their catalytic effectiveness is generally not known. For this reason, it is usually assumed that the surface of the foreign substrate is homogeneous and that the extent of wetting of the solid by the liquid determines the wetting angle of the clusters on the solid surface.

The classical theory of heterogeneous nucleation was formulated in the same way as the traditional approach to the homogeneous process [1,2]. It employs the detailed balance principle and a hypothetical equilibrium distribution of clusters to calculate the rate constant of evaporation from clusters in terms of the rate constant of condensation. The equilibrium distribution is calculated on the basis of macroscopic thermodynamics. In particular, the macroscopic surface tension is used to calculate the surface free energy of a cluster. The extension of macroscopic properties to the small clusters constitutes the most questionable feature of the classical theory.

Macroscopic thermodynamics was avoided in the lately developed kinetic theory of homogeneous nucleation [3,4]. In that approach, the rate of evaporation from a cluster was obtained independent of the rate of condensation on the basis of the solution of the equation governing the motion of a molecule interacting with the cluster. In this sense, the latter approach can be considered more kinetic than the conventional treatment. The kinetic theory was applied to liquid/solid [3] and vapor/liquid [4] phase transitions.

In this paper, we extend the formalism developed for homogeneous nucleation in gases [4] to the heterogeneous nucleation on a flat, homogeneous surface. The properties of the surface are assumed uniform, in the sense that the solid/liquid $\sigma_{ls}$ and solid/gas $\sigma_{sg}$ interfacial tensions have constant values over the entire surface. The nuclei formed are assumed to have the shape of a spherical cap, with a wetting angle $\theta$ at the edge of the nuclei given by the Young equation [5]

$$\sigma_{ls} + \sigma_{lg} \cos\theta = \sigma_{sg}, \tag{2.228}$$

where $\sigma_{lg}$ is the liquid/gas interfacial tension.

In the next section, the theory [4] for the rate of evaporation of molecules from a cluster is extended to heterogeneous nucleation by taking into account the interactions between the evaporating molecule and the system consisting of the cluster and the supporting solid. The obtained results are used, in a subsequent section, to calculate the rate of nucleation from the steady-state solution of the equations describing the population of clusters. Finally, numerical results are presented for the rate of heterogeneous nucleation. A comparison with the predictions of the classical nucleation theory concludes the paper.

### 2.7.2 Rate of Evaporation from the Cluster

In the layer over the free surface of the cluster, a potential well is generated due to the attractive intermolecular interactions. The form of the pair interaction potential $\phi(r_{12})$ is assumed to combine the London–van der Waals attraction with the rigid core repulsion:

$$\begin{aligned} \phi(r_{12}) &= \infty & r_{12} &< \sigma \\ &= -\epsilon\left(\sigma/r_{12}\right)^6 & r_{12} &\geq \sigma, \end{aligned} \tag{2.229}$$

* *J. Phys. Chem.*, 96, 2313 (1992). Republished with permission.

where $r_{12}$ is the distance between the centers of the molecules, $\epsilon$ is an interaction constant, and $\sigma$ is the molecular diameter (assumed the same for all the species). The vapor molecule in the well interacts with the cluster as well as with the solid. Assuming pairwise additivity, the effective interaction potential $\Phi(r)$ in the position $\mathbf{r}$ in the well is calculated by integrating over the interactions with the molecules of the cluster and with those of the solid substrate. The interaction constant with the cluster, $\epsilon_c$, is assumed to be greater than that with the solid, $\epsilon_s$. The integration over the volume of the semi-infinite solid yields the following expression for the interactions $\Phi_s$ of the vapor molecule in the well with the solid:

$$\Phi_s = -\frac{\pi\epsilon_s\rho_s\sigma^3}{6}\left(\frac{\sigma}{x}\right)^3 \tag{2.230}$$

where $x$ is the distance of the molecule from the surface of the solid and $\rho_s$ is the number density of molecules in the solid.

The volume integral over the spherical cap, involved in the calculation of the interaction potential $\Phi_c$ with the cluster, can be performed analytically for two coordinates; the resulting formula has a complicated form, and for this reason the integration with respect to the third variable was carried out numerically. The details are presented in Appendix 2E.

As long as the motion of the vapor molecule is confined within the well, one can consider that it belongs to the cluster. In this case, its energy $E$ lies in the range $\Phi_0 < E < 0$, where $\Phi_0$ is the potential energy at the bottom of the well. The energy of the molecule is a stochastic variable, because of the exchange of energy during the collisions with the molecules of the gaseous medium. Eventually, the molecule in the well can acquire a sufficient amount of energy to overcome the potential well, and hence to leave the well. Therefore, the rate of evaporation of molecules from the cluster is related to the rate by which their energy passes over the upper energetic boundary of the well, $E = 0$.

In gases, the collisions with the medium are relatively infrequent. As a result, the energy of the molecules in the well changes slowly in comparison with the fast changes in their impulses and positions. Consequently, the time evolution in the energy space is governed by the diffusion-like equation for the probability density $P(E,t)$ of energy [4,6]:

$$\frac{\partial P(E,t)}{\partial t} = \frac{\partial}{\partial E} D(E) P_{eq}(E) \frac{\partial}{\partial E}\frac{P(E,t)}{P_{eq}(E)}. \tag{2.231}$$

Here, $D(E)$ is the diffusion coefficient in the energy space, $P_{eq}(E)$ is the equilibrium distribution of energy given by the equation

$$P_{eq}(E) = \Omega(E)\exp(-E/kT) \Big/ \int_{\Phi_0}^{0} \mathrm{d}E\,\Omega(E)\exp(-E/kT) \tag{2.232}$$

and $\Omega(E)$ is the number of states of energy $E$. For a single molecule with the Hamiltonian $H(\mathbf{r},\mathbf{p})$, $\Omega(E)$ is given by

$$\Omega(E) = \int_{\Gamma} \mathrm{d}\mathbf{r}\,\mathrm{d}\mathbf{p}\,\delta(E - H(\mathbf{r},\mathbf{p})), \tag{2.233}$$

where $\mathbf{r}$ and $\mathbf{p}$ are the position and impulse of the molecule, respectively.

The flux $j(E,t)$ in the energy space is defined by

$$\frac{\partial P(E,t)}{\partial t} = -\frac{\partial}{\partial E} j(E,t). \tag{2.234}$$

Using Equation 2.231, one can write

$$j(E,t) = -D(E)P_{eq}(E)\frac{\partial}{\partial E}\frac{P(E,t)}{P_{eq}(E)}. \tag{2.235}$$

The rate of evaporation is provided by the steady-state solution of Equation 2.235, i.e., by the time-independent flux $j_E$. Assuming that every molecule that reaches the upper energy of the well separates from the cluster, one obtains the absorption boundary condition, $P(E) = 0$ at $E = 0$. The stationary solution of Equation 2.235, which satisfies this boundary condition, is given by

$$P(E) = j_E P_{eq}(E)\int_E^0 \frac{dE'}{D(E')P_{eq}(E')}. \tag{2.336}$$

The normalization constraint, $\int P(E)dE = 1$, yields an expression for the rate of evaporation per molecule, $j_E = \alpha^1$:

$$\alpha^1 = \left[\int_{\Phi_0}^0 dEP_{eq}(E)\int_E^0 \frac{dE'}{D(E')P_{eq}(E')}\right]^{-1}. \tag{2.237}$$

For Brownian dynamics, $D(E)$ is given by [7]

$$D(E) = \frac{kT\xi d}{m\Omega(E)}\int_{\Phi_0}^E dE'\Omega(E'), \tag{2.238}$$

where $m$ is the mass of the molecule, $\xi$ is the friction coefficient, and $d$ is the dimensionality of the geometrical space in which the Brownian motion takes place (equal to 3 in the system considered).

The overall evaporation rate per cluster $\alpha$ is given by the product between $\alpha^1$ and the number of molecules in the potential well over the cluster. Denoting by $n_w$ the number density of the vapor molecules in the well, the total number of molecules in a well of thickness $\lambda$ is $N_w = 2\pi(1 - \cos\theta)R^2\lambda n_w$. The overall evaporation rate is therefore given by

$$\alpha = \frac{2\pi(1-\cos\theta)R^2\lambda n_w}{\int_{\Phi_0}^0 dEP_{eq}(E)\int_E^0 \frac{dE'}{D(E')P_{eq}(E')}}. \tag{2.239}$$

Explicit values for the quantities $n_w$ and $\lambda$ are not needed because they are not present in the final equations.

One may note that the above treatment implicitly assumes equipartition over the degrees of freedom, an assumption that should break down for large clusters.

### 2.7.3 Growth of Cluster

The rate of change of the size of the cluster is determined by the difference between the rates of evaporation and condensation. In gaseous media, the latter rate is given by the number of impingements of the vapor molecules on the surface of the cluster

$$\beta = \frac{\pi}{2}(1-\cos\theta)R^2 n_1\bar{\upsilon}, \tag{2.240}$$

where $\bar{\upsilon}$ is the mean thermal velocity of the vapor molecules and $n_1$ is the number density of the vapor at a large distance from the cluster. For a given value of $n_1$, there exists a size, called critical cluster, for which the rates of evaporation and condensation are equal. Clusters smaller than the critical one decay (statistically), whereas those that are larger grow. Hence, the critical cluster is in unstable equilibrium with the surrounding vapor. Using Equations 2.239 and 2.240, one can obtain the following relation between the vapor number density $n_1$ and the radius $R_\star$ of the critical cluster:

$$n_1 = n_w \frac{4\lambda}{\bar{\upsilon}} \alpha^1(R_\star). \tag{2.241}$$

The number density in the vapor that coexists with an infinite cluster is the saturation number density, $n_s$. The supersaturation $s = n_1/n_s$ constitutes the thermodynamic driving force for nucleation. Using Equation 2.241, it can be expressed as

$$s = \alpha^1(R_\star)/\alpha^1(\infty). \tag{2.242}$$

In the macroscopic thermodynamics, the supersaturation is related to the critical radius via the Kelvin equation

$$s = \exp\left(\frac{2\sigma_{lg}}{kT\rho R_\star}\right), \tag{2.243}$$

where $\sigma_{lg}$ is the macroscopic surface tension of the liquid and $\rho$ is its number density.

The Kelvin equation plays an important role in the classical theory of nucleation. The macroscopic thermodynamics is expected to be valid for large clusters; hence, Equation 2.243 should be recovered in the present theory for large nuclei. Consequently, comparing Equations 2.242 and 2.243 for $R_\star/\sigma$ @ 1, one can calculate the macroscopic surface tension in terms of molecular parameters.

### 2.7.4 Rate of Nucleation

The evolution of the ensemble of nuclei can be expressed in terms of the net flux of clusters passing from those consisting of $i$ molecules, $n_i$, to those containing $(i + 1)$ molecules, $n_{i+1}$,

$$I_i = \beta_i n_i - \alpha_{i+1} n_{i+1}. \tag{2.244}$$

$I_i$ vanishes at equilibrium, and Equation 2.244 then yields the detailed balance condition. Since the kinetic theory of nucleation provides independent results for $\alpha$ and $\beta$, the equilibrium distribution can be explicitly obtained from the detailed balance equation in the form

$$n_i^{eq} = n_1^{eq} \prod_{j=2}^{i} \frac{\beta_{j-1}}{\alpha_j} \equiv n_i^{eq} \exp(h(i)). \tag{2.245}$$

For convenience, Equation 2.244 is converted to a continuum form, which, using the detailed balance equation, can be written as [8,9]

$$I(g,t) = \beta(g)\, n^{eq}(g) \frac{\partial}{\partial g} \frac{n(g,t)}{n^{eq}(g)}. \tag{2.246}$$

The continuum variable $g$ replaces in Equation 2.246 the discrete variable $i$. The rate of nucleation, calculated as the stationary solution of Equation 2.246, is given by

$$I = \left[\int_1^\infty \frac{dg}{\beta(g) n^{eq}(g)}\right]^{-1}. \tag{2.247}$$

The equilibrium distribution $n^{eq}(g)$ exhibits a very sharp minimum at the critical size $g_*$. This allows us to calculate the integral in Equation 2.247 by the steepest descent method. One obtains finally

$$I \cong n_1\beta_1\left(\frac{R_*}{\sigma}\right)^2\left(\frac{h''(g_*)}{2\pi}\right)^{1/2}\exp(h(g_*)). \tag{2.248}$$

In the classical theory of nucleation, the hypothetical equilibrium distribution $n_{eq}$ is calculated in terms of the macroscopic surface free energy of clusters. Equation 2.247 then provides the following expression for the steady-state nucleation

$$I_{cl} = \frac{n_1\beta_1}{4\pi\rho\sigma^3}\left(\frac{\gamma\sigma^2}{kT}\right)^{1/2}\exp\left(-\frac{16\pi\gamma^3}{3(kT)^3\rho^2\ln^2 s}\right), \tag{2.249}$$

where

$$\gamma = \sigma_{lg}\left(\frac{2-3\cos\theta+\cos^3\theta}{4}\right)^{1/3}$$

The rates of nucleation were calculated using both Equations 2.248 and 2.249. The surface tension $\sigma_{lg}$ needed in the classical expression was calculated as indicated in the previous section. This procedure enabled us to compare the predictions of the kinetic theory of nucleation with those of the classical theory.

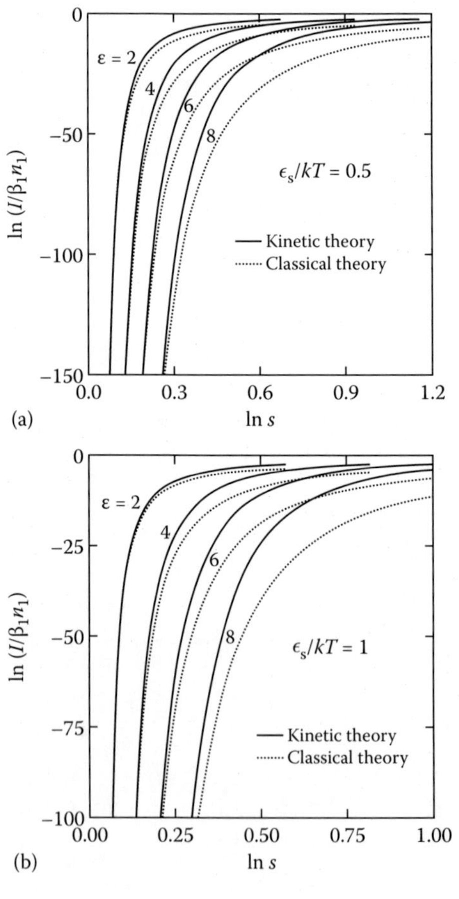


Figure 2.21

Rate of nucleation ln $(I/\beta_1 n_1)$ against supersaturation (ln $s$) for various values of $\varepsilon = \epsilon_c/kT$ and for (a) $\epsilon_s/kT = 0.5$ and (b) $\epsilon_s/kT = 1$ provided by the kinetic (—) and classical (...) theories.

### 2.7.5 Results of Calculations and Discussion

The numerical calculations were performed using for the interaction constant with the cluster the values $\epsilon_c/kT = 2$, 4, 6, and 8, and for the interaction constant with the solid the values $\epsilon_s/kT = 0.5$ and 1. The diameters of the molecules of all the species involved were assumed equal, and the number densities of molecules in the cluster and in the solid were calculated using the relation $\rho\sigma^3 = 1.2$ (valid for a random packing of spheres [10]). The wetting angle for the liquid forming the nucleus was taken $\theta = \pi/6$. The rate of nucleation based on the kinetic theory of nucleation was calculated numerically using Equation 2.248, and the predictions of the classical theory were obtained using Equation 2.249. Figure 2.21a and b presents the results obtained in a ln–ln scale, namely $-\ln(I/\beta_1 n_1)$ versus the logarithm of supersaturation, ln $s$, for $\epsilon_s/kT$ equal to 0.5 and 1, respectively. The continuous lines are based on the kinetic theory and the dashed lines on the classical theory. The values of the surface tension used in the classical formula (Equation 2.249) were calculated on the basis of the results of kinetic theory (Equation 2.242) for clusters greater than $R_*/\sigma > 17$. The results obtained have been fitted to the Kelvin Equation 2.243, and the surface tension was calculated by the least-squares method. The values of $\sigma_{lg}\sigma^2/kT$ are listed in Table 2.6.

Figure 2.22 compares the results provided by the kinetic theory of nucleation for two interaction constants with the solid substrate, namely $\epsilon_s/kT = 0.5$ and 1. The rate of nucleation is higher for $\epsilon_s/kT = 1$ than for $\epsilon_s/kT = 0.5$, because of the stronger "catalytic effect" of the solid. The stronger the attraction of the molecules in the well by the solid, the greater the difficulty to leave the well and evaporate from the cluster. Hence, the critical cluster is smaller for higher values of $\epsilon_s$, and as a result the nucleation rate is higher.

The predictions of both theories coincide in the range of small supersaturations, corresponding to large critical clusters. The classical theory is expected to be valid for large clusters, because it is based on macroscopic thermodynamics that is asymptotically correct in this range. At the other end, large supersaturations, the present theory provides much higher rates of nucleation than the classical theory. High supersaturations lead to small critical clusters, to which the macroscopic thermodynamics is no longer applicable. The macroscopic surface tension exhibited by bulk, planar interfaces overestimates the surface energy of small clusters. Consequently, according to Equation 2.249 the classical theory underpredicts the population of critical clusters and the rate of nucleation. Similar relations between the results of the classical and the kinetic theories of nucleation were also observed in the case of homogeneous nucleation [3,4].

Table 2.6 Values of the Surface Tension $\sigma_{lg}\sigma^2/kT$ for Various Interaction Parameters $\varepsilon_c/kT$

| $\varepsilon_c/kT$ | $\sigma_{lg}\sigma^2/kT$ |
|---|---|
| 2 | 1.791 ± 29 |
| 4 | 2.589 ± 51 |
| 6 | 3.357 ± 75 |
| 8 | 4.182 ± 86 |

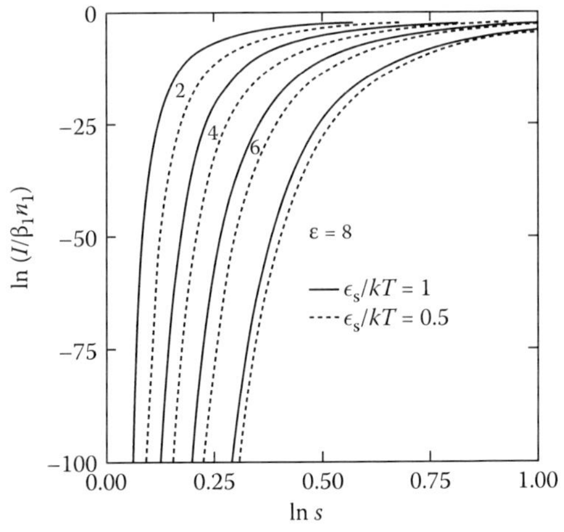


Figure 2.22

Comparison between the rates of nucleation obtained on the basis of the kinetic theory for various values of $\varepsilon = \epsilon_c/kT$ and for $\epsilon_s/kT = 0.5$ (...) and $\epsilon_c/kT = 1$ (—).

The macroscopic thermodynamics on which the classical theory of nucleation is based was avoided in the kinetic approach. Another assumption of the classical theory that can be reconsidered is the use of the simple formula (Equation 2.240) of the kinetic theory of gases for the number of collisions of the vapor molecules with the cluster. Equation 2.240 does not include the interaction of the condensing molecule with the substrate and the cluster. Such an interaction was shown to have some effect in the case of spherical particles on which condensation occurs [11].

## Appendix 2E: Interaction Potential between a Molecule and a Spherical Cap Cluster

Assuming pairwise additivity, the effective interaction potential acting on a molecule in the well can be calculated using the integral

$$\Phi(\mathbf{r}) = \int_V \rho_c \phi(|\mathbf{r}-\mathbf{r}'|)d\mathbf{r}', \tag{2E.1}$$

where $\mathbf{r}$ is the position of the molecule, $\mathbf{r}'$ is the position of a molecule in the cluster, and $\rho_c$ is the number density of molecules in the cluster (assumed uniform).

To calculate the volume integral over the spherical cap of the cluster, we locate the origin of the coordinate system $\mathbf{r}'$ in the center of the sphere containing the cap, and switch to cylindrical coordinates, in which $\mathbf{r}' = (\rho \cos \phi, \rho \sin \phi, z)$. Because of the cylindrical symmetry, the position of the molecule in the well can always be written in the form $\mathbf{r} = (r \sin \psi, 0, r \cos \psi)$, where $\psi$ is the azimuthal angle of the position $\mathbf{r}$. The square of the distance between the molecule and the point of integration $\mathbf{r}'$ is

$$(\mathbf{r} - \mathbf{r}')^2 = (r \cos \psi - z)^2 + \rho^2 + (r \sin \psi)^2 - 2\rho r \cos \phi \sin \psi. \tag{2E.2}$$

Using Equations 2E.2 and 2.229, the integral 2E.1 over the spherical cap acquires the form

$$\begin{aligned}\Phi_c = -\epsilon_c \rho_c \sigma^6 \int_{R\cos\theta}^{R} dz \int_0^{(R^2-z^2)^{1/2}} \rho d\rho \\ \times \int_0^{2\pi} \frac{d\phi}{[(r\cos\psi - z)^2 + \rho^2 + (r\sin\psi)^2 - 2\rho r \sin\psi \cos\phi]^3}.\end{aligned} \tag{2E.3}$$

The integration over the polar angle # can be easily performed, and one obtains

$$\Phi_c = -\epsilon_c \rho_c \sigma^6 \int_{R\cos\theta}^{R} dz \int_0^{(R^2-z^2)^{1/2}} \rho d\rho \frac{\pi(2a^2 + b^2)}{(a^2 - b^2)^{5/2}}, \tag{2E.4}$$

where

$$a(\rho, z) = (r \cos \psi - z)^2 + \rho^2 + (r \sin \psi)^2 \tag{2E.5}$$

and

$$b(\rho, z) = 2\rho r \sin \psi. \tag{2E.6}$$

The denominator of the function under the integral in Equation 2E.4 can be recast in the form

$$a^2 - b^2 = [\rho^2 + (r \cos \psi - z)^2 - (r \sin \psi)^2]^2 + 4(r \cos \psi - z)^2(r \sin \psi)^2. \tag{2E.7}$$

Except for the unphysical case when $\mathbf{r} = \mathbf{r}'$, expression 2E.7 is always different from zero (being the sum of squares). Hence, the function under the integral Equation 2E.4 has no singularities. The integration of Equation 2E.4 can be further performed by changing to the variable $q = \rho^2$. This yields

$$\Phi_c = -\epsilon_c \rho_c \sigma^6 \int_{R\cos\theta}^{R} dz (Q_1(q) + Q_2(q) + Q_3(q))|_{q_1}^{q_2}, \tag{2E.8}$$

where the functions $Q_i$, have the form

$$Q_1(q) = \frac{q}{4(r\cos\psi - z)^2 (r\sin\psi)^2 [q^2 + 4(r\cos\psi - z)^2 (r\sin\psi)^2]^{1/2}} \tag{2E.9}$$

$$Q_2(q) = \frac{-2(r\sin\psi)^2}{[q^2 + 4(r\cos\psi - z)^2 (r\sin\psi)^2]^{3/2}} \tag{2E.10}$$

$$\begin{aligned} Q_3(q) &= 1/2(r\sin\psi)^2[(r\sin\psi)^2 - (r\cos\psi - z)^2] \\ &\times \frac{q}{(r\cos\psi - z)^2 (r\sin\psi)^2 [q^2 + 4(r\cos\psi - z)^2 (r\sin\psi)^2]^{3/2}} \\ &\times \left\{ 1 + \frac{q^2 + 4(r\cos\psi - z)^2 (r\sin\psi)^2}{2(r\cos\psi - z)^2 (r\sin\psi)^2} \right\}. \end{aligned} \tag{2E.11}$$

The functions $Q_i$, appear in the integral 2E.8 for the two limits of $q$

$$q_1 = (r\cos\psi - z)^2 - (r\sin\psi)^2 \tag{2E.12}$$

$$q_2 = (r\cos\psi - z)^2 + R^2 - z^2 - (r\sin\psi)^2. \tag{2E.13}$$

The third integral was calculated numerically.

## References


1. Abraham, F.F., *Homogeneous Nucleation Theory*, Academic Press, New York, 1974.
2. Zettlemoyer, A.C., Ed., *Nucleation*, Dekker, New York, 1969.
3. Ruckenstein, E., Nowakowski, B., *J. Colloid Interface Sci.* 137, 583 (1990). (Section 1.3 of this volume.)
4. Nowakowski, B., Ruckenstein, E., *J. Chem. Phys.* 94, 1397, 8487 (1991). (Section 2.1 of this volume.)
5. de Gennes, P.G., *Rev. Mod. Phys.* 57, 827 (1985).
6. Keck, J., Carrier, G. *J. Chem. Phys.* 43, 2284 (1965).
7. Borkovec, M., Berne, B.J., *J. Chem. Phys.* 82, 797 (1985).
8. Zeldovich, J.B., *Acta Physicochim. USSR* 18, 1 (1943).
9. Shizgal, B., Barrett, J.C., *J. Chem. Phys.* 91, 6505 (1989).
10. Kittel, C., *Introduction to Solid State Physics*, Wiley, New York, 1974.
11. Marlow, W.H., *J. Chem. Phys.* 73, 6284 (1980).

# 2.8 New Approach to the Kinetics of Heterogeneous Unary Nucleation on Liquid Aerosols of a Binary Solution

*Yuri Djikaev and Eli Ruckenstein**

The formation of a droplet on a hygroscopic center may occur either in a barrierless way via Köhler activation or via nucleation by overcoming a free energy barrier. Unlike the former, the latter mechanism of this process has been studied very little and only in the framework of the classical nucleation theory based on the capillarity approximation whereby a nucleating droplet behaves like a bulk liquid. In this paper, the authors apply another approach to the kinetics of heterogeneous nucleation on liquid binary aerosols, based on a first passage time analysis that avoids the concept of surface tension for tiny droplets involved in nucleation. Liquid aerosols of a binary solution containing a nonvolatile solute are considered. In addition to modeling aerosols formed through the deliquescence of solid soluble particles, the considered aerosols constitute a rough model of "processed" marine aerosols. The theoretical results are illustrated by numerical calculations for the condensation of water vapor on binary aqueous aerosols with nonvolatile nondissociating solute molecules using Lennard–Jones potentials for the molecular interactions. (© *2006 American Institute of Physics.* doi: 10.1063/1.2423012.)

## 2.8.1 Introduction

Heterogeneous condensation constitutes the most probable way of formation of atmospheric droplets. Droplets are formed in a homogeneous way only at high enough supersaturations of water vapor. The minimum supersaturation of water vapor required for droplet formation by homogeneous nucleation is about 300% [1], whereas the condensation of water vapor leading to cloud formation in the atmosphere occurs when the supersaturation of water is 1–2% [2]. In the atmosphere, large supersaturations are not required because of the presence of foreign nucleating centers.

There are several kinds of nucleating centers among which soluble nuclei are believed to be most efficient and hence most important for nucleation and condensation in the atmosphere. Among all kinds of soluble nuclei, the preexisting liquid aerosols are recognized as most widespread in the atmosphere. Therefore, it is important to develop an adequate theory of heterogeneous condensation on preexisting aerosols.

Many theoretical aspects of heterogeneous unary condensation on different kinds of heterogeneous centers have been well investigated (for short reviews, see, e.g., Ref. [3]). The thermodynamics of heterogeneous binary condensation on soluble aerosols was extensively examined [3–17], whereas the kinetics of the process has been studied to a lesser extent [18]. Moreover, the existing theory of nucleation on heterogeneous centers (including liquid aerosols) is based on the capillarity approximation (CA), whereby the droplet is considered as a particle with a sharp boundary and with the properties of a bulk liquid phase [19]. This approximation uses the macroscopic concept of surface tension for tiny droplets.

As an alternative to the kinetic theory of homogeneous nucleation in single component systems, Ruckenstein and coworkers [20–25] proposed a new approach to the kinetics of nucleation that takes into account the molecular interactions and does not employ the traditional thermodynamics, thus avoiding the use of the concept of surface tension for tiny droplets involved in nucleation. In the new kinetic theory, the rate of emission of molecules by a new-phase particle is determined through a mean first passage time analysis. This time is calculated by solving a single-molecule master equation for the probability distribution function of a surface layer molecule moving in a potential field generated by the droplet. Ruckenstein's theory was developed for both liquid-to-solid and vapor-to-liquid phase transitions. In the former case [20–22], the single-molecule master equation is a Fokker–Planck equation in the phase space that can be reduced to the Smoluchowski equation owing to the hierarchy of the characteristic time scales. In the latter case [23–25] the starting master equation was a Fokker–Planck equation for the probability distribution function of a surface layer molecule with respect to both its energy and phase coordinates. Unlike the case of liquid-to-solid nucleation, this Fokker–Planck equation cannot be reduced to the Smoluchowski equation, but the hierarchy of time scales does allow one to reduce it to a Fokker–Planck equation in the energy space.

A further development of that kinetic theory for vapor-to-liquid nucleation has been carried out by combining it with the density functional theory [26,27]. Recently [28], the Ruckenstein approach was extended to the kinetic theory of deliquescence, which represents the initial stage of heterogeneous condensation on a solid soluble particle during which the solid nucleus dissolves in a liquid film formed by the vapor

* *J. Chem. Phys.*, 125, 244707 (2006). Republished with permission.

molecules condensed on the nucleus. Previously, the kinetics of nucleational deliquescence was examined in the framework of the CA that involves the interfacial tensions of the films even at the early stages of deliquescence when they are very thin. Moreover, in the previous theory of deliquescence, the distribution of solute molecules within the film was assumed to be uniform. A new approach to the kinetics of deliquescence [28] does not employ the classical thermodynamics and the above approximations. The rates of emission of molecules from a liquid film into the vapor and from the solid core into the film were determined through a mean first passage time analysis. This time was calculated by solving the single-molecule master equation for the probability distribution of a "surface" molecule moving in a potential field created by the cluster. The temporal evolution of a liquid film around the solid core was described by a couple of first order ordinary differential equations involving the rates of absorption and emission of molecules by the liquid film.

After the deliquescence stage is over, the emerging aerosol particle represents a liquid droplet of a binary solution. Under favorable external conditions, this aerosol can serve as a heterogeneous center for further condensation of the vapor (or vapor mixture). Usually, in the atmosphere the water vapor is the dominant component that participates in droplet formation and therefore the condensation usually has a unary character. If the solute in the aerosol in nonvolatile, the state of the aerosol can be characterized by a single variable, e.g., its radius or its number of solvent (water) molecules.

The formation of a droplet on a hygroscopic center may occur either via the Köhler activation or via nucleation. The Köhler activation of the particle occurs when the vapor saturation ratio is equal to or higher than a threshold value at which the height of the activation barrier becomes zero (see Figure 2.23). In atmospheric models it is conventionally assumed that the droplet formation on hygroscopic particles occurs only via the Köhler activation. However, if the vapor saturation ratio is just slightly lower than its threshold value, the droplet formation may occur via the nucleation mechanism. Although, as shown below, the range of saturation ratios in which this mechanism is important is rather narrow, under some conditions it can contribute significantly to the number of droplets formed in the atmosphere, even though the traditional atmospheric models consider that in that range none will form.

So far, the nucleation mechanism of droplet formation on hygroscopic aerosols has been studied only on the basis of the classical thermodynamics in the framework of the CA. In this approximation the droplet is characterized by the surface tension considered equal to the surface tension of a bulk liquid. In this paper,

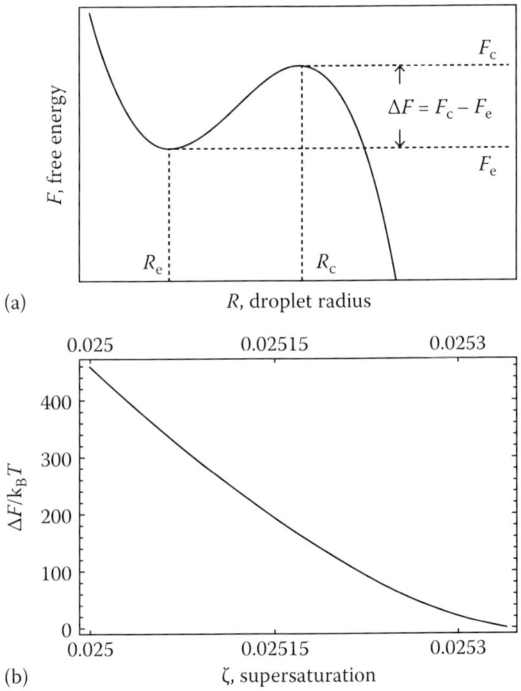


Figure 2.23

(a) Free energy $F$ of heterogeneous formation of a droplet on a foreign nucleating center as a function of the droplet radius $R$. The difference between $F_c$ and $F_e$ determines the height $\Delta F$ of the activation barrier. (b) The height of the activation barrier $\Delta F$ versus $\zeta_1$, for the heterogeneous condensation of water vapor on a spherical quartz particle of $R_0 = 30$ nm at $T = 293$ K (see Refs. [3–5]).

we apply a new approach, based on a first passage time analysis, to the kinetics of heterogeneous nucleation of water vapor on a liquid aerosol of an aqueous binary solution "water + nonvolatile solute." The solute molecules will be assumed to be nondissociating and interacting with the water molecules via a Lennard–Jones potential, which is also used for the "water–water" interactions. In addition to modeling an aerosol that results from the total deliquescence of a solid soluble particle, such an aerosol constitutes a rough model for a "processed" marine aerosol. Marine aerosols have been the subject of many *in situ* measurements and experimental studies (for short reviews, see, e.g., Refs. [29–31]). It was established that in the marine boundary layer the organic matter constitutes a significant fraction of the mass of sea salt aerosols. A widely known model of a marine aerosol was proposed by Gill et al. [32] and called an "inverted micelle." That model consists of an aqueous "brine" core encapsulated by a monolayer of organic surfactants with the hydrophobic tails exposed to air and the hydrophilic head groups immersed in water [29,30,32]. The chain reactions triggered by the atmospheric OH radicals oxidize the hydrophobic coating of the marine aerosol transforming it into a hydrophilic layer, which considerably enhances the ability of aerosols to nucleate.

It should be emphasized that the Köhler theory [33] provides only the threshold of the saturation ratio at or above which the droplet formation occurs as a barrierless process. It does not address the issues regarding the possibility of droplet formation under conditions when there exists an activation barrier [34]. The present paper is concerned with the nucleation mechanism of heterogeneous droplet formation that occurs under conditions in which the Köhler activation cannot take place.

The paper is structured as follows. Section 2.8.2 briefly outlines the derivation of the kinetic equation of heterogeneous nucleation on a liquid binary aerosol starting with the discrete equation of balance for the size distribution of droplets. Section 2.8.3 presents the new approach to the kinetics of heterogeneous nucleation based on a first passage time analysis. The theory is developed for an aqueous aerosol containing a nonvolatile solute in a water vapor. In Section 2.8.4, the new theory is illustrated with numerical calculations. Section 2.8.5 summarizes the conclusions.

## 2.8.2 Kinetics of Nucleation

Consider a liquid aerosol of a binary solution of component 1 (solvent) and component 2 (nonvolatile solute) at equilibrium with the vapor of component 1. At a given temperature $T$, the state of the vapor can be characterized by its supersaturation $\zeta$ defined as

$$\zeta = P/P_{\infty} - 1, \tag{2.250}$$

where $P$ is the pressure of the vapor (of component 1) and $P_{\infty}$ is the pressure of the vapor in equilibrium with its pure bulk liquid.

The number of molecules of component $i$ in the initial aerosol will be denoted by $n_{i0}$ ($i = 1, 2$). Clearly, the number of solute molecules $n_{20}$ in the aerosol does not change. However, once the supersaturation $\zeta$ becomes large enough, the aerosol can grow via vapor condensation onto it. The initial absorption of vapor molecules by the aerosol may be thermodynamically favorable (see, e.g., Refs. [7,18]), as long as this reduces the free energy of the system compared to its initial state (vapor + aerosol). However, at some droplet size this condition no longer holds, and the incorporation of a new molecule into the droplet begins to demand work, i.e., becomes thermodynamically unfavorable. The growth can still occur owing to fluctuations until the droplet attains a critical size, after which its growth becomes thermodynamically favorable and occurs irreversibly. Thus, $F(n_1)$, the free energy of heterogeneous formation of a droplet, depends on the droplet size as follows [7,18] (Figure 2.23a): with increasing size the free energy first decreases, attains a minimum $F_e$, increases again until it attains a maximum $F_c$, and after which the free energy decreases monotonically.

The physical mechanism underlying the initial decrease of $F$ depends on the nature of the heterogeneous nucleus. For an insoluble nucleus, it involves the disjoining pressure in the thin liquid film around the nucleus which reduces the chemical potential of the condensate within the film compared to the bulk liquid. For liquid aerosols of binary solutions, the mechanism consists of the decrease of the solute chemical potential in the droplet because the binary solution therein becomes increasingly dilute due to water condensation. In each case, the increase in the interfacial contribution to the free energy of the system, which occurs because of the increase in the interfacial area, is dominated by the decrease in the free energy due to one of the aforementioned mechanisms. The difference $\Delta F = F_c - F_e$ provides the height of the activation barrier for heterogeneous nucleation and is extremely sensitive to the vapor supersaturation.

When $\Delta F \lesssim kT$, the phase transition becomes practically barrierless and occurs as spinodal decomposition. The formation of droplets under such conditions is often referred to as the Köhler activation of heterogeneous centers [33]. The nucleation mechanism of heterogeneous droplet formation is important only in a very narrow range of supersaturations slightly just below the threshold for the Köhler activation [34]. This is demonstrated in Figure 2.23b, which presents the dependence of $\Delta F$ on $\zeta$ for heterogeneous condensation of a pure water vapor on an insoluble spherical quartz particle (for details, see figure caption). The curve was obtained by using a CA-based theory of binary nucleation on an insoluble particle. It should be noted that Köhler himself did not calculate the free energy of formation $F$. Instead, he obtained an expression for the threshold supersaturation $\zeta_{\text{th}}$ by examining the chemical potential of the water molecules in the droplet as a function of its radius. The maximum value of that chemical potential provided the threshold value of the water chemical potential in the vapor phase (hence its supersaturation), because if the latter is higher than the former, the vapor condensation into the droplet becomes thermodynamically favorable and occurs in a barrierless way. The result can be presented in the form $\ln(\zeta_{\text{th}}+1)=2\left(8\pi d_1^2\sigma/9k_BT\right)^{3/2}/n_{20}$, where $d_1$ is the radius of a sphere having the same volume as a water molecule in the liquid phase and $\sigma$ is the surface tension of the droplet. The super saturation at which the Köhler activation of most atmospheric aerosols occurs depends on the nature of the aerosols and is of the order of $10^{-2}$ for most aerosols [2]. While both aforementioned methods for determining the threshold for the Köhler activation are based on the classical thermodynamics (and hence on the concept of surface tension), in the following discussion we will present an alternative approach thereto that is based on a first passage time analysis. The application of both previous methods for a given temperature requires the knowledge of the surface tension of the droplet for that temperature, whereas the application of our method requires the knowledge of the temperature independent parameters of intermolecular interactions. Besides, the use of the surface tension concept for small droplets may lead to a large inaccuracy in theoretical predictions.

The number of molecules of component 1 passed into the aerosol from the vapor will be denoted by $n_1$. The total number of condensate molecules in the droplet is $n_{10}+n_1$, and its radius $R$ is related to $n_1$, $n_{10}$, and $n_2$ by $R=[3((n_1+n_{10})\upsilon_1+n_2\upsilon_2)/4\pi]^{1/3}$, where $\upsilon_i$ $(i=1, 2)$ is the volume per molecule of component $i$ in the binary droplet.

In the framework of the CA, it is assumed that the characteristic time of the internal relaxation processes in the droplet is very small in comparison with the characteristic time between two successive elementary interactions of the droplet with the vapor. This allows us to assume that the droplet is in internal thermodynamic equilibrium at any time, i.e., can be characterized by a uniform composition (solute mole fraction) $x=n_2/(n_1+n_{10}+n_{20})$ and temperature $T$.

Let us denote the number of initial aerosols (all assumed to be of the same size and composition) in the vapor by $N_0$. If the vapor saturation ratio is not too high, the formation of regularly growing droplets occurs in a fluctuational mode via nucleation as opposed to the barrierless way via Köhler activation) [33]. Therefore, the size distribution of growing droplets, $g(n_1,t)$, with respect to the variable $n_1$ at time $t$ will not be monodisperse. Denoting by $W_1^+(n_1)$ and $W_1^-(n_1)$ the numbers of molecules of component 1 that a droplet absorbs and emits, respectively, per unit time, one can write the following discrete equation of conservation:

$$\frac{\partial g(n_1,t)}{\partial t}=W_1^+(n_1-1)g(n_1-1,t)-W_1^+(n_1)g(n_1,t)+W_1^-(n_1+1)g(n_1+1,t)-W_1^-(n_1)g(n_1,t). \tag{2.251}$$

This equation assumes that the evolution of droplets occurs through the absorption and emission of a molecule of component 2.250. For the sake of simplicity, we neglect the multimer absorption and emission.

Assuming $n_1 \gg 1$, one can rewrite Equation 2.251 in a continuous form by using Taylor series expansions for its right-hand side (RHS) and retaining only the lowest order terms. The resulting partial differential equation has the Fokker–Planck form

$$\frac{\partial g(n_1,t)}{\partial t}=\frac{\partial}{\partial n_1}J_1(n_1,t), \tag{2.252}$$

where

$$J_1(n_1,t)=-A_1(n_1)g(n_1,t)-\frac{\partial}{\partial n_1}(B_1(n_1)g(n_1,t)). \tag{2.253}$$

Here, the coefficients $A_1(n_1)$ and $B_1(n_1)$ depend on the emission and absorption rates $W_1^-$ and $W_1^+$ of the liquid cluster. The steady-state solution of the kinetic equation (Equation 2.252) is in the vicinity of the critical point subject to the conventional boundary conditions:

$$\frac{g(n_1)}{g_e(n_1)} \to 1 \quad (n_1 \to 0), \quad \frac{g(n_1)}{g_e(n_1)} \to 0 \quad (n_1 \to \infty), \tag{2.254}$$

where $g_e(n_1)$ is the equilibrium distribution, provides the steady-state nucleation rate.

To solve Equation 2.252 subject to the boundary conditions (Equation 2.254), it is necessary to know its coefficients $A_1(n_1)$ and $B_1(n_1)$, i.e., the emission and absorption rates $W_1^-(n_1)$ and $W_1^+(n_1)$. The rate of absorption $W_1^+(n_1)$ of condensate molecules is usually found by using the kinetic theory of gases. The emission rate $W_1^-(n_1)$ in the classical nucleation theory (CNT) is assumed to be provided by the absorption rate that the droplet would have if it were in equilibrium with an imaginary mother phase. In this case, the determination of $W_1^-(n_1)$ is based (via the Gibbs–Kelvin relation) on the interfacial tension of the droplet. Because the concept of interfacial tension may not be adequate for too small droplets (such as those involved in nucleation), not to mention the assumption (of the CA) that it is equal to the interfacial tension of a bulk liquid, its use constitutes a drawback of CNT.

### 2.8.3 A New Approach to Kinetics of Heterogeneous Nucleation

Equation 2.252 can be rewritten as

$$\frac{\partial g(n_1,t)}{\partial t} = W_{1c}^+ \frac{\partial}{\partial n_1}\left[\frac{\partial}{\partial n_1} + \frac{\partial G}{\partial n_1}\right] g(n_1,t), \tag{2.255}$$

(see Appendix 2F), where $G = -\ln g_e(n_1)$. The quantity $G$ in the present theory plays a role similar to that of the difference $F(n_1) - F_e$ for the free energy of cluster formation in CNT and has a shape similar to the latter (Figure 2.24a). The essence of the new method, as an alternative to the CA-based theory, consists of constructing the equilibrium distribution of clusters, $g_e(n_1)$ [and function $G(n_1)$], without applying the classical thermodynamics to droplets, hence *without* using the macroscopic interfacial tension.

The above differential equation describes the time evolution of heterogeneously formed droplets. In order to proceed to its solution, it is necessary to determine the condensate absorption rate $W_{1c}^+$ for the critical droplet and the equilibrium distribution $g_e(n_1)$, which depend on the emission and absorption rates $W_1^+$ and $W_1^-$.

One usually assumes that the exchange of molecules between the liquid droplet and the vapor occurs in a free-molecular regime (this exchange may occur in a diffusion regime as well if the density of the vapor is high enough). In this regime, the condensate absorption rate $W_1^+$ is given by the well-known expression [35] $W_1^+ = \frac{1}{4}\alpha_1 \upsilon_{T1} \rho_1^{\upsilon} S$, where $\alpha_1$ is the sticking coefficient of component 1, $\upsilon_{T1}$ is the mean thermal velocity of the vapor molecules, $\rho_1^{\upsilon}$ is the number density of the water vapor, and $S = 4\pi R^2$ is the surface area of the droplet. If the exchange of molecules occurs in the diffusion regime, the rate $W_1^+$ can be easily calculated as well.

In order to avoid the use of macroscopic thermodynamics in the unary nucleation theory, a kinetic theory based on a first passage time analysis was developed by Ruckenstein and coworkers for unary nucleation [20–25]. The key idea of that approach consists of calculating the emission rates $W_1^-(n_1)$ of a binary cluster directly (rather than indirectly by using the Gibbs–Kelvin relation, as usually done in the classical theory) by solving the Fokker–Planck or Smoluchowski equation governing the motion of a single molecule in a potential well around the cluster.

Using that method (for details, see Appendix 2G) [20–25], one can obtain the following expression for the rate of emission of molecules of component 1 from the droplet:

$$W_1^- = \frac{N_1}{\bar{\tau}_1} = N_1 D_1 \omega_1, \tag{2.256}$$

where $N_1$ is the number of water molecules in the potential well around the cluster, $\bar{\tau}_1$ is the mean first passage time of a molecule over the outer "boundary" of the well (i.e., average dissociation time), $D_1$ is its diffusion coefficient in the well, and $\omega_1$ is defined as $\omega_1 \equiv 1/D_1\bar{\tau}_1$. Clearly, $\omega_1$ and $W_1^-$ are functions of $R$.

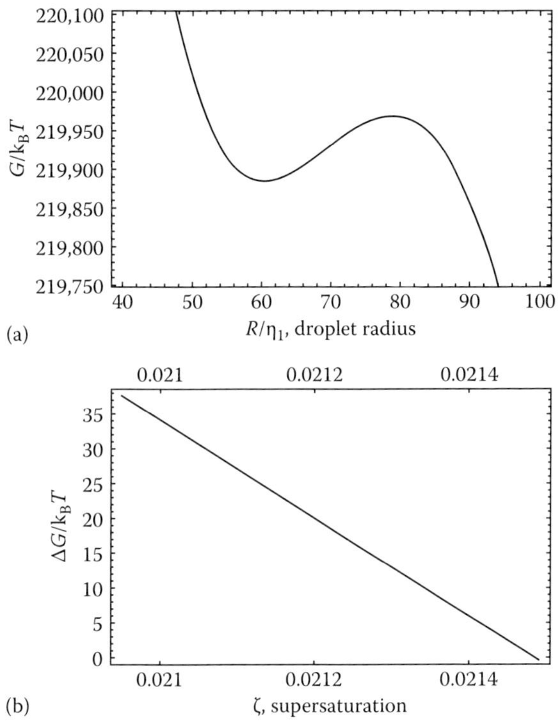


Figure 2.24

Heterogeneous condensation of water on a model aerosol of aqueous binary solution of $n_{20} = 6 \times 10^3$ at the salinity of seawater (corresponding to $R_0 = 35\eta$ and $T = 293$ K). (a) Function $G$ versus the droplet radius $R$ for $\epsilon_{11}/\epsilon_{12} = 4.47$. The difference between $G_c$ and $G_e$ between the maximum and minimum of $G$ determines the height $\Delta G$ of the "kinetic" activation barrier. (b) The height of the activation barrier $\Delta G$ versus $\zeta$, for $\epsilon_{11}/\epsilon_{12} = 4.47$.

At a given saturation ratio, the rates of emission and absorption are equal to each other for a critical droplet, the radius and composition of which will be denoted by $R_c$ and $x_c$, respectively. The equality $W_{1c}^- = W_{1c}^+$ allows one to obtain the following expressions for the absorption rate $W_1^+(R)$ and saturation ratio $\zeta$:

$$W_1^+(R) = N_1 D_1 \frac{R_2}{R_c^2} \omega_1(R_c), \tag{2.257}$$

$$\zeta = \omega_1(R_c)/\omega_{1\infty} - 1, \tag{2.258}$$

where the subscript $\infty$ marks the quantity $\omega_1$ for a solution with a flat interface and of concentration $x_c$.

For $R_c$ much larger than the average molecular diameter, the results of this theory are expected to coincide with those of the CA-based theory. In particular, the classical Kelvin relations between the saturation ratio and the critical radius $R_c$ are recovered and can be employed to evaluate the macroscopic surface tension [20–25].

The quantity $N_1$ in Equation 2.256 for $W_1^-$ depends on $\lambda_1$. However, owing to Equation 2.257, neither $N_1$ nor $D_1$ enters into the final theoretical expressions since they depend on $W_1^-$ only via the ratios $W_1^-/W_1^+$ in the quantity $G = -\ln g_e(n_1)$ (see Equation 2.255). One can obtain the following expression for the equilibrium distribution $g_e(n_1)$ (see Appendix 2G):

$$g_e(n_1) = N_0 \tilde{P}(n_1) \Big/ \sum_{n_1=0}^{\infty} \tilde{P}(n_1),$$
$$\tilde{P}(n_1) = \frac{\omega_{1c}^{n_1}}{R_c^{2n_1}} \frac{R^2(0)}{R^2(n_1)} \prod_{i=1}^{n_1} \frac{R^2(n_1 - i + 1)}{\omega_1(n_1 - i + 1)}. \tag{2.259}$$

Thus, knowing $\omega_1$ and $R$ as functions of $n_1$, one can find $G = -\ln g_e(n_1)$ from Equation 2.259. Clearly, $G$ in the present theory plays a role similar to that of the free energy $F$ of a cluster formation in the CA-based theory. As a function of $n_1$ (or, equivalently, $R$) it has the shape shown in Section 2.8.4 (Figure 2.24a). The crucial advantage of the present theory is that it does not employ the macroscopic interfacial tension for small droplets.

The steady-state nucleation rate is then obtained by solving Equation 2.255 subject to the boundary conditions (Equation 2.254). It can be written in the form

$$J_s = Zg_e(n_{1c}), \tag{2.260}$$

where the subscript c marks quantities at the maximum of the function $G = G(n_1)$ and $Z = W_{1c}^{+}\sqrt{|G_c''|/2\pi}$ is usually referred to as the Zeldovich factor ($G_c''$ is the second derivative of $G$ with respect to $n_1$ at $n_1 = n_{1c}$).

## 2.8.4 Numerical Evaluations

To illustrate the above theory with numerical results, we carried out numerical evaluations for heterogeneous condensation of water vapor on liquid aerosols of an aqueous binary solution with a nonvolatile solute. Such an aerosol may emerge as a result of deliquescence of a solid particle in a water vapor and can serve as a heterogeneous center for further condensation of vapor (or vapor mixture), leading to the formation of a stable droplet either via Köhler activation or via nucleation, depending on the saturation ratio of the vapor and the characteristics of the aerosol. Besides, such an aerosol can serve as a rough model of a marine aerosol whereof the surfactant coating has been processed and rendered hydrophilic by atmospheric hydroxyl radicals. The temperature of the water vapor was taken $T = 293.15$ K and the number density of initial aerosols therein was $N_0 = 10^5$ cm$^{-3}$. The interactions between a pair of molecules of components $i$ and $j$ were modeled via the Lennard–Jones (LJ) potentials

$$\phi_{ij}(r) = 4\epsilon_{ij}\left[\left(\frac{\eta_{ij}}{r}\right)^{12} - \left(\frac{\eta_{ij}}{r}\right)^{6}\right] \quad (i, j = 1, 2), \tag{2.261}$$

where $\epsilon_{ij}$ and $\eta_{ij}$ are energy and length parameters, respectively, satisfying the simple mixing rules $\epsilon_{ij} = \sqrt{\epsilon_{ii}\epsilon_{jj}}$ and $\eta_{ij} = \frac{1}{2}(\eta_i + \eta_j)$ with $\eta_i$ ($i$ = 1, 2) being the length parameter of the interaction potential between two molecules of component $i$. The following values of the system parameters were used in the calculations:

$$\eta_1 = 3.15\times10^{-8}\text{cm}, \quad \eta_2/\eta_1 = 30.1, \quad \epsilon_{11}/kT = 0.26,$$

$$R_0/\eta_1 = 35, \quad \upsilon_1/\eta_1^3 = 0.96, \quad \upsilon_2/\eta_1^3 = 28.7,$$

$$D_1 = 1\times10^{-5}\text{cm}^2/\text{s}.$$

This set of parameters was chosen as representative for water (component 1) and some large nondissociating molecules (component 2). The results of the numerical calculations are presented in Figures 2.24 through 2.27.

In Figure 2.24a, the quantity $G$, which *can be interpreted* as (but *is not equivalent* to) the free energy of formation of a droplet (see Section 2.8.3), is plotted as a function of $R$ for $\epsilon_{11}/\epsilon_{12} = 4.7$ at $\zeta = 0.02134$. In the present work the threshold saturation ratio of the Köhler activation is evaluated using the condition $\Delta G \simeq 0$, where $\Delta G \equiv G_c - G_e$ is the difference between the values of $G$ at the maximum and at the minimum, respectively. To emphasize that the nucleation mechanism of heterogeneous droplet formation is important only in a very narrow range of supersaturations just slightly below the threshold of the Köhler activation, in Figure 2.24b $\Delta G$ is plotted as a function of $\zeta$ for $\epsilon_{11}/\epsilon_{12} = 4.47$. The threshold of the Köhler activation is found to be approximately $\zeta = 0.02148$. As clearly shown in Figure 2.24b, our approach predicts that the nucleation mechanism of heterogeneous droplet formation can occur only in a narrow range of supersaturations just below the Köhler activation threshold. Nevertheless, under some conditions (saturation ratios below the Köhler activation threshold and sufficiently long times), the nucleation mechanism can contribute significantly to the number of droplets formed, even though the traditional models assume that in that range none will form.

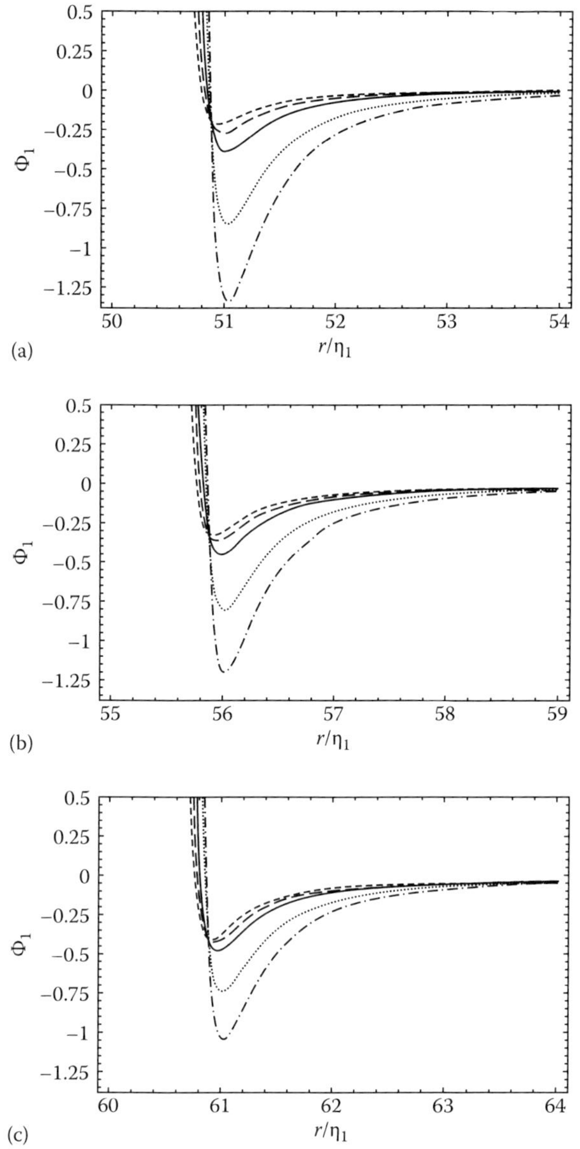


Figure 2.25

The potential well $\Phi_1$ around a droplet as a function of the distance $r$ from the droplet center, for (a) $R/\eta_1 = 35$, (b) $R/\eta_1 = 40$, and (c) $R/\eta_1 = 45$ and fixed number of solute molecules $n_{20} = 6 \times 10^3$. The short-dashed curves are for $\epsilon_{11}/\epsilon_{12} = 4.47$, the long-dashed curve for $\epsilon_{11}/\epsilon_{12} = 2$, the solid for $\epsilon_{11}/\epsilon_{12} = 1$, the dotted for $\epsilon_{11}/\epsilon_{12} = 0.34$, and the dot-dashed for $\epsilon_{11}/\epsilon_{12} = 0.2$.

Figure 2.25 presents the potential well $\Phi_1$ around the droplet as a function of the distance from the droplet center, for $R/\eta_1 = 35$ (Figure 2.25a), $R/\eta_1 = 40$ (Figure 2.25b), and $R/\eta_1 = 45$ (Figure 2.25c) and fixed number of solute molecules $n_{20} = 6 \times 10^3$. Different curves in the same panel correspond to different ratios $\epsilon_{11}/\epsilon_{12}$ (see figure caption for details). Physically, the ratio $\epsilon_{11}/\epsilon_{12}$ can be interpreted as an indicator of the solute solubility in water: the lower this ratio, the higher the solubility. As expected, the depth of the potential well is more sensitive to the ratio $\epsilon_{11}/\epsilon_{12}$ at the initial stage of the formation of the droplet on the aerosol. As the droplet grows, the solution becomes more dilute and the contribution of the solute molecules to $\Phi_1(r)$ gradually decreases.

In Figure 2.26, the ratio $W_1^-/N_1D_1$ (i.e., $\omega_1(R)$) for the emission rate of the droplet is plotted as a function of droplet size, $R/\eta_1$, for different ratios $\epsilon_{11}/\epsilon_{12}$ (for details, see figure caption). The number of solute molecules in the droplet remains constant, $n_{20} = 6 \times 10^3$. Remarkably, the larger the ratio $\epsilon_{11}/\epsilon_{12}$, the sharper is the decrease of the emission rate of the droplet with increasing droplet radius. For $\epsilon_{11}/\epsilon_{12} = 4.47$, the function $\omega(R)$ attains its asymptotic value (presumably corresponding to a flat interface) at $R/\eta_1 \simeq 150$, whereas

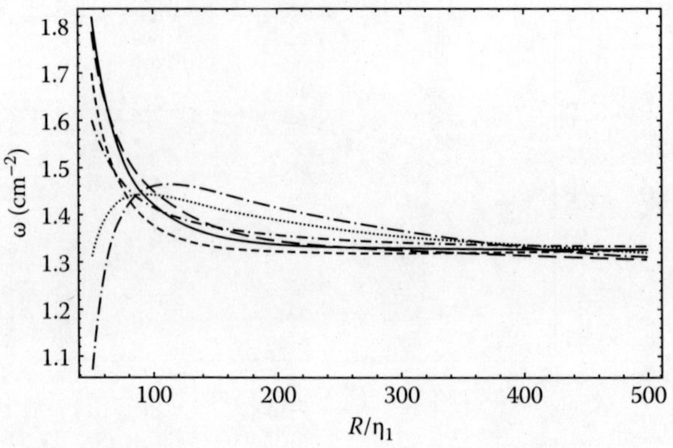


Figure 2.26

The emission characteristic function $\omega_1(R)$ for $\epsilon_{11}/\epsilon_{12} = 4.47$ (short-dashed curve), $\epsilon_{11}/\epsilon_{12} = 2$ (long-dashed curve), $\epsilon_{11}/\epsilon_{12} = 1$ (solid curve), $\epsilon_{11}/\epsilon_{12} = 0.34$ (dot-dashed curve), $\epsilon_{11}/\epsilon_{12} = 0.2$ (dotted curve), and $\epsilon_{11}/\epsilon_{12} = 0.14$ (long-dashed-dotted curve).

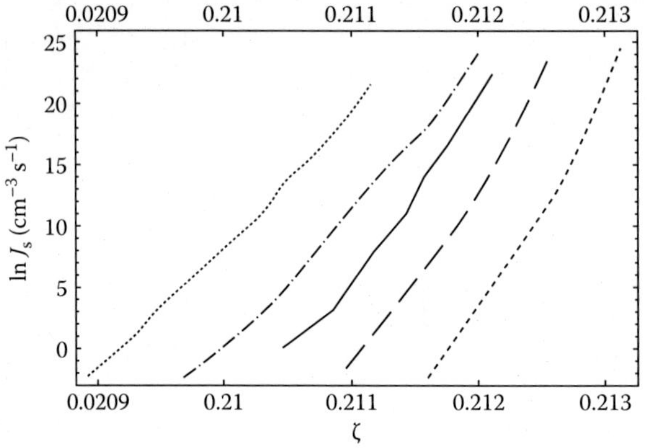


Figure 2.27

The steady-state rate $J_s$ of heterogeneous nucleation of water vapor on binary aerosols as a function of water vapor saturation ratio $\zeta_1$. The short-dashed curve corresponds to $\epsilon_{11}/\epsilon_{12} = 4.47$, the long-dashed curve to $\epsilon_{11}/\epsilon_{12} = 2$, the solid to $\epsilon_{11}/\epsilon_{12} = 1$, the dot-dashed to $\epsilon_{11}/\epsilon_{12} = 0.34$, and the dotted to $\epsilon_{11}/\epsilon_{12} = 0.2$. The aerosols were assumed to have the initial radius $R_0 = 35\eta_1$ and contain $n_{20} = 6 \times 10^3$ solute molecules each. The number density of aerosols in the system was taken to be $N_0 = 10^3$ cm$^{-3}$.

for $\epsilon_{11}/\epsilon_{12} = 0.14$ this occurs only for droplets with $R/\eta_1 > 450$. The function $\omega(R)$ is not monotonic for sufficiently small values of the ratio $\epsilon_{11}/\epsilon_{12}$. For $\epsilon_{11}/\epsilon_{12} = 0.14$, the function $\omega_1(R)$ increases as $R$ decreases from $R = \infty$, but at some droplet radius it attains a maximum and begins to decrease upon further decrease in $R$. Since the number of solute molecules in the droplet does not change, this nonmonotonic behavior of $\omega_1(R)$ can be explained as follows. On one hand, an increase in the concentration of solute with decreasing $R$ favors a decrease in the emission rate of the droplet because in this case the solute–water interaction is much stronger than the water–water interaction. On the other hand, the total number of molecules in the droplet decreases with decreasing $R$ and this is expected to favor an increase in the droplet emission rate because the potential well becomes shallower. The former effect dominates when the droplet size is sufficiently small.

Figure 2.27 presents $J_s$, the steady-state rate of heterogeneous nucleation of water vapor on a binary aerosol as a function of water vapor supersaturation $\zeta$, for different values of the ratio $\epsilon_{11}/\epsilon_{12}$ (see figure caption for details). The aerosols were assumed to have the initial radius $R_0 = 35\eta_1$ and to contain $n_{20} = 6 \times 10^3$ solute molecules each. The number density of aerosols in the system was taken to be $N_0 = 10^3$ cm$^{-3}$. The nucleation rate changes significantly with $\epsilon_{11}/\epsilon_{12}$. We evaluated that the threshold supersaturation for the Köhler activation is about 0.02–0.025. These values are consistent with the previous observations [2] that the condensation of water vapor leading to cloud formation in the atmosphere occurs when the relative humidity is 1–2%.

The CA-based theory of heterogeneous nucleation on soluble particles also predicts that the Köhler activation of particles [6,18] with $R_0 \simeq 10^{-6}$ cm, $n_{20} \simeq 10^5$, and $\upsilon_2 = 3 \times 10^{-23}$ cm$^3$ in a water vapor at $T = 293.15$ K occurs at a supersaturation of about 0.019.

### 2.8.5 Conclusions

We have studied the nucleation stage of heterogeneous condensation of a unary vapor on liquid aerosols of a binary solution containing a nonvolatile solute. Heterogeneous condensation of water vapor plays a major role in the formation of atmospheric droplets. Among various kinds of nucleating centers in the atmosphere, the liquid aerosols of aqueous solutions with nonvolatile solutes are most widespread. The formation of a droplet on a hygroscopic center may occur either by the Köhler activation or by nucleation. Unlike the Köhler activation, the nucleation mechanism of this process has been studied very little and only in the framework of the CNT based on classical thermodynamics and the CA. In that approximation, the droplet is characterized by the surface tension considered equal to the surface tension of a bulk liquid. In order to avoid the use of thermodynamics in the kinetic theory of nucleation, an alternative approach was previously proposed by Ruckenstein and coworkers. The new approach involves molecular interactions and does not employ the traditional thermodynamics, thus avoiding the use of the concept of surface tension for tiny clusters. The rate of emission of molecules by a droplet is determined using a mean first passage time analysis. This time is calculated by solving the single-molecule master equation for the probability distribution function of a surface layer molecule moving in a potential field created by the droplet.

In this paper, we have applied that new approach to the kinetics of heterogeneous nucleation on aqueous aerosols containing a nonvolatile solute. In addition to modeling the aerosols that result from the deliquescence of solid soluble particle, such aerosols can serve as a rough model for "processed" marine aerosols (whereof the surfactant coating has been processed and rendered hydrophilic by atmospheric hydroxyl radicals).

The theoretical results are illustrated by numerical calculations for the condensation of water vapor on aqueous binary aerosols with nonvolatile nondissociating solute molecules. The interactions between all pairs of molecules were modeled by Lennard–Jones (LJ) potentials. The set of system parameters was chosen to represent water (component 1) and some large nonvolatile nondissociating molecules (component 2). As expected, the new theory predicts that the Köhler activation of liquid aerosols occurs at saturation ratios much lower than those required for homogeneous nucleation of water vapor. In addition, it predicts a substantial droplet formation via the nucleation mechanism at saturation ratios slightly below the threshold of Köhler activation.

## Acknowledgments

This work was supported by the National Science Foundation through the Grant No. CTS-0000548.

## Appendix 2F: Kinetic Equation of Heterogeneous Nucleation

To derive Equation 2.255, let us first transform the right-hand side of Equation 2.251 by using Taylor series expansions. One thus obtains $B_1(n_1) = -\frac{1}{2}\left(W_1^+(n_1) + W_1^-(n_1)\right)$. On the other hand, according to the principle of detailed balance (see Equation 2G.1), $W_1^-(n_1+1)/W_1^+(n_1) = \exp[-G(n_1) + G(n_1+1)] \simeq \exp[\partial G(n_1)/\partial n_1]$. The first derivatives $\partial G(n_1)/\partial n_1$ is equal to zero at the critical point (i.e., at the maximum of $G(n_1)$) and is close to zero in its vicinity where, in addition, $W_1^+(n_1+1) \simeq W_1^+(n_1)$. Therefore, one can write

$$W_1^-(n_1)/W_1^+(n_1) \simeq 1 + o(\partial G(n_1)/\partial n_1), \tag{2F.1}$$

where $o(x)$ denotes a quantity of the same order of magnitude as $x$. Neglecting the terms of the order $o(\partial G(n_1)/\partial n_1)$, one can assume that the sum $W_1^+(n_1) + W_1^-(n_1)$ in the vicinity of the critical point (where the kinetic equation [2.255] has to be solved) is equal to $2W_1^+(n_1)$. Furthermore, in this vicinity $W_1^+(n_1) \simeq W_1^+(n_{1c}) \equiv W_{1c}^+$; hence

$$B_1 \simeq -W_{1c}^+. \tag{2F.2}$$

Because at equilibrium $J_1 = 0$, Equation 2.253 leads to

$$A_1(n_1)g_e(n_1)+B_1(n_1)\frac{\partial g_e(n_1)}{\partial n_1}=0, \tag{2F.3}$$

hence one obtains

$$A_1(n_1)=-B_1(n_1)\frac{1}{g_e(n_1)}\frac{\partial g_e(n_1)}{\partial n_1}\simeq W_{1c}^{+}\frac{\partial G(n_1)}{\partial n_1}, \tag{2F.4}$$

where $G(n_1) = -\ln g_e(n_1)$. The coefficients $A_1(n_1)$ and $B_1(n_1)$ being the same for equilibrium and nonequilibrium states, $J_1$ in Equation 2.253 acquires the form

$$J_1(n_1,t)=-W_{1c}^{+}\left(\frac{\partial}{\partial n_1}+\frac{\partial G(n_1)}{\partial n_1}\right). \tag{2F.5}$$

The substitution of this equation into Equation 2.252 leads to Equation 2.255.

## Appendix 2G: Emission Rate via a First Passage Time Analysis

Consider a spherical binary cluster immersed in a single component initial phase. A molecule located in the surface layer of the cluster performs thermal chaotic motion in a spherically symmetric potential well $\phi_1(r)$ created in the surface layer by its interactions with the cluster and surrounding medium. Neglecting the internal structure of molecules and assuming pairwise additivity of the intermolecular interactions, the potential $\phi_1(r)$ is given by

$$\phi_1(r)=\sum_{j=1}^{2}\int_V d\mathbf{r}'\rho_j(r')\phi_{1j}(|\mathbf{r}'-\mathbf{r}|). \tag{2G.1}$$

Here, $\mathbf{r}$ is the coordinate of the surface molecule of component 1, $\rho_j(r')$ $(j = 1, 2)$ is the number density of molecules of component $j$ at point $\mathbf{r}'$ (spherical symmetry is assumed), and $\phi_{1j}(|\mathbf{r}'-\mathbf{r}|)$ is the interaction potential between two molecules of components 1 and $j$ located at points $\mathbf{r}$ and $\mathbf{r}'$, respectively. The integration in Equation 2G.1 is over the whole volume of the system, but the contribution of the medium in which the cluster is immersed can be assumed to be negligible because of the small density. The molecule is considered as belonging to the cluster as long as it remains in the potential well and as dissociated from the cluster when it passes over the outer boundary of the well. The rate of emission $W_1^-$ is determined by the mean first passage time necessary for this threshold point to be reached.

The mean first passage time of a molecule dissociating from the cluster is calculated on the basis of a kinetic equation governing the chaotic motion of the molecule escaping from the potential well generated by the cluster. The chaotic motion of a molecule is governed by the Fokker–Planck equation for the single-molecule distribution function with respect to its coordinates and momenta, i.e., in the phase space [36–38]. Prior to a dissociation event, the evolution of a molecule (belonging to the cluster) occurs in the surface layer of a condensed phase cluster, where the relaxation time for its velocity distribution is extremely short and negligible compared to the characteristic time scale of the dissociation process. Under these conditions, the Fokker–Planck equation reduces to the Smoluchowski equation, which involves diffusion in an external field [37,38]. In the case of spherical symmetry, it can be written as [20–22]

$$\frac{\partial p_1(r,t|r_0)}{\partial t}=D_1r^{-2}\frac{\partial}{\partial r}\left(r^2e^{-\Phi_1(r)}\frac{\partial}{\partial r}e^{\Phi_1(r)}p_1(r,t|r_0)\right), \tag{2G.2}$$

where $p_1(r,t|r_0)$ is the probability of observing a condensate molecule between $r$ and $r + dr$ at time $t$ given that initially it was at a radial distance $r_0$, $D_1$ is its diffusion coefficient in the well, and $\Phi_1(r) = \phi_1(r)/kT$.

The mean passage time depends on the initial position (distance from the center of the cluster) $r_0$ of the molecule. It is convenient to use the backward Smoluchowski equation [20–22] that expresses the dependence of the transition probability $p_1(r,t|r_0)$ on $r_0$:

$$\frac{\partial p_1(r,t|r_0)}{\partial t} = D_1 r_0^{-2} e^{\Phi_1(r_0)} \frac{\partial}{\partial r_0}\left( r_0^2 e^{-\Phi_1(r_0)} \frac{\partial}{\partial r_0} p_1(r,t|r_0) \right). \tag{2G.3}$$

The probability that a condensate molecule, initially at a distance $r_0$ within the surface layer, will remain in this region after time $t$ is given by the survival probability [20–22]

$$q_1(t|r_0) = \int_R^{R+\lambda_1} \mathrm{d}r r^2 p_1(r,t|r_0) \quad (R < r_0 < R + \lambda_1), \tag{2G.4}$$

where $R$ is the radius of the cluster and $\lambda_1$ is the width of the potential well defined such that a condensate molecule can be assumed to have dissociated from the cluster at the distance $R + \lambda_1$ from its center. The width $\lambda_1$ should be chosen such that $\lambda_1/R \lesssim 1$ and $\Phi_1(R+\lambda_1)/\Phi_1^0 \ll 1$, where $\Phi_1^0$ is the depth of the well. Having chosen $\lambda_1$, the probability for the dissociation time to be between 0 and $t$ is equal to $1 - q_1(t|r_0)$, and the probability density for the dissociation time is given by $-\partial q_1/\partial t$. The first passage time is provided by [20–22]

$$\tau_1(r_0) = -\int_0^\infty t \frac{\partial q_1(t|r_0)}{\partial t} \mathrm{d}t = \int_0^\infty q_1(t|r_0) \mathrm{d}t. \tag{2G.5}$$

The equation for the first passage time is obtained by integrating the backward Smoluchowski equation (Equation 2G.3) with respect to $r$ and $t$ over the entire range and using the boundary conditions $q_1(0|r_0) = 1$ and $q_1(t|r_0) \to 0$ as $t \to \infty$ for any $r_0$. This yields [20–22]

$$-D_1 r_0^{-2} e^{\Phi(r_0)} \frac{\partial}{\partial r_0}\left( r_0^2 e^{-\Phi_1(r_0)} \frac{\partial}{\partial r_0} \tau_1(r_0) \right) = 1. \tag{2G.6}$$

The solution of Equation 2G.6 subject to the reflecting boundary condition at the inner boundary of the well ($\mathrm{d}\tau_1/\mathrm{d}r_0 = 0$ at $r_0 = R$) and the radiation boundary condition at the outer boundary [$\tau_1(r_0) = 0$ at $r_0 = R + \lambda_1$] has the form

$$\tau_1(r_0) = \frac{1}{D_1} \int_0^{R+\lambda_1} \mathrm{d}y y^{-2} e^{\Phi_1(y)} \int_R^y \mathrm{d}r' r'^2 e^{-\Phi_1(r')}. \tag{2G.7}$$

The mean first passage time (i.e., average dissociation time) $\overline{\tau}_1$ is obtained by averaging $\tau_1(r_0)$ with the Boltzmann factor over all possible initial positions $r_0$:

$$\overline{\tau}_1 = \frac{1}{Z} \int_R^{R+\lambda_1} \mathrm{d}r_0 r_0^2 e^{-\Phi_1(r_0)} \tau_1(r_0), \tag{2G.8}$$

with

$$Z = \int_R^{R+\lambda_1} \mathrm{d}r_0 r_0^2 e^{-\Phi_1(r_0)}. \tag{2G.9}$$

Clearly, $\overline{\tau}_1 \equiv \overline{\tau}_1(R) \equiv \overline{\tau}_1(n_1)$.

The rate of emission of molecules of component 1 from the droplet can be expressed as

$$W_1^- = \frac{N_1}{\bar{\tau}_1} = N_1 D_1 \omega_1, \tag{2G.10}$$

where $N_1$ denotes the number of molecules in the well and $\omega_1$ is defined as $\omega_1 \equiv 1/D_1\bar{\tau}_1$.

## Appendix 2H: Equilibrium Distribution of Heterogeneously Formed Droplets

Let us derive Equation 2.259. According to the principle of detailed balance,

$$W_1^+(n_1-1)g_e(n_1-1) = W_1^{-1}(n_1)g_e(n_1), \tag{2H.1}$$

which can be rewritten as

$$\frac{g_e(n_1)}{g_e(n_1-1)} = \frac{W_1^+(n_1-1)}{W_1^-(n_1)}. \tag{2H.2}$$

By applying Equation 2H.2 to the droplets $(n_1 - i)$ with $i = 1, 2, 3, \ldots, n_1$, multiplying the right- and left-hand sides of all equations, one obtains

$$\frac{g_e(n_1)}{g_e(0)} = \prod_{i=1}^{n_1} \frac{W_1^+(n_1-i)}{W_1^-(n_1-i+1)}. \tag{2H.3}$$

Using the normalization condition $\Sigma_{n_1=0}^{\infty} g_e(n_1) = N_0$, where $N_0$ is the total number of initial aerosols in unit volume of the system, one obtains

$$g_e(0) = N \Big/ \sum_{n_1=0}^{\infty} \prod_{i=1}^{n_1} \frac{W_1^+(n_1-i)}{W_1^-(n_1-i+1)}, \tag{2H.4}$$

so that the equilibrium distribution of droplets acquires the form

$$g_e(n_1) = N_0 \tilde{p}(n_1) \Big/ \sum_{n_1'=0}^{\infty} \tilde{p}(n_1'), \tag{2H.5}$$

where

$$\tilde{p}(n_1) = \prod_{i=1}^{n_1} \frac{W_1^+(n_1-i)}{W_1^-(n_1-i+1)}, \tag{2H.6}$$

which can be rewritten as

$$\tilde{p}(n_1) = \frac{W_1^+(0)}{W_1^+(n_1)} \prod_{i=1}^{n_1} \frac{W_1^+(n_1-i+1)}{W_1^-(n_1-i+1)}. \tag{2H.7}$$

Finally, substituting Equations 2.256 and 2.257 into the right-hand side of Equation 2H.7 and then substituting the result into Equation 2H.5, one obtains Equation 2.259.

## References


1. Miller, R.C., Anderson, R.J., Kassner, J.L., Jr., Hagen, D.E., *J. Chem. Phys.* 78, 3204 (1983).
2. N.H. Fletcher, *The Physics of Rainclouds*, Cambridge University Press, New York, 1962.
3. Djikaev, Y.S., Donaldson, D.J., *J. Geophys. Res.* 104, 14283 (1999).
4. Djikaev, Y.S., Donaldson, D.J., *J. Chem. Phys.* 113, 6822 (2000).
5. Djikaev, Y.S., Donaldson, D.J., *J. Geophys. Res.* 104, 14447 (2001).
6. Kuni, F.M., Shchekin, A.K., Rusanov, A.I., *Colloid J. Russ. Acad. Sci. (Engl. Transl.)* 55, 174 (1993).
7. Kuni, F.M., Shchekin, A.K., Rusanov, A.I., *Colloid J. Russ. Acad. Sci. (Engl. Transl.)* 55, 202 (1993).
8. Kuni, F.M., Shchekin, A.K., Rusanov, A.I., *Colloid J. Russ. Acad. Sci. (Engl. Transl.)* 55, 211 (1993).
9. Kuni, F.M., Shchekin, A.K., Rusanov, A.I., *Colloid J. Russ. Acad. Sci. (Engl. Transl.)* 55, 686 (1993).
10. Shchekin, A.K., Rusanov, A.I., Kuni, F.M., *Colloid J. Russ. Acad. Sci. (Engl. Transl.)* 55, 227 (1993).
11. Shchekin, A.K., Rusanov, A.I., Kuni, F.M., *Colloid J. Russ. Acad. Sci. (Engl. Transl.)* 55, 776 (1993).
12. MacKenzie, A.R., Kulmala, M., Laaksonen, A., Vesala, T., *J. Geophys. Res.* 100, 11275 (1995).
13. Carslaw, K.S., Luo, B.P., Clegg, S.L., Peter, T., Brimblecombe, P., Crutzen, P.J., *Geophys. Res. Lett.* 21, 2479 (1994).
14. Meilinger, S.K., Koop, T., Luo, B.P., Huthwelker, T., Carslaw, K.S., Krieger, P.J., Crutzen, U., Peter, T., *Geophys. Res. Lett.* 22, 3031 (1995).
15. Ruckenstein, E., *J. Colloid Interface Sci.* 45, 115 (1973).
16. Konopka, P., *J. Aerosol Sci.* 28, 1411 (1997).
17. Gorbunov, B., *J. Chem. Phys.* 110, 10035 (1999).
18. Kuni, F.M., Shchekin, A.K., Rusanov, A.I., *Colloid J. Russ. Acad. Sci. (Engl. Transl.)* 55, 184 (1993).
19. Lothe, J., Pound, G.M.J., in: *Nucleation*, Zettlemoyer, A.C., Ed., Dekker, New York, 1969.
20. Narsimhan, G., Ruckenstein, E., *J. Colloid Interface Sci.* 128, 549 (1989). (Section 1.2 of this volume.)
21. Ruckenstein, E., Nowakowski, B., *J. Colloid Interface Sci.* 137, 583 (1990). (Section 1.3 of this volume.)
22. Nowakowski, B., Ruckenstein, E., *J. Colloid Interface Sci.* 139, 500 (1990). (Section 1.4 of this volume.)
23. Nowakowski, B., Ruckenstein, E., *J. Chem. Phys.* 94, 1397 (1991). (Section 2.1 of this volume.)
24. Nowakowski, B., Ruckenstein, E., *J. Chem. Phys.* 94, 8487 (1991). (Section 2.3 of this volume.)
25. Ruckenstein, E., Nowakowski, B., *Langmuir* 7, 1537 (1991). (Section 2.2 of this volume.)
26. Djikaev, Y.S., Ruckenstein, E., *J. Chem. Phys.* 123, 214503 (2005). (Section 2.5 of this volume.)
27. Djikaev, Y.S., Ruckenstein, E., *J. Chem. Phys.* 124, 124521 (2006). (Section 2.6 of this volume.)
28. Djikaev, Y.S., Ruckenstein, E., *J. Chem. Phys.* 124, 194709 (2006). (Section 1.6 of this volume.)
29. Ellison, G.B., Tuck, A.F., Vaida, V., *J. Geophys. Res.* 104, 11633 (1999).
30. Vaida, V., Tuck, A.F., Ellison, G.B., *Phys. Chem. Earth, Part C, Sol.-Terr. Planet. Sci.* 25, 195 (2000).
31. Murphy, D.M., Thomson, D.S., Middlebrook, A.M., *Geophys. Res. Lett.* 24, 3197 (1997).
32. Gill, P.S., Graedel, T.E., Wesher, C.J., *Rev. Geophys. Space Phys.* 21, 903 (1983).
33. Köhler, H., *Trans. Faraday Soc.* 32, 1152 (1936).
34. Mirabel, P., Reiss, H., Bowles, R., *J. Chem. Phys.* 113, 8194 (2000).
35. Abraham, F., *Homogeneous Nucleation Theory*, Academic, New York, 1974.
36. Chandrasekhar, S., *Rev. Mod. Phys.* 15, 1 (1949).
37. Gardiner, C.W., *Handbook of Stochastic Methods*, Springer, New York, 1983.
38. Agmon, N., *J. Chem. Phys.* 81, 3644 (1984).

## 2.9 Kinetics of Heterogeneous Nucleation on a Rough Surface: Nucleation of a Liquid Phase in Nanocavities

*Eli Ruckenstein and Gersh Berim**

The kinetic theory of heterogeneous nucleation of a liquid developed by Nowakowski and Ruckenstein (*J. Phys. Chem.* 96 (1992) 2313 [Section 2.7 of this volume]) for a planar surface is extended to the nucleation of liquid clusters in spherical nanocavities of a solid surface. It is shown that for all considered contact angles and radii of the cavity, the nucleation rate is higher than that on a planar surface and that it increases with increasing strength of the fluid–solid interactions. The difference is small at high supersaturations (small critical clusters), but is orders of magnitude higher at lower supersaturations when the cluster has a size comparable with the size of the cavity. At constant contact angle, the nucleation rate increases with increasing curvature of the cavity. The dependence of the nucleation rate on the contact angle is not monotonous.

### 2.9.1 Introduction

The increased nucleation rate of crystallization in the presence of a scratched surface compared to that over a smooth surface was noted a long time ago in a paper of Gay-Lussac [1] published in 1813. Since that time, the understanding of the peculiarities of heterogeneous nucleation remained an important issue both for practical (e.g., semiconductor film deposition, weather modification) and theoretical (understanding of the role of the interaction potentials, of substrate geometry, external parameters such as temperature, acoustic fields, etc.) reasons (see Refs. [2–10]).

Because of the random nature of the roughness that can appear as cavities, asperities, grooves, etc., it is practically impossible to provide a complete treatment of the problem and most studies are devoted to a specific kind of roughness. In Ref. [2], the nucleation of a crystalline phase in wedges was treated by Monte Carlo simulations and it was shown that the wedge angle affected the nucleation rate, which can become orders-of-magnitude higher in a wedge than on a flat surface. The same conclusion was drawn with respect to the nucleation of a magnetically ordered phase in a rectangular cavity within a two dimensional spin lattice model [11]. The classical approach developed in Refs. [12–16] was extended in Ref. [6] to the nucleation rate in nanocavities. It was found that at high supersaturations the nucleation rate increases with increasing contact angle and cavity radius. However, at low supersaturations the nucleation rate decreases with increasing cavity radius and/or contact angle. In Ref. [4], the classical nucleation theory was used to examine the effect of the geometry of the cavity on the nucleation of bubbles in liquids. Spherical, cylindrical, and conical cavities were considered and it was shown that the cavity geometry affected the threshold for unstable growth of the gas–liquid interface inside the cavity leading to its emergence outside the cavity.

Note that the main shortcoming of the classical nucleation theory is the assumption that macroscopic thermodynamics, in particular, the concept of surface tension, can be satisfactorily used for small clusters consisting of only a few molecules. The critical cluster size is determined in that theory from the maximum of the cluster free energy.

A kinetic approach to the nucleation theory that does not use the concept of surface tension was developed in Refs. [17–21], in which the critical cluster size was determined by equating the rate of evaporation of the molecules from the surface of the cluster to the rate of condensation. The rate of evaporation of molecules from a cluster was obtained on the basis of a kinetic equation for the motion of a molecule in the potential well generated by the interactions of the molecule with the cluster and the substrate on which the cluster is located.

In the papers of Nowakowski and Ruckenstein [21,22], the kinetic approach to the nucleation theory was applied to the heterogeneous nucleation in the presence of a smooth planar solid surface and the results were compared with the predictions of the classical heterogeneous nucleation theory [21]. The kinetic approach provides much higher nucleation rates than the classical one.

In the present paper, the nucleation theory of Nowakowski and Ruckenstein is applied to the nucleation of a liquid phase in the nanocavities of a surface, which have the shape of a spherical segment of radius $R_s$ and depth $h$ (see Figure 2.28).

In Sections 2.9.2 and 2.9.3, the description of the system and the presentation of the basic equations of the kinetic theory are provided. The results and their discussion are presented in Sections 2.9.4 and 2.9.5.

* *J. Colloid Interface Sci.*, 351, 277 (2010). Republished with permission.

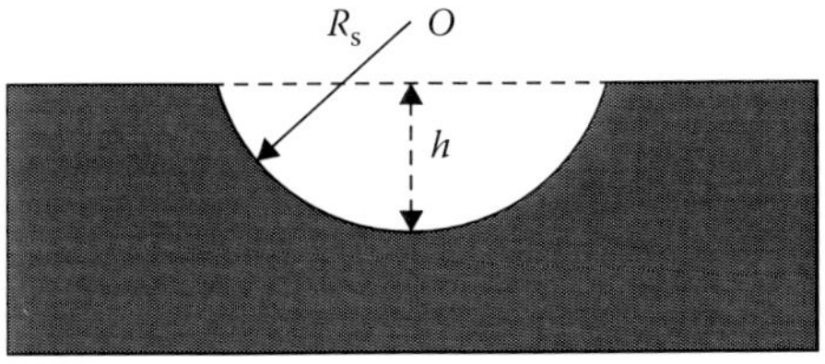


Figure 2.28

Spherical cavity of radius $R_s$ with the center in the point $O$ considered in the present paper.

## 2.9.2 The System and the Interaction Potentials

The considered system consists of a semi-infinite solid possessing cavities of the kind shown in Figure 2.28, a cluster of liquid phase being present inside the cavity in contact with a vapor phase of the same fluid. It is assumed that the liquid cluster makes with the solid a contact angle $\theta$ that is assumed to be independent of the curvature of the solid and cluster size. The free surface of the cluster is considered spherical, its radius being dependent on the size of the cluster (number of molecules $N$ in the cluster) and on the contact angle $\theta$.

Depending on the values of $N$ and $\theta$, both convex and concave vapor–liquid interfaces can form in the cavity. The analysis of possible geometries of the cluster indicates that for $\theta \geq \theta_c$, where $\cos \theta_c = 1 - h/R_s$, the free surface of the cluster is convex for any cluster with a contact line located on the surface of the cavity (see Figure 2.29a).

For $\theta < \theta_c$, the free surface of an initially convex cluster (the darker cluster in Figure 2.29b) is transformed with increasing size of the cluster first into a planar one (dashed line) and then in a concave surface. The transition between convex and concave interfaces occurs at a cluster size for which the distance $d$ in Figure 2.29b is equal to $d = R_s(1 - \cos \theta)$.

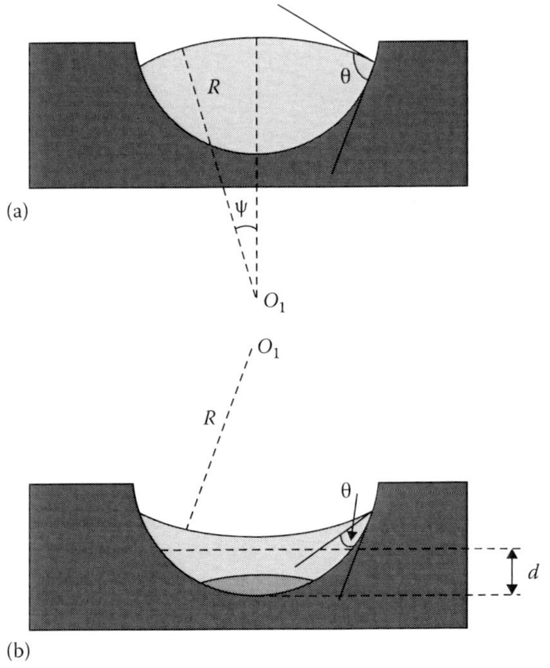


Figure 2.29

(a) Convex and (b) concave free spherical surfaces of a cluster of radius $R$ inside the cavity. For $\theta > \theta_c = \cos^{-1}(1 - h/R_s)$, where $R_s$ and $h$ are the cavity radius and depth (see Figure 2.28), the surface of the cluster is always convex. For $\theta < \theta_c$, the surface of the small cluster (the darker one in panel b) is convex but for larger clusters it can become concave. The transformation occurs through a planar surface at the distance $d = R_s(1 - \cos \theta)$ from the bottom of the cavity (dashed line in panel b). The center of the surface of the cluster is located at point $O_1$. The angle $\psi$ determines the position of a molecule on the cluster surface.

The interaction potential between two molecules of vapor combines the London–van der Waals attraction with the hard core repulsion

$$\phi(r_{12}) = \begin{cases} -4\epsilon\left(\dfrac{\sigma}{r_{12}}\right)^6, & r_{12} \geq \sigma, \\ \infty, & r_{12} < \sigma, \end{cases} \tag{2.262}$$

where $r_{12} = |\mathbf{r} - \mathbf{r}'|$ denotes the distance between the centers of the interacting molecules, $\sigma$ being the diameter of the hard core repulsion (molecular diameter), and $\epsilon$ is the energy parameter of the London–van der Waals attraction. The interaction of a fluid molecule of the cluster with a molecule of the solid substrate is selected in the same form as Equation 2.262, the energy parameter $\epsilon$ in the latter being replaced by $\epsilon_s$.

A fluid molecule near the liquid–vapor interface is in a potential well caused by the interactions between that molecule with all the molecules of the cluster and the solid. The effective potential $\Phi(\mathbf{r})$ of interaction with the cluster ($\mathbf{r}$ being the position of a fluid molecule in the well) and the supporting solid can be calculated by integrating the corresponding potentials over the volumes of the cluster and the solid. Because of the axial symmetry of the system, both integrations can be performed analytically for two coordinates but only numerically for the third one. The procedure is similar to that used in Ref. [21] for a planar smooth surface. Note that the depth of the potential well depends on the position of the fluid molecule with respect to the cluster and the supporting solid.

### 2.9.3 Basic Equations of the Kinetic Theory

The kinetic theory developed in Refs. [19–21] is based on the fact that in vapors the collisions of molecules are relatively infrequent and, consequently, the energy of the molecules changes slowly compared to the changes in their linear momentum and position. Therefore, the time evolution in the energy space can be described by a diffusion-like equation for the probability density $P(E,t)$ of the energy $E$ [19,23]:

$$\frac{\partial P(E,t)}{\partial t} = \frac{\partial}{\partial E} D(E) P_{eq}(E) \frac{\partial}{\partial E} \frac{P(E,t)}{P_{eq}(E)}, \tag{2.263}$$

where $D(E)$ is the diffusion coefficient in the energy space, $P_{eq}(E)$ is the equilibrium distribution of the energy at temperature $T$ provided by the equation

$$P_{eq}(E) = \Omega(E)\exp(-E/kT)/Z. \tag{2.264}$$

In Equation 2.264, $Z$ is the partition function

$$Z = \int_{\Phi_0}^{\Phi_1} \Omega(E)\exp(-E/kT)\mathrm{d}E, \tag{2.265}$$

$\Omega(E)$ is the density of states of energy $E$, $\Phi_0$, and $\Phi_1$ are the energies at the lower and the upper boundaries of the potential well. For a single molecule with the Hamiltonian $H(\mathbf{r},\mathbf{p})$, where $\mathbf{r}$ and $\mathbf{p}$ are the position and the linear momentum of the molecule, respectively, $\Omega(E)$ is provided by the equation

$$\Omega(E) = \int_{\Gamma} \mathrm{d}\Gamma \delta(E - H(\mathbf{r},\mathbf{p})), \tag{2.266}$$

where $\Gamma$ is the region of the phase space available to the molecule and $\delta(E - H(\mathbf{r},\mathbf{p}))$ is the Dirac $\delta$ function.

The boundary condition for Equation 2.263 is selected as [21]

$$P(E)|_{E=\Phi_1} = 0 \tag{2.267}$$

and involves the assumption that any molecule that acquires the energy $\Phi_1$ separates from the cluster. With this boundary condition, the stationary solution of Equation 2.263 has the form

$$P(E) = j_E P_{eq}(E) \int_E^{\Phi_1} \frac{dE}{D(E')P_{eq}(E')}, \tag{2.268}$$

where $j_E$ is the steady-state solution for the flux $j(E, t)$ in the energy space, the latter being defined as

$$\frac{\partial P(E,t)}{\partial t} = \frac{\partial}{\partial E} j(E,t) \tag{2.269}$$

and is given by the expression [19,21]

$$j(E,t) = -D(E)P_{eq}(E)\frac{\partial}{\partial E}\frac{P(E,t)}{P_{eq}(E)}. \tag{2.270}$$

Using the normalization condition, $\int P(E)dE = 1$, the following expression for the rate of evaporation per molecule, $\alpha^1$, is obtained [21]:

$$\alpha^1 = \left[\int_{\Phi_0}^{\Phi_1} dE P_{eq}(E) \int_E^{\Phi_1} \frac{dE'}{D(E')P_{eq}(E')}\right]^{-1}. \tag{2.271}$$

Note that $\alpha^1$ depends on the cluster size $N$ (number of molecules in the cluster), i.e., $\alpha^1 \equiv \alpha^1(N)$. The applicability of the steady-state approximation used here to calculate the evaporation rate was examined in detail in Ref. [21]. The lag time in which the steady state is achieved was calculated by Nowakowski and Ruckenstein [24].

Assuming Brownian dynamics, the diffusion coefficient $D(E)$ in the energy space is given by [25]

$$D(E) = \frac{3kT\xi}{m\Omega(E)} \int_{\Phi_0}^{E} dE'\Omega(E'), \tag{2.272}$$

where $m$ is the mass of a molecule and $\xi$ is a friction coefficient.

The net evaporation rate, $\alpha$, is proportional to $\alpha^1$ and the number of molecules in the potential well. Assuming that the number density $n_w$ of the fluid molecules in the well is constant, the total number of molecules in a well of thickness $\lambda$ is

$$\alpha = \alpha^1 \lambda n_w S_c, \tag{2.273}$$

where $S_c \equiv S_c(N)$ is the free surface of the cluster that depends on the cluster size. In the considered case,

$$S_c = 2\pi R^2 f_c(x), \tag{2.274}$$

where $R$ is the cluster radius, $f_c(x)$ is a geometrical factor given by

$$f_c(x) = 1 - \frac{1 + x\cos\theta}{\sqrt{1 + 2x\cos\theta + x^2}} \tag{2.275}$$

and $x = (R_s - \sigma)/R$. The subtraction of $\sigma$ in the nominator of the expression of $x$ accounts for the hard core repulsion between the surface of the cavity and the fluid molecules that leads to an effective decrease of the radius of the cavity. The radius of a cluster of a given size $N$ can be determined from the equation

$$V_c(R)\rho = N, \tag{2.276}$$

where $\rho$ is the fluid density in the cluster and the cluster volume $V_c(R)$ is given by

$$V_c(R) = \frac{1}{3}\pi R^3 \left\{ f_c^2(x)[3 - f_c(x)] + f_s^2(x)[3x - f_s(x)] \right\} \tag{2.277}$$

with

$$f_s(x) = x\left[1 - \frac{x + \cos\theta}{\sqrt{1 + 2x\cos\theta + x^2}}\right]. \tag{2.278}$$

For the planar surface considered in Ref. [21], $f_c(x) = 1 - \cos\theta$ and $f_s(x) = 0$.

The critical cluster size, $N_*$, is determined by equating the evaporation and condensation rates. The latter rate, $\beta \equiv \beta(N)$, is provided by the number of impingements of the vapor molecules on the free surface $S_c$ of the cluster and is given by [26]

$$\beta = \frac{\bar{\upsilon}}{4} S_c n_1, \tag{2.279}$$

where $\bar{\upsilon}$ is the mean velocity of the vapor molecules and $n_1$ is the number density of the vapor. Combining Equations 2.273 and 2.279, the following relation between $n_1$ and the size $N_*$, of the critical cluster is obtained [19,21]:

$$n_1 = n_w \frac{4\lambda}{\bar{\upsilon}} \alpha^1(N_*). \tag{2.280}$$

The supersaturation $s$ is defined as $s = n_1/n_s$, where $n_s$ is the vapor number density in equilibrium with the bulk condensed phase. Formally, $n_s$ can be determined from Equation 2.280 as the limiting value of $n_1$ for $N_* \to \infty$ [19,21]; consequently, the supersaturation can be expressed as

$$s = \frac{\alpha^1(N_*)}{\alpha^1(\infty)}. \tag{2.281}$$

Note that the specific values of $\bar{\upsilon}$, $\lambda$, and $n_w$ are not needed because these quantities are not present in the final equations.

In the kinetic approach developed in Refs. [19,21], the nucleation rate is calculated on the basis of the equation for the evolution of the population $n_i$ of clusters consisting of $n_i$ molecules

$$\frac{dn_i}{dt} = I_{i-1} - I_i, \tag{2.282}$$

where the time-dependent flux $I_i$ of clusters passing from the $i$ population to the $(i + 1)$ population is given by

$$I_i = \beta_i n_i - \alpha_{i+1} n_{i+1}. \tag{2.283}$$

At equilibrium, $I_i$ vanishes providing the detailed balance equation from which the equilibrium distribution can be obtained [21]:

$$n_i^{eq} = n_1^{eq} \prod_{j=2}^{i} \frac{\beta_{j-1}}{\alpha_j} \equiv n_1^{eq} \exp(H(i)). \tag{2.284}$$

After the conversion of Equation 2.283 to a continuum form the rate of nucleation, $I$, calculated as the stationary solution of that equation, can be written in the form [21]

$$I = \left[ \int_1^{\infty} \frac{dg}{\beta(g) n^{eq}(g)} \right]^{-1}. \qquad (2.285)$$

At the critical size $g = N_*$, the equilibrium distribution $n^{eq}(g)$ has a sharp minimum and this allows us to calculate $I$ using the steepest descent approximation for the integral in Equation 2.285 [21]:

$$I \simeq n_1 \beta_1 \frac{S_c(N_*)}{S_c(1)} \left( \frac{H''(N_*)}{2\pi} \right)^{1/2} \exp(H(N_*)). \qquad (2.286)$$

For calculating the area $S_c(1)$ of the free surface of the cluster consisting of one molecule, Equation 2.274 for $S_c(N)$ is used below. Although not absolutely correct, this approach is consistent with the consideration of the fluid as a continuous medium.

### 2.9.4 Results

Numerical calculations were performed using for the energy parameter of the fluid–fluid interactions the value $\epsilon/kT = 0.5$ and for the fluid–solid interactions the values $\epsilon_s/kT = 1$ and 0.25 ($\epsilon/\epsilon_s = 0.5$ and 2.0). For the contact angle the following values were selected: $\theta = 55°$, 65°, and 85°. Three different cavity radii $R_s = 10\sigma$, $15\sigma$, and $20\sigma$ were considered, with the same parameter $h$ (see Figure 2.28), $h = 5\sigma$, for all cavities. Only clusters with convex free surfaces and contact lines on the surface of the cavity were considered. The number densities of molecules in the cluster and in the solid were taken $\rho_L\sigma^3 = 1.2$ and $\rho_s\sigma^3 = 1.44$, respectively. The upper bound, $\Phi_1$ of the potential well is provided by the potential $\Phi(\mathbf{r})$ at infinite distance between a molecule and the solid surface, i.e., $\Phi_1 = 0$. However, in the numerical calculations it was taken equal $\Phi_1 = 0.02\Phi_0$ to avoid the numerical integration over an infinite range when calculating $\Omega(E)$ (see Equation 2.266).

Note that in the presence of a solid surface $\Phi_0$ depends on the location of the molecule on the surface of the cluster. As example, in Figure 2.30, $\Phi_0$ is plotted as function of the angle $\psi$ (see Figure 2.29) for $\theta = 65°$, $\epsilon/kT = 0.5$, $\epsilon_s/kT = 1.0$, $R_s = 15\sigma$, $R = 10\sigma$, where $R$ is the radius of the free surface of the cluster. However, Equation 2.273 is valid if the depth $\Phi_0$ of the potential well is constant along the free surface of the cluster. To account for the variation of $\Phi_0$, the range of values of $\Phi_0$ was divided into several parts, and in each of them $\Phi_0$ was taken equal to its average value over that part. The evaporation rate was calculated for each $\Phi_0$ separately and then averaged with weights equal to the fractions of the corresponding parts with respect to the entire range of variation of $\Phi_0$.

The value of a $\alpha^1(\infty)$ in Equation 2.281, which is necessary to calculate the supersaturation $s$, does not depend on the presence of the solid surface. It was determined by extrapolating $\alpha^1(N)$ calculated for a

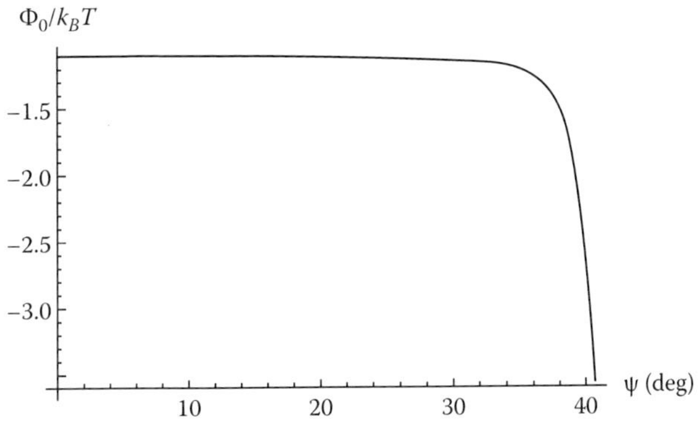


Figure 2.30

Dependence of the potential well depth $\Phi_0$ on the position of the molecule on the cluster surface for $\theta = 65°$, $\epsilon/kT = 0.5$, $\epsilon_s/kT = 1.0$, $R_s = 15\sigma$, $R = 10\sigma$. This position is provided by the angle $\psi$ presented in Figure 2.29a.

cluster on a planar surface for large values of $N$. For the considered fluid ($\epsilon/k_B T = 0.5$) this procedure provides $\alpha^1(\infty) = 0.667$.

In Figure 2.31, the nucleation rate is presented for a cavity of radius $R_s = 15\sigma$ (solid lines) and for a planar surface (dashed lines) for various contact angles as function of supersaturation $s$. In all cases, the nucleation rate in the cavity is larger than on a planar surface. As expected, this difference is small at high supersaturations when the critical cluster has a small size and while present in a cavity it can be considered as located on a planar surface.

For lower supersaturation ($\log(s) \sim 0.3$–$0.4$), the critical cluster is comparatively large and the nucleation rate in the cavity can be 5–6 orders of magnitude higher than on a planar surface, as one can see in Figure 2.31 for $\theta = 55°$, $\theta = 65°$, and $\theta = 85°$, respectively.

The value of the contact angle affects the nucleation rate both in a cavity and on a planar surface. For a cavity, the change in nucleation rate due to the change of $\theta$ is relatively small (does not exceed 2–3 orders of magnitude at low supersaturations ($\log(s) < 0.4$)). In contrast, on a planar surface the difference in nucleation rates for $\log(s) < 0.4$; $\theta = 65°$ and $\theta = 85°$ can reach 10 orders of magnitude. The $s$ dependence of the nucleation rate for various contact angles is presented for a cavity in Figure 2.32a and for a planar surface in Figure 2.32b. One can see from this figure that at the same supersaturation, the dependence of nucleation

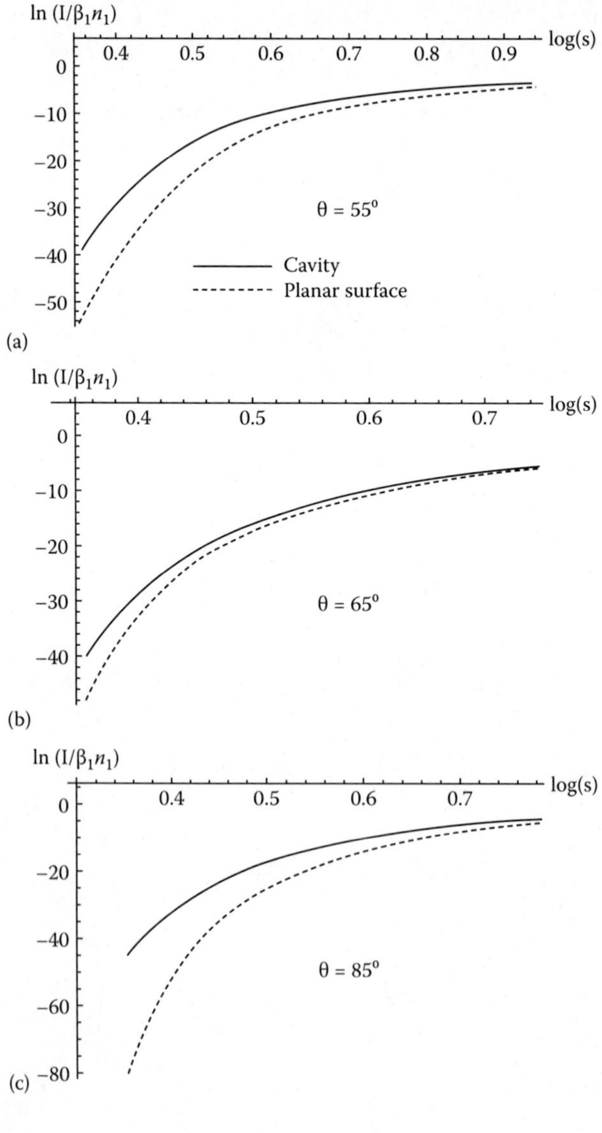


Figure 2.31

Dependence of nucleation rate on supersaturation for a cavity (solid curves) and for a planar surface (dashed curves) for various values of the contact angle $\theta$. The results are presented for $R_s = 15\sigma$, $h = 5\sigma$, $\epsilon_s = 0.5\epsilon$, and $\epsilon/k_B T = 0.5$.

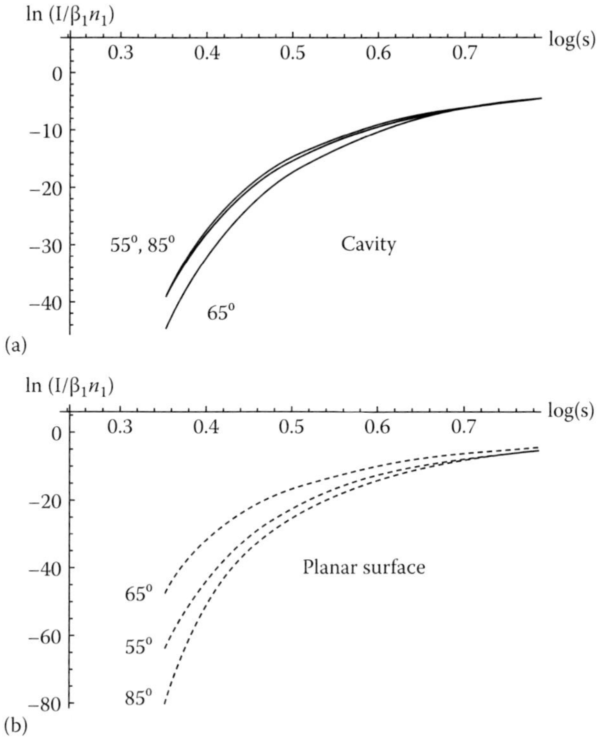


Figure 2.32

Dependence of the nucleation rate on supersaturation for various angles θ in a cavity (solid curves) and on a planar surface (dashed curves). The values of the contact angle are marked on each of the curves. Other parameters are the same as in Figure 2.31.

rate on the contact angle is not monotonous both for a cavity and a planar surface. For a cavity, as the angle θ increases from 55° to 65°, the nucleation rate decreases. A further increase of θ to 85° increases the nucleation rate. For a planar surface, as the angle increases from 55° to 65°, the nucleation rate increases, whereas an increase of the contact angle from 65° to 85° decreases the nucleation rate. Consequently, for the considered values of θ, the nucleation rate passes through a minimum for a cavity and a maximum for a planar surface. The difference between the nucleation rates for $\theta = 55°$ and $\theta = 85°$ is small in the cavity for any supersaturation, but is about 6 orders of magnitude on the planar surface for $\log(s) < 0.4$. Note that the nonmonotonous dependence of nucleation rate on contact angle is also predicted by the classical nucleation theory [6]. It is difficult to provide a simple physical explanation of such a behavior of the nucleation rate as function of θ because it is a result of a complicated dependence of the geometrical characteristic of a drop (free surface area, volume, etc.) on the contact angle.

The $s$-dependence of nucleation rate in the cavity as function of the cavity radius is presented by the solid lines in Figure 2.33 for $\theta = 85°$. For comparison, the same dependence for a planar surface ($R_s = \infty$) is presented by the dashed line. The nucleation rate in the cavity increases monotonously with decreasing cavity radius. A similar behavior occurs for the other contact angles. The examination of the nucleation rate dependence on the cavity depth $h$ has shown that this dependence is very weak for all considered cavity radii and can be neglected.

The above-mentioned dependence of nucleation rate on cavity radius differs qualitatively from that predicted by the classical theory of nucleation [6] in the high supersaturation range. In the latter case, the nucleation rate decreases with decreasing cavity radius.

The influence of the strength of the fluid–solid interaction on the nucleation rate in a cavity of radius $R_s = 15\sigma$ and on a planar surface is illustrated in Figure 2.34, where the s-dependences of the nucleation rate in the cavity (solid lines) and on the planar surface (dashed lines) are plotted for $\theta = 65°$ and two values of the energy parameter $\epsilon_s$, ($\epsilon_s = 2\epsilon$ and $\epsilon_s = 0.5\epsilon$). Both for the cavity and planar surface the nucleation rate increases with increasing $\epsilon_s$. Such a behavior is a result of the decrease of the evaporation rate as a result of the increase in the depth of the potential well that leads to a decrease of the critical cluster size at the same supersaturation and, consequently, an increase of the nucleation rate.

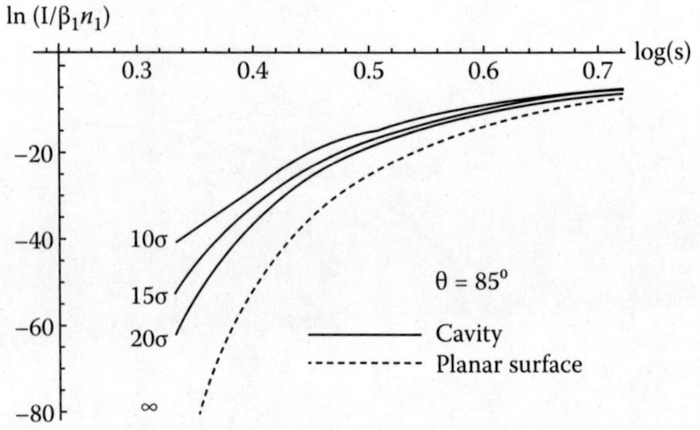


Figure 2.33

Dependence of the nucleation rate on supersaturation for cavities of various radii and $h = 5\sigma$ for a contact angle $\theta = 85°$. The cavity radius is provided on each curve. For comparison, the same dependence for a cluster on a planar surface ($R_s \to \infty$) is presented as a dashed line. $\epsilon/k_BT = 0.5$ and $\epsilon_s = 0.5\epsilon$.

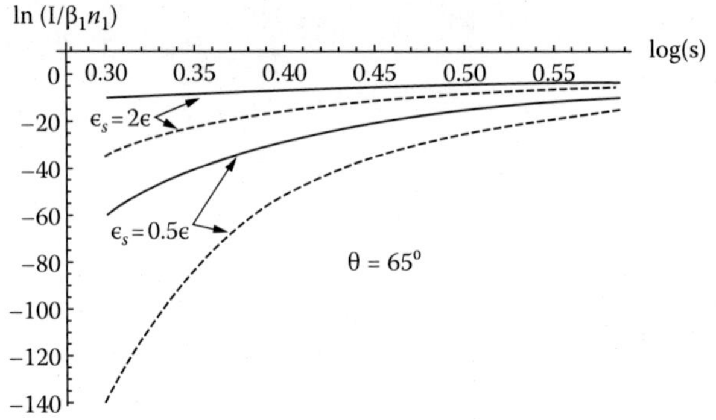


Figure 2.34

Dependence of the nucleation rate on supersaturation for various values of the energy parameter $\epsilon_s$ of the fluid–solid interaction and $\theta = 65°$. The solid curves are for nucleation in a cavity of $R_s = 15\sigma$ and $h = 5\sigma$ and the dashed curves for a planar surface.

### 2.9.5 Discussion

The above results reveal that the nucleation rate in cavities is much larger in a wide range of supersaturation than on a planar surface. The former rate is affected by the cavity radius, i.e., by the curvature of the solid surface at the place of formation of the critical cluster. The larger the curvature, the larger is the nucleation rate. Qualitatively, the increase of the nucleation rate due to cavities can be understood as a result of the increase of the interaction strength between a molecule on the surface of the cluster and the solid supporting the cluster. Indeed, the liquid–solid contact area for clusters of equal volume on a planar surface and in the cavity is larger for those in the cavity. For instance, for a cavity with $R_s = 15\sigma$ and a cluster volume $V_c = 230.0\sigma^3$, the contact areas for $\theta = 65°$ are equal to $138.0\sigma^2$ and $108.3\sigma^2$ for clusters in the cavity and on the planar surface, respectively. Such a difference leads to a smaller average distance of the free surface of the cluster in the cavity from the solid surface compared to that for a cluster on a planar surface. As a result, the average depth of the potential well in the cavity is larger for a cluster in the cavity. This causes a decrease in the evaporation rate and an increase in the nucleation rate.

For supersaturation smaller than those considered in the present paper, the leading edge of the critical cluster can be located outside the cavity and can have the shape presented in Figure 2.35 by the solid line. To achieve such a shape, the growing cluster should overcome the discontinuous transformation from a cluster with the leading edge on the surface of the cavity (dashed line in Figure 2.35) to one with the leading edge immediately near to the opening of the cavity (dotted dashed line in Figure 2.35). Such a transformation is accompanied by a discontinuous change in the cluster size that should be taken into account in the theoretical treatment. Note that in real situations, the opening of the cavity is not sharp and the above transformation may occur continuously.

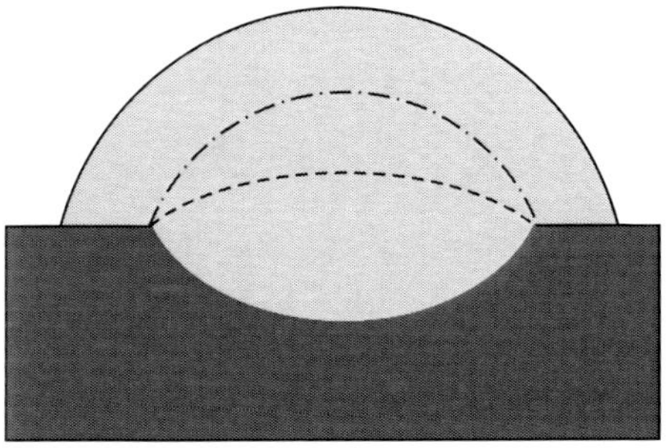

Figure 2.35

Transformation of a cluster surface when the contact line leaves the cavity. The dashed line represents the surface when the contact line of the cluster coincides with the cavity opening. The dotted-dashed line is that of the surface immediately after the contact line is located on the planar part of the solid surface.

For the sake of simplicity, in the present considerations, the values of the contact angle $\theta$ were selected in the range $\theta > \theta_c = \cos^{-1}(1 - h/R_s)$. However, the treatment of the problem on the basis of microscopic interaction potentials should provide this angle along with the nucleation rate. The appropriate approach for calculating the contact angles were developed in Refs. [27,28], where the minimization of the total potential energy of a drop on a solid surface [27] or the density functional theory [28] were used. Incorporation of these treatments into the kinetics of nucleation constitutes a task for future investigation.

## Acknowledgments

The authors are grateful to Dr. Nowakowski for useful discussions.

## References

1. Gay-Lussac, J.L., *Ann. Chim.* 87, 225 (1813).
2. Page, A.J., Sear, R.P., *J. Am. Chem. Soc.* 131, 17550 (2009).
3. Page, A.J., Sear, R.P., *Phys. Rev. E* 80, 031605 (2009).
4. Chappell, M.A., Payne, S.J., *J. Acoust. Soc. Am.* 121, 853 (2007).
5. Qian, M., Ma, J., *J. Chem. Phys.* 130, 214709 (2009).
6. Liu, Q.X., Zhu, Y.J., Yang, G.W., Yang, Q.B., *J. Mater. Sci. Technol.* 24, 183 (2008).
7. Shevkunov, S.V., *J. Exp. Theor. Phys.* 107, 965 (2008).
8. Cooper, S.J., Nicholson, C.E., Liu, J., *J. Chem. Phys.* 129, 124715 (2008).
9. Kimura, T., Maruyama, S., *Microsc. Thermophys. Eng.* 6, 3 (2002).
10. Eisenmenger-Sittner, C., *J. Crystal Growth* 205, 441 (1999).
11. Page, A.J., Sear, R.P., *Phys. Rev. Lett.* 97, 065701 (2006).
12. Farkas, L., *Z. Phys. Chem. A* 125, 236 (1927).
13. Volmer, M., *Z. Elektrochem.* 35, 555 (1929).
14. Becker, R., Doering, W., *Ann. Phys.* 24, 719 (1935).
15. Becker, R., *Discuss. Faraday Soc.* 5, 55 (1949).
16. Fletcher, N.H., *J. Chem. Phys.* 29, 572 (1958).
17. Narsimhan, G., Ruckenstein, E., *J. Colloid Interface Sci.* 128, 549 (1989). (Section 1.2 of this volume.)
18. Ruckenstein, E., Nowakowski, B., *J. Colloid Interface Sci.* 137, 583 (1990). (Section 1.3 of this volume.)
19. Nowakowski, B., Ruckenstein, E., *J. Chem. Phys.* 94, 1397 (1991). (Section 2.1 of this volume.)
20. Nowakowski, B., Ruckenstein, E., *J. Chem. Phys.* 94, 8487 (1991). (Section 2.3 of this volume.)
21. Nowakowski, B., Ruckenstein, E., *J. Phys. Chem.* 96, 2313 (1992). (Section 2.7 of this volume.)
22. Ruckenstein, E., Nowakowski, B., *Langmuir* 8, 1470 (1992).
23. Keck, J., Carrier, G., *J. Chem. Phys.* 43, 2284 (1965).
24. Nowakowski, B., Ruckenstein, E., *J. Colloid Interface Sci.* 145, 182 (1991).
25. Borkovec, M., Berne, B.J., *J. Chem. Phys.* 82, 794 (1985).
26. Abraham, F.F., *Homogeneous Nucleation Theory*, Academic, New York, 1974.
27. Berim, G.O., Ruckenstein, E., *J. Phys. Chem. B* 108, 19339 (2004).
28. Berim, G.O., Ruckenstein, E., *J. Chem. Phys.* 129, 014708 (2008).

## 2.10 Kinetic Theory of Heterogeneous Nucleation: Effect of Nonuniform Density in the Nuclei

*Gersh Berim and Eli Ruckenstein**

The heterogenous nucleation of a liquid from a vapor in contact with a planar solid surface or a solid surface with cavities is examined on the basis of the kinetic theory of nucleation developed by Nowakowski and Ruckenstein (*J. Phys. Chem.* 96 [1992] 2313) which is extended to nonuniform fluid density distribution (FDD) in the nucleus. The latter is determined under the assumption that at each moment the FDD in the nucleus is provided by the density functional theory (DFT) for a nanodrop. As a result of this assumption, the theory does not require users to consider that the contact angle that the nucleus makes with the solid surface and the density of the nucleus are independent parameters since they are provided by the DFT. For all considered cases, the nucleation rate is higher in the cavities than on a planar surface and increases with increasing strength of the fluid–solid interactions and decreasing cavity radius. The difference is small at high supersaturations (small critical nuclei), but becomes larger at low supersaturations when the critical nucleus has a size comparable with the size of the cavity. The nonuniformity of the FDD in the nucleus decreases the nucleation rate when compared to the uniform FDD.

### 2.10.1 Introduction

The formation of a critical nucleus of a new phase in the body of the initial phase (for instance, a liquid nucleus in the vapor phase) is a complicated nonequilibrium process occurring because of local density fluctuations. Because of this complexity, the theories of nucleation usually describe the nucleus in a simplified manner by using well developed theories valid at macroscopic scale.

The classical nucleation theory (CNT) [1–8] calculates the free energy of a nucleus using the concept of surface tension. As a result, this theory provides the critical size of a nucleus, above which it grows indefinitely, from the maximum free energy with respect to its size.

The kinetic nucleation theory (KNT) [9–16] does not involve the macroscopic concepts of surface tension and free energy and determines the critical nucleus from the balance of evaporation and condensation rates on the surface of the nucleus. For this reason, the latter theory appears to be more adequate to describe the nucleation phenomenon. Note that both KNT and CNT use the assumption of uniform density in the nucleus and, in the case of heterogeneous nucleation, the assumption of a constant contact angle that the nucleus makes with the surface. (Both assumptions can be considered to be valid for large critical nuclei, i.e., at low supersaturations.) Using such assumptions in the framework of KNT, the rate of heterogeneous nucleation of the liquid phase in a spherical cavity was calculated in Ref. [16] for various cavity radii and contact angles that the nucleus makes with the solid surface. The comparison of the obtained results with those for a planar surface has shown that roughness (cavities) always increases the nucleation rate, in agreement with the experimental and simulations results [17–19].

However, as shown previously on the basis of the density functional theory (DFT) and numerical simulations, the fluid density distribution (FDD) in nuclei (nanodrops) in contact with a solid is not uniform [20–23]; in addition the contact angle $\theta$ of the nucleus in nanocavities can vary in a wide range with the change in the size of the nucleus [24]. Even for a nucleus on a planar surface, $\theta$ in not constant, but its change with the size of the nucleus is much smaller than that in a cavity. These results suggest that a nucleus of the liquid phase on a solid surface has most likely a nonuniform density, which should be taken into account in a theoretical calculation.

The purpose of the present paper is to extend the kinetic theory of heterogeneous nucleation (using as an example the nucleation of a liquid phase from vapor in the presence of a solid surface) by including the nonuniformity of the FDD inside the nucleus. The assumption that the FDD of the nonhomogeneous nucleus of the liquid phase can be obtained at each moment via DFT is employed.

### 2.10.2 The System, the Interaction Potentials, and the Definition of the Nucleus

The considered system consists of a one-component fluid of fixed average density $\rho_{av}$ in contact with a semi-infinite solid of uniform density $\rho_s$, possessing a cylindrical cavity (groove) like that presented in Figure 2.36. The system has the finite dimensions $L_x$ and $L_h$ in the $x$ and $h$ directions, respectively, and an infinite dimension in the y direction (normal to the plane of the figure). Cavities with radii $R_s = 15\sigma_{ff}$ and $R_s = 11\sigma_{ff}$

* *J. Colloid Interface Sci.*, 355, 259 (2011). Republished with permission.

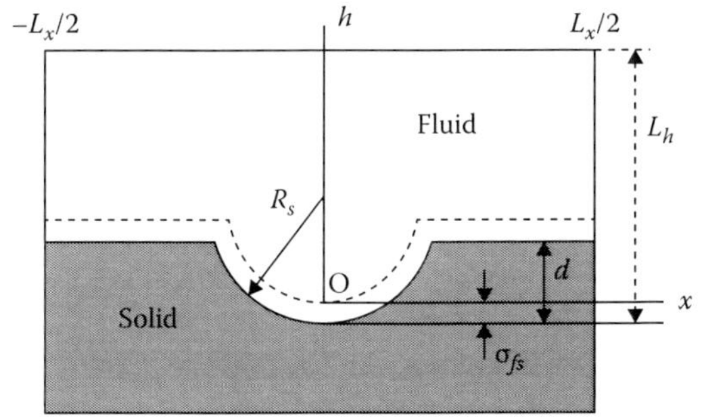


Figure 2.36

Schematic presentation of the considered system that is infinite in the $y$ direction (normal to the plane of the figure) and periodic with period $L_x$ in the $x$ direction. $R_s$ and $d$ are the radius and depth of the cavity, respectively, $\sigma_{fs}$ is the hard core diameter of the fluid–solid interaction potential. The dashed line indicates the location of the centers of the fluid molecules in the first (nearest to the solid) layer.

with depths $d = 5\sigma_{ff}$ and $d = 10\sigma_{ff}$, respectively, where $\sigma_{ff}$ is the length parameter in the fluid–fluid interaction potential were considered. The width $L_x$ was taken large enough for the effect of the presence of the cavity at $x = \pm L_x/2$ to be negligible. As a reference system, a planar surface (without cavities) is also considered. In all cases, a periodic boundary condition is employed in the $x$ direction and the upper boundary of the box is treated as a hard wall (without attractive interactions with the fluid molecules). The FDD $\rho(\mathbf{r})$ in such a system is assumed to be uniform in the $y$ direction and nonuniform in the $x$ and $h$ directions, i.e., $\rho(\mathbf{r}) = \rho(x,h)$. In what follows, all extensive quantities (e.g., free energy, number of molecules in cylindrical nuclei) will be calculated per unit length in the $y$ direction.

The interaction between the fluid molecules is considered of the Lennard–Jones type with a hard core repulsion:

$$\phi_{ff}\left(|\mathbf{r}-\mathbf{r}'|\right)=\begin{cases} 4\epsilon_{ff}\left[\left(\dfrac{\sigma_{ff}}{r}\right)^{12}-\left(\dfrac{\sigma_{ff}}{r}\right)^{6}\right], & r\geq\sigma_{ff}, \\ \infty, & r<\sigma_{ff}, \end{cases} \tag{2.287}$$

where the coordinates $\mathbf{r}$ and $\mathbf{r}'$ provide the locations of the fluid molecules, $r = |\mathbf{r} - \mathbf{r}'|$, $\sigma_{ff}$ and $\epsilon_{ff}$ are the fluid hard core diameter and the energy parameter, respectively.

The potential $\phi_{fs}(|\mathbf{r} - \mathbf{r}'|)$ for the fluid–solid interactions is selected of the same form as $\phi_{ff}(|\mathbf{r} - \mathbf{r}'|)$, with the energy parameter $\epsilon_{ff}$ and the hard core diameter $\sigma_{ff}$ in Equation 2.287 replaced by $\epsilon_{fs}$ and $\sigma_{fs}$, respectively, $r$ being the distance between a molecule of fluid and that of the solid walls. Because of hard core repulsion, the fluid molecules cannot be located at distances smaller than $\sigma_{fs}$ from the solid surface (marked by the dashed line in Figure 2.36).

The net potential generated by the solid is provided by the expression

$$U_{fs}(\mathbf{r})=\rho_s\int_{V_s}\phi_{fs}\left(|\mathbf{r}-\mathbf{r}'|\right)d\mathbf{r}', \tag{2.288}$$

where $V_s$ is the volume occupied by the solid. The integration over $y$ and $x$ coordinates in the three-dimensional integral (Equation 2.288) can be carried out in analytical form, whereas the integration with respect to $h$ can be performed only numerically.

The total Helmholtz free energy $F[\rho(\mathbf{r})]$ of the fluid in the external potential $U_{fs}(\mathbf{r})$ is expressed as the sum of an ideal gas free energy, $F_{id}[\rho(\mathbf{r})]$, a free energy $F_{hs}[\rho(\mathbf{r})]$ of a reference system of hard spheres, a free energy $F_{attr}[\rho(\mathbf{r})]$ due to the attractive interactions between the fluid molecules (in the mean-field approximation), and a free energy $F_{fs}[\rho(\mathbf{r})]$ due to the interactions between the fluid and the solid [25,26]. Explicit expressions for all these contributions are provided in Appendix 2I.

Minimizing $F[\rho(\mathbf{r})]$ with respect to the FDD leads to the Euler–Lagrange equation for $\rho(\mathbf{r})$ (see Appendix 2I), which can be solved numerically by iterations to provide the equilibrium two-dimensional FDD. A typical example of FDD is presented in Figure 2.37a.

(a)

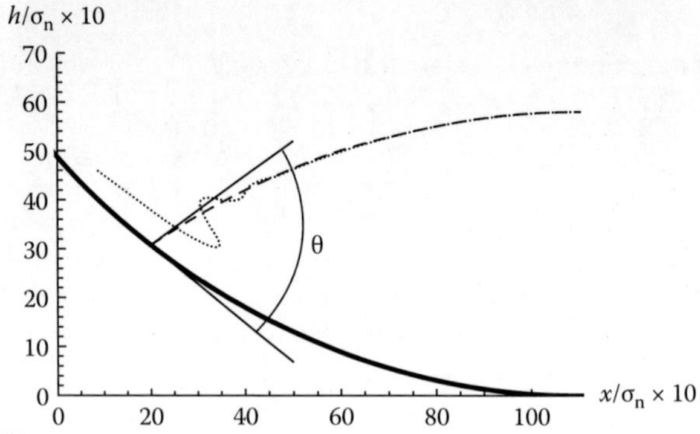


(b)

Figure 2.37

(a) Example of a two-dimensional fluid density distribution (FDD) in a cavity of radius $R_s = 15\sigma_{ff}$ and depth $d = 5\sigma_{ff}$. The lighter areas correspond to higher fluid densities. (b) Half of the drop profile (dotted line) extracted from FDD of panel a. The solid line is the surface of the cavity, the dashed line represents the circular approximation of the upper part of the drop profile, and θ is the contact angle.

To determine the profile of a nucleus on the basis of the obtained FDD, a procedure similar to that used in Ref. [23] was used. This procedure is based on the selection of a dividing density, $\rho_{div}$, and extraction of the drop profile from the two-dimensional FDD as a curve along which the local fluid density is $\rho_{div}$. To follow the KNT, a liquid like value of $\rho_{div}$ $\left(\rho_{div}\sigma_{ff}^3 \simeq 0.60\right)$ was selected and a sharp-kink interface assumed between a liquid like and a vapor like phases. An example of a nucleus profile based on the selected dividing density is presented in Figure 2.37b. Near the solid surface this profile has a complicated structure. For this reason, its upper part was approximated by a circle that was extended up to the solid surface. The angle that this circle makes with the surface was considered the contact angle.

The size of the cylindrical drop is defined as the number of molecules $N$ it contains per unit length along the $y$ direction of the drop. This number is provided by the expression

$$N = \int_S \rho(x,h)\mathrm{d}x\,\mathrm{d}h, \tag{2.289}$$

where $S$ is the area between the solid surface and the circle that approximates the drop profile.

Our numerical calculations have indicated that droplike solutions of the Euler–Lagrange equation can be obtained only if the average fluid density $\rho_{av}$ of the system is bounded by the densities $\rho_{av,min}$ and $\rho_{av,max}$ ($\rho_{av,min} < \rho_{av} < \rho_{av,max}$) both dependent on the interaction potentials, the geometry of the system and temperature. For $\rho_{av} < \rho_{av,min}$ and $\rho_{av} > \rho_{av,max}$, there are no droplike solutions and the fluid generates a film on the solid surface. Typically, $\rho_{av,min}\sigma_{ff}^3$ is between 0.04 and 0.05 and $\rho_{av,max}\sigma_{ff}^3$ is between 0.18 and 0.22. The number of molecules in the considered nuclei is restricted between $N_{min}$ and $N_{max}$, which can be calculated using Equation 2.289.

## 2.10.3 Basic Equations of KNT

The kinetic theory of nucleation developed in Refs. [11–15] considers the change of nucleus size as a result of the competition between the evaporation of the molecules from the surface of the nucleus and the condensation of vapor molecules on that surface, the rates of both processes being functions of the number of molecules, $N$, in the nucleus. When the rate of evaporation $\alpha(N)$ is larger than the rate of condensation $\beta(N)$, the nucleus is shrinking. In the opposite case ($\alpha(N) < \beta(N)$), the nucleus grows. The critical size $N_*$ of the nucleus is provided by the equation $\alpha(N_*) = \beta(N_*)$. Whereas the rate of condensation is considered to be dependent only on the properties of the vapor molecules, the rate of evaporation is determined by the interactions of a fluid molecule at the nucleus surface with all the molecules of the nucleus and of the solid. Hence, $\alpha(N)$ depends on the interaction potential, FDD inside the nucleus, and the shape of the nucleus.

As a result, the nucleation rate is provided by the expression [14,16]

$$I \simeq n_1\beta(1)\frac{S_c(N_*)}{S_c(1)}\left(\frac{H''(N_*)}{2\pi}\right)^{1/2}\exp(H(N_*)), \tag{2.290}$$

where $n_1$ is the vapor number density, $S_c(N)$ is the free surface of the nucleus containing $N$ molecules and $H(N)$ is provided by the following equation [14]:

$$\exp(H(N)) = \prod_{j=2}^{N}\frac{\beta(j-1)}{\alpha(j)}. \tag{2.291}$$

Explicit expressions for $\beta(N)$ and $\alpha(N)$ and some other details of KNT are presented in Appendix 2J. Here we only emphasize that the calculation of the functions $H(N)$ and $H''(N)$ present in Equations 2.290 and 2.291, respectively, and of the supersaturation $s$ provided by Equation 2J.14 involves the evaporation rate per molecule $\alpha^1(N)$ (see Equation 2J.9), which has to be known for all values of $N$ between one and infinity. In turn, the calculation of this function requires the determination of the potential generated by a nucleus containing $N$ molecules and by the solid. However, as previously mentioned, nucleuslike solutions of the Euler–Lagrange equation could be obtained only in a restricted range of the average fluid density of the system. As a result, the size of the nucleus can be varied only between $N_{min}$ and $N_{max}$, the numbers of molecules in the smallest and largest possible nuclei, respectively, and hence the function $\alpha^1(N)$ could be determined only in the range $N_{min} < N < N_{max}$. Whereas $\alpha^1(N)$ has a clear asymptotic limit $\alpha^1(\infty)$ its behavior for $1 < N < N_{min}$ cannot be obtained in the framework of the present model. To estimate the latter behavior, the small nuclei with $N \leq N_{min}$ were modeled to have the same uniform density for all $N$ and to have circular profile that, because of their smallness, make a contact angle equal to that of a nanodrop on a planar surface. This uniform fluid density was determined by matching the evaporation rates of a uniform and nonuniform nucleus containing $N_{min}$ molecules. After the evaporation rate was calculated for small nuclei for sizes between $N = 1$ and $N = N_{min}$, the obtained data for small and large nuclei were approximated by a fitting function that could be used to calculate the nucleation rate.

Let us illustrate the procedure to obtain the fitting function for a solid substrate with $\epsilon_{fs} = 1.0\epsilon_{ff}$. First, the best fitting for the evaporation rate $\alpha^1(x)$ as function of the nucleus size $x$ was determined for a planar surface using a built-in fitting function of MATHEMATICA. In this case $\alpha^1(x)$ has the form $\alpha_p^1(x) = 0.07266 + 2.5297/(1 + x^{0.3})^{3.4} - 0.03574x^{-0.13}\exp(-0.035x)$, where $x$ is the number of molecules in the nucleus. The determination of the fitting functions for nuclei in cavities was restricted to functions that have the same asymptotic values at $x \to \infty$ as for planar surfaces, because the evaporation rate for very large nuclei is expected to be the same for planar and rough surfaces. In this way the functions $\alpha_{15}^1(x) = 0.07266 + 1.1282/(0.645 + x^{0.3})^{3.4} - 0.0005839x^{-0.13}\exp(-0.008x)$ and $\alpha_{11}^1(x) = 0.07266 + 2.5297/(1 + x^{0.3})^{3.4} - 0.03574x^{-0.13}\exp(-0.035x)$ were obtained for cavities with radii $R_s = 15\sigma_{ff}$ and $R_s = 11\sigma_{ff}$, respectively.

### 2.10.4 Results and Conclusion

In the calculations, argon was selected as the fluid with the interaction parameters $\epsilon_{ff}/k_B = 119.76$ K and $\sigma_{ff} = 3.405$ Å [27]. As supporting solid, substrates with $\epsilon_{fs} = \epsilon_{ff}$, and $\epsilon_{fs} = 0.8\epsilon_{ff}$ and $\sigma_{fs} = 3.727$ Å were considered. Such substrates provided the largest numbers of nucleus sizes in the DFT calculations. For larger and smaller $\epsilon_{fs}$, droplike solutions of the Euler–Lagrange equation for the FDD are present in a smaller range of nucleus sizes. All calculations have been carried out at a temperature $T = 85$ K.

In Figure 2.38, the nucleation rate is plotted as a function of supersaturation $s$ for a planar surface (thick solid line) and for cavities of radii $R_s = 15\sigma_{ff}$ (dashed dotted line) and $R_s = 11\sigma_{ff}$ (dashed line) at $\epsilon_{fs} = 0.8\epsilon_{ff}$. As expected, the nucleation rate at higher supersaturations (small critical nuclei) is practically the same in all considered cases. For smaller supersaturation, the nucleation rate increases with increasing curvature of the substrate. For the considered range of supersaturations this difference is of about 2 orders of magnitude when the nucleation rates for a cavity with $R_s = 11\sigma_{ff}$ and a planar surface (zero curvature) are compared.

Similar results were obtained for $\epsilon_{fs} = 1.0\epsilon_{ff}$ (see Figure 2.39).

In Figure 2.40, the dependence of the nucleation rate on the energy parameter $\epsilon_{fs}$ of the fluid–solid interaction is presented for a planar surface (panel a) and cavities with $R_s = 15\sigma_{ff}$ (panel *b*) and $R_s = 11\sigma_{ff}$ (panel c). In all cases, the larger is $\epsilon_{fs}$ the larger is the nucleation rate. This result is expected because an increase in the fluid–solid interactions decreases the evaporation rate and hence the size of the critical nucleus at the same supersaturation.

To estimate the effect of the nonuniformity of the fluid density on the nucleation rate, we calculated the latter by substituting the inhomogeneous nucleus obtained via the DFT calculations with another one of the same shape and number of molecules, but with uniform density. The results obtained in this manner for $\epsilon_{fs} = 0.8\epsilon_{ff}$ are presented in Figure 2.41. Qualitatively, they coincide with those for nonuniform distribution.

A typical difference between nucleation rates calculated for nonuniform and uniform FDDs for the same other conditions is presented in Figure 2.42 for a cavity with radius $R_s = 11\sigma_{ff}$ and $\epsilon_{fs} = 0.8\epsilon_{ff}$. The nucleation rate calculated for uniform FDD is somewhat larger than that calculated for nonuniform FDD. The difference is almost negligible at high supersaturations $s$ and increases with decreasing $s$. For all considered cases, this difference does not exceed 2 orders of magnitude.

In conclusion, the accounting of the nonuniform internal structure of a nucleus on the basis of the DFT provides for the nucleation rate results which, for small supersaturations, are 2 orders of magnitude smaller that those obtained by considering a uniform nucleus. In both cases, the results predict an increase of the nucleation rate, $I$, in the cavity with increasing cavity curvature (decreasing cavity radius) and increasing

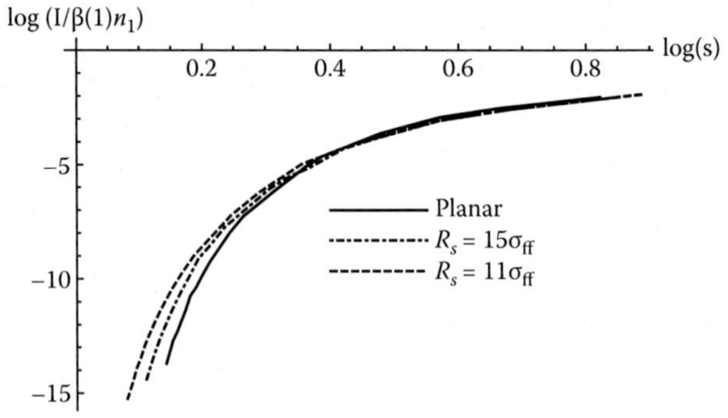


Figure 2.38

Nucleation rates as functions of supersaturation $s$ for $\epsilon_{fs} = 0.8\epsilon_{ff}$ and a planar surface as well as for cavities with radii equal to $15\sigma_{ff}$ and $11\sigma_{ff}$.

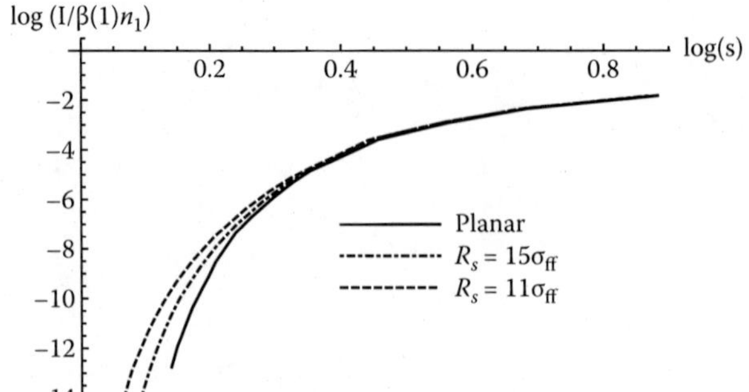


Figure 2.39

As in Figure 2.38 but for $\epsilon_{fs} = 1.0\epsilon_{ff}$.

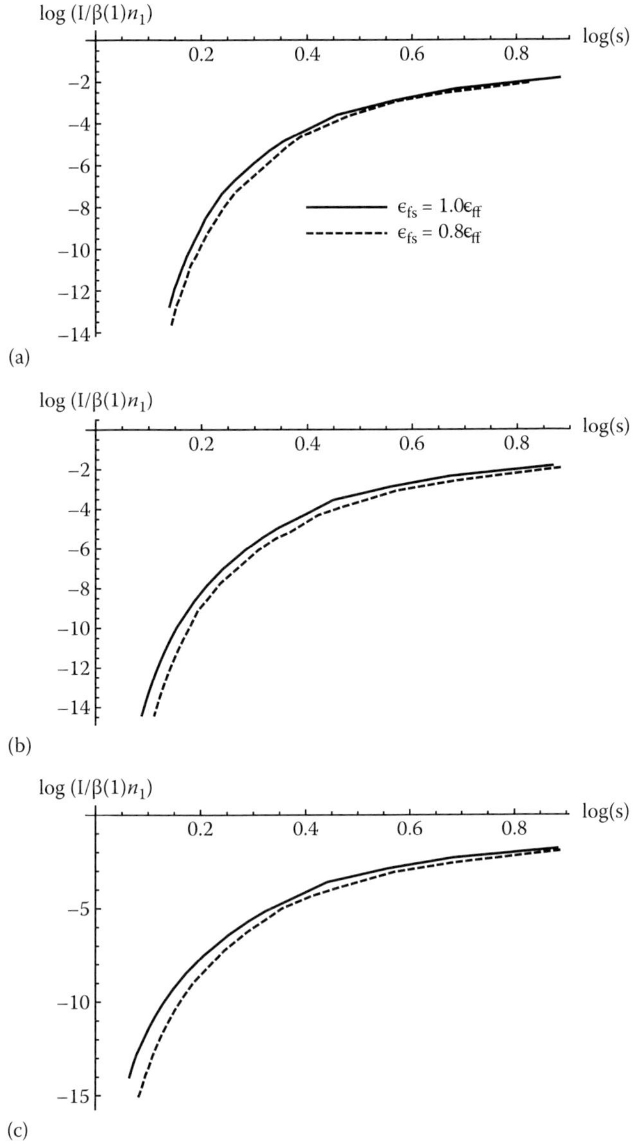


Figure 2.40

Comparison of nucleation rates for $\epsilon_{fs} = 0.8\epsilon_{ff}$ and $\epsilon_{fs} = 1.0\epsilon_{ff}$. (a) Planar surface, (b) cavity with radius $R_s = 15\sigma_{ff}$, and (c) cavity with radius $R_s = 11\sigma_{ff}$.

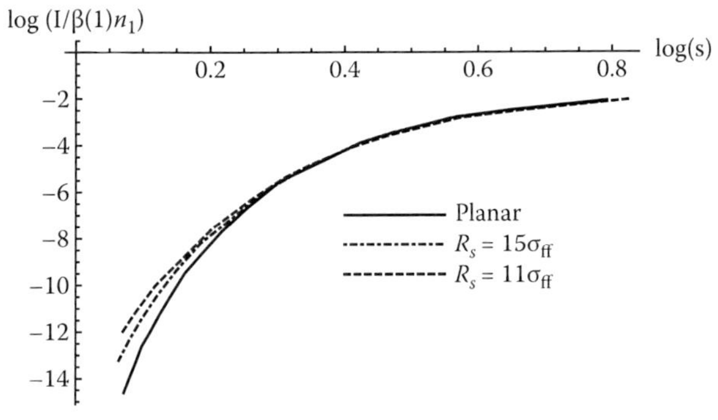


Figure 2.41

Nucleation rates calculated under the assumption of uniform density in the nucleus for $\epsilon_{fs} = 0.8\epsilon_{ff}$ and a planar surface and for cavities with radii equal $15\sigma_{ff}$ and $11\sigma_{ff}$.

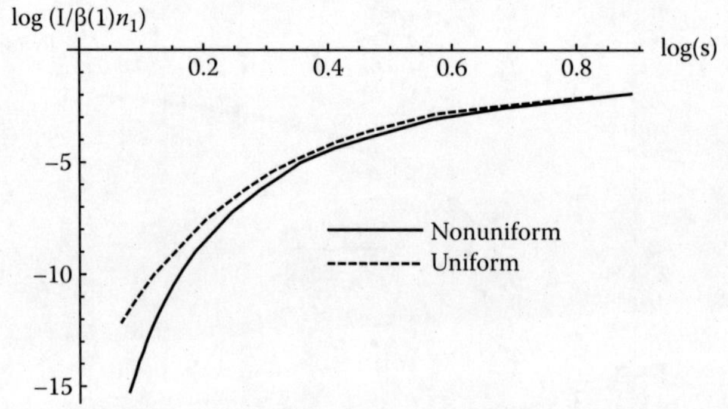


Figure 2.42

Cavity with radius $r_s = 11\sigma_{ff}$. Comparison of nucleation rates for $\epsilon_{fs} = 0.8\epsilon_{ff}$ calculated with and without the assumption of uniform density in the nucleus.

strength of the fluid–solid interactions. The major difference between the present and previous (see Refs. [14–16]) considerations is that previously the contact angle and the uniform density were independent parameters that remained the same for all sizes of the nucleus, whereas in the present considerations the contact angle $\theta$ and nonuniform density are provided by the calculated FDD. The angle $\theta$ depends on the size $N$ of the nucleus and increases with increasing $N$. For example, for a cavity with $R_s = 11\sigma_{ff}$ and $\epsilon_{fs} = 0.8\epsilon_{ff}$, this angle changes between 107° and 162° with increasing size of the nucleus from $N\sigma_{ff} = 31.2$ to $N\sigma_{ff} = 135.0$.

## Appendix 2I: Contributions to the Free Energy of the System and Derivation of Euler–Lagrange Equation for FDD

The contributions to the free energy listed in Section 2.10.2 can be expressed as follows [25,26]:

$$F_{id}[\rho(\mathbf{r})] = k_B T \int d\mathbf{r}\rho(\mathbf{r})\{\log[\Lambda^3\rho(\mathbf{r})] - 1\}, \tag{2I.1}$$

$$F_{hs}[\rho(\mathbf{r})] = \int d\mathbf{r}\rho(\mathbf{r})\Delta\psi_{hs}(\mathbf{r}), \tag{2I.2}$$

where $\Lambda = h_P/(2\pi m k_B T)^{1/2}$ is the thermal de Broglie wavelength, $k_B$ and $h_P$ are the Boltzmann and Planck constants, respectively, $m$ is the mass of a fluid molecule,

$$\Delta\psi_{hs}(\mathbf{r}) = k_B T \eta_{\bar{\rho}} \frac{4 - 3\eta_{\bar{\rho}}}{(1 - \eta_{\bar{\rho}})^2}, \tag{2I.3}$$

$\eta_{\bar{\rho}} = \frac{1}{6}\pi\bar{\rho}(\mathbf{r})\sigma_{ff}^3$ is the packing fraction of the fluid molecules, $\sigma_{ff}$ is the fluid hard core diameter, and $\bar{\rho}(\mathbf{r})$ is the smoothed density defined as

$$\bar{\rho}(\mathbf{r}) = \int d\mathbf{r}'\rho(\mathbf{r}')W(|\mathbf{r} - \mathbf{r}'|). \tag{2I.4}$$

The weighting function $W(|\mathbf{r} - \mathbf{r}'|)$ is selected in the form [28]

$$W(|\mathbf{r} - \mathbf{r}'|) = \begin{cases} \dfrac{3}{\pi\sigma_{ff}^3}\left(1 - \dfrac{r}{\sigma_{ff}}\right), & r \le \sigma_{ff} \\ 0, & r > \sigma_{ff}, \end{cases}$$

where $r = |\mathbf{r} - \mathbf{r}'|$.

The contribution to the free energy due to the attraction between the fluid–fluid molecules is calculated in the mean-field approximation

$$F_{\mathrm{attr}}[\rho(\mathbf{r})]=\frac{1}{2}\iint \mathrm{d}\mathbf{r}\,\mathrm{d}\mathbf{r}'\rho(\mathbf{r})\rho(\mathbf{r}')\phi_{\mathrm{ff}}\left(|\mathbf{r}-\mathbf{r}'|\right) \tag{2I.5}$$

and the contribution due to fluid–solid interactions is given by

$$F_{\mathrm{fs}}[\rho(\mathbf{r})]=\int \mathrm{d}\mathbf{r}\rho(\mathbf{r})U_{\mathrm{fs}}(\mathbf{r}). \tag{2I.6}$$

In Equations 2I.5 and 2I.6, the integration is over the volume occupied by the fluid.

Minimizing the total free energy with respect to $\rho(\mathbf{r})$ provides the following Euler–Lagrange equation for the two-dimensional FDD $\rho(x,h)$:

$$\log[\Lambda^3\rho(x,h)]-Q(x,h)=\frac{\lambda}{k_{\mathrm{B}}T} \tag{2I.7}$$

with

$$Q(x,h)=-\frac{1}{k_{\mathrm{B}}T}[\Delta\psi_{\mathrm{hs}}(x,h)+\overline{\Delta\psi'}_{\mathrm{hs}}(x,h)+U_{\mathrm{ff}}(x,h)+U_{\mathrm{fs}}(x,h)], \tag{2I.8}$$

where $U_{\mathrm{fs}}(x,h)$ is provided by Equation 2.288,

$$U_{\mathrm{ff}}(x,h)=\int\!\!\int \mathrm{d}x'\mathrm{d}h'\rho(x',h')\phi_{\mathrm{ff},y}(|x-x'|,|h-h'|), \tag{2I.9}$$

$$\overline{\Delta\psi'}_{\mathrm{hs}}(x,h)=\int\!\!\int \mathrm{d}x'\mathrm{d}h'\rho(x',h')W_y(|x-x'|,|h-h'|)\times\frac{\partial}{\partial\bar{\rho}}\Delta\psi_{\mathrm{hs}}(\bar{\rho})|_{\bar{\rho}=\bar{\rho}(x',h')}, \tag{2I.10}$$

$\phi_{\mathrm{ff},y}(|x-x'|,|h-h'|)$ and $W_y(|x-x'|,|h-h'|)$ are obtained by integrating the potential $\phi_{\mathrm{ff}}(|\mathbf{r}-\mathbf{r}'|)$ and the weighted function $W(|\mathbf{r}-\mathbf{r}'|)$ with respect to the $y$ direction, from $-\infty$ to $+\infty$, respectively. In Equation 2I.7, $\lambda$ is a Lagrange multiplier, which can be found from the constraint of fixed average density of the fluid.

When calculating the term $U_{\mathrm{ff}}(x,h)$ of the Euler–Lagrange equation arising due to the long-range fluid–fluid interactions, a cutoff at a distance equal to four molecular diameters $\sigma_{\mathrm{ff}}$ for the range of Lennard–Jones attraction was employed.

## Appendix 2J: Basic Equations of KNT

The kinetic theory developed in Refs. [11–15] starts from the fact that in vapors the collisions between molecules are relatively infrequent and, consequently, the energy of the molecules changes slowly compared to the changes in their linear momentum and position. Therefore, the time evolution in the energy space can be described by a diffusion-like equation for the probability density $P(E,t)$ of the energy $E$ [11,29]:

$$\frac{\partial P(E,t)}{\partial t}=\frac{\partial}{\partial E}D(E)P_{\mathrm{eq}}(E)\frac{\partial}{\partial E}\frac{P(E,t)}{P_{\mathrm{eq}}(E)}, \tag{2J.1}$$

where $D(E)$ is the diffusion coefficient in the energy space, $P_{eq}(E)$ is the equilibrium distribution of the energy at temperature $T$ provided by the equation

$$P_{eq}(E) = \Omega(E)\exp(-E/kT)/Z. \tag{2J.2}$$

In Equation 2J.2, $Z$ is the partition function

$$Z = \int_{\Phi_0}^{\Phi_1} \Omega(E)\exp(-E/kT)\mathrm{d}E, \tag{2J.3}$$

$\Omega(E)$ is the density of states of energy $E$, and $\Phi_0$ and $\Phi_1$ are the energies at the lower and the upper boundaries of the potential well, respectively. For a single molecule with the Hamiltonian $H(\mathbf{r},\mathbf{p})$, where $\mathbf{r}$ and $\mathbf{p}$ are the position and the linear momentum of the molecule, respectively, $\Omega(E)$ is provided by the equation

$$\Omega(E) = \int_{\Gamma} \mathrm{d}\Gamma\delta(E - H(\mathbf{r},\mathbf{p})), \tag{2J.4}$$

where $\Gamma$ is the region of the phase space available to the molecule and $\delta(E - H(\mathbf{r},\mathbf{p}))$ is the Dirac $\delta$ function.

The boundary condition for Equation 2J.1 is selected as [14]

$$P(E)|_{E=\Phi_1} = 0 \tag{2J.5}$$

and involves the assumption that any molecule that acquires the energy $\Phi_1$ separates from the nucleus. With this boundary condition, the stationary solution of Equation 2J.1 has the form

$$P(E) = j_E P_{eq}(E)\int_{E}^{\Phi_1} \frac{\mathrm{d}E}{D(E')P_{eq}(E')}, \tag{2J.6}$$

where $j_E$ is the steady-state solution for the flux $j(E, t)$ in the energy space, the latter being defined as

$$\frac{\partial P(E,t)}{\partial t} = \frac{\partial}{\partial E} j(E,t) \tag{2J.7}$$

and is given by the expression [11,14]

$$j(E,t) = -D(E)P_{eq}(E)\frac{\partial}{\partial E}\frac{P(E,t)}{P_{eq}(E)}. \tag{2J.8}$$

Using the normalization condition, $\int P(E)\mathrm{d}E = 1$, the following expression for the rate of evaporation per molecule, $\alpha^1$, is obtained [14]:

$$\alpha^1 = \left[\int_{\Phi_0}^{\Phi_1} \mathrm{d}EP_{eq}(E)\int_{E}^{\Phi_1} \frac{\mathrm{d}E'}{D(E')P_{eq}(E')}\right]^{-1}. \tag{2J.9}$$

It should be noted that $\alpha^1$ depends on the nucleus size $N$ (number of molecules in the nucleus), i.e., $\alpha^1 = \alpha^1(N)$. The applicability of the steady-state approximation used here to calculate the evaporation rate was examined in detail in Ref. [14]. The lag time in which the steady state is achieved was calculated by Nowakowski and Ruckenstein [13].

Assuming Brownian dynamics, the diffusion coefficient $D(E)$ in the energy space is given by [30]

$$D(E)=\frac{3kT\xi}{m\Omega(E)}\int_{\Phi_0}^{E}\mathrm{d}E'\Omega(E'), \tag{2J.10}$$

where $m$ is the mass of a molecule and $\xi$ is a friction coefficient.

The net evaporation rate, $\alpha$, is proportional to $\alpha^1$ and the number of molecules in the potential well. Assuming that the number density $n_w$ of the fluid molecules in the well is constant, the total number of molecules in a well of thickness $l$ is

$$\alpha = \alpha^1 n_w l S_c, \tag{2J.11}$$

where $S_c \equiv S_c(N)$ is the free surface of the nucleus that depends on the nucleus size.

The critical nucleus size, $N_*$, is determined by equating the evaporation and condensation rates. The latter rate, $\beta \equiv \beta(N)$, is provided by the number of impingements of the vapor molecules on the free surface $S_c$ of the nucleus and is given by [6]

$$\beta=\frac{\bar{\upsilon}}{4}S_c n_1, \tag{2J.12}$$

where $\bar{\upsilon}$ is the mean velocity of the vapor molecules and $n_1$ is the number density of the vapor. Combining Equations 2J.11 and 2J.12, the following relation between $n_1$ and the size $N_*$ of the critical nucleus is obtained [11,14]:

$$n_1 = n_w \frac{4l}{\bar{\upsilon}}\alpha^1(N_*). \tag{2J.13}$$

The supersaturation $s$ is defined as $s = n_1/n_s$, where $n_s$ is the vapor number density in equilibrium with the bulk condensed phase. Formally, $n_s$ can be determined from Equation 2J.13 as the limiting value of $n_1$ for $N_* \to \infty$ [11,14], consequently the supersaturation can be expressed as

$$s=\frac{\alpha^1(N_*)}{\alpha^1(\infty)}. \tag{2J.14}$$

Note that the specific values of $\bar{\upsilon}$, $l$, and $n_w$ are not needed because these quantities are not present in the final equations.

In the kinetic approach developed in Refs. [11,14], the nucleation rate is calculated on the basis of the equation for the evolution of the population $n_i$ of nuclei consisting of $i$ molecules:

$$\frac{\mathrm{d}n_i}{\mathrm{d}t} = I_{i-1} - I_i, \tag{2J.15}$$

where the time-dependent flux $I_i$ of nuclei passing from the $i$ population to the $(i + 1)$ population is given by

$$I_i = \beta(i)n_i - \alpha(i + 1)n_{i+1}. \tag{2J.16}$$

At equilibrium, $I_i$ vanishes providing the detailed balance equation from which the equilibrium distribution can be obtained [14]:

$$n_i^{\mathrm{eq}} = n_1^{\mathrm{eq}}\prod_{j=2}^{i}\frac{\beta(j-1)}{\alpha(j)} \equiv n_1^{\mathrm{eq}}\exp(H(i)). \tag{2J.17}$$

After the conversion of Equation 2J.16 to a continuum form the rate of nucleation, $I$, calculated as the stationary solution of that equation, can be written in the form [14]

$$I = \left[ \int_1^{\infty} \frac{dg}{\beta(g) n^{eq}(g)} \right]^{-1}. \tag{2J.18}$$

At the critical size $g = N_*$, the equilibrium distribution $n^{eq}(g)$ has a sharp minimum and this allows us to calculate I using the steepest descent approximation for the integral in Equation 2J.18 [14]:

$$I \simeq n_1 \beta_1 \frac{S_c(N_*)}{S_c(1)} \left( \frac{H''(N_*)}{2\pi} \right)^{1/2} \exp(H(N_*)). \tag{2J.19}$$

## References


1. Farkas, L., *Z. Phys. Chem. A* 125, 236 (1927).
2. Volmer, M., *Z. Elektrochem.* 35, 555 (1929).
3. Becker, R., Doering, W., *Ann. Phys.* 24, 719 (1935).
4. Becker, R., *Discuss. Faraday Soc.* 5, 55 (1949).
5. Fletcher, N.H., *J. Chem. Phys.* 29, 572 (1958).
6. Abraham, F.F., *Homogeneous Nucleation Theory*, Academic, New York, 1974.
7. Liu, Q.X., Zhu, Y.J., Yang, G.W., Yang, Q.B., *J. Mater. Sci. Technol.* 24, 183 (2008).
8. Qian, M., Ma, J., *J. Chem. Phys.* 130, 214709 (2009).
9. Narsimhan, G., Ruckenstein, E., *J. Colloid Interface Sci.* 128, 549 (1989). (Section 1.2 of this volume.)
10. Ruckenstein, E., Nowakowski, B., *J. Colloid Interface Sci.* 137, 583 (1990). (Section 1.3 of this volume.)
11. Nowakowski, B., Ruckenstein, E., *J. Chem. Phys.* 94, 1397 (1991). (Section 2.1 of this volume.)
12. Nowakowski, B., Ruckenstein, E., *J. Chem. Phys.* 94, 8487 (1991). (Section 2.3 of this volume.)
13. Nowakowski, B., Ruckenstein, E., *J. Colloid Interface Sci.* 145, 182 (1991). (Section 2.4 of this volume.)
14. Nowakowski, B., Ruckenstein, E., *J. Phys. Chem.* 96, 2313 (1992). (Section 2.7 of this volume.)
15. Ruckenstein, E., Nowakowski, B., *Langmuir* 8, 1470 (1992).
16. Ruckenstein, E., Berim, G.O., *J. Colloid Interface Sci.* 351, 277 (2010). (Section 2.9 of this volume.)
17. Gay-Lussac, J.L., *Ann. Chim.* 87, 225 (1813).
18. Page, A.J., Sear, R.P., *Phys. Rev. Lett.* 97, 065701 (2006).
19. Page, A.J., Sear, R.P., *J. Am. Chem. Soc.* 131, 17550 (2009).
20. de Ruijter, M.J., Blake, T.D., De Coninck, J., *Langmuir* 15, 7836 (1999).
21. Porcheron, F., Monson, P.A., *Langmuir* 22, 1595 (2006).
22. Giovambattista, N., Debenedetti, P.G., Rossky, P.J., *J. Phys. Chem. B* 111, 9581 (2007).
23. Berim, G.O., Ruckenstein, E., *J. Chem. Phys.* 129, 014708 (2008).
24. Berim, G.O., Ruckenstein, E., unpublished results.
25. Tarazona, P., *Phys. Rev. A* 31, 2672 (1985).
26. Tarazona, P., Marconi, U.M.B., Evans, R., *Mol. Phys.* 60, 573 (1987).
27. Evans, R., Tarazona, P., *Phys. Rev. A* 28, 1864 (1983).
28. Nilson, R.H., Griffits, S.K., *J. Chem. Phys.* 111, 4281 (1999).
29. Keck, J., Carrier, G., *J. Chem. Phys.* 43, 2284 (1965).
30. Borkovec, M., Berne, B.J., *J. Chem. Phys.* 82, 794 (1985).

# 3

# Application of the Kinetic Nucleation Theory to Protein Folding

Numerous processes in the living body are dependent on protein molecules that must have a specific (native) structure to be biologically active. The formation and loss of this structure are known as protein folding and denaturation. Many thermodynamic and kinetic aspects of these processes remain unclear [1–10].

It is known that protein folding occurs through the formation of a critical number of tertiary (native) contacts [3,4,11–14]. As soon as this number is attained, the native structure is formed without passing through any detectable intermediate. In Sections 3.1 through 3.5, this process (protein folding) is modeled using the ideas of kinetic nucleation theory. The protein is treated as a heteropolymer chain consisting of neutral, hydrophobic, and hydrophilic beads with all the bonds and bond angles equal and constant [3,4,12]. The model takes into account the pairwise repulsive and attractive "molecular" interactions, configurational potentials due to protein residues, and confining potential that arises because of the finite size of the protein. The dihedral potential was averaged over all possible configurations of all neighboring residues along the protein chain. Combined with an effective pairwise and confining potentials, it gives rise to an overall potential around the cluster. This allows one to evaluate the protein folding time.

The reverse process, protein denaturation (unfolding), if occurs, results in the loss of biological activity that may lead to catastrophic consequences for a living organism. Denaturation can be caused by heating or cooling, changes in pH, changes in the dielectric constant, in the ionic strength of the solution, etc. In Section 3.5, the model for protein denaturation is presented with the focus on the role of temperature in this process. One of the main features of this model is the barrierless way of protein denaturation similar to spinodal decomposition.

In Section 3.6, the original model is made more realistic by including explicitly the hydrogen bonding into consideration. The disruption of hydrogen bond networks around the interacting particles provides a contribution in the overall potential field around a cluster. The hydrogen bond contribution plays a dominant role in the total potential field around the cluster.

## References


1. Stryer, L., *Biochemistry*, W.H. Freeman and Co, 3rd ed., 1988.
2. Nölting, B., *Protein Folding Kinetics*, Springer-Verlag, Berlin, 2006.
3. Honeycutt, J.D., Thirumalai, D., Metastability of the folded states of globular-proteins. *Proc. Natl. Acad. Sci U.S.A.* 87, 3526–3529 (1990).

4. Honeycutt, J.D., Thirumalai, D., The nature of folded states of globular-proteins. *Biopolymers* 32, 695–709 (1992).
5. Nölting, B.J., Structural resolution of the folding pathway of a protein by correlation of Phi-values with inter-residue contacts. *Theor. Biol.* 194, 419–428 (1998).
6. Nölting, B.J., Analysis of the folding pathway of chymotrypsin inhibitor by correlation of Phi-values with inter-residue contacts. *Theor. Biol.* 197, 113–121 (1999).
7. Weissman, J.S., Kim, P.S., Reexamination of the folding of BPTI—Predominance of native intermediates. *Science* 253, 1386–1393 (1991).
8. Creighton, T.E., Protein folding—Up the kinetic pathway. *Nature* 356, 194–195 (1992).
9. Gursky, O., Atkinson, D., High- and low-temperature unfolding of human high-density apolipoprotein A-2. *Protein Sci.* 5, 1874–1882 (1996).
10. Paci, E., Karplus, M., Unfolding proteins by external forces and temperature: The importance of topology and energetic. *Proc. Natl. Acad. Sci. U.S.A.* 97, 6521–6526 (2000).
11. Abkevich, V.I., Gutin, A.M., Shakhnovich, E.I., Specific nucleus as the transition-state for protein-folding—Evidence from the lattice model. *Biochemistry* 33, 10026–10036 (1994).
12. Guo, Z., Thirumalai, D., Kinetics of protein-folding—Nucleation mechanism, time scales, and pathways. *Biopolymers* 36, 83–102 (1995).
13. Fersht, A.R., Optimization of rates of protein-folding—The nucleation–condensation mechanism and its implications. *Proc. Natl. Acad. Sci. U.S.A.* 92, 10869–10873 (1995).
14. Fersht, A.R., Nucleation mechanisms in protein folding. *Curr. Opin. Struct. Biol.* 7, 3–9 (1997).

# 3.1 Model for the Nucleation Mechanism of Protein Folding

*Yuri Djikaev and Eli Ruckenstein**

A nucleation-like pathway of protein folding involves the formation of a cluster containing native residues that grows by including residues from the unfolded part of the protein. This pathway is examined by using a heteropolymer as a protein model. The model heteropolymer consists of hydrophobic and hydrophilic beads with fixed bond lengths and bond angles. The total energy of the heteropolymer is determined by the pairwise repulsive/attractive interactions between nonlinked beads and by the contribution from the dihedral angles involved. The parameters of these interactions can be rigorously defined, unlike the ill-defined surface tension of a cluster of protein residues that constitutes the basis of a previous nucleation model. The main idea underlying the new model consists of averaging the dihedral potential of a selected residue over all possible configurations of all neighboring residues along the protein chain. The resulting average dihedral potential depends on the distance between the selected residue and the cluster center. Its combination with the average pairwise potential of the selected residue and with a confining potential caused by the bonds between the residues leads to an overall potential around the cluster that has a double-well shape. Residues in the inner (closer to the cluster) well are considered as belonging to the folded cluster, whereas those in the outer well are treated as belonging to the unfolded part of the protein. Transitions of residues from the inner well into the outer one, and vice versa, are considered as elementary emission and absorption events, respectively. The double-well character of the potential well around the cluster allows one to determine the rates of both emission and absorption of residues by the cluster using a first passage time analysis. Once these rates are found as functions of the cluster size, one can develop a self-consistent kinetic theory for the nucleation mechanism of folding of a protein. The model allows one to evaluate the size of the nucleus and the protein folding time. The latter is evaluated as the sum of the times necessary for the first nucleation event to occur and for the nucleus to grow to the maximum size (of the folded protein). Depending on the diffusion coefficients of the native residues in the range from $10^{-6}$ to $10^{-8}$ $cm^2/s$, numerical calculations for a protein of 2500 residues suggest that the folding time ranges from several seconds to several hundreds of seconds.

## 3.1.1 Introduction

Proteins play an overwhelmingly dominant role in life. If a specific job has to be performed in a living organism, then it is almost always a protein that does it. Life depends on thousands of different proteins whose structures are fashioned so that the individual protein molecules combine, with exquisite precision, with other molecules. For a protein molecule to carry out a specific biological function, it has to adopt a well-defined (native) three-dimensional structure [1,2]. The formation of this structure (of a biologically active globular protein) constitutes the core of the so-called "protein folding problem" [3]. Many thermodynamic and kinetic aspects of the process remain obscure, and its mechanism is elusive [4–9].

Experiment and simulation suggest that there are multiple pathways for protein folding [5–25]. It is believed that initially an unfolded protein transforms very quickly into a compact (but not native) configuration with a few, insignificant number of tertiary contacts. The transition from such a compact configuration to the native one has been suggested to occur via two distinct mechanisms. One of them can be referred to as a "transition state mechanism" whereby the tertiary contacts of the native structure are formed as the protein passes through a sequence of intermediate states thus gradually achieving its unique spatial configuration [7–24]. The protein in the intermediate states has a nativelike overall topology but is stabilized by incorrect hydrophobic contacts. These states correspond to misfolded forms of the native protein. The transition from these intermediate, misfolded states to the correctly folded, native structure is a slow process (compared to the formation of the compact configuration) that occurs on a relatively large timescale (because it involves a large-scale rearrangement of the molecule) [15,16]. Alternatively, the transition from the compact "amorphous" configuration to the native state occurs immediately following the formation of a number of tertiary contacts [15,16]. This mechanism is similar to nucleation—i.e., once a critical number of (native) tertiary contacts is established the native structure is formed without passing through any detectable intermediates [15].

So far, most of the work on protein folding has been carried out by using either Monte Carlo (MC) or molecular dynamics (MD) simulations. A rigorous theoretical treatment of protein folding by means of statistical mechanics is hardly practicable because of the extreme complexity of the system, although

* *J. Phys. Chem. B* 111, 886 (2007). Republished with permission.

some approximate treatments have been already reported [26–29]. Moreover, a theoretical model for the nucleation mechanism of the process has so far remained underdeveloped [15,30–32]. The model, a thermodynamic one, considers the formation of a cluster of protein residues and calculates its free energy change much like the classical nucleation theory (CNT) does. The cluster is characterized by the number of residues $\nu$ (with the mole fraction of the hydrophobic ones assumed to be known). As usual in CNT, the critical cluster (nucleus) corresponds to the maximum of the free energy of formation as a function of $\nu$. Such an approach involves the use of the concept of surface tension for a cluster consisting of protein residues. This quantity is an intrinsically ill-defined physical quantity and can be considered only as an adjustable parameter; clearly, no direct experimental measurement thereof is possible. After the formation of a nucleus (a critical size cluster of residues), the protein quickly reaches its native state.

In this paper, we present a new, microscopic model for the nucleation mechanism of protein folding. The new model is based on "molecular" interactions both long-range (i.e., pairwise repulsion/attraction) and configurational (bond and dihedral angles) in which protein residues are involved. These parameters can be rigorously defined, and it should be possible (although not straightforward) to determine them theoretically, computationally, or experimentally. The ill-defined surface tension of a cluster of protein residues (within a protein in a compact but nonnative configuration) does not enter into the new model, which is thus more reasonable than the old one. The main idea underlying the new model consists of averaging the dihedral potential of a selected residue over all possible configurations of neighboring residues. The resulting average dihedral potential depends on the distance between the residue and the cluster center. Its combination with the average long-range potential between the cluster and the residue and with a confining potential due to the bonds between the residues generates a pair of potential wells around the cluster with a barrier between them. The residues in the inner well are considered to belong to the cluster (part of the protein with correct tertiary contacts) while those in the outer well are treated as belonging to the mother phase (amorphous part of the protein with incorrect tertiary contacts). Transitions of residues from the inner well into the outer one and vice versa are considered as elementary emission and absorption events, respectively. The rates of emission and absorption of residues by the cluster are determined by using a first passage time analysis [33–38]. Once these rates are found as functions of the cluster size, one can develop a self-consistent kinetic theory for the nucleation mechanism of folding of a protein. For example, the size of the critical cluster (nucleus) is then found as the one for which these rates are equal. The time necessary for the protein to fold can be evaluated as the sum of the times necessary for the appearance of the first nucleus and the time necessary for the nucleus to grow to the maximum size (of the folded protein in the native state). Both can be obtained if the emission and absorption rates are known as functions of the cluster size.

The paper is structured as follows. In Section 3.1.2, we describe a random heteropolymer chain whereby a protein molecule is often modeled [7,15] and outline a CNT-based model for the nucleation mechanism for protein folding. A new, microscopic model is proposed in Section 3.1.3, and the results of numerical calculations are presented in Section 3.1.4. A brief discussion and conclusions are summarized in Section 3.1.5.

## 3.1.2 Heteropolymer Chain as a Protein Model and a CNT-Based Model for the Nucleation Mechanism of Protein Folding

### *3.1.2.1 Heteropolymer as a Protein Model*

A heteropolymer was introduced as a simple model of a protein in the MD and MC simulations of protein folding dynamics [7,15]. The polypeptide chain of a protein was modeled as a heteropolymer consisting of $N$ connected beads that can be thought of as representing the α-carbons of various amino acids. The heteropolymer may consist of hydrophobic (b), hydrophilic (l), and neutral (n) beads. Two adjacent beads are connected by a covalent bond of fixed length $\eta$. This model (and its variants), completed with the appropriate potentials described below, has been shown [7,15,26,27,30] to be able to capture the essential characteristics of protein folding even though it contains only some of the features of a real polypeptide chain. For example, this model ignores the side groups although they are known to be crucial for intramolecular hydrogen bonding [1]. Besides, the presence of a solvent (water) in a real physical system has been usually accounted for too simplistically, although protein dynamics were reported to become more realistic in those MD simulations where the solvent molecules were explicitly taken into account [39]. Despite these limitations, various modifications of the heteropolymer model [6,26,27,30,40–44] shed light on some important details regarding the transition of a protein from its unfolded state to the native one [7,15].

The total energy of the heteropolymer (polypeptide chain) can contain three different types of contributions. First, the contribution from repulsive/attractive forces between pairs of nonadjacent beads (these can be, e.g., of Lennard–Jones or other types). The next contribution can arise from the harmonic forces due to the oscillations of bond angles. Finally, there is a contribution from the dihedral angle potential due to the rotation around the peptide bonds. There are various ways to model these three types of contributions [7,15,40–43].

A pair interaction between two nonadjacent beads $i$ and $j$ at a distance $d$ away from each other can be considered to be of the form [7,15]

$$\phi_{ij}(d)=\begin{cases} 4\epsilon_b[(\eta/d)^{12}-(\eta/d)^6] & (i,j=b) \\ 4\epsilon_l[(\eta/d)^{12}+(\eta/d)^6] & (i=l,\ j=b,l) \\ 4\epsilon_n(\eta/d)^{12} & (i=n,\ j=b,l,n), \end{cases} \tag{3.1}$$

where $\eta$ is the bond length (fixed) and $\epsilon_b$, $\epsilon_l$, and $\epsilon_n$ are energy parameters.

The angle $\beta$ between two successive bonds (in the heteropolymer) can be regarded to be subjected to the harmonic potential

$$\phi_\beta=\frac{k_\beta}{2}(\beta-\beta_0)^2, \tag{3.2}$$

where the spring constant $k_\beta$ is sufficiently large for the deviation of the bond angles from the average value $\beta_0$ to be small. As argued in Refs. [7,15], the bond angle forces play a minor role in protein folding/unfolding; hence, all bond angles can be set to be equal to $\beta_0$.

The dihedral angle potential arises due to the rotation of three successive peptide bonds connecting four successive beads and is related to the dihedral angle $\delta$ via

$$\phi_\delta=\epsilon_\delta'(1+\cos\delta)+\epsilon_\delta''(1+\cos 3\delta), \tag{3.3}$$

where $\epsilon_\delta'$ and $\epsilon_\delta''$ are independent energy parameters that depend on the nature and sequence of the four beads involved in the dihedral angle $\delta$. This potential has three minima, one in the *trans* configuration at $\delta = 0$ and two others in the gauche configurations at $\delta=\pm\arccos\sqrt{(3\epsilon_\delta''-3\epsilon_\delta')/12\epsilon_\delta''}$ (the former one being the lowest).

The above structure of the potential functions of a heteropolymer was suggested by Honeycutt and Thirumalai [6]. Discrete analogs (for a protein on a diamond lattice) of Equations 3.1 and 3.3 completed with a "cooperativity potential" were also proposed [40–43], and random energy models were used [30]. It was shown [7,15] via MD simulations, which employed low friction Langevin dynamics, that a proper balance between the above three contributions (Equations 3.1 through 3.3) to the total energy of the heteropolymer ensures that the heteropolymer folds into a well-defined $\beta$-barrel structure. The balancing between these contributions was achieved by adjusting the energy parameters $\epsilon_b$, $\epsilon_l$, $\epsilon_n$, $\epsilon_\delta'$, and $\epsilon_\delta''$ for each of the types of beads. It was also found [7,15] that the balance between the dihedral angle potential, which tends to stretch the molecule into a state with all bonds in a *trans* configuration, and the attractive hydrophobic potential is crucial to induce folding into a $\beta$-barrel-like structure upon cooling. If attractive forces are excessively dominant, then they make the heteropolymer fold into a globulelike structure, while an overwhelming dihedral angle potential forces the chain to remain in an unfolded (elongated) state (even at low temperatures) with bonds mainly in the *trans* configuration.

The possibility that a nucleation-like mechanism constitutes a viable pathway for protein folding was first suggested in Refs. [15,30]. The formalism of the nucleation theory was used to evaluate the size of a critical cluster (nucleus) of native protein residues, the formation of which leads to a rapid transition of the whole protein to its native state. Let us denote the total number of residues in the protein by $N_0$ and consider the formation of a cluster having a correct tertiary structure in an unfolded protein. The free energy of formation of such a cluster of $\nu$ native residues (i.e., residues which are in the same state as they are in the native protein) can be written in the framework of CNT as

$$W=-\nu\Delta\mu+\sigma 4\pi\lambda^2\nu^{2/3}, \tag{3.4}$$

where $\Delta\mu \equiv \mu_u - \mu_f$ is the difference between the free energies per residue in the unfolded and folded (native) states, respectively (marked with the subscripts "u" and "f"), $\sigma$ is the "surface" tension of the boundary between the cluster (having a native structure) and the unfolded part of the protein, $\lambda = (3v/4\pi)^{1/3}$, and $v$ is the volume of a protein residue in its native state.

It was argued [15] that the initial stage of protein folding is driven by the hydrophobic attractive forces so that the volume term (i.e., the first one) in Equation 3.4 is determined by the number of hydrophobic contacts in the cluster and hence can be written as $-(1/2)\epsilon_b\chi\nu(\chi\nu-1)$, where $\chi$ is the mole fraction of hydrophobic residues in the cluster (assumed to be the same as in the whole protein). As a result, the number of residues in the critical cluster is given by $\nu_c - (8\pi\sigma\lambda^2/3\chi^2\epsilon_b)^{3/4}$, which for typical values of $\lambda$, $\sigma$, and $\epsilon_b$ was estimated [15] to be on the order of 10. In Ref. [30], $\Delta\mu$, which appears in the volume term of Equation 3.4, was evaluated to be on the order of 0.1 $k_BT$ (where $k_B$ is the Boltzmann constant and $T$ is the absolute temperature). The surface tension was argued to arise because the amino acid residues located at the cluster surface interact more strongly with the cluster interior than with the unfolded part of the protein. Since the interaction energies in protein folding are on the order of $k_BT$, the surface tension $\sigma$ could be obtained from $\sigma 4\pi\lambda^2 \approx k_BT$ and the number of residues in the critical cluster was evaluated [30] to be on the order of 100 (for $N_0 = 150$). Despite the significant difference between them, both estimates corroborate the idea that the nucleation mechanism constitutes a viable pathway for protein folding [18–22,45–49].

### 3.1.3 A Kinetic Approach to the Nucleation Mechanism of Protein Folding

The above nucleation model of protein folding involves thermodynamics within the framework of CNT. However, it was argued that the concept of surface tension may not be adequate for too small clusters (such as those of interest in nucleation) [33–35], not to mention the assumption (of CNT) that it is equal to the surface tension of a planar interface. Although CNT produces reasonable agreement with experiment for unary nucleation, its application to multicomponent nucleation leads to several inconsistencies and large discrepancies with experiment [50–54] that are blamed on the use of the concept of surface tension. In the case of protein folding this problem is even more complex because $\sigma$ in Equation 3.4 is an ill-defined quantity that cannot be determined experimentally due to the nonexistence of bulk "folded" and "unfolded" proteins as real physical phases, not to mention a flat interface between them.

To avoid the use of macroscopic thermodynamics in the kinetic theory of unary nucleation, an alternative approach was proposed by Ruckenstein and coworkers [33–35] on the basis of a mean first passage time analysis. Unlike CNT, the new theory [33–35] is built on molecular interactions and does not make use of the surface tension for the tiny clusters involved in nucleation. Instead, the theory [33–35] exploits the fact that one can derive and solve the kinetic equation of nucleation (hence find the nucleation rate) if the emission and absorption rates of a cluster are determined as functions of its size. For the rate of absorption of molecules by the cluster, the new theory uses (as CNT does) a standard gas-kinetic expression [55], but (and this is the main idea of the new approach [33–35]) the rate of emission of molecules by the cluster is determined via a mean first passage time analysis. This time is calculated by solving a single-molecule master equation for the probability distribution function of a surface layer molecule moving in a potential well around the cluster. The master equation is a Fokker–Planck equation in the phase space that can be reduced to the Smoluchowski equation owing to the hierarchy of the characteristic timescales involved in the evolution of the single-molecule distribution function with respect to coordinates and momenta [33–35]. Recently, a further development of that kinetic theory was proposed by combining it [36,37] with the density functional theory (DFT) and extending it to binary [37] and heterogeneous [38] systems.

Although the emission rate of the cluster in Refs. [33–38] was found via a first passage time analysis, the absorption rate there was calculated using an expression derived in the framework of the gas-kinetic theory [55], which assumes a Maxwellian distribution of the velocities of the mother phase molecules. Although this assumption is unquestionably valid for vapor-to-liquid nucleation in dilute (if not ideal) gases, it becomes increasingly inaccurate as the density of the mother phase increases and molecular interactions therein become nonnegligible. Clearly, this assumption (hence the absorption rate based thereupon) is inadequate in the treatment of cluster formation during protein folding. Indeed, the amino acid residues of the protein are successively linked by bonds of virtually fixed length and fixed angle between each pair.

In this section, we present a new, kinetic model for the nucleation mechanism of protein folding based on a first passage analysis that will be used to determine not only the rate of emission (of native residues from the cluster) but also (unlike Refs. [33–38]) the rate of absorption (of nonnative residues by the cluster). The general formalism of our model is a mean first passage time analysis, but a major difference, compared to Refs. [33–38], is the double-well potential generated in this case around the cluster.

#### *3.1.3.1 Potential Well around a Cluster of Native Residues*

A heteropolymer chain as a protein model, proposed in Refs. [7,15] and described above, consists of three types of beads: neutral, hydrophobic, and hydrophilic. The neutral beads ($n$) play an important role in that model. Their interaction with each other is purely repulsive, and the dihedral angle forces are considered to be weaker for the bonds involving them so that the bending is enhanced where they are present. Indeed, MD simulations [7,15] show that such a heteropolymer acquires a β-barrel shape in the lowest energy configuration, with neutral residues appearing mostly in the bend regions. For the sake of simplicity, we will consider a heteropolymer consisting of only hydrophobic (b) and hydrophilic (l) beads rather than three kinds of beads. However, this simplification requires us to rebalance the $\epsilon$'s in Equation 3.1 and particularly in Equation 3.3 to allow the formation of loops and turns in a folded heteropolymer chain.

Consequently, we will adopt a two-component analog of a heteropolymer chain with the pair interaction, bond angle, and dihedral angle potentials given by Equations 3.1 through 3.3 as a model for a protein. Besides, one must introduce a confining potential that excludes the volume free of protein residues. Therefore, the formation of a cluster consisting of native residues during protein folding can be regarded as binary nucleation. We shall therefore present a nucleation mechanism model of protein folding in terms of binary nucleation by using a first passage time analysis [33–38] that accounts for the particular shape of the present potential.

Consider a binary cluster of spherical shape (with sharp boundaries and radius $R$) immersed in a binary fluid mixture. In the original papers regarding nucleation [33–35] and recent developments [36–38], a molecule of component $i$ ($i$ = b, l) located in the surface layer of the cluster was considered to perform a thermal chaotic motion in a spherically symmetric potential well $\phi_i(r)$ resulting from the pair interactions of this molecule with those in the cluster.

Assuming pairwise additivity of the interactions between beads, $\phi_i(r)$ is provided by

$$\phi_i(r) = \sum_j \int_v d\mathbf{r}' \rho_j(r')\phi_{ij}(|\mathbf{r}' - \mathbf{r}|). \tag{3.5}$$

Here, $\mathbf{r}$ is the coordinate of the surface bead $i$, $\rho_j(r')$ ($j$ = 1, 2) is the number density of beads of component $j$ at point $\mathbf{r}'$ (spherical symmetry is assumed with the cluster center chosen as the origin of the coordinate system), and $\phi_{ij}(|\mathbf{r}' - \mathbf{r}|)$ is the interaction potential between two beads of components $i$ and $j$ at points $\mathbf{r}$ and $\mathbf{r}'$, respectively. The integration in Equation 3.5 has to be carried out over the whole volume of the system, but the contribution from the unfolded part can be assumed to be small and accounted for by particular choices of $\epsilon_b$ and $\epsilon_l$.

For nucleation in proteins the potential well $\psi_i(r)$ for a residue of type $i$ around the cluster is determined not only by the potential $\phi_i(r)$ but also by three other constituents, $\phi_\beta(r)$, $\bar{\phi}_\delta(r)$, and $\phi_{co}$, which represent the bond angle, dihedral angle, and confining potentials, respectively:

$$\psi_i(r) = \phi_i(r) + \phi_\beta(r) + \bar{\phi}_\delta(r) + \phi_{co}.$$

Without affecting the generality of the model, one can significantly simplify the algebra and eventual numerical calculations by assuming that all bond angles are fixed and equal to $\beta_0 = 105°$. Under this assumption, the contribution to the potential energy of the protein arising from the bond angle potential is constant and does not depend on the distance $r$ between the selected bead and the center of the cluster. Therefore, the term $\phi_\beta(r)$ on the right-hand side of the above equation can be disregarded (or, equivalently, be chosen as a reference level for the potential energy), i.e.,

$$\psi_i(r) = \phi_i(r) + \bar{\phi}_\delta(r) + \phi_{co}. \tag{3.6}$$

Because of the bonds between the residues, none of them can be located outside a volume that depends on the densities of the folded and unfolded regions. Assuming that volume to be spherical, its radius $r_{co}$ is given by the expression $r_{co} = [R^3 + 3(N_0 - v)4\pi\rho_u]^{1/3}$ with $v = (4\pi/3)R^3\rho_f$, where $\rho_u$ and $\rho_f$ denote the unfolded and folded densities, respectively. Consequently, the confining potential has the form

$$\phi_{co} = \begin{cases} 0 & (r < r_{co}) \\ \infty & (r \geq r_{co}) \end{cases} \tag{3.7}$$

The term $\bar{\phi}_\delta(r)$ in $\psi_i(r)$ is due to the dihedral angle potential of the whole protein. Consider a bead 1 (of type b or l) at a distance $\tilde{d}$ from the surface of the cluster, so that $r = R + \tilde{d}$ (Figure 3.1). The total dihedral

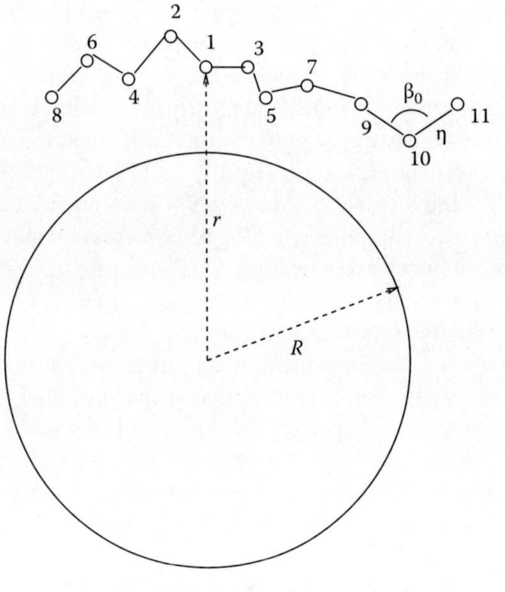


Figure 3.1

Scheme of a piece of a heteropolymer chain around a spherical cluster of radius $R$. Bead 1 is in the plane of the figure, whereas other beads may all lie in different planes, but all bond angles are equal to 105°, and their lengths are equal to $\eta$. The distance between the selected bead 1 and the center of the cluster is $r$.

angle potential $\phi'_\delta$ of the whole protein chain for a given configuration of beads 1, 2, 3, …, $N_0$ is a function of $r$, $\mathbf{r}_2$, $\mathbf{r}_3$, $\mathbf{r}_4$, $\mathbf{r}_5$, $\mathbf{r}_6$, and $\mathbf{r}_7$ and can be written in the form (for simplicity, the arguments $\mathbf{r}_2$, $\mathbf{r}_3$, $\mathbf{r}_4$, $\mathbf{r}_5$, $\mathbf{r}_6$, and $\mathbf{r}_7$ of $\phi'_\delta$ are not explicitly noted)

$$\phi'_\delta(r)=\phi_\delta\left(\delta^{642}_{421}(r)\right)+\phi_\delta\left(\delta^{421}_{213}(r)\right)+\phi_\delta\left(\delta^{213}_{135}(r)\right)+\phi_\delta\left(\delta^{135}_{357}(r)\right)+\text{constant},$$

where $\delta^{ijk}_{jkl}$ is a dihedral angle between two planes, one of which is determined by beads $i$, $j$, and $k$ and the other by beads $j$, $k$, and $l$. The last term on the right-hand side of the above equation, representing the contributions from the dihedral angles involving beads 8, 9, …, $N_0$, does not depend on $r$ because these dihedral angles remain unaffected as the position of bead 1 changes. Hence, this contribution (constant) is omitted hereafter without being specified since it can be regarded as affecting only the reference level for $\psi_i(r)$

$$\phi'_\delta(r)=\phi_\delta\left(\delta^{642}_{421}(r)\right)+\phi_\delta\left(\delta^{421}_{213}(r)\right)+\phi_\delta\left(\delta^{213}_{135}(r)\right)+\phi_\delta\left(\delta^{135}_{357}(r)\right). \tag{3.8}$$

For a given location of bead 1, various configurations of beads 2, 3, …, $N_0$ (subject to the constraints on the bond length and bond angle as well as to the constraint of excluded cluster volume) lead to various sets of dihedral angles. However, variations in the locations of beads 8, 9, …, $N_0$ lead to variations in the dihedral potential that are independent of $r$. Thus, the dihedral term $\bar{\phi}_\delta(r)$ (less a term independent of $r$ and omitted hereafter) on the right-hand side of Equation 3.6 can be obtained by averaging Equation 3.8 with the Boltzmann factor $\exp[-\phi'_\delta(r)/k_BT]$ (and appropriate normalization constant) over all possible configurations of beads 2, 3, …, 7 and assigning the result to a selected bead with fixed coordinates, i.e., bead 1:

$$\bar{\phi}_\delta(r)=f^{-1}\int_{\Omega_{18}} d\mathbf{r}_2\,d\mathbf{r}_3\,d\mathbf{r}_4\,d\mathbf{r}_5\,d\mathbf{r}_6\,d\mathbf{r}_7\,\phi'_\delta(r)\exp[-\phi'_\delta(r)/k_BT], \tag{3.9}$$

where $f$ is a normalization constant,

$$f^{-1}\int_{\Omega_{18}} d\mathbf{r}_2\,d\mathbf{r}_3\,d\mathbf{r}_4\,d\mathbf{r}_5\,d\mathbf{r}_6\,d\mathbf{r}_7\exp[-\phi'_\delta(r)/k_BT], \tag{3.10}$$

and $\Omega_{18}$ is the integration region in an 18-dimensional space. The 18-fold integrals in Equations 3.9 and 3.10 can be reduced to sevenfold integrals by taking into account the constraints of fixed bond length and fixed bond angle (see Appendix 3A):

$$\bar{\phi}_\delta(r) = f^{-1} \sum_{i,j,k,m,n=-}^{+} \int_0^{2\pi} d\varphi_2 \int_{L_2} dr_2 \int_{L_3^i} dx_3 \int_{L_4^j} dx_4 \int_{L_5^k} dx_5 \times \int_{L_6^m} dx_6 \int_{L_7^n} dx_7 r_2^2 \sin\Theta_2(r,r_2)\tilde{\phi}_{ijkmn}(r,\varphi_2,r_2,x_3,\ldots,x_7) \times \exp[-\tilde{\phi}_{ijkmn}(\varphi_2,r_2,x_3,\ldots,x_7)/k_BT]. \tag{3.11}$$

$$f = \sum_{i,j,k,m,n=-}^{+} \int_0^{2\pi} d\varphi_2 \int_{L_2} dr_2 \int_{L_3^i} dx_3 \int_{L_4^j} dx_4 \int_{L_5^k} dx_5 \times \int_{L_6^m} dx_6 \int_{L_7^n} dx_7 r_2^2 \sin\Theta_2(r,r_2) \times \exp[-\tilde{\phi}_{ijkmn}(r,\varphi_2,r_2,x_3,\ldots,x_7)/k_BT]. \tag{3.12}$$

Each of the summation indices in Equations 3.11 and 3.12 takes on two values, denoted (see Appendix 3A) + and −, so that there are $5^2$ terms in the sum differing by the integrand as well as by the integration ranges (except for $L_2$, which is independent of $i$, $j$, $k$, $m$, and $n$). The definitions of the coordinates $\varphi_2$, $r_2$, $x_3$, $x_4$, $x_5$, $x_6$, and $x_7$, the explicit forms of the integration ranges $L_2$, $L_3^i$, $L_4^j$, $L_5^k$, $L_6^m$, and $L_7^n$, and the rules for transforming the function $\phi'_\delta(r)$ into the function $\phi_{ijkmn}$ $(\varphi_2, r_2, x_3, \ldots, x_7)$ are provided in Appendix 3A.

Figure 3.2 presents typical shapes of the constituents $\phi_i(r)$ and $\bar{\phi}_\delta(r)$ of the potential well as functions of the distance from the cluster center as well as the overall potential well $\psi_i(r)$ itself. (For details regarding the numerical calculations, see Section 3.1.4.) The contribution $\phi_i(r)$, arising from the pairwise interactions, has

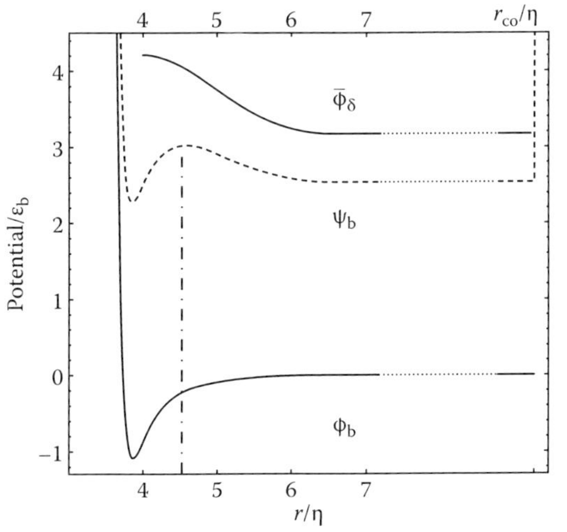


Figure 3.2

Typical shapes of the potentials $\phi_b(r)$ (lower solid curve), $\bar{\phi}_\delta(r)$ (upper solid curve), and $\psi_b(r)$ (dashed curve) for a hydrophobic bead around the cluster as functions of the distance $r$ from the center of the cluster of radius $R = 3\eta$. The vertical dashed line at $r \simeq 4.53\eta$ indicates the location of the maximum of the barrier between the ipw and the opw that determines the widths $\lambda_b^{iw}$ and $\lambda_b^{ow}$ of both potential wells. The outer boundary of the opw ($r_{co} \simeq 12.99\eta$) was assumed to coincide with the outer boundary of the volume wherein the whole protein is encompassed.

a familiar form [33–38] reminiscent of the underlying Lennard–Jones potential, whereas the contribution from the average dihedral potential has a rather remarkable behavior. Starting with its maximum value at the cluster surface, it monotonically decreases with increasing $r$ until it becomes constant for $r \geq \tilde{r}$ (Section 3.1.4, Equation 3.32). Thus, except very short distances from the cluster surface where $\phi_i(r)$ sharply decreases from $\infty$ to its global minimum, the potential $\psi_i(r)$ is shaped by two competing terms, $\phi_i(r)$ and $\bar{\phi}_\delta(r)$, which increase and decrease, respectively, with increasing $r$ and by the confining potential $\phi_{co}$. As a result, the overall potential $\psi_i(r)$ has a double-well shape: The inner well is separated by a potential barrier from the outer well. This shape of $\psi_i(r)$ is of crucial importance to our nucleation mechanism model of protein folding because it allows one to use the mean first passage time analysis for the determination of the rates of absorption and emission of beads by the cluster.

### *3.1.3.2 Determination of Emission and Absorption Rates*

Formulating our model in the spirit of mean first passage time analysis [33–38], a bead (residue) is considered as belonging to a cluster as long as it remains in the inner potential well (hereinafter referred to as "ipw") and as dissociated from the cluster when it passes over the barrier between the ipw and the outer potential well (hereinafter referred to as "opw"). The rate of emission, $W^-$, is determined by the mean time necessary for the passage of the bead from the ipw over the barrier into the opw. Likewise, a bead is considered as belonging to the unfolded part of the heteropolymer (protein) as long as it remains in the opw and as absorbed by the cluster when it passes over the barrier between the opw and the ipw. The rate of emission, $W^+$, is determined by the mean time necessary for the passage of the bead from the opw over the barrier into the ipw.

The mean first passage time of a bead escaping from a potential well is calculated on the basis of a kinetic equation governing the chaotic motion of the bead in that potential well. The chaotic motion of the bead is assumed to be governed by the Fokker–Planck equation for the single-particle distribution function with respect to its coordinates and momenta, i.e., in the phase space [56–58]. Under favorable conditions, the Fokker–Planck equation reduces to the Smoluchowski equation, which involves diffusion in an external field [57,58]. Solving that equation [33–38], one can obtain the following expressions for $W^-$ and $W^+$, the emission and absorption rates (see Appendix 3B for details), respectively:

$$W^- = n^{\text{iw}} D^{\text{iw}} \omega^{\text{iw}} \tag{3.13}$$

$$W^+ = n^{\text{ow}} D^{\text{ow}} \omega^{\text{ow}}. \tag{3.14}$$

Here, the superscripts "iw" and "ow" mark the quantities for the inner and outer potential wells, respectively; $n$ is the number of beads in the well, $D$ is the diffusion coefficient of the bead, and $\omega^{\text{iw}}$ and $\omega^{\text{ow}}$ are defined as

$$\omega^{\text{iw}} \equiv 1/(D^{\text{iw}} \bar{\tau}^{\text{iw}}),\ \omega^{\text{ow}} \equiv 1/(D^{\text{ow}} \bar{\tau}^{\text{ow}}), \tag{3.15}$$

where $\bar{\tau}^{\text{iw}}$ and $\bar{\tau}^{\text{iw}}$ are the mean first passage times for a bead in the ipw to cross over the barrier into the opw, and vice versa, respectively. Note that in the original work [33–35] on the mean first passage time analysis in the nucleation theory and its recent development [36–38], this method was used only for determining $W^-$ but not $W^+$. In the present model, the double-well shape of the overall potential $\psi(r)$ around the cluster allows one to use the first passage time analysis to determine $W^+$ as well. Clearly, the quantities $W^-$, $W^+$, $\omega^{\text{iw}}$, $\omega^{\text{iw}}$, $\bar{\tau}^{\text{iw}}$, and $\bar{\tau}^{\text{iw}}$ are functions of the cluster size and composition (for their explicit forms, see Appendix 3B). However, since the overall composition of the protein (heteropolymer) is fixed, one can assume that the cluster that forms during folding has a constant composition equal to the overall protein composition, which reduces the theory to a unary nucleation theory. (Strictly speaking, one can develop a theoretical model for the nucleation mechanism of protein folding without this assumption, which would lead to a binary nucleation theory, but this would drastically complicate the problem computationally.) Under these assumptions, the aforementioned quantities are functions of only the size of the cluster (say, its radius $R$ or the total number of beads $\nu$ therein).

### *3.1.3.3 Equilibrium Distribution and Steady-State Nucleation Rate*

During protein folding, clusters of various sizes may emerge and exist simultaneously with different probabilities. Let us denote the distribution of clusters with respect to the number of beads in a cluster at time $t$ by $g(\nu, t)$. Once the emission and absorption rates $W^- = W^-(\nu)$ and $W^+ = W^+(\nu)$ are known as functions of cluster size, one can find the equilibrium distribution of clusters $g_e(\nu)$ and solve the kinetic equation of nucleation to find the steady-state nucleation rate.

According to the principle of detailed balance,

$$W^{+}(v-1)g_{e}(v-1) = W^{-}(v)g_{e}(v), \tag{3.16}$$

which can be rewritten as

$$\frac{g_{e}(v)}{g_{e}(v-1)} = \frac{W^{+}(v-1)}{W^{-}(v)}. \tag{3.17}$$

By applying Equation 3.17 to $(v - i)$ with $i = 2, 3 \ldots, v - 1$ and multiplying the right-hand and left-hand sides of all equalities, one obtains

$$\frac{g_{e}(v)}{g_{e}(1)} = \prod_{i=1}^{v-i} \frac{W^{+}(v-i)}{W^{-}(v-i+1)}. \tag{3.18}$$

The equilibrium distribution of clusters $v = 1$ is just the number density of residues in a compact (but unfolded) protein, i.e., $g_{e}(1) = \rho_{u}$, so that Equation 3.18 can be rewritten as

$$g_{e}(v) = \rho_{u} \frac{W^{+}(1)}{W^{+}(v)} \prod_{i=1}^{v-1} \frac{W^{+}(v-i+1)}{W^{-}(v-i+1)}. \tag{3.19}$$

Let us introduce the function $G(v) = -k_{B}T \ln[g_{e}(v)/\rho_{u}]$. Clearly, $G(v)$ in the present theory plays a role similar to the free energy of cluster formation in CNT [59–61] and can be expected to have a shape similar to the latter with a global maximum attained at some critical $v_{c}$ (hereinafter the subscript "c" marks quantities at the critical point). In the vicinity of the critical size, the function $G(v)$ can be accurately represented by its bilinear form $G(v) = G_{c} + (1/2)G_{c}''(v - v_{c})^{2}$, where $G_{c}''$ is the second derivative of $G(v)$ at $v = v_{c}$. The essence of our model, as an alternative to the CA-based theory, consists of constructing the equilibrium distribution of clusters, $g_{e}(v)$, and the function $G(v)$ without employing classical thermodynamics.

The kinetic equation of nucleation in the vicinity of the critical point can be derived from the discrete equation governing the temporal evolution of the distribution $g(v, t)$:

$$\begin{aligned} \frac{\partial g(v,t)}{\partial t} &= W^{+}(v-1)g(v-1,t) - W^{+}(v)g(v,t) \\ &\quad + W^{-}(v+1)g(v+1,t) - W^{-}(v)g(v,t). \end{aligned} \tag{3.20}$$

This equation assumes that the evolution of clusters occurs through the absorption and emission of a bead. For the sake of simplicity, we neglect the multimer absorption and emission. Assuming $v \gg 1$, one can rewrite Equation 3.20 in a continuous form by using Taylor series expansions for its right-hand side and retaining only the lowest order terms. The resulting partial differential equation has the Fokker–Planck form

$$\frac{\partial g(v,t)}{\partial t} = -\frac{\partial}{\partial v} J(v,t), \tag{3.21}$$

where

$$J(v,t) = -A(v)g(v,t) - \frac{\partial}{\partial v}(B(v)g(v,t)). \tag{3.22}$$

Here, the coefficients $A(v)$ and $B(v)$ depend on the emission and absorption rates $W^-$ and $W^+$ of the cluster

$$A(v)=W^+(v)-W^-(v) \quad B(v)=-\frac{1}{2}[W^+(v)+W^-(v)].$$

However, according to the principle of detailed balance, $W^-(v+1)/W^+(v) = \exp[-G(v)+G(v+1)] \simeq \exp[\partial G(v)/\partial v]$. The first derivative $\partial G(v)/\partial v$ is equal to zero at the critical point $v = v_c$ and is close to zero in its vicinity where, in addition, $W^+(v+1) \simeq W^+(v)$. Therefore, $W^-(v)/W^+(v) \simeq 1 + o(\partial G(v)/\partial v)$, where $o(x)$ denotes a quantity on the same order of magnitude as $x$. Neglecting the terms of order, $o(\partial G(v)/\partial v)$, one can assume that $W^+(v) + W^-(v) \simeq 2W^+(v)$ and $W^+(v) \simeq W^+(v_c) \equiv W_c^+$; hence $B \simeq -W_c^+$. Because at equilibrium $J = 0$, Equation 3.22 leads to

$$A(v)g_e(v)+B(v)\frac{\partial g_e(v)}{\partial v}=0,$$

whence one obtains

$$A(v)=-B(v)\frac{1}{g_e(v)}\frac{\partial g_e(v)}{\partial v}\simeq W_c+\frac{\partial G(v)}{\partial v}$$

and $J$ in Equation 3.22 acquires the form

$$J(v,t)=-W_c^+\left(\frac{\partial}{\partial v}+\frac{\partial G(v)}{\partial v}\right)g(v,t)=-W_c^+g_e(v)\frac{\partial}{\partial v}\frac{g(v,t)}{g_e(v)}. \tag{3.23}$$

The substitution of this equation into Equation 3.21 leads to the equation

$$\frac{\partial g(v,t)}{\partial t}=W_c^+\frac{\partial}{\partial v}\left[\frac{\partial}{\partial v}+\frac{\partial G}{\partial v}\right]g(v,t). \tag{3.24}$$

The steady-state solution of Equation 3.24 in the vicinity of $v_c$, subject to the conventional boundary conditions

$$\begin{aligned} &\frac{g(v,t)}{g_e(v)}\to 1\ (v\to 0) \\ &\frac{g(v,t)}{g_e(v)}\to 0\ (v\to\infty), \end{aligned} \tag{3.25}$$

where $g_e(v)$ is the equilibrium distribution and provides the steady-state nucleation rate [59–61] $J(v,t) = J_s$. Indeed, the rightmost equality in Equation 3.23 in the steady state takes the form

$$J_s=-W_c^+g_e(v)\frac{\partial}{\partial v}\frac{g(v)}{g_e(v)}. \tag{3.26}$$

Using the boundary conditions (Equation 3.25) and the bilinear form of $G(v)=G_c+(1/2)G_c''(v-v_c)^2$ in the vicinity of the critical point, one can easily integrate the above equation and obtain for $J_s$ the expression

$$J_s=\frac{W_c^+}{\sqrt{\pi}\Delta v_c}\rho_u e^{-G_c/k_BT}, \tag{3.27}$$

where $\Delta v_c = \left|\partial^2 G/\partial v^2\right|_c^{-1/2}$.

### 3.1.3.4 Evaluation of Protein Folding Time

Knowing the emission and absorption rates as functions of $v$ as well as the nucleation rate $J_s$, one can estimate the time $t_f$ necessary for the protein to fold via nucleation. Roughly speaking, protein folding (via nucleation) consists of two stages. During the first stage, a critical cluster of native residues is formed (nucleation proper). During this stage, i.e., for $v < v_c$, the emission rate $W^-$ is larger than $W^+$, but the cluster still can attain the critical size by means of fluctuations. At the second stage, the nucleus grows via regular absorption of native residues, which dominates their emission, $W^- < W^+$ for $v > v_c$. Thus, the folding time is given by

$$t_f \simeq t_n + t_g, \tag{3.28}$$

where $t_n$ is the time necessary for one critical cluster to nucleate within a compact (but still unfolded) protein and $t_g$ is the time necessary for the nucleus to grow up to the maximum size, i.e., attain the size of the folded protein.

The time $t_n$ of the first nucleation event can be estimated as

$$t_n \simeq 1/J_s V_0, \tag{3.29}$$

where $V_0$ is the volume of the unfolded protein in a compact state. The growth time $t_g$ can be found by solving the differential equation

$$\frac{dv}{dt} = W^+(v) - W^-(v) \tag{3.30}$$

subject to the initial conditions $v = v_c$ at $t = 0$ and $v = N_0$ at $t = t_g$. The solution of Equation 3.30 is given by the integral

$$t_g \simeq \int_{v_c}^{N_0} \frac{dv}{W^+(v) - W^-(v)}. \tag{3.31}$$

## 3.1.4 Numerical Evaluations

In this section, we will present some numerical results of the application of our model to the folding of a model protein, namely, a heteropolymer consisting of a total of 2500 hydrophobic and hydrophilic residues, with the mole fraction of hydrophobic residues $\chi_0 = 0.75$. The interactions between a pair of nonlinked beads were modeled by Lennard–Jones (LJ) type potentials (Equation 3.1), whereas the potential due to the dihedral angle $\delta$ was modeled according to Equation 3.3. The presence of water molecules was not taken into account explicitly but was assumed to be included into the model via the potential parameters.

All numerical calculations were carried out for the following values of the interaction parameters:

$$\eta = 5.39 \times 10^{-8}\ \text{cm}$$
$$\epsilon_1 = (2/700)\epsilon_b$$
$$\epsilon'_\delta = \epsilon''_\delta = 0.3\epsilon_b$$
$$\epsilon_b / kT = 1.$$

The typical density of a folded protein was evaluated using the data from Refs. [62,63] and was set to $\rho_f\eta^3 = 1.05$, whereas the typical density of an unfolded protein in the compact configuration was set to be $\rho_u = 0.25\rho_f$. (Note that comparable values for $\rho_f$ and $\rho_d$ were also suggested in Ref. [30].) Taking into account the results presented in Refs. [64,65], the diffusion coefficients in the ipw and the opw were assumed to be related as $D^{iw}\rho_f = D^{ow}\rho_u$. Because of the lack of reliable data on the diffusion coefficient of a residue in a protein chain, $D^{iw}$ was assumed to vary between $10^{-6}$ and $10^{-8}$ cm $^2$/s.

Figure 3.2 presents typical shapes of the potentials $\phi_b(r)$ (lower solid curve), $\bar{\phi}_\delta(r)$ (upper solid curve), and $\psi_b(r)$ (dashed curve) for a hydrophobic bead around the cluster as functions of the distance $r$ from the center of a cluster of radius $R = 3\eta$. The potential $\phi_b(r)$ is due to the pairwise interactions of the Lennard–Jones type and has a shape reminiscent thereof. Previous applications [33–38] of a mean first passage time analysis to nucleation employed this kind of potential well around a cluster.

The average dihedral potential (assigned to a selected bead) $\bar{\phi}_\delta(r)$ has a maximum value at the cluster surface and decreases monotonically with increasing $r$ until it becomes constant for $r \geq \tilde{r}$, with $\tilde{r} - R$ being the maximum distance between beads 1 and 6 (or beads 1 and 7), which depends on $R$, $\eta$, and $\beta_0$:

$$\tilde{r} = R + \eta\sqrt{3 - 2\cos\beta_0 + 2\sqrt{2(1-\cos\beta_0)}\sin\frac{\beta_0}{2}}. \tag{3.32}$$

Such a behavior of $\bar{\phi}_\delta(r)$ can be thought to be a consequence of an increase in the entropy of the heteropolymer chain as the selected bead approaches the cluster surface for $r < \tilde{r}$. It occurs because the configurational space available to the neighboring beads (2–7) becomes increasingly restricted. This piece of the heteropolymer chain (beads 1–7) would not feel the presence of the cluster any more at $r > \tilde{r}$ if there were no pairwise interactions.

As a result of the combination of $\phi_b(r)$, $\bar{\phi}_\delta(r)$, and $\phi_{co}$, the overall potential $\psi_b(r)$ has a double-well shape: the inner well is separated by a potential barrier from the outer well. The geometric characteristics of the wells (widths, depths, etc.) and the height and location of the barrier between them depend on the interaction parameters $\epsilon_b$, $\epsilon_1$ and $\epsilon_\delta = \epsilon'_\delta = \epsilon''_\delta$. The larger the ratio $\epsilon_\delta/\epsilon_b$, the higher the barrier between the wells, the wider the ipw, and the narrower the opw. Note that the barrier has different heights for the ipw and opw beads. For a given $N_0$ the location $r_{co}$ of the outer boundary of the opw is determined by the size of the cluster and densities $\rho_f$ and $\rho_u$ (with the above choices thereof $r_{co} \simeq 12.99\eta$ for a cluster $R = 3\eta$). The existence of an opw allows one to consider the absorption of a bead by the cluster as an escape of the bead from the opw by crossing over the barrier into the ipw. One can therefore use a mean first passage time analysis for the determination not only of the emission rate but also of the rate of absorption of beads by the cluster. Since the use of the traditional expression for the absorption rate (based on the gas-kinetic theory) [50] is rather inadequate for the cluster growth within the protein, the double well shape of $\psi_i(r)$ is of crucial importance in our model for the nucleation mechanism of protein folding.

In Figure 3.3, the average dihedral potential (assigned to a selected bead) $\bar{\phi}_\delta(r)$ is plotted as a function of $r$ for three clusters of sizes (1) $R = 3\eta$, (2) $R = 6\eta$, and (3) $R = 9\eta$. The points represent the actual numerical results obtained by using MC integration with $1 \times 10^6$ to $2 \times 10^6$ points in calculating the sevenfold integrals in Equations 3.11 and 3.12. The vertical dashed lines correspond to $\tilde{r}$ (which was defined above), and $\bar{\phi}_\delta(r)$ is expected to be constant for $r > \tilde{r}$. The solid lines are analytical fits by an expression of the form $a + b\exp[-c(r - d)^2]$. Within the accuracy of our calculations, the parameters $a$, $b$, and $c$ of this fit do not change with $R$, while the parameter $d$ is roughly $R + \eta$. Clearly, with an increased accuracy we may eventually find some dependence of $a$, $b$, and $c$ on $R$. Undoubtedly, the calculation of $\bar{\phi}_\delta(r)$ constitutes the most time-consuming step in applying our model to real proteins. It took approximately 24 h to obtain one value of $\bar{\phi}_\delta(r)$ (one point in Figure 3.3) by using MATHEMATICA on a Dell computer with a 3-GHz Pentium 4 processor and 512 Mb of memory.

Figure 3.4 presents $W^-$ and $W^+$, the emission and absorption rates, respectively, as functions of the cluster size $R$. The location of the intersection of these functions determines the size of the critical cluster, $R_c$. The emission rate is greater than the absorption rate, $W^-(r) > W^+(r)$, for clusters with $R < R_c$, whereas for clusters larger than the nucleus absorption dominates emission, $W^-(r) < W^+(r)$ for $R > R_c$. Note that both $W^-$ and $W^+$ increase with increasing $R$, but $W^-$ increases roughly linearly with $R$ whereas $W^+$ rapidly increases by several orders of magnitude after the cluster becomes supercritical. This occurs because the width of the opw quickly decreases as the cluster grows, whereas the outer height of the barrier between the ipw and the opw virtually does not change. For this reason, it becomes increasingly easier for a bead located in the opw to cross the barrier and to fall into the ipw.

The behavior of $W^-$ and $W^+$ also explains our estimates for the characteristic times of the first nucleation event $t_n$, growth time $t_g$, and total folding time $t_f$ provided by Equations 3.28, 3.29, and 3.31. Although these values depend strongly on $R_c$ and on $D^{iw}$ ($R_c$ itself does not depend on $D^{iw}$ but only on the ratio $D^{ow}/D^{iw}$), we have always $t_n \gg t_g$, i.e., the time of protein folding is mainly determined by the time necessary for the first nucleation event to occur. Physically, this happens because the increase of the cluster size from $\nu = 1$ to $\nu = \nu_c$ takes place only via fluctuations that have to overcome the natural tendency of a small cluster to decay ($W^+ < W^-$ for $\nu < \nu_c$). For supercritical clusters, $W^+$ overwhelms $W^-$ so quickly that the fluctuations are unable to impede the natural tendency of the cluster to grow. (Strictly speaking, this is true only for $\nu > \nu_c + \Delta\nu_c$, but

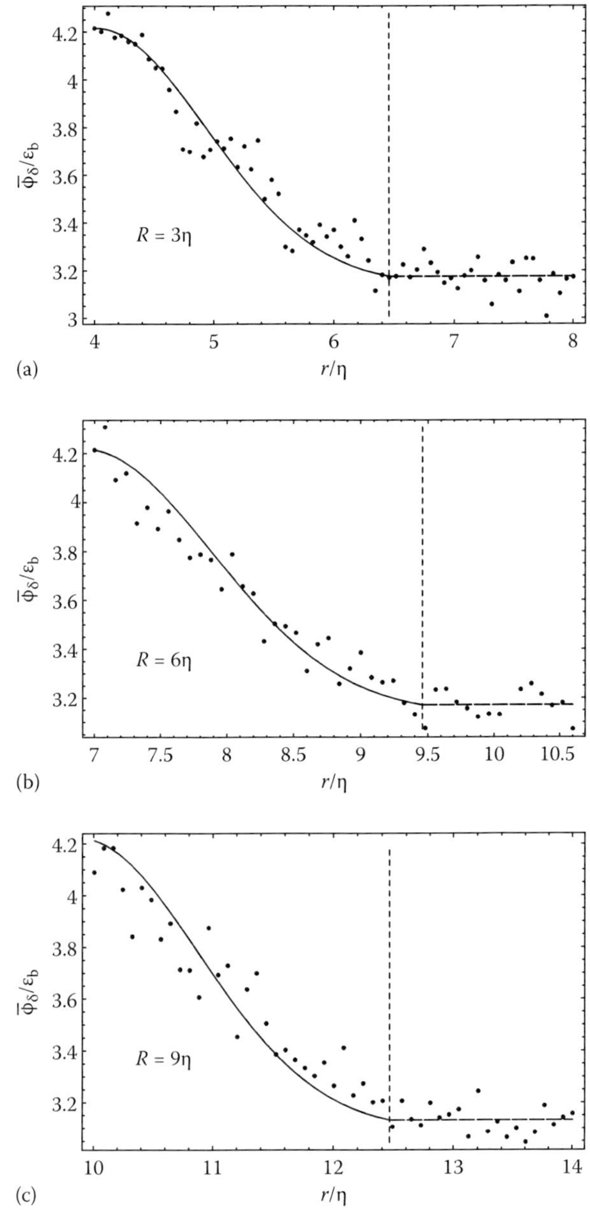


Figure 3.3

Average dihedral potential (assigned to a selected bead) $\bar{\phi}_\delta(r)$ as a function of $r$ for three clusters of sizes (a) $R = 3\eta$, (b) $R = 6\eta$, and (c) $R = 9\eta$. The points represent the actual numerical results obtained by using Monte Carlo integration in Equations 3.11 and 3.12. The vertical dashed lines correspond to $\tilde{r}$. The solid lines are analytical fits by the expression $a + b\exp[-c(r - d)^2]$.

for rough estimates Equation 3.31 is acceptable.) With the above choice of the system parameters and with $D_{iw}$ in the range from $10^{-6}$ to $10^{-8}$ cm$^2$/s, for $N_0 = 2500$ our model predicts $R_c \simeq 3.7\eta$, $\nu_c \simeq 220$, and characteristic times for protein folding in the range from several seconds to several hundreds of seconds, which is in a good agreement with experimental data [5].

### 3.1.5 Conclusions

So far, most of the work on protein folding has been carried out by using either MC or MD simulations. A rigorous theoretical treatment of protein folding by means of statistical mechanics is hardly practicable because of the extreme complexity of the system. Simulations have suggested that there are multiple pathways for a protein to fold, and one of them has been identified as reminiscent of nucleation. However, a theoretical model involving the nucleation mechanism was based on CNT, hence on classical thermodynamics

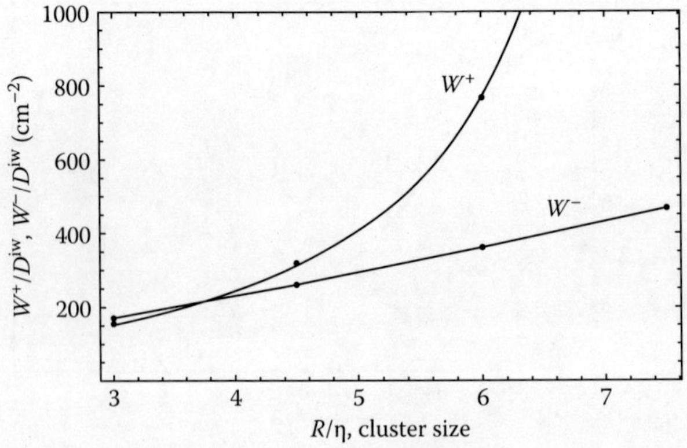


Figure 3.4

$W^-$ and $W^+$, the emission and absorption rates, respectively, as functions of cluster size $R$. The location of the intersection of these functions determines the size of the critical cluster, $R_c$.

[15,30]. In that approach the size of the critical cluster (nucleus) was provided by the location of the maximum of the free energy of cluster formation as a function of the cluster size. In such a model, the free energy of cluster formation depends on the surface tension of a cluster of protein residues. This quantity is an ill-defined physical quantity and can be considered only as an adjustable parameter. According to the nucleation mechanism, after the formation of a nucleus (critical size cluster of residues), the protein quickly reaches its native state.

In this paper, we have presented a new, microscopic model for the nucleation mechanism of protein folding. A protein is considered as a heteropolymer consisting of two types of beads (hydrophobic and hydrophilic) linked with bonds of fixed length. All bond angles are also assumed to be fixed and equal to 105°. All nonadjacent beads are assumed to interact via a Lennard–Jones-like potential. Besides these interactions, the total energy of the heteropolymer contains a contribution from dihedral angles of all triads of successive links. Unlike the old model, the present one is developed without recurring to the CNT approach but is based on the above "molecular" interactions, both pairwise and configurational. The parameters of these potentials can be rigorously defined, unlike the ill-defined surface tension of a cluster of protein residues within a protein.

The main idea underlying the new model consists of averaging the dihedral potential in which a selected residue is involved over all possible configurations of neighboring residues. The resulting average dihedral potential depends on the distance between the residue and the cluster center. It has a maximum at the cluster surface and monotonically decreases with increasing distance therefrom. Its combination with the average potential due to pairwise interactions between the selected residue and those in the cluster and with a confining potential provides a double potential well around the cluster with a barrier between the two wells. Residues in the inner well are considered to belong to the cluster (part of the protein with correct tertiary contacts) while those in the outer well are treated as belonging to the mother phase (amorphous part of the protein with incorrect tertiary contacts). Transitions of residues from the inner well into the outer one and vice versa are considered as elementary emission and absorption events, respectively. The rates of these processes are determined by using a mean first passage time analysis. Once these rates are found as functions of the cluster size, one can develop a self-consistent kinetic theory for the nucleation mechanism of protein folding. For example, the size of the critical cluster (nucleus) is then found as the one for which these rates are equal. The time necessary for the protein to fold can be evaluated as the sum of the times necessary for the appearance of the first nucleus and the time necessary for the nucleus to grow to the maximum size (of the folded protein in the native state).

For numerical illustration, we have considered a model protein consisting of 2500 beads with the mole fraction of hydrophobic beads equal to 0.75. The composition of the cluster during its formation and growth was assumed to be constant and equal to the composition of the whole protein. This allows one to consider the model as a single-component one. The size of the critical cluster and the folding time predicted by the model depend very much on the parameters of the system, such as the interaction parameters, densities of the protein in the unfolded (but compact) and folded states, and diffusion coefficients therein. With appropriate choices of interaction parameters and densities, the size of the critical cluster predicted by our model is approximately 220 residues, the free energy of nucleus formation being approximately 20 $k_BT$. These results suggest that the quantity equivalent to the "surface tension" in the old model of nucleation in a protein should be smaller than the previous estimates of the latter [15,30] by approximately 1 order of magnitude. Considering the diffusion coefficient of native residues in the range from $10^{-6}$ to $10^{-8}$ cm$^2$/s, the characteristic time of protein folding was estimated to be in the range of several seconds to several hundreds of seconds. These values are consistent with the experimental data [5] on typical folding times of proteins as well as with estimates provided by other theoretical models [30] and simulations [7,15].

A further development of our model will require the removal of several simplifying assumptions that we have employed in the present paper. For example, it would be more appropriate to model a protein as a three-component heteropolymer (including not only hydrophobic and hydrophilic residues, but also neutral ones). This will result in longer numerical calculations of the average dihedral potential because the dihedral potential involving neutral beads is expected to be weaker and requires separate calculations. Next, the cluster composition during its formation and growth can quite significantly depend on the cluster size, particularly in the vicinity of the critical size; hence the assumption that it is constant can lead to inaccuracies in the results. By including neutral beads in the model and allowing the cluster composition to differ from that of the protein will result in a binary or even ternary nucleation mechanism of protein folding. However, besides some increase in the computational efforts, there is no conceptual difficulty in developing the model in those directions.

## Appendix 3A: Derivation of Equation 3.11 from Equation 3.9

Consider beads 1–7 of the unfolded part of the heteropolymer (Figure 3.1) around a cluster of radius $R$ and denote the distance between bead 1 and the cluster center by $r$. It is convenient to choose a Cartesian system of coordinates with the origin in the cluster center in such a way that the coordinates of bead 1 are $x_1 = 0$, $y_1 = 0$, $z_1 = r$. The Cartesian coordinates of other beads will be denoted by $x_i, y_i, z_i$ ($i = 2, \ldots, 7$). The Cartesian coordinates of bead 2 are related to its spherical ones $r_2, \Theta_2, \varphi_2$:

$$x_2 = r_2 \sin\Theta_2 \cos\varphi \quad y_2 = r_2 \sin\Theta_2 \sin\varphi \quad z_2 = r_2 \cos\Theta_2. \tag{3A.1}$$

At a given $r$ (since the location of bead 1 is fixed), the polar angle $\Theta_2$ of bead 2 is uniquely determined by $r_2$ due to the constant bond length constraint

$$\Theta_2(r,r_2) = \arccos\left[\left(r^2 + r_2^2 - \eta^2\right)/(2r_2r)\right] \; (0 \le \Theta_2 \le \pi), \tag{3A.2}$$

whereas the azimuthal angle $0 \le \phi_2 \le 2\pi$. The distance $r_2$ varies in the range $r_{2\min} \le r_2 \le r + \eta$, where $r_{2\min} = \max(R, r - \eta)$, which determines the integration range $L_2$ in Equations 3.11 and 3.12. Thus, the integration with respect to $r_2$ in Equations 3.9 and 3.10 reduces to the integration with respect to $\varphi_2$ and $r_2$ with fixed $\Theta_2$ (given by Equation 3A.2) in Equations 3.11 and 3.12.

For given locations of beads 1 and 2, the possible locations of beads 3 and 4 lie on circles of radius $r_1 = \eta \sin\beta_0$ with their location and orientation completely determined by the coordinates of beads 1 and 2. This is attributable to the constraints that all bond angles are equal to $\beta_0$ and all bond lengths are equal to $\eta$. Due to the same constraints, if the locations of beads 2 and 4 are given, then the possible locations of bead 6 lie on a circle of radius $r_1$, whereof the location and orientation are completely determined by the coordinates of beads 2 and 4. Furthermore, for given locations of beads 1 and 3, the possible locations of bead 5 lie on a circle of radius $r_1$ with the position and orientation completely determined by the coordinates of beads 1 and 3. Finally, for the given locations of beads 3 and 5, the possible locations of bead 7 are on a circle of radius $r_1$, with the location and orientation completely determined by the coordinates of beads 3 and 5.

Let us consider bead $s$ with unknown coordinates and two other beads, $c$ and $n$ (closest to and next to the closest to bead $s$), with fixed (known) coordinates $x_c, y_c, z_c$ and $x_n, y_n, z_n$. For example, if $s = 7$, then $c = 5$ and $n = 3$; if $s = 4$, then $c = 2$ and $n = 1$. Bead $s$ lies on a circle of radius $r_1$ with the coordinates of the center

$$x_0 \equiv x_0(x_n, y_n, z_n, x_c, y_c, z_c)$$
$$= x_n + (x_c - x_n)\frac{\eta\left(1+\left|\cos\beta_0\right|\right)}{\sqrt{(x_c - x_n)^2 + (y_c - y_n)^2 + (z_c - z_n)^2}} \tag{3A.3}$$

$$y_0 \equiv y_0(x_n, y_n, z_n, x_c, y_c, z_c)$$
$$= y_n + (y_c - y_n)\frac{\eta\left(1+\left|\cos\beta_0\right|\right)}{\sqrt{(x_c - x_n)^2 + (y_c - y_n)^2 + (z_c - z_n)^2}} \tag{3A.4}$$

$$z_0 \equiv z_0(x_n, y_n, z_n, x_c, y_c, z_c)$$
$$= z_n + (z_c - z_n)\frac{\eta\left(1+\left|\cos\beta_0\right|\right)}{\sqrt{(x_c - x_n)^2 + (y_c - y_n)^2 + (z_c - z_n)^2}}. \tag{3A.5}$$

The coordinate $x_s$ of bead $s$ can change in the range

$$x_c - x_b \leq x_s \leq x_c + x_b, \tag{3A.6}$$

where

$$x_b \equiv x_b(x_n, y_n, z_n, x_c, y_c, z_c)$$
$$= \eta \sin\beta_0 \sqrt{\frac{\left((y_c - y_0)^2 + (z_c - z_0)^2\right)}{(x_c - x_n)^2 + (y_c - y_n)^2 + (z_c - z_n)^2}}. \tag{3A.7}$$

For a given $x_s$, the coordinate $y_s$ of bead $s$ can have only one of two values:

$$y_s^{\pm} = y_s^{\pm}(x_s, x_n, y_n, z_n, x_c, y_c, z_c)$$
$$= y_0 + \frac{-(x_c - x_0)(y_c - y_0)(x_s - x_0)}{(y_c - y_0)^2 + (z_c - z_0)^2} \tag{3A.8}$$
$$\pm \frac{\left|z_c - z_0\right| \sqrt{\begin{array}{c}[(y_c - y_0)^2 + (z_c - z_0)^2]\eta^2 \sin^2\beta_0 \\ -\left[(x_c - x_n)^2 + (y_c - y_n)^2 + (z_c - z_n)^2\right](x_s - x_0)^2\end{array}}}{(y_c - y_0)^2 + (z_c - z_0)^2}.$$

For given $x_s$ and $y_s^{\pm}$, the coordinate $z_s$ of bead $s$ can have only a single value:

$$z_s^+ \equiv z_s\left(x_s, y_s^+, x_n, y_n, z_n, x_c, y_c, z_c\right)$$
$$= z_0 - \frac{(x_c - x_0)(x_s - x_0) + (y_c - y_0)\left(y_s^+ - y_0\right)}{z_c - z_0}. \tag{3A.9}$$

Since the volume occupied by the cluster is unavailable to beads 2, 3, …7, an additional constraint is imposed on $x_s$, $y_s$, and $z_s$, namely, $x_s^2 + y_s^2 + z_s^2 > R^2$. Subject to this constraint, the double inequality in Equation 3A.6 and Equations 3A.7 through 3A.9 completely determine the integration ranges in Equations 3.11 and 3.12. To transform the function $\phi_\delta'(r) \equiv \phi_\delta'(r, \mathbf{r}_2, \mathbf{r}_3, \mathbf{r}_4, \mathbf{r}_5, \mathbf{r}_6, \mathbf{r}_7)$ into the function $\tilde{\phi}_{ijkmn}(r, \varphi_2, r_2, x_3, \ldots, x_7)$, the variables $y_s$ ($s = 3, \ldots, 7$) and $z_s$ ($s = 3, \ldots, 7$) in the former must be replaced by $y_s^\pm$ and $z_s^\pm$ in all possible combinations each of which gives rise to a term in the sums in Equations 3.11 and 3.12. (Note that $z_s^+ = z_s(x_s, y_s^+, \ldots)$ and $z_s^- = z_s(x_s, y_s^-, \ldots)$.) The polar angle of bead 2 must be replaced by $\Theta_2(r, r_2)$.

For example, consider the term with $i = +, j = -, k = +, m = +, n = -$ in the sum in Equations 3.11 and 3.12. In the corresponding integrands, the function $\tilde{\phi}_{+-++-}(\varphi_2, r_2, x_3, \ldots, x_7)$ is obtained from $\phi_\delta'(r, \mathbf{r}_2, \mathbf{r}_3, \mathbf{r}_4, \mathbf{r}_5, \mathbf{r}_6, \mathbf{r}_7)$ as follows.

$y_3$ must be replaced by $y_3^+ \equiv y_3^+(x_3, x_2, y_2, z_2, x_1, y_1, z_1)$ (defined by Equation 3A.8) and $z_3$ by $z_3^+ \equiv z_3(x_3, y_3^+, x_2, y_2, z_2, x_1, y_1, z_1)$ (defined by Equation 3A.9), $y_4$ must be replaced by $y_4^- \equiv y_4^-(x_4, x_1, y_1, z_1, x_2, y_2, z_2)$ and $z_4$ by $z_4^- \equiv z_4(x_4, y_4^-, x_1, y_1, z_1, x_2, y_2, z_2)$, $y_5$ must be replaced by $y_5^+ \equiv y_5^+(x_5, x_1, y_1, z_1, x_3, y_3^+, z_3^+)$ and $z_5$ by $z_5^+ \equiv z_5(x_5, y_5^+, x_1, y_1, z_1, x_3, y_3^+, z_3^+)$, $y_6$ must be replaced by $y_6^+ \equiv y_6^+(x_6, x_2, y_2, z_2, x_4, y_4^-, z_4^-)$ and $z_6$ by $z_6^+ = z_6(x_6, y_6^+, x_2, y_2, z_2, x_4, y_4^-, z_4^-)$, $y_7$ must be replaced by $y_7^- \equiv y_7^-(x_7, x_3, y_3^+, z_3^+, x_5, y_5^+, z_5^+)$ and $z_7$ by $z_7^- \equiv z_7(x_7, y_7^-, x_3, y_3^+, z_3^+, x_5, y_5^+, z_5^+)$. The coordinates $x_2$, $y_2$, and $z_2$ are functions of $r_2$ and $\varphi_2$ provided by Equations 3A.1 and 3A.2.

## Appendix 3B: Derivation of Expressions for Emission and Absorption Rates by a Mean First Passage Time Analysis

Let us use the mean first passage time analysis [33–38] to calculate the emission and absorption rates of a cluster during protein folding. The mean first passage time of a bead escaping from a potential well is calculated on the basis of a kinetic equation governing the chaotic motion of the bead in that potential well. The chaotic motion of the bead is assumed to be governed by the Fokker–Planck equation for the single-particle distribution function with respect to its coordinates and momenta, i.e., in the phase space [56–58]. Prior to the passage event, the evolution of a bead in both the ipw and the opw occurs in a dense enough medium (cluster of folded residues or unfolded but compact part of the protein), where the relaxation time for its velocity distribution function is very short and negligible compared to the characteristic timescale of the passage process.

Under these conditions, the Fokker–Planck equation reduces to the Smoluchowski equation, which involves diffusion in an external field [57,58]. In the case of spherical symmetry, it can be written in the form [33–35]

$$\frac{\partial p_i\left(r,t|r_0\right)}{\partial t} = D_i r^{-2} \frac{\partial}{\partial r}\left(r^2 e^{-\Psi_i(r)} \frac{\partial}{\partial r} e^{\Psi_i(r)} p_i\left(r,t|r_0\right)\right), \tag{3B.1}$$

where $p_i\,(r,t|r_0)$ is the probability of observing a bead of species $i$ ($i$ = b, l) between $r$ and $r + dr$ at time $t$ given that initially it was at a radial distance $r_0$, $D_i$ is its diffusion coefficient in the well, and $\Psi_i(r) = \psi_i(r)/kT$.

The mean passage time depends on the initial position (distance from the center of the cluster) $r_0$ of the bead. It is convenient to use the backward Smoluchowski equation [33–35], which expresses the dependence of the transition probability $p_i\,(r,t|r_0)$ on $r_0$:

$$\frac{\partial p_i\left(r,t|r_0\right)}{\partial t} = D_i r_0^{-2} e^{\Psi_i(r_0)} \frac{\partial}{\partial r_0}\left(r_0^2 e^{-\Psi_i(r_0)} \frac{\partial}{\partial r_0} p_i\left(r,t|r_0\right)\right). \tag{3B.2}$$

Let us first consider the ipw and find the emission rate of beads therefrom. The probability that a bead $i$, initially at a distance $r_0$ within the surface layer, will remain in this region after time $t$ is given by the so-called survival probability [33–35]:

$$q_i\left(t\middle|r_0\right)=\int_R^{R+\lambda_i^w} drr^2 p_i\left(r,t\middle|r_0\right)\ \left(R<r_0<R+\lambda_i^w\right), \tag{3B.3}$$

where $R$ is the radius of the cluster and $\lambda_i^w$ is the width of the potential well determined by the location of the barrier between the ipw and the opw (Figure 3.2). (Previously [33–38], $\lambda_i^w$ was defined so that a bead $i$ could be assumed to be dissociated from the cluster at the distance $R+\lambda_i^w$ from its center. The width $\lambda_i^w$ had to be chosen such that $\lambda_i^w/R \ll 1$ and $\Psi_i\left(R+\lambda_i^w\right)/\Psi_i^w \ll 1$, where $\Psi_i^w$ was the depth of the well.) The probability for the dissociation time to be between 0 and $t$ is equal to $1 - q_i(t|r_0)$, and the probability density for the dissociation time is given by $-\partial q_i/\partial t$. The first passage time is provided by [56–58]

$$\tau_i(r_0)=-\int_0^\infty t\frac{\partial q_i(t\,|\,r_0)}{\partial t}dt=\int_0^\infty q_i(t\,|\,r_0)dt. \tag{3B.4}$$

The equation for the first passage time is obtained by integrating the backward Smoluchowski Equation 3B.2 with respect to $r$ and $t$ over the entire range and using the boundary conditions $q_i(0|r_0) = 1$ and $q_i(t|r_0) \to 0$ as $t \to \infty$ for any $r_0$. This yields [33–35]

$$-D_i r_0^{-2} e^{\Psi(r_0)}\frac{\partial}{\partial r_0}\left(r_0^2 e^{-\Psi_i(r_0)}\frac{\partial}{\partial r_0}\tau_i(r_0)\right)=1. \tag{3B.5}$$

One can solve Equation 3B.5 by assuming a reflecting inner boundary of the well ($d\tau_i/dr_0 = 0$ at $r_0 = R$) and the radiation boundary condition at the outer boundary ($\tau_i(r_0) = 0$ at $r_0 = R+\lambda_i^w$). One thus obtains for the first passage time

$$\tau_i(r_0)=\frac{1}{D_i}\int_{r_0}^{R+\lambda_i^w} dyy^{-2}e^{\Psi_i(y)}\int_R^y dxx^2 e^{-\Psi_i(x)}. \tag{3B.6}$$

The average dissociation time $\bar{\tau}_i$, or the mean first passage time, is obtained by averaging $\tau_i(r_0)$ with the Boltzmann factor over all possible initial positions $r_0$:

$$\bar{\tau}_i=\frac{1}{Z}\int_R^{R+\lambda_i^w} dr_0 r_0^2\, e^{-\Psi_i(r_0)}\tau_i(r_0) \tag{3B.7}$$

with

$$Z=\int_R^{R+\lambda_i^w} dr_0 r_0^2 e^{-\Psi_i(r_0)}. \tag{3B.8}$$

Defining the quantity $\omega_i$ as

$$\omega_i \equiv 1/(D_i\bar{\tau}_i), \tag{3B.9}$$

the emission rate $W_i^-$, can be expressed as

$$W_i^- = \frac{N_i^w}{\bar{\tau}_i} = N_i^w D_i w_i, \quad (3B.10)$$

where $N_i^w$ denotes the number of molecules in the well. Clearly, $\bar{\tau}_i$, $\omega_i$, and $W_i^-$ are functions of $R$ and $\chi = \nu_b/(\nu_b + \nu_l)$, where $\nu_i$ ($i$ = b, l) is the number of beads of type $i$ in the cluster.

Considering the opw, a similar procedure can be used to find the rate of absorption of beads by the cluster, $W_i^+$.

## Acknowledgment

This work was supported by the National Science Foundation through Grant No. CTS-0000548.

## References

1. Creighton, T.E., *Proteins: Structure and Molecular Properties*, W.H. Freeman, San Francisco, 1984.
2. Stryer, L., *Biochemistry*, 3rd ed., W.H. Freeman, New York, 1988.
3. Ghelis, C., Yan, J., *Protein Folding*, Academic Press, New York, 1982.
4. Anfinsen, C.B., *Science* 181, 223–230 (1973).
5. Nölting, B., *Protein Folding Kinetics: Biophysical Methods*, 2nd ed., Springer-Verlag: Berlin, 2006.
6. Honeycutt, J.D., Thirumalai, D., *Proc. Natl. Acad. Sci. U.S.A.* 87, 3526–3529 (1990).
7. Honeycutt, J.D., Thirumalai, D., *Biopolymers* 32, 695–709 (1992).
8. Weissman, J.S., Kim, P.S., *Science* 253, 1386–1393 (1991).
9. Creighton, T.E., *Nature* (*London*) 356, 194–195 (1992).
10. Kim, P.S., Baldwin, R.I., *Annu. Rev. Biochem.* 51, 459 (1982).
11. Kim, P.S., Baldwin, R.I., *Annu. Rev. Biochem.* 59, 631 (1990).
12. Creighton, T.E., *Biochem. J.* 240, 1 (1990).
13. Creighton, T.E., *Prog. Biophys. Mol. Biol.* 33, 231 (1978).
14. Guo, Z., Thirumalai, D., Honeycutt, J.D., *J. Chem. Phys.* 97, 525–535 (1992).
15. Guo, Z., Thirumalai, D., *Biopolymers* 36, 83–102 (1995).
16. Thirumalai, D., Guo, Z., *Biopolymers* 35, 137–140 (1995).
17. Nölting, B., Golbik, R., Neira, J.L., Soler-Gonzalez, A.S., Schreiber, G., Fersht, A.R., *Proc. Natl. Acad. Sci. U.S.A.* 84, 826 (1997).
18. Nölting, B., *J. Theor. Biol.* 194, 419 (1998).
19. Nölting, B., *J. Theor. Biol.* 197, 113 (1999).
20. Fersht, A.R., *Proc. Natl. Acad. Sci. U.S.A.* 92, 10869 (1995).
21. Fersht, A.R., *Curr. Opin. Struct. Biol.* 7, 3 (1997).
22. Shakhnovich, E.I., *Curr. Opin. Struct. Biol.* 7, 29 (1997).
23. Shakhnovich, E.I., Abkevich, V., Ptitsyn, O., *Nature* 379, 96 (1996).
24. Abkevich, V.I., Gutin, A.M., Shakhnovich, E.I., *J. Chem. Phys.* 101, 6052 (1994).
25. Finkelstein, A.V., Badretdinov, A.Y., *Folding Des.* 2, 115 (1997).
26. Bryngelson, J.D., Wolynes, P.G., *Proc. Natl. Acad. Sci. U.S.A.* 84, 7524 (1987).
27. Bryngelson, J.D., Wolynes, P.G., *J. Phys. Chem.* 93, 6902 (1998).
28. Shakhnovich, E.I., Gutin, A.M., *Nature* 346, 773 (1990).
29. Shakhnovich, E.I., Gutin, A.M., *J. Phys. A Math. Gen.* 22, 1647 (1989).
30. Bryngelson, J.D., Wolynes, P.G., *Biopolymers* 30, 177 (1990).
31. Dill, K.A., *Biochemistry* 24, 1501 (1985).
32. Dill, K.A., *Biochemistry* 29, 7133 (1990).
33. Narsimhan, G., Ruckenstein, E., *J. Colloid Interface Sci.* 128, 549 (1989). (Section 1.2 of this volume.)
34. Ruckenstein, E., Nowakowski, B., *J. Colloid Interface Sci.* 137, 583 (1990). (Section 1.3 of this volume.)
35. Nowakowski, B., Ruckenstein, E., *J. Colloid Interface Sci.* 139, 500 (1990). (Section 1.4 of this volume.)
36. Djikaev, Y.S., Ruckenstein, E., *J. Chem. Phys.* 123, 214503 (2005). (Section 2.5 of this volume.)
37. Djikaev, Y.S., Ruckenstein, E., *J. Chem. Phys.* 124, 124521 (2006). (Section 2.6 of this volume.)
38. Djikaev, Y.S., Ruckenstein, E., *J. Chem. Phys.* 124, 194709 (2006). (Section 1.6 of this volume.)

39. Levitt, M., Sharon, R., *Proc. Natl. Acad. Sci. U.S.A.* 85, 8557 (1988).
40. Skolnick, J., Kolinski, A., Yaris, R., *Biopolymers* 28, 1059 (1989).
41. Kolinski, A., Skolnick, J., Yaris, R., *Biopolymers* 26, 937 (1987).
42. Sikorski, A., Skolnick, J., *Biopolymers* 28, 1097 (1989).
43. Sikorski, A., Skolnick, J., *J. Mol. Biol.* 215, 183 (1990).
44. Skolnick, J., Kolinski, A., *Science* 250, 1121 (1990).
45. Levinthal, C., in: *Mossbauer Spectroscopy in Biological Systems*, Debrunner, P., Tsibris, J.C.M., Mönck, E., Eds., University of Illinois Press, Urbana, IL, 1968.
46. Wetlaufer, D., *Proc. Natl. Acad. Sci. U.S.A.* 70, 697 (1973).
47. Tsong, T.Y., Baldwin, R., McPhie, P., *J. Mol. Biol.* 63, 453 (1972).
48. Moult, J., Unger, R., *Biochemistry* 30, 3816 (1991).
49. Dill, K., Fiebig, K., Chan, H.S., *Proc. Natl. Acad. Sci. U.S.A.* 90, 1942 (1993).
50. Flageollet, C., Dihn Cao, M., Mirabel, P., *J. Chem. Phys.* 72, 544 (1980).
51. Wilemski, G., *J. Phys. Chem.* 91, 2492 (1987).
52. Wilemski, G., *J. Chem. Phys.* 80, 1370 (1984).
53. Laaksonen, A., *J. Chem. Phys.* 97, 1983 (1992).
54. Djikaev, Y.S., Napari, I., Laaksonen, A., *J. Chem. Phys.* 97, 9752 (2004).
55. Abraham, F.F., *Homogeneous Nucleation Theory*, Academic Press, New York, 1974.
56. Chandrasekhar, S., *Rev. Mod. Phys.* 15, 1 (1949).
57. Gardiner, C.W., *Handbook of Stochastic Methods*, Springer Series in Synergetics 13, Springer, New York, 1983.
58. Agmon, N., *J. Chem. Phys.* 81, 3644 (1984).
59. Lothe, J., Pound, G.M.J., in: *Nucleation*, Zettlemoyer, A.C., Ed., Marcel-Dekker, New York, 1969.
60. Schmelzer, J.W.P., Röpke, G., Priezzhev, V.B., Eds., *Nucleation Theory and Applications*, Joint Institute for Nuclear Research, Dubna, Russia, 1999.
61. Kashchiev, D., *Nucleation: Basic Theory with Applications*, Butterworth-Heinemann: Oxford, U.K., 2000.
62. Harpaz, Y., Gerstein, M., Chothia, C., *Structure* 2, 641–649 (1994).
63. Huang, D.M., Chandler, D., *Proc. Natl. Acad. Sci. U.S.A.* 15, 8324–8327 (2000).
64. Hirshfelder, J.O., Curtiss, C.F., Bird, R.B., *Molecular Theory of Gases and Liquids*, Wiley, New York, 1964.
65. Wakeham, W.A., *J. Phys. B At. Mol. Phys.* 6, 372 (1973).

## 3.2 A Ternary Nucleation Model for the Nucleation Pathway of Protein Folding

*Yuri Djikaev and Eli Ruckenstein**

Recently (Y. S. Djikaev and E. Ruckenstein, *J. Phys. Chem. B* 111, 886 [2007] [Section 3.1 of this volume]), the authors proposed a kinetic model for the nucleation mechanism of protein folding where a protein was modeled as a heteropolymer consisting of hydrophobic and hydrophilic beads and the composition of the growing cluster of protein residues was assumed to be constant and equal to the overall protein composition. Here, they further develop the model by considering a protein as a three-component heteropolymer and by allowing the composition of the growing cluster of protein residues to vary independently of the overall one. All the bonds in the heteropolymer (now consisting of hydrophobic, hydrophilic, and neutral beads) have the same constant length, and all the bond angles are equal and fixed. As a crucial idea of the model, an overall potential around the cluster wherein a residue performs a chaotic motion is considered to be a combination of the average dihedral and average pairwise potentials assigned to the bead. The overall potential as a function of the distance from the cluster center has a double well shape that allows one to determine its emission and absorption rates by using a first passage time analysis. Knowing these rates as functions of three independent variables of a ternary cluster, one can develop a self-consistent kinetic theory for the nucleation mechanism of folding of a protein using a ternary nucleation formalism and evaluate the size and composition of the nucleus and the protein folding time. As an illustration, the model is applied to the folding of bovine pancreatic ribonuclease consisting of 124 amino acids whereof 40 are hydrophobic, 81 hydrophilic, and 3 neutral. With a reasonable choice of diffusion coefficients of the residues in the native state and potential parameters, the model predicts folding times in the range of 1–100 s. (© *2007 American Institute of Physics.*)

### 3.2.1 Introduction

The formation of the native structure of a biologically active protein constitutes the core of the so-called "protein folding problem" [1,2]. A well-defined three-dimensional structure [3,4] is necessary for a protein molecule to carry out a specific biological function. Many thermodynamic and kinetic aspects of the protein folding process remain unexplained [5–9].

An unfolded protein would have to explore an astronomical number of configurations if it searched for its native one at random [10]. Many mechanisms have been proposed to model the protein folding phenomenon. At present, it is believed that, initially, an unfolded protein transforms quickly into a compact (but not native) configuration. Both experiment and simulation suggest [5–28] that there are multiple pathways for the further evolution of the protein. The transition from a compact configuration to the native one may occur via two distinct mechanisms. One of them can be referred to as a "transition state mechanism," whereby the tertiary contacts of the native structure are formed as the protein passes through a sequence of intermediate states, thus gradually achieving its unique spatial configuration [7–30]. In the intermediate states the protein has a nativelike structure but is stabilized by incorrect hydrophobic contacts. These states correspond to misfolded forms of the native protein. The transition from these misfolded states to the native structure involves a large-scale rearrangement of the molecule and hence is a slow process (compared to the formation of the compact configuration) [16,17]. Alternatively, the transition from the compact "amorphous" configuration to the native state occurs immediately following the formation of a number of tertiary contacts [16,17,21]. This mechanism is similar to nucleation, i.e., once a critical number of (native) tertiary contacts is established the native structure is formed without passing through any detectable intermediates [16].

Computer simulations are an essential tool for studying molecular systems and processes which are too complicated for analytical methods, as is the case of protein folding. No wonder that, so far, most of the work on protein folding has been carried out by using either Monte Carlo (MC) or molecular dynamics (MD) simulations. Few theoretical models for the nucleation mechanism of folding have been proposed. Initially [16,31–33], the problem was approached by using classical thermodynamics and calculating the free energy of formation of a cluster of protein residues in the framework of the capillarity approximation [34] much like the classical nucleation theory (CNT) does. Such an approach involves the concept of surface tension for a cluster consisting of protein residues which is an intrinsically ill-defined physical quantity and can be considered only as an adjustable parameter; clearly, no direct experimental measurement thereof is possible.

* *J. Chem. Phys.* 126, 175103 (2007). Republished with permission.

Recently [35], we have presented a new microscopic model for the nucleation mechanism of protein folding. The model is based on pairwise "molecular" interactions (i.e., repulsion/attraction) and configurational (dihedral angle) potentials in which protein residues are involved. The parameters of the potentials can be rigorously defined. The rates of emission and absorption of residues by the cluster are determined by using a first passage time analysis [36–41]. Once these rates are found as functions of the cluster size, one can develop a self-consistent kinetic theory for the nucleation mechanism of folding of a protein [35]. The time necessary for the protein to fold can be evaluated as the sum of the times necessary for the appearance of the first nucleus and the time necessary for the nucleus to grow to the maximum size (of the folded protein in the native state). The ill-defined surface tension of a cluster of protein residues (within a protein in a compact but nonnative configuration) does not enter into the new model at all.

Some protein models treated it as a heteropolymer chain consisting of three types of beads—neutral, hydrophobic, and hydrophilic [7,16] (with all the bonds of the same constant length and all the bond angles equal and fixed). However, for the sake of simplicity and (mostly) to reduce the computational effort, in the previous paper [35] on the nucleation mechanism of protein folding we considered a protein as a heteropolymer consisting of only hydrophobic and hydrophilic beads without neutral ones. Even with this simplification the numerical calculations remained too lengthy and we had to recur to an additional assumption that the composition of the cluster remained constant and equal to the overall composition of the protein at all times.

In the present paper, we remove these simplifications and assumptions. A protein is modeled as a heteropolymer consisting of three kinds of beads (hydrophobic, hydrophilic, and neutral) and the composition of the cluster is allowed to vary. We shall therefore present a model for the nucleation mechanism of protein folding in terms of ternary nucleation by using a first passage time analysis [36–41] that accounts for the particular shape of the potential around the cluster of native residues.

The paper is structured as follows. In Section 3.2.2, we outline the previous models for the nucleation mechanism of protein folding. A ternary nucleation model for protein folding is presented in Section 3.2.3. The results of numerical calculations are presented in Section 3.2.4, and a brief discussion and conclusions are summarized in Section 3.2.5.

## 3.2.2 Previous Models for the Nucleation Mechanism of Protein Folding

The previous models [16,31,35] for the nucleation mechanism of protein folding considered a protein as a heteropolymer chain. Hence, we briefly describe this protein model before outlining the models for the nucleation mechanism themselves.

### *3.2.2.1 Heteropolymer as a Protein Model*

A heteropolymer was introduced as a simple model of a protein in the MD and MC simulations of protein folding dynamics [7,16]. The polypeptide chain of a protein was modeled as a heteropolymer consisting of $N$ connected beads which can be thought of as representing the $\alpha$ carbons of various amino acids. The heteropolymer may consist of hydrophobic (b), hydrophilic (l), and neutral (n) beads. Two adjacent beads are connected by a covalent bond of fixed length $\eta$. This model (and its variants), completed with the appropriate potentials described below, has been shown [7,16,27,28,31] to be able to capture the essential characteristics of protein folding even though it contains only some of the features of a real polypeptide chain. For example, this model ignores the side groups although they are known to be crucial for the intramolecular hydrogen bonding [3]. Besides, the presence of a solvent (water) in a real physical system has been usually accounted for too simplistically, although the protein dynamics was reported to become more realistic in those MD simulations where the solvent molecules were explicitly taken into account [42]. Despite these limitations, various modifications of the heteropolymer model [6,7,27,28,31,43–48] shed light on some important details regarding the folding transition of a protein [7,16].

The total energy of the heteropolymer (polypeptide chain) can contain three contributions of different types: first, the contribution from repulsive/attractive forces between pairs of nonadjacent beads (these can be, e.g., of Lennard–Jones or other types); next, a contribution from the harmonic forces due to the oscillations of the bond angles; finally, a contribution from the dihedral angle potential due to the rotation around the peptide bonds. There are various ways to model these three types of contributions [7,16,44–47]. Such a structure of the potential functions of a heteropolymer was suggested by Honeycutt and Thirumalai [7] and further explored by Guo and Thirumalai [16]. Discrete analogs (for a protein on a diamond lattice) thereof completed with a "cooperativity potential" were also proposed [44–47] and random energy models used [27,28,31]. It was shown [7,16] via MD simulations, which employed low friction Langevin dynamics, that a proper balance between the above three contributions to the total energy of the heteropolymer ensures that the heteropolymer folds into a well-defined β barrel structure. The balancing between these contributions was achieved by adjusting the energy parameters for each type of beads. It was also found [7,16] that the balance between the dihedral angle potential, which tends to stretch the molecule into a state with all bonds in a *trans* configuration, and the attractive hydrophobic potential is crucial to induce folding into a β barrel-like structure upon cooling. If attractive forces are excessively dominant, they make the heteropolymer fold into a globulelike structure, while an overwhelming dihedral angle potential forces the chain to remain in an elongated state (even at low temperatures) with bonds mainly in the *trans* configuration.

#### *3.2.2.2 A CNT-Based Model for Nucleation Mechanism of Protein Folding*

The possibility that a nucleation-like mechanism constitutes a viable pathway for protein folding was suggested in Refs. [16,31], where it was treated by means of a CNT-like formalism. A cluster of protein residues stabilized with correct tertiary contacts was treated in the framework of the capillarity approximation [34], i.e., assumed to be a sphere with sharp boundaries with uniform distribution of native residues (which are in the same state as they are in the native protein) therein. As in CNT, the free energy of formation of such a cluster was determined as the difference between the free energies of the protein in a partly folded state (with a cluster therein) and in a completely unfolded (but compact) one. In doing so, another assumption of capillarity approximation comes into play, namely, that the cluster can be characterized by a surface tension, equal to the interfacial tension of a flat interface between bulk phase "folded protein" and "unfolded (but compact) protein." Clearly, this is rather an ill-defined physical quantity, which cannot be determined experimentally in principle. The cluster was modeled as consisting of residues of a single type which allowed one to represent the free energy of formation of such a cluster of $\nu$ native residues as

$$\Delta F = -\nu\Delta\mu + \sigma 4\pi\lambda^2\nu^{2/3}, \tag{3.33}$$

where $\Delta\mu \equiv \mu_u - \mu_f$ is the difference between the free energies per residue in the unfolded and folded (native) states, respectively (marked with the subscripts "u" and "f"); $\sigma$ is the "surface" tension of the cluster, $\lambda = (3\upsilon/4\pi)^{1/3}$; and $\upsilon$ is the volume of a protein residue in its native state. This expression allows one to find the size of the critical cluster (nucleus) whereof the formation is immediately followed by a quick transition of the protein into the native state (whence the analogy with nucleation). So far, two ways [16,31] have been suggested for evaluating the volume and surface contributions to $\Delta F$, i.e., the first and second terms in Equation 3.33.

In Ref. [16], the initial stage of protein folding was considered to be driven by hydrophobic attractive forces. Consequently, the first term in Equation 3.33 was taken to be equal to $-(1/2)\epsilon_b\chi\nu(\chi\nu-1)$, where $\epsilon_b$ is the energy parameter of the Lennard–Jones potential between hydrophobic residues and $\chi$ denotes their mole fraction in the cluster (assumed to be the same as in the whole protein). The number of residues in the nucleus is thus given by $\nu_c = (8\pi\sigma\lambda^2/3\chi^2\epsilon_b)^{3/4}$, which for typical values of $\lambda$, $\sigma$, and $\epsilon_b$ was estimated [16] to be of the order of 10.

In Ref. [31], it was argued that the surface tension arises because the surface located residues interact with the cluster interior stronger than with the unfolded part of the protein. Since the interaction energy characteristic to protein folding is of the order of $k_BT$ (where $k_B$ is the Boltzmann constant and $T$ is the absolute temperature), the surface tension $\sigma$ was assumed to be given by $\sigma 4\pi\lambda^2 \sim k_BT$. The difference between the chemical potentials in the folded and unfolded parts of the protein, $\Delta\mu$, was evaluated to be of the order of 0.1 $k_BT$, so that the number of residues in the critical cluster was evaluated [31] to be of the order of 100 for a protein with the total number of residues $N_0 = 150$.

Despite a significant difference between them, both estimates corroborate the idea that the nucleation mechanism constitutes a viable pathway for protein folding [21–25,49–52]. However, the CNT-based model is riddled with flaws and is very difficult to be improved. In a more realistic (heteropolymer) model the protein is considered to consist of more than one type of residues. Using these protein models in a capillarity approximation-based nucleation model would give rise to a multicomponent (binary, ternary, etc.) CNT, which is widely known to contain several inconsistencies and to provide a poor agreement with experiment for various reasons [53–57], such as the inability to take account of adsorption effects and of the composition dependence of the surface tension (whereof the concept itself is questioned when it comes to clusters as small as those usually involved in nucleation). In the case of protein folding, this problem is even more complex because $\sigma$ in Equation 3.33 is an ill-defined quantity that cannot be determined experimentally due to the nonexistence of bulk folded and unfolded proteins as real physical phases, not to mention a flat interface between them.

### *3.2.2.3 A Model for Nucleation Mechanism of Protein Folding Based on a First Passage Time Analysis*

In order to avoid the use of macroscopic thermodynamics in the kinetic theory of unary nucleation, an alternative approach was proposed by Ruckenstein and coworkers [36–38] on the basis of a mean first passage time analysis. Unlike CNT, the new approach [36–38] is built on molecular interactions and does not make use of the surface tension for the tiny clusters involved in nucleation. Instead, it exploits the fact that one can derive and solve the kinetic equation of nucleation (hence find the nucleation rate) if the emission and absorption rates of a cluster are known as functions of its size. For the rate of absorption of molecules by the cluster, the new theory uses (as CNT does) a standard gas-kinetic expression [58], but (and this constitutes the main idea of the new approach [36–38]) the rate of emission of molecules by the cluster is determined via a mean first passage time analysis. This time is calculated by solving a single-molecule master equation for the probability distribution function of a surface layer molecule moving in a potential well around the cluster. Recently, a further development of that kinetic theory was proposed by combining it [39,40] with the density functional theory and extending it to binary [40] and heterogeneous [41] systems.

In Ref. [35], we presented a new microscopic model for the nucleation mechanism of protein folding. A protein was considered as a heteropolymer consisting of two types of beads (hydrophobic and hydrophilic) linked with bonds of fixed length. The bond angles were all assumed to be fixed and equal to 105° and all nonadjacent beads (at least three links apart) to interact only via pairwise potentials. Besides these interactions, the total energy of the heteropolymer contains contributions from the bond angles and from dihedral angles of all triads of successive links. Unlike the CNT-based model, which is based on the ill-defined surface tension of a cluster of protein residues within a protein, the microscopic model [35] was developed without treating the cluster by means of classical thermodynamics. Instead, it is based on the molecular interactions, pairwise and configurational, whereof the parameters can be rigorously defined.

The main idea underlying our model [35] consists of averaging the dihedral potential in which a selected residue is involved over all possible configurations of neighboring residues. The resulting average dihedral potential depends on the distance between the residue and the cluster center. It is maximum at the cluster surface and monotonically decreases with increasing distance therefrom. Strictly speaking, the total potential $\psi_i(r)$ for a residue of type $i$ around the cluster is determined by four constituents—$\phi_i(r)$, $\phi_\beta(r)$, $\bar{\phi}_\delta(r)$, and $\phi_{cp}$—which represent the average pairwise (for interactions of the selected residue with those in the cluster), bond angle, dihedral angle, and confining (representing the external boundary of the volume available to the unfolded residues) potentials, respectively, $\psi_i(r)=\phi_i(r)+\phi_\beta(r)+\bar{\phi}_\delta(r)+\phi_{cp}$ (for more details, see the next section). With some assumptions that did not affect the generality of the model and significantly simplified the algebra and numerical calculations, we showed [35] that the above combination of potentials gives rise to a double potential well around the cluster with a barrier between the two wells. Residues in the inner well are considered to belong to the cluster (part of the protein with correct tertiary contacts), while those in the outer well are treated as belonging to the mother phase (amorphous part of the protein with incorrect tertiary contacts). Transitions of residues from the inner well into the outer one and vice versa are considered as elementary emission and absorption events, respectively. The rates of these processes are determined by using a mean first passage time analysis. This time is calculated by solving a single-molecule master equation for the probability distribution function of a surface layer molecule moving in a potential well around the cluster. The master equation is a Fokker–Planck equation in the phase space which can be reduced to the Smoluchowski equation owing to the hierarchy of the characteristic timescales involved in the evolution of the single-molecule distribution function with respect to coordinates and momenta [36–38]. Once the rates of emission and absorption are found as functions of the cluster size, one can develop a self-consistent kinetic theory for the nucleation mechanism of protein folding [35]. For example, the size of the critical cluster (nucleus) is then found as the one for which these rates are equal. The time necessary for the protein to fold can be evaluated as the sum of the times necessary for the appearance of the first nucleus and the time necessary for the nucleus to grow to the maximum size (of the folded protein in the native state).

Due to computational constraints, our previous model was developed in the framework of the binary nucleation theory (the neutral beads were neglected) with an additional assumption that the composition of clusters is the same as that of the mother phase which essentially reduced the theory to a unary one. Below, we present a full-fledged ternary nucleation model for the nucleation mechanism of protein folding which takes into account that all three different kinds of amino acids (hydrophobic, hydrophilic, and neutral) can be present in the protein, and no assumption is made regarding the cluster composition.

As the previous (unary) model [35], a ternary nucleation model of protein folding will be developed by using a first passage analysis to determine both the rate of emission of native residues from the cluster and the rate of absorption of nonnative residues by the cluster. This is possible owing to the double well shape of the potential generated in this case around the cluster. In the earlier applications [36–41] of the method to nucleation in conventional first order phase transitions, the potential generated around the cluster had a single-well shape. Hence, only the emission rate of the cluster could be determined via a first passage time analysis, whereas the absorption rate there was calculated using an expression derived in the framework of the gas-kinetic theory [58] that assumes a Maxwellian distribution of the velocities of the mother phase molecules.

## 3.2.3 A Ternary Nucleation Mechanism of Protein Folding

### 3.2.3.1 Potential Well around a Cluster of Native Residues

A heteropolymer chain as a protein model, proposed in Refs. [7,16] and briefly described earlier, consists of three types of beads—neutral, hydrophobic, and hydrophilic. A pair interaction between two nonadjacent beads $i$ and $j$ at a distance $d$ away from each other can be considered to be of the Lennard–Jones type,

$$\phi_{ij}(d) = 4\epsilon_{ij}[(\eta/d)^{12} - (\eta/d)^{6}] \quad (i, j = \mathrm{b}, \mathrm{l}, \mathrm{n}), \tag{3.34}$$

where $\eta$ is the bond length (fixed) and $\epsilon_{ij}(i, j = \mathrm{b}, \mathrm{l}, \mathrm{n})$ is an energy parameter satisfying the mixing rules $\epsilon_{ij} = \sqrt{\epsilon_{ii}\epsilon_{jj}}$. The neutral beads (n) play an important role in this model. Their short-range repulsion from each other is assumed to be stronger, and the dihedral angle forces for the bonds involving them are considered to be weaker so that the bending is enhanced where they are present. As MD simulations [7,16] showed, such a heteropolymer acquires a β-barrel shape in the lowest energy configuration, with neutral residues appearing mostly in the bend regions.

The angle $\beta$ between two successive bonds (in the heteropolymer) can be regarded to be subjected to the harmonic potential $\phi_\beta = (k_\beta/2)(\beta - \beta_0)^2$, where the spring constant $k_\beta$ is sufficiently large for the deviation of the bond angles from the average value $\beta_0$ to be small. The bond angle forces are believed [7,16] to play a minor role in the protein folding/unfolding. Hence, all bond angles can be set to be equal to $\beta_0$.

The dihedral angle potential arises due to the rotation of three successive peptide bonds connecting four successive beads and is related to the dihedral angle $\delta$ via

$$\phi_\delta = \epsilon'_\delta(1 + \cos\delta) + \epsilon''_\delta(1 + \cos 3\delta), \tag{3.35}$$

where $\epsilon'_\delta$ and $\epsilon''_\delta$ are independent energy parameters that depend on the nature and sequence of the four beads involved in the dihedral angle $\delta$. This potential has three minima, one in the *trans* configuration at $\delta = 0$ and two others in the *gauche* configurations at $\delta = \pm\arccos\sqrt{(3\epsilon''_\delta - \epsilon''_\delta)/12\epsilon''_\delta}$ (the former one being the lowest).

In addition to the above potentials, one must introduce a confining potential which arises because all the residues of the protein are linked and there is an external boundary of the volume available to them. The formation of a cluster consisting of native residues during protein folding can be regarded as a ternary nucleation. We shall therefore present a nucleation mechanism model of protein folding in terms of ternary nucleation by using a first passage time analysis [36–41] that accounts for the particular shape of the present potential around the cluster.

Consider a ternary cluster of spherical shape (with sharp boundaries and radius $R$) immersed in a ternary fluid mixture. The number of beads of component $i$ in the cluster will be denoted by $\nu_i$ ($i = \mathrm{b}, \mathrm{l}, \mathrm{n}$), whereas the total number of beads of component $i$ in the protein is denoted by $N_i$. In the previous applications of a first passage time analysis to nucleation [36–41], a molecule of component $i$ (in our model $i = \mathrm{b}, \mathrm{l}, \mathrm{n}$) located in the surface layer of the cluster was considered to perform a thermal chaotic motion in a spherically symmetric potential well $\phi_i(r)$ resulting from the pair interactions of this molecule with those in the cluster.

Assuming pairwise additivity of the interactions between nonadjacent beads, $\phi_i(r)$ is provided by

$$\phi_i(r) = \sum_j \int_V d\mathbf{r}' \rho_j(r') \phi_{ij}(|\mathbf{r}' - \mathbf{r}|). \tag{3.36}$$

Here, $\mathbf{r}$ is the coordinate of the surface bead of type $i$, $\rho_j(r')$ ($j = \mathrm{b}, \mathrm{l}, \mathrm{n}$) is the number density of beads of component $j$ at point $\mathbf{r}'$ (spherical symmetry is assumed with the cluster center chosen as the origin of the coordinate system), and $\phi_{ij}$ ($|\mathbf{r}' - \mathbf{r}|$) is the interaction potential between two beads of components $i$ and $j$ at points $\mathbf{r}$ and $\mathbf{r}'$, respectively. The integration in Equation 3.36 has to be carried out over the whole volume of the system, but the contribution from the unfolded part can be assumed to be small and accounted for by particular choices of the energy parameters.

For nucleation in proteins the potential well $\psi_i(r)$ for a residue of type $i$ around the cluster is determined not only by the potential $\phi_i(r)$, but also by three other constituents, $\phi_\beta(r)$, $\bar{\phi}_i^\delta(r)$, and $\phi_{cp}$, which represent the bond angle, dihedral angle, and confining potentials, respectively,

$$\psi_i(r) = \phi_i(r) + \phi_\beta(r) + \bar{\phi}_i^\delta(r) + \phi_{cp}.$$

Without affecting the generality of the model, one can significantly simplify the algebra and eventual numerical calculations by assuming that all bond angles are fixed and equal to $\beta_0 = 105°$. Under this assumption, the contribution to the potential energy of the protein arising from the bond angle potential is constant and does not depend on the distance $r$ between the selected bead and the center of the cluster. Therefore, the term $\phi_\beta(r)$ on the right-hand side of the above equation can be disregarded (or, equivalently, be chosen as a reference level for the potential energy), i.e.,

$$\psi_i(r) = \phi_i(r) + \bar{\phi}_i^\delta(r) + \phi_{cp}(r) \quad (i = \mathrm{b, l, n}). \tag{3.37}$$

Because of the bonds between the residues, none of them can be located outside a volume which depends on the densities of the folded and unfolded regions. Assuming that volume to be spherical, its radius $r_{cp}$ is given by the expression $r_{cp} = [R^3 + 3(N_0 - \nu)/4\pi\rho_u]^{1/3}$, where $N_0$ and $\nu$ are the total numbers of beads in the whole protein and in the cluster, respectively, and $\rho_u = \rho_{bu} + \rho_{lu} + \rho_{nu}$ is the overall number density of beads (of all three components) in the unfolded part of the protein. Clearly, $\nu = (4\pi/3)R^3\rho_f$, with $\rho_f = \rho_{bf} + \rho_{lf} + \rho_{nf}$ being the overall number density of beads in the folded part of the protein. Consequently, the confining potential $\phi_{cp}(r) = 0$ for $r < r_{cp}$ and $\phi_{cp} = \infty$ for $r \geq r_{cp}$. The term $\bar{\phi}_i^\delta(r) \equiv \bar{\phi}_i^\delta(r, \mathbf{r}_2, \mathbf{r}_3, \ldots, \mathbf{r}_7)$ in $\psi_i(r)$ is due to the $r$-dependent part of the dihedral angle potential of the whole protein. Consider a bead 1 (of type b, l, or n) at a distance $r$ from the (center of the) cluster, $r > R$ (see Figure 3.5). The total dihedral angle potential of the whole protein chain for a given configuration of beads 1, 2, 3, ..., $N_0$ is a function of $r, \mathbf{r}_2, \mathbf{r}_3, \ldots, \mathbf{r}_{N_0}$ and can be written as

$$\phi_{i_1}^\delta\left(\delta_{421}^{642}(r)\right) + \phi_{i_1}^\delta\left(\delta_{213}^{421}(r)\right) + \phi_{i_1}^\delta\left(\delta_{135}^{213}(r)\right) + \phi_{i_1}^\delta\left(\delta_{357}^{135}(r)\right) + \text{const},$$

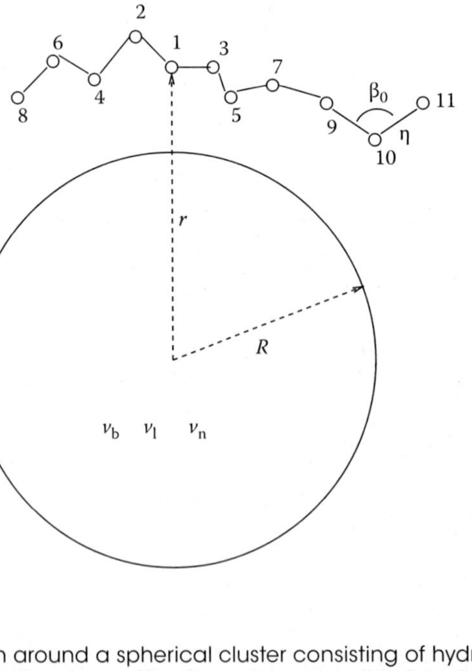


Figure 3.5

A piece of a heteropolymer chain around a spherical cluster consisting of hydrophobic ($\nu_b$), hydrophilic ($\nu_l$), and neutral beads ($\nu_n$). Bead 1 is in the plane of the figure, whereas other beads may all lie in different planes. All bond angles are equal to $\beta_0$, and their lengths are equal to $\eta$. The radius of the cluster is $R$, and the distance from the selected bead 1 to the cluster center is $r$.

where $\delta_{jkl}^{ijk}$ is a dihedral angle between two planes, one of which is determined by beads $i$, $j$, and $k$ and the other by beads $j$, $k$, and l [an occasionally used subscript $i_s$ ($i_s$ = b, l, n) indicates the type of bead $s$ ($s = 1, \ldots, 7$)]. The last term on the right-hand side of the above equation, representing the contributions from the dihedral angles involving beads 8, 9, …, $N_0$, does not depend on $r$ because those dihedral angles remain unaffected as the position of bead 1 changes. Hence, this contribution (const) can be omitted hereafter without being specified since it can be regarded as affecting only the reference level for $\psi_{i_1}(r)$,

$$\begin{aligned}\phi_{i_1}^{\delta}(r) &\equiv \phi_{i_1}^{\delta}(r, \mathbf{r}_2, \mathbf{r}_3, \ldots, \mathbf{r}_7)\\ &= \phi_{i_1}^{\delta}\left(\delta_{421}^{642}(r)\right) + \phi_{i_1}^{\delta}\left(\delta_{213}^{421}(r)\right)\\ &\quad + \phi_{i_1}^{\delta}\left(\delta_{135}^{213}(r)\right) + \phi_{i_1}^{\delta}\left(\delta_{357}^{135}(r)\right).\end{aligned} \tag{3.38}$$

For a given location of bead 1, various configurations of beads 2, 3, …, $N_0$ (subject to the constraints on the bond length and bond angle as well as to the constraint of the excluded cluster volume) lead to various sets of dihedral angles. However, variations in the location of beads 8, 9, …, $N_0$ give rise to variations in the dihedral potential which are independent of $r$. Thus, the dihedral term $\bar{\phi}_i^{\delta}(r)$ (less a term independent of $r$ and omitted hereafter) on the right-hand side of Equation 3.37 can be obtained by averaging Equation 3.38 with the Boltzmann factor $\exp\left[-(\phi_{i_1}^{\delta 27}(r)/k_B T\right]$ with

$$\phi_{i_1}^{\delta 27}(r) \equiv \phi_{i_1}^{\delta 27}(r, \mathbf{r}_2, \ldots, \mathbf{r}_7) = \phi_{i_1}^{\delta}(r) + \sum_{s=2}^{7} \phi_{i_s}(\mathbf{r}_s), \tag{3.39}$$

over all possible configurations of beads 2, 3, …, 7 and assigning the result to a selected bead with fixed coordinates, i.e., bead 1,

$$\bar{\phi}_{i_1}^{\delta}(r) = f^{-1} \int_{\Omega_{18}} d\mathbf{r}_2\, d\mathbf{r}_3\, d\mathbf{r}_4\, d\mathbf{r}_5\, d\mathbf{r}_6\, d\mathbf{r}_7 \phi_{i_1}^{\delta}(r) \times \exp[-\phi_{i_1}^{\delta 27}(r)/k_B T], \tag{3.40}$$

where the normalization constant

$$f = \int_{\Omega_{18}} d\mathbf{r}_2\, d\mathbf{r}_3\, d\mathbf{r}_4\, d\mathbf{r}_5\, d\mathbf{r}_6\, d\mathbf{r}_7 \exp\left[-\phi_{i_1}^{\delta 27}(r)/k_B T\right], \tag{3.41}$$

where $\Omega_{18}$ is the integration region in an 18-dimensional space. Note that (unlike in Ref. [35]) the exponent in the Boltzmann factor is a sum of two terms, the first representing the dihedral angle potential of beads 1–7 and the second the sum of "bead–cluster" interactions for beads 2–7. Clearly, both contributions affect the probability of configurations of beads 1–7, but the second one was neglected in the previous paper [35] to increase the efficiency of numerical calculations. In calculating $\phi_i^{\delta 27}$ in Equation 3.39, only the type of bead 1 is supposed to be known, whereas the types of beads 2–7 are treated in a probabilistic way. For example, if bead 1 is of kind b, then any of the beads 2–7 is either of kind b with the probability $(N_b - \nu_b - 1)/(N_0 - \nu - 1)$ or of kind l with the probability $(N_l - \nu_l)/(N_0 - \nu - 1)$ or of kind n with the probability $(N_n - \nu_n)/(N_0 - \nu - 1)$ (where $N_0 = N_b + N_l + N_n$ and $\nu = \nu_b + \nu_l + \nu_n$).

The 18-fold integrals in Equations 3.40 and 3.41 can be reduced to sevenfold ones by taking into account the constraints of fixed bond length and fixed bond angle,

$$\begin{aligned}\bar{\phi}_{i_1}^{\delta}(r) = f^{-1} \sum_{i,j,k,m,n=-}^{+} \int_0^{2\pi} d\varphi_2 \int_{L_2} dr_2 \int_{L_3^i} dx_3 \int_{L_4^j} dx_4 \int_{L_5^k} dx_5 \int_{L_6^m} dx_6 \int_{L_7^n} dx_7 r_2^2 \sin\Theta(r, r_2) \tilde{\phi}_{ijkmn}(r, \varphi_2, r_2, x_3, \ldots, x_7)\\ \times \exp[-\tilde{\phi}_{ijkmn}(\varphi, r_2, x_3, \ldots, x_7)/k_B T],\end{aligned} \tag{3.42}$$

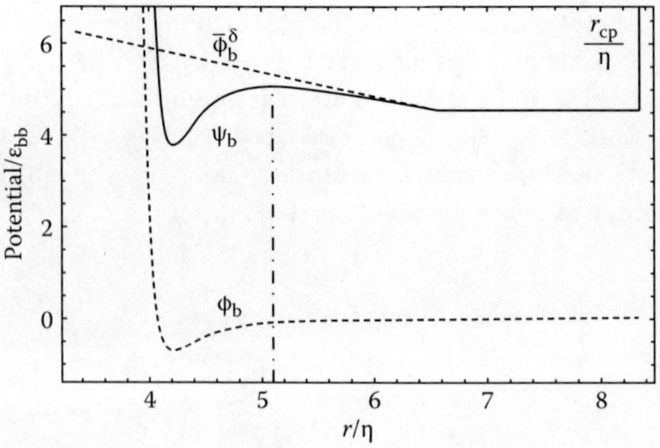


Figure 3.6

The potentials $\phi_b(r)$ (lower dashed curve), $\bar{\phi}_b^{\delta}(r)$ (upper dashed curve), and $\psi_b(r)$ (solid curve) for a hydrophobic bead around the cluster as functions of the distance $r$ from the (center of the) cluster consisting of $v_b = 29$, $v_l = 50$, and $v_n = 2.4$ hydrophobic, hydrophilic, and neutral beads, respectively. The vertical dashed line indicates the location of the maximum of the barrier between the ipw and the opw which determines the widths $\lambda_b^{iw}$ and $\lambda_b^{ow}$ of both potential wells. The outer boundary of the opw is indicated by $r_{cp}$.

$$f = \sum_{i,j,k,m,n=-}^{+} \int_0^{2\pi} d\varphi_2 \int_{L_2} dr_2 \int_{L_3^i} dx_3 \int_{L_4^j} dx_4 \int_{L_5^k} dx_5 \int_{L_6^m} dx_6 \int_{L_7^n} dx_7 r_2^2 \sin\Theta_2(r, r_2)$$
$$\exp[-\tilde{\phi}_{ijkmn}(r, \varphi_2, r_2, x_3, \ldots, x_7)/k_B T]. \tag{3.43}$$

Each of the summation indices in Equations 3.42 and 3.43 takes on two values, denoted + and −, so that there are $2^5$ terms in the sum differing by the integrand as well as by the integration ranges (except for $L_2$ which is independent of $i, j, k, m$, and $n$). The definitions of the coordinates $\varphi_2, r_2, x_3, x_4, x_5, x_6, x_7$, the explicit forms of the integration ranges $L_2, L_3^i, L_4^j, L_5^k, L_6^m, L_7^n$, and the rules for transforming the function $\phi_{i_1}^{\delta 27}(r)$ into the function $\tilde{\phi}_{ijkmn}(\varphi_2, r_2, x_3, \ldots, x_7)$ are provided in Appendix 3C.

Figure 3.6 presents typical shapes of the constituents $\phi_i(r)$ and $\bar{\phi}_i^{\delta 27}(r)$ of the potential well as functions of the distance from the cluster center, as well as the overall potential well $\psi_i(r)$ itself (for details regarding the numerical calculations, see Section 3.2.4). The contribution $\phi_i(r)$, arising from the pairwise interactions, has a familiar form [36–41] reminiscent of the underlying Lennard–Jones potential, whereas the contribution from the average dihedral potential has a rather remarkable behavior. Indeed, starting with a maximum value at the cluster surface, it monotonically decreases with increasing $r$ until it becomes constant for $r \geq R + \tilde{d}$ (see Section 3.2.4, Equation 3.64). Thus, except for very short distances from the cluster surface where $\phi_i(r)$ sharply decreases from $\infty$ to its global minimum, the potential $\psi_i(r)$ is shaped by two competing terms, $\phi_i(r)$ and $\bar{\phi}_i^{\delta 27}(r)$, which increase and decrease, respectively, with increasing $r$, and by the confining potential $\phi_{cp}$. As a result, the overall potential $\psi_i(r)$ has a double well shape: the inner well is separated by a potential barrier from the outer well. This shape of $\psi_i(r)$ is of crucial importance to our nucleation mechanism model of protein folding because it allows one to use a mean first passage time analysis for the determination of the rates of absorption and emission of beads by the cluster.

### 3.2.3.2 Determination of Emission and Absorption Rates

Formulating our model in the spirit of a mean first passage time analysis [36–41], a bead (residue) is considered as belonging to a cluster as long as it remains in the inner potential well (hereinafter referred to as "ipw"), and as dissociated from the cluster when it passes over the barrier between the ipw and the outer potential well (hereinafter referred to as "opw"). The rate of emission, $W^-$, is determined by the mean time necessary for the passage of the bead from the ipw over the barrier into the opw. Likewise, a bead is considered as belonging to the unfolded part of the heteropolymer (protein) as long as it remains in the opw, and as absorbed by the cluster when it passes over the barrier between the opw and the ipw. The rate of emission, $W^+$, is determined by the mean time necessary for the passage of the bead from the opw over the barrier into the ipw.

The mean first passage time of a bead escaping from a potential well is calculated on the basis of a kinetic equation governing the chaotic motion of the bead in that potential well. The chaotic motion of the bead is assumed to be governed by the Fokker–Planck equation for the single-particle distribution function with respect to its coordinates and momenta, i.e., in the phase space [59–61]. When the relaxation with respect to momenta occurs on a timescale much shorter than that for the evolution with respect to coordinates, the Fokker–Planck equation reduces to the Smoluchowski equation, which involves diffusion in an external field [60,61]. Solving that equation [36–41], one can obtain the following expressions for $W^-$ and $W^+$, the emission and absorption rates (see Appendix 3D for details), respectively:

$$W_i^- = n_i^{\text{iw}} D_i^{\text{iw}} \omega_i^{\text{iw}}, \ W_i^+ = n_i^{\text{ow}} D_i^{\text{ow}} \omega_i^{\text{ow}} \ (i = \text{b}, \text{l}, \text{n}). \tag{3.44}$$

Here, the superscripts "iw" and "ow" mark the quantities for the inner and outer potential wells, respectively; $n$ is the number of beads in the well, $D$ is the diffusion coefficient of the bead, and $\omega^{\text{iw}}$ and $\omega^{\text{ow}}$ are defined as

$$\omega_i^{\text{iw}} \equiv 1/\left(D_i^{\text{iw}} \overline{\tau}_i^{\text{iw}}\right), \ \omega_i^{\text{ow}} \equiv 1/\left(D_i^{\text{ow}} \overline{\tau}_i^{\text{ow}}\right) \ (i = \text{b}, \text{l}, \text{n}), \tag{3.45}$$

where $\overline{\tau}_i^{\text{iw}}$ and $\overline{\tau}_i^{\text{ow}}$ are the mean first passage times for a bead in the ipw to cross over the barrier into the opw, and vice versa, respectively. Note that in the original work [36–38] on the mean first passage time analysis in the nucleation theory and its recent development [39–41], this method was used only for determining $W^-$ but not $W^+$. In the present model, the double well shape of the overall potential $\psi(r)$ around the cluster allows one to use the first passage time analysis to determine $W^+$ as well.

Clearly, the state of a three-component cluster is determined by three independent variables; hence, the quantities $W^-$, $W^+$, $\omega^{\text{iw}}$, $\omega^{\text{ow}}$, $\overline{\tau}^{\text{iw}}$ and $\overline{\tau}^{\text{ow}}$ are functions of three variables, which can be, for example, the numbers $\nu_{\text{b}}$, $\nu_{\text{l}}$, and $\nu_{\text{n}}$ of hydrophobic, hydrophilic, and neutral beads, respectively, or, alternatively, the cluster radius $R$ and the mole fractions $\chi_{\text{b}} = \nu_{\text{b}}/(\nu_{\text{b}} + \nu_{\text{l}} + \nu_{\text{n}})$ and $\chi_{\text{l}} = \nu_{\text{l}}/(\nu_{\text{b}} + \nu_{\text{l}} + \nu_{\text{n}})$ of the hydrophobic and hydrophilic beads therein (see Appendix 3D for more details). In the framework of the present model, the nucleation mechanism of protein folding must be treated in terms of ternary nucleation.

#### *3.2.3.3 Equilibrium Distribution and Steady-State Nucleation Rate*

During protein folding, clusters of various sizes emerge with different probabilities. Let us denote the distribution of clusters with respect to the number of beads in a cluster at time $t$ by $g(\nu_{\text{b}}, \nu_{\text{l}}, \nu_{\text{n}}, t)$. The temporal evolution of the distribution function $g(\nu_{\text{b}}, \nu_{\text{l}}, \nu_{\text{n}}, t)$ is governed by the balance equation

$$\frac{\partial g(\nu_{\text{b}}, \nu_{\text{l}}, \nu_{\text{n}}, t)}{\partial t} = \Theta_{\text{b}} + \Theta_{\text{l}} + \Theta_{\text{n}}, \tag{3.46}$$

where

$$\begin{aligned} \Theta_{\text{b}} = {} & W_{\text{b}}^+(\nu_{\text{b}} - 1, \nu_{\text{l}}, \nu_{\text{n}}) g(\nu_{\text{b}} - 1, \nu_{\text{l}}, \nu_{\text{n}}, t) \\ & - W_{\text{b}}^+(\nu_{\text{b}}, \nu_{\text{l}}, \nu_{\text{n}}) g(\nu_{\text{b}}, \nu_{\text{l}}, \nu_{\text{n}}, t) + W_{\text{b}}^-(\nu_{\text{b}} + 1, \nu_{\text{l}}, \nu_{\text{n}}) \\ & \times g(\nu_{\text{b}} + 1, \nu_{\text{l}}, \nu_{\text{n}}, t) - W_{\text{b}}^-(\nu_{\text{b}}, \nu_{\text{l}}, \nu_{\text{n}}) g(\nu_{\text{b}}, \nu_{\text{l}}, \nu_{\text{n}}, t), \end{aligned} \tag{3.47}$$

$$\begin{aligned} \Theta_{\text{l}} = {} & W_{\text{l}}^+(\nu_{\text{b}}, \nu_{\text{l}} - 1, \nu_{\text{n}}) g(\nu_{\text{b}}, \nu_{\text{l}} - 1, \nu_{\text{n}}, t) \\ & - W_{\text{l}}^+(\nu_{\text{b}}, \nu_{\text{l}}, \nu_{\text{n}}) g(\nu_{\text{b}}, \nu_{\text{l}}, \nu_{\text{n}}, t) \\ & + W_l^-(\nu_{\text{b}}, \nu_{\text{l}}, +1, \nu_{\text{n}}) g(\nu_{\text{b}}, \nu_{\text{l}} + 1, \nu_{\text{n}}, t) \\ & - W_l^-(\nu_{\text{b}}, \nu_{\text{l}}, \nu_{\text{n}}) g(\nu_{\text{b}}, \nu_{\text{l}}, \nu_{\text{n}}, t), \end{aligned} \tag{3.48}$$

$$\Theta_{\text{n}} = W_{\text{n}}^{+}(\nu_{\text{b}},\nu_{\text{l}},\nu_{\text{n}}-1)g(\nu_{\text{b}},\nu_{\text{l}},\nu_{\text{n}},-1,t)$$
$$-W_{\text{n}}^{+}(\nu_{\text{b}},\nu_{\text{l}},\nu_{\text{n}})g(\nu_{\text{b}},\nu_{\text{l}},\nu_{\text{n}},t)$$
$$+W_{\text{n}}^{-}(\nu_{\text{b}},\nu_{\text{l}},\nu_{\text{n}}+1)g(\nu_{\text{b}},\nu_{\text{l}},\nu_{\text{n}}+1,t) \tag{3.49}$$
$$-W_{\text{n}}^{-}(\nu_{\text{b}},\nu_{\text{l}},\nu_{\text{n}})g(\nu_{\text{b}},\nu_{\text{l}},\nu_{\text{n}},t),$$

where $W_i^-(\nu_{\text{b}},\nu_{\text{l}},\nu_{\text{n}})$ and $W_i^+(\nu_{\text{b}},\nu_{\text{l}},\nu_{\text{n}})$ $(i=\text{b},\text{l},\text{n})$ are the numbers of molecules of component $i$ that a cluster $\nu_{\text{b}}$, $\nu_{\text{l}}$, $\nu_{\text{n}}$ emits and absorbs, respectively, per unit time. These equations assume that the evolution of clusters occurs through the absorption and emission of single beads (the multimer absorption and emission are neglected).

The equilibrium distribution $g_{\text{e}}(\nu_{\text{b}}, \nu_{\text{l}}, \nu_{\text{n}})$ can be obtained once the emission and absorption rates $W_i^- = W_i^-(\nu_{\text{b}},\nu_{\text{l}},\nu_{\text{n}})$ and $W_i^+ = W_i^+(\nu_{\text{b}},\nu_{\text{l}},\nu_{\text{n}})$ are known as functions of $\nu_{\text{b}}$, $\nu_{\text{l}}$, $\nu_{\text{p}}$. Indeed, according to the principle of detailed balance [62,63] in ternary nucleation:

$$W_{\text{b}}^{+}(\nu_{\text{b}}-1,\nu_{\text{l}},\nu_{\text{n}})g_{\text{e}}(\nu_{\text{b}}-1,\nu_{\text{l}},\nu_{\text{n}}) = W_{\text{b}}^{-}(\nu_{\text{b}},\nu_{\text{l}},\nu_{\text{n}})g_{\text{e}}(\nu_{\text{b}},\nu_{\text{l}},\nu_{\text{n}}), \tag{3.50}$$

$$W_{\text{l}}^{+}(\nu_{\text{b}},\nu_{\text{l}}-1,\nu_{\text{n}})g_{\text{e}}(\nu_{\text{b}},\nu_{\text{l}}-1,\nu_{\text{n}}) = W_{\text{l}}^{-}(\nu_{\text{b}},\nu_{\text{l}},\nu_{\text{n}})g_{\text{e}}(\nu_{\text{b}},\nu_{\text{l}},\nu_{\text{n}}), \tag{3.51}$$

$$W_{\text{n}}^{+}(\nu_{\text{b}},\nu_{\text{l}},\nu_{\text{n}}-1)g_{\text{e}}(\nu_{\text{b}},\nu_{\text{l}},\nu_{\text{n}}-1) = W_{\text{n}}^{-}(\nu_{\text{b}},\nu_{\text{l}},\nu_{\text{n}})g_{\text{e}}(\nu_{\text{b}},\nu_{\text{l}},\nu_{\text{n}}). \tag{3.52}$$

The application of these equations allows one to derive the expression

$$
\begin{aligned}
g_{\text{e}}(\nu_{\text{b}},\nu_{\text{l}},\nu_{\text{n}}) = \frac{1}{3}(\rho_{\text{bu}}+\rho_{\text{lu}}+\rho_{\text{nu}})\Bigg[ & \frac{W_{\text{b}}^{+}(1,\nu_{\text{l}},\nu_{\text{n}})W_{\text{l}}^{+}(1,0,\nu_{\text{n}})W_{\text{n}}^{+}(1,0,0)}{W_{\text{b}}^{+}(\nu_{\text{b}},\nu_{\text{l}},\nu_{\text{n}})W_{\text{l}}^{+}(1,\nu_{\text{l}},\nu_{\text{n}})W_{\text{n}}^{+}(1,0,\nu_{\text{n}})} \\
& \times \prod_{i=1}^{\nu_{\text{b}}-1}\frac{W_{\text{b}}^{+}(\nu_{\text{b}}-i+1,\nu_{\text{l}},\nu_{\text{n}})}{W_{\text{b}}^{-}(\nu_{\text{b}}-i+1,\nu_{\text{l}},\nu_{\text{n}})}\prod_{j=1}^{\nu_{\text{l}}}\frac{W_{\text{l}}^{+}(1,\nu_{\text{l}}-j+1,\nu_{\text{n}})}{W_{\text{l}}^{-}(1,\nu_{\text{l}},-j+1,\nu_{\text{n}})}\prod_{k=1}^{\nu_{\text{n}}}\frac{W_{\text{n}}^{+}(1,0,\nu_{\text{n}}-k+1)}{W_{\text{n}}^{-}(1,0,\nu_{\text{n}}-k+1)} \\
& + \frac{W_{\text{b}}^{+}(0,1,\nu_{\text{n}})W_{\text{l}}^{+}(\nu_{\text{b}},1,\nu_{\text{n}})W_{\text{n}}^{+}(0,1,0)}{W_{\text{b}}^{+}(\nu_{\text{b}},1,\nu_{\text{n}})W_{\text{l}}^{+}(\nu_{\text{b}},\nu_{\text{l}},\nu_{\text{n}})W_{\text{n}}^{+}(0,1,\nu_{\text{n}})} \\
& \times \prod_{i=1}^{\nu_{\text{b}}}\frac{W_{\text{b}}^{+}(\nu_{\text{b}}-i+1,1,\nu_{\text{n}})}{W_{\text{b}}^{-}(\nu_{\text{b}}-i+1,\nu_{\text{l}},\nu_{\text{n}})}\prod_{j=1}^{\nu_{\text{l}}-1}\frac{W_{\text{l}}^{+}(\nu_{\text{b}},\nu_{\text{l}}-j+1,\nu_{\text{n}})}{W_{\text{l}}^{-}(\nu_{\text{b}},\nu_{\text{l}}-j+1,\nu_{\text{n}})}\prod_{k=1}^{\nu_{\text{n}}}\frac{W_{\text{n}}^{+}(0,1,\nu_{\text{n}}-k+1)}{W_{\text{n}}^{-}(0,1,\nu_{\text{n}}-k+1)} \\
& + \frac{W_{\text{b}}^{+}(0,\nu_{\text{l}},1)W_{\text{l}}^{+}(0,0,1)W_{\text{n}}^{+}(\nu_{\text{b}},\nu_{\text{l}},1)}{W_{\text{b}}^{+}(\nu_{\text{b}},\nu_{\text{l}},1)W_{\text{l}}^{+}(0,\nu_{\text{l}},1)W_{\text{n}}^{+}(\nu_{\text{b}},\nu_{\text{l}},\nu_{\text{n}})} \\
& \times \prod_{i=1}^{\nu_{\text{b}}}\frac{W_{\text{b}}^{+}(\nu_{\text{b}}-i+1,\nu_{\text{l}},1)}{W_{\text{b}}^{-}(\nu_{\text{b}}-i+1,\nu_{\text{l}},1)}\prod_{j=1}^{\nu_{\text{l}}}\frac{W_{\text{l}}^{+}(0,\nu_{\text{l}}-j+1,1)}{W_{\text{b}}^{-}(0,\nu_{\text{l}}-j+1,1)}\prod_{k=1}^{\nu_{\text{n}}-1}\frac{W_{\text{n}}^{+}(\nu_{\text{b}},\nu_{\text{l}},\nu_{\text{n}}-k+1)}{W_{\text{n}}^{-}(\nu_{\text{b}},\nu_{\text{l}},\nu_{\text{n}}-k+1)}\Bigg]
\end{aligned} \tag{3.53}
$$

(for details, see Appendix 3E).

In the present theory, the function $G(\nu_{\text{b}}, \nu_{\text{l}}, \nu_{\text{n}}) = -\ln[g_{\text{e}}(\nu_{\text{b}}, \nu_{\text{l}}, \nu_{\text{n}})/(\rho_{\text{bu}} + \rho_{\text{lu}} + \rho_{\text{nu}})]$ (with $g_{\text{e}}(\nu_{\text{b}}, \nu_{\text{l}}, \nu_{\text{n}})$ defined by Equation 3.53) plays a role similar to the free energy of cluster formation in $k_{\text{B}}T$ units in multicomponent CNT [64,65]. It determines a surface in a four-dimensional space which, under appropriate conditions (i.e., high enough metastability of the initial phase), is expected to have the shape of a hyperbolic paraboloid with at least one "saddle" point at the coordinates $\nu_{\text{bc}}$, $\nu_{\text{lc}}$, and $\nu_{\text{nc}}$ (hereinafter the subscript c marks quantities at the saddle point). In the vicinity of the saddle point the function $G(\nu_{\text{b}}, \nu_{\text{l}}, \nu_{\text{n}})$ can be accurately represented by the bilinear form

$$G(\nu_{\text{b}},\nu_{\text{l}},\nu_{\text{n}}) = G_{\text{c}} + \frac{1}{2}\sum_{i,j=\text{b},\text{l},\text{n}} G_{ij}''(\nu_i-\nu_{ic})(\nu_j-\nu_{jc}), \tag{3.54}$$

where

$$G''_{ij} = \partial^2 G/\partial v_i \partial v_j |_c \quad (i, j = \text{b}, \text{l}, \text{n}). \tag{3.55}$$

In the vicinity of the saddle point, the discrete master equation (Equation 3.46) can be reduced to the differential kinetic equation of nucleation,

$$\frac{\partial g(v_\text{b}, v_\text{l}, v_\text{n}, t)}{\partial t} = \sum_{i=\text{b,l,n}} W_{ic}^{+} \frac{\partial}{\partial v_i}\left[\frac{\partial}{\partial v_i} + \frac{\partial G}{\partial v_i}\right] g(v_\text{b}, v_\text{l}, v_\text{n}, t) \tag{3.56}$$

(see Appendix 3F).

To find the steady-state nucleation rate, Equation 3.56 has to be solved in the vicinity of the saddle point subject to boundary conditions, one expressing the assumption that small clusters are in equilibrium and the other the nonexistence of too large clusters. Using the method proposed by Trinkaus [64], one can obtain the steady-state rate of ternary nucleation,

$$J_\text{s} = \frac{|\lambda_0|/2\pi k_\text{B} T}{\sqrt{-\det(G''/2\pi k_\text{B} T)}} g_\text{e}(v_\text{bc}, v_\text{lc}, v_\text{nc}), \tag{3.57}$$

where $\mathbf{G}''$ is the matrix of second derivatives (defined by Equation 3.55) of the function $G(v_\text{b}, v_\text{l}, v_\text{n})$ and $\lambda_0$ is a negative eigenvalue of the matrix $\mathbf{A}\cdot\mathbf{G}''$, with the elements of the diagonal matrix $\mathbf{A}$ of the absorption rates given by $A_{ij} = \delta_{ij} W_i^{+}$ $(i, j = \text{b}, \text{l}, \text{n})$ and $\delta_{ij}$ being the Kronecker delta. Note that although the form of the expression for $J_\text{s}$ in Equation 3.57 is identical to that of Trinkaus [64], the latter was obtained (and applied to binary and ternary nucleation) in the framework of CNT. The crucial difference is hidden in the method for obtaining the equilibrium distribution $g_\text{e}(v_\text{b}, v_\text{l}, v_\text{n})$. In our method, it is obtained by using a first passage time analysis (and the principle of detailed balance) and is given by Equation 3.53, while in the CNT (whereupon the use of Trinkaus' expression had been previously based) the equilibrium distribution would have the form $(\rho_\text{bu} + \rho_\text{lu} + \rho_\text{nu})\exp[-F(v_\text{b}, v_\text{l}, v_\text{n})]$, where $F(v_\text{b}, v_\text{l}, v_\text{n})$ is the free energy of formation of the cluster in $k_\text{B}T$ units, derived by using the concept of surface tension.

#### 3.2.3.4 Evaluation of Protein Folding Time

Knowing the emission and absorption rates as functions of $v$ as well as the nucleation rate $J_\text{s}$, one can estimate the time $t_\text{f}$ necessary for the protein to fold via nucleation. Roughly speaking, the protein folding (via nucleation) consists of two stages. During the first stage, a critical cluster (nucleus) of native residues is formed (nucleation proper). Until the nucleus forms, the emission rate $W^-$ is larger than $W^+$, but the cluster can still attain the critical size and composition by means of fluctuations. At the second stage the nucleus grows via regular absorption of native residues which dominates their emission, $W^- < W^+$. Thus, the folding time is given by

$$t_\text{f} \simeq t_\text{n} + t_\text{g}, \tag{3.58}$$

where $t_\text{n}$ is the time necessary for one critical cluster to nucleate within a compact (but still unfolded) protein and $t_\text{g}$ is the time necessary for the nucleus to grow up to the maximum size, i.e., the size of the entirely folded protein.

The time $t_\text{n}$ of the first nucleation event can be estimated as

$$t_\text{n} \simeq 1/J_\text{s} V_0, \tag{3.59}$$

where $V_0$ is the volume of the unfolded protein in a compact state. The growth time $t_\text{g}$ can be found by solving the differential equation

$$\frac{dv_\text{b}}{dt} = W_\text{b}^{+}(v_\text{b}, v_\text{l}(v_\text{b}), v_\text{n}(v_\text{b})) - W_\text{b}^{-}(v_\text{b}, v_\text{l}(v_\text{b}), v_\text{n}(v_\text{b})) \tag{3.60}$$

subject to the initial conditions $\nu_b = \nu_{bc}$ at $t = 0$ and $\nu_b = N_b$ at $t = t_g$. The functions $\nu_l = \nu_l(\nu_b)$ and $\nu_n = \nu_n(\nu_b)$ determine the growth path in parametric form and can be found as the solution of the simultaneous differential equations

$$\frac{d\nu_l}{d\nu_b} = \frac{W_l^+(\nu_b, \nu_l, \nu_n) - W_l^-(\nu_b, \nu_l, \nu_n)}{W_b^+(\nu_b, \nu_l, \nu_n) - W_b^-(\nu_b, \nu_l, \nu_n)}, \tag{3.61}$$

$$\frac{d\nu_n}{d\nu_b} = \frac{W_n^+(\nu_b, \nu_l, \nu_n) - W_n^-(\nu_b, \nu_l, \nu_n)}{W_b^+(\nu_b, \nu_l, \nu_n) - W_b^-(\nu_b, \nu_l, \nu_n)}. \tag{3.62}$$

The solution of Equation 3.60 is given by the integral

$$t_g \simeq \int_{\nu_{bc}}^{N_b} \frac{d\nu}{W_b^+(\nu_b, \nu_l(\nu_b), \nu_n(\nu_b)) - W_b^-(\nu_b, \nu_l(\nu_b), \nu_n(\nu_b))}. \tag{3.63}$$

Note that $t_g$ provided by Equation 3.63 can be considered only as a rough estimate for two reasons. First, the actual most probable evolution path of clusters during nucleation and growth may not pass through the saddle point (as well known [64,65] in the binary CNT). Second, in the vicinity of the saddle point $d\nu_i/dt$ ($i$ = $b$, $l$, $n$) is not equal to $W_i^+ - W_i^-$ but contains also stochastic terms reflecting the fluctuations. Nevertheless, Equation 3.63 is expected to provide a reasonable estimate for $t_g$.

### 3.2.4 Numerical Evaluations

In order to illustrate the above theory by numerical calculations we have applied it to the folding of the bovine pancreatic ribonuclease (BPR). According to Elliott and Elliott [66], the 20 main amino acids can be roughly classified as neutral (glycine), hydrophobic (alanine, valine, leucine, isoleucine, methionine, phenylalanine, tyrosine, tryptophan), and hydrophylic (aspartic acid, glutamic acid, lysine, arginine, histidine, asparagine, glutamine, serine, threonine, cysteine, and proline). According to this classification, BPR consists of 40 hydrophobic, 81 hydrophilic, and 3 neutral amino acids, thus totaling 124 beads. Consequently, we have considered a model heteropolymer consisting of the same numbers of hydrophobic, hydrophilic, and neutral beads as they are present in the BPR. The presence of water molecules was not taken into account explicitly but was assumed to be included into the model via the potential parameters.

The interactions between any two nonadjacent (i.e., at least three links apart) beads were modeled by Lennard–Jones (LJ) potentials (see Equation 3.34), whereas the potential due to the dihedral angle $\delta$ was represented by Equation 3.35. Using the optimized potentials for liquid simulations for proteins developed by Jorgensen and Tirado-Rives [67], the parameters in the LJ potential in Equation 3.33 were chosen as $\epsilon_{bb}/kT = 1$, $\epsilon_{ll} = (0.137/0.143)\epsilon_{bb}$, $\epsilon_{nn} = (0.151/0.143)\epsilon_{bb}$, and $\eta$ = 4.3 Å. As suggested by Honeycutt and Thirumalai [7] and Guo and Thirumalai [16], the potential parameters for the dihedral angle potential were selected so that $\epsilon'_\delta = 0$ and $\epsilon''_\delta = 0.2\epsilon_{bb}$ when there are two successive neutral beads in a quartet (forming a dihedral angle); otherwise, $\epsilon'_\delta = \epsilon''_\delta = 1.2\epsilon_{bb}$. The typical density of the folded protein was evaluated using data from Refs. [68,69], and was set to $\rho_f\eta^3 = 0.57$, whereas the typical density of the unfolded protein in the compact configuration was set to be $\rho_u = 0.2\rho_f$. Taking into account the results of Refs. [70,71], the diffusion coefficients in the ipw and the opw were assumed to be related as $D^{iw}\rho_f = D^{ow}\rho_u$. Because of the lack of reliable data on the diffusion coefficient of a residue in a protein chain, $D^{iw}$ was assumed to vary between $10^{-6}$ and $10^{-8}$ cm$^2$/s (the diffusion coefficients for all three components were assumed to be equal, i.e., $D^{iw} = D_i^{iw}$ $(i = \text{b, l, n})$ and $D^{ow} = D_i^{ow}$ $(i = \text{b, l, n})$).

The numerical calculations for evaluating the folding time of the protein have been carried out as follows. First, using Equation 3.36, the average pairwise potential $\phi_i(r)$ for a bead of each type ($i$ = b, l, n) is calculated as a function of $r$ (distance from the center of a cluster) for a cluster with the numbers of beads therein, $\nu_b$, $\nu_l$, and $\nu_n$, taken on a three-dimensional grid which must be fine enough to be able to construct $\phi_i(r|\nu_b, \nu_l, \nu_n)$, the average pairwise potential at fixed $r$ as a function of three continuous variables, $\nu_b$, $\nu_l$, and $\nu_n$ [note that the integration in Equation 3.36 can be carried out analytically if the densities $\rho_i$ ($i$ = b, l, n) inside the cluster are assumed to be uniform]. This function represents the first term in Equation 3.37 for the potential field around a cluster. It is also needed (see Equations 3.39 through 3.41) for calculating the average dihedral potential $\bar{\phi}_i^{\delta}(r)$ as a function of $r$ for given $\nu_b$, $\nu_l$, and $\nu_n$, using a three-dimensional grid. The grid must be fine enough to construct $\bar{\phi}_i^{\delta}(r\,|\,\nu_b,\nu_l,\nu_n)$, the average dihedral potential as a function of three continuous variables, $\nu_b$, $\nu_l$, and $\nu_n$, at fixed $r$. This function represents the second term in Equation 3.37 for the potential field around a cluster. Thus, one can numerically construct $\psi_i(r, \nu_b, \nu_l, \nu_n)$, the potential field around a cluster as a function of four variables, $r$, $\nu_b$, $\nu_l$, and $\nu_n$. Knowing this function, one can then calculate the emission and absorption rates of a cluster (as outlined in Section 3.2.3.2 and Appendix 3D) as functions of three variables, $\nu_b$, $\nu_l$, and $\nu_n$. (The accuracy in building these functions is extremely important because thereupon depends the accuracy with which the function $G(\nu_b, \nu_l, \nu_n)$ and its "saddle point" (i.e., the parameters of the critical cluster) are determined. This, in turn, has a major impact on the accuracy of the prediction of the nucleation rate and hence folding time.) Knowing the emission and absorption rates $W_i^+(\nu_b,\nu_l,\nu_n)$ and $W_i^-(\nu_b,\nu_l,\nu_n)$, the equilibrium distribution $g_e(\nu_b, \nu_l, \nu_n)$ is constructed via Equation 3.53 and the nucleation rate is calculated with Equation 3.57, which contains the function $G(\nu_b, \nu_l, \nu_n) = -\ln[g_e(\nu_b, \nu_l, \nu_n)/(\rho_{bu} + \rho_{lu} + \rho_{nu})]$. The time $t_n$, necessary for the first nucleation event to occur and the time $t_g$ necessary for the nucleus to grow to its maximum size are evaluated with Equations 3.59 and 3.63, respectively. Finally, the folding time $t_f$ is evaluated as $t_f = t_n + t_g$.

Figure 3.6 presents the shapes of the potentials $\phi_i(r)$ (lower dashed curve), $\bar{\phi}_i^{\delta}(r)$ (upper dashed curve), and $\psi_i(r)$ (solid curve) for a bead around a cluster (with $\nu_b = 29$, $\nu_l = 50$, and $\nu_n = 2$) as functions of the distance $r$ from the cluster center [the curves are shown only for a hydrophobic bead ($i$ = b) because for the hydrophilic ($i$ = l) and neutral ($i$ = n) ones they have a very similar shape]. The potential $\phi_i(r)$ is due to the pairwise interactions of Lennard–Jones type and has a shape reminiscent thereof. Previous applications [36–41] of the mean first passage time analysis to nucleation employed this kind of potential well around a cluster.

The average dihedral potential (assigned to a selected bead) $\bar{\phi}_i^{\delta}(r)$ has a maximum value at the cluster surface and decreases monotonically with increasing $r$ until it becomes constant for $r \geq R+\tilde{d}$, where $\tilde{d}$ is the maximum distance between beads 1 and 6 (or beads 1 and 7) that depends on $\eta$ and $\beta_0$:

$$\tilde{d} \simeq \eta\sqrt{3-2\cos\beta_0+2\sqrt{2(1-\cos\beta_0)}\,\sin\frac{\beta_0}{2}}. \tag{3.64}$$

Such a behavior of $\bar{\phi}_i^{\delta}(r)$ can be thought to be a consequence of an increase in the entropic contribution to the free energy of the heteropolymer chain as the selected bead (1) approaches the cluster surface for $r < R+\tilde{d}$. It occurs because the configurational space available to the neighboring beads (2–7) becomes increasingly restricted. This piece of the heteropolymer chain (beads 1–7) would not feel the presence of the cluster any more at $r > R+\tilde{d}$ if there were no pairwise interactions.

As a result of the combination of $\phi_i(r)$, $\bar{\phi}_i^{\delta}(r)$, and $\phi_{cp}$, the overall potential $\psi_i(r)$ has a double well shape: the inner well is separated by a potential barrier from the outer well. The geometric characteristics of the wells (widths, depths, etc.) and the height and location of the barrier between them depend on the interaction parameters $\epsilon_{bb}$, $\epsilon_{ll}$, $\epsilon_{nn}$, $\epsilon_{\delta}'$ and $\epsilon_{\delta}''$. The larger the ratio $\epsilon_{\delta}/\epsilon_{bb}$, the higher the barrier between the wells, the wider the ipw, and the narrower the opw. Note that the barrier has different heights for the ipw and opw beads. For a given $N_i$, the location $r_{cp}$ of the outer boundary of the opw is determined by the size of the cluster and densities $\rho_{if}$ and $\rho_{iu}$. The existence of an opw allows one to consider the absorption of a bead by the cluster as an escape of the bead from the opw by crossing over the barrier into the ipw. One can therefore use the mean first passage time analysis for the determination not only of the emission rate but also of the rate of absorption of beads by the cluster. Since the use of the traditional expression for the absorption rate (based on the gas-kinetic theory) [58] is rather inadequate for the cluster growth within the protein, the double well shape of $\psi_i(r)$ is of crucial importance in our model for the nucleation mechanism of protein folding.

In Figures 3.7 and 3.8, the average dihedral potential (assigned to a selected bead) $\bar{\phi}_i^{\delta}(r)$ is plotted as a function of $r$ for all three kinds of beads [hydrophobic (a), hydrophilic (b), and neutral (c)] for a cluster characterized by $\nu_b = 29$, $\nu_l = 50$, and $\nu_n = 2$. The points represent the actual numerical results obtained by using $1 \times 10^6$ point MC integration in calculating the sevenfold integrals in Equations 3.42 and 3.43. The results in Figure 3.7 correspond to the Boltzmann factor $\exp\left[-\left(\phi_i^{\delta 27}(r)\right)/k_BT\right]$ (see Equation 3.39) in Equations 3.40 and 3.41, whereas the results in Figure 3.8 are obtained with the Boltzmann factor $\exp\left[-\left(\phi_i^{\delta}(r)\right)/k_BT\right]$ in Equations 3.40 and 3.41. The vertical dashed lines correspond to $R+\tilde{d}$, and $\tilde{\phi}_i^{\delta}(r)$ is expected to be constant for $r > R+\tilde{d}$. The comparison of Figures 3.7 and 3.8 suggests that the use of the truncated Boltzmann factor $\exp\left[-\tilde{\phi}_i^{\delta}(r)/k_BT\right]$ instead of the physically more appropriate $\exp\left[-\left(\phi_i^{\delta 27}(r)\right)/k_BT\right]$ slightly affects (decreases)

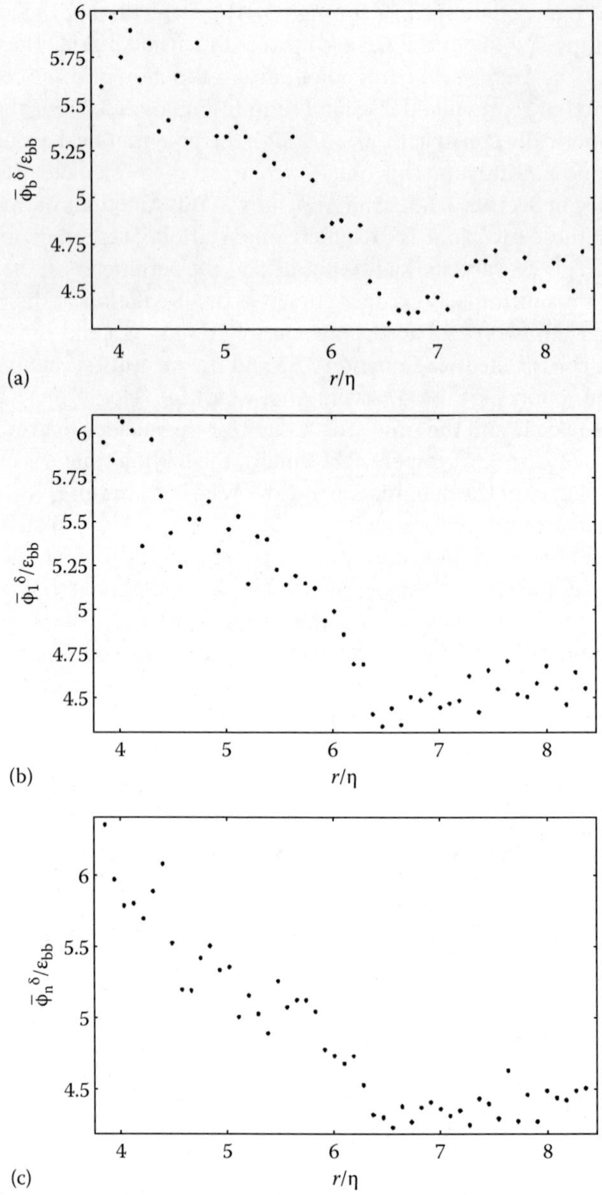


Figure 3.7

Average dihedral potential (assigned to a single bead of type $i$) $\bar{\phi}_i^{\delta}(r)$ ($i$ = b, l, n) as a function of $r$ for a cluster consisting of $v_b = 29$, $v_l = 50$, and $v_n = 2.4$ hydrophobic, hydrophilic, and neutral beads, respectively. The points represent the actual numerical results obtained by using Monte Carlo integration in Equations 3.40 and 3.41 with the Boltzmann factor $\exp\left[-\left(\phi_i^{\delta 27}/k_B T\right)\right]$ (see Equation 3.39) for the (a) hydrophobic, (b) hydrophilic, and (c) neutral beads.

the average dihedral potential only in the close vicinity of the cluster (by about 5%). The uncertainty in the results (due to the MC integration), however, makes this effect rather insignificant.

The emission and absorption rates of a cluster, $W^-$ and $W^+$, respectively, are functions of three independent variables of its state. Figure 3.9 presents the dependence of $W^-$ and $W^+$ (as calculated by the mean first passage time analysis, Equation 3.44) on the total number of beads in a cluster, at constant mole fractions of the hydrophobic and hydrophylic beads therein. Figure 3.9a presents the case of $\chi_b = 0.48$ and $\chi_l = 0.49$, whereas Figure 3.9b is for $\chi_b = 0.37$ and $\chi_l = 0.61$. The dashed (marked with b), dash-dotted (marked with l), and solid curves are for the hydrophobic, hydrophilic, and neutral beads, respectively. Note that, for the case shown in Figure 3.9a, the emission rates (for all three components) are always greater than the absorption rates for all cluster sizes, i.e., the clusters with such a composition ($\chi_b = 0.48$ and $\chi_l = 0.49$) have a strong

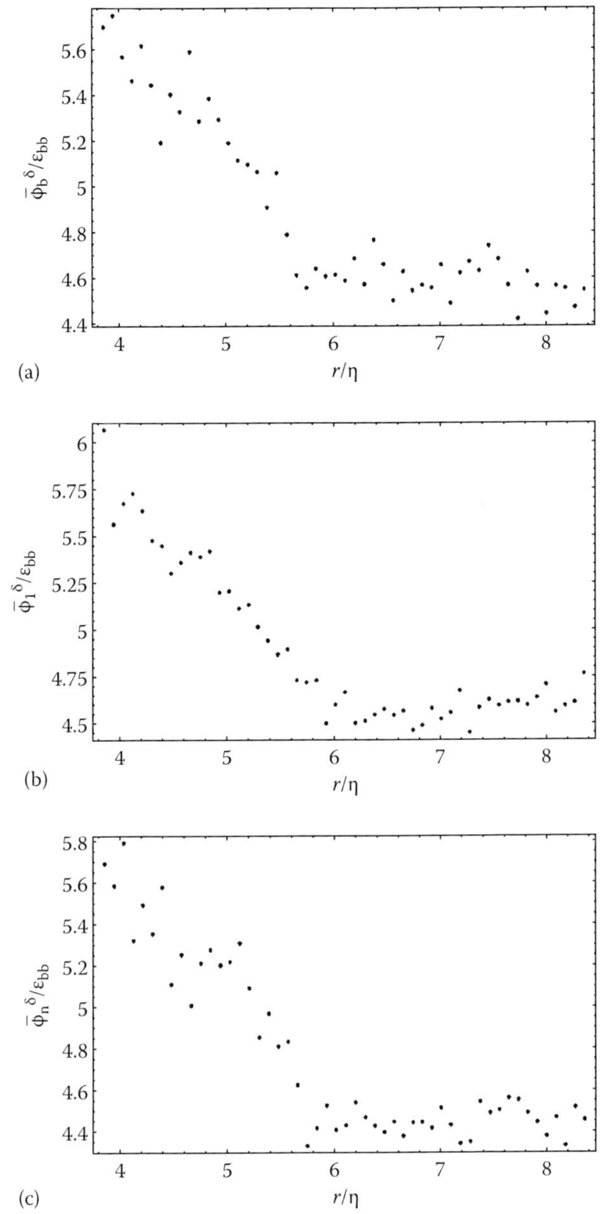


Figure 3.8

Same as in Figure 3.3, except for the Boltzmann factor $\exp\left[-\left(\phi_i^{\delta}/k_BT\right)\right]$ which was used in the Monte Carlo integration in Equations 3.40 and 3.41 instead of $\exp\left[-\left(\phi_i^{\delta 27}/k_BT\right)\right]$.

tendency to disaggregate. However, for the case shown in Figure 3.9b ($\chi_b = 0.37$ and $\chi_l = 0.61$), the emission rates are greater than the absorption rates, $W^- > W^+$, for smaller clusters, whereas for larger clusters absorption dominates emission, $W^- < W^+$. The location of the simultaneous intersection of the three pairs of functions $W_i^-$ and $W_i^+$ ($i = $ b, l, n) determines the characteristics of the critical cluster, $\nu_{bc}$, $\nu_{lc}$, and $\nu_{nc}$. Note that both $W^-$ and $W^+$ increase with increasing $\nu$, but $W^-$ increases roughly linearly with $\nu$ whereas $W^+$ shoots up by orders of magnitude after the intersection. This occurs because the width of the opw quickly decreases as the cluster grows, whereas the outer height of the barrier between the ipw and opw remains virtually unchanged. For this reason it becomes increasingly easier for a bead located in the opw to cross the barrier and to fall into the ipw.

The behavior of $W^-$ and $W^+$ also explains our estimates for the characteristic times of the first nucleation event $t_n$, growth time $t_g$, and total folding time $t_f$ provided by Equations 3.60, 3.61, and 3.63 (nucleation rate $J_s$ was evaluated by using Equation 3.57). These values are very sensitive to $D^{iw}$, $D^{ow}$, and coordinates of the

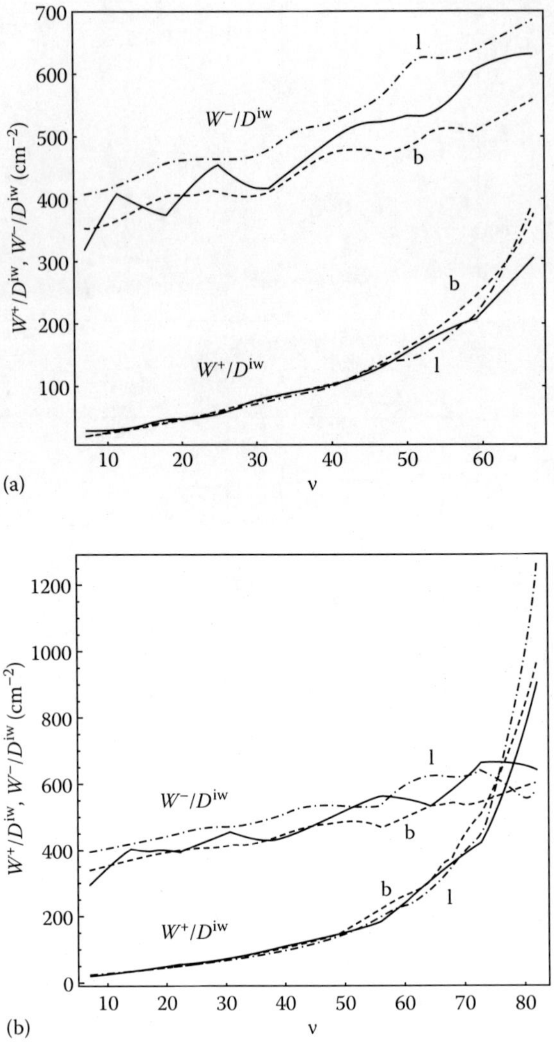


Figure 3.9

The dependence of $W_i^-$ and $W_i^+ (i = \text{b, l, n})$ on the total number of beads in a cluster, $\nu$, at constant mole fractions of the hydrophobic and hydrophylic beads therein: (a) $\chi_b = 0.48$, $\chi_l = 0.49$; (b) $\chi_b = 0.37$, $\chi_l = 0.61$. The dashed curves (marked with "b") are for hydrophobic beads, the dash-dotted ones (marked with "l") for hydrophilic, and the solid curves for neutral ones. The lower series is for $W_i^+$, and the upper one for $W_i^-$. The location of the simultaneous intersection of the three pairs of functions $W_i^-$ and $W_i^+$ $(i = \text{b, l, n})$ determines the characteristics of the critical cluster.

saddle point (which themselves depend on the diffusion coefficients only through the ratio $D^{ow}/D^{iw}$), but the time of protein folding is determined mainly by the time necessary for the first nucleation event to occur, i.e., $t_n \gg t_g$. This occurs because the increase of the cluster size from $\nu = 1$ to the saddle point takes place only via fluctuations which have to overcome the natural tendency of a small cluster to decay (because $W^+ < W^-$). For supercritical clusters, $W^+$ quickly overwhelms $W^-$ and the fluctuations are unable to impede the natural tendency of the cluster to grow (strictly speaking, this is true only sufficiently far away from the saddle point, but for rough estimates Equation 3.63 is acceptable). The characteristics of the critical cluster can be evaluated by numerically solving three simultaneous equations, $\left(W_i^-(\nu_b,\nu_l,\nu_n) = W_i^+(\nu_b,\nu_l,\nu_n)\ (i = \text{b, l, n})\right)$. Alternatively, $\nu_{bc}$, $\nu_{lc}$, and $\nu_{nc}$ can be found as the solution of three simultaneous equations $\partial G(\nu_b, \nu_l, \nu_n)/\partial \nu_i = 0$ $(i = \text{b, l, n})$ satisfying the condition $\det(\mathbf{G}'') < 0$. We used the former method since it requires less computational effort. For the protein considered (BPR with $N_0 = 124$) with the above choices of the system parameters, our model predicts $\nu_{bc} \simeq 23$, $\nu_{lc} \simeq 51$, and $\nu_{nc} \simeq 2$ (corresponding to $R_c \simeq 3.3\eta$), the free energy of nucleus formation being about $15k_BT$. According to the classical Gibbs' expression, the free energy of nucleus formation is equal to 1/3 × surface tension × surface area. Thus, our results indicate that the quantity equivalent to the "surface tension" in the CNT-based model of nucleation in a protein should be about $0.8k_BT/(4\pi\lambda^2)$, i.e., somewhat smaller than the previous estimates of the latter [16,31] [roughly $k_BT/(4\pi\lambda^2)$]. Finally, for the diffusion coefficient $D^{iw}$ in the range of $10^{-6}$ to $10^{-8}$ cm$^2$/s, our model estimates the characteristic time of protein folding to be in the range of a second to hundred seconds, which is in agreement with experimental data [5].

### 3.2.5 Conclusions

So far, most of the work on protein folding has been carried out by using either MC or MD simulations. Both experiments and simulations have suggested that there are multiple pathways for protein folding, one of which involves a nucleation mechanism. Its theory was first approached [14,31–33] by means of the classical thermodynamics, calculating the free energy of formation of a cluster of protein residues in the framework of the capillarity approximation, much like the CNT does. Such an approach involves the use of the concept of surface tension for a cluster consisting of protein residues which is an intrinsically ill-defined physical quantity and can be considered at best as an adjustable parameter.

Recently [35], we have presented a new, microscopic model for the nucleation mechanism of protein folding. The model is based on pairwise interactions and configurational potentials (in which the protein residues are involved) whereof the parameters can be clearly defined. The rates of emission and absorption of residues by the cluster are determined by using a first passage time analysis [36–41]. Once these rates are found as functions of the cluster size, one can develop a self-consistent kinetic theory for the nucleation mechanism of folding of a protein. The time necessary for the protein to fold can be evaluated as the sum of the times necessary for the appearance of the first nucleus and the time necessary for the nucleus to grow to the maximum size (of the folded protein in the native state).

However, for the sake of simplicity in our previous work [35] we represented a protein as a heteropolymer consisting of only hydrophobic and hydrophilic beads without neutral ones. In addition, we assumed that the composition of a cluster remained constant and equal to the overall composition of the protein at all times. In the present paper, we have removed these simplifying assumptions. A protein has been modeled as a heteropolymer consisting of three kinds of beads (hydrophobic, hydrophilic, and neutral), and the composition of a cluster has been allowed to vary during folding. We have thus developed a model for the nucleation mechanism of protein folding in terms of ternary nucleation by using a first passage time analysis [36–41] that accounts for the particular shape of the potential around the cluster of native residues. As in our previous model, the main idea underlying the new model consists of averaging the dihedral potential in which a selected residue is involved over all the possible configurations of all neighboring residues along the protein chain. The combination of the average dihedral potential with the average pairwise potential of the selected residue and with a confining potential caused by the bonds between the residues leads to an overall potential around the cluster that has a double well shape. Residues in the inner (closer to the cluster) well are considered as belonging to the folded cluster, whereas those in the outer well are treated as belonging to the unfolded part of the protein. Transitions of residues from the inner well into the outer one and vice versa are considered as elementary emission and absorption events, respectively. The double well character of the potential field around the cluster allows one to determine the rates of both emission and absorption of residues by the cluster using a first passage time analysis. Once these rates are found as functions of the cluster size, one can develop a self-consistent kinetic theory for the nucleation mechanism of folding of a protein. The model allows one to evaluate the size and composition of the nucleus and the protein folding time. The latter is evaluated as the sum of the times necessary for the first nucleation event to occur and for the nucleus to grow to the maximum size of the folded protein.

For numerical illustration we have considered the folding of BPR protein consisting of 124 residues whereof 40 can be classified as hydrophobic, 81 as hydrophilic, and 3 as neutral. The composition of the cluster during its formation and growth was allowed to vary independently (to a limited degree, of course) of the composition of the whole protein, so that the model is developed by using the formalism of ternary nucleation. The size and composition of the critical cluster and the folding time predicted by the model depend very much on the parameters of the system, such as the interaction parameters, densities of the protein in the unfolded (but compact) and folded states, diffusion coefficients therein. With appropriate choices of the interaction parameters and densities and the diffusion coefficient of native residues in the range of $10^{-6}$ to $10^{-8}$ cm$^2$/s, the characteristic time of protein folding was estimated to be in the range of 1–100 s. These values are consistent with the experimental data [5] on typical folding times of proteins as well as with estimates provided by other theoretical models [27,28,31] and simulations [7,16]. Note that the predictive power of our model is mostly determined by the accuracy of MC integration (in Equations 3.42 and 3.43) and the availability of reliable data on the diffusion coefficients and other parameters of the protein.

As a next step in developing our model, it seems to be appropriate to take into account that some protein residues may carry charges which presumably play an important role in the formation of the tertiary structure of a protein under *in vivo* conditions. Besides an increase in the computational efforts there seem to be no conceptual difficulty in developing the model in this direction.

## Appendix 3C: Derivation of Equations 3.42 and 3.43 from Equations 3.40 and 3.41

Let us consider a cluster of radius $R$ consisting of protein residues stabilized by correct tertiary contacts (see Figure 3.5). This will constitute the folded part of the protein. In the unfolded part of the heteropolymer, let us consider beads 1–7 around a cluster and denote their Cartesian coordinates by $x_i$, $y_i$, and $z_i$ ($i = 2, \ldots, 7$). It is convenient to choose the origin of the Cartesian system of coordinates in such a way that $x_1 = 0$, $y_1 = 0$, and $z_1 = r$, where $r$ is the distance between bead 1 and the cluster center.

For bead 2, it is also convenient to introduce the spherical coordinates $r_2$, $\Theta_2$, and $\varphi_2$, which are related to its Cartesian ones by

$$x_2 = r_2 \sin\Theta_2 \cos\varphi,\ y_2 = r_2 \sin\Theta_2 \sin\varphi,\ z_2 = r_2 \cos\Theta_2. \tag{3C.1}$$

Clearly, since the location of bead 1 is fixed (hence $r$ given), the polar angle $\Theta$ of bead 2 is uniquely determined by $r_2$ due to the constant bond length constraint,

$$\Theta_2(r,r_2) = \arccos\left[\left(r^2 + r_2^2 - \eta^2\right)/2r_2 r\right] \quad (0 \le \Theta_2 \le \pi), \tag{3C.2}$$

whereas the azimuthal angle $0 \le \phi_2 \le 2\pi$. The distance $r_2$ varies in the range $r_{2\min} \le r_2 \le r + \eta$, where $r_{2\min} = \max(R, r - \eta)$, which determines the integration range $L_2$ in Equations 3.42 and 3.43. Thus, the integration with respect to $r_2$ in Equations 3.40 and 3.41 reduces to the integration with respect to $\varphi_2$ and $r_2$ at fixed $\Theta_2$ (given by Equation 3C.2) in Equations 3.42 and 3.43.

For given locations of beads 1 and 2, the possible locations of beads 3 and 4 lie on circles of radius $r_1 = \eta \sin\beta_0$, with their location and orientation completely determined by the coordinates of beads 1 and 2. This is due to the constraints that all bond angles are equal to $\beta_0$ and all bond lengths are equal to $\eta$. Due to the same constraints, if the locations of beads 2 and 4 are given, the possible locations of bead 6 lie on a circle of radius $r_1$, whereof the location and orientation are completely determined by the coordinates of beads 2 and 4. Furthermore, for given locations of beads 1 and 3, the possible locations of bead 5 lie on a circle of radius $r_1$ with the position and orientation completely determined by the coordinates of beads 1 and 3. Finally, for the given locations of beads 3 and 5, the possible locations of bead 7 are on a circle of radius $r_1$, with the location and orientation completely determined by the coordinates of beads 3 and 5.

Let us consider bead $s$ ($s = 3, \ldots, 7$) with unknown coordinates and two other beads, $c$ and $n$ (closest to and next to the closest to bead $s$), with fixed (known) coordinates $x_c$, $y_c$, and $z_c$ and $x_n$, $y_n$, and $z_n$. For example, if $s = 7$, then $c = 5$ and $n = 3$; if $s = 4$, then $c = 2$ and $n = 1$. Bead $s$ lies on a circle of radius $r_1$ with the coordinates of the center,

$$\begin{aligned} x_0 &\equiv x_0(x_n, y_n, z_n, x_c, y_c, z_c,) \\ &= x_n + (x_c - x_n)\frac{\eta\left(1 + |\cos\beta_0|\right)}{\sqrt{(x_c - x_n)^2 + (y_c - y_n)^2 + (z_c - z_n)^2}}, \end{aligned} \tag{3C.3}$$

$$\begin{aligned} y_0 &\equiv y_0(x_n, y_n, z_n, x_c, y_c, z_c) \\ &= y_n + (y_c - y_n)\frac{\eta\left(1 + |\cos\beta_0|\right)}{\sqrt{(x_c - x_n)^2 + (y_c - y_n)^2 + (z_c - z_n)^2}}, \end{aligned} \tag{3C.4}$$

$$z_0 \equiv z_0(x_n, y_n, z_n, x_c, y_c, z_c)$$

$$= z_n + (z_c - z_n)\frac{\eta\left(1+\left|\cos\beta_0\right|\right)}{\sqrt{(x_c - x_n)^2 + (y_c - y_n)^2 + (z_c - z_n)^2}}. \tag{3C.5}$$

The coordinate $x_s$ of bead $s$ can change in the range

$$x_c - x_b \le x_s \le x_c + x_b \;(s = 3,\ldots,7) \tag{3C.6}$$

where

$$x_b \equiv x_b(x_n, y_n, z_n, x_c, y_c, z_c)$$

$$= \eta \sin \beta_0 \sqrt{\frac{\left((y_c - y_0)^2 + (z_c - z_0)^2\right)}{(x_c - x_n)^2 + (y_c - y_n)^2 + (z_c - z_n)^2}}. \tag{3C.7}$$

For a given $x_s$, and, the coordinate $y_s$ of bead $s$ can have only one of two values:

$$y_s^{\pm} \equiv y_s^{\pm}(x_s, x_n, y_n, z_n, x_c, y_c, z_c)$$

$$= y_0 + \frac{-(x_c - x_0)(y_c - y_0)(x_s - x_0)}{(y_c - y_0)^2 + (z_c - z_0)^2} \tag{3C.8}$$

$$\pm \frac{\left|z_c - z_0\right|\sqrt{[(y_c - y_0)^2 + (z_c - z_0)^2]\eta^2 \sin^2\beta_0 - [(x_c - x_n)^2 + (y_c - y_n)^2 + (z_c - z_n)^2](x_s - x_0)^2}}{(y_c - y_0)^2 + (z_c - z_0)^2}.$$

For a given $x_s$, and $y_s^{\pm}$, the coordinate $z_s$ of bead $s$ can have only a single value:

$$z_s^{\pm} \equiv z_s\left(x_s, y_s^{\pm}, x_n, y_n, z_n, x_c, y_c, z_c\right)$$

$$= z_0 - \frac{(x_c - x_0)(x_s - x_0) + (y_c - y_0)\left(y_s^{\pm} - y_0\right)}{z_c - z_0}. \tag{3C.9}$$

Since the volume occupied by the cluster is unavailable to beads 2, 3, …, 7, an additional constraint is imposed on $x_s$, $y_s$, and $z_s$, namely, $x_s^2 + y_s^2 + z_s^2 > R^2$. Subject to this constraint, the double inequality in Equation 3C.6 and Equations 3C.7 through 3C.9 completely determine the integration ranges in Equations 3.42 and 3.43.

In order to transform the function $\phi_{i_1}^{\delta 27}(r) \equiv \phi_{i_1}^{\delta}(r, \mathbf{r}_2, \mathbf{r}_3, \ldots, \mathbf{r}_7)$ into the function $\bar{\phi}_{ijkmn}(r, \varphi_2, r_2, x_3, \ldots, x_7)$, the variables $y_s$ ($s$ = 3, …,7) and $z_s$ ($s$ = 3, …, 7) in the former must be replaced by $y_s^{\pm}$ and $z_s^{\pm}$ in all possible combinations, each of which gives rise to a term in the sums in Equations 3.42 and 3.43 [note that $z_s^{+} = z_s\left(x_s, y_s^{+}, \ldots\right)$ and $z_s^{-} = z_s\left(x_s, y_s^{-}, \ldots\right)$]. The polar angle of bead 2 must be replaced by $\Theta_2(r, r_2)$.

Indeed, at fixed $\mathbf{r}_2$, $\mathbf{r}_3$, $\mathbf{r}_4$, $\mathbf{r}_5$, $\mathbf{r}_6$ (i.e., fixed locations of beads 2–6), the trajectory of bead 7 represents a circle; hence, equivalently, the integrand in Equation 3.40, $I(\mathbf{r}_2, \mathbf{r}_3, \mathbf{r}_4, \mathbf{r}_5, \mathbf{r}_6, x_7, y_7, z_7)$, contains a factor $[\delta(y_7 - y_7^{+})\,\delta(z_7 - z_7^{+}) + \delta(y_7 - y_7^{-})\,\delta(z_7 - z_7^{-})]$, where $\delta(x)$ is Dirac's delta function and $y_7^{\pm}$ and $z_7^{\pm}$ are given by Equations

3C.8 and 3C.9. Therefore, the integration with respect to $y_7$ and $z_7$ in Equation 3.40 results in the replacement of the integrand $I(\mathbf{r}_2, \mathbf{r}_3, \mathbf{r}_4, \mathbf{r}_5, \mathbf{r}_6, x_7, y_7, z_7)$ by the sum $J(\mathbf{r}_2, \mathbf{r}_3, \mathbf{r}_4, \mathbf{r}_5, \mathbf{r}_6, x_7) \equiv I(\mathbf{r}_2, \mathbf{r}_3, \mathbf{r}_4, \mathbf{r}_5, \mathbf{r}_6, x_7, y_7^+, z_7^+) + I(\mathbf{r}_2, \mathbf{r}_3, \mathbf{r}_4, \mathbf{r}_5, \mathbf{r}_6, x_7, y_7^-, z_7^-)$. Next, at fixed $\mathbf{r}_2, \mathbf{r}_3, \mathbf{r}_4, \mathbf{r}_5$ the trajectory of bead 6 is a circle; hence, equivalently, the integrand $J(\mathbf{r}_2, \mathbf{r}_3, \mathbf{r}_4, \mathbf{r}_5, \mathbf{r}_6, x_7)$ contains a factor $[\delta(y_6 - y_6^+)\,\delta(z_6 - z_6^+) + \delta(y_6 - y_6^-)\,\delta(z_7 - z_6^-)]$, where functions $y_6^\pm$ and $z_6^\pm$ are given by Equations 3C.8 and 3C.9. Therefore, the integration with respect to $y_6$ and $z_6$ in Equation 3.40 results in the replacement of the integrand $J(\mathbf{r}_2, \mathbf{r}_3, \mathbf{r}_4, \mathbf{r}_5, x_6, y_6, z_6, x_7)$ by the sum

$$
\begin{aligned}
&K(\mathbf{r}_2,\mathbf{r}_3,\mathbf{r}_4,\mathbf{r}_5,x_6,x_7)\\
&\equiv J\left(\mathbf{r}_2,\mathbf{r}_3,\mathbf{r}_4,\mathbf{r}_5,x_6,y_6^+,z_6^+,x_7\right)\\
&\quad+J\left(\mathbf{r}_2,\mathbf{r}_3,\mathbf{r}_4,\mathbf{r}_5,x_6,y_6^-,z_6^-,x_7\right)\\
&=I\left(\mathbf{r}_2,\mathbf{r}_3,\mathbf{r}_4,\mathbf{r}_5,x_6,y_6^+,z_6^+,x_7,y_7^+,z_7^+\right)\\
&\quad+I\left(\mathbf{r}_2,\mathbf{r}_3,\mathbf{r}_4,\mathbf{r}_5,x_6,y_6^+,z_6^+,x_7,y_7^-,z_7^-\right)\\
&\quad+I\left(\mathbf{r}_2,\mathbf{r}_3,\mathbf{r}_4,\mathbf{r}_5,x_6,y_6^-,z_6^-,x_7,y_7^+,z_7^+\right)\\
&\quad+I\left(\mathbf{r}_2,\mathbf{r}_3,\mathbf{r}_4,\mathbf{r}_5,x_6,y_6^-,z_6^-,x_7,y_7^-,z_7^-\right).
\end{aligned}
$$

Furthermore, at fixed $\mathbf{r}_2, \mathbf{r}_3, \mathbf{r}_4$ the trajectory of bead 5 is a circle; hence, equivalently, the integrand $K(\mathbf{r}_2, \mathbf{r}_3, \mathbf{r}_4, \mathbf{r}_5, x_6, x_7)$ contains a factor $[\delta(y_5 - y_5^+)\,\delta(z_5 - z_5^+) + \delta(y_5 - y_5^-)\,\delta(z_5 - z_5^-)]$, where functions $y_5^\pm$ and $z_5^\pm$ are given by Equations 3C.8 and 3C.9. Therefore, the integration with respect to $y_6$ and $z_6$ in Equation 3.40 results in the replacement of the integrand $K(\mathbf{r}_2, \mathbf{r}_3, \mathbf{r}_4, x_5, y_5, z_5, x_6, x_7)$ by the sum

$$
\begin{aligned}
&L(\mathbf{r}_2,\mathbf{r}_3,\mathbf{r}_4,x_5,x_6,x_7)\equiv K\left(\mathbf{r}_2,\mathbf{r}_3,\mathbf{r}_4,x_5,y_5^+,z_5^+,x_6,x_7\right)+K\left(\mathbf{r}_2,\mathbf{r}_3,\mathbf{r}_4,x_5,y_5^-,z_5^-,x_6,x_7\right)\\
&=I\left(\mathbf{r}_2,\mathbf{r}_3,\mathbf{r}_4,x_5,y_5^+,z_5^+,x_6,y_6^+,z_6^+,x_7,y_7^+,z_7^+\right)\\
&+I\left(\mathbf{r}_2,\mathbf{r}_3,\mathbf{r}_4,x_5,y_5^+,z_5^+,x_6,y_6^+,z_6^+,x_7,y_7^-,z_7^-\right)+I\left(\mathbf{r}_2,\mathbf{r}_3,\mathbf{r}_4,x_5,y_5^+,z_5^+,x_6,y_6^-,z_6^-,x_7,y_7^+,z_7^+\right)\\
&+I\left(\mathbf{r}_2,\mathbf{r}_3,\mathbf{r}_4,x_5,y_5^+,z_5^+,x_6,y_6^-,z_6^-,x_7,y_7^-,z_7^-\right)+I\left(\mathbf{r}_2,\mathbf{r}_3,\mathbf{r}_4,x_5,y_5^-,z_5^-,x_6,y_6^+,z_6^+,x_7,y_7^+,z_7^+\right)\\
&+I\left(\mathbf{r}_2,\mathbf{r}_3,\mathbf{r}_4,x_5,y_5^-,z_5^- x_6,y_6^+,z_6^+,x_7,y_7^-,z_7^-\right)+I\left(\mathbf{r}_2,\mathbf{r}_3,\mathbf{r}_4,x_5,y_5^-,z_5^-,x_6,y_6^-,z_6^-,x_7,y_7^+,z_7^+\right)\\
&+I\left(\mathbf{r}_2,\mathbf{r}_3,\mathbf{r}_4,x_5,y_5^-,z_5^-,x_6,y_6^-,z_6^-,x_7,y_7^-,z_7^-\right).
\end{aligned}
$$

Repeating this procedure twice more (for the integration with respect to $y_4$, $z_4$, and $y_3$, $z_3$), the integrand of Equation 3.40 is reduced to a sum of $32 = 2^5$ terms corresponding to all possible combinations of pairs $y_s^\pm$ and $z_s^\pm$ ($s = 3, \ldots, 7$) in $I(\mathbf{r}_2, x_3, y_3, z_3, x_4, y_4, z_4, x_5, y_5, z_5, x_6, y_6, z_6, x_7, y_7, z_7)$, and Equation 3.40 is reduced to Equation 3.42. The same procedure reduces Equation 3.41 to Equation 3.43.

To further clarify the above procedure, consider, for example, the term with $i = +, j = -, k = +, m = +, n = -$ in the sum in Equations 3.42 and 3.43. In the corresponding integrands, the function $\tilde{\phi}_{+-++-}(\varphi_2, r_2, x_3, \ldots, x_7)$ is obtained from $\phi_i^\delta(r, \mathbf{r}_2, \mathbf{r}_3, \ldots, \mathbf{r}_7)$ as follows: $y_3$ must be replaced by $y_3^- \equiv y_3^-(x_3, x_2, y_2, z_2, x_1, y_1, z_1)$ and $z_3$ by $z_3^- \equiv z_3\left(x_3, y_3^-, x_2, y_2, z_2, x_1, y_1, z_1\right)$, $y_4$ must be replaced by $y_4^+ \equiv y_4^+(x_4, x_1, y_1, z_1, x_2, y_2, z_2)$ and $z_4$ by $z_4^+ \equiv z_4\left(x_4, y_4^+, x_1, y_1, z_1, x_2, y_2, z_2\right)$, $y_5$ must be replaced by $y_5^+ \equiv y_5^+\left(x_5, x_1, y_1, z_1, x_3, y_3^-, z_3^-\right)$ and $z_5$, by $z_5^+ \equiv z_5\left(x_5, y_5^+, x_1, y_1, z_1, x_3, y_3^-, z_3^-\right)$, $y_6$ must be replaced by $y_6^- \equiv y_6^-(x_6, x_2, y_2, z_2, x_4, y_4^+, z_4^+)$ and $z_6$ by $z_6^- \equiv z_6\left(x_6, y_6^-, x_2, y_2, z_2, x_4, y_4^+, z_4^+\right)$, and $y_7$ must be replaced by $y_7^+ \equiv y_7^+\left(x_7, x_3, y_3^-, z_3^-, x_5, y_5^+, z_5^+\right)$ and $z_7$ by $z_7^+ \equiv z_7\left(x_7, y_7^+, x_3, y_3^-, z_3^-, x_5, y_5^+, z_5^+\right)$. The coordinates $x_2$, $y_2$, and $z_2$ are functions of $r_2$ and $\varphi_2$ given by Equations 3C.1 and 3C.2.

## Appendix 3D: Derivation of Expressions for Emission and Absorption Rates by a Mean First Passage Time Analysis

A mean first passage time analysis [36–38] can be used to obtain the rates of emission and absorption of beads by a cluster during protein folding [35]. To this end, a bead in the vicinity of the cluster is considered to perform a chaotic motion in the potential wells arising as a combination of the bead–cluster interactions with the dihedral angle potential. The mean first passage time of a bead escaping from a potential well is calculated on the basis of a kinetic equation governing the chaotic motion of the bead in that potential well. The chaotic motion of the bead is assumed to be governed by the Fokker–Planck equation for the single-particle distribution function with respect to its coordinates and momenta, i.e., in the phase space [59–61]. Prior to the passage event, the evolution of a bead in both the ipw and opw occurs in a dense enough medium (cluster of folded residues or unfolded, but compact part of the protein), where the relaxation time for its velocity distribution function is very short and negligible compared to the characteristic timescale of the passage process. Under these conditions, the Fokker–Planck equation reduces to the Smoluchowski equation, which involves diffusion in an external field [60,61]. In the case of spherical symmetry, it can be written in the form [36–38]

$$\frac{\partial p_i(r,t\,|\,r_0)}{\partial t} = D_i r^{-2}\frac{\partial}{\partial r}\left(r^2 e^{-\Psi_i(r)}\frac{\partial}{\partial r}e^{\Psi_i(r)}p_i(r,t\,|\,r_0)\right), \tag{3D.1}$$

where $p_i(r,t|r_0)$ is the probability of observing a bead of species $i$ ($i$ = b, l, n) between $r$ and $r + dr$ at time $t$ given that initially it was at a radial distance $r_0$, $D_i$ is its diffusion coefficient in the well, and $\Psi_i(r) = \psi_i(r)/kT$.

The mean passage time depends on the initial position (distance from the center of the cluster) $r_0$ of the bead. It is convenient to use the backward Smoluchowski equation [36–38] which expresses the dependence of the transition probability $p_i(r,t|r_0)$ on $r_0$:

$$\frac{\partial p_i(r,t\,|\,r_0)}{\partial t} = D_i r_0^{-2} e^{\Psi_i(r_0)}\frac{\partial}{\partial r_0}\left(r_0^2 e^{-\Psi_i(r_0)}\frac{\partial}{\partial r_0}p_i(r,t\,|\,r_0)\right). \tag{3D.2}$$

Let us first consider the ipw and find the emission rate of beads therefrom. The probability that a bead $i$, initially at a distance $r_0$ within the surface layer, will remain in this region after time $t$ is given by the so-called survival probability [36–38,72],

$$q_i(t\,|\,r_0) = \int_R^{R+\lambda_i^{w}} \mathrm{d}r r^2 p_i(r,t\,|\,r_0)\ \left(R < r_0 < R+\lambda_i^{w}\right), \tag{3D.3}$$

where $R$ is the radius of the cluster and $\lambda_i^{w}$ is the width of the potential well determined by the location of the barrier between the ipw and the opw (see Figure 3.6). The probability for the dissociation time to be between 0 and $t$ is equal to $1 - q_i(t|r_0)$ and the probability density for the dissociation time is given by $-\partial q_i/\partial t$ The first passage time is provided by [59–61]

$$\tau_i(r_0) = -\int_0^{\infty} t\frac{\partial q_i(t\,|\,r_0)}{\partial t}\mathrm{d}t = \int_0^{\infty} q_i(t\,|\,r_0)\mathrm{d}t. \tag{3D.4}$$

The equation for the first passage time is obtained by integrating the backward Smoluchowski equation (Equation 3D.2) with respect to $r$ and $t$ over the entire range and by using the boundary conditions $q_i(0|r_0) = 1$ and $q_i(t|r_0) \to 0$ as $t \to \infty$ for any $r_0$. This yields [36–38]

$$-D_i r_0^{-2} e^{\Psi(r_0)}\frac{\partial}{\partial r_0}\left(r_0^2 e^{-\Psi_i(r_0)}\frac{\partial}{\partial r_0}\tau_i(r_0)\right) = 1. \tag{3D.5}$$

One can solve Equation 3D.5 by assuming a reflecting inner boundary of the well ($d\tau_i/dr_0 = 0$ at $r_0 = R$) and the radiation boundary condition at the outer boundary ($\tau_i(r_0) = 0$ at $r_0 = R + \lambda_i^{\mathrm{w}}$). One thus obtains for the first passage time,

$$\tau_i(r_0) = \frac{1}{D_i} \int_{r_0}^{R+\lambda_i^{\mathrm{w}}} dy\, y^{-2} e^{\Psi_i(y)} \int_R^y dx\, x^2 e^{-\Psi_i(x)}. \tag{3D.6}$$

The average dissociation time, $\bar{\tau}_i$, or the mean first passage time, is obtained by averaging $\tau_i(r_0)$ with the Boltzmann factor over all possible initial positions $r_0$:

$$\bar{\tau}_i = \frac{1}{Z} \int_R^{R+\lambda_i^{\mathrm{w}}} dr_0 r_0^2 e^{-\Psi_i(r_0)} \tau_i(r_0) \tag{3D.7}$$

with

$$Z = \int_R^{R+\lambda_i^{\mathrm{w}}} dr_0 r_0^2 e^{-\Psi_i(r_0)}. \tag{3D.8}$$

Defining the quantity $\omega_i$ as

$$\omega_i \equiv 1/(D_i \bar{\tau}_i), \tag{3D.9}$$

the emission rate, $W_i^-$, can be expressed as

$$W_i^- = \frac{N_i^{\mathrm{w}}}{\bar{\tau}_i} = N_i^{\mathrm{w}} D_i \omega_i, \tag{3D.10}$$

where $N_i^{\mathrm{w}}$ denotes the number of molecules in the well. Clearly, $\bar{\tau}_i$, $\omega_i$, and $W_i^-$ are functions of $R$, $\chi_{\mathrm{b}}$, and $\chi_{\mathrm{l}}$.

Considering the opw, a similar procedure can be used to find the rate of absorption of beads by the cluster, $W_i^+$.

## Appendix 3E: Equilibrium Distribution of Ternary Clusters

Equation 3.53 can be derived by applying the principle of detailed balance to ternary nucleation (Equations 3.50 through 3.52). According to Equations 3.50 through 3.52,

$$\frac{g_{\mathrm{e}}(\nu_{\mathrm{b}}, \nu_{\mathrm{l}}, \nu_{\mathrm{n}})}{g_{\mathrm{e}}(\nu_{\mathrm{b}} - 1, \nu_{\mathrm{l}}, \nu_{\mathrm{n}})} = \frac{W_{\mathrm{b}}^+(\nu_{\mathrm{b}} - 1, \nu_{\mathrm{l}}, \nu_{\mathrm{n}})}{W_{\mathrm{b}}^-(\nu_{\mathrm{b}}, \nu_{\mathrm{l}}, \nu_{\mathrm{n}})}, \tag{3E.1}$$

$$\frac{g_{\mathrm{e}}(\nu_{\mathrm{b}}, \nu_{\mathrm{l}}, \nu_{\mathrm{n}})}{g_{\mathrm{e}}(\nu_{\mathrm{b}}, \nu_{\mathrm{l}} - 1, \nu_{\mathrm{n}})} = \frac{W_{\mathrm{l}}^+(\nu_{\mathrm{b}}, \nu_{\mathrm{l}} - 1, \nu_{\mathrm{n}})}{W_{\mathrm{l}}^-(\nu_{\mathrm{b}}, \nu_{\mathrm{l}}, \nu_{\mathrm{n}})}, \tag{3E.2}$$

$$\frac{g_{\mathrm{e}}(\nu_{\mathrm{b}}, \nu_{\mathrm{l}}, \nu_{\mathrm{n}})}{g_{\mathrm{e}}(\nu_{\mathrm{b}}, \nu_{\mathrm{l}}, \nu_{\mathrm{n}} - 1)} = \frac{W_{\mathrm{n}}^+(\nu_{\mathrm{b}}, \nu_{\mathrm{l}}, \nu_{\mathrm{n}} - 1)}{W_{\mathrm{n}}^-(\nu_{\mathrm{b}}, \nu_{\mathrm{l}}, \nu_{\mathrm{n}})}. \tag{3E.3}$$

By applying Equation 3E.1 to the clusters ($\nu_b - i, \nu_l, \nu_n$), with $i = 2, 3, \ldots, \nu_b - 1$, and multiplying the right-hand sides and left-hand sides of all equations, one obtains

$$\frac{g_e(\nu_b,\nu_l,\nu_n)}{g_e(1,\nu_l,\nu_n)} = \prod_{i=1}^{\nu_b-1} \frac{W_b^+(\nu_b - i,\nu_l,\nu_n)}{W_b^-(\nu_b - i + 1,\nu_l,\nu_n)}. \tag{3E.4}$$

By applying Equation 3E.2 to the clusters ($1, \nu_l - j, \nu_n$), with $j = 1, 2, \ldots, \nu_l$ and multiplying the right-hand sides and left-hand sides of all equations, one obtains

$$\frac{g_e(1,\nu_l,\nu_n)}{g_e(1,0,\nu_n)} = \prod_{j=1}^{\nu_l} \frac{W_l^+(1,\nu_l - j,\nu_n)}{W_l^-(1,\nu_l - j + 1,\nu_n)}. \tag{3E.5}$$

Finally, applying Equation 3E.3 to the clusters ($1, 0, \nu_n - k$) with $k = 1, 2, \ldots, \nu_n$ and multiplying the right-hand sides and left-hand sides of all equations, one obtains

$$\frac{g_e(1,0,\nu_n)}{g_e(1,0,0)} = \prod_{k=1}^{\nu_n} \frac{W_n^+(1,0,\nu_n - k)}{W_n^-(1,0,\nu_n - k + 1)}. \tag{3E.6}$$

The equilibrium distribution of clusters (1,0,0) is just the number density of beads of component b in the unfolded part of the protein, i.e., $g_e(1,0,0) = \rho_{bu}$. Therefore, the combination of Equations 3E.4 through 3E.6 provides

$$\begin{aligned} g_e(\nu_b,\nu_l,\nu_n) = \rho_{bu} & \frac{W_b^+(1,\nu_l,\nu_n)W_l^+(1,0,\nu_n)W_n^+(1,0,0)}{W_b^+(\nu_b,\nu_l,\nu_n)W_l^+(1,\nu_l,\nu_n)W_n^+(1,0,\nu_n)} \\ & \times \prod_{i=1}^{\nu_b-1} \frac{W_b^+(\nu_b - i + 1,\nu_l,\nu_n)}{W_b^-(\nu_b - i + 1,\nu_l,\nu_n)} \\ & \times \prod_{j=1}^{\nu_l} \frac{W_l^+(1,\nu_l - j + 1,\nu_n)}{W_l^-(1,\nu_l - j + 1,\nu_n)} \\ & \times \prod_{k=1}^{\nu_n} \frac{W_n^+(1,0,\nu_n - k + 1)}{W_n^-(1,0,\nu_n - k + 1)}. \end{aligned} \tag{3E.7}$$

In a similar way one can derive two other expressions for $g_e(\nu_b,\nu_l,\nu_n)$:

$$\begin{aligned} g_e(\nu_b,\nu_l,\nu_n) = \rho_{lu} & \frac{W_b^+(0,1,\nu_n)W_l^+(\nu_b,1,\nu_n)W_n^+(0,1,0)}{W_b^+(\nu_b,1,\nu_n)W_l^+(\nu_b,\nu_l,\nu_n)W_n^+(0,1,\nu_n)} \\ & \times \prod_{i=1}^{\nu_b} \frac{W_b^+(\nu_b - i + 1,1,\nu_n)}{W_b^-(\nu_b - i + 1,1,\nu_n)} \\ & \times \prod_{j=1}^{\nu_l-1} \frac{W_l^+(\nu_b,\nu_l - j + 1,\nu_n)}{W_l^-(\nu_b,\nu_l - j + 1,\nu_n)} \\ & \times \prod_{k=1}^{\nu_n} \frac{W_n^+(0,1,\nu_n - k + 1)}{W_n^-(0,1,\nu_n - k + 1)}, \end{aligned} \tag{3E.8}$$

$$g_e(\nu_b,\nu_l,\nu_n)=\rho_{nu}\frac{W_b^+(0,\nu_l,1)W_l^+(0,0,1)W_n^+(\nu_b,\nu_l,1)}{W_b^+(\nu_b,\nu_l,1)W_l^+(0,\nu_l,1)W_n^+(\nu_b,\nu_l,\nu_n)}$$
$$\times\prod_{i=1}^{\nu_b}\frac{W_b^+(\nu_b-i+1,\nu_l,1)}{W_b^-(\nu_b-i+1,\nu_l,1)}$$
$$\times\prod_{j=1}^{\nu_l}\frac{W_l^+(0,\nu_l-j+1,1)}{W_l^-(0,\nu_l-j+1,1)}$$
$$\times\prod_{k=1}^{\nu_n-1}\frac{W_n^+(\nu_b,\nu_l,\nu_n-k+1)}{W_n^-(\nu_b,\nu_l,\nu_n-k+1)},\tag{3E.9}$$

which (along with Equation 3E.7) allow one to represent $g_e(\nu_b,\nu_l,\nu_n)$,in the form given by Equation 3.53.

## Appendix 3F: Kinetic Equation of Ternary Nucleation

The goal of this appendix is to derive Equation 3.56 from Equations 3.46 through 3.49. Transforming $\theta_i$ ($i$ = b, l, n) (Equations 3.47 through 3.49) into a continuous form by using Taylor series expansions and retaining only the lowest order terms, one obtains the following partial differential equation of the Fokker–Planck form:

$$\frac{\partial g(\nu_b,\nu_l,\nu_n,t)}{\partial t}=-\sum_{i=b,l,n}\frac{\partial}{\partial\nu}J_i(\nu_b,\nu_l,\nu_n,t),\tag{3F.1}$$

where

$$J_i(\nu_b,\nu_l,\nu_n,t)=-A_i(\nu_b,\nu_l,\nu_n)g(\nu_b,\nu_l,\nu_n,t)$$
$$-\frac{\partial}{\partial\nu_i}(B_i(\nu_b,\nu_l,\nu_n)g(\nu_b,\nu_l,\nu_n,t))\ (i=\text{b, l, n}).\tag{3F.2}$$

Here, the coefficients $A_i(\nu_b, \nu_l, \nu_n)$ and $B_i(\nu_b, \nu_l, \nu_n)$ depend on the emission and absorption rates $W^-$ and $W^+$ of the cluster:

$$A_i(\nu_b,\nu_l,\nu_n)=W_i^+(\nu_b,\nu_l,\nu_n)-W_i^-(\nu_b,\nu_l,\nu_n),$$
$$B_i(\nu_b,\nu_l,\nu_n)=-\frac{1}{2}\left[W_i^+(\nu_b,\nu_l,\nu_n)+W_i^-(\nu_b,\nu_l,\nu_n)\right].$$

On the other hand, according to the principle of detailed balance (see Equations 3E.1 through 3E.3 in Appendix 3E), $W_b^-(\nu_b+1,\nu_l,\nu_n)/W_b^+(\nu_b,\nu_l,\nu_n)=\exp[-G(\nu_b,\nu_l,\nu_n)+G(\nu_b+1,\nu_l,\nu_n)]\simeq\exp[\partial G(\nu_b,\nu_l,\nu_n)/\partial\nu_b]$, $W_l^-(\nu_b,\nu_l+1,\nu_n)/W_l^+(\nu_b,\nu_l,\nu_n)=\exp[-G(\nu_b,\nu_l,\nu_n)+G(\nu_b,\nu_l+1,\nu_n)]\simeq\exp[\partial G(\nu_b,\nu_l,\nu_n)/\partial\nu_l]$, $W_n^-(\nu_b,\nu_l,\nu_n+1)/W_n^+(\nu_b,\nu_l,\nu_n)=\exp[-G(\nu_b,\nu_l,\nu_n)+G(\nu_b,\nu_l,\nu_n+1)]\simeq\exp[\partial G(\nu_b,\nu_l,\nu_n)/\partial\nu_n]$. The first derivatives $\partial G(\nu_b,\nu_l,\nu_n)/\partial\nu_i$ ($i$ = b, l, n) are equal to zero at the saddle point and are close to zero in its vicinity where, in addition, $W_b^+(\nu_b+1,\nu_l,\nu_n)\simeq W_b^+(\nu_b,\nu_l,\nu_n)$ and $W_l^+(\nu_b,\nu_l+1,\nu_n)\simeq W_l^+(\nu_b,\nu_l,\nu_n)$. Therefore, one can write

$$W_i^-(\nu_b,\nu_l,\nu_n)/W_l^+(\nu_b,\nu_l,\nu_n)\simeq 1+o(\partial G(\nu_b,\nu_l,\nu_n)/\partial\nu_i),\tag{3F.3}$$

where $o(x)$ denotes a quantity of the same order of magnitude as $x$. Neglecting the terms of the order $o(\partial G(\nu_b,\nu_l,\nu_n)/\partial\nu_i)$, one can thus assume that the sum $W_i^+(\nu_b,\nu_l,\nu_n)+W_i^-(\nu_b,\nu_l,\nu_n)$ in the vicinity of the saddle point [where the kinetic equation (Equation 3F.1) has to be solved] is equal to $2W_i^+(\nu_b,\nu_l,\nu_n)$. Furthermore, in the vicinity of the saddle point $W_i^+(\nu_b,\nu_l,\nu_n) \simeq W_i^+(\nu_{bc},\nu_{lc},\nu_{nc}) \equiv W_{ic}^+$; hence,

$$B_i \simeq -W_{ic}^+ \quad (i = \text{b, l, n}). \tag{3F.4}$$

Because at equilibrium $J_b = J_l = J_n = 0$, Equation 3F.2 leads to

$$A_i(\nu_b,\nu_l,\nu_n)g_e(\nu_b,\nu_l,\nu_n) + B_i(\nu_b,\nu_l,\nu_n)\frac{\partial g_e(\nu_b,\nu_l,\nu_n)}{\partial \nu_i} = 0 \ (i = \text{b, l, n}), \tag{3F.5}$$

whence one obtains

$$\begin{aligned} A_i(\nu_b,\nu_l,\nu_n) &= -B_i(\nu_b,\nu_l,\nu_n)\frac{1}{g_e(\nu_b,\nu_l,\nu_n)}\frac{\partial g_e(\nu_b,\nu_l,\nu_n)}{\partial \nu_i} \\ &\simeq W_{ic}^+\frac{\partial G(\nu_b,\nu_l,\nu_n)}{\partial \nu_i}. \end{aligned} \tag{3F.6}$$

Thus, $J_i$ ($i$ = b, l, n) in Equation 3F.2 acquires the form

$$J_i(\nu_b,\nu_l,\nu_n,t) = -W_{ic}^+\left(\frac{\partial}{\partial \nu_i} + \frac{\partial G(\nu_b,\nu_l,\nu_n)}{\partial \nu_i}\right). \tag{3F.7}$$

The substitution of this equation into Equation 3F.1 leads to Equation 3.56.

## References


1. Anfinsen, C.B., *Science* 181, 223 (1973).
2. Ghelis, C., Yan, J., *Protein Folding*, Academic, New York, 1982.
3. Creighton, T.E., *Proteins: Structure and Molecular Properties*, Freeman, San Francisco, 1984.
4. Stryer, L., *Biochemistry*, 3rd ed., Freeman, New York, 1988.
5. Nölting, B., *Protein Folding Kinetics*, Springer-Verlag, Berlin, 2006.
6. Honeycutt, J.D., Thirumalai, D., *Proc. Natl. Acad. Sci. U.S.A.* 87, 3526 (1990).
7. Honeycutt, J.D., Thirumalai, D., *Biopolymers* 32, 695 (1992).
8. Weissman, J.S., Kim, P.S., *Science* 253, 1386 (1991).
9. Creighton, T.E., *Nature (London)* 356, 194 (1992).
10. Levinthal, C., *J. Chim. Phys. Phys.-Chim. Biol.* 65, 44 (1968).
11. Kim, P.S., Baldwin, R.I., *Annu. Rev. Biochem.* 51, 459 (1982).
12. Kim, P.S., Baldwin, R.I., *Annu. Rev. Biochem.* 59, 631 (1990).
13. Creighton, T.E., *Biochem. J.* 240, 1 (1990).
14. Creighton, T.E., *Prog. Biophys. Mol. Biol.* 33, 231 (1978).
15. Guo, Z., Thirumalai, D., Honeycutt, J.D., *J. Chem. Phys.* 97, 525 (1992).
16. Guo, Z., Thirumalai, D., *Biopolymers* 36, 83 (1995).
17. Thirumalai, D., Guo, Z., *Biopolymers* 35, 137 (1995).
18. Nölting, B., Golbik, R., Neira, J.L., Soler-Gonzalez, A.S., Schreiber, G., Fersht, A.R., *Proc. Natl. Acad. Sci. U.S.A.* 94, 826 (1997).
19. Nölting, B., *J. Theor. Biol.* 194, 419 (1998).
20. Nölting, B., *J. Theor. Biol.* 197, 113 (1999).
21. Fersht, A.R., *Proc. Natl. Acad. Sci. U.S.A.* 92, 10869 (1995).
22. Fersht, A.R., *Curr. Opin. Struct. Biol.* 7, 3 (1997).
23. Shakhnovich, E.I., *Curr. Opin. Struct. Biol.* 7, 29 (1997).
24. Shakhnovich, E.I., Abkevich, V., Ptitsyn, O., *Nature (London)* 379, 96 (1996).

25. Abkevich, V.I., Gutin, A.M., Shakhnovich, E.I., *J. Chem. Phys.* 101, 6052 (1994).
26. Finkelstein, A.V., Badretdinov, A.Y., *Folding Des.* 2, 115 (1997).
27. Bryngelson, J.D., Wolynes, P.G., *Proc. Natl. Acad. Sci. U.S.A.* 84, 7524 (1987).
28. Bryngelson, J.D., Wolynes, P.G., *J. Phys. Chem.* 93, 6902 (1998).
29. Shakhnovich, E.I., Gutin, A.M., *Nature (London)* 346, 773 (1990).
30. Shakhnovich, E.I., Gutin, A.M., *J. Phys. A* 22, 1647 (1989).
31. Bryngelson, J.D., Wolynes, P.G., *Biopolymers* 30, 177 (1990).
32. Dill, K.A., *Biochemistry* 24, 1501 (1985).
33. Dill, K.A., *Biochemistry* 29, 7133 (1990).
34. Lothe, J., Pound, G.M.J., in: *Nucleation*, Zettlemoyer, A., Ed., Marcel-Dekker, New York, 1969.
35. Djikaev, Y.S., Ruckenstein, E., *J. Phys. Chem. B* 111, 886 (2007). (Section 3.1 of this volume.)
36. Narsimhan, G., Ruckenstein, E., *J. Colloid Interface Sci.* 128, 549 (1989). (Section 1.2 of this volume.)
37. Ruckenstein, E., Nowakowski, B., *J. Colloid Interface Sci.* 137, 583 (1990). (Section 1.3 of this volume.)
38. Nowakowski, B., Ruckenstein, E., *J. Colloid Interface Sci.* 139, 500 (1990). (Section 1.4 of this volume.)
39. Djikaev, Y.S., Ruckenstein, E., *J. Chem. Phys.* 123, 214503 (2005). (Section 2.5 of this volume.)
40. Djikaev, Y.S., Ruckenstein, E., *J. Chem. Phys.* 124, 124521 (2006). (Section 2.6 of this volume.)
41. Djikaev, Y.S., Ruckenstein, E., *J. Chem. Phys.* 124, 194709 (2006). (Section 1.6 of this volume.)
42. Levitt, M., Sharon, R., *Proc. Natl. Acad. Sci. U.S.A.* 85, 8557 (1988).
43. Skolnik, J., Kolinski, A., Yaris, R., *Biopolymers* 28, 937 (1989).
44. Kolinski, A., Skolnik, J., Yaris, R., *Biopolymers* 28, 937 (1989).
45. Sikorski, A., Skolnik, J., *Biopolymers* 28, 1097 (1989).
46. Sikorski, A., Skolnik, J., *J. Mol. Biol.* 213, 183 (1990).
47. Skolnik, J., Kolinski, A., *Science* 250, 1121 (1990).
48. Levinthal, C., in: *Mössbauer Spectroscopy in Biological Systems*, Debrunner, P., Tsibris, J.C., Mönck, E., Eds., University of Illinois Press, Urbana, 1968.
49. Wetlaufer, D., *Proc. Natl. Acad. Sci. U.S.A.* 70, 697 (1973).
50. Tsong, T.Y., Baldwin, R., McPhie, P., *J. Mol. Biol.* 63, 453 (1972).
51. Moult, J., Unger, R., *Biochemistry* 30, 3816 (1991).
52. Dill, K., Fiebig, K., Chan, H.S., *Proc. Natl. Acad. Sci. U.S.A.* 90, 1942 (1993).
53. Flageollet, C., Dihn Cao, M., Mirabel, P., *J. Chem. Phys.* 72, 544 (1980).
54. Wilemski, G., *J. Phys. Chem.* 91, 2492 (1987).
55. Wilemski, G., *J. Chem. Phys.* 80, 1370 (1984).
56. Laaksonen, A., *J. Chem. Phys.* 97, 1983 (1992).
57. Djikaev, Y.S., Napari, I., Laaksonen, A., *J. Chem. Phys.* 120, 9752 (2004).
58. Abraham, F.F., *Homogeneous Nucleation Theory*, Academic, New York, 1974.
59. Chandrasekhar, S., *Rev. Mod. Phys.* 15, 1 (1949).
60. Gardiner, C.W., *Handbook of Stochastic Methods*, Springer, New York, 1983.
61. Agmon, N., *J. Chem. Phys.* 81, 3644 (1984).
62. Schmelzer, J., Röpke, G., Priezzhev, V.B., Eds., *Nucleation Theory and Applications*, JINR, Dubna, 1999.
63. Kashchiev, D., *Nucleation: Basic Theory with Applications*, Butterworth-Heinemann, Oxford, 2000.
64. Trinkaus, H., *Phys. Rev. B* 27, 7372 (1983).
65. Kuni, F.M., Melikhov, A.A., Novozhilova, T.Yu., Terent'ev, I.A., *Theor. Math. Phys.* 83, 530 (1990).
66. Elliott, W.H., Elliott, D.C., *Biochemistry and Molecular Biology*, Oxford University Press, New York, 2003.
67. Jorgensen, W.L., Tirado-Rives, J., *J. Am. Chem. Soc.* 110, 1657 (1988).
68. Harpaz, Y., Gerstein, M., Chothia, C., *Structure (London)* 2, 641 (1994).
69. Huang, D.M., Chandler, D., *Proc. Natl. Acad. Sci. U.S.A.* 15, 8324 (2000).
70. Hirshfelder, J.O., Curtiss, C.F., Bird, R.B., *Molecular Theory of Gases and Liquids*, Wiley, New York, 1964.
71. Wakeham, W.A., *J. Phys. B* 6, 372 (1973).
72. Sano, H., Tachia, M., *J. Chem. Phys.* 71, 1276 (1979).

## 3.3 Effect of Ionized Protein Residues on the Nucleation Pathway of Protein Folding

*Yuri Djikaev and Eli Ruckenstein**

Using a ternary nucleation formalism, we have recently (Y. S. Djikaev and E. Ruckenstein, *J. Chem. Phys.* 126, 175103 [2007]) proposed a kinetic model for the nucleation mechanism of protein folding. A protein was considered as a heteropolymer consisting of hydrophobic, hydrophilic, and neutral beads with all the bonds having the same constant length and all the bond angles equal and fixed. In this paper, we further develop that model by taking into account the ionizability of some of the protein residues. As previously, an overall potential around the cluster wherein a protein residue performs a chaotic motion is considered to be a combination of the average dihedral and average pairwise potentials (the latter now including an electrostatic contribution for ionized residues) assigned to the residue and the confining potential due to the polymer connectivity constraint. The overall potential as a function of the distance from the cluster has a double well shape (even for ionized beads) that allows one to determine the rates of emission and absorption of residues by the cluster by using a first passage time analysis. Assuming the equality of the ratios of the numbers of negatively and positively ionized residues in the cluster and in the entire protein, one can keep the modified model within the framework of the ternary nucleation formalism and evaluate the size and composition of the nucleus and the protein folding time as in the previous model. As an illustration, the model is again applied to the folding of bovine pancreatic ribonuclease consisting of 124 amino acids, whereof 40 are hydrophobic, 81 hydrophilic (of which 10 are negatively and 18 positively ionizable), and 3 neutral. Numerical calculations at pH = 6.3, pH = 7.3, and pH = 8.3 show that for this protein the time of folding via nucleation is significantly affected by electrostatic interactions only for the unusually low pH of 6.3 and that among all pH's considered pH = 7.3 provides the lowest folding time. (© *2008 American Institute of Physics.* doi: 10.1063/1.2820771.)

### 3.3.1 Introduction

A well-defined three-dimensional structure [1,2] is necessary for a protein molecule to carry out a specific biological function. The formation of the native structure of a biologically active protein constitutes the core of the so-called protein folding problem [3,4]. Many thermodynamic and kinetic aspects of the protein folding process remain unexplained [5–8].

Although free amino acids in aqueous solution have zwitterionic structures in which the α-amino and α-carboxyl groups are ionized, these groups are no longer ionizable (except for the terminal ones) when incorporated into a polypeptide chain. The ionized state of a protein is therefore almost entirely due to the amino acid residues of aspartic and glutamic acids, lysine, arginine, and histidine since their ionizable side chain groups are not blocked by peptide bond formation. Charges of the same sign repel one another, whereas positive and negative charges attract one another. Although hydrophobic interactions provide the major driving force in initiating protein folding and stabilizing protein structure [9–13], electrostatic interactions and the distribution of charged amino acid residues in a protein have an important effect on the spatial conformations that a polypeptide can adopt and hence they can play a role in protein folding [15–19].

It is believed that initially an unfolded protein transforms quickly into a compact (but not native) configuration [20,21]. The transition from a compact configuration to the native one can occur either via the gradual (relatively slow) formation of tertiary contacts of the native structure as the protein passes through a sequence of intermediate states [20–32] or, alternatively, immediately following the formation of a number of tertiary contacts [6,20–22,33–35]. The latter mechanism is similar to nucleation, i.e., once a critical number of (native) tertiary contacts is established the native structure is formed without passing through any detectable intermediates [20–22].

Few theoretical models for the nucleation mechanism of folding have been proposed. Initially [11,12,21,35], the problem was approached by using classical thermodynamics in the framework of the capillarity approximation of the classical nucleation theory (CNT). Such an approach involves the concept of surface tension for a cluster consisting of protein residues which is an intrinsically ill-defined physical quantity and can be considered only as an adjustable parameter; clearly, no direct experimental measurement thereof is possible.

* *J. Chem. Phys.* 128, 025103 (2008). Republished with permission.

We have recently [36,37] developed an alternative, microscopic model for the nucleation mechanism of protein folding which is based on pairwise "molecular" interactions (i.e., repulsion/attraction) and configurational (dihedral angle) potentials in which the protein residues are involved and whereof the parameters can be rigorously defined. The rates of emission and absorption of residues by the cluster are determined by using a first passage time analysis [38–43]. Once these rates are found as functions of the cluster size, one can develop a self-consistent kinetic theory for the nucleation mechanism of folding of a protein. The time necessary for the protein to fold can be evaluated as the sum of the times necessary for the appearance of the first nucleus (consisting of protein residues stabilized by native tertiary contacts) and the time necessary for the nucleus to grow to the maximum size (of the folded protein in the native state). The ill-defined surface tension of a cluster of protein residues (within a protein in a compact but nonnative configuration) does not enter into the new model at all.

The model in Refs. [36,37] was developed by considering a protein as a heteropolymer with all the bonds of the same constant length and all the bond angles equal and fixed. In Ref. [36], the heteropolymer consisted of hydrophobic and hydrophilic beads and the cluster composition were considered to remain constant and equal to the overall composition of the protein at all times, which allowed us to use a unary nucleation formalism in modeling the nucleation mechanism of protein folding. In Ref. [37], the protein consisted of three kinds of beads (hydrophobic, hydrophilic, and neutral), and the composition of the cluster (of native residues) was allowed to vary so that a ternary nucleation formalism was adopted. Neither Ref. [36] nor Ref. [37] considered the possibility that five amino acids (aspartic and glutamic acids, lysine, arginine, and histidine) are ionizable. Thus, the effect of electrostatic interactions on protein folding was not taken into account.

To clarify the impact of charged residues on the nucleation mechanism of protein folding, in the present paper we modify the previous model [37] by considering the hydrophilic amino acid residues to be of three subtypes, namely, those which cannot be ionized at all and those which can (depending on the pH of the surrounding solution) be either positively or negatively ionized. Strictly speaking, a rigorous treatment of this modified model requires the use of a five-component nucleation formalism which would lead to an enormous increase in computations involved in its applications. To avoid this increase, it will be assumed that the ratio of negatively and positively charged residues is the same in the cluster and in the entire protein. This will allow us to treat the model in the framework of the ternary nucleation formalism which requires manageable (although still very extensive) numerical computations.

The paper is structured as follows. In Section 3.3.2, we briefly outline the model for the nucleation mechanism of protein folding developed previously in Ref. [37]. The modifications of that model which are necessary for taking into account the effect of electrostatic interactions on the nucleation mechanism of protein folding are presented in Section 3.3.3. Sections 3.3.4 and 3.3.5 describe the procedures to calculate the emission and absorption rates and the folding time, respectively. Section 3.3.6 concerns the results of numerical calculations, and Section 3.3.7 summarizes the conclusions.

## 3.3.2 Ternary Nucleation Model of Protein Folding Based on a First Passage Time Analysis

### *3.3.2.1 Heteropolymer as a Protein Model*

Previous models [21,35–37] for the nucleation mechanism of protein folding (as well as numerous computer simulations of protein folding dynamics [6,21,22]) considered a protein as a heteropolymer chain of connected beads than can be thought of as representing the α-carbons of various amino acids. Two adjacent beads are connected by a covalent bond of fixed length $\eta$. This model (and its variants), completed with the appropriate potentials described below, has been shown [6,21,35] to be able to capture the essential characteristics of protein folding even though it involves only some of the features of a real polypeptide chain. For example, this model ignores the side groups although they are known to be crucial for the intramolecular hydrogen bonding [1]. Besides, the presence of a solvent (water) in a real physical system has been usually accounted for too simplistically, although the protein dynamics was reported to become more realistic in those molecular dynamics simulations where the solvent molecules were explicitly taken into account [44,45]. Despite these limitations, various modifications of the heteropolymer model [6,21,33–35,45–47] have shed light on some important details regarding the folding transition of a protein [6,23]. The approach presented in Refs. [36,37] is based on a protein model, wherein the sequence of amino acids is treated in a probabilistic way, with their uniform distribution along the polypeptide chain, so that our model for the nucleation mechanism of folding does not take into account the detailed sequence of amino acids. To some extent, this is similar to the so-called random (disordered) heteropolymer models [33,34,48–50] which provided valuable insight into many

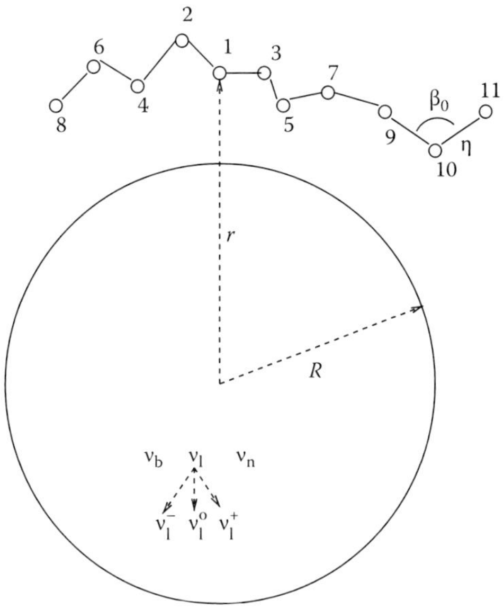


Figure 3.10

A piece of a heteropolymer chain around a spherical cluster consisting of $\nu_b$ hydrophobic, $\nu_l$ hydrophilic, and $\nu_n$ neutral beads. Among the hydrophilic beads themselves $\nu_0$ are uncharged, $\nu_-$ are negatively charged, and $\nu_+$ are positively charged. Bead 1 is in the plane of the figure, whereas other beads may all lie in different planes. All bond angles are equal to $\beta_0$ and their lengths are equal to $\eta$. The radius of the cluster is $R$ and the distance from the selected bead 1 to the cluster center is $r$.

aspects of protein folding. Clearly, sequence-dependent models of a protein are more realistic, and they have been extensively studied by computer simulations. However, at this point they are hardly tractable (when applied to the nucleation mechanism) by simple analytical methods like the one proposed in this work and previously in Refs. [36,37]. Nevertheless, one can expect the proposed model to provide a reasonable estimate for the folding time of a protein with a given composition, an estimate which would be representative for folding times of proteins with various sequences of (roughly) given composition.

The total potential field to which a bead is subjected in the vicinity of the cluster contains three contributions of different types: the contribution from repulsive/attractive forces between pairs of nonadjacent beads (these can be, e.g., of Lennard–Jones [LJ] or other types); a contribution from the harmonic forces due to the oscillations of the bond angles; and finally, a contribution from the dihedral angle potential due to the rotation around the peptide bonds. There are various ways to model these three types of contributions [6,21,33–35,45–47].

In our previous model [37], we considered the polypeptide chain of a protein as a three-component heteropolymer consisting of hydrophobic (b), hydrophilic (l), and neutral (n) connected beads (Figure 3.10). A pair interaction between two nonadjacent beads $i$ and $j$ was considered to be of the LJ type. The bond angle forces are believed [6,21] to play a minor role in the protein folding/unfolding, hence all bond angles are most often set to be equal to $\beta_0$. The dihedral angle potential arises due to the rotation of three successive peptide bonds connecting four successive beads and depends [6,21,36,37] not only on the dihedral angle $\delta$ (which is an angle between two planes each of which is determined by a sequence of three beads) but also on the nature and sequence of the four beads involved. In addition to these potentials, one must introduce a confining potential which arises because all the residues of the protein are linked and there is an external boundary of the volume available to them.

In Ref. [37], the formation of a cluster consisting of native residues during protein folding was treated as ternary nucleation. A nucleation mechanism model of protein folding was developed [37] in terms of ternary nucleation by using a first passage time analysis [38–43].

### *3.3.2.2 Potential Field for a Protein Residue around a Cluster (Folded Part of the Protein)*

A mean first passage time analysis as an alternative approach to the kinetics of unary nucleation was first proposed by Ruckenstein and coworkers [38–40]. Unlike CNT, that approach [38–40] avoids the use of macroscopic thermodynamics and does not make use of the surface tension for the tiny clusters involved in nucleation. It is built on molecular interactions and exploits the fact that one can derive and solve the kinetic equation of nucleation (hence find the nucleation rate) if the emission and absorption rates of a cluster are known as functions of its size. For the rate of absorption of molecules by the cluster, the new theory uses (as CNT does) a standard gas-kinetic expression [51], but (and this constitutes the main idea of the new approach [38–40]) the rate of emission of molecules by the cluster is determined via a mean first passage time analysis. This time is calculated by solving a single-molecule master equation for the probability distribution function of a surface layer molecule moving in a potential well around the cluster. The master equation is a Fokker–Planck equation in the phase space which can be reduced to the Smoluchowski equation owing to the hierarchy of the characteristic timescales involved in the evolution of the single-molecule distribution function with respect to coordinates and momenta.

Consider a ternary cluster of spherical shape (with sharp boundaries and radius $R$) immersed in a ternary fluid mixture. The number of beads of component $i$ in the cluster will be denoted by $\nu_i$ ($i$ = b, l, n), whereas the total number of beads of component $i$ in the protein by $N_i$. In the previous applications of a first passage time analysis to nucleation [38–43], a molecule of component $i$ (in our model $i$ = b, l, n) located in the surface layer of the cluster was considered to perform a thermal chaotic motion in a spherically symmetric potential well $\phi_i(r)$ resulting from the pair interactions (say, of LJ type) of this molecule with those in the cluster ($r$ is the distance between the centers of the cluster and bead 1; see Figure 3.10).

In Ref. [37], the total potential $\psi_i(r)$ for a residue of type $i$ around the cluster was determined by three constituents, $\phi_i(r)$, $\bar{\phi}_i^{\delta}(r)$, and $\phi_{\text{cp}}$, which represent the effective LJ (for interactions of the selected residue with those in the cluster), dihedral angle, and confining (representing the external boundary of the volume available to the unfolded residues) potentials, respectively, $\psi_i(r) = \phi_i(r) + \bar{\phi}_i^{\delta}(r) + \phi_{\text{cp}}(r)$ ($i$ = b, l, n).

It was shown [37] that this combination of potentials gives rise to a double potential well around the cluster with a barrier between the two wells. Residues in the inner well are considered to belong to the cluster (part of the protein with correct tertiary contacts), whereas those in the outer well are treated as belonging to the mother phase (amorphous part of the protein with incorrect tertiary contacts). Transitions of residues from the inner well into the outer one and vice versa are considered as elementary emission and absorption events, respectively. The rates of these processes are determined by using a mean first passage time analysis. Knowing the rates of emission and absorption as functions of the cluster size, one can develop a self-consistent kinetic theory for the nucleation mechanism of protein folding [36,37]. For example, the size of the critical cluster (nucleus) is then found as the one for which these rates are equal. The time necessary for the protein to fold can be evaluated as the sum of the times necessary for the appearance of the first nucleus and the time necessary for the nucleus to grow to the maximum size (of the folded protein in the native state).

The pairwise (LJ) contribution $\phi_i(r)$ to $\psi_i(r)$ is calculated by integrating the pair interactions of the selected bead of type $i$ with all the elements of the cluster (folded part of the protein). A contribution to $\phi_i(r)$, from the interaction of the selected bead with the unfolded part can be assumed to be small and accounted for by particular choices of the energy parameters.

Because of the bonds between the residues, none of them can be located outside some confining boundary. We have assumed [37] it to be spherical, so that the confining potential $\phi_{\text{cp}}(r) = 0$ for $r < r_{\text{cp}}$ and $\phi_{\text{cp}}(r) = \infty$ for $r \geq r_{\text{cp}}$. The radius of the confining boundary $r_{\text{cp}}$ is assumed to be determined [37] by the total number of residues in the protein, densities of residues in its unfolded and folded parts, and the radius of the cluster (folded part). Clearly, the polymer connectivity imposes strict constraints on $r_{\text{cp}}$. For every bead in the unfolded part of the protein, the maximum distance from the cluster at which it can be found is determined by many factors, such as the location of the bead in the chain and the location of the beads that have already formed a cluster. Taking all these details into account would significantly complicate any analytical model. Moreover, this approach would be appropriate only at the early stages of the evolution of the protein, immediately following its synthesis when the sequence of connected amino acids has an elongated shape. However, such an approach seems to be inadequate in developing a model for the nucleation mechanism of folding. Indeed, as suggested by both experiment and simulations (see, e.g., Refs. [20,21] and references therein), the initial transformation of a protein from an extended shape into a compact (but nonnative) state occurs extremely quickly. The nucleation mechanism is one of the possible pathways for the transformation of this compact configuration of the protein into the native one. Therefore, the location of the confining potential $r_{\text{cp}}$ is hardly determined by pure polymer connectivity constraints (appropriate for the protein in the extended state). Although somewhat rough, the approach used in Ref. [37] for determining $r_{\text{cp}}$ seems to be more appropriate in the case of a protein in a compact state (and certainly much more amenable analytically), leading to $r_{\text{cp}} = (3/4\pi)^{1/3}(\nu/\rho_{\text{f}} + (N_0 - \nu)/\rho_{\text{u}})^{1/3}$, where $N$ and $\nu$ are the total numbers of beads in the whole protein and in the cluster, respectively, and $\rho_{\text{u}}$ and $\rho_{\text{f}}$ are the overall number densities of beads in the unfolded and folded parts of the protein, respectively.

The term $\overline{\phi}_i^{\delta}(r) \equiv \overline{\phi}_i^{\delta}(r, \mathbf{r}_2, \mathbf{r}_3, \ldots, \mathbf{r}_7)$ in $\psi_i(r)$ is due to the $r$-dependent part of the dihedral angle potential of the whole protein and is determined from

$$\overline{\phi}_{i_1}^{\delta}(r) = f^{-1} \int_{\Omega_{18}} d\mathbf{r}_2\, d\mathbf{r}_3\, d\mathbf{r}_4\, d\mathbf{r}_5\, d\mathbf{r}_6\, d\mathbf{r}_7 \phi_{i_1}^{\delta}(r) \exp\left[-\phi_{i_1}^{\delta 27}(r)/k_B T\right], \tag{3.65}$$

where $\mathbf{r}_s$ ($s = 2, \ldots, 7$) is the radius vector of bead $s$ (with the origin in the cluster center), $f$ is a normalization constant, and $\Omega_{18}$ is the integration region (of all possible configurations of beads 2, 3, …, 7) in an 18-dimensional space and the subscript $i_s$ ($i_s$ = b, l, n) indicates the type of bead $s$ ($s = 1, \ldots, 7$). The exponent in the Boltzmann factor is the sum of two terms, the first representing the dihedral angle potential of beads 1–7 and the second the sum of "bead–cluster" interactions for beads 2–7:

$$\phi_{i_1}^{\delta 27}(r) \equiv \phi_{i_1}^{\delta 27}(r, \mathbf{r}_2, \ldots, \mathbf{r}_7) = \phi_{i_1}^{\delta}(r) + \sum_{s=2}^{7} \phi_{i_s}(\mathbf{r}_s), \tag{3.66}$$

where

$$\begin{aligned} \phi_{i_1}^{\delta}(r) &\equiv \phi_{i_1}^{\delta}(r, \mathbf{r}_2 \mathbf{r}_3, \ldots, \mathbf{r}_7) \\ &= \phi_{i_1}^{\delta}\left(\delta_{421}^{642}(r)\right) + \phi_{i_1}^{\delta}\left(\delta_{213}^{421}(r)\right) + \phi_{i_1}^{\delta}\left(\delta_{135}^{213}(r)\right) + \phi_{i_1}^{\delta}\left(\delta_{357}^{135}(r)\right) \end{aligned} \tag{3.67}$$

and $\phi_{i_1}^{\delta}\left(\delta_{jkl}^{ijk}(r)\right)$ is the dihedral angle potential corresponding to the dihedral angle $\delta_{jkl}^{ijk}$ between two planes, one of which is determined by beads $i, j, k$ and the other by beads $j, k, l$. The 18-fold integral in Equation 3.65 can be reduced to a sevenfold one by taking into account the constraints of fixed bond length and fixed bond angle (see Ref. [37] for all the details on calculating $\overline{\phi}_i^{\delta}(r)$).

### 3.3.3 Taking Account of the Ionizability of Protein Residues

According to Elliott and Elliott [52], the 20 main amino acids can be roughly classified as neutral (glycine), hydrophobic (alanine, valine, leucine, isoleucine, methionine, phenylalanine, tyrosine, tryptophan), and hydrophilic (aspartic acid, glutamic acid, lysine, arginine, histidine, asparagine, glutamine, serine, threonine, cysteine, and proline). Five out of the 11 hydrophilic amino acids are ionizable even when they are incorporated into a protein. The acidic residues (aspartic and glutamic acids) have an extra –COOH on the side chains which can be negatively charged. On the other hand, the basic residues (lysine, arginine, and histidine) have extra $-NH_2$ groups on the side chains which can become positively charged.

The ionizability of these five hydrophilic beads can be easily taken into account in a modified version of our previous model [37] (whereof some of the elements are given above). Indeed, let $N_b$, $N_l$, $N_n$ be the total numbers of hydrophobic, hydrophilic, and neutral beads, respectively, in the whole protein (the corresponding quantities for the cluster were denoted by $\nu_b$, $\nu_l$, $\nu_n$). As a step forward in our model, let us assume that some of the $N_l$ hydrophylic beads are negatively ionizable (the aspartic and glutamic acids), some are positively ionizable (lysine, arginine, and histidine), and all the others are nonionizable. Denote the numbers of these beads by $\tilde{N}_l^-$, $\tilde{N}_l^+$ and $\tilde{N}_l^0$, respectively. Clearly, not all of $\tilde{N}_l^-$ and $\tilde{N}_l^+$ will be actually ionized. The overall actual numbers of positively and negatively ionized protein residues will be denoted by $N_l^-$ and $N_l^+$, respectively, such that

$$N_l^- = k^- \tilde{N}_l^-, \quad N_l^+ = k^+ \tilde{N}_l^-, \tag{3.68}$$

where $k^-$ and $k^+$ are the effective dissociation coefficients of acidic and basic residues in the protein which depend on the pH of the medium, wherein the protein is present and on the p$K_a$ and p$K_b$, respectively: $k^- = 1/(1 + 10^{-\mathrm{pH}+\mathrm{p}K_a})$, $k^+ = 1/(1 + 10^{-14+\mathrm{pH}-\mathrm{p}K_b})$, where p$K_a = -\log K_a$, p$K_b = -\log K_b$ with $K_a$ and $K_b$ being the effective dissociation constants of acidic and basic residues, respectively. These expressions can easily be derived from the definitions of $K_a = [H^+][A^-]/[HA]$ and $K_b = [BH^+][OH^-]/[B]$, where HA and B indicate the acidic and basic species, respectively. Obviously, $N_l = \tilde{N}_l^- + \tilde{N}_l^+ + \tilde{N}_l^0 = N_l^- + N_l^+ + N_l^0$, where $N_l^0 = (1-k^-)\tilde{N}_l^- + (1-k^+)\tilde{N}_l^+ + \tilde{N}_l^0$ is the actual number of *uncharged* hydrophilic residues in the whole protein.

Let us denote the numbers of negatively and positively charged residues in the cluster by $\nu_l^-$ and $\nu_l^+$, respectively, and the number of uncharged hydrophilic residues therein by $\nu_l^0$ so that $\nu_l = \nu_l^- + \nu_l^+ + \nu_l^0$. Strictly speaking, a cluster of native residues involved in the nucleation mechanism of protein folding should be characterized by five independent variables $\nu_b$, $\nu_l^-$, $\nu_l^+$, $\nu_l^0$, $\nu_n$, and it would be necessary to use the formalism of a five-component nucleation to develop a model for this process. The application of such a model to real proteins would, however, require extremely lengthy numerical calculations (by an order of magnitude longer than the already very extensive ones required by our previous, ternary nucleation model [37]). To avoid this difficulty, one can assume that, as the protein folds via nucleation, the "hydrophilic" mole fractions of positive and negative residues in the cluster remain equal to those in the whole protein, i.e.,

$$\frac{\nu_l^-}{\nu_l} = \frac{N_l^-}{N_l}, \quad \frac{\nu_l^+}{\nu_l} = \frac{N_l^+}{N_l}.$$

Under these assumptions,

$$\nu_l^- = \nu_l \frac{N_l^-}{N_l}, \; \nu_l^+ = \nu_l \frac{N_l^+}{N_l}, \; \nu_l^0 = \nu_l \left(1 - \frac{N_l^- + N_l^+}{N_l}\right), \tag{3.69}$$

and, as noted previously [37], the cluster can be characterized by only three independent variables $\nu_b$, $\nu_l$, $\nu_n$. Thus, a model (taking into account electrostatic interactions between charged residues) for the nucleation mechanism of protein folding can be again developed in terms of a ternary nucleation theory.

A major difference between the present model and the previous one [37] is in the rates with which the hydrophilic residues are emitted into and absorbed from the unfolded part of the protein by a cluster of native residues (itself containing $\nu_l$ hydrophilic residues of which $\nu_l^+ + \nu_l^-$ are charged).In the previous model [37], all the hydrophilic residues interacted identically with the cluster, hence they had identical emission and absorption rates. In the present model, the cluster has a total charge $q_\nu = e(\nu_l^+ - \nu_l^-)$ (each of the ionized residues can carry only one elementary charge, $+e$ or $-e$), which will be assumed to be uniformly distributed over its volume). Thus, positively and negatively charged residues interact electrostatically with the cluster in opposite (to one another) ways. The interactions of uncharged hydrophilic residues (as well as hydrophobic and neutral ones) with the cluster are not affected by $q_\nu$, so their emission and absorption rates are determined as in the previous model [37]. Thus, the hydrophilic beads are emitted from and absorbed by the cluster with the total rates

$$W_l^e = W_-^e + W_+^e + W_0^e, \; W_l^a = W_-^a + W_+^a + W_0^a, \tag{3.70}$$

respectively. Here, $W \equiv W(\nu_b, \nu_l, \nu_n)$ is the rate of emission or adsorption (marked with the superscripts e or a, respectively) of negatively charged, positively charged, and uncharged hydrophilic residues (marked with the subscripts −, +, and 0, respectively) by a cluster $\nu_b$, $\nu_l$, $\nu_n$.

As previously noted [37], the total potential $\psi_i(r)$ for residues of type b, n, and subtype 0 (uncharged hydrophilic) around the cluster is provided by the equation

$$\psi_i(r) = \phi_i(r) + \overline{\phi}_i^\delta(r) + \phi_{cp}(r) \;\; (i = \text{b, n, 0}). \tag{3.71}$$

However, the total potentials $\psi_-(r)$ and $\psi_+(r)$ for negatively and positively charged hydrophilic residues around the cluster have additional (electrostatic) contributions, $u_-(r)$ and $u_+(r)$, respectively, equal in absolute values but with opposite signs: $u_+(r) = -u_-(r)$,

$$\psi_\pm(r) = \phi_\pm(r) + \overline{\phi}_\pm^\delta(r) + \phi_{cp}(r) + u_\pm(r). \tag{3.72}$$

As a reasonable assumption, for all hydrophilic residues the effective LJ potential can be considered to be the same, i.e., $\phi_l(r) \equiv \phi_0(r) = \phi_-(r) = \phi_+(r)$. Similarly, one can assume for the average dihedral angle potential that $\overline{\phi}_l^\delta(r) \equiv \overline{\phi}_0^\delta(r) = \overline{\phi}_-^\delta(r) = \overline{\phi}_+^\delta(r)$. The total electrostatic potential $u_\pm(r)$ of a residue carrying a charge $q_\pm = \pm e$ around the cluster (at a distance $r$ from its center) can be represented in the form

$$u_{\pm}(r)=\int_{v} d\mathbf{r}'\rho_q(r')w_{\pm}(|\mathbf{r}'-\mathbf{r}|). \tag{3.73}$$

Here, $\mathbf{r}$ is the coordinate of the selected bead with a charge $q_{\pm}$ and the "number density" of charged residues at point $\mathbf{r}'$ within the cluster is assumed to be uniform, i.e.,

$$\rho_q(r')=\frac{v_1^+-v_1^-}{4\pi R^3/3}.$$

Note that $\rho_q(r')$ can be negative, in which case the total charge of the cluster is negative. The electrostatic interactions between the selected charge $q_{\pm}$ at $\mathbf{r}$ and the elementary charge at point $\mathbf{r}'$ is given in the Debye–Hückel approximation for the screened Coulomb potential by

$$w_{\pm}(|\mathbf{r}'-\mathbf{r}|)=\frac{k}{\varepsilon}\frac{eq_{\pm}}{|\mathbf{r}'-\mathbf{r}|}\mathbf{e}^{-\kappa|\mathbf{r}'-\mathbf{r}|}, \tag{3.74}$$

where $k$ is the electrostatic constant (1 in cgs units and $1/4\pi\varepsilon_0$ in SI units, with $\varepsilon_0 = 8.854 \times 10^{-12}$ F/m being the dielectric permittivity of vacuum), $\varepsilon$ is the relative permittivity of the medium wherein the protein is present, and $\kappa$ is the inverse Debye length (in our numerical calculations, $1/\kappa \simeq 0.3$ nm). The integration in Equation 3.73 has to be carried out over the whole volume of the system, but the contribution from the unfolded part is assumed to be small owing to the smaller density.

Clearly, for hydrophobic, neutral, and uncharged hydrophilic beads, the potential fields $\psi_i(r)$ ($i$ = b, n, 0) around the cluster are not affected by the charged residues in the cluster, so they will have a double well shape with a barrier between the two wells. Therefore, the same procedure [37] (based on a first passage time analysis) can be used to determine their adsorption and emission rates $W_i^{\mathrm{a}}$ and $W_i^{\mathrm{e}}$ ($i$ = b, n, 0).

Figure 3.11 presents typical shapes of the overall potential well $\psi_i(r)$ ($i$ = 0, −, + and of its different constituents as functions of the distance from the cluster center for the uncharged and negatively and positively charged hydrophilic beads when the total charge of the cluster is positive (for details regarding the numerical calculations, see Section 3.3.6). The contributions $\phi_1(r)$, $\phi_1(r) + u_-(r)$, $\phi_1(r) + u_+(r)$, arising from the pairwise interactions (LJ and electrostatic), have a form reminiscent of the LJ potential. For a positively charged cluster $\phi_1(r) + u_-(r) < \phi_1(r) < \phi_1(r) + u_+(r)$. Consequently, the potential well (due to the well of the LJ potential) becomes shallower for positively charged hydrophilic residues and deeper for negatively charged ones as compared to uncharged ones. The contribution from the average dihedral potential starts with a maximum value at the cluster surface and monotonically decreases with increasing $r$ until it becomes constant for large enough $r$ (see Section 3.3.6). Thus, except very short distances from the cluster surface where $\phi_1(r)$, sharply decreases from $\infty$ to its global minimum, the potential $\psi_{\pm}(r)$ for all charged hydrophilic beads (as well as uncharged ones) is shaped by two competing terms, $\phi_1(r) + u_{\pm}(r)$ and $\bar{\phi}_l^{\delta}(r)$, which increase and decrease, respectively, with increasing $r$, and by the confining potential $\phi_{\mathrm{cp}}$ As a result, the overall potential $\psi_{\pm}(r)$ has a double well shape for charged residues, with the inner well separated by a potential barrier from the outer well. This shape of $\psi_{\pm}(r)$ allows one to use a mean first passage time analysis for the determination of the rates of absorption and emission of charged hydrophilic beads by the cluster exactly as done for hydrophobic, neutral, and uncharged hydrophilic ones.

### 3.3.4 Determination of Emission and Absorption Rates

As previously noted [36,37], a bead (residue) is considered as belonging to a cluster as long as it remains in the inner potential well (hereinafter referred to as "ipw"), and as dissociated from the cluster when it passes over the barrier between the ipw and the outer potential well (hereinafter referred to as "opw"). The rate of emission $W^{\mathrm{e}}$ is determined by the mean time necessary for the passage of the bead from the ipw over the barrier into the opw. Likewise, a bead is considered as belonging to the unfolded part of the heteropolymer (protein) as long as it remains in the opw, and as absorbed by the cluster when it passes over the barrier between the opw and the ipw. The rate of absorption $W^{\mathrm{a}}$ is determined by the mean time necessary for the passage of the bead from the opw over the barrier into the ipw.

The mean first passage time of a bead escaping from a potential well is calculated on the basis of a kinetic equation governing the chaotic motion of the bead in that potential well which is a Fokker–Planck equation for the single-particle distribution function with respect to its coordinates and momenta, i.e., in the phase

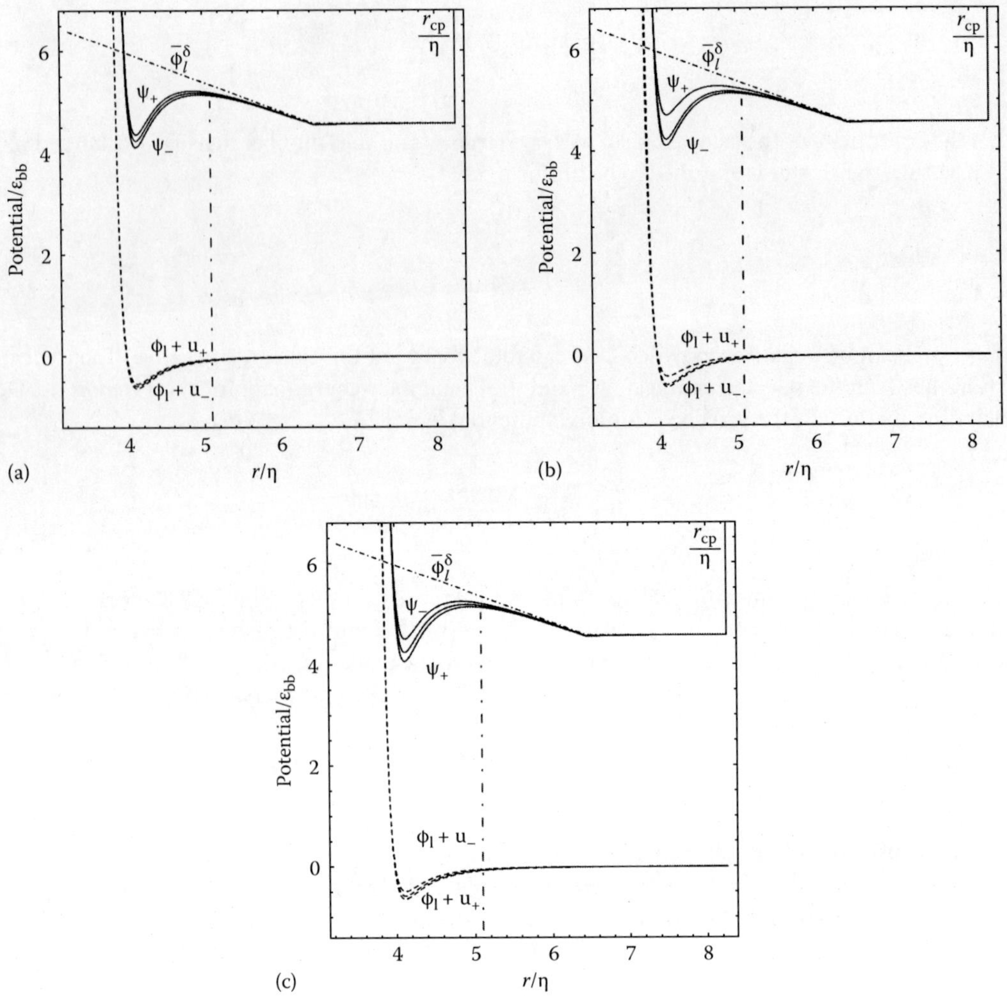


Figure 3.11

The potentials $\phi_l(r) + u_+(r)$ [upper dashed curves in (a) and (b) and lower dashed curve in (c)], $\phi_l(r)$ (middle dashed curve), and $\phi_l(r) + u_-(r)$) [lower dashed curves in (a) and (b) and upper dashed curve in (c)], $\bar{\phi}_l^\delta(r)$ (dot-dashed curve), $\psi_+(r)$ [upper solid curve in (a) and (b) and lower solid curve in (c)], $\psi_0(r)$ (middle solid curve), and $\psi_-(r)$ [lower solid curves in (a) and (b) and upper solid curve in (c)] for a bead around a cluster with: (a) $\nu_b = 23$, $\nu_l = 41$, $\nu_- = 4.5$, $\nu_+ = 6.5$, $\nu_n = 2$ at pH = 7.3; (b) $\nu_b = 23$, $\nu_l = 41$, $\nu_- = 3$, $\nu_+ = 8$, $\nu_n = 2$ at pH = 6.3; (c) $\nu_b = 23$, $\nu_l = 40$, $\nu_- = 7.5$, $\nu_+ = 4.5$, $\nu_n = 2$ at pH = 8.3 as functions of the distance $r$ from the cluster center. The vertical dashed line indicates the location of the maximum of the barrier between the ipw and the opw which determines the widths $\lambda_l^{iw}$ and $\lambda_l^{ow}$ of both potential wells. The outer boundary of the opw is indicated by $r_{cp}$. The potentials are expressed in units of $\epsilon_{bb}$, energy parameter of the LJ potential for the "hydrophobic-hydrophobic" interactions (see Ref. [28]).

space [53,54]. Solving that equation [36,37], one can obtain the following expressions for $W^e$ and $W^a$, the emission and absorption rates (for details, see Appendix B in Ref. [37]), respectively:

$$W_i^e = n_i^{iw} D_i^{iw} \omega_i^{iw}, \; W_i^a = n_i^{ow} D_i^{ow} \omega_i^{ow} \;\; (i = b, n, 0, -, +). \tag{3.75}$$

Here the superscripts "iw" and "ow" mark the quantities for the inner and outer potential wells, respectively; $n$ is the number of beads in the well, $D$ is the diffusion coefficient of the bead, and $\omega^{iw}$ and $\omega^{ow}$ are defined as

$$\omega_i^{iw} \equiv 1/\left(D_i^{iw}\bar{\tau}_i^{iw}\right), \; \omega_i^{ow} \equiv 1/\left(D_i^{ow}\bar{\tau}_i^{ow}\right) \;\; (i = b, n, 0, -, +), \tag{3.76}$$

where $\bar{\tau}_i^{iw}$ and $\bar{\tau}_i^{ow}$ are the mean first passage times for a bead in the ipw to cross over the barrier into the opw and vice versa, respectively. Note that in the original work [38–40] on the mean first passage time analysis in the nucleation theory and its recent development [41–43] this method was used only for determining $W^e$, but not $W^a$. For nucleation in proteins [36,37], the double well shape of the overall potential $\psi(r)$ around the cluster allows one to use the first passage time analysis to determine $W^a$ as well.

The quantities $W^e$, $W^a$, $\omega^{iw}$, $\omega^{ow}$, $\bar{\tau}^{iw}$ and $\bar{\tau}^{ow}$ are functions of three independent variables of state of a cluster. These variables can be, for example, the numbers $\nu_b$, $\nu_l$, and $\nu_n$ of hydrophobic, hydrophilic, and neutral beads, respectively, or, alternatively, the cluster radius $R$ and the mole fractions $\chi_b = \nu_b/(\nu_b + \nu_l + \nu_n)$ and $\chi_l = \nu_l/(\nu_b + \nu_l + \nu_n)$ of the hydrophobic and hydrophilic beads therein (see Appendix B in Ref. [37]). Thus, the nucleation mechanism of protein folding can be treated as previously [37] in terms of ternary nucleation.

## 3.3.5 Evaluation of Protein Folding Time

Knowing the emission and absorption rates as functions of $\nu_b$, $\nu_l$, and $\nu_n$, one can estimate the time $t_f$ necessary for the protein to fold via nucleation. Roughly speaking, the process of folding via nucleation consists of two stages. During the first stage, a critical cluster (nucleus) of native residues is formed (nucleation proper). Until the nucleus forms, the emission rate $W^e$ is larger than $W^a$, but the cluster still can attain the critical size and composition by means of fluctuations. At the second stage, the nucleus grows via regular absorption of native residues which dominates their emission, $W^e < W^a$. Thus, the folding time is given by

$$t_f \simeq t_n + t_g, \tag{3.77}$$

where $t_n$ is the time necessary for one critical cluster to nucleate within a compact (but still unfolded) protein and $t_g$ is the time necessary for the nucleus to grow up to the maximum size, i.e., the size of the entirely folded protein.

The time $t_n$ of the first nucleation event can be estimated as

$$t_n \simeq 1/(J_s, V_0), \tag{3.78}$$

where $J_s$ is the steady-state nucleation rate and $V_0$ is the volume of the unfolded protein in a compact state. The steady-state nucleation rate can be found by solving the kinetic equation of ternary nucleation, subject to the appropriate boundary conditions, in the vicinity of the saddle point of the function $G(\nu_b, \nu_l, \nu_n)$, $= -\ln[g_e(\nu_b, \nu_l, \nu_n)/(\rho_{bu} + \rho_{lu} + \rho_{nu})]$, where $g_e(\nu_b, \nu_l, \nu_n)$, is the equilibrium distribution of clusters with respect to $\nu_b$, $\nu_l$, $\nu_n$, and $\rho_{bu}$, $\rho_{lu}$, and $\rho_{nu}$ are the number densities of hydrophobic, hydrophilic, and neutral beads, respectively, in the unfolded (but compact) protein (for more details, see Ref. [37]). One can show that

$$J_s = \frac{|\lambda_0| 2\pi k_B T}{\sqrt{-\det(\mathbf{G}''/2\pi k_B T)}} g_e = (\nu_{bc}, \nu_{lc}, \nu_{nc}), \tag{3.79}$$

where $\mathbf{G}''$ is the matrix of the second derivatives of the function $G(\nu_b, \nu_l, \nu_n)$ at its "saddle point" (at which all quantities are marked with the subscript $c$) and $\lambda_0$ is a negative eigenvalue of the matrix $\mathbf{A}\cdot\mathbf{G}''$, with the elements of the diagonal matrix $\mathbf{A}$ of the absorption rates given by $A_{ij} = \delta_{ij} W_i^a$ $(i, j = \mathrm{b, l, n})$ and $\delta_{ij}$ being the Kronecker delta (for more details, see Ref. [37]).

The growth time $t_g$ can be found by solving the differential equation

$$\frac{d\nu_b}{dt} = W_b^a(\nu_b, \nu_l(\nu_b)\nu_n(\nu_b)) - W_b^e(\nu_b, \nu_l(\nu_b), \nu_n(\nu_b)) \tag{3.80}$$

subject to the initial conditions $\nu_b = \nu_{bc}$ at $t = 0$ and $\nu_b = N_b$ at $t = t_g$. One can thus find

$$t_g \simeq \int_{\nu_{bc}}^{N_b} \frac{d\nu}{W_b^a(\nu_b, \nu_l(\nu_b), \nu_n(\nu_b)) - W_b^e(\nu_b, \nu_l(\nu_b), \nu_n(\nu_b)))}, \tag{3.81}$$

where the functions $\nu_l = \nu_l(\nu_b)$ and $\nu_n = \nu_n(\nu_b)$ determine the growth path in parametric form and can be found as the solution of two simultaneous differential equations (see Ref. [37] for more details).

### 3.3.6 Numerical Evaluations

As an illustration of the above theory, we have carried out numerical calculations for the folding of the bovine pancreatic ribonuclease (BPR). According to Elliott and Elliott [52], BPR consists of $N_b = 40$ hydrophobic, $N_l = 81$ hydrophilic, and $N_n = 3$ neutral amino acids, thus totaling $N = 124$ beads. Consequently, we have considered a model heteropolymer consisting of the same numbers of hydrophobic, hydrophilic, and neutral beads as they are present in BPR. Out of 81 hydrophilic beads, only 28 are ionizable, of which $\tilde{N}_l^- = 10$ and $\tilde{N}_l^+ = 18$ can be charged negatively and positively, respectively. The presence of water and solute molecules and ionized beads counterions was not taken into account explicitly but was assumed to be included into the model via the potential parameters in LJ and dihedral angle potentials as well as the inverse Debye length and dielectric permittivity in Equation 3.74.

The actual numbers $N_-$ and $N_+$ of negatively and positively ionized residues in the protein can be smaller than $\tilde{N}_l^-$ and $\tilde{N}_l^+$, respectively, with the average dissociation coefficients $k_- = N_l^-/\tilde{N}_l^-$ and $k_+ = N_l^+/\tilde{N}_l^+$ depending on the pH of the surrounding medium and on the average dissociation coefficients $K_a$ and $K_b$ of the acidic and basic residues in the protein. In the BPR, there are five residues of aspartic acid (asp), five of glutamic acid (glu), ten of lysine (lys), four of arginine (arg), and four of histidine (his). Thus, the average dissociation constant of acidic residues $K_a$ in BPR can be determined through the dissociation coefficients of the corresponding amino acids as $K_a = (5K_{asp} + 5K_{glu})/(5 + 5)$, whereas for the basic amino acids we have $K_b = (10K_{lys} + 4K_{arg} + 4K_{his})/(10 + 4 + 4)$. Data on $K_{asp}$, $K_{glu}$, $K_{lys}$, $K_{arg}$, and $K_{his}$ are available in the literature [52,55] (albeit they exhibit some scatter); hence, $K_a$ and $K_b$ can be estimated. At any given pH, the dissociation coefficients $k_-$ and $k_+$ are completely determined by p$K_a$ and p$K_b$, respectively. We have carried out numerical calculations for pH = 7.3 (roughly that of living cells) at which $N_l^-/N_l^+ \simeq 0.69$, as well as for pH = 6.3 at which $N_l^-/N_l^+ \simeq 0.38$ and for pH = 8.3 at which $N_l^-/N_l^+ \simeq 1.59$.

The nonelectrostatic interactions between any two nonadjacent (i.e., at least three links apart) beads were modeled by LJ potentials, Equation 3.34, whereas the potential due to the dihedral angle δ was represented by Equation 3.35. All the parameters of these two potentials were chosen as in Ref. [37] in order to be able to make a meaningful comparison in clarifying the effect of ionized residues on the nucleation pathway. Likewise, the typical total densities $\rho_f$ and $\rho_u$ of protein residues in the folded and unfolded (but compact) states and the diffusion coefficients of residues $D^{iw}$ and $D^{ow}$ in the ipw and the opw were taken the same as in Ref. [37]: $\rho_f\eta^3 = 0.57$, $\rho_u = 0.2\rho_f$, $D^{iw}\rho_f = D^{ow}\rho_u$, $D^{iw} = D_i^{iw}$ $(i = \text{b, l, n})$, $D^{ow} = D_i^{ow}$ $(i = \text{b, l, n})$, with $D^{iw}$ assumed to vary between $10^{-6}$ and $10^{-8}$ cm$^2$/s. The electrostatic interactions between nonadjacent beads were modeled by the Debye–Hückel approximation for the screened Coulomb potential, Equation 3.74, where the inverse Debye length was assumed to be approximately constant and equal to 1/0.3 nm$^{-1}$, which roughly corresponds to an electrolyte concentration of 0.3 M (typical for living cells [52]) in the solution surrounding the protein. The numerical calculations are presented in Figures 3.11 through 3.13.

Figure 3.11 presents the shapes of the potentials $\phi_l(r) + u_-(r)$ [upper dashed curve in (a) and (b) and lower dashed curve in (c)], $\phi_l(r$ (middle dashed curve), and $\phi_l(r) + u_+(r)$ [lower dashed curves in (a) and (b) and upper dashed curve in (c)], $\bar{\phi}_i^{\delta}(r)$ (dotted–dashed curve), $\psi_+(r)$ [upper solid curves in (a) and (b) and lower solid curve in (c)], $\psi_0(r)$ (middle solid curve) $\psi_-(r)$ [lower solid curves in (a) and (b) and upper solid curve in (c)] for the hydrophilic beads around a cluster as functions of the distance $r$ from the cluster center [the curves $\phi_i(r)$, $\bar{\phi}_i^{\delta}(r)$, and $\psi_i(r)$ $(i = \text{b, n})$ for the hydrophobic and neutral beads have a shape very similar to those presented in the figure]. Figure 3.11a is for pH = 7.3 and a cluster $\nu_b = 23$, $\nu_l^0 = 41$, $\nu_l^- = 4.5$, $\nu_l^+ = 6.5$, $\nu_n = 2$ carrying a total charge $q_\nu = +2e$, Figure 3.11b is for pH = 6.3 and a cluster $\nu_b = 23$, $\nu_l^0 = 41$, $\nu_l^- = 3$, $\nu_l^+ = 8$, $\nu_n = 2$ with $q_\nu = +5e$, and Figure 3.11c is for pH = 8.3 and a cluster $\nu_b = 23$, $\nu_l^0 = 40$, $\nu_l^- = 7.5$, $\nu_l^+ = 4.5$, $\nu_n = 2$ with $q_\nu = -3e$.

The potentials $\phi_l(r)$, $\phi_l(r) + u_-(r)$, $\phi_l(r) + u_+(r)$ arising from the pairwise interactions (LJ and electrostatic) have a form reminiscent of the LJ potential. As one could expect for a positively charged cluster, $\phi_l(r),+ u_-(r) < \phi_l(r) < \phi_l(r) + u_+(r)$ so that the potential well (due to the well of the LJ potential) becomes shallower for positively charged hydrophilic residues and deeper for negatively charged ones as compared to uncharged ones. The effect is not symmetric and, the asymmetry is greater in Figure 3.11b, corresponding to a larger $|q_\nu|$, i.e., to a larger asymmetry in the presence of ionized residues in the system (and hence in the cluster).

The average dihedral potential (assigned to a selected bead) $\bar{\phi}_i^{\delta}(r)$ has a maximum value at the cluster surface and decreases monotonically with increasing $r$ until it becomes constant for $r \geq R + \tilde{d}$, where $\tilde{d}$ is the maximum [37] distance between beads 1 and 6 (or beads 1 and 7) which depends on η and $\beta_0$ (=105°). Such a behavior of $\bar{\phi}_i^{\delta}(r)$ can be interpreted as a consequence of the decrease in entropy (hence an increase in the free energy) of the heteropolymer chain as the selected bead 1 approaches the cluster surface for $r < R + \tilde{d}$, which, in turn, is due to a decrease in the configurational space available to the neighboring beads (2–7).

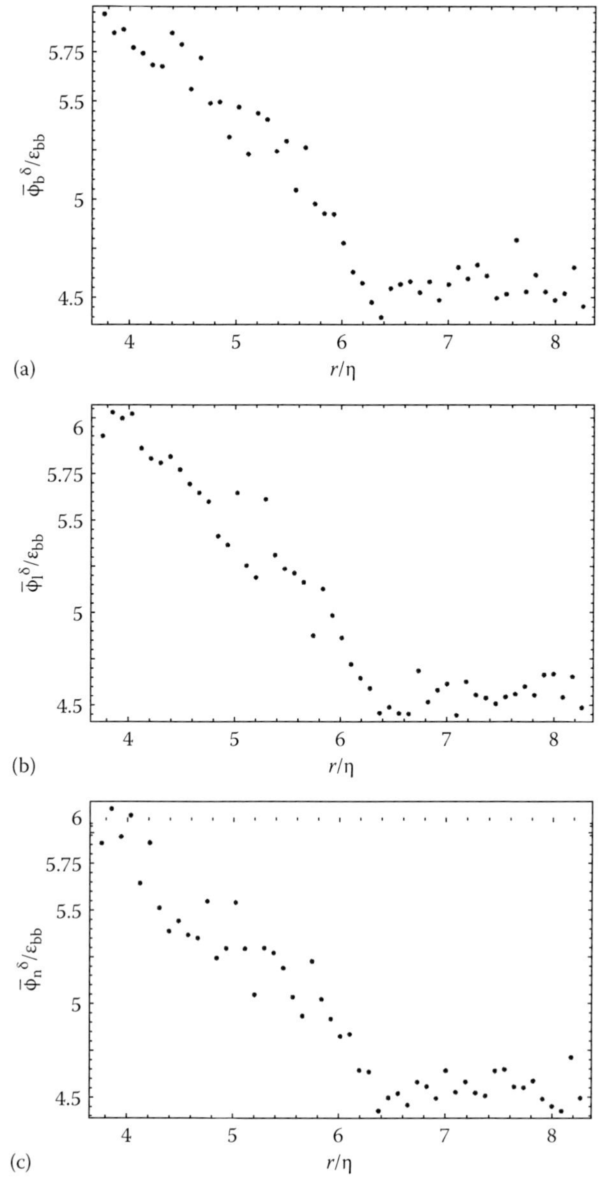


Figure 3.12

Average dihedral potential (assigned to a single bead of type $i$) $\bar{\phi}_i^{\delta}(r)$ $(i = \text{b, l, n})$ as a function of $r$ for a cluster consisting of $\nu_b = 23$, $\nu_l = 52$, and $\nu_n = 2$ hydrophobic, hydrophilic, and neutral beads, respectively. The points represent the actual numerical results obtained by using Monte Carlo integration in Equations 10 and 11 in Ref. [28] for the hydrophobic (a), hydrophilic (b), and neutral (c) beads. The potential $\bar{\phi}_i^{\delta}(r)$ is expressed in units of $\epsilon_{bb}$, energy parameter of the LJ potential for the hydrophobic–hydrophobic pairwise interactions (see Ref. [28]).

As a result of the combination of pairwise interactions-based potential [i.e., either $\phi_l(r)$ or $\phi_l(r) + u_-(r)$ or $\phi_l(r) + u_+(r)$], dihedral angle potential [i.e., $\bar{\phi}_i^{\delta}(r)$], and confining potential (i.e., $\phi_{cp}$) contributions, the overall potential $\psi_l(r)$ has a double well shape: the inner well is separated by a potential barrier from the outer well. The geometric characteristics of the wells (widths, depths, etc.) and the height and location of the barrier between them depend on the interaction parameters. Note that the barrier has different heights for the ipw and opw beads. For a given $N_i$, the location $r_{cp}$ of the outer boundary of the opw is determined by the size of the cluster and densities $\rho_{if}$ and $\rho_{iu}$. The existence of an opw allows one to consider the absorption of a bead by the cluster as an escape of the bead from the opw by crossing over the barrier into the ipw. One can therefore use the mean first passage time analysis for the determination not only of the emission rate but also of the rate of absorption of beads by the cluster.

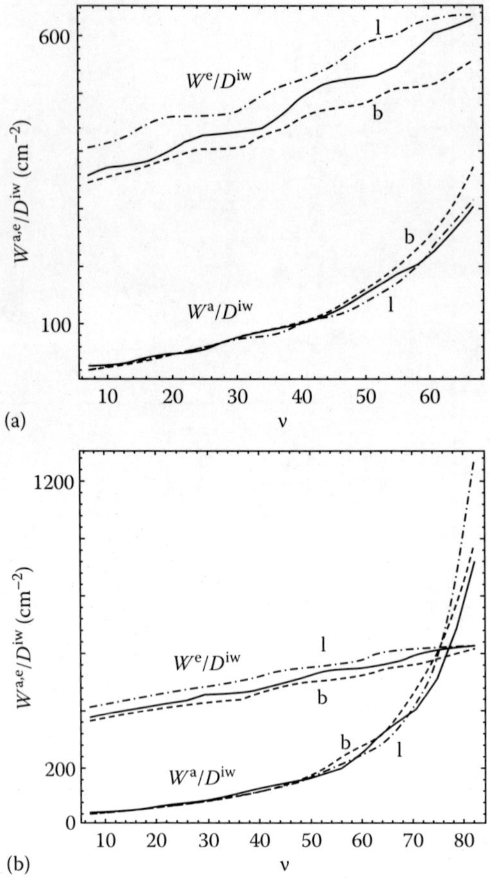


Figure 3.13

The dependence of $W_i^e$ and $W_i^a$ ($i$ = b, l, n) on the total number of beads in a cluster, $\nu = \nu_b + \nu_l + \nu_n$ for pH = 7.3 at constant mole fractions of the hydrophobic and hydrophylic beads therein: (a) $\chi_b = 0.48$, $\chi_l = 0.49$; (b) $\chi_b = 0.31$, $\chi_l = 0.67$. The dashed curves (marked with b) are for hydrophobic beads, the dashed-dotted ones (marked with l) for hydrophilic, and the solid curves for neutral ones. The lower series is for $W_i^a$ and the upper one for $W_i^e$. The location of the simultaneous intersection of the three pairs of functions $W_i^e$ and $W_i^a$ ($i$ = b, l, n) determines the characteristics of the critical cluster, $\nu_{bc} \simeq 25$, $\nu_{lc} \simeq 55$, $\nu_{nc} \simeq 2.3$.

Since the cluster (in both Figure 3.11a and b) is positively charged, both potential wells are shallower (as compared to uncharged beads) for positively ionized beads and deeper for negatively ionized ones which is naturally accounted for by the electrostatic interactions between the selected bead and the cluster. This effect is much weaker for the opw. Furthermore, the depth and width of the opw are hardly affected by changing pH, but those of the ipw are quite sensitive to the absolute value of the overall cluster charge $|q_\nu|$, which, in turn, is a function of pH of the surrounding medium. An increase (decrease) in $|q_\nu|$ results from an increase (decrease) in the ionization ratio $\xi = \max\{N_-, N_+\}/\min\{N_-, N_+\}$. Therefore, an increase in $|q_\nu|$ is accompanied by an increase in the rate of emission of *increasingly larger fraction* of hydrophilic beads in the protein, which leads to a decrease in the ternary nucleation rate and hence to an increase in the folding time of the protein.

In Figure 3.12, the average dihedral potential (assigned to a selected bead) $\bar{\phi}_i^\delta(r)$ is plotted as a function of $r$ for all three kinds of beads [hydrophobic (a), hydrophilic (b), and neutral (c)] for a cluster characterized by $\nu_b = 23$, $\nu_l = 52$, $\nu_n = 2$. The points represent the actual numerical results obtained by using $1 \times 10^6$ point Monte Carlo integration in calculating the sevenfold integrals of Equations 3.42 and 3.43 (note that the exponent of the Boltzmann factor in Equation 3.65 now contains a contribution from the electrostatic interactions of beads 2–7 with the cluster). The vertical dashed lines correspond to $R+\tilde{d}$ and $\bar{\phi}_i^\delta(r)$ is expected to be constant for $r > R+\tilde{d}$.

The emission and absorption rates of a cluster, $W^e$ and $W^a$, respectively, are functions of three independent variables of its state. Figure 3.13 presents the dependence of $W^e$ and $W^a$ (calculated by the mean first passage time analysis, Equations 3.70, 3.75, and 3.76) on the total number of beads in a cluster, $\nu = \nu_b + \nu_l + \nu_n$, at constant mole fractions of the hydrophobic and hydrophylic beads therein. Figure 3.13a presents the case of $\chi_b = 0.48$, $\chi_l = 0.49$, whereas Figure 3.13b is for $\chi_b = 0.31$ $\chi_l = 0.67$. The dashed (marked with b), dashed-dotted (marked with l), and solid curves are for the hydrophobic, hydrophilic, and neutral beads, respectively. Note that, for the case shown in Figure 3.13a, the emission rates (for all three components) are always greater than the absorption rates for all cluster sizes, i.e., the clusters with such a composition ($\chi_b = 0.48$ $\chi_l = 0.49$) have a strong tendency to disaggregate. However, for the case shown in Figure 3.13b ($\chi_b = 0.31$, $x_l = 0.67$), the emission rates are greater than the absorption rates, $W^e > W^a$, for smaller clusters, whereas for larger clusters absorption dominates emission, $W^e < W^a$. The location of the simultaneous intersection of the three pairs of functions $W_i^e$ and $W_i^a$ $(i = \mathrm{b, l, n})$ determines the characteristics of the critical cluster, $\nu_{bc}$, $\nu_{lc}$, $\nu_{nc}$. Note that both $W^e$ and $W^a$ increase with increasing $\nu$, but $W^e$ increases roughly linearly with $\nu$, whereas $W^a$ shoots up by orders of magnitude after the intersection. This occurs because the width of the opw quickly decreases as the cluster grows, whereas the outer height of the barrier between the ipw and opw remains virtually unchanged. For this reason, it becomes increasingly easier for a bead located in the opw to cross the barrier and to fall into the ipw. The characteristics of the critical cluster $\nu_{bc}$, $\nu_{lc}$, and $\nu_{nc}$ can be evaluated by numerically solving three simultaneous equations $W_i^e(\nu_b, \nu_l, \nu_n) = W_i^a(\nu_b, \nu_l, \nu_n)$ $(i = \mathrm{b, l, n})$. Alternatively, $\nu_{bc}$, $\nu_{lc}$, and $\nu_{nc}$ can be found as the solution of three simultaneous equations $\partial G(\nu_b, \nu_l, \nu_n)/d\nu_i, = 0$ $(i = \mathrm{b, l, n})$ satisfying the condition $\det(\mathbf{G}'') < 0$. We used the former method since it involves less computational effort. For the protein considered (BPR with $N_0 = 124$) with the above choices of the system parameters, our model predicts $\nu_{bc} \simeq 25$, $\nu_{lc} \simeq 55$, $\nu_{nc} \simeq 2.3$. (corresponding to $R_c \simeq 3.7\eta$) at pH = 7.3. Our estimates show that the time of protein folding is determined mainly by the time necessary for the first nucleation event to occur because $t_n \gg t_g$. For the diffusion coefficient $D^{iw}$ in the range from $10^{-6}$ to $10^{-8}$ $cm^2/s$, our model estimates the characteristic time of BPR folding to be in the ranges from about 3 to about 300 s at pH = 8.3, from about 2 to about 200 s at pH = 7.3, and from about 7 to about 700 for pH = 6.3. Thus, when one takes into account the ionizability of some of the hydrophilic beads in BPR, an increase in the folding time occurs [37]. However, the effect is significant only for rather unusually low pH = 6.3, whereas for a physiological pH = 7.3 of living cells the effect is just slightly higher than the statistical uncertainty due to the Monte Carlo integration in calculating the average dihedral angle potential and to the numerical procedure of determining $\nu_{bc}$, $\nu_{lc}$, $\nu_{nc}$. Among all the pH values considered, the physiological pH provides the lowest folding time.

### 3.3.7 Conclusions

One of the multiple pathways for protein folding involves a nucleation mechanism. Its theory was first approached [11,12,21,35] by means of the classical thermodynamics, calculating the free energy of formation of a cluster of protein residues in the framework of the capillarity approximation, much like the CNT does. In this approach, a crucial role is played by the surface tension for a cluster of protein residues which is an intrinsically ill-defined physical quantity.

To avoid this difficulty, we have recently developed [37] a model for the nucleation mechanism of protein folding in terms of ternary nucleation by using a first passage time analysis. In that model, a protein was considered as a random heteropolymer consisting of hydrophobic, hydrophilic, and neutral beads with all the bonds in the heteropolymer having the same constant length and all the bond angles equal and fixed. The main idea underlying the model consists of averaging the dihedral potential in which a selected residue is involved over all the possible configurations of all neighboring residues along the protein chain. The combination of the average dihedral potential with the effective pairwise potential of the selected residue and with a confining potential caused by the bonds between the residues leads to an overall potential around the cluster that has a double well shape. Residues in the inner (closer to the cluster) well are considered as belonging to the folded cluster, whereas those in the outer well are treated as belonging to the unfolded part of the protein. Transitions of residues from the inner well into the outer one and vice versa are considered as elementary emission and absorption events, respectively. The double well character of the potential field around the cluster allows one to determine the rates of both emission and absorption of residues by the cluster using a first passage time analysis. Once these rates are found as functions of the cluster size, one can develop a self-consistent kinetic theory for the nucleation mechanism of folding of a protein. The model allows one to evaluate the size and composition of the nucleus and the protein folding time. The latter is evaluated as the sum of the times necessary for the first nucleation event to occur and for the nucleus to grow to the maximum size of the folded protein.

It should be noted that computer simulations reported the nucleus during protein folding to be a specific fragment of a native protein [22]. In our model, the specific sequence of beads is averaged out by considering them uniformly distributed along the polymer which is similar to modeling the protein as a variant of a random heteropolymer [48–50]. However, important details are still preserved upon this averaging because the nucleus is determined not only by its size but also by its composition, i.e., by three independent variables of state.

In this work, we further develop that model by taking into account the ionizability of some of protein residues (aspartic and glutamic acids, lysine, arginine, and histidine). As a result, the effective pairwise potential around the cluster of native residues now includes an electrostatic contribution if the selected residue is ionized. As noted previously, an overall potential around the cluster wherein a protein residue performs a chaotic motion as a function of the distance from the cluster center has a double well shape even for ionized beads. Strictly speaking, taking into account the ionizability of some residues required the development of a model in terms of five-component nucleation. This would lead to a huge increase in the amount of numerical calculations needed for the application of the model. However, one can avoid this by assuming that the ratio of the numbers of negatively and positively ionized residues in the cluster remains the same as in the entire protein during its folding. With this assumption, the modified model remains within the framework of the ternary nucleation formalism, and one can evaluate the size and composition of the nucleus and the protein folding time as in the previous model.

For a numerical illustration, we have again applied the model to the folding of BPR, protein consisting of 124 residues, whereof 40 can be classified as hydrophobic, 81 as hydrophilic (of which 10 are negatively and 18 positively ionizable), and 3 as neutral. Numerical calculations have been performed at pH = 8.3, pH = 7.3, and pH = 6.3. They show that taking into account the ionizability of some of the hydrophilic beads of BPR leads to an increase in the folding time [37]. However, the effect of electrostatic interactions leads to a significant increase in the folding time roughly by a factor of 7 at pH = 6.3, by a factor of 2 at pH = 7.3, and by a factor of 3 at pH = 8.3. Thus, a more realistic model, taking into account the ionizability of some protein residues, may predict protein folding times which may differ by as much as an order of magnitude from the predictions of the previous version of the model where this effect was neglected. However, these conclusions should not be generalized because they are specific to the particular protein considered.

Clearly, the model we have proposed is coarse grained and does not implement many details which can be important for the folding phenomenon (such as the sequence of amino acids in a protein, effect of solvent, hydrogen bonding). However, a ternary heteropolymer model and the corresponding pairwise and configurational potentials have been successful in studying and elucidating many aspects of folding (as briefly described in Refs. [36,37]). Developing a theoretical approach for protein folding is much more complicated than doing simulations due to the notorious complexity of the problem. The model for the nucleation mechanism of folding presented above is significantly more realistic than the previous ones. Although its application involves extensive numerical calculations, one can hardly expect to develop a theoretical model of protein folding which would be more or less realistic and at the same time amenable on the "back of the envelope." Clearly, the model requires thorough testing and development, but this is a gradual process and will be the subject of our future reports.

## References

1. Creighton, T.E., *Proteins: Structure and Molecular Properties*, Freeman, San Francisco, 1984.
2. Stryer, L., *Biochemistry*, 3rd ed., Freeman, San Francisco, 1988.
3. Anfinsen, C.B., *Science* 181, 223 (1973).
4. Ghelis, C., Yan, J., *Protein Folding*, Academic, New York, 1982.
5. Nölting, B., *Protein Folding Kinetics*, Springer-Verlag, Berlin, 2006.
6. Honeycutt, J.D., Thirumalai, D., *Dev. Mech.* 32, 695 (1992).
7. Weissman, J.S., Kim, P.S., *Science* 253, 1386 (1991).
8. Creighton, T.E., *Nature (London)* 356, 194 (1992).
9. DeGrado, W.F., Wasserman, Z.R., Lear, J.D., *Science* 243, 622 (1989).
10. Pace, C.N., *J. Mol. Biol.* 226, 29 (1992).
11. Dill, K.A., *Biochemistry* 24, 1501 (1985).
12. Dill, K.A., *Biochemistry* 29, 7133 (1990).
13. Creighton, T. E., *Proteins: Structure and Molecular Properties*, Freeman, New York, 1993.
14. Kohn, W.D., Kay, C.M., Hodges, R.S., *Protein Sci.* 4, 237 (1995).
15. Kohn, W.D., Kay, C.M., Hodges, R.S., *J. Mol. Biol.* 237, 500 (1994).
16. Kohn, W.D., Kay, C.M., Hodges, R.S., *Protein Eng.* 7, 1365 (1994).
17. Kohn, W.D., Kay, C.M., Hodges, R.S., *J. Mol. Biol.* 283, 993 (1998).

18. Kohn, W.D., Monera, O.D., Kay, C.M., Hodges, R.S., *J. Biol. Chem.* 270, 25495 (1995).
19. Horovitz, A., Serrano, L., Avron, B., Bycroft, M., Fersht, A.R., *J. Mol. Biol.* 216, 1031 (1990).
20. Fersht, A.R., *Curr. Opin. Struct. Biol.* 7, 3 (1997).
21. Guo, Z., Thirumalai, D., *Biopolymers* 36, 83 (1995).
22. Abkevich, V.I., Gutin, A.M., Shakhnovich, E.I., *Biochemistry* 33, 10026 (1994).
23. Levinthal, C., *J. Chim. Phys. Phys.-Chim. Biol.* 85, 44 (1968).
24. Kim, P.S., Baldwin, R.I., *Annu. Rev. Biochem.* 51, 459 (1982).
25. Kim, P.S., Baldwin, R.I., *Annu. Rev. Biochem.* 59, 631 (1990).
26. Creighton, T.E., *Biochem. J.* 240, 1 (1990).
27. Creighton, T.E., *Prog. Biophys. Mol. Biol.* 33, 231 (1978).
28. Nölting, B., *J. Theor. Biol.* 194, 419 (1998).
29. Nölting, B., *J. Theor. Biol.* 197, 113 (1999).
30. Shakhnovich, E.I., *Curr. Opin. Struct. Biol.* 7, 29 (1997).
31. Shakhnovich, E.I., Abkevich, V., Ptitsyn, O., *Nature (London)* 379, 96 (1996).
32. Finkelstein, A.V., Badretdinov, A.Y., *Folding Des.* 2, 115 (1997).
33. Bryngelson, J.D., Wolynes, P.G., *Proc. Natl. Acad. Sci. U.S.A.* 84, 7524 (1987).
34. Bryngelson, J.D., Wolynes, P.G., *J. Phys. Chem.* 93, 6902 (1998).
35. Bryngelson, J.D., Wolynes, P.G., *Biopolymers* 30, 177 (1990).
36. Djikaev, Y.S., Ruckenstein, E., *J. Phys. Chem. B* 111, 886 (2007). (Section 3.1 of this volume.)
37. Djikaev, Y.S., Ruckenstein, E., *J. Chem. Phys.* 126, 175103 (2007). (Section 3.2 of this volume.)
38. Narsimhan, G., Ruckenstein, E., *J. Colloid Interface Sci.* 128, 549 (1989). (Section 1.2 of this volume.)
39. Ruckenstein, E., Nowakowski, B., *J. Colloid Interface Sci.* 137, 583 (1990). (Section 1.3 of this volume.)
40. Nowakowski, B., Ruckenstein, E., *J. Colloid Interface Sci.* 139, 500 (1990). (Section 1.4 of this volume.)
41. Djikaev, Y.S., Ruckenstein, E., *J. Chem. Phys.* 123, 214503 (2005). (Section 2.5 of this volume.)
42. Djikaev, Y.S., Ruckenstein, E., *J. Chem. Phys.* 124, 124521 (2006). (Section 2.6 of this volume.)
43. Djikaev, Y.S., Ruckenstein, E., *J. Chem. Phys.* 124, 194709 (2006). (Section 1.6 of this volume.)
44. Levitt, M., Sharon, R., *Proc. Natl. Acad. Sci. U.S.A.* 85, 8557 (1988).
45. Sikorski, A., Skolnik, J., *Biopolymers* 28, 1097 (1989).
46. Sikorski, A., Skolnik, J., *J. Mol. Biol.* 213, 183 (1990).
47. Skolnik, J., Kolinski, A., *Science* 250, 1121 (1990).
48. Shakhnovich, E.I., Gutin, A.M., *J. Phys. A* 22, 1647 (1989).
49. Bratko, D., Chakraborty, A.K., Shakhnovich, E.I., *J. Chem. Phys.* 106, 1264 (1997).
50. Konkoli, Z., Hertz, J., Franz, S., *Phys. Rev. E* 64, 051910 (2001).
51. Abraham, F.F., *Homogeneous Nucleation Theory*, Academic, New York, 1974.
52. Elliott, W.H., Elliott, D.C., *Biochemistry and Molecular Biology*, Oxford University Press, New York, 2003.
53. Chandrasekhar, S., *Rev. Mod. Phys.* 15, 1 (1949).
54. Gardiner, C.W., *Handbook of Stochastic Methods*, Springer, New York, 1983.
55. Schmidt, C.L.A., Kirk, P.L., Appleman, W.K., *J. Biol. Chem.* 88, 285 (1930).

## 3.4 Temperature Effects on the Nucleation Mechanism of Protein Folding and on the Barrierless Thermal Denaturation of a Native Protein

*Yuri Djikaev and Eli Ruckenstein**

Recently, the authors proposed a kinetic model for the nucleation mechanism of protein folding where a protein was treated as a heteropolymer with all the bonds and bond angles equal and constant. As a crucial idea of the model, an overall potential around a cluster of native residues in which a protein residue performs a chaotic motion is considered to be a combination of three potentials: effective pairwise, average dihedral, and confining. The overall potential as a function of the distance from the cluster center has a double well shape that allows one to determine the rates with which the cluster emits and absorbs residues by using a first passage time analysis. One can then develop a kinetic theory for the nucleation mechanism of protein folding and evaluate the protein folding time. In the present paper, we evaluate the optimal temperature at which the protein folding time is the shortest. A method is also proposed to determine the temperature dependence of the folding time without carrying out the time consuming calculations for a series of temperatures. Using Taylor series expansions in the formalism of the first passage time analysis, one can calculate the temperature dependence of the cluster emission and absorption rates in the vicinity of some temperature $T_0$ if they are known at $T_0$. Thus, one can evaluate the protein folding time $t_f$ at any other temperature $T$ in the vicinity of $T_0$ at which the folding time $t_f$ is known. We also present a model for the thermal denaturation of a protein occurring via the decay of the native structure of the protein. Because of a sufficiently large temperature increase or decrease, the rate with which a cluster of native residues within a protein emits residues becomes larger than the absorption rate in the whole range of cluster sizes up to the size of the whole protein. This leads to the unfolding of the protein in a barrierless way, i.e., as spinodal decomposition. Knowing the cluster emission and absorption rates as functions of temperature and cluster size, one can find the threshold temperatures of cold and hot barrierless denaturation as well as the corresponding unfolding times. Both proposed methods are illustrated by numerical calculations for two model proteins, one consisting of 124 amino acids, the other consisting of 2500 residues. The first one roughly mimics a bovine pancreatic ribonuclease whereas the second one is a representative of the largest proteins that are extremely difficult to study by straightforward Monte Carlo or molecular dynamics simulations.

### 3.4.1 Introduction

For a protein molecule to carry out a specific biological function, it has to adopt a well-defined (native) three-dimensional structure [1,2]. The formation and stability of this structure (of a biologically active protein) constitute the core of two exciting (and intrinsically related) topics of modern biophysics, namely, "protein folding" and "protein denaturation" (i.e., unfolding) problems [2–5]. Many thermodynamic and kinetic aspects of these processes remain obscure [6–17].

Computer simulations, based either on Monte Carlo (MC) or on molecular dynamics (MD) methods, have been an essential tool for examining these phenomena which are extremely complicated for analytical methods. Few theoretical models have been proposed for the nucleation mechanism of protein folding which is believed to be one of the possible pathways for the transition of a protein from a compact "amorphous" configuration (into which an initially unfolded protein quickly transforms) to the native state. Such a transition occurs immediately following the formation of a number of tertiary (native) contacts [8,9,18–21]; once a critical number thereof is attained, the native structure is formed without passing through any detectable intermediates. Analytically the problem was initially [15,22–24] approached by using classical thermodynamics and calculating the free energy of formation of a cluster of protein residues (stabilized by correct, native tertiary contacts) in the framework of the classical nucleation theory (CNT) based on the capillarity approximation [25]. Such an approach involves the concept of surface tension for a cluster consisting of protein residues which has no clear physical meaning and can be considered at best as an adjustable parameter.

With the goal of improving the theory and bringing it to a level more adequate to the current state of simulations and experiments [26–30], we have recently [31–33] presented a new, microscopic model for the nucleation mechanism of protein folding. The model is based on the first passage time analysis [34–37] first applied to nucleation phenomena in vapor-to-liquid and liquid-to-solid phase transitions by Ruckenstein

* *Phys. Chem. Chem. Phys.* 10, 6281 (2008). Republished with permission.

and coworkers [38–40]. Our model [31–33] for nucleation in protein folding involves pairwise "molecular" interactions (i.e., repulsion/attraction), configurational (dihedral angle) potentials in which protein residues are involved, and confining potential which arises because of the finite size of the protein. The parameters of these potentials can be rigorously defined. The protein itself is treated as a heteropolymer chain consisting of three types of beads—neutral, hydrophobic, and hydrophilic [8,9,19,32] (with all the bonds and bond angles equal and constant; see Figure 3.14a). The ionizability of the latter type of residues was also taken into account [33]. The rates of emission and absorption of residues by the cluster of native residues (within a protein) are determined by using a first passage time analysis [34–37]. Once these rates are found as functions of cluster size, one can develop a self-consistent kinetic theory for the nucleation mechanism of folding of a protein. The time necessary for the protein to fold can be evaluated as the sum of the times necessary for the appearance of a nucleus and the time necessary for the nucleus to grow to the maximum size (of the folded protein in the native state). The ill-defined surface tension of a cluster of protein residues (within a protein in a compact but nonnative configuration) does not enter into the new model at all.

Both experiments and simulations have demonstrated that a change in temperature significantly affects the process of protein folding [41–45] as well as the stability of a native protein (which, having folded at temperature $T_0$, can unfold upon both cooling and heating [14–17]). In the present paper we first examine the effect of temperature on protein folding. We propose a method for obtaining the temperature dependence

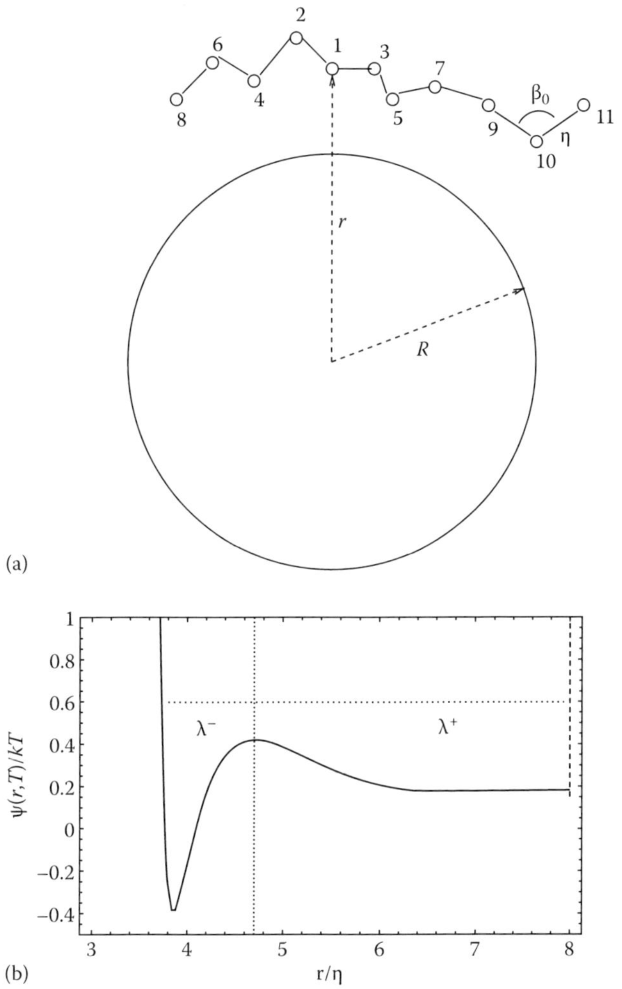


Figure 3.14

(a) A piece of a heteropolymer chain around a spherical cluster consisting of $v$ beads. Bead 1 is in the plane of the figure, whereas other beads may all lie in different planes. All bond angles are equal to $\beta_0$ and their lengths are equal to $\eta$. The radius of the cluster is $R$ and the distance from the selected bead 1 to the cluster center is $r$. (b) A schematic representation of the total potential field around the cluster of radius $R = 3\eta$, which a selected bead (bead 1 in panel a) is subjected to, as a function of the distance $r/\eta$ from the cluster center. The widths of the inner and outer potential wells are $\lambda^-$ and $\lambda^+$, respectively.

of the protein folding time without carrying out long and tedious calculations for a series of temperatures. Instead, the method makes use of Taylor series expansions in the formalism of the first passage time analysis whereupon our model for the nucleation mechanism of protein folding is based [31–33].

Once the protein is in its native state, the change in temperature may lead to its unfolding (denaturation). The unfolding of a protein results in the loss of its biological functionality which may lead to catastrophic consequences for a living organism.

Protein denaturation can be defined as a process (or sequence of processes) whereby the spatial arrangement of the polypeptide chains within the molecule changes from that of the native protein to a more disordered one [4]. This change may alter the secondary, tertiary or quaternary structure of the protein. Note that what constitutes denaturation is largely dependent upon its cause. Changes in the structure of proteins can be caused by a variety of factors, such as heating or cooling, changes in pH, contact with liquid–vapor or liquid–liquid interfaces, changes in the dielectric constant (upon addition of a cosolvent that is miscible with water but is less/more polar), changes in the ionic strength of the solution. In this paper, we will be concerned mainly with the thermal mode of protein denaturation.

As evidenced by both experiments and simulations, a native protein, stable at temperature $T_0$, can unfold upon cooling as well as heating [4,5,14–17]. Protein unfolding (denaturation) is easier understood at high temperatures where it can be at least partly accounted for by a decreased stability of hydrogen bonding between water molecules and between water and protein residues. When proteins are exposed to increasing temperature, losses of solubility or enzymatic activity occur over a relatively narrow range of temperatures. Depending on the protein and the severity of heating, these changes may or may not be reversible.

With increasing temperature, some bonds in a protein molecule are weakened. The first to be affected are the long-range interactions responsible for the stability of the tertiary structure. As these are weakened and broken, the protein becomes more flexible and its (previously) internal groups become more exposed to the solvent. If heating ceases at this stage, the protein is able to readily refold to the native structure. Upon further heating, some of the hydrogen bonds that stabilize the helical structure begin to break, so that water can interact with and form new hydrogen bonds with the amide nitrogen and carbonyl oxygens of peptide bonds. Thus, as the helical structure is broken, new hydrogen bonding groups and hydrophobic groups are exposed to the solvent, which leads to an increase in the amount of water bound to the protein molecules. The net result will be an attempt by the protein to minimize its free energy by burying as many hydrophobic groups as possible while exposing as many polar groups as possible to the solvent. Although this is analogous to what occurred when the protein initially folded, it now occurs at a much higher temperature. This greatly weakens the short-range interactions that initially direct protein folding, and the resulting structures are often very different from those of the native protein.

The widespread existence of protein unfolding at low temperatures is surprising, particularly because it is unexpectedly accompanied by a decrease in entropy [46]. It was shown [5] that the cold denaturation is a general phenomenon caused by the very specific and strongly temperature-dependent interactions of the protein nonpolar groups with water. In contrast to expectations, the hydration of these groups is favorable thermodynamically, i.e., the Gibbs energy of hydration is negative and increases in magnitude as the temperature decreases [5]. As a result, the polypeptide chain, tightly packed in a compact native structure, unfolds at sufficiently low temperatures by exposing internal nonpolar groups to water.

So far, all theoretical treatments of protein unfolding have been mostly intended for extracting information from computer simulations of the process. So far, virtually no analytical description of protein denaturation has been proposed that would be similar to those [19,22,23,31–33] for the nucleation mechanism of protein folding [which is believed to be one of the possible pathways for the transition of a protein from the compact "amorphous" configuration (into which an initially unfolded protein quickly transforms) to the native state].

In this paper, we present an analytical model for the process of protein denaturation occurring in a barrierless way (i.e., as spinodal decomposition). The approach used is similar to that of the model for the nucleation mechanism of protein folding. We consider a native protein stable at temperature $T_0$ and examine the temperature dependence of the rates with which it can emit and absorb residues. At $T_0$, the absorption rate is greater than the emission rate for large enough clusters (of native residues). The unfolding temperature is determined as the temperature at which the emission rate becomes greater than the absorption rate in the whole range of cluster sizes (including the one corresponding to the entire protein). As in studying the temperature effect on protein folding, the temperature dependence of the emission and absorption rates is examined by using Taylor series expansions in the formalism of a first passage time analysis [31–33].

### 3.4.2 Temperature Expansions for Protein Folding Time

#### 3.4.2.1 First Passage Time Analysis-Based Model for the Nucleation Mechanism of Protein Folding

In our previous work [31–33], we considered a protein as a random heteropolymer [47–51] consisting of $N$ connected beads which can be thought of as representing the $\alpha$-carbons of various amino acids (Figure 3.14). The study of simplified random heteropolymer models (in both "random bond" and "random energy" versions [47,50–54]) may provide a useful insight into the phenomenon of protein folding. For simplicity, let us consider the case where the heteropolymer consists only [31] of hydrophobic (b) and hydrophilic (l) beads sequentially (and randomly) connected to one another by covalent bonds of equal and constant length $\eta$ with equal and constant angles $\beta_0$ between two neighboring bonds.

It is believed (and assumed hereafter) that any proposed protein folding pathway must start with the so called "random coil" state and finish with the unique folded conformation with all the native protein contacts. The folded state is stable and well defined, but the nature of the unfolded state is less obvious. Definitions for the characteristics of a random coil vary [55–57]. Moreover, owing to a number of recent studies [58–63,] it appears that the random coil state is far less random than previously suggested.

Assuming that the composition of a cluster of native residues within a protein (which is in a compact, but not-native state) remains constant and equal to the overall composition of the protein at all times, one can [31] treat the nucleation mechanism of protein folding in terms of unary nucleation and characterize the cluster by $\nu$, the total number of residues therein. In that model [31] (as well as in its more detailed versions [32,33]) the protein folding time $t_{\mathrm{f}}$ is determined as the sum

$$t_{\mathrm{f}} \simeq t_{\mathrm{n}} + t_{\mathrm{g}}, \tag{3.82}$$

where $t_{\mathrm{n}}$ is the time necessary for a critical cluster to nucleate within a compact (but still unfolded) protein and $t_{\mathrm{g}}$ is the time necessary for the nucleus to grow up to the maximum size, i.e., the size of the entirely folded protein.

The time $t_{\mathrm{n}}$ of the first nucleation event can be estimated as the inverse of the frequency of nucleation events in the whole volume of the protein in a compact (but unfolded) state. In the steady state, this frequency is equal to the product $J_{\mathrm{s}}V_0$, where $J_{\mathrm{s}} \equiv J_{\mathrm{s}}(T)$ is the steady-state nucleation rate at temperature $T$ and $V_0$ is the volume of the unfolded protein in a compact state. Thus,

$$t_{\mathrm{n}}(T) \simeq 1/[J_{\mathrm{s}}(T)V_0]. \tag{3.83}$$

The steady-state nucleation rate $J_{\mathrm{s}}(T)$ is given by

$$J_{\mathrm{s}}(T) = \frac{W_{\mathrm{c}}^{+}}{\sqrt{\pi \Delta \nu_{\mathrm{c}}}} \rho_{\mathrm{u}} e^{-G_{\mathrm{c}}}. \tag{3.84}$$

Here, the subscript "c" marks quantities for the critical cluster, $W^{+} \equiv W^{+}(\nu)$ is the overall rate with which a cluster containing $\nu$ residues absorbs residues from the unfolded part of the protein, $\Delta \nu_{\mathrm{c}} = \left|\partial^2 G/\partial \nu^2\right|_{\mathrm{c}}^{-1/2}$, where the function $G \equiv G(\nu) = -\ln[g_{\mathrm{e}}(\nu)/\rho_{\mathrm{u}}]$, $g_{\mathrm{e}}(\nu)$ being the equilibrium distribution of clusters with respect to $\nu$ and $\rho_{\mathrm{u}}$ the number density of residues in the unfolded (although compact) protein (for more details, see Ref. [31]). The equilibrium distribution $g_{\mathrm{e}}(\nu)$ has the form [31]

$$g_{\mathrm{e}}(\nu) = \rho_{\mathrm{u}} \frac{W^{+}(1)}{W^{+}(\nu)} \prod_{i=1}^{\nu-1} \frac{W^{+}(\nu-i+1)}{W^{-}(\nu-i+1)}, \tag{3.85}$$

where $W^{-} \equiv W^{-}(\nu)$ is the overall rate with which a cluster of size $\nu$ emits residues into the unfolded part of the protein. Note that on the right-hand side of Equation 3.84, the quantities $W_{\mathrm{c}}^{+}$, $\Delta \nu_{\mathrm{c}}$, and $G_{\mathrm{c}}$ depend on the temperature $T$: $W_{\mathrm{c}}^{+} = W_{\mathrm{c}}^{+}(T)$, $\Delta \nu_{\mathrm{c}} = \Delta \nu_{\mathrm{c}}(T)$, $G_{\mathrm{c}} = G_{\mathrm{c}}(T)$. The equilibrium distribution $g_{\mathrm{e}}(\nu)$ is also a function of $T$ because the rates of both absorption and emission, $W^{+}(\nu)$ and $W^{-}(\nu)$, depend on $T$.

The total rates of emission and absorption of residues by the cluster are defined as

$$W^- = W_b^- + W_l^-, \; W^+ = W_b^+ + W_l^+,$$

where $W_i^- \equiv W_i^-$ $(\nu_b,\nu_l)$ and $W_i^+ \equiv W_i^+$ $(\nu_b,\nu_l)$ $(i = \text{b, l})$ are the rates of emission and absorption, respectively, of beads of type $i$ by a cluster $(\nu_b,\nu_l)$, $\nu_b$ and $\nu_l$ being the numbers of hydrophobic and hydrophilic residues, respectively, in the cluster. Owing to the approximation that the composition of a cluster of native residues within a protein (which is in a compact but not-native state) remains constant and equal to the overall composition of the protein at all times, $\nu_b$ and $\nu_l$ are related to $\nu = \nu_b + \nu_l$ as $\nu_b = \chi_0\nu$ and $\nu_l = (1 - \chi_0)\nu$, where $\chi_0 = N_b/(N_b + N_l)$ is the mole fraction of hydrophobic residues in the entire protein ($N_i$ $(i = \text{b, l})$ is the total number of residues of type $i$ in the protein). Thus, one can [31] characterize the cluster by $\nu$, the total number of residues therein, so the above definitions of $W^+$ and $W^-$ can be rewritten

$$\begin{aligned} W^w(\nu) &= W_b^w(\chi_0\nu,(1-\chi_0)\nu) + W_l^w(\chi_0\nu,(1-\chi_0)\nu) \\ &\equiv W_b^w(\nu_b,\nu_l) + W_l^w(\nu_b,\nu_l) \quad (w = +,-). \end{aligned} \tag{3.86}$$

The growth time $t_g$ can be evaluated by solving the differential equations, governing the temporal evolution of the variables $\nu_b$ and $\nu_l$ for supercritical clusters, $d\nu_i/dt = W_i^+ - W_i^-$, $(i = \text{b, l})$ (where $t$ is the time) subject to the initial conditions $\nu_i = \nu_{ic}$ $(i = \text{b, l})$ at $t = 0$ and $\nu_i = N_i$ $(i = \text{b, l})$ at $t = t_g$. Because of the approximation $\nu_b/(\nu_b + \nu_l) = \text{const} = \chi_0$, this couple of simultaneous equations can be reduced to a single independent equation which, due to Equation 3.86, can be written in the form $d\nu/dt = W^+(\nu) - W^-(\nu)$ and is subject to the initial conditions $\nu = \nu_c$ at $t = 0$ and $\nu = N$ at $t = t_g$ (with $N = N_b + N_l$). One thus obtains

$$t_g \simeq \int_{\nu_c}^{N} \frac{d\nu}{W^+(\nu) - W^-(\nu)}.$$

Note that the right-hand side of this equation is positive because $W^+(\nu) > W^-(\nu)$ for $\nu > \nu_c$, according to the definition of a nucleus [31–33].

Numerical evaluations showed [31–33] that the protein folding time is determined mainly by the time necessary for the first nucleation event to occur because $t_n \gg t_g$ so that, for a given temperature $T$, $t_f(T) \simeq t_n(T)$ (when important and convenient, the temperature dependence of some of the quantities involved will be explicitly emphasized). Thus,

$$t_f(T) \simeq 1/[J_s(T)V_0] \tag{3.87}$$

and, as clearly seen in Equations 3.83 through 3.85 and 3.87, one can estimate the folding time $t_f$ of a protein if the emission and absorption rates $W^+(\nu)$ and $W^-(\nu)$ are known as functions of the cluster variable $\nu$.

#### 3.4.2.2 Determination of $W^+ = W^+(\nu)$ and $W^- = W^-(\nu)$ via a First Passage Time Analysis

At any given temperature, both functions $W^+ = W^+(\nu)$ and $W^- = W^-(\nu)$ can be determined by using a first passage time analysis [31–33]. Indeed, let us consider a spherical cluster of native residues (with the structure stabilized by native tertiary contacts) within a protein and denote the distance from its center by $r$. In our model [31–33], the total potential $\psi_i(r)$ of a selected bead of type $i$ around the cluster contains three terms: the effective pairwise potential $\phi_i(r)$, confining potential $\phi_{cp}$, and average dihedral potential $\bar{\phi}_i^\delta(r)$:

$$\psi_i(r) = \phi_i(r) + \phi_{cp} + \bar{\phi}_i^\delta(r) \tag{3.88}$$

(for details, see Appendix 3G).

The term $\phi_i(r)$ on the right-hand side of Equation 3.88 represents the effective pairwise potential due to the interactions of the selected residue with all the residues in the cluster. The confining potential $\phi_{cp}$ is due to the polymer connectivity constraint: because of the bonds between the residues, none of them can be located outside some confining boundary. Finally, the term $\bar{\phi}_i^\delta(r)$ in $\psi_i(r)$ is due to the $r$-dependent part of the dihedral angle potential of the whole protein. This is a configurational potential consisting of contributions from every dihedral angle formed by four consecutive beads in the heteropolymer. The potential due to the dihedral angle δ can be modeled, e.g., as [19]

$$\phi_\delta = \varepsilon'_\delta(1+\cos\delta)+\varepsilon''_\delta(1+\cos 3\delta). \tag{3.89}$$

Here, $\varepsilon'_\delta$ and $\varepsilon''_\delta$ are independent energy parameters that depend on the nature and sequence of the four beads involved in the dihedral angle δ. This potential has three minima, one in the *trans* configuration at δ = 0 and two others in the *gauche* configurations at $\delta = \pm \arccos\sqrt{(3\varepsilon''_\delta - \varepsilon'_\delta)/12\varepsilon''_\delta}$ (the former one being the lowest).

Let us consider a bead 1 (of type *b* or *l*) at a distance *r* from the (center of the) cluster, $r > R$ (see Figure 3.14), assign the total dihedral angle potential of the whole protein chain for a given configuration of beads 1, 2, 3, …, *N* to bead 1, and denote it by $\phi^\delta_{i_1}(r)$ [the subscript $i_s$ indicates the type of bead *s* ($s = 1,\ldots, N$; $i_s$ = b, l)]. The potential $\phi^\delta_{i_1}(r)$ contains contributions from all the dihedral angles in the heteropolymer each of which is given by Equation 3.89. Clearly, $\phi^\delta_{i_1}(r)$ is a function not only of *r*, but also of the coordinates of beads 2 through *N*, i.e., $r_2, r_3, \ldots, r_N$.

The contribution to $\phi^\delta_{i_1}(r)$ from six nearest neighbors of bead 1 (three on each side of bead 1 along the heteropolymer), i.e., from beads 2 through 7, depends on *r*. However, the contribution to $\phi^\delta_{i_1}(r)$ from the dihedral angles involving beads 8, 9, …, *N* does not depend on *r* because those dihedral angles remain unaffected as the position of bead 1 changes because bead 1 participates in the formation of dihedral angles only with three nearest beads on either side of the heteropolymer (beads 2, 4, and 6 on one side and beads 3, 5, and 7 on the other; see Figure 3.14). Hence, this contribution (independent of *r*) can be omitted because it can be regarded as affecting only the reference level for $\bar{\phi}^\delta_{i_1}(r)$.

For a given location of bead 1, various configurations of beads 2, 3, …, *N* (subject to the bond length and bond angle constraints as well as to the constraint of excluded cluster volume) lead to various sets of dihedral angles. However, variations in the location of beads 8, 9, …, *N* give rise to variations in the dihedral potential which are independent of the position of bead 1, characterized by *r*. The dihedral term $\bar{\phi}^\delta_i(r)$ on the right-hand side of Equation 3.88 can be obtained by averaging $\phi^\delta_{i_1}(r)$ with the Boltzmann factor exp $\left[-\left(\phi^{\delta 27}_{i_1}(r)/k_BT\right)\right]$, where $\phi^{\delta 27}_{i_1}(r) \equiv \phi^{\delta 27}_{i_1}$ $(r, \mathbf{r}_2, \ldots, \mathbf{r}_7)$ $= \phi^\delta_{i_1}(r) + \sum_{s=2}^{7} \phi_{i_s}(r_s)$, over all the possible configurations of beads 2, 3, …, 7 and assigning the result to the selected bead with fixed coordinates, i.e., to bead 1 (for more details, see Refs. [31–33] or Appendix 3G). Because of this averaging, the mean dihedral potential depends on the temperature *T*, and so does $\psi_i(r)$ in Equation 3.88, i.e., $\bar{\phi}^\delta_i(r) \equiv \bar{\phi}^\delta_i(r,T)$, $\psi_i(r) \equiv \psi_i(r,T)$. Therefore, $\Psi_i(r) \equiv \Psi_i(r,T) = \psi_i(r,T)/k_BT$. It should be emphasized that in this model the polymer connectivity is taken into account not only through the confining potential $\phi_{cp}$, but also through the average dihedral potential $\bar{\phi}^\delta_i(r,T)$. Both of them are constituents of the overall potential field $\psi_i(r,T)$ whereto a selected bead is subjected.

It was shown [31–33] that the combination of potentials on the right-hand side of Equation 3.88 gives rise to a double potential well around the cluster with a barrier between the two wells. Residues in the inner (closer to the cluster) potential well (ipw) are considered as belonging to the folded cluster, whereas those in the outer potential well (opw) are treated as belonging to the unfolded part of the protein. Transitions of residues from the inner well into the outer one and vice versa are considered as elementary emission and absorption events, respectively. The double well character of the potential well around the cluster allows one to determine the rates of both emission and absorption of residues by the cluster using a first passage time analysis.

Let us consider a heteropolymer bead (i.e., a protein residue) of type *i* (*i* = b, l) performing a chaotic motion in a spherically symmetric potential well $\psi_i(r)$ with one boundary infinitely high, say, at $r = R$ (where *R* is the radius of the cluster) and the other one of finite height, say, at $r = R + \lambda_i^-$ (where $\lambda_i^-$ is the width of the well). Such a choice of the well boundaries and their locations correspond to the ipw ($\lambda_i^-$ is the distance from the cluster surface and the location of the barrier between the ipw and the opw) for which all quantities will be marked with the superscript "–". According to a first passage time analysis [31–33], the mean first passage time $\tau_i^-$ necessary for the molecule to escape from the ipw is given by

$$\tau_i^- = \frac{1}{Z_i^-}\frac{1}{D_i^-}\int_R^{R+\lambda_i^-} dr_0 r_0^2 e^{-\psi_i(r_0)} \int_{r_0}^{R+\lambda_i^-} dy\, y^{-2} e^{\psi_i(y)} \times \int_R^y dx\, x^2 e^{-\psi_{i(x)}}, \tag{3.90}$$

where $D_i^-$ is the diffusion coefficient of a residue of type $i$ in the ipw,

$$Z_i^- = \int_R^{R+\lambda_i^-} dr_0 r_0^2 e^{-\psi_i(r_0)}. \tag{3.91}$$

The expression for $\tau_i$ is derived by solving a single-molecule master equation for the probability distribution function of a surface layer molecule (residue/bead) moving in a potential field $\psi_i(r)$. The thermal motion of the bead is assumed to be governed by the Fokker–Planck equation which reduces [34–37] to the Smoluchowski equation (involving diffusion in an external field) if the relaxation time for the velocity distribution function of the molecule is very short and negligible compared to the characteristic timescale of the passage process (for more details, see Appendix 3H).

The rate of emission of beads of type $i$ by the cluster (i.e., the number of residues of type $i$ escaping from the ipw into the opw per unit time) is thus provided by

$$W_i^- = \frac{N_i^-}{\tau_i^-}, \tag{3.92}$$

where $N_i^-$ denotes the number of molecules in the ipw. In a similar fashion, one can determine the rate with which the cluster absorbs beads of type $i$ (i.e., the number of residues of type $i$ passing from the opw into the ipw per unit time) as

$$W_i^+ = \frac{N_i^+}{\tau_i^+}, \tag{3.93}$$

where $N_i^+$ is the number of beads in the opw and $\tau_i^+$ is the mean first passage time for the transition of a bead of type $i$ from the opw into the ipw (all quantities for the opw are marked with the superscript "+"). The total rates of emission and absorption of residues by the cluster are determined by Equation 3.86. The quantities $N_i^-$ and $N_i^+$, numbers of molecules in the ipw and opw, respectively, are given by

$$N_i^- = \frac{4\pi}{3}\left[(R+\lambda^-)^3 - R^3\right]\rho_f,$$
$$N_i^+ = \frac{4\pi}{3}\left[(R+\lambda^- +\lambda^+)^3 - (R+\lambda^-)^3\right]\rho_u,$$

where $\lambda^-$ and $\lambda^+$ are the widths of the inner and outer potential wells, $\rho_f$ and $\rho_u$ are the overall number densities of residues in the folded and unfolded parts of the protein.

### 3.4.2.3 Optimal Temperature of Protein Folding

Every protein has optimal conditions for its fastest folding. For instance, if only the temperature of the protein system can vary, there exists such an optimal temperature $T^o$, at which the folding time of the protein is the shortest [42]. The model developed in Refs. [31–33] and outlined above allows one to obtain an estimate for the optimal temperature of a given model protein.

Indeed, according to Equations 3.84 and 3.87, the temperature dependence of the folding time is due to the temperature dependence of $W_c^+ = W_c^+(T)$, $\Delta\nu_c = \Delta\nu_c(T)$, and $g_e(\nu_c) = g_e(\nu_c,T)$:

$$t_f(T) = \frac{\sqrt{\pi}}{V_0}\frac{\Delta\nu_c(T)}{W_c^+(T)g_e(\nu_c,T)}. \tag{3.94}$$

The extremum of this function can be found by solving the equation $dt_f(T)/dT = 0$ with respect to $T$. The folding time is known to become infinitely large at both very high and very low temperatures. As $t_f(T)$ is not expected to have any oscillations, it is clear that if the function $t_f(T)$ has an extremum, it should be its minimum. Thus, in the framework of our approach the optimal temperature $T^o$ of protein folding is the solution of the equation

$$\frac{d}{dT}\frac{\Delta v_c(T)}{W_c^+(T)g_e(v_c,T)}\bigg|_{T^o}=0, \tag{3.95}$$

which can be rewritten as

$$\begin{aligned}&g_e(v_c,T)\Delta v_c[dW_c^+(T)/dT]+W_c^+(T)\Delta v_c(T)[dg_e(v_c,T)dT]\\&\quad -g_e(v_c,T)W_c^+(T)[d\Delta v_c/dT]=0.\end{aligned} \tag{3.96}$$

The quantity $\Delta v_c$ characterizes the width of the near-critical region of the function $G(v)$. According to the definition of this vicinity around $v_c$, the function $G(v)$ decreases by $k_BT$, compared to $G(v_c)$, when $v$ deviates from $v_c$ by $\pm\Delta v_c$. In the kinetics of nucleation based on the first passage time analysis, the function $G(v)$ plays a role equivalent to the free energy of cluster formation whereof the width of the near-critical region is known [64,65] to be rather insensitive to $T$. Thus, a rough estimate for the optimal temperature $T^o$ of protein folding can be obtained by solving the approximate equation

$$g_e(v_c,T)[dW_c^+(T)/dT]+W_c^+(T)[dg_e(v_c,T)/dT]=0 \tag{3.97}$$

to which Equation 3.96 is reduced if the last term on its right-hand side is neglected. Strictly speaking, it is impossible to solve Equation 3.97 analytically because at any given temperature all the components of the equation (i.e., $v_c$, $g_e(v_c)$, and $W_c^+$) are found numerically. However, one can suggest the following semianalytical procedure to evaluate $T^o$ with the least possible computational effort.

Assume that $g_e(v_c,T)$ and $W_c^+(T)$ are linear functions of $T$. This approximation is acceptable for any function if used in a temperature range $T_a \le T \le T_b$ narrow enough for the nonlinearities to be negligible. The functions $g_e(v_c,T)$ and $W_c^+(T)$ can then be written as

$$g_e(v_c,T)\simeq g_0+g_1T,\ W_c^+(T)=w_0+w_1T. \tag{3.98}$$

The coefficients $g_0$, $g_1$ and $w_0$, $w_1$ can be determined by the values of $g_e(v_c,T)$ and $W_c^+(T)$ at the two temperatures, $T_a$ and $T_b$:

$$\begin{aligned}g_0&=g_e(v_c,T_a)-\frac{T_a}{T_a-T_b}[g_e(v_c,T_a)-g_e(v_c,T_b)],\\ g_1&=\frac{g_e(v_c,T_a)-g_e(v_c,T_b)}{T_a-T_b},\end{aligned} \tag{3.99}$$

$$\begin{aligned}w_0&=W_c^+(T_a)-\frac{T_a}{T_a-T_b}\left[W_c^+(T_a)-W_c^+(T_b)\right],\\ w_1&=\frac{W_c^+(T_a)-W_c^+(T_b)}{T_a-T_b}.\end{aligned} \tag{3.100}$$

It should be noted that $\nu_c$ is a function of the temperature $T$. Substituting Equation 3.98 into Equation 3.97 and solving with respect to $T$, one obtains for the optimal temperature of protein folding

$$T^0 = -\frac{1}{2}\left(\frac{g_0}{g_1} + \frac{w_0}{w_1}\right). \tag{3.101}$$

Clearly, the accuracy of this estimate will be largely dependent on the accuracy of the linear approximations in Equation 3.98.

#### 3.4.2.4 Taylor Series Expansions for Protein Folding Time

Now that all temperature-dependent quantities in our model have been identified, let us develop a procedure for obtaining the temperature dependence of the protein folding time $t_f$ on the basis of calculations carried out at a single temperature $T_0$ when $T$ is in the vicinity of $T_0$. As already mentioned, the procedure is based on a Taylor series expansion in the vicinity of $T_0$. Using linear expansions in some vicinity $T_0 - \Delta T \le T \le T_0 + \Delta T$ of $T_0$, Equations 3.85 and 3.87 lead to the following relation between the folding times at temperatures $T$ and $T_0$ (for details, see Appendix 3I):

$$\frac{t_f(T)}{t_f(T_0)} \simeq \varepsilon_g(\nu_c, T_0, T)\left[1 + \varepsilon_w^+(\nu_c, T_0)(T - T_0)\right] \times [1 + \varepsilon_\nu(\nu_c, T_0)(T - T_0)], \tag{3.102}$$

where

$$\varepsilon_w^{\pm}(\nu, T) = \frac{d\tau^{\pm}(\nu, T)/dT}{\tau^{\pm}(\nu, T)}, \tag{3.103}$$

$$\varepsilon_\nu(\nu, T) = \frac{1}{2}(\Delta\nu_c(T))^2 \frac{\partial^2}{\partial \nu^2}\left(\varepsilon_w^+(\nu, T) + \varepsilon_w(\nu, T)\right), \tag{3.104}$$

$$\varepsilon_w(\nu, T) = \sum_{i=1}^{\nu-1}\left(-\varepsilon_w^+(\nu - i + 1, T) + \varepsilon_w^-(\nu - i + 1, T)\right), \tag{3.105}$$

$$\varepsilon_g(\nu, T_0, T) = \frac{1 - \varepsilon_w^+(1, T_0)(T - T_0)}{1 - \varepsilon_w^+(\nu, T_0)(T - T_0)} \times \prod_{i=1}^{\nu-1} \frac{1 - \varepsilon_w^+(\nu - i + 1, T_0)(T - T_0)}{1 - \varepsilon_w^-(\nu - i + 1, T_0)(T - T_0)}. \tag{3.106}$$

In Equation 3.103, $\tau^+$ and $\tau^-$ are the mean first passage times for the transitions of a bead from the outer potential well into the inner one (absorption of a bead) and from the inner potential well into the outer one (emission of a bead), respectively (remind that the potential field around the cluster has a double well shape).

Numerical calculations (involving MC integration in Equations 3G.10, 3G.11, 3G.14, and 3G.15 in Appendix 3G) that are necessary to construct the functions $\varepsilon_w^{\pm}(\nu, T)$, $\varepsilon_\nu(\nu, T)$, $\varepsilon_w(\nu, T)$, and $\varepsilon_g(\nu, T_0, T)$ for a given cluster at a selected temperature $T_0$, require approximately one day on a Dell (Pentium 4, 3 GHz, 1 GB) computer. On the other hand, on the same computer, the full procedure for the evaluation of the folding time $t_f(T)$ at a selected temperature $T$ requires about (or more than) 20 days. First, the effective pairwise potential $\phi(r)$ for a bead in the vicinity of the cluster is calculated as a function of $r$ (distance from the center of the cluster) for a cluster containing $v$ beads taken on a grid which must be fine enough to be able to construct $\phi(r|v)$, the effective pairwise potential at a fixed $r$ as a function of the continuous variable $v$ (note that the integration in calculating $\phi(r)$ can be carried out analytically if the density $\rho$ inside the cluster is assumed to be uniform). This function represents the first term in $\psi(r) = \phi(r) + \bar{\phi}_\delta(r) + \phi_{cp}$, the potential field around a cluster. Next, the average dihedral potential $\bar{\phi}^\delta(r)$ as a function of $r$ (for given $v$) is calculated by using the same grid which must be fine enough to construct $\bar{\phi}^\delta(r\,|\,v)$, the average dihedral potential as a function of the continuous variables $v$ at fixed $r$. This function represents the second term in $\psi(r,v)$. Thus, one can numerically calculate $\psi(r,v)$, the potential field around a cluster, as a function of two variables $r$, $v$. Knowing this function, one can then calculate the emission and absorption rates of a cluster (as outlined in Refs. [31–33]) as functions of $v$. (The accuracy in building these functions is extremely important because thereupon depends the accuracy with which the function $G(v)$ and its "saddle point" (i.e., the parameters of the critical cluster) are determined. This, in turn, has a major impact on the accuracy of the prediction of the nucleation rate and hence folding time.) Knowing the emission and absorption rates $W^+(v)$ and $W^-(v)$, the equilibrium distribution $g_e(v)$ is obtained via Equation 3.85 and the nucleation rate is calculated with Equation 3.84, which involves the function $G(v) = -\ln[g_e(v)/\rho_u)]$. The time $t_f$ necessary for the protein to fold is evaluated with Equation 3.87.

Thus, in order to examine the temperature dependence of the protein folding time $t_f$, the straightforward application of the method proposed in Refs. [31–33] would require extremely long calculations. On the other hand, if $t_f$ is known at some temperature $T_0$, the Taylor series expansions, Equation 3.102, allow one to evaluate the ratio $t_f(T)/t_f(T_0)$ for any $T$ in a close enough vicinity of $T_0$ on a timescale by 1 or 2 orders of magnitude shorter than the time necessary for the evaluation of $t_f(T_0)$.

If the range of temperatures (i.e., the width $2\Delta T$ of the interval centered at $T_0$), in which Equation 3.102 is valid, is not large enough to examine the behavior of $t_f$ as a function of $T$, one can increase that range by retaining higher order terms (quadratic, cubic, etc., in $T - T_0$) in the Taylor series expansions of $W^+$, $W^-$, and $\Delta v_c$ in Equation 3.104. As a result, the right-hand side of Equation 3.102 will become more complicated, but all the changes in the procedure will just slightly affect the timescale of the calculations.

### 3.4.3 Time of Protein Denaturation via Spinodal Decomposition

As mentioned earlier, the stability of a folded (native) protein can be disrupted by various factors. Depending on the strength of the destabilizing factor(s), the denaturation (unfolding) of the protein can occur either fluctuationally as nucleation (at weak destabilization) or, alternatively, in a barrierless way as spinodal decomposition (at strong destabilization). In the former case (i.e., unfolding via nucleation), the initial unfolding of the protein is thermodynamically unfavorable, but can still occur owing to fluctuations. After the protein unfolded to some critical extent, further denaturation becomes thermodynamically favorable and occurs with the decrease in the free energy of the system (protein + medium). In the latter case (i.e., unfolding via spinodal decomposition), it is thermodynamically favorable for a protein to unfold starting from its native state down to its full denaturation.

Being interested in thermal denaturation, let us consider a protein consisting of $N_b$ hydrophobic and $N_l$ hydrophilic residues. Initially folded at some temperature $T_0$, assume that it can unfold via spinodal decomposition at either $T \geq T_u^+ > T_0$, which corresponds to hot denaturation, or $T \leq T_u^- < T_0$, which corresponds to cold denaturation. The temperature $T_u^-$ represents the highest temperature at which cold denaturation can occur, whereas $T_u^+$ represents the lowest temperature of hot denaturation.

The threshold temperatures of hot and cold denaturation ($T_u^+$ and $T_u^-$, respectively) can be determined by solving the equation $W^-(N,T) = W^+(N,T)$ with respect to $T$. If the equation has two solutions, $T_u^+$ and $T_u^-$, the protein can unfold upon both heating and cooling. If the equation has just one solution, either $T_u^+$ or $T_u^-$, protein denaturation can occur just upon either heating or cooling. Plotting the difference $W^-(N,T) - W^+(N,T)$ as a function of $T - T_0$, Figure 3.15a illustrates the former situation qualitatively, whereas Figure 3.15b and c illustrates the two latter situations quantitatively (for more details, see Section 3.4.4). In general, $|T_u^+ - T_0|$ does not have to be equal to $|T_u^- - T_0|$ because denaturation upon heating and cooling is due to several factors some of which are of different nature in the two cases and even those which are of the same nature may be nonlinear functions of temperature.

The emission and absorption rates of a cluster (of native residues) of size $v$ ($0 \leq v \leq N$) as functions of $v$ are shown in Figure 3.16a (for a protein at $T = T_0$), Figure 3.16b (for a protein at $T = T_u^+$ or $T = T_u^-$), and Figure 3.16c (for a protein at $T > T_u^+$ or $T < T_u^-$). If the protein is initially unfolded at $T_0$ (Figure 3.16a), it can fold via nucleation [8,9,18,20,21,31–33]. In principle, the native protein at this temperature can also unfold via nucleation, but in this work we assume that the thermodynamic barrier for the transition of a cluster of size

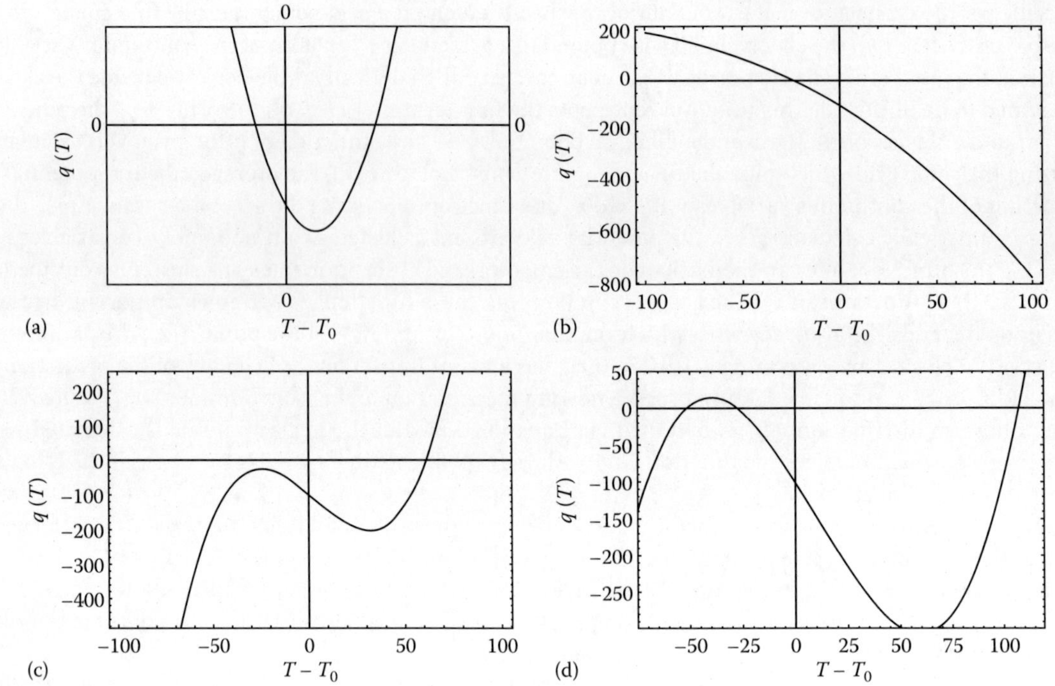


Figure 3.15

The function $q(T) = W^-(N,T) - W^+(N,T)$ plotted versus $T - T_0$ for a protein (with $N = 2500$) exhibiting: (a) both cold and hot denaturation; (b) only cold denaturation; (c) only hot denaturation; (d) both cold and hot denaturation, but, unlike case (a), as predicted in the approximation of the third-order series expansions for $W^+(N,T)$ and $W^-(N,T)$ (see the text).

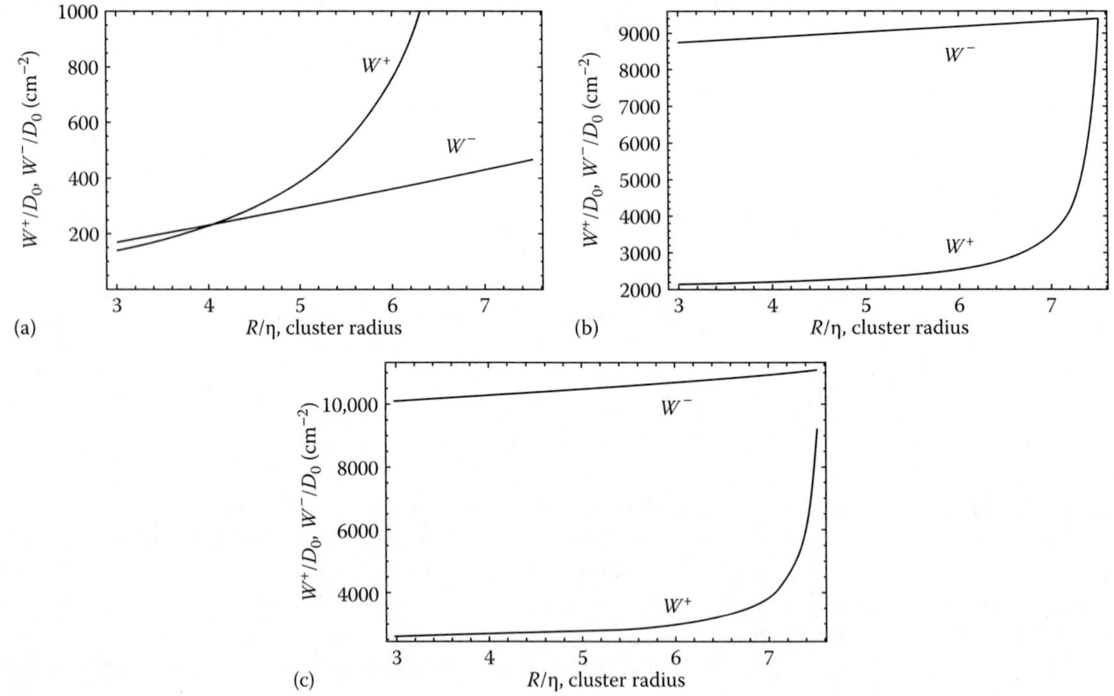


Figure 3.16

The emission and absorption rates of a cluster (of native residues), $W^+(\nu)$ and $W^-(\nu)$, as functions of $\nu$ at $T = T_0 = 293.15$ K (protein in the native state, panel a), $T = T_u^+ \simeq 358$ K (threshold temperature of hot denaturation, panel b), and $T = T_u^+ + 5$ K (temperature above the threshold one of hot denaturation, panel c). The data are for the large protein model of $N = 2500$.

$N$ (the whole protein in a native state) to the size $\nu_c$ (size of the critical cluster) is so high that the unfolding time is virtually infinity. On the other hand (see Figure 3.16b), $W^+(\nu) \le W^-(\nu)$ for $\nu \le N$ at temperature $T_u^{\pm}$ (the equality $W^+ = W^-$ holds only for the largest cluster possible, i.e., at $\nu = N$). Therefore, if a protein is initially unfolded at $T_u^{\pm}$, it will never fold, whereas, if it is initially folded at this temperature, it will immediately start unfolding via regular loss of residues from its native structure. The unfolding will occur even faster at $T > T_u^+$ (or $T < T_u^-$), as shown in Figure 3.16c, because in this case $W^+(\nu) < W^-(\nu)$ for all $\nu \le N$ (including $\nu = N$) and the difference $W^-(\nu) - W^-(\nu)$ is greater than for $T = T_u^{\pm}$ (see Section 3.4.4 for more details).

The full unfolding time $t_u$ can be evaluated by solving a couple of simultaneous differential equations governing the temporal evolution of the variables $\nu_b$ and $\nu_l$ for $\nu_b \le N_b$ and $\nu_l, \le N_l$:

$$d\nu_i/dt = W_i^+ - W_i^-, \ (i = b, l) \tag{3.107}$$

subject to the initial conditions $\nu_i = N_i$ $(i = b, l)$ at $t = 0$ and $\nu_i = 0$ $(i = b, l)$ at $t = t_g$. However, by virtue of the chosen approximation $\nu_b/(\nu_b + \nu_l) = \text{const} = \chi_0$, these simultaneous equations reduce to a single independent equation

$$\frac{d\nu}{dt} = -[W^-(\nu) - W^+(\nu)], \tag{3.108}$$

which has to be solved subject to the initial conditions $\nu = N$ at $t = 0$ and $\nu = 0$ at $t = t_u$. One can thus obtain

$$t_u \simeq \int_0^N \frac{d\nu}{W^-(\nu) - W^+(\nu)}. \tag{3.109}$$

Note that the right-hand side of this equation is positive because $W^+(\nu) < W^-(\nu)$ in the whole range $0 \le \nu \le N$ at $T = T_u^{\pm}$ (see Figure 3.16b).

#### 3.4.3.1 Taylor Series Expansions for Emission and Absorption Rates

As clearly shown in the previous section, in order to evaluate $T_u^+$ and $T_u^-$ the minimum and maximum temperatures of hot and cold denaturation, respectively, via spinodal decomposition, it is necessary to know the temperature dependence of the functions $W^- = W^-(\nu)$ and $W^+ = W^+(\nu)$, i.e., the functions $W_i^w = W_i^w(\nu, T)$ $(w = +, -; i = b, l)$, however, even determining $W_i^w$ $(w = +, -; i = b, l)$ as functions of only $\nu$ at a given $T$ requires lengthy computations. The straightforward application of the method proposed in Refs. [31–33] for constructing $W_i^w$ $(w = +, -; i = b, l)$ as functions of both $\nu$ and $T$ and determining the temperatures of hot and cold denaturations $T_u^+$ and $T_u^-$ and the corresponding unfolding times $t_u^+$ and $t_u^-$ would require extremely long calculations. In order to avoid this problem, one can use the method proposed above where the temperature dependence of the protein folding time was determined by using Taylor series expansions.

Indeed, if $W_i^w$ is known as a function of $\nu$ at some temperature $T_0$, one can evaluate $W_i^w(\nu, T)$ for any $T$ in a close enough vicinity of $T_0$ using the series expansion

$$\begin{aligned} W_i^w(\nu,T) = W_i^w(\nu,T_0)[1 - \alpha_i^w(\nu,T_0)(T-T_0) - \beta_i^w(\nu,T_0) \\ \times (T-T_0)^2 - \gamma_i^w(\nu,T_0)(T-T_0)^3 + \cdots] \\ (w = +,-;\ i = b, l), \end{aligned} \tag{3.110}$$

where

$$\begin{aligned} \alpha_i^w(\nu,T) &= \frac{d\tau_i^w(\nu,T)/dT}{\tau_i^w(\nu,T)}, \\ \beta_i^w(\nu,T) &= \frac{1}{2}\frac{d^2\tau_i^w(\nu,T)/dT^2}{\tau_i^w(\nu,T)}, \\ \gamma_i^w(\nu,T) &= \frac{1}{6}\frac{d^3\tau_i^w(\nu,T)/dT^3}{\tau_i^w(\nu,T)} \end{aligned} \tag{3.111}$$

(for details, see Appendix 3J). Here, $\tau_i^w$ ($w = +,-$) is the mean first passage time necessary for the bead of type $i$ ($i$ = b, l) to escape from the well $w$. Equation 3.110 is derived from Equations 3.92 and 3.93 by assuming that $N_b^-$, $N_b^+$, $N_l^-$ and $N_l^+$, the numbers of beads of types b and l in the inner and outer potential wells, do not depend on $T$. Clearly, the range of validity of the expansions in Equation 3.110 becomes larger when higher order (in $T - T_0$) terms are retained on its right-hand side.

#### 3.4.3.2 Threshold Temperature and Time of Thermal Denaturation

If a protein is subject to both hot and cold denaturation, the corresponding threshold temperatures at which its unfolding can occur via spinodal decomposition ($T_u^+$ and $T_u^-$, respectively), can be determined by solving the equation $W^-(N,T) = W^+(N,T)$ with respect to $T$, i.e.,

$$W^-(N,T_u^+) = W^+(N,T_u^+), W^-(N,T_u^-) = W^+(N,T_u^-) \tag{3.112}$$

(note that both $W^-(\nu,T)$ and $W^+(\nu,T)$ are monotonic functions of $\nu$). The first of these equalities means that the threshold temperature of hot denaturation is the lowest temperature at which the emission rate of a cluster is greater than or equal to its absorption rate for clusters of any size, i.e., $W^-(\nu,T) \geq W^+(\nu,T)$ for any $\nu \leq N$ and $T \geq T_u^+$. The second of the equalities in Equation 3.112 means that the threshold temperature of cold denaturation is the highest temperature at which the emission rate of a cluster is greater than or equal to its absorption rate for clusters of any size, i.e., $W^-(\nu,T) \geq W^+(\nu,T)$ for any $\nu \leq N$ and $T \leq T_u^-$.

The estimates of the times necessary for the protein to unfold at these temperatures via spinodal decomposition are given by the equations

$$\begin{aligned} t_u^- &\simeq \int_0^{N-0} \frac{d\nu}{W^-(\nu,T_u^-) - W^+(\nu,T_u^-)} \\ t_u^+ &\simeq \int_0^{N-0} \frac{d\nu}{W^-(\nu,T_u^+) - W^+(\nu,T_u^+)}. \end{aligned} \tag{3.113}$$

Clearly, the unfolding time of hot denaturation at a temperature $T > T_u^+$ is smaller than $t_u^+$, whereas the unfolding time of cold denaturation at a temperature $T < T_u^-$ is smaller than $t_u^-$. The divergence of the integrands at the upper limit of integration at $T = T_u^+$ and $T = T_u^-$ complicates the application of these expressions. The character of the divergence determines whether the $t_u$'s are finite or infinite, but this issue cannot be rigorously addressed unless the $\nu$ dependence of $W^-(\nu) - W^+(\nu)$ is explicitly known. Reasonable estimates for $t_u^-$ and $t_u^+$ can be obtained either by evaluating the integrals at temperatures very close to the threshold ones, $T_u^{\pm}(1 \pm \delta)$ with a positive $\delta \ll 1$ or by slightly decreasing the upper limit of integration. The latter method could be substantiated by the effect of fluctuations whereof the physical impact is to decrease the size of the folded cluster of size $N$ but which are not formally included in the model.

### 3.4.4 Numerical Evaluations

In order to illustrate the above methods we applied them to the folding and unfolding of two model proteins composed of 124 and 2500 amino acid residues. The first roughly mimics a bovine pancreatic ribonuclease and was previously studied in Refs. [32,33]. In this work, however (unlike Refs. [32,33]), the neutral residues were treated as hydrophobic so that the whole heteropolymer was considered to consist of hydrophilic and hydrophobic residues with the mole fraction of the latter being $\chi_0 = 43/124 \simeq 0.347$. The second model protein is a representative of large proteins. As in Ref. [31], it was a heteropolymer consisting of a total of 2500 hydrophobic and hydrophilic residues, with the mole fraction of hydrophobic residues $\chi_0 = 0.75$. We intentionally modeled such a large protein (relatively rare in nature) to demonstrate that our method provides a reasonable temperature dependence of the folding/unfolding times on a reasonable timescale even for the largest proteins. Straightforward MC or MD simulations of folding/unfolding of such large proteins are currently impossible within a sensible timeframe.

The interactions between a pair of nonadjacent (at least three links apart) beads were modeled by Lennard–Jones (LJ) type potentials (Equation 3.1). The interactions of nearest and next nearest neighbor beads are completely taken into account by the constant bond length and constant bond angles constraints (whereof the contribution to the total energy of the system is constant hence can be omitted). The potential due to the dihedral angle $\delta$ was modeled according to Equation 3.89. The presence of water molecules was not taken into account explicitly but was assumed to be included into the model via the potential parameters and diffusion coefficients. Clearly, hydrophilic residues are readily solvated by water molecules.

The formation of hydrogen bonds between water molecules and amino acid residues (particularly hydrophilic ones) has various effects on the behavior of the polypeptide chain. The pairwise interactions of protein residues are affected by the water molecules hydrating them. Dihedral angle potential may also be affected due to changes in the properties of the hydrated residues involved. The direct interaction between water molecules hydrogen-bound to protein residues also contributes to residue–residue interactions. All these effects are included into the model *via* the choice of not only potential parameters of both pairwise and dihedral angle potentials, but also the temperature dependence of diffusion coefficients. Indeed, the destruction of hydrogen bonds between water molecules and protein residues with increasing temperature provides an additional physical mechanism for the increase of diffusion coefficients with increasing temperature which was implemented in the numerical calculations. (Currently, we are developing a model which takes into account the effect of water–water hydrogen bonds on the total potential well around the cluster. The disruption of the hydrogen bond network in the first hydration shell around a cluster occurring when a protein residue [which is itself surrounded by a network of hydrogen bound water molecules] approaches the cluster from the unfolded part of the protein will lead to the decrease in the depth of the inner potential well, which will likely lead to an increase in the protein folding time and a decrease in the unfolding time.)

Clearly, such a heteropolymer is a two-component system and, strictly speaking, the nucleation mechanism of its folding must be treated in terms of binary nucleation. However, because the overall composition of the protein (heteropolymer) is fixed, in Ref. [31] the cluster which is formed during folding was assumed to have a constant composition equal to the overall protein composition. This assumption reduces the theory to a unary nucleation theory and substantially simplifies the numerical calculations. Note that in Refs. [32,33] this assumption was removed from the model. In the present work, however, for the sake of simplicity, we have carried out the Taylor series expansions in the framework of the unary nucleation mechanism. The generalization of this method to the ternary nucleation mechanism of protein folding does not involve any major difficulties.

All numerical calculations were carried out for the following values of the interaction parameters: (1) $\beta = 105°$, $\eta = 4.3 \times 10^{-8}$ cm, $\varepsilon_{bb} = 4 \times 10^{-14}$ erg, $\varepsilon_{ll} = (0.137/0.143)\ \varepsilon_{bb}$, $\varepsilon_{bl} = \sqrt{\varepsilon_{bb}\varepsilon_{ll}}$, $\varepsilon'_{\delta} = \varepsilon''_{\delta} = 1.2\varepsilon_{bb}$ for the short protein model (124 residues); (2) $\beta = 105°$, $\eta = 5.39 \times 10^{-8}$ cm, $\varepsilon_{ll} = (2/700)\ \varepsilon_{bb}$, $\varepsilon_{bl} = \sqrt{\varepsilon_{bb}\varepsilon_{ll}}$, $\varepsilon'_{\delta} = \varepsilon''_{\delta} = 0.3\varepsilon_{bb}$ for the large protein model (2500 residues). In both cases, $\varepsilon_{bb} = 4 \times 10^{-14}$ (note that for the shorter protein the parameters in the LJ potential were chosen in consistency with the optimized potentials for liquid simulations for proteins developed by Jorgensen and Tirado-Rives [66]). Here, $\beta$ is the bond angle, $\eta$ is the length parameter in the LJ potential, whereas $\varepsilon_{bb}$, $\varepsilon_{ll}$, and $\varepsilon_{bl}$ are the energy parameters in the LJ potentials for the hydrophobic–hydrophobic, hydrophilic–hydrophilic, and hydrophobic–hydrophilic interactions, respectively (the mole fraction of hydrophobic beads was taken $\chi = 0.75$ both in the whole protein and in the cluster), and $\varepsilon'_{\delta} = \varepsilon''_{\delta}$ are the energy parameters in the dihedral angle potential (for more details, see Refs. [31]). The typical density of the folded protein was evaluated using data from Refs. [67,68] and was set to $\rho_f\eta^3 = 1.05$, whereas the typical density of the unfolded protein in the compact configuration was set to be $\rho_u = 0.25\rho_f$. The temperature dependence of the diffusion coefficients $D^{iw}$ and $D^{ow}$ was modeled as a power law $D_0(1 - T_c/T)^{\gamma}$ as predicted [69,70] by the mode coupling theory. Taking into account the results of Refs. [71,72], the diffusion coefficients in the ipw and the opw were assumed to be related by $D_0^{iw}\rho_f = D_0^{ow}\rho_u$. Because of the lack of reliable data on the diffusion coefficient of a residue in a native protein, $D_0^{iw}$ was assumed to vary between $10^{-7}$ and $10^{-9}$ cm$^2$ s$^{-1}$ (much smaller than typical diffusion coefficients for gases but somewhat larger than typical values for solids). The parameters $T_c$ and $\gamma$ were chosen to be $T_c = 240$ K and $\gamma = 2.1$ (for the short protein) and $T_c = 250$ K and $\gamma = 2.3$ (for the large protein). The parameter $T_c$ determines the temperature at which the "freezing" of the protein residues occurs. Our choice of $T_c$ is reasonable because, although water can be supercooled down to 235 K, the system "water molecules"–"protein residues" can be considered a binary solution in which small particles are known [69,70] to remain mobile even when bigger particles have already stopped their motion. Note that the $T$-dependence of $D^{iw}$ arises not only from the direct temperature effect on the chaotic motion of protein residues but also from the indirect effect of either weakening (with increasing $T$) or strengthening (with decreasing $T$) of their hydrogen bonds with water molecules and between water molecules themselves.

#### 3.4.4.1 Results for Protein Folding

At temperature $T_0 = 293$ K, the model proteins are predicted [31–33] to fold into native states on a timescale of 1 to 100 s (depending on what numerical value is chosen for the factor $D_0^{iw}$ in the diffusion coefficient $D^{iw}$). In order to evaluate the optimal temperature of folding as described in Section 3.4.2.2, calculations were carried out also at $T = 298$ K. Using the numerical results for $T_a = 293$ K and $T_b = 298$ K, we estimated the optimal folding temperatures for the above model protein to be $T^\circ \simeq 291.5$ K (for the short protein) and $T^\circ \simeq 290.5$ K (for the large protein). The accuracy of these estimates is rather unclear because of the uncertainty related to the use of the linear approximations for $W_c^+(T)$ and $g_e(\nu_c,T)$ (as functions of temperature) in the temperature interval from 293 K to 298 K.

The results of numerical calculations are presented in Figures 3.17 and 3.18. Figure 3.18 shows the ratio $t_f(T)/t_f(T_0)$, given by Equation 3.102, as a function of temperature $T$ at fixed $T_0$. For the large protein (2500 residues), the calculations for $T_0 = 293$ K were carried out in Ref. [31], whereas for the short protein (124 residues) all the calculations were carried out in the present paper. In both cases, the above parameters were chosen in order to obtain reasonable folding times in the range between 1 s (for $D_0^{iw} = 10^{-6}$ cm$^2$ s$^{-1}$) to 100 s (for $D_0^{iw} = 10^{-8}$ cm$^2$ s$^{-1}$). [In principle, the interaction potential parameters for the protein model considered could be determined by performing MD simulations and adjusting the parameters so that the proteins should fold on timescales observed experimentally (say, of the order of 1 s). In reality, such long simulations are not yet feasible.] With the interaction potential parameters given, as clear from Figure 3.15, the protein folding time $t_f(T)$ sharply increases (compared to $t_f(T_0)$) with relatively small deviations of the temperature $T$ from $T_0$: the ratio $t_f(T)/t_f(T_0)$ increases by several orders of magnitude as $T$ changes from $T_0$ to $T_0 \pm 5$ (for the larger protein) and from $T_0$ to $T_0 \pm 10$ (for the shorter protein). Thus, our model predicts that although the considered proteins (initially unfolded) are very likely to fold on a reasonable timescale (of 1 to 100 s) at $T_0 = 293$ $K$, this becomes absolutely unlikely upon either relatively slight heating or relatively slight cooling of the system. Note that the folding time of the 2500-residue protein is much more sensitive to temperature changes than that of the 124-residue one.

A very similar, U (or V)-shaped dependence of the folding time on temperature was obtained by discontinuous MD simulations of an all-atom model of the second α-hairpin fragment of protein G by Zhou et al. [41] They suggested that for a given protein there is an optimal (i.e., minimum) folding temperature because at high temperatures the native state is less stable, whereas at low temperatures there is an onset of low-energy nonnative intermediate states where the protein gets trapped during folding. Likewise, Faisca and Ball [42] reported a U (or V)-shaped dependence of the folding time obtained by MC simulations of a simple lattice model of protein folding. Studying polypeptide chains of various lengths, they also pointed out that, as the length of the chain increases, the temperature range where folding can be observed becomes narrower. This corroborates the predictions of our model that for a protein of $N = 2500$ residues the range of temperatures (around the optimal temperature $T_0$), where the folding can occur on a biologically reasonable timescale, should be significantly narrower than for a protein of $N = 124$ residues. A similar, U

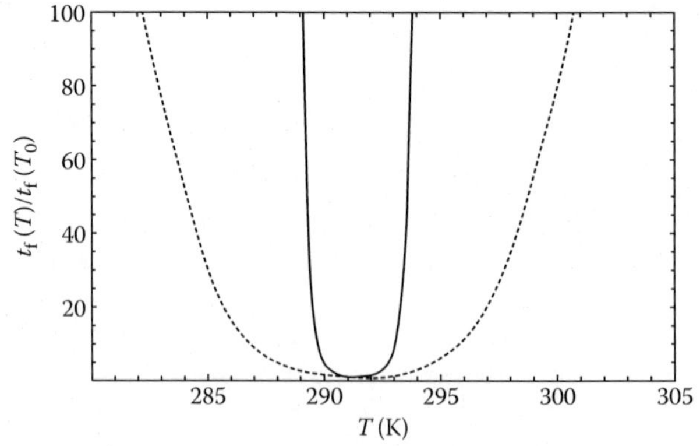


Figure 3.17

The ratio of the protein folding times at temperatures $T$ and $T_0$ as a function of temperature $T$ for $T_0 = 293$ K, i.e., $t_f(T)/t_f(T_0)$ versus $T$ as given by Equation 3.102. The solid curve is for the larger protein ($N = 2500$ residues) and the dashed curve is for the smaller one ($N = 124$ residues). The optimal temperature of protein folding was evaluated to be $T \simeq 290.5$ K (for the larger protein) and $T \simeq 291.5$ K (for the smaller protein).

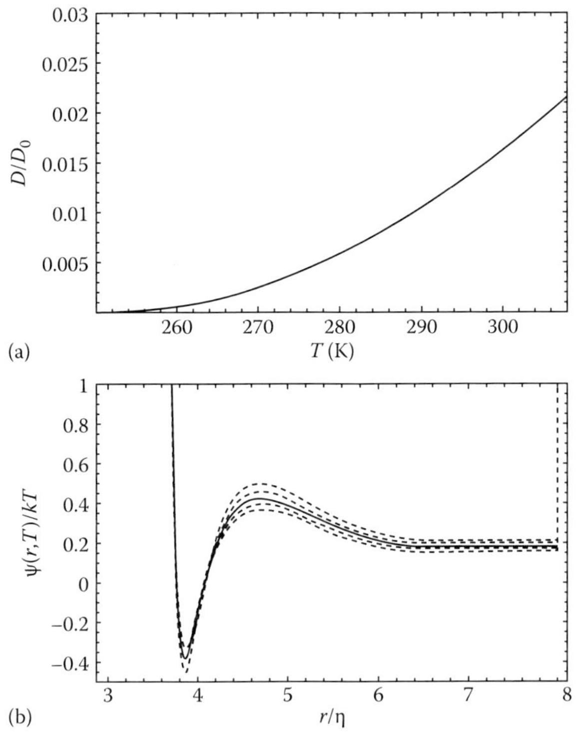


Figure 3.18

(a) The temperature dependence of the dimensionless diffusion coefficients $D^{iw}(T)/D_0^{iw}$ and $D^{ow}(T)/D_0^{ow}$, i.e., $(1\text{-}T_c/T)^{\gamma}$ versus $T$ (for the protein of $N$ = 2500 residues). (b) The temperature variations of the potential field around the cluster in $k_BT$ units presented as $\psi(r,T)/k_BT$ versus $r$ at various $T$ for the cluster of radius $R = 3\eta$. The curves are for $T = T_0 = 20$, $T_0 - 10$, $T_0$, $T_0 + 10$, and $T_0 + 20$ from top to bottom (the solid curve is for $T = T_0$, and the confining potential is shown as a vertical dashed line (for the protein of $N$ = 2500 residues).

(or V)-shaped dependence of $t_f$ on $T$ was obtained by Li et al., who studied the folding kinetics of three-dimensional lattice Go models with side chains by using MC simulations. The results of all the aforementioned simulations [41–43] are consistent with the experimental data on the folding (and unfolding) kinetics of a 16-residue β-hairpin that undergoes cold denaturation at ambient temperatures and which were investigated by time-resolved infrared spectroscopy coupled with the laser-induced temperature jump initiation method [44].

The underlying physics that would explain the above behavior of $t_f(T)/t_f(T_0)$ as a function of $T$ is not immediately clear. The temperature dependence of the related folding rate is strongly non-Arrhenius, a known fact that has been discussed in the literature. Three possible explanations for this non-Arrhenius dependence were previously [73] suggested: (1) super-Arrhenius dependence associated to a rough energy landscape; (2) temperature dependence of the hydrophobic interaction, as reflected in a large activation heat capacity change; (3) temperature dependence of protein stability. In our model [31–33], for the nucleation mechanism of protein folding the temperature affects the folding time through two major factors. One is that of the diffusion coefficients in both inner and outer potential wells and the other is that of the depths of these wells in $k_BT$ units, i.e., $D^{iw}(T)$, $D^{ow}(T)$ and $\Psi_{iw}(r,T)$, $\Psi_{ow}(r,T)$, respectively. The temperature dependence of the diffusion coefficients is shown in Figure 3.17a as $(1 - T/T)^{\gamma}$ versus $T$ (note that $D^{iw}(T)/D_0^{iw} = D^{ow}(T)D_0^{ow} = (1 - T_c/T)^{\gamma}$). The temperature variations of the potential field around the cluster in $k_BT$ units is presented in Figure 3.17b as $\psi(r,T)/k_BT$ versus $r$ at various temperatures $T$ for a cluster of radius $R = 3\eta$. The curves are for $T = T_0 - 45$, $T_0 - 20$, $T_0$, $T_0 + 20$, and $T_0 - 45$ from top to bottom (the solid curve is for $T = T_0$, and the location of the confining potential is the same for all temperatures and given by a vertical dashed line at around $r/\eta \simeq 8$). Thus, with changing temperature all these quantities behave monotonically. However, the protein folding time is a complex function of them and the interplay of their changes (with changing temperature) leads to predictions expected from experiments, namely, that a given protein acquires its native structure on a reasonable timescale under relatively narrow external conditions (temperature, pressure, etc.).

#### 3.4.4.2 Results for Protein Unfolding

In the numerical calculations for the protein unfolding model, the parameters $T_c$ and $\gamma$ were chosen to be: (1) $T_c = 240$ K and $\gamma = 2.1$; (2) $T_c = 270$ K and $\gamma = 2.9$. The results of numerical calculations are shown in Figures 3.15, 3.16, and 3.19.

Because the temperature dependence of $W^+(\nu,T)$ and $W^-(\nu,T)$ was obtained by using third-order Taylor series expansions (see Equation 3.110), the equation $W^-(N,T) = W^+(N,T)$ was cubic in $T - T_0$, i.e., had the form $q(T) \equiv a(T - T_0)^3 + b(T - T_0)^2 + c(T - T_0) + d = 0$, and hence can have up to three different real solutions. The discriminant of this equation is $\Delta = 4b^3d - b^2c^2 + 4ac^3 - 18abcd + 27a^2d^2$. If $\Delta = 0$, the equation $q(T) = 0$ has three real roots of which at least two coincide. It may be that all three roots coincide yielding a triple real root or, alternatively, the equation has a double real root and another distinct single real root. If $\Delta > 0$, the equation has one real root and a pair of complex conjugate roots, the real root corresponding to either cold or hot denaturation depending on whether the root is smaller than or greater than $T_0$. The latter two situations were found in our calculations for two different sets of the diffusion coefficient parameters. Figure 3.15 presents results for the large protein model (2500 residues). In Figure 3.15b, the function $q(T)$ is plotted versus $T - T_0$ for $T_c = 240$ K and $\gamma = 2.1$ (with $\Delta \simeq 0.6$), whereas Figure 3.15c is for $T_c = 270$ K and $\gamma = 3.9$ (with $\Delta \simeq 1.5$). As shown clearly, the protein corresponding to Figure 3.15b is subject to cold denaturation at $T = T_u^- \simeq T_0 - 23$ K, whereas Figure 3.15c shows hot denaturation of another protein at $T = T_u^+ \simeq T_0 + 60$ K. Finally, if $\Delta < 0$, the equation has three distinct real roots (as shown in Figure 3.15d for $T_c = 260$ K and $\gamma = 3.2$): one at $T = T_u^+ > T_0$ corresponding to the threshold temperature of hot denaturation, another one at $T = T_u^- < T_0$ corresponding to the threshold temperature of cold denaturation, and the third one at some $T' > T_u^+$ or $T' < T_u^-$, which can be reasonably considered to be physically meaningless and hence disregarded as arising only due to the approximation of $W^+(\nu,T)$ and $W^-(\nu,T)$ by third-order Taylor series expansions.

Figure 3.16 presents $W^+(\nu,T)$ and $W^-(\nu,T)$, the emission and absorption rates of a cluster (of native residues), as functions of $\nu$ for the large protein model (2500 residues) at three temperatures: $T = T_0 = 293.15$ K (Figure 3.16a), $T = T_u^+ \simeq 358$ K (Figure 3.16b), and $T = T_u^+ + 5$ K (Figure 3.16c). At $T = T_0$ (Figure 3.16a),

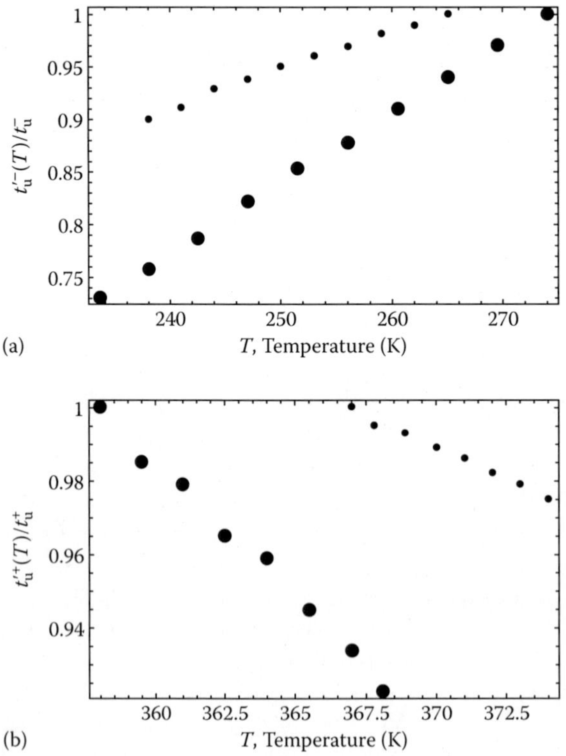


Figure 3.19

The temperature dependence of $t_u'^-/t_u^-$ (a) and $t_u'^+/t_u^+$ (b), the time of protein denaturation (via spinodal decomposition) upon cooling and heating, respectively, divided by their values at the corresponding threshold temperatures. In panels a and b, the large points (lower series) are for the large protein model ($N = 2500$), whereas small points (upper series) are for the small protein model ($N = 124$).

$W^+(\nu) < W^-(\nu)$ for $\nu < \nu_c$ and $W^+(\nu) > W^-(\nu)$ for $\nu_c < \nu < N$. Clearly, if the protein is initially unfolded at this temperature, it cannot fold by a regular growth of the folded cluster, but it can fold via nucleation. The intersection of $W^+(\nu)$ and $W^-(\nu)$ determines the size $\nu_c$ of the critical cluster (nucleus). After the cluster of native residues attains the size $\nu_c$, the whole native structure forms immediately without passing through any detectable intermediates states, because the time $t_g$ of growth of the cluster from size $\nu_c$ to size $N$ is much smaller than the time $t_n$ of the formation of the nucleus [31–33]. On the other hand, at temperature $T_u^+$ (Figure 3.16b), $W^+(\nu) < W^-(\nu)$ for $0 < \nu < N$. Therefore, if a protein is initially unfolded at $T_u^+$ it will never fold, whereas, if it is initially folded at this temperature, it will immediately start unfolding via the loss of residues from its native structure, i.e., in a barrierless way. If the temperature is increased beyond $T_u^+$ (Figure 3.16c), the time necessary for the protein to unfold *via* spinodal decomposition becomes smaller (compared to that time at $T = T_u^+$), because the higher $T$ is above $T_u^+$, the greater the positive difference $W^-(\nu)-W^+(\nu)$ is. Similar behavior is observed when the temperature is decreased down from $T_u^-$.

Let us denote the times of protein denaturation (via spinodal decomposition) upon heating and cooling at some temperature $T$ by $t_u'^+ \equiv t_u'^+(T)$ and $t_u'^- \equiv t_u'^-(T)$, respectively. By definition, $t_u'^+ = t_u'^+(T_u^+)$ and $t_u^- = t_u'^-(T_u^-)$. The threshold temperatures of cold and hot denaturation were determined (as outlined in Section 3.4.3) to be $T_u^- \simeq 274$ K $= T_0 - 19$ K and $T_u^+ \simeq 358$ K $= T_0 + 65$ K, respectively (for the larger protein), and $T_u^- \simeq 265$ K $= T_0 - 28$ K and $T_u^+ \simeq 366$ K $= T_0 + 73$ K, respectively (for the smaller protein). The temperature dependence of $t_u'^-$ and $t_u'^+$ is presented in Figure 3.19a and b, respectively. In this figure, the large points are for the large protein model, whereas the small points are for the small protein model. The third-order Taylor series expansions (see Equation 3.110), used for determining $W^+$ and $W^-$ as functions of temperature, are valid in the range $|T - T_0| \leq 80$ K, so that we were able to study the dependence of $t_u'^+$ on $T$ only in the range from $T = T_u^+$ to $T \simeq T_0 + 80$ K and the dependence of $t_u'^-$ on $T$ in the range from $T \simeq T_0 - 80$ K to $T = T_u^-$ K. The unfolding times are presented as the ratios $t_u'^- / t_u^-$ (Figure 3.19a) and $t_u'^+ / t_u^+$ (Figure 3.19b). As is clear from this figure, the temperature sensitivity of the time of barrierless denaturation is predicted to be higher for the larger protein than for the smaller one. Indeed, the unfolding time of hot denaturation linearly decreases by about 5% as the temperature increases from $T = T_u^+$ to $T = T_u^+ + 10$ for the large protein and by about 2.5% as the temperature increases from $T = T_u^+$ to $T = T_u^+ + 7$ for the small protein. The unfolding time of cold denaturation linearly decreases by about 30% as the temperature decreases from $T = T_u^-$ to $T = T_u^+ - 46$ for the large protein and by about 10% as the temperature decreases from $T = T_u^-$ to $T = T_u^+ - 27$ for the small protein.

The unfolding times $t_u^-$ and $t_u^+$ are inversely proportional to the coefficient $D_0 \equiv D_0^{iw}$ in the temperature dependence of the diffusion coefficient $D^{iw}$. Because of the lack of reliable data on the diffusion coefficient of a residue in a native protein, $D_0^{iw}$ was assumed to vary between $10^{-7}$ and $10^{-9}$ cm$^2$ s$^{-1}$. With $W_c^+$ the unfolding times of hot and cold denaturation would be $t_u^+ \simeq 0.05$ ms and $t_u^+ \simeq 0.7$ ms, respectively, for the large protein model and $t_u^+ \simeq 0.003$ ms and $t_u^- \simeq 0.06$ ms, respectively, for the small protein model. With $D_0^{iw} = 10^{-9}\,\text{cm}^2\text{s}^{-1}$, the unfolding times of hot and cold denaturation would be $t_u^+ \simeq 5$ ms and $t_u^- \simeq 70$ ms, respectively, for the large protein model and $t_u^+ \simeq 0.3$ ms and $t_u^- \simeq 6$ ms, respectively, for the small protein model. These estimates are consistent with experimental data on the unfolding times of hot and cold denaturation [74–78]. Many MD simulations report [79,80] much shorter timescales for hot denaturation, but those simulations are usually concerned with the sequence of structural changes occurring upon protein unfolding and hence use artificial means to speed up the process and make unfolding occur on a timescale amenable to MD simulations, which is typically less than 100 ns.

### 3.4.5 Conclusions

So far, most of the work on protein folding has been carried out by using either MC or MD simulations. Both experiments and simulations have suggested that there are multiple pathways for protein folding, one of which involves a nucleation mechanism. Its theory was first approached [19,22–24] by means of the classical thermodynamics, calculating the free energy of formation of a cluster of protein residues in the framework of the capillarity approximation, much like the CNT does. Such an approach involves the use of the concept of surface tension for a cluster consisting of protein residues which is an intrinsically ill-defined physical quantity and can be considered at best as an adjustable parameter.

Recently [31–33], we have presented a new, microscopic model for the nucleation mechanism of protein folding. A protein was treated as a random heteropolymer with all the bonds having the same constant length and all the bond angles equal and fixed. As a crucial idea of the model, an overall potential around the cluster of native residues wherein a protein residue performs a chaotic motion is considered to be a combination of three terms representing the average dihedral, effective pairwise, and confining potentials. The overall potential as a function of the distance from the cluster center has a double well shape which allows one to determine both its emission and absorption rates by using a first passage time analysis. One can thus develop a self-consistent kinetic theory for the nucleation mechanism of protein folding and evaluate the size and composition of the nucleus and the protein folding time.

In the present paper, we have proposed a method for examining the temperature dependence of protein folding time without carrying out the time-consuming calculations for a series of temperature. To evaluate the folding time just for one temperature by using our model [31–33] requires about 20 days of intensive computational work. Thus, straightforwardly applying that model to the same protein at six temperatures would require half a year. To avoid such lengthy calculations, in this paper we have proposed a method based on Taylor series expansions in the formalism of the first passage time analysis. Once the folding time $t_f$ is evaluated at one temperature $T_0$, the proposed method allows one to evaluate $t_f$ at any other temperature $T$ in the vicinity of $T_0$. The width of the vicinity in which the method is valid can be increased by retaining incrementally higher order terms in the Taylor series expansions with respect to $T - T_0$.

Besides, we have presented a model for the process of protein denaturation occurring in a barrierless way (i.e., as spinodal decomposition). The key elements of the proposed model are the temperature-dependent rates with which a cluster of native residues within a compact (but not native) protein emits and absorbs residues. Considering a native protein, stable at temperature $T_0$, we have examined the temperature dependence of the emission and absorption rates of the cluster (which are also functions of the cluster size). To numerically construct the emission and absorption rates as functions of the cluster size for just one temperature, the straightforward computations would require many days. To numerically construct these rates as functions of both cluster size and temperature would thus require months of straightforward computations. To avoid such lengthy calculations, we have applied a method based on Taylor series expansions with respect to $T - T_0$ in the formalism of the first passage time analysis. Once the emission and absorption rates are known as functions of cluster size at one temperature $T_0$, i.e., $W^-(\nu,T_0)$ and $W^+(\nu,T_0)$, one can use the Taylor series representation of $W^-(\nu,T)$ and $W^+(\nu,T)$ to evaluate them at any other temperature $T$ in the vicinity of $T_0$. The width of the vicinity in which the method is valid depends on how many terms are retained in the Taylor series expansions.

To illustrate the proposed methods by numerical calculations, we have applied them to two model proteins previously studied in Refs. [31–33] at room temperature $T_0 = 293$ K, namely, heteropolymers consisting of 2500 and 124 beads. The short protein model was supposed to mimic the bovine pancreatic ribonuclease. The larger heteropolymer was considered to demonstrate that the method provides a reasonable temperature dependence of the folding time on a reasonable timescale even for the largest proteins, whereas straightforward MC/MD simulations of such proteins would require extremely lengthy computations.

The model for the temperature dependence of the protein folding time predicts that this time $t_f(T)$ sharply increases (compared to $t_f(T_0)$) with relatively small deviations of the temperature $T$ from $T_0$ and this increase is sharper for the larger protein. Thus, our model predicts that while the considered proteins (initially unfolded) reasonably quickly fold at $T_0 = 293$ K, their folding becomes highly unlikely upon either slight heating or slight cooling. These results are consistent with those predicted by a model (also based on the first passage time analysis) of protein denaturation which can predict both cold and hot denaturation, subject to an appropriate choice of diffusion coefficients (and their temperature dependence) in the folded and unfolded protein. The estimates for the unfolding times of hot and cold denaturation are essentially in the range of typical times of protein denaturation. One can notice that the unfolding time $t_u^-$ at the threshold temperature of cold denaturation is predicted to be an order-of-magnitude greater compared to that of hot denaturation, $t_u^+$.

## Appendix 3G: Three Constituents of the Potential Field around the Cluster for a Selected Bead

Let us consider a cluster of radius $R$ consisting of protein residues stabilized by correct tertiary contacts (see Figure 3.14). This will constitute the folded part of the protein.

### Effective Pairwise Potential

A pair interaction between two nonadjacent beads $i$ and $j$ at a distance $d$ away from each other can be considered to be of the LJ type

$$\phi_{ij}(d) = 4\varepsilon_{ij}\,[(\eta/d)^{12} - (\eta/d)^{6}]\ (i, j = \mathrm{b}, \mathrm{l}), \tag{3G.1}$$

where $\eta$ is the bond length (fixed) and $\varepsilon_{ij}$ ($i,j$ = b, l) is an energy parameter satisfying the mixing rules $\varepsilon_{ij} = \sqrt{\varepsilon_{ii}\varepsilon_{jj}}$.

Assuming pairwise additivity of the interactions between nonadjacent beads, the effective pairwise potential for a bead of type $i$ at a distance $r$ from the cluster center is $\phi_i(r)$ is provided by

$$\phi_i(r) = \sum_j \int_V d\mathbf{r}'\rho_j(r')\phi_{ij}(|\mathbf{r}' - \mathbf{r}|), \quad j = \text{b, l}. \tag{3G.2}$$

Here, $\boldsymbol{r}$ is the coordinate of the surface bead of type $i$, $\rho_j(r')$ ($j$ = b, l) is the number density of beads of component $j$ at point $\boldsymbol{r}'$ (spherical symmetry is assumed with the cluster center chosen as the origin of the coordinate system), and $\phi_{ij}(|\boldsymbol{r}' - \boldsymbol{r}|)$ is the interaction potential between two beads of components $i$ and $j$ at points $\boldsymbol{r}$ and $\boldsymbol{r}'$, respectively. The integration in Equation 3G.2 has to be carried out over the whole volume of the system, but the contribution from the unfolded part can be assumed to be small and accounted for by particular choices of the energy parameters.

## Confining Potential

Because of the bonds between the residues, none of them can be located outside a volume which depends on the densities of the folded and unfolded regions. Assuming that volume to be spherical, its radius $r_{\text{cp}}$ is given by the expression $r_{\text{cp}} = [R^3 + 3(N - \nu)/4\pi\rho_{\text{u}}]^{1/3}$, where $N$ and $\nu$ are the total numbers of beads in the whole protein and in the cluster, respectively, and $\rho_{\text{u}} = \rho_{\text{bu}} + \rho_{\text{lu}}$ is the overall number density of beads (of both components) in the unfolded part of the protein. Clearly, $\nu = (4\pi/3)R^3\rho_{\text{f}}$ with $\rho_{\text{f}} = \rho_{\text{bf}} + \rho_{\text{lf}}$ being the overall number density of beads in the folded part of the protein. Consequently, the confining potential $\phi_{\text{cp}}(r) = 0$ for $r < r_{\text{cp}}$ and $\phi_{\text{cp}}(r) = \infty$ for $r \geq r_{\text{cp}}$.

## Average Dihedral Angle Potential

The term $\bar{\phi}_i^{\delta}(r) \equiv \bar{\phi}_i^{\delta}(r, \mathbf{r}_2, \mathbf{r}_3, \ldots, \mathbf{r}_7)$ in $\psi_i(r)$ is due to the $r$-dependent part of the dihedral angle potential of the whole protein. The dihedral angle potential arises due to the rotation of three successive peptide bonds connecting four successive beads, and is related to the dihedral angle $\delta$ via

$$\phi_\delta = \varepsilon'_\delta(1 + \cos\delta) + \varepsilon''_\delta(1 + \cos 3\delta), \tag{3G.3}$$

where $\varepsilon'_\delta$ and $\varepsilon''_\delta$ are independent energy parameters which depend on the nature and sequence of the four beads involved in the dihedral angle $\delta$. This potential has three minima, one in the *trans* configuration at $\delta = 0$ and two others in the *gauche* configurations at $\delta = \pm\arccos\sqrt{(3\varepsilon''_\delta - \varepsilon'_\delta)/12\varepsilon''_\delta}$ (the former one being the lowest).

Consider a bead 1 (of type b or l) at a distance $r$ from the (center of the) cluster, $r > R$ (see Figure 3.14). The total dihedral angle potential of the whole protein chain for a given configuration of beads 1, 2, 3, …, $N$ is a function of $r$, $\mathbf{r}_2$, …, $\boldsymbol{r}_N$ and can be written as

$$\phi_{i_1}^{\delta}\left(\delta_{421}^{642}(r)\right) + \phi_{i_1}^{\delta}\left(\delta_{213}^{421}(r)\right) + \phi_{i_1}^{\delta}\left(\delta_{135}^{213}(r)\right) + \phi_{i_1}^{\delta}\left(\delta_{357}^{135}(r)\right) + \text{const},$$

where $\delta_{jkl}^{ijk}$ is a dihedral angle between two planes, one of which is determined by beads $i$, $j$, $k$ and the other by beads $j$, $k$, $l$. The last term on the right-hand side of the above equation, representing the contributions from the dihedral angles involving beads 8, 9, …, $N$, does not depend on $r$ because those dihedral angles remain unaffected as the position of bead 1 changes. Hence, this contribution (const) can be omitted hereafter without being specified since it can be regarded as affecting only the reference level for $\psi_{i1}(r)$:

$$\begin{aligned}\phi_{i_1}^{\delta} &\equiv \phi_{i_1}^{\delta}(r, \mathbf{r}_2, \ldots, \mathbf{r}_7) \\ &= \phi_{i_1}^{\delta}\left(\delta_{421}^{642}(r)\right) + \phi_{i_1}^{\delta}\left(\delta_{213}^{421}(r)\right) + \phi_{i_1}^{\delta}\left(\delta_{135}^{213}(r)\right) + \phi_{i_1}^{\delta}\left(\delta_{357}^{135}(r)\right)\end{aligned} \tag{3G.4}$$

(an occasionally used subscript $i_s$ ($i_s$ = b, l) indicates the type of bead $s$ ($s$ = 1, …, 7)).

For a given location of bead 1, various configurations of beads 2, 3, …, $N$ (subject to the constraints on the bond length and bond angle as well as to the constraint of excluded cluster volume) lead to various sets of dihedral angles. However, variations in the location of beads 8, 9, …, $N$ give rise to variations in the dihedral potential which are independent of $r$. Thus, the dihedral term $\bar{\phi}_i^{\delta}(r)$ (less a term independent of $r$ and omitted hereafter) on the right-hand side of Equation 3G.5 can be obtained by averaging Equation 3G.4 with the Boltzmann factor $\exp\left[-\left(\phi_{i_1}^{\delta 27}(r)/k_B T\right)\right]$ with

$$\phi_{i_1}^{\delta 27}(r) \equiv \phi_{i_1}^{\delta 27}(r, \mathbf{r}_2, \ldots, \mathbf{r}_7) = \phi_{i_1}^{\delta}(r) + \sum_{s=2}^{7} \phi_{i_s}(\mathbf{r}_s), \tag{3G.5}$$

over all possible configurations of beads 2, 3, …, 7 and assigning the result to a selected bead with fixed coordinates, i.e., bead 1:

$$\bar{\phi}_{i_1}^{\delta}(r) = f^{-1} \int_{\Omega_{18}} d\mathbf{r}_2 \ldots d\mathbf{r}_7\, \phi_{i_1}^{\delta}(r) \exp\left[-\phi_{i_1}^{\delta 27}(r)/k_B T\right], \tag{3G.6}$$

where the normalization constant

$$f = \int_{\Omega_{18}} d\mathbf{r}_2 \ldots d\mathbf{r}_7 \exp\left[-\phi_{i_1}^{\delta 27}(r)/k_B T\right], \tag{3G.7}$$

where $\Omega_{18}$ is the integration region in an 18-dimensional space. Note that (unlike Ref. [31]) the exponent in the Boltzmann factor is a sum of two terms, the first representing the dihedral angle potential of beads 1 through 7 and the second the sum of "bead–cluster" interactions for beads 2 through 7. Clearly, both contributions affect the probability of configurations of beads 1 through 7 during integration, but the second one was neglected in Ref. [31] to increase the efficiency of numerical calculations. In calculating $\phi_{i_1}^{\delta 27}$ in Equation 3G.5, only the type of bead 1 is supposed to be known, whereas the types of beads 2 through 7 are treated in a probabilistic way. For example, if bead 1 is of kind b, then any of the beads 2 through 7 is either of kind b with the probability $(N_b - \nu_b - 1)/(N - \nu - 1)$ or of kind l with the probability $(N_l - \nu_l)/(N - \nu - 1)$ (remind that $N = N_b + N_l$ and $\nu = \nu_b + \nu_l$).

The 18-fold integrals in Equations 3G.6 and 3G.7 can be reduced to sevenfold ones by taking into account the constraints of fixed bond length and fixed bond angle:

$$\begin{aligned} \bar{\phi}_{i_j}^{\delta} = f^{-1} \sum_{i,j,k,m,n=-}^{+} & \int_0^{2\pi} d\varphi_2 \int_{L_2} dr_2 \int_{L_3^i} dx_3 \int_{L_4^j} dx_4 \\ & \times \int_{L_5^k} dx_5 \int_{L_6^m} dx_6 \int_{L_7^n} dx_7 \\ & \times r_2^2 \sin\Theta_2(r, r_2) \tilde{\phi}_{ijkmn}(r, \varphi_2, r_2, x_3, \ldots, x_7) \\ & \times \exp[-\tilde{\phi}_{ijkmn}(\varphi_2, r_2, x_3, \ldots, x_7)/k_B T], \end{aligned} \tag{3G.8}$$

$$f = \sum_{i,j,k,m,n=-}^{+} \int_{0}^{2\pi} d\varphi_2 \int_{L_2} dr_2 \int_{L_3^i} dx_3 \int_{L_4^j} dx_4 \times \int_{L_5^k} dx_5 \int_{L_6^m} dx_6 \int_{L_7^n} dx_7 \times r_2^2 \sin\Theta_2(r, r_2) \times \exp[-\tilde{\phi}_{ijkmn}(r, \varphi_2, r_2, x_3, \ldots, x_7)/k_B T]. \tag{3G.9}$$

Each of the summation indices in Equations 3G.8 and 3G.9 takes on two values, denoted + and –, so that there are $2^5$ terms in the sum differing by the integrand as well as by the integration ranges (except for $L_2$, which is independent of $i$, $j$, $k$, $m$, $n$). The definitions of the coordinates $\varphi_2$, $r_2$, $x_3$, $x_4$, $x_5$, $x_6$, $x_7$, the explicit forms of the integration ranges $L_2$, $L_3^i, L_4^j, L_5^k, L_6^m, L_7^n$, and the rules for transforming the function $\phi_{i_1}^{\delta 27}(r)$ into the function $\bar{\phi}_{ijkmn}(\varphi_2, r_2, x_3, \ldots, x_7)$ are given below.

Let us consider beads 1 through 7 around a cluster in the unfolded part of the heteropolymer and denote their Cartesian coordinates by $x_i$, $y_i$, $z_i$ ($i$ = 2, ..., 7). It is convenient to choose the origin of the Cartesian system of coordinates in such a way that $x_1 = 0$, $y_1 = 0$, $z_1 = r$ where $r$ is the distance between bead 1 and the cluster center.

For bead 2, it is also convenient to introduce the spherical coordinates $r_2$, $\Theta_2$, $\varphi_2$, which are related to its Cartesian ones by

$$x_2 = r_2 \sin\Theta_2 \cos\varphi, \quad y_2 = r_2 \sin\Theta_2 \sin\varphi, \quad z_2 = r_2 \cos\Theta_2. \tag{3G.10}$$

Clearly, since the location of bead 1 is fixed (hence, $r$ given), the polar angle $\Theta$ of bead 2 is uniquely determined by $r_2$ due to the constant bond length constraint:

$$\Theta_2(r, r_2) = \arccos\left[\left(r^2 + r_2^2 - \eta^2\right)/(2r_2 r)\right] \quad (0 \le \Theta_2 \le \pi), \tag{3G.11}$$

whereas the azimuthal angle $0 \le \phi_2 \le 2\pi$. The distance $r_2$ varies in the range $r_{2\min} \le r_2 \le r + \eta$, where $r_{2\min} = \max(R, r - \eta)$, which determines the integration range $L_2$ in Equations 3G.8 and 3G.9. Thus, the integration with respect to $r_2$ in Equations 3G.6 and 3G.7 reduces to the integration with respect to $\varphi_2$ and $r_2$ at fixed $\Theta_2$ (given by Equation 3G.11) in Equations 3G.8 and 3G.9.

For given locations of beads 1 and 2, the possible locations of beads 3 and 4 lie on circles of radius $r_1 = \eta \sin \beta_0$ with their location and orientation completely determined by the coordinates of beads 1 and 2. This is due to the constraints that all bond angles are equal to $\beta_0$ and all bond lengths are equal to $\eta$. Owing to the same constraints, if the locations of beads 2 and 4 are given, the possible locations of bead 6 lie on a circle of radius $r_1$, where the location and orientation are completely determined by the coordinates of beads 2 and 4. Furthermore, for given locations of beads 1 and 3, the possible locations of bead 5 lie on a circle of radius $r_1$ with the position and orientation completely determined by the coordinates of beads 1 and 3. Finally, for the given locations of beads 3 and 5, the possible locations of bead 7 are on a circle of radius $r_1$, with the location and orientation completely determined by the coordinates of beads 3 and 5.

Let us consider bead $s$ ($s$ = 3, …, 7) with unknown coordinates and two other beads, $c$ and $n$ (closest to and next to the closest to bead $s$) with fixed (known) coordinates $x_c$, $y_c$, $z_c$ and $x_n$, $y_n$, $z_n$. For example, if $s = 7$, then $c = 5$, $n = 3$; if $s = 4$, then $c = 2$, $n = 1$. Bead $s$ lies on a circle of radius $r_1$ with the coordinates of the center

$$x_0 \equiv x_0(x_n, y_n, z_n, x_c, y_c, z_c) = x_n + (x_c - x_n) \times \frac{\eta\left(1+|\cos\beta_0|\right)}{\sqrt{(x_c - x_n)^2 + (y_c - y_n)^2 + (z_c - z_n)^2}}, \tag{3G.12}$$

$$y_0 \equiv y_0(x_n, y_n, z_n, x_c, y_c, z_c) = y_n + (y_c - y_n) \times \frac{\eta\left(1+|\cos\beta_0|\right)}{\sqrt{(x_c - x_n)^2 + (y_c - y_n)^2 + (z_c - z_n)^2}}, \tag{3G.13}$$

$$z_0 \equiv z_0(x_n, y_n, z_n, x_c, y_c, z_c) = z_n + (z_c - z_n) \times \frac{\eta\left(1+|\cos\beta_0|\right)}{\sqrt{(x_c - x_n)^2 + (y_c - y_n)^2 + (z_c - z_n)^2}}. \tag{3G.14}$$

The coordinate $x_s$ of bead $s$ can change in the range

$$x_c - x_b \le x_s \le x_c + x_b \ (s = 3, \ldots, 7), \tag{3G.15}$$

where

$$x_b \equiv x_b(x_n, y_n, z_n, x_c, y_c, z_c) = \eta \sin\beta_0 \sqrt{\frac{\left((y_c - y_0)^2 + (z_c - z_0)^2\right)}{(x_c - x_n)^2 + (y_c - y_n)^2 + (z_c - z_n)^2}}. \tag{3G.16}$$

For a given $x_s$, the coordinate $y_s$ of bead $s$ can have only one of two values.
For given $x_s$ and $y_s^{\pm}$, the coordinate $z_s$ of bead $s$ can only have a single value:

$$y_s^{\pm} \equiv y_s^{\pm}(x_s, x_n, y_n, z_n, x_c, y_c, z_c) = y_0 + \frac{-(x_c - x_0)(y_c - y_0)(x_s - x_0)}{(y_c - y_0)^2 + (z_c - z_0)^2} \pm \frac{|z_c - z_0|\sqrt{[(y_c - y_0)^2 + (z_c - z_0)^2]\eta^2 \sin^2\beta_0 - [(x_c - x_n)^2 + (y_c - y_n)^2 + (z_c - z_n)^2](x_s - x_0)^2}}{(y_c - y_0)^2 + (z_c - z_0)^2}, \tag{3G.17}$$

$$z_s^{\pm} \equiv z_s\left(x_s, y_s^{\pm}, x_n, y_n, z_n, x_c, y_c, z_c\right)$$

$$= z_0 - \frac{(x_c - x_0)(x_s - x_0) + (y_c - y_0)\left(y_s^{\pm} - y_0\right)}{z_c - z_0}. \qquad (3G.18)$$

Since the volume occupied by the cluster is unavailable to beads 2, 3, …, 7, an additional constraint is imposed on $x_s$, $y_s$, and $z_s$, namely, $x_s^2 + y_s^2 + z_s^2 > R^2$. Subject to this constraint, the double inequality Equation 3G.15 and Equations 3G.16 through 3G.18 completely determine the integration ranges in Equations 3G.8 and 3G.9.

In order to transform the function $\phi_{i_1}^{\delta 27}(r) \equiv \phi_{i_1}^{\delta}(r, r_2, r_3, \ldots, r_7)$ into the function $\bar{\phi}_{ijkmn}(r, \varphi_2, r_2, x_3, \ldots, x_7)$, the variables $y_s$ ($s = 3, \ldots, 7$) and $z_s$ ($s = 3, \ldots, 7$) in the former must be replaced by $y_s^{\pm}$ and $z_s^{\pm}$ in all possible combinations each of which gives rise to a term in the sums in Equations 3G.8 and 3G.9 (note that $z_s^{+} = z_s\left(x_s, y_s^{+}, \ldots\right)$ and $z_s^{-} = z_s\left(x_s, y_s^{-}, \ldots\right)$). The polar angle of bead 2 must be replaced by $\Theta_2(r, r_2)$.

Indeed, at fixed $\boldsymbol{r}_2$, $\boldsymbol{r}_3$, $\boldsymbol{r}_4$, $\boldsymbol{r}_5$, $\boldsymbol{r}_6$ (i.e., fixed locations of beads 2 through 6), the trajectory of bead 7 represents a circle hence, equivalently, the integrand in Equation 3G.6, $\mathrm{I}(\boldsymbol{r}_2, \boldsymbol{r}_3, \boldsymbol{r}_4, \boldsymbol{r}_5, \boldsymbol{r}_6, x_7, y_7, z_7)$, contains a factor $\left[\delta\left(y_7 - y_7^{+}\right)\delta\left(z_7 - z_7^{+}\right) + \delta\left(y_7 - y_7^{-}\right)\delta\left(z_7 - z_7^{-}\right)\right]$, where $\delta(x)$ is Dirac's delta-function and $y_7^{\pm}$ and $z_7^{\pm}$ are given by Equations 3G.17 and 3G.18. Therefore, the integration with respect to $y_7$ and $z_7$ in Equation 3G.6 results in the replacement of the integrand $\mathrm{I}(\boldsymbol{r}_2, \boldsymbol{r}_3, \boldsymbol{r}_4, \boldsymbol{r}_5, \boldsymbol{r}_6, x_7, y_7, z_7)$ by the sum $\mathrm{J}(\boldsymbol{r}_2, \boldsymbol{r}_3, \boldsymbol{r}_4, \boldsymbol{r}_5, \boldsymbol{r}_6, x_7) \equiv \mathrm{I}(\boldsymbol{r}_2, \boldsymbol{r}_3, \boldsymbol{r}_4, \boldsymbol{r}_5, \boldsymbol{r}_6, x_7, y_7^{+}, z_7^{+}) + \mathrm{I}(\boldsymbol{r}_2, \boldsymbol{r}_3, \boldsymbol{r}_4, \boldsymbol{r}_5, \boldsymbol{r}_6, x_7, y_7^{-}, z_7^{-})$. Next, at fixed $\boldsymbol{r}_2$, $\boldsymbol{r}_3$, $\boldsymbol{r}_4$, $\boldsymbol{r}_5$, the trajectory of bead 6 is a circle, hence, equivalently, the integrand $\mathrm{J}(\boldsymbol{r}_2, \boldsymbol{r}_3, \boldsymbol{r}_4, \boldsymbol{r}_5, \boldsymbol{r}_6, x_7)$ contains a factor $\left[\delta\left(y_6 - y_6^{+}\right)\delta\left(z_6 - z_6^{+}\right) + \delta\left(y_6 - y_6^{-}\right)\delta\left(z_6 - z_6^{-}\right)\right]$, where functions $y_6^{\pm}$ and $z_6^{\pm}$ are given by Equations 3G.17 and 3G.18. Therefore, the integration with respect to $y_6$ and $z_6$ in Equation 3G.6 results in the replacement of the integrand $\mathrm{J}(\boldsymbol{r}_2, \boldsymbol{r}_3, \boldsymbol{r}_4, \boldsymbol{r}_5, x_6, y_6, z_6, x_7)$ by the sum

$$\begin{aligned}
\mathrm{K}(\boldsymbol{r}_2, \boldsymbol{r}_3, \boldsymbol{r}_4, \boldsymbol{r}_5, x_6, x_7) &\equiv \mathrm{J}\left(\boldsymbol{r}_2, \boldsymbol{r}_3, \boldsymbol{r}_4, \boldsymbol{r}_5, x_6, y_6^{+}, z_6^{+}, x_7\right) + \mathrm{J}\left(\boldsymbol{r}_2, \boldsymbol{r}_3, \boldsymbol{r}_4, \boldsymbol{r}_5, x_6, y_6^{-}, z_6^{-}, x_7\right) \\
&= \mathrm{I}\left(\boldsymbol{r}_2, \boldsymbol{r}_3, \boldsymbol{r}_4, \boldsymbol{r}_5, x_6, y_6^{+}, z_6^{+}, x_7, y_7^{+}, z_7^{+}\right) + \mathrm{I}(\boldsymbol{r}_2, \boldsymbol{r}_3, \boldsymbol{r}_4, \boldsymbol{r}_5, x_6, y_6^{+}, z_6^{+}, x_7, y_7^{-}, z_7^{-}) \\
&\quad + \mathrm{I}\left(\boldsymbol{r}_2, \boldsymbol{r}_3, \boldsymbol{r}_4, \boldsymbol{r}_5, x_6, y_6^{-}, z_6^{-}, x_7, y_7^{+}, z_7^{+}\right) + \mathrm{I}(\boldsymbol{r}_2, \boldsymbol{r}_3, \boldsymbol{r}_4, \boldsymbol{r}_5, x_6, y_6^{-}, z_6^{-}, x_7, y_7^{-}, z_7^{-}).
\end{aligned}$$

Furthermore, at fixed $\boldsymbol{r}_2$, $\boldsymbol{r}_3$, $\boldsymbol{r}_4$ the trajectory of bead 5 is a circle, hence, equivalently, the integrand $\mathrm{K}(\boldsymbol{r}_2, \boldsymbol{r}_3, \boldsymbol{r}_4, \boldsymbol{r}_5, x_6, x_7)$ contains a factor $\left(\left[\delta\left(y_5 - y_5^{+}\right)\delta\left(z_5 - z_5^{+}\right) + \delta\left(y_5 - y_5^{-}\right)\delta\left(z_5 - z_5^{-}\right)\right]\right)$, where functions $y_5^{\pm}$ and $z_5^{\pm}$ are given by Equations 3G.17 and 3G.18. Therefore, the integration with respect to $y_6$ and $z_6$ in Equation 3G.6 results in the replacement of the integrand $\mathrm{K}(\boldsymbol{r}_2, \boldsymbol{r}_3, \boldsymbol{r}_4, x_5, y_5, z_5, x_6, x_7)$ by the sum

$$\begin{aligned}
&L(\boldsymbol{r}_2, \boldsymbol{r}_3, \boldsymbol{r}_4, x_5, x_6, x_7) \\
&\equiv \mathrm{K}\left(\boldsymbol{r}_2, \boldsymbol{r}_3, \boldsymbol{r}_4, x_5, y_5^{+}, z_5^{+}, x_6, x_7\right) + \mathrm{K}\left(\boldsymbol{r}_2, \boldsymbol{r}_3, \boldsymbol{r}_4, x_5, y_5^{-}, z_5^{-}, x_6, x_7\right) \\
&= \mathrm{I}\left(\boldsymbol{r}_2, \boldsymbol{r}_3, \boldsymbol{r}_4, x_5, y_5^{+}, z_5^{+}, x_6, y_6^{+}, z_6^{+}, x_7, y_7^{+}, z_7^{+}\right) \\
&\quad + \mathrm{I}\left(\boldsymbol{r}_2, \boldsymbol{r}_3, \boldsymbol{r}_4, x_5, y_5^{+}, z_5^{+}, x_6, y_6^{+}, z_6^{+}, x_7, y_7^{-}, z_7^{-}\right) \\
&\quad + \mathrm{I}\left(\boldsymbol{r}_2, \boldsymbol{r}_3, \boldsymbol{r}_4, x_5, y_5^{+}, z_5^{+}, x_6, y_6^{-}, z_6^{-}, x_7, y_7^{+}, z_7^{+}\right) \\
&\quad + \mathrm{I}\left(\boldsymbol{r}_2, \boldsymbol{r}_3, \boldsymbol{r}_4, x_5, y_5^{+}, z_5^{+}, x_6, y_6^{-}, z_6^{-}, x_7, y_7^{-}, z_7^{-}\right) \\
&\quad + \mathrm{I}\left(\boldsymbol{r}_2, \boldsymbol{r}_3, \boldsymbol{r}_4, x_5, y_5^{-}, z_5^{-}, x_6, y_6^{+}, z_6^{+}, x_7, y_7^{+}, z_7^{+}\right) \\
&\quad + \mathrm{I}\left(\boldsymbol{r}_2, \boldsymbol{r}_3, \boldsymbol{r}_4, x_5, y_5^{-}, z_5^{-}, x_6, y_6^{+}, z_6^{+}, x_7, y_7^{-}, z_7^{-}\right) \\
&\quad + \mathrm{I}\left(\boldsymbol{r}_2, \boldsymbol{r}_3, \boldsymbol{r}_4, x_5, y_5^{-}, z_5^{-}, x_6, y_6^{-}, z_6^{-}, x_7, y_7^{+}, z_7^{+}\right) \\
&\quad + \mathrm{I}\left(\boldsymbol{r}_2, \boldsymbol{r}_3, \boldsymbol{r}_4, x_5, y_5^{-}, z_5^{-}, x_6, y_6^{-}, z_6^{-}, x_7, y_7^{-}, z_7^{-}\right).
\end{aligned}$$

Repeating this procedure twice more (for the integration with respect to $y_4$, $z_4$ and $y_3$, $z_3$, the integrand of Equation 3G.6 is reduced to a sum of 32 = $2^5$ terms corresponding to all possible combinations of pairs $y_s^{\pm}$, $z_s^{\pm}$ ($s = 3,\ldots,7$) in $I(\mathbf{r}_2, x_3, y_3, z_3, x_4, y_4, z_4, x_5, y_5, z_5, x_6, y_6, z_6, x_7, y_7, z_7)$ and Equation 3G.6 is reduced to Equation 3G.8. The same procedure reduces Equation 3G.7 to Equation 3G.9.

To further clarify the above procedure, consider, for example, the term with $i = +, j = -, k = +, m = +, n = -$ in the sum in Equations 3G.8 and 3G.9. In the corresponding integrands, the function $\bar{\phi}_{+-++-}(\varphi_2, r_2, x_3, \ldots, x_7)$ is obtained from $\phi_i^{\delta}(r, \mathbf{r}_2, \mathbf{r}_3, \ldots, \mathbf{r}_7)$ as follows.

$y_3$ must be replaced by $y_3^- \equiv y_3^-(x_3, x_2, y_2, z_2, x_1, y_1, z_1)$ and $z_3$ by $z_3^- \equiv z_3(x_3, y_3^-, x_2, y_2, z_2, x_1, y_1, z_1)$, $y_4$ must be replaced by $y_4^+ \equiv y_4^+(x_4, x_1, y_1, z_1, x_2, y_2, z_2)$ and $z_4$ by $z_4^+ \equiv z_4(x_4, y_4^+, x_1, y_1, z_1, x_2, y_2, z_2)$, $y_5$ must be replaced by $y_5^+ \equiv y_5^+(x_5, x_1, y_1, z_1, x_3, y_3^-, z_3^-)$ and $z_5$ by $z_5^+ \equiv z_5(x_5, y_5^+, x_1, y_1, z_1, x_3, y_3^-, z_3^-)$, $y_6$ must be replaced by $y_6^- \equiv y_6^-(x_6, x_2, y_2, z_2, x_4, y_4^-, z_4^-)$ and $z_6$ by $z_6^- \equiv z_6(x_6, y_6^-, x_2, y_2, z_2, x_4, y_4^+, z_4^+)$, $y_7$ must be replaced by $y_7^+ \equiv y_7^+(x_7, x_3, y_3^-, z_3^-, x_5, y_5^+, z_5^+)$ and $z_7$ by $z_7^+ \equiv z_7(x_7, y_7^+, x_3, y_3^-, z_3^-, x_5, y_5^+, z_5^+)$. The coordinates $x_2$, $y_2$, and $z_2$ are functions of $r_2$ and $\varphi_2$ given by Equations 3G.10 and 3G.11.

## Appendix 3H: Derivation of Expressions for Emission and Absorption Rates by a Mean First Passage Time Analysis

A mean first passage time analysis [34–37] can be used to obtain the rates of emission and absorption of beads by a cluster of native residues within a protein [31–33]. To this end, a bead in the vicinity of the cluster is considered to perform a chaotic motion in the potential wells arising as a combination of the bead–cluster interactions with the dihedral angle potential. The mean first passage time of a bead escaping from a potential well is calculated on the basis of a kinetic equation governing the chaotic motion of the bead in that potential well. The chaotic motion of the bead is assumed to be governed by the Fokker–Planck equation for the single-particle distribution function with respect to its coordinates and momenta, i.e., in the phase space [34–36]. Prior to the passage event, the evolution of a bead in both the ipw and opw occurs in a dense enough medium (cluster of folded residues or unfolded, but compact part of the protein), where the relaxation time for its velocity distribution function is very short and negligible compared to the characteristic timescale of the passage process. Under these conditions, the Fokker–Planck equation reduces to the Smoluchowski equation, which involves diffusion in an external field [35–37]. In the case of spherical symmetry, it can be written in the form [34–37]

$$\frac{\partial p_i(r,t\,|\,r_0)}{\partial t} = D_i r^{-2}\frac{\partial}{\partial r}\left(r^2 e^{-\psi_i(r)}\frac{\partial}{\partial r}e^{\psi_i(r)}p_i(r,t\,|\,r_0)\right), \tag{3H.1}$$

where $p_i(r,t|r_0)$ is the probability of observing a bead of species $i$ ($i$ = b, l) between $r$ and $r + dr$ at time $t$ given that initially it was at a radial distance $r_0$, $D_i$ is its diffusion coefficient in the well, and $\Psi_i(r) = \Psi_i(r)/kT$.

The mean passage time depends on the initial position (distance from the center of the cluster) $r_0$ of the bead. It is convenient to use the backward Smoluchowski equation [34–37] which expresses the dependence of the transition probability $P_i(r,t|r_0)$ on $r_0$:

$$\frac{\partial p_i(r,t\,|\,r_0)}{\partial t} = D_i r_0^{-2} e^{\psi_i(r_0)}\frac{\partial}{\partial r_0}\left(r_0^2 e^{-\psi_i(r_0)}\frac{\partial}{\partial r_0}p_i(r,t\,|\,r_0)\right). \tag{3H.2}$$

Let us first consider the ipw and find the emission rate of beads therefrom. The probability that a bead $i$, initially at a distance $r_0$ within the surface layer, will remain in this region after time $t$ is given by the so-called survival probability [38–40],

$$q_i(t\,|\,r_0) = \int_R^{R+\lambda_i^-} \mathrm{d}r r^2 p_i(r,t\,|\,r_0)\left(R < r_0 < R+\lambda_i^-\right), \tag{3H.3}$$

where $R$ is the radius of the cluster and $\lambda_i^-$ is the width of the potential well determined by the location of the barrier between the ipw and the opw. The probability for the dissociation time to be between 0 and $t$ is equal to $1 - q_i(t|r_0)$, and the probability density for the dissociation time is given by $-\partial q_i/\partial t$. The first passage time is provided by [38–40]

$$\tau_i(r_0) = \int_0^\infty t \frac{\partial q_i(t|r_0)}{\partial t} dt = \int_0^\infty q_i(t|r_0) dt. \tag{3H.4}$$

The equation for the first passage time is obtained by integrating the backward Smoluchowski Equation 3H.2 with respect to $r$ and $t$ over the entire range and using the boundary conditions $q_i(0|r_0) = 1$ and $q_i(t|r_0) \rightarrow 0$ as $t \rightarrow \infty$ for any $r_0$. This yields [38–49]

$$-D_i^- r_0^{-2} e^{\psi(r_0)} \frac{\partial}{\partial r_0}\left( r_0^2 e^{-\psi_i(r_0)} \frac{\partial}{\partial r_0} \tau_i(r_0) \right) = 1. \tag{3H.5}$$

One can solve Equation 3H.5 by assuming a reflecting inner boundary of the well ($d\tau_i/dr_0 = 0$ at $r_0 = R$) and the radiation boundary condition at the outer boundary ($\tau_i(r_0) = 0$ at $r_0 = R + \lambda_i^-$). One thus obtains for the first passage time,

$$\tau_i(r_0) = \frac{1}{D_i} \int_r^{R+\lambda_i^-} dy\, y^{-2} e^{\psi_i(y)} \int_R^y dx\, x^2 e^{-\psi_i(x)}. \tag{3H.6}$$

The average dissociation time, $\tau_i^-$, or the mean first passage time for the transition from the ipw into the opw, is obtained by averaging $\tau_i(r_0)$ with the Boltzmann factor over all possible initial positions $r_0$:

$$\tau_i^- = \frac{1}{Z_i^-} \int_R^{R+\lambda_i^-} dr_0 r_0^2 e^{-\psi_i(r_0)} \tau_i(r_0) \tag{3H.7}$$

with

$$Z_i^- = \int_R^{R+\lambda_i^-} dr_0 r_0^2 e^{-\psi_i(r_0)}. \tag{3H.8}$$

The substitution of Equation 3H.6 into Equation 3H.7 leads to Equations 3.90 and 3.91.

Considering the opw, a similar procedure can be used to find $\tau_i^+$, the mean first passage time of an absorption event (for a bead of type $i$), i.e., mean first passage time for the transition of a bead of type $i$ from the opw into the ipw. It will be given by an expression almost identical to Equation 3.90, except for different integration limits corresponding to the outer potential well.

## Appendix 3I: Derivation of Equation 3.102 and Relevant Details

Equation 3.102 can be easily derived by first expanding in Taylor series the temperature dependent quantities on the right-hand side of Equation 3.87:

$$W^{\pm}(v,T) = W^{\pm}(v,T_0)\left[1 - \varepsilon_w^{\pm}(v,T_0)(T - T_0)\right], \tag{3I.1}$$

$$\Delta v_c(T) = \Delta v_c(T_0)\, [1 + \varepsilon_v(v_c, T_0)(T - T_0)], \tag{3I.2}$$

$$g_e(v, T) = g_e(v, T_0)\, \varepsilon_g(v, T_0, T), \tag{3I.3}$$

where the functions $\varepsilon_w^{\pm}(v,T)$, $\varepsilon_v(v,T)$, $\varepsilon_w(v,T)$, and $\varepsilon_g(v,T_0,T)$ are defined by Equations 3.103 through 3.106. Taking into account the definition of $G(v,T)$ and Equations 3.84, 3.85, 3.87, and 3I.1 through 3I.3, one obtains Equation 3.102.

Because the diffusion coefficient $D$ depends on temperature ($D = D(T)$), Equation 3.90 for the mean first passage time can be rewritten, by virtue of (10), as

$$\tau(T) = \frac{Q(T)}{Z(T)D(T)}, \tag{3I.4}$$

where

$$\begin{aligned} Q(T) &= \int_{r_a}^{r_b} \mathrm{d}r r^2 \int_{r}^{r_b} \mathrm{d}y y^{-2} \int_{r_a}^{y} \mathrm{d}x x^2 e^{-\psi(r,T)/k_BT} \times e^{\psi(y,T)/k_BT} e^{-\psi(x,T)/k_BT} \\ &= \int_{r_a}^{r_b} \mathrm{d}r \int_{r}^{r_b} \mathrm{d}y \int_{r_a}^{y} \mathrm{d}x r^2 y^{-2} x^2 e^{-\psi_0(r,y,x,T)/k_BT}, \end{aligned} \tag{3I.5}$$

with $\Psi(u_1, u_2, u_3, T) = -(\Psi(u_1, T) - \psi(u_2, T) + \psi(u_3, T))$ and

$$Z(T) = \int_{r_a}^{r_b} \mathrm{d}r r^2 e^{-\psi(r,T)/k_BT}. \tag{3I.6}$$

Thus, the first derivative $\mathrm{d}\tau^{+}(\nu,T)/\mathrm{d}T$ on the right-hand side of Equation 3.103 can be expressed through the functions $Q(T)$, $Z(T)$, $D(T)$ and their first derivatives with respect to $T$.

While the first derivative of $D(T)$ depends on the particular form of this function (assumed to be known), the first derivatives of $Q(T)$ and $Z(T)$ can be easily obtained from Equations 3I.5 and 3I.6:

$$\begin{aligned} \frac{\mathrm{d}Q(T)}{\mathrm{d}T} &= \frac{1}{T}\int_{r_a}^{r_b} \mathrm{d}r \int_{r}^{r_b} \mathrm{d}y \int_{r_a}^{y} \mathrm{d}x r^2 y^{-2} x^2 \frac{\psi_0(r,y,x,T)}{k_BT} \times e^{-\psi_0(r,y,x,T)/k_BT} \\ &- \int_{r_a}^{r_b} \mathrm{d}r \int_{r}^{r_b} \mathrm{d}y \int_{r_a}^{y} \mathrm{d}x r^2 y^{-2} x^2 \frac{1}{k_BT} \\ &\times \left( \frac{\mathrm{d}\bar{\phi}_\delta(r,T)}{\mathrm{d}T} - \frac{\mathrm{d}\bar{\phi}_\delta(y,T)}{\mathrm{d}T} + \frac{\mathrm{d}\bar{\phi}_\delta(x,T)}{\mathrm{d}T} \right) \\ &\times e^{-\psi_0(r,y,x,T)/k_BT}, \end{aligned} \tag{3I.7}$$

$$\frac{\mathrm{d}Z(T)}{\mathrm{d}T} = \frac{1}{T}\int_{r_a}^{r_b} \mathrm{d}r r^2 \frac{\psi_o(r,T)}{k_BT} e^{-\psi_0(r,T)/k_BT} - \int_{r_a}^{r_b} \mathrm{d}r r^2 \frac{1}{k_BT} \frac{\mathrm{d}\bar{\phi}_\delta(r,T)}{\mathrm{d}T} e^{-\psi_0(r,T)/k_BT}. \tag{3I.8}$$

The average dihedral potential $\bar{\phi}_\delta(r)$ attributed to the selected bead, 1, at the distance $r$ from the cluster center is determined from

$$\bar{\phi}_\delta(r) = n(T)/m(T) \tag{3I.9}$$

with

$$\begin{aligned} n(T) &\equiv n(r,T) \\ &= \int_{\Omega_{18}} \mathrm{d}r_2 \ldots \mathrm{d}r_7 \phi_\delta^1(r) \exp\left[-\phi_\delta^{27}(r)/k_BT\right], \end{aligned} \tag{3I.10}$$

$$m(T) \equiv m(r,T) = \int_{\Omega_{18}} \mathrm{d}\boldsymbol{r}_2 \ldots \mathrm{d}\boldsymbol{r}_7 \exp\left[-\phi_\delta^{27}(r)/k_\mathrm{B}T\right], \tag{3I.11}$$

where $\boldsymbol{r}_i$ ($i = 2, \ldots, 7$) is the radius-vector of bead $i$ (with the origin in the cluster center) and $\Omega_{18}$ is the integration region (of all possible configurations of beads 2, 3, …, 7) in an 18-dimensional space. The exponent in the Boltzmann factor is the sum of two terms, the first representing the dihedral angle potential of beads 1 through 7 and the second the sum of "bead–cluster" interactions for beads 2 through 7,

$$\phi_\delta^{27}(r) \equiv \phi_\delta^{27}(r, \boldsymbol{r}_2, \ldots, \boldsymbol{r}_7) = \phi_\delta^1(r) + \sum_{s=2}^{7} \phi_s(\boldsymbol{r}_s), \tag{3I.12}$$

where

$$\begin{aligned}\phi_\delta^1(r) \equiv \phi_\delta(r, \boldsymbol{r}_2, \boldsymbol{r}_3, \ldots, \boldsymbol{r}_7) &= \phi_\delta^1\left(\delta_{421}^{642}(r)\right) + \phi_1^\delta\left(\delta_{213}^{421}(r)\right) \\ &+ \phi_1^\delta\left(\delta_{135}^{213}(r)\right) + \phi_1^\delta\left(\delta_{357}^{135}(r)\right)\end{aligned} \tag{3I.13}$$

and $\phi^\delta\left(\delta_{jkl}^{ijk}(r)\right)$ is the dihedral angle potential corresponding to the dihedral angle $\delta_{jkl}^{ijk}$ between two planes, one of which is determined by beads $i, j, k$ and the other by beads $j, k, l$. There are four terms on the right-hand side of this equation because bead 1 is involved in four dihedral angles. The 18-fold integrals in Equations 3I.10 and 3I.11 can be reduced to sevenfold ones by taking into account the constraints of fixed bond length and fixed bond angle (see Refs. [31–33] for all the details on calculating $\bar{\phi}^\delta(r)$).

According to Equation 3I.9, the first derivative of $\bar{\phi}^\delta(r)$ with respect to $T$ is $\mathrm{d}\bar{\phi}_\delta(r)/\mathrm{d}T = [m(T)\mathrm{d}n(T)/\mathrm{d}T - n(T)\mathrm{d}m(T)/\mathrm{d}T]/m^2(T)$, where

$$\frac{\mathrm{d}n(T)}{\mathrm{d}T} = \frac{1}{T} \int_{\Omega_{18}} \mathrm{d}\boldsymbol{r}_2 \ldots \mathrm{d}\boldsymbol{r}_7 \phi_\delta^1(r) \frac{\phi_\delta^{27}(r)}{k_\mathrm{B}T} \exp\left[-\phi_\delta^{27}(r)/k_\mathrm{B}T\right], \tag{3I.14}$$

$$\frac{\mathrm{d}m(T)}{\mathrm{d}T} = \frac{1}{T} \int_{\Omega_{18}} \mathrm{d}\boldsymbol{r}_2 \ldots \mathrm{d}\boldsymbol{r}_7 \frac{\phi_\delta^{27}(r)}{k_\mathrm{B}T} \exp\left[-\phi_\delta^{27}(r)/k_\mathrm{B}T\right], \tag{3I.15}$$

and $n(T)$ and $m(T)$ are given by Equations 3I.10 and 3I.11.

## Appendix 3J: Coefficients in Taylor Series Expansion (Equation 3.110)

The coefficient functions $\alpha_i^\mathrm{w}(\nu, T_0)$, $\beta_i^\mathrm{w}(\nu, T_0)$, and $\gamma_i^\mathrm{w}(\nu, T_0)$ in Equation 3.110 are determined through the derivatives of the mean first passage time $\tau$ with respect to temperature $T$. Because the diffusion coefficient $D$ depends on temperature ($D = D(T)$), the expression for the mean first passage time [32] can be rewritten as

$$\tau(T) = \frac{Q(T)}{Z(T)D(T)}, \tag{3J.1}$$

where

$$\begin{aligned}Q(T) &= \int \mathrm{d}r r^2 \int \mathrm{d}y y^{-2} \int \mathrm{d}x x^2 e^{-\psi(r,T)/k_\mathrm{B}T} e^{\psi(y,T)/k_\mathrm{B}T} e^{-\psi(x,T)/k_\mathrm{B}T} \\ &= \int \mathrm{d}r \int \mathrm{d}y \int \mathrm{d}x r^2 y^{-2} x^2 e^{-\psi_0(r,y,x,T)/k_\mathrm{B}T},\end{aligned} \tag{3J.2}$$

with $\psi_0(u_1,u_2,u_3,T) = -(\Psi(u_1,T) - \Psi(u_2,T) + \Psi(u_3,T))$ and

$$Z(T) = \int dr r^2 e^{-\psi(r,T)/k_B T}. \tag{3J.3}$$

Henceforth, we omit the integration limits and the subscript for the bead type, so that the same expression can be used for the mean first passage times $\tau^+$ and $\tau^-$ and their derivatives for beads of both types.

Thus, the derivatives of $\tau(v,T)$ with respect to $T$ in Equation 3.111 can be expressed through the functions $Q(T)$, $Z(T)$, $D(T)$ and their derivatives with respect to $T$. While the derivatives of $D(T)$ depend on a particular form of this function (assumed to be known), the derivatives of $Q(T)$ and $Z(T)$ can be easily obtained from Equations 3J.2 and 3J.3.

For the first derivatives, we have

$$\frac{d\tau(T)}{dT} = \frac{Z(T)D(T)dQ(T)/dT - Q(T)[D(T)dZ(T)/dT + Z(T)dD(T)/T]}{[Z(T)D(T)]^2} \tag{3J.4}$$

$$\begin{aligned}\frac{dQ(T)}{dT} &= \frac{1}{T}\int dr \int dy \int dx r^2 y^{-2} x^2 \frac{\psi_0(r,y,x,T)}{k_B T} e^{-\psi_0(r,y,x,T)/k_B T} \\ &- \int dr \int dy \int dx r^2 y^{-2} x^2 \frac{1}{k_B T} \\ &\times \left( \frac{d\bar{\phi}^\delta(r,T)}{dT} - \frac{d\bar{\phi}^\delta(y,T)}{dT} + \frac{d\bar{\phi}^\delta(x,T)}{dT} \right) e^{-\psi_0(r,y,x,T)/k_B T}\end{aligned} \tag{3J.5}$$

$$\frac{dZ(T)}{dT} = \frac{1}{T}\int dr r^2 \frac{\psi(r,T)}{k_B T} e^{-\psi(r,T)/k_B T} - \int dr r^2 \frac{1}{k_B T} \frac{d\bar{\phi}^\delta(r,T)}{dT} e^{-\psi_0(r,T)/k_B T}, \tag{3J.6}$$

where $\bar{\phi}_i^\delta(r,T)$ is the average dihedral potential of the selected bead of type $i$ at the distance $r$ from the cluster center at temperature $T$.

Let us select bead 1 (its type is not indicated as agreed above). The average dihedral potential $\bar{\phi}^\delta \equiv \bar{\phi}^\delta(r,T)$ attributed to this bead can be written as (see Refs. [31–33])

$$\bar{\phi}^\delta(r) = n(T)/m(T) \tag{3J.7}$$

with

$$n(T) \equiv n(r,T) = \int_{\Omega_{18}} d\mathbf{r}_2 \ldots d\mathbf{r}_7 \phi_\delta^1(r) \exp[-\phi^{\delta 27}(r)/k_B T], \tag{3J.8}$$

$$m(T) \equiv m(r,T) = \int_{\Omega_{18}} d\mathbf{r}_2 \ldots d\mathbf{r}_7 \exp[-\phi^{\delta 27}(r)/k_B T], \tag{3J.9}$$

where $r_s$ ($s = 2, \ldots, 7$) is the radius-vector of bead $s$ (with the origin in the cluster center) and $\Omega_{18}$ is the integration region (of all possible configurations of beads 2, 3, …, 7) in an 18-dimensional space. The exponent in the Boltzmann factor is the sum of two terms, the first being $\phi^\delta(r)$, the total dihedral angle potential in which bead 1 is involved (for a given configuration of all neighboring beads), and the second the sum of effective "bead–cluster" interactions for beads 2 through 7 (see Refs. [31–33]).

According to Equation 3J.7, the first derivative of $\bar{\phi}_\delta(r)$ with respect to $T$ is $\mathrm{d}\bar{\phi}_\delta(r)/\mathrm{d}T = [m(T)\mathrm{d}n(\mathrm{T})/\mathrm{d}T - n(T)\mathrm{d}m(T)/\mathrm{d}T]/m^2(T)$, where

$$\frac{\mathrm{d}n(T)}{\mathrm{d}T} = \frac{1}{T}\int_{\Omega_{18}} \mathrm{d}\boldsymbol{r}_2\ldots\mathrm{d}\boldsymbol{r}_7\phi_\delta^1(r)\frac{\phi^{\delta 27}(r)}{k_\mathrm{B}T}\exp\left[-\phi_\delta^{27}(r)/k_\mathrm{B}T\right] \tag{3J.10}$$

$$\frac{\mathrm{d}m(T)}{\mathrm{d}T} = \frac{1}{T}\int_{\Omega_{18}} \mathrm{d}\boldsymbol{r}_2\ldots\mathrm{d}\boldsymbol{r}_7\frac{\phi^{\delta 27}(r)}{k_\mathrm{B}T}\exp\left[-\phi_\delta^{27}(r)/k_\mathrm{B}T\right], \tag{3J.11}$$

and $n(T)$ and $m(T)$ are given by Equations 3J.8 and 3J.9.

Using the single, double, and triple primes to indicate the corresponding derivatives with respect ot $T$ and omitting the argument $T$ of the temperature dependent $Q(T)$, $Z(T)$, and $T$, the second derivative of $\tau$ with respect to $T$ can be written as

$$\frac{\mathrm{d}^2\tau(T)}{\mathrm{d}T^2} = \frac{Z^2\mathrm{d}P(T)/\mathrm{d}T - P(T)2ZZ'}{Z^4}, \tag{3J.12}$$

where

$$P(T) = \frac{(Q'D - QD')Z}{D^2} - \frac{QZ'}{D}, \tag{3J.13}$$

$$\frac{\mathrm{d}P(T)}{T} = \frac{[(Q''D - QD'')D^2 - (Q'D - QD')2DD']Z}{D^4} - \frac{QZ''}{D} \tag{3J.14}$$

$$\begin{aligned} Q'' = &\int \mathrm{d}r \int \mathrm{d}y \int \mathrm{d}x r^2 y^{-2} x^2 e^{-\psi_0(r,y,x,T)/k_\mathrm{B}T} \\ &\times\left[\left(\frac{1}{T}\frac{\psi_0(r,y,x,T)}{k_\mathrm{B}T} - \frac{1}{k_\mathrm{B}T}\frac{\mathrm{d}\bar{\phi}^\delta(r,y,x,T)}{\mathrm{d}T}\right)^2\right. \\ &\left.+\left(-\frac{2}{T^2}\frac{\psi_0(r,y,x,T)}{k_\mathrm{B}T} + \frac{2}{T}\frac{1}{k_\mathrm{B}T}\frac{\mathrm{d}\bar{\phi}_0^\delta(r,y,x,T)}{\mathrm{d}T} - \frac{1}{k_\mathrm{B}T}\frac{\mathrm{d}^2\bar{\phi}_0^\delta(r,y,x,T)}{\mathrm{d}T^2}\right)\right], \end{aligned} \tag{3J.15}$$

$$\begin{aligned} Z'' = &\int \mathrm{d}r r^2 e^{-\psi(r,T)/K_\mathrm{B}T} \times \left[\left(\frac{1}{T}\frac{\psi(r,T)}{k_\mathrm{B}T} - \frac{1}{k_\mathrm{B}T}\frac{\mathrm{d}\bar{\phi}_i^\delta(r,T)}{\mathrm{d}T}\right)^2\right. \\ &\left.+\left(-\frac{2}{T^2}\frac{\psi(r,T)}{k_\mathrm{B}T} + \frac{2}{T}\frac{1}{k_\mathrm{B}T}\frac{\mathrm{d}\bar{\phi}^\delta(r,T)}{\mathrm{d}T} - \frac{1}{k_\mathrm{B}T}\frac{\mathrm{d}^2\bar{\phi}^\delta(r,T)}{\mathrm{d}T^2}\right)\right], \end{aligned} \tag{3J.16}$$

where $\bar{\phi}_0^\delta(u_1,u_2,u_3,T) = -(\bar{\phi}^\delta(u_1,T) - \bar{\phi}^\delta(u_2,T) + \phi^{-\delta}(u_3,T))$.

The third derivative of $\tau$ with respect to $T$ can be written as

$$\frac{\mathrm{d}^3\tau(T)}{\mathrm{d}T^3} = \frac{Z^2P'' + 2P(3Z'^2 - ZZ'')}{Z^4}, \tag{3J.17}$$

where

$$P'' = [(Q'''D + Q''D' - Q'D'' - QD''')D^2 - (Q'D - QD')2(D'^2 + DD'')]\frac{Z}{D^4}$$
$$+ [(Q''D - QD'')D^2 - (Q'D - QD')2DD']\frac{Z'D - Z4D'}{D^5} \tag{3J.18}$$
$$-\frac{(Q'Z'' + QZ''')D - QZ''D'}{D^2}.$$

## References


1. Anfinsen, C.B., *Science* 181, 223–230 (1973).
2. Ghelis, C., Yan, J., *Protein Folding*, Academic Press, New York, 1982.
3. Creighton, T.E., *Proteins: Structure and Molecular Properties*, W.H. Freeman, San Francisco, 1984.
4. Kauzmann, W., *Adv. Protein Chem.* 14, 1 (1959).
5. Privalov, P.L., *Crit. Rev. Biochem. Mol. Biol.* 25, 281 (1990).
6. Stryer, L., *Biochemistry*, 3rd ed., W.H. Freeman, New York, 1988.
7. Nölting, B., *Protein Folding Kinetics*, Springer-Verlag, Berlin, 2006.
8. Honeycutt, J.D., Thirumalai, D., *Proc. Natl. Acad. Sci. U.S.A.* 87, 3526 (1990).
9. Honeycutt, J.D., Thirumalai, D., *Biopolymers* 32, 695 (1992).
10. Nölting, B., *J. Theor. Biol.* 194, 419 (1998).
11. Nölting, B., *J. Theor. Biol.* 197, 113 (1999).
12. Weissman, J.S., Kim, P.S., *Science* 253, 1386–1393 (1991).
13. Creighton, T.E., *Nature* 356, 194–195 (1992).
14. Pastore, A., Martin, S.R., Politou, A., Kondapalli, K.C., Stemmler, T., Temussi, P.A., *J. Am. Chem. Soc.* 129, 5374 (2007).
15. Marqués, M.I., *Phys. Status Solidi A* 203, 1487 (2006).
16. Gursky, O., Atkinson, D., *Protein Sci.* 5, 1874 (1996).
17. Paci, E., Karplus, M., *Proc. Natl. Acad. Sci. U.S.A.* 97, 6521 (2000).
18. Abkevich, V.I., Gutin, A.M., Shakhnovich, E.I., *Biochemistry* 33, 10026 (1994).
19. Guo, Z., Thirumalai, D., *Biopolymers* 36, 83–102 (1995).
20. Fersht, A.R., *Proc. Natl. Acad. Sci. U.S.A.* 92, 10869 (1995).
21. Fersht, A.R., *Curr. Opin. Struct. Biol.* 7, 3 (1997).
22. Dill, K.A., *Biochemistry* 24, 1501 (1985).
23. Dill, K.A., *Biochemistry* 29, 7133 (1990).
24. Bryngelson, J.D., Wolynes, P.G., *Biopolymers* 30, 177 (1990).
25. Lothe, J., Pound, G.M.J., in: *Nucleation*, Zettlemoyer, A.C., Ed., Marcel-Dekker, New York, 1969.
26. Nölting, B., Golbik, R., Neira, J., Soler-Gonzales, A.S., Schreiber, G., Fersht, A., *Proc. Natl. Acad. Sci. U.S.A.* 94, 826 (1997).
27. Otzen, D.E., Itzhaki, L.S., ElMasry, N.F., Jackson, S.E., Fersht, A., *Proc. Natl. Acad. Sci. U.S.A.* 91, 10422 (1994).
28. Koizumi, M., Hirai, H., Onai, T., Inoue, K., Hirai, M., *J. Appl. Crystallogr.* 40, 175–178 (2007).
29. Brovchenko, I., Krukau, A., Smolin, N., Oleinikova, A., Geiger, A., Winter, R., *J. Chem. Phys.* 123, 224905 (2005).
30. Scheraga, H.A., Khalili, M., Liwo, A., *Annu. Rev. Phys. Chem.* 58, 57–83 (2007).
31. Djikaev, Y.S., Ruckenstein, E., *J. Phys. Chem. B* 111, 886 (2007). (Section 3.1 of this volume.)
32. Djikaev, Y.S., Ruckenstein, E., *J. Chem. Phys.* 126, 175103 (2007). (Section 3.2 of this volume.)
33. Djikaev, Y.S., Ruckenstein, E., *J. Chem. Phys.* 128, 025103 (2008). (Section 3.3 of this volume.)
34. Chandrasekhar, S., *Rev. Mod. Phys.* 15, 1 (1943).
35. Gardiner, C.V., *Handbook of Stochastic Methods*, Springer, New York, 1983.
36. Szabo, A., Schulten, K., Schulten, Z., *J. Chem. Phys.* 139, 500 (1980).
37. Park, S., Sener, M.K., Lu, Schulten, K., *J. Chem. Phys.* 119, 1313 (2003).
38. Narsimhan, G., Ruckenstein, E., *J. Colloid Interface Sci.* 128, 549 (1989). (Section 1.2 of this volume.)
39. Ruckenstein, E., Nowakowski, B., *J. Colloid Interface Sci.* 137, 583 (1990). (Section 1.3 of this volume.)

40. Nowakowski, B., Ruckenstein, E., *J. Colloid Interface Sci.* 139, 500 (1990) (Section 1.4 of this volume).
41. Zhou, Y., Zhang, C., Stell, G., Wang, J., *J. Am. Chem. Soc.* 125, 6300 (2003).
42. Faisca, P.F.N., Ball, R.C., *Phys. Rev. Lett.* 116, 7231 (2002).
43. Li, M.S., Klimov, D.K., Thirumalai, D., *Comput. Phys. Commun.* 147, 625–628 (2002).
44. Xu, Y., Wang, T., Gai, F., *Chem. Phys.* 323, 21–27 (2006).
45. Gutin, A.M., Abkevich, V.I., Shakhnovich, E.I., *Phys. Rev. Lett.* 77, 5433 (1996).
46. Tsai, C.-J., Maizel, J.V., Nussinov, R., *Crit. Rev. Biochem. Mol. Biol.* 37, 55–69 (2002).
47. Shakhnovich, E.I., Gutin, A.M., *J. Phys. A: Math. Gen.* 22, 1647 (1989).
48. Bratko, D., Chakraborty, A.K., Shakhnovich, E.I., *J. Chem. Phys.* 106, 1264 (1997).
49. Konkoli, Z., Hertz, J., Franz, S., *Phys. Rev. E: Stat., Nonlinear, Soft Matter Phys.* 64, 051910 (2001).
50. Bryngelson, J.D., Wolynes, P.G., *Proc. Natl. Acad. Sci. U.S.A.* 1987, 84, 7524.
51. Bryngelson, J.D., Wolynes, P.G., *J. Phys. Chem.* 93, 6902 (1998).
52. Garel, T., Orland, H., Pitard, E., in: *Spin Glasses and Random Fields*, Young, A.P., Ed., World Scientific, Singapore, 1998.
53. Derrida, B., *Phys. Rev. B: Condens. Matter Mater. Phys.* 24, 2613 (1981).
54. Moskalenko, A., Szamel, G., *Europhys. Lett.* 48(6), 641–647 (1999).
55. Tanford, C., *Adv. Protein Chem.* 24, 1–95 (1970).
56. Shortle, D., *FASEB J.* 10(1), 27–34 (1996).
57. Smith, L.J., Fiebig, K.M., Schwalbe, H., Dobson, C.M., *Folding Des.* 1(5), R95–R106 (1996).
58. Bernstein, F.C., Koetzle, T.F., Williams, G.J.B., Meyer, J.E.F., Brice, M.D., Rodgers, J.R., Kennard, O., Shimanouchi, T., Tasumi, M., *J. Mol. Biol.* 112, 535–542 (1977).
59. Kanesaka, S., Kamiguchi, K., Kanekiyo, M., Kurok, S., Ando, I., *Biomacromolecules* 7, 1323–1328 (2006).
60. Swindells, M.B., Macarthur, M.W., Thornton, J.M., *Nat. Struct. Biol.* 2(7), 596–603 (1995).
61. Neri, D., Billeter, M., Wider, G., Wathrich, K., *Science* 257(5076), 1559–1563 (1992).
62. Bolin, K.A., Pitkeathly, M., Miranker, A., Smith, L.J., Dobson, C.M., *J. Mol. Biol.* 261(3), 443–453 (1996).
63. Fiebig, K.M., Schwalbe, H., Buck, M., Smith, L.J., Dobson, C.M., *J. Phys. Chem.* 100(7), 2661–2666 (1996).
64. Kuni, F.M., *Colloid J. USSR* 46, 595–601 (1984).
65. Kuni, F.M., *Colloid J. USSR* 46, 791–797 (1984).
66. Jorgensen, W.L., Tirado-Rives, J., *J. Am. Chem. Soc.* 110, 1657 (1988).
67. Harpaz, Y., Gerstein, M., Chothia, C., *Structure* 2, 641–649 (1994).
68. Huang, D.M., Chandler, D., *Proc. Natl. Acad. Sci. U.S.A.* 15, 8324–8327 (2000).
69. Mukherjee, A., Bhattacharyya, S., Bagchi, B., *J. Chem. Phys.* 116, 4577–4586 (2002).
70. Bosse, J., Kaneko, Y., *Phys. Rev. Lett.* 74, 4023 (1995).
71. Hirshfelder, J.O., Curtiss, C.F., Bird, R.B., *Molecular Theory of Gases and Liquids*, Wiley, New York, 1964.
72. Wakeham, W.A., *J. Phys. B: Atom. Mol. Phys.* 6, 372 (1973).
73. Scalley, M.L., Barker, D., *Proc. Natl. Acad. Sci. U.S.A.* 94, 10636 (1997).
74. Georg, H., Wharton, C.W., Siebert, F., *Laser Chem.* 19, 233–235 (1999).
75. Mayor, U., Johnson, C.M., Daggett, V., Fersht, A.R., *Proc. Natl. Acad. Sci. U.S.A.* 97, 13518 (2000).
76. Jackson, S.E., Fersht, A.R., *Biochemistry* 30, 10428–10435 (1991).
77. Spector, S., Raleigh, D.P., *J. Mol. Biol.* 293, 763–768 (1999).
78. Tang, X., Pikal, M.J., *Pharm. Res.* 22, 1176 (2005).
79. Williams, M.A., Thornton, J.M., Goodfellow, J.M., *Protein Eng.* 10, 895 (1997).
80. Day, R., Daggett, V., *Proc. Natl. Acad. Sci. U.S.A.* 102, 13445 (2005).

## 3.5 First Passage Time Analysis of Protein Folding via Nucleation and of Barrierless Protein Denaturation

*Yuri Djikaev and Eli Ruckenstein**

A review of the kinetic models, recently developed by the authors for the nucleation mechanism of protein folding and for the barrierless thermal denaturation, is presented. Both models are based on the mean first passage time analysis. A protein is treated as a random heteropolymer consisting of hydrophobic, hydrophilic, or neutral beads. As a crucial idea of the model, an overall potential around the cluster of native residues wherein a residue performs a chaotic motion is considered as the combination of the average dihedral, effective pairwise, and confining potentials. The overall potential as a function of the distance from the cluster center has a double well shape that allows one to determine its emission and absorption rates by the first passage time analysis. One can thus develop a theory for the nucleation mechanism of protein folding and calculate the temperature dependence of the folding time. A kinetic model for protein denaturation occurring in a barrierless way has been also developed by using the same approach. The numerical calculations for two model proteins (one consisting of 124 amino acids and the other of 2500 amino acids) demonstrate that the models can predict folding and unfolding times consistent with experimental data.

### 3.5.1 Introduction

Life depends on thousands of various proteins whose structures are fashioned so that the individual protein molecules combine, with amazing precision, with other molecules. If a specific job has to be performed in a living organism, it is almost always a protein that does it. For a protein molecule to carry out a specific biological function, it has to adopt and maintain a well-defined three-dimensional (conventionally referred to as "native") structure [1,2]. The formation and stability of this structure (of a biologically active protein) constitute the core of two exciting (and intrinsically related) topics of modern biophysics, namely, "protein folding" and "protein denaturation" (i.e., unfolding) [2–5]. Many thermodynamic and kinetic aspects of these processes remain obscure [6–16].

Recently [17–19], we have presented a microscopic model for the nucleation mechanism of protein folding which is believed to be one of the possible pathways for the transition of a protein from a compact "amorphous" configuration (into which an initially unfolded protein quickly transforms) to the native state. Such a transition occurs immediately following the formation of a number of tertiary (native) contacts [8,9,20–23]; once a critical number thereof is attained, the native structure is formed without passing through any detectable intermediates. Our model of this process is based on the mean first passage time analysis (MFPTA) [24–27] previously applied [28–30] to nucleation phenomena in vapor-to-liquid and liquid-to-solid phase transitions. The model involves pairwise "molecular" interactions (i.e., repulsion/attraction), configurational (dihedral angle) potentials in which protein residues are involved, and confining potential which arises because of the finite size of the protein. The protein itself is treated as a heteropolymer chain consisting of three types of beads—neutral, hydrophobic, and hydrophilic [8,9,17,18,21] (with all the bonds and bond angles equal and constant; see Figure 3.20). The ionizability of the latter type of residues was also taken into account [19]. A cluster of native residues within a protein emits and absorbs residues with rates determined by means of MFPTA [24–27]. Knowing these rates as functions of cluster size, one can develop a self-consistent kinetic theory for the nucleation mechanism of protein folding. The time necessary for the protein to fold can be evaluated as the sum of the times necessary for the appearance of a nucleus and the time necessary for the nucleus to grow to the maximum size (of the folded protein in the native state).

Both experiments and simulations have demonstrated that a change in temperature significantly affects the process of protein folding [31–35]. The MFPTA-based method allows one to examine the effect of temperature on protein folding on reasonable timescales (much shorter than those required by conventional computer simulations). These timescales can be further shortened [36] by using Taylor series expansions in the MFPTA formalism.

Once the protein is in its native state, the change in temperature may affect the stability of a native protein and, in fact, lead to its unfolding (denaturation) upon both cooling and heating [13–16]. The unfolding of a protein results in the loss of its biological functionality which may lead to catastrophic consequences for a living organism.

Protein denaturation can be defined as a process (or sequence of processes) whereby the spatial arrangement of the polypeptide chains within the molecule changes from that of the native protein to a more disordered

* *Adv. Colloid Interface Sci.* 146, 18 (2009). Republished with permission.

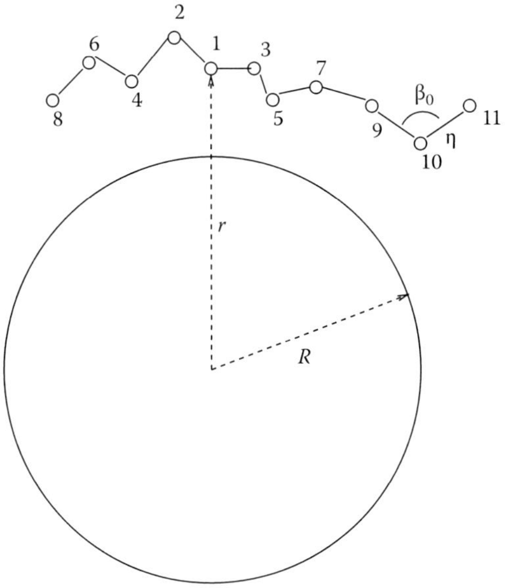


Figure 3.20

A piece of a heteropolymer chain around a spherical cluster consisting of $v$ beads. Bead 1 is in the plane of the figure, whereas other beads may all lie in different planes. All bond angles are equal to $\beta_0$, and their lengths are equal to $\eta$. The radius of the cluster is $R$ and the distance from the selected bead 1 to the cluster center is $r$.

one [4]. This change may alter the secondary, tertiary or quaternary structure of the protein. What constitutes denaturation is largely dependent upon its cause (such as heating or cooling, changes in pH, changes in the dielectric constant [upon addition of a cosolvent miscible with water but less/more polar], changes in the ionic strength of the solution). In this paper, we will be mainly concerned with the thermal mode of protein denaturation.

Protein unfolding (denaturation) is easier understood at high temperatures where it can be at least partly accounted for by a decreased stability of hydrogen bonding between water molecules and between water and protein residues. When proteins are exposed to increasing temperatures, losses of solubility or enzymatic activity occur over a relatively narrow range of temperatures. Depending upon the protein and the severity of heating, these changes may or may not be reversible.

The widespread existence of protein unfolding at low temperatures is surprising, particularly because it is unexpectedly accompanied by a decrease in entropy [37]. It was shown [5] that the cold denaturation is a general phenomenon caused by the very specific and strongly temperature-dependent interactions of the protein nonpolar groups with water. In contrast to expectations, the hydration of these groups is favorable thermodynamically, i.e., the Gibbs energy of hydration is negative and increases in magnitude as the temperature decreases [5]. As a result, the polypeptide chain, tightly packed in a compact native structure, unfolds at sufficiently low temperatures by exposing internal nonpolar groups to water.

Recently [36,38], we have presented a model for protein denaturation occurring in a barrierless way (similar to spinodal decomposition). The approach used is similar to that of the model for the nucleation mechanism of protein folding. We consider a native protein stable at temperature $T_0$ and examine the temperature dependence of the rates with which it can emit and absorb residues. At $T_0$ the absorption rate is greater than the emission rate for large enough clusters (of native residues). The unfolding temperature is determined as the temperature at which the emission rate becomes greater than the absorption rate in the whole range of cluster sizes (including the one corresponding to the entire protein). As in studying the temperature effect on protein folding, the temperature dependence of the emission and absorption rates can be examined [38] by using Taylor series expansions in the MFPTA formalism.

In this paper, we present a brief description of both theoretical models, namely, for the nucleation mechanism of protein folding and for barrierless protein denaturation. The two processes occur via different mechanisms but the models we have developed for them are both based on the MFPTA. The presentation of the models is not technically very detailed, because we would like to put the emphasis on the basic underlying ideas and concepts as well as on the predictive power of the models. For more details readers are referred to the original papers [17–19,36,38].

## 3.5.2 MFPTA in Protein Folding/Unfolding Problems

### 3.5.2.1 Heteropolymer as a Protein Model

Both the model for the nucleation mechanism of protein folding [17–19] and the model for barrierless thermal denaturation [36,38] considered the polypeptide chain of a protein as a heteropolymer chain which was first introduced as a simple model of a protein in the MD and MC simulations of protein folding dynamics [7,15]. The heteropolymer consists of N connected beads that can be thought of as representing the $\alpha$ carbons of various amino acids. The heteropolymer may consist of hydrophobic (b), hydrophilic (l), and neutral (n) beads. Two adjacent beads are connected by a covalent bond of fixed length $\eta$. This model (and its variants), completed with the appropriate potentials described below, has been shown [7,15,26,28] to be able to capture the essential characteristics of protein folding even though it contains only some of the features of a real polypeptide chain. For example, this model ignores the side groups although they are known to be crucial for the intramolecular hydrogen bonding [3]. Besides, the presence of a solvent (water) in a real physical system has been usually accounted for too simplistically, although the protein dynamics was reported to become more realistic in those MD simulations where the solvent molecules were explicitly taken into account [31]. Despite these limitations, various modifications of the heteropolymer model [6,7,32–35,37,39] shed light on some important details regarding the folding transition of a protein [7,15].

The total energy of the heteropolymer (polypeptide chain) can contain three contributions of different types: the contribution from repulsive/attractive forces between pairs of nonadjacent beads (these can be, e.g., of Lennard–Jones (LJ) or other types); a contribution from the harmonic forces due to the oscillations of the bond angles; a contribution from the dihedral angle potential due to the rotation around the peptide bonds. The bond angle forces are believed [7,15] to play a minor role in the protein folding/unfolding, hence all bond angles can be set to be equal to $\beta_0$. It was shown [7,15] by MD simulations, which employed low friction Langevin dynamics, that a proper balance between the remaining two contributions to the total energy of the heteropolymer ensures that the heteropolymer folds into a well defined $\beta$-barrel structure. It was also found [7,15] that the balance between the dihedral angle potential, which tends to stretch the molecule into a state with all bonds in a *trans* configuration, and the attractive hydrophobic potential is crucial to induce folding into a $\beta$-barrel like structure upon cooling. If attractive forces are excessively dominant they make the heteropolymer fold into a globule-like structure, while an overwhelming dihedral angle potential forces the chain to remain in an elongated state (even at low temperatures) with bonds mainly in the *trans* configuration.

The physical process of protein folding is modeled to occur via the formation and evolution of a cluster of native residues within the protein. The cluster is assumed to have a spherical shape all the time. Protein denaturation is pictured to occur in the inverse fashion, with the protein residues (initially in the globular, native state) escaping from the surface of a spherical cluster of native residues. In the framework of the three-component heteropolymer representation, the protein folding/unfolding process can be regarded as a ternary phase transition with the ternary cluster (of native residues) within a ternary mother phase (unfolded part of the protein).

### 3.5.2.2 Potential Well around a Cluster of Native Residues

Consider a ternary cluster of spherical shape (with sharp boundaries and radius $R$) immersed in a ternary fluid mixture. The number of molecules (beads) of component $i$ in the cluster will be denoted by $\nu_i$ ($i$ = b, l, n), whereas the total number of beads of component $i$ in the protein by $N_i$. In the previous applications of the first passage time analysis to nucleation [28–30] a molecule of component $i$ located in the surface layer of the cluster was considered to perform a thermal chaotic motion in a spherically symmetric potential well $\phi_i(r)$ resulting from the pair interactions (say, of LJ type) of this molecule with those in the cluster.

In the case of protein folding, the total potential $\psi_i(r)$ for a residue of type $i$ around the cluster has three constituents, $\phi_i(r)$, $\overline{\phi}_i^{\delta}(r)$ and $\phi_{\text{cp}}$, which represent the effective LJ (for pairwise interactions of the selected residue with those in the cluster), average dihedral angle, and confining (representing the external boundary of the volume available to the unfolded residues) potentials, respectively,

$$\psi_i(r) = \phi_i(r) + \phi_{\text{cp}}(r) + \overline{\phi}_i^{\delta}(r) \quad (i = \text{b, l, n}). \tag{3.114}$$

Complete details concerning the physical nature and calculation of $\phi_i(r)$, $\overline{\phi}_i^{\delta}(r)$, and $\phi_{\text{cp}}$ are given in Refs. [17–19,36,38]. Only a brief description of these contributions to $\psi_i(r)$ is given below.

The pairwise (LJ) contribution $\phi_i(r)$ to $\psi_i(r)$ is calculated by integrating the pair interactions of the selected bead of type $i$ with all the elements of the cluster (folded part of the protein). The contribution to $\phi_i(r)$ from the interaction of the selected bead with the unfolded part can be assumed to be small and accounted for by particular choices of the energy parameters.

Because of the bonds between the residues, none of them can be located outside some confining boundary. We have assumed [17–19] it to be spherical so that the confining potential $\phi_{cp}(r) = 0$ for $r < r_{cp}$ and $\phi_{cp}(r) = \infty$ for $r \geq r_{cp}$. The radius of the confining boundary $r_{cp}$ is assumed to be determined [17–19] by the total number of residues in the protein, densities of residues in its unfolded and folded parts, and the radius of the cluster (folded part).

The term $\bar{\phi}_i^{\delta}(r)$ in $\psi_i(r)$ is due to the $r$-dependent part of the dihedral angle potential of the whole protein. The dihedral angle potential arises due to the rotation of three successive peptide bonds connecting four successive beads, and is related to the dihedral angle δ via $\phi_{\delta} = \epsilon'_{\delta}(1+\cos\delta)+\epsilon''_{\delta}(1+\cos 3\delta)$, where $\epsilon'_{\delta}$ and $\epsilon''_{\delta}$ are energy parameters that depend on the nature and sequence of the four beads involved in the dihedral angle δ. The main idea underlying our model [17–19] consists of averaging the dihedral potential in which a selected residue is involved over all possible configurations of neighboring residues. The resulting average dihedral potential $\bar{\phi}_i^{\delta}(r)$ depends on the distance between the residue and the cluster center.

The combination of potentials in Equation 3.114 gives rise to a double potential well around the cluster with a barrier between the two wells [17–19]. Figure 3.21a presents typical shapes of the constituents $\phi_i(r)$ and $\bar{\phi}_i^{\delta}(r)$ of the potential well as functions of the distance from the cluster center, as well as the overall potential well $\psi_i(r)$ itself (for details regarding the numerical calculations, see Section 3.5.5). The contribution $\phi_i(r)$, arising from the pairwise interactions, has a form reminiscent of the underlying LJ potential, whereas the contribution from the average dihedral potential has a rather remarkable behavior. Indeed, starting with a maximum value at the cluster surface, it monotonically decreases with increasing $r$ until it becomes constant for large enough $r$'s. Thus, except very short distances from the cluster surface where $\phi_i(r)$

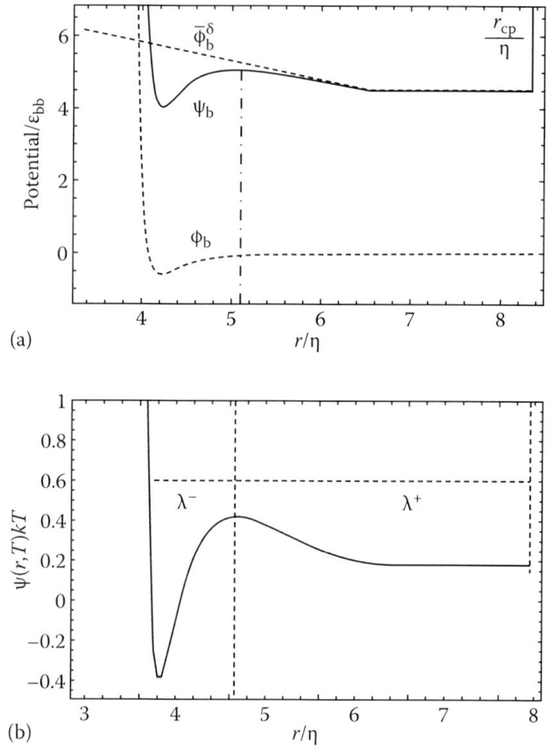


Figure 3.21

(a) The potentials $\phi_b(r)$ (lower dashed curve), $\bar{\phi}_b^{\delta}(r)$ (upper dashed curve), and $\psi_b(r)$ (solid curve) for a hydrophobic bead around the cluster as functions of the distance $r$ from the (center of the) cluster consisting of $\nu_b = 29$, $\nu_l = 50$, and $\nu_n = 2.4$ hydrophobic, hydrophilic, and neutral beads, respectively. The vertical dashed line indicates the location of the maximum of the barrier between the ipw and the opw which determines the widths $\lambda_b^-$ and $\lambda_b^+$ of both potential wells. The outer boundary of the opw is determined by the location of the confining potential indicated by $r_{cp}$. (b) A schematic representation of the total potential field around the cluster of radius $R = 3\eta$, which a selected bead (bead 1 in Figure 3.20) is subjected to, as a function of the distance $r/\eta$ from the cluster center. The widths of the inner and outer potential wells are $\lambda^-$ and $\lambda^+$, respectively.

sharply decreases from ∞ to its global minimum, the potential $\psi_i(r)$ is shaped by two competing terms, $\phi_i(r)$ and $\bar{\phi}_i^{\delta}(r)$, which increase and decrease, respectively, with increasing $r$, and by the confining potential $\phi_{cp}$. As a result, the overall potential $\psi_i(r)$ has a double well shape: the inner potential well (ipw) is separated by a barrier from the outer potential well (opw). This shape of $\psi_i(r)$ is of crucial importance to our nucleation mechanism model of protein folding because it allows one to use the MFPTA for the determination of the rates of both absorption and emission of beads by the cluster.

The residues in the inner well are considered to belong to the cluster (part of the protein with correct tertiary contacts) while those in the outer well are treated as belonging to the mother phase (amorphous part of the protein with incorrect tertiary contacts). Transitions of residues from the inner well into the outer one and vice versa are considered as elementary emission and absorption events, respectively. The rates of these processes are determined by using an MFPTA. This time is calculated by solving a single-molecule master equation for the probability distribution function of a surface layer molecule moving in a potential well around the cluster. The master equation is a Fokker–Planck equation in the phase space which can be reduced to the Smoluchowski equation owing to the hierarchy of the characteristic timescales involved in the evolution of the single-molecule distribution function with respect to coordinates and momenta [24–27]. Once the rates of emission and absorption are found as functions of the cluster variables, one can develop self-consistent kinetic theories for the nucleation mechanism of protein folding [17–19,36] and for the barrierless thermal denaturation of a native protein [36,38].

Note that, on the right-hand side of Equation 3.114 for the total potential around the cluster, a contribution due to hydrogen bonding is not included, although hydrogen bonding is believed to play an important role in both protein folding and protein unfolding. In addition to hydrogen bonds between the atoms of the polypeptide backbone (main chain-main chain N–H...O bonds) which are responsible for the formation of the secondary structure of proteins (both α-helices and β-sheets), many hydrogen bonded interactions are provided by the polar groups of side chains. These provide important stabilizing interactions on the protein surface and sometimes in the molecular interior, through side chain–side chain and side chain–main chain hydrogen bonds, or, for surface groups, through interactions with the external solvent. Moreover, the biological activity of proteins appears to depend on the formation of a two-dimensional (2D) hydrogen-bonded network spanning most of the protein surface and connecting all the surface hydrogen-bonded water clusters [40,41]. Such a water network is critical, not only for the correct folding of proteins but also for the maintenance of the native structure. Heat denaturation and loss of biological activity has been linked to the breakup of the 2D-spanning water network around the protein [39] (due to increasing hydrogen bond breakage with temperature), which otherwise acts restrictively on protein vibrational dynamics [42].

Clearly, in order to render our kinetic models [17–19,36,38] for protein folding and unfolding more realistic, it is necessary to find a way for taking into account hydrogen bonding "water–water" and "water–protein residue" during these processes. As a first step to this goal, we are currently developing a probabilistic method for studying the effect of hydrogen bond networks of water molecules around two large particles in water on their interaction potential.

### 3.5.2.3 Rates of Emission and Absorption of Residues by a Cluster

Let us use $W_i^- \equiv W_i^-(\nu_b, \nu_l, \nu_n)$ and $W_i^+ \equiv W_i^+(\nu_b, \nu_l, \nu_n)$ ($i$ = b, l, n) to denote the rates of emission and absorption, respectively, of beads of type $i$ by a cluster containing hydrophobic ($\nu_b$), hydrophilic ($\nu_l$), and neutral ($\nu_n$) residues. These rates represent the fundamental kinetic characteristics of the protein folding/unfolding processes. At any given temperature both functions $W^-$ and $W^+$ can be determined by using a first passage time analysis (the method was first [28–30] applied to calculating $W^- = W^-(\nu)$ in unary nucleation and later extended [17–19,36,38] to both $W^-$ and $W^+$ in protein folding/unfolding).

Consider a heteropolymer bead (i.e., a protein residue) of type $i$ ($i$ = b, l, n) performing a chaotic motion in a spherically symmetric potential well $\psi(r)$ with one boundary infinitely high (say, at $r = r_a$) and the other one of finite height (say, at $r = r_b$) (see Figure 3.21b). The mean first passage time τ necessary for the molecule to escape from the well is

$$\tau = \frac{1}{Z}\frac{1}{D}\int_{r_a}^{r_b} \mathrm{d}r r^2 e^{-\psi(r)} \int_{r}^{r_b} \mathrm{d}y y^{-2} e^{\psi(y)} \int_{r_a}^{y} \mathrm{d}x x^2 e^{-\psi(x)}, \quad (3.115)$$

where $D$ is the diffusion coefficient of a residue, $\Psi(r) = \psi(r)/k_BT$ ($k_B$ is Boltzmann's constant and $T$ is the temperature), and

$$Z = \int_{r_a}^{r_b} dr r^2 e^{-\psi(r)}. \tag{3.116}$$

The expression for $\tau$ was derived by solving a single-molecule master equation for the probability distribution function of a surface layer molecule (residue/bead) moving in a potential field $\psi(r)$ [17–19,24–27]. The diffusive motion of the bead is assumed to be governed by the Fokker–Planck equation. The Fokker–Planck equation reduces to the Smoluchowski equation (which involves diffusion in an external field) if the relaxation time for the velocity distribution function of the molecule is very short and negligible compared to the characteristic timescale of the passage process.

The rates of emission and absorption of beads of type $i$ by the cluster (i.e., the numbers of residues of type $i$ escaping from the ipw into the opw and from the opw into the ipw, respectively, per unit time) are provided by

$$W_i^- = \frac{N_i^-}{\tau_i^-}, \; W_i^+ = \frac{N_i^+}{\tau_i^+}, \tag{3.117}$$

where $N_i^-$ and $N_i^+$ denote the numbers of molecules in the ipw and opw, respectively, and $\tau_i^-$ and $\tau_i^+$ are the mean first passage times for the transition of a bead of type $i$ from the ipw into the opw and from the opw into the ipw, respectively. Applying Equations 3.115 and 3.116 to calculate $W^-$, the locations of the boundaries of the ipw must be used, that is, $r_a = R$ and $r_b = R + \lambda^-$, where $R$ is the radius of the cluster and $\lambda^-$ is the width of the ipw (Figure 3.21b). One the other hand, calculating $W^+$, the locations of the boundaries of the opw must be used in Equations 3.115 and 3.116, that is, $r_a = R + \lambda^- + \lambda^+$ and $r_b = R + \lambda^-$, with $\lambda^+$ the width of the opw. The quantities $N_i^-$ and $N_i^+$ can be calculated as the product of the "volume × number density"

$$N_i^- = \frac{4\pi}{3}\left[(R+\lambda^-)^3 - R^3\right]\rho_f, \; N_i^+ \frac{4\pi}{3}\left[(R+\lambda^- + \lambda^+)^3 - (R+\lambda^-)^3\right]\rho_u,$$

where $\rho_f$ and $\rho_u$ are the number densities of residues in the folded and unfolded parts of the protein.

### 3.5.3 Protein Folding

#### 3.5.3.1 Evaluation of Protein Folding Time in the Framework of Ternary Nucleation Formalism

Knowing the emission and absorption rates as functions of $v$ as well as the nucleation rate $J_s$, one can estimate the time $t_f$ necessary for the protein to fold via nucleation. Roughly speaking, the protein folding (via nucleation) consists of two stages. During the first stage, a critical cluster (nucleus) of native residues is formed (nucleation proper). Until the nucleus forms, the emission rate $W^-$ is larger than $W^+$, but the cluster still can attain the critical size and composition by means of fluctuations. At the second stage the nucleus grows via regular absorption of native residues which dominates their emission, $W^- < W^+$. Thus, the folding time is given by

$$t_f \simeq t_n + t_g, \tag{3.118}$$

where $t_n$ is the time necessary for one critical cluster to nucleate within a compact (but still unfolded) protein and $t_g$ is the time necessary for the nucleus to grow up to the maximum size, i.e., the size of the entirely folded protein.

The time $t_n$ of the first nucleation event can be estimated as

$$t_n \simeq 1/[J_s V_0], \tag{3.119}$$

where $J_s$ is the steady-state rate of ternary nucleation and $V_0$ is the volume of the unfolded protein in a compact configuration.

The nucleation rate $J_s$ is found by solving the steady-state version of the kinetic equation of ternary nucleation governing the temporal evolution of $g(\nu_b, \nu_l, \nu_n, t)$, the distribution of clusters with respect to their three independent variables of state at time $t$. This equation has to be solved in the vicinity of the saddle point of the function $G(\nu_b,\nu_l, \nu_n) = -k_BT \ln[g_e(\nu_b, \nu_l, \nu_n)/(\rho_{bu} + \rho_{lu} + \rho_{nu})]$, where $g_e(\nu_b, \nu_l, \nu_n)$ is the equilibrium distribution of clusters which can be constructed once the emission and absorption rates $W_i^- = W_i^-(\nu_b,\nu_l,\nu_n)$ and $W_i^+ = W_i^+(\nu_b,\nu_l,\nu_n)$ are known as functions of $\nu_b$, $\nu_l$, $\nu_n$ (for more details, see Refs. [17–19]). Note that in the MFPTA-based approach to the kinetics of multicomponent nucleation the function $G(\nu_b, \nu_l, \nu_n)$ plays a role similar to the free energy of cluster formation in the classical nucleation theory (CNT). It determines a surface in a four-dimensional space which, under appropriate conditions (i.e., high enough metastability of the initial phase), is expected to have the shape of a hyperbolic paraboloid with at least one "saddle" point at the coordinates $\nu_{bc}$, $\nu_{lc}$, $\nu_{nc}$ (hereinafter the subscript "c" marks quantities at the saddle point). The steady-state kinetic equation of nucleation has to be solved subject to two boundary conditions, one expressing the assumption that small clusters are in equilibrium and the other the absence of too large clusters:

$$J_s = \frac{|\lambda_0| / 2\pi k_B T}{\sqrt{-\det(\mathbf{G}''/2\pi k_B T)}} g_e(\nu_{bc}, \nu_{lc}, \nu_{nc}), \tag{3.120}$$

where $\mathbf{G}''$ is the matrix of second derivatives of the function $G(\nu_b, \nu_l, \nu_n)$ with respect to $\nu_b$, $\nu_l$, $\nu_n$, and $\lambda_0$ is a negative eigenvalue of the matrix $\mathbf{A}{\cdot}\mathbf{G}''$, with the elements of the diagonal matrix $\mathbf{A}$ of the absorption rates given by $A_{ij} = \delta_{ij} W_i^+$ $(i, j = \mathrm{b, l, n})$, $\delta_{ij}$ being the Kronecker delta.

Note that although the form of the expression for $J_s$ in Equation 3.120 is identical to that given by Trinkaus [43], the latter was obtained (and applied to binary and ternary nucleation) in the framework of CNT. The crucial difference is hidden in the method for obtaining the equilibrium distribution $g_e(\nu_b, \nu_l, \nu_n)$. In our method it is obtained by using the MFPTA and the principle of detailed balance, while in CNT (whereupon the use of Trinkaus' expression had been previously based) the equilibrium distribution would have the form $(\rho_{bu} + \rho_{lu} + \rho_{nu})\exp[-F(\nu_b, \nu_l, \nu_n)/kT]$, where $F(\nu_b, \nu_l, \nu_n)$ is the free energy of formation of the cluster, derived by using the concept of surface tension.

The growth time $t_g$ is provided by the integral

$$t_g \simeq \int_{\nu_{bc}}^{N_b} \frac{d\nu}{W_b^+(\nu_b, \nu_l(\nu_b), \nu_n(\nu_b)) - W_b^-(\nu_b, \nu_l(\nu_b), \nu_n(\nu_b))}. \tag{3.121}$$

Here, the functions $\nu_l = \nu_l(\nu_b)$ and $\nu_n = \nu_n(\nu_b)$ determine the growth path in parametric form and can be found as the solution of a couple of simultaneous differential equations (see Refs. [18,19]):

$$\frac{d\nu_l}{d\nu_b} = \frac{W_l^+(\nu_b,\nu_l,\nu_n) - W_l^-(\nu_b,\nu_l,\nu_n)}{W_b^+(\nu_b,\nu_l,\nu_n) - W_b^-(\nu_b,\nu_l,\nu_n)}, \quad \frac{d\nu_n}{d\nu_b} = \frac{W_n^+(\nu_b,\nu_l,\nu_n) - W_n^-(\nu_b,\nu_l,\nu_n)}{W_b^+(\nu_b,\nu_l,\nu_n) - W_b^-(\nu_b,\nu_l,\nu_n)}.$$

### 3.5.3.2 Temperature Dependence of Protein Folding Time

*3.5.3.2.1 A Simplified Model* For simplicity, the temperature dependence of the protein folding time was examined in the case where the heteropolymer consists only [17] of hydrophobic (b) and hydrophilic (l) beads sequentially (and randomly) connected to one another by covalent bonds of equal and constant length $\eta$ with equal and constant angles $\beta_0$ between two neighboring bonds.

Assuming that the composition of a cluster of native residues within a protein (which is in a compact, but not-native state) remains constant and equal to the overall composition of the protein at all times, one can [17] treat the nucleation mechanism of protein folding in terms of unary nucleation and characterize the cluster by $\nu$, the total number of residues therein. Thus, the steady-state nucleation rate $J_s(T)$ is given by

$$J_s(T) = \frac{W_c^+}{\sqrt{\pi}\,\Delta\nu_c} \rho_u e^{-G_c}. \tag{3.122}$$

Here, the subscript "c" marks quantities for the critical cluster, $W^+ \equiv W^+(\nu)$ is the overall rate with which a cluster containing $\nu$ residues absorbs residues from the unfolded part of the protein, $\Delta\nu_c = \left|\delta^2 G/\delta\nu^2\right|_c^{-1/2}$, where the function $G \equiv G(\nu) = -\ln[g_e(\nu)/\rho_u]$, where $g_e(\nu)$ is the equilibrium distribution of clusters with respect to $\nu$ and $p_u$ is the number density of residues in the unfolded (although compact) protein (for more details, see Ref. [17]). Note that on the right-hand side of Equation 3.122 the quantities $W_c^+$, $\Delta\nu_c$, and $G_c$ depend on the temperature $T: W_c^+ = W_c^+(T)$, $\Delta\nu_c = \Delta\nu_c(T)$, $G_c = G_c(T)$. The equilibrium distribution $g_e(\nu)$ is also a function of $T$ because the rates of both absorption and emission, $W^+(\nu)$ and $W^-(\nu)$, depend on $T$.

The total rates of emission and absorption of residues by the cluster are defined as

$$W^- = W_b^- + W_l^-, \quad W^+ = W_b^+ + W_l^+,$$

where $W_i^- \equiv W_i^-(\nu_b, \nu_l)$ and $W_i^+ \equiv W_i^+(\nu_b, \nu_l) (i = \mathrm{b}, \mathrm{l})$ are the rates of emission and absorption, respectively, of beads of type $i$ by a cluster $(\nu_b, \nu_l)$, $\nu_b$ and $\nu_l$ being the numbers of hydrophobic and hydrophilic residues, respectively, in the cluster. Owing to the approximation that the composition of a cluster of native residues within a protein (which is in a compact but not-native state) remains constant and equal to the overall composition of the protein at all times, $\nu_b$ and $\nu_l$ are related to $\nu = \nu_b + \nu_l$ by $\nu_b = \chi_0\nu$ and $\nu_l = (1 - \chi_0)\nu$, where $\chi_0 = N_b/N$ (with $N = N_b + N_l$ and $N_i$ ($i$ = b, l) the total number of residues of type $i$ in the protein) is the mole fraction of hydrophobic residues in the entire protein. The above definitions of $W^+$ and $W^-$ can be rewritten

$$\begin{aligned} W^w(\nu) &= W_b^w(\chi_0\nu, (1-\chi_0)\nu) + W_l^w(\chi_0\nu, (1-\chi_0)\nu) \\ &\equiv W_b^w(\nu_b, \nu_l) + W_l^w(\nu_b, \nu_l) \quad (w = +, -). \end{aligned} \tag{3.123}$$

The growth time $t_g$ can be evaluated by solving the differential equations, governing the temporal evolution of the variables $\nu_b$ and $\nu_l$ for supercritical clusters, $d\nu_i/dt = W_i^+ - W_i^-$, $(i = \mathrm{b}, \mathrm{l})$ (where $t$ is the time) subject to the initial conditions $\nu_i = \nu_{ic}$ ($i$ = b, l) at $t = 0$ and $\nu_i = N_i$ ($i$ = b, l) at $t = t_g$. Because of the approximation $\nu_b/(\nu_b + \nu_l) = \text{const} = \chi_0$, this couple of simultaneous equations can be reduced to a single independent equation that, due to Equation 3.123, can be written in the form $d\nu/dt = W^+(\nu) - W^-(\nu)$ and is subject to the initial conditions $\nu = \nu_c$ at $t = 0$ and $\nu = N$ at $t = t_g$ (with $N = N_b + N_l$). One thus obtains

$$t_g \simeq \int_{\nu_c}^{N} \frac{d\nu}{W^+(\nu) - W^-(\nu)}.$$

Note that the right-hand side of this equation is positive because $W^+(\nu) > W^-(\nu)$ for $\nu > \nu_c$, according to the definition of a nucleus [17,36].

Numerical evaluations showed [17–19] that the protein folding time is determined mainly by the time necessary for the first nucleation event to occur because $t_n \gg t_g$ so that, for a given temperature $T$, $t_f(T) \simeq t_n(T)$ (when important and convenient, the temperature dependence of some of the quantities involved will be explicitly emphasized). Thus,

$$t_f(T) \simeq 1/[J_s(T)V_0] \tag{3.124}$$

and one can thus estimate the folding time $t_f$ of a protein if the emission and absorption rates $W^+(\nu)$ and $W^-(\nu)$ are known as functions of the cluster variable $\nu$.

*3.5.3.2.2 Optimal Temperature of Protein Folding* Every protein has optimal conditions for its fastest folding. For instance, if only the temperature of the protein system can vary, there exists such an optimal temperature $T^0$, at which the folding time of the protein is the shortest [32]. The model developed in Refs. [17–19] and outlined above allows one to obtain an estimate for the optimal temperature of a given model protein [36].

Indeed, according to Equations 3.122 and 3.124, the temperature dependence of the folding time is due to the temperature dependence of $W_c^+ = W_c^+(T)$, $\Delta\nu_c = \Delta\nu_c(T)$, and $g_e(\nu_c) = g_e(\nu_c, T)$:

$$t_f(T) = \frac{\sqrt{\pi}}{V_0} \frac{\Delta\nu_c(T)}{W_c^+(T) g_e(\nu_c, T)}. \tag{3.125}$$

Thus, in the framework of our approach the optimal temperature $T^0$ of protein folding is the solution of the equation

$$\frac{\mathrm{d}}{\mathrm{d}T}\frac{\Delta v_c(T)}{W_c^+(T)g_e(v_c,T)}\bigg|_{T^0}=0. \tag{3.126}$$

A rough estimate for the optimal temperature $T^0$ of protein folding can be obtained by neglecting the $T$-dependence of $\Delta v_c$ and solving the approximate equation

$$g_e(v_c,T)\left[\mathrm{d}W_c^+(T)/\mathrm{d}T\right]+W_c^+(T)\mathrm{d}g_e(v_c,T)/\mathrm{d}T]=0. \tag{3.127}$$

Strictly speaking, it is impossible to solve Equation 3.127 analytically because at any given temperature all the components of the equation (i.e., $v_c$, $g_e(v_c)$, and $W_c^+$) are found numerically. However, one can suggest the following semianalytical procedure to evaluate $T^0$ with the least possible computational effort.

Assume that $g_e(v_c,T)$ and $W_c^+(T)$ are linear functions of $T$ (this approximation is acceptable for any function if used in a temperature range $T_a \le T \le T_b$ narrow enough for the nonlinearities to be negligible):

$$g_e(v_c,T)\simeq g_0+g_1T,\ W_c^+(T)=w_0+w_1T. \tag{3.128}$$

The coefficients $g_0$, $g_1$ and $w_0$, $w_1$ can be determined by the values of $g_e(v_c, T)$ and $W_c^+(T)$ at the two temperatures, $T_a$ and $T_b$ (see Ref. [36]). Substituting Equation 3.128 into Equation 3.127 and solving with respect to $T$, one obtains for the optimal temperature of protein folding:

$$T^0=-\frac{1}{2}\left(\frac{g_0}{g_1}+\frac{w_0}{w_1}\right). \tag{3.129}$$

Clearly, the accuracy of this estimate is largely dependent on the accuracy of the linear approximations in Equation 3.128.

*3.5.3.2.3 Taylor Series Expansions for Protein Folding Time* Since all temperature dependent quantities in our model have been identified, one can develop a procedure for obtaining the temperature dependence of the protein folding time $t_f$ on the basis of calculations carried out at a single temperature $T_0$ when $T$ is in the vicinity of $T_0$. As already mentioned, the procedure is based on a Taylor series expansion in the vicinity of $T_0$. Using linear expansions in some vicinity $T_0 - \Delta T \le T \le T_0 + \Delta T$ of $T_0$, Equation 3.125 leads to the following relation between the folding times at temperatures $T$ and $T_0$:

$$\frac{t_f(T)}{t_f(T_0)}\simeq\varepsilon_g(v_c,T_0,T)\left[1+\varepsilon_w^+(v_c,T_0)(T-T_0)\right]\left[1+\varepsilon_v(v_c,T_0)(T-T_0)\right], \tag{3.130}$$

where the functions $\varepsilon_w^+(v,T)$, $\varepsilon_v(v,T)$, and $\varepsilon_g(v, T_0, T)$ are expressed through $\mathrm{dln}(\tau^{\pm}(v,T))/\mathrm{d}T$, with $\tau^+$ and $\tau^-$ the mean first passage times for the transitions of a bead from the outer potential well into the inner one (absorption of a bead) and from the inner potential well into the outer one (emission of a bead), respectively (see Ref. [38]). At $T = T_0$ the function $\varepsilon_g(v, T_0, T_0) = 1$.

In order to examine the temperature dependence of the protein folding time $t_f$, the straightforward application of the method proposed in Refs. [17–19] would require very long calculations. On the other hand, if $t_f$ is known at some temperature $T_0$, the Taylor series expansions, Equation 3.130, can be used as an alternative way for studying the temperature dependence of the protein folding time. Indeed, they allow one to evaluate the ratio $t_f(T)/t_f(T_0)$ for any $T$ in a close enough vicinity of $T_0$ on a timescale by 1 or 2 orders of magnitude shorter than the time necessary for the evaluation of $t_f(T_0)$.

## 3.5.4 Barrierless Denaturation

### 3.5.4.1 Time of Barrierless Protein Denaturation

As mentioned earlier, the stability of a folded (native) protein can be disrupted by various factors. Depending on the strength of the destabilizing factor(s), the denaturation (unfolding) of the protein can occur either fluctuationally as nucleation (at weak destabilization) or, alternatively, in a barrierless way similar to spinodal decomposition. In the former case (i.e., unfolding via nucleation), the initial unfolding of the protein is thermodynamically unfavorable, but can still occur owing to fluctuations. After the protein unfolded to some critical extent, further denaturation becomes thermodynamically favorable and occurs with the decrease in the free energy of the system (protein + medium). In the latter case (i.e., barrierless unfolding), it is thermodynamically favorable for a protein to unfold starting from its native state down to its full denaturation.

Examining thermal denaturation, for simplicity we again consider a protein that consists of $N_b$ hydrophobic and $N_l$ hydrophilic residues. Initially folded at some temperature $T_0$, assume that it can unfold via spinodal decomposition at either $T \geq T_u^+ > T_0$, which corresponds to hot denaturation, or $T \leq T_u^- < T_0$, which corresponds to cold denaturation. The temperature $T_u^-$ represents the highest temperature at which cold denaturation can occur, whereas $T_u^+$ represents the lowest temperature of hot denaturation.

The threshold temperatures of hot and cold denaturation ($T_u^+$ and $T_u^-$, respectively) can be determined by solving the equation $W^-(N,T) = W^+(N,T)$ with respect to T. If the equation has two solutions, $T_u^+$ and $T_u^-$, the protein can unfold upon both heating and cooling. If the equation has just one solution, either $T_u^+$ or $T_u^-$, protein denaturation can occur just upon either heating or cooling.

The emission and absorption rates of a cluster (of native residues) of size $\nu$ $(0 \leq \nu \leq N)$ as functions of $\nu$ are shown in Figure 3.22a (for a protein at $T = T_0$), Figure 3.22b (for a protein at $T = T_u^+$ or $T = T_u^-$), and Figure 3.22c (for a protein at $T > T_u^+$ or $T < T_u^-$). If the protein is initially unfolded at $T_0$ (Figure 3.22a), it can fold via nucleation [8,9,17–23]. In principle, the native protein at this temperature can also unfold via nucleation, but in this work we assume that the thermodynamic barrier for the transition of a cluster of size $N$ (the whole protein in a native state) to the size $\nu_c$ (size of the critical cluster) is so high that the unfolding time is virtually

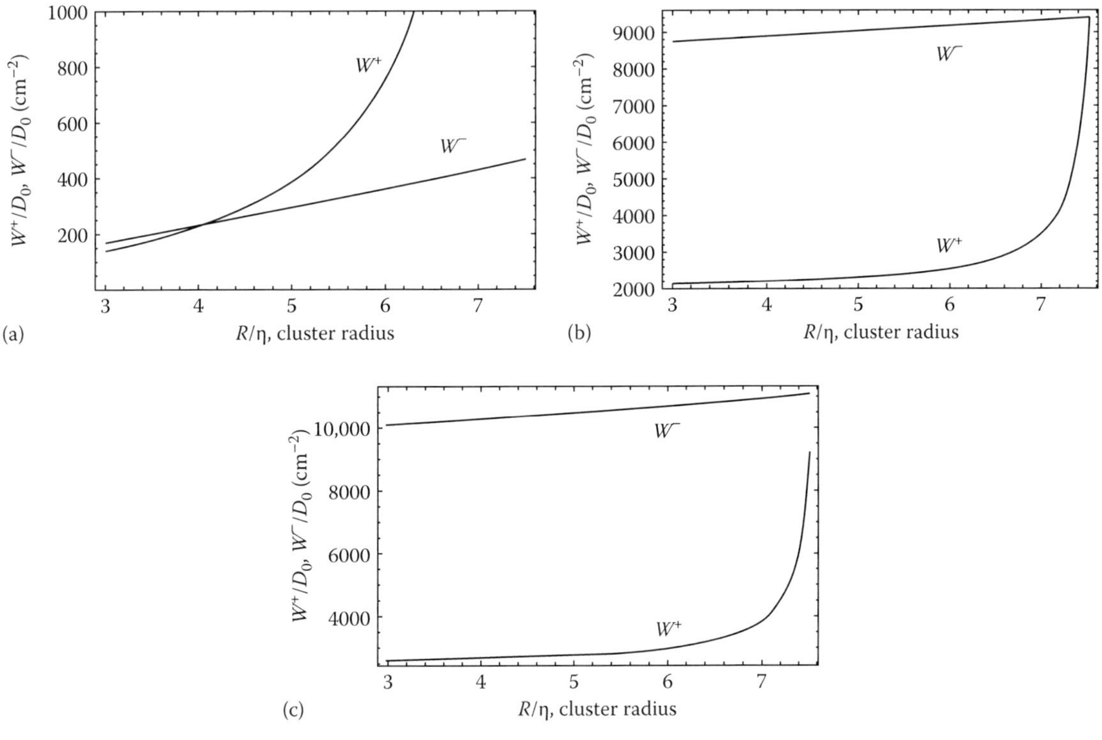


Figure 3.22

The emission and absorption rates of a cluster (of native residues), $W^+(\nu)$ and $W^-(\nu)$, as functions of $\nu$ at $T = T_0 = 293.15$ K (protein in the native state, panel a), $T = T^+ \simeq 358$ K (threshold temperature of hot denaturation, panel b), and $T = T_u^+ + 5$ K (temperature above the threshold one of hot denaturation, panel c). The data are for the large protein model of $N = 2500$.

infinity. On the other hand (see Figure 3.22b), $W^{+}(\nu) \leq W^{-}(\nu)$ for $\nu \leq N$ at the temperature $T_{\mathrm{u}}^{\pm}$ (the equality $W^{+} = W^{-}$ holds only for the largest cluster possible, i.e., at $\nu = N$). Therefore, if a protein is initially unfolded at $T_{\mathrm{u}}^{\pm}$, it will never fold, whereas, if it is initially folded at this temperature, it will immediately start unfolding via regular loss of residues from its native structure. The unfolding will occur even faster at $T > T_{\mathrm{u}}^{+}$ (or $T < T_{\mathrm{u}}^{-}$), as shown in Figure 3.22c, because in this case $W^{+}(\nu) < W^{-}(\nu)$ for all $\nu \leq N$ (including $\nu = N$) and the difference $W^{-}(\nu) - W^{+}(\nu)$ is greater than at $T = T_{\mathrm{u}}^{-}$ (see Section 3.5.5 for more details).

The full unfolding time $t_{\mathrm{u}}$ can be evaluated by solving a couple of simultaneous differential equations governing the temporal evolution of the variables $\nu_{\mathrm{b}}$ and $\nu_{\mathrm{l}}$ for $\nu_{\mathrm{b}} \leq N_{\mathrm{b}}$ and $\nu_{\mathrm{l}} \leq N_{\mathrm{l}}$, $d\nu_i/dt = W_i^{+} - W_i^{-}$, $(i = \mathrm{b}, \mathrm{l})$ subject to the initial conditions $\nu_i = N_i$ $(i = \mathrm{b}, \mathrm{l})$ at $t = 0$ and $\nu_i = 0$ $(i = \mathrm{b}, \mathrm{l})$ at $t = t_{\mathrm{g}}$. However, by virtue of the chosen approximation $\nu_{\mathrm{b}}/(\nu_{\mathrm{b}} + \nu_{\mathrm{l}}) = \mathrm{const} = \chi_0$, these simultaneous equations reduce to a single independent equation $\frac{d\nu}{dt} = -\left[W^{-}(\nu) - W^{+}(\nu)\right]$ which has to be solved subject to the initial conditions $\nu = N$ at $t = 0$ and $\nu = 0$ at $t = t_{\mathrm{u}}$. One can thus obtain

$$t_{\mathrm{u}} \simeq \int_{0}^{N} \frac{d\nu}{W^{-}(\nu) - W^{+}(\nu)}. \tag{3.131}$$

Note that the right-hand side of this equation is positive because $W^{+}(\nu) < W^{-}(\nu)$ in the whole range $0 \leq \nu \leq N$ at $T = T_{\mathrm{u}}^{\pm}$ (see Figure 3.22b).

### 3.5.4.2 Threshold Temperatures $T_{\mathrm{u}}^{-}$ and $T_{\mathrm{u}}^{+}$ of Cold and Hot Denaturation and the Unfolding Times

In order to evaluate $T_{\mathrm{u}}^{+}$ and/or $T_{\mathrm{u}}^{-}$, the minimum and maximum temperatures of hot and cold denaturation, respectively, that occur in a barrierless way, it is necessary to know the temperature dependence of the functions $W^{-}(N)$ and $W^{+}(N)$, i.e., the functions $W_i^{w} = W_i^{w}(\nu, T)$ $(w = +, -;\ i = \mathrm{b}, \mathrm{l})$.

If a protein is subject to both hot and cold denaturation, the corresponding threshold temperatures at which its barrierless unfolding can occur ($T_{\mathrm{u}}^{+}$ and $T_{\mathrm{u}}^{-}$, respectively), can be determined by solving the equation $W^{-}(N, T) = W^{+}(N, T)$ with respect to $T$, i.e.,

$$W^{-}\left(N, T_{\mathrm{u}}^{+}\right) = W^{+}\left(N, T_{\mathrm{u}}^{+}\right), \quad W^{-}\left(N, T_{\mathrm{u}}^{-}\right) = W^{+}\left(N, T_{\mathrm{u}}^{-}\right) \tag{3.132}$$

(note that both $W^{-}(\nu, T)$ and $W^{+}(\nu, T)$ are monotonic functions of $\nu$). The first of these equalities means that the threshold temperature of hot denaturation is the lowest temperature at which the emission rate of a cluster is greater than or equal to its absorption rate for clusters of any size, i.e., $W^{-}(\nu, T) \geq W^{+}(\nu, T)$ for any $\nu \leq N$ and $T \geq T_{\mathrm{u}}^{+}$. The second of the equalities in Equation 3.132 means that the threshold temperature of cold denaturation is the highest temperature at which the emission rate of a cluster is greater than or equal to its absorption rate for clusters of any size, i.e., $W^{-}(\nu, T) \geq W^{+}(\nu, T)$ for any $\nu \leq N$ and $T \leq T_{\mathrm{u}}^{-}$.

Estimates for the times necessary for the protein to unfold at temperatures $T = T_{\mathrm{u}}^{-}$ or/and $T = T_{\mathrm{u}}^{-}$ via spinodal decomposition are provided by the equations

$$t_{\mathrm{u}}^{-} \simeq \int_{0}^{N-0} \frac{d\nu}{W^{-}\left(\nu, T_{\mathrm{u}}^{-}\right) - W^{+}\left(\nu, T_{\mathrm{u}}^{-}\right)}, \quad t_{\mathrm{u}}^{+} \simeq \int_{0}^{N-0} \frac{d\nu}{W^{-}\left(\nu, T_{\mathrm{u}}^{+}\right) - W^{+}\left(\nu, T_{\mathrm{u}}^{+}\right)}. \tag{3.133}$$

Clearly, the unfolding time of hot denaturation at a temperature $T > T_{\mathrm{u}_{+}}$ is smaller than $t_{\mathrm{u}}^{+}$, while the unfolding time of cold denaturation at a temperature $T < T_{\mathrm{u}}^{-}$ is smaller than $t_{\mathrm{u}}^{-}$. The divergence of the integrands at the upper limit of integration at $T = T_{\mathrm{u}}^{+}$ and $T = T_{\mathrm{u}}^{-}$ complicates the application of these expressions. The character of the divergence determines whether the $t_{\mathrm{u}}$'s are finite or infinite, but this issue cannot be rigorously addressed unless the $\nu$ dependence of $W^{-}(\nu) - W^{+}(\nu)$ is explicitly known. Reasonable estimates for $t_{\mathrm{u}}^{-}$ and $t_{\mathrm{u}}^{+}$ can be obtained either by evaluating the integrals at temperatures very close to the threshold ones, $T_{\mathrm{u}}^{\pm}(1 \pm \delta)$ with a positive $\delta \ll 1$, or by slightly decreasing the upper limit of integration which is indicated by writing it as $N - 0$. The latter method could be substantiated by the effect of fluctuations whereof the physical impact is to decrease the size of the folded cluster of size $N$ but which are not formally included in the model.

### 3.5.4.3 Taylor Series Expansions for Emission and Absorption Rates

As clear from the previous subsection, even determining $W_i^{\mathrm{w}}(w=+,-;\ i=\mathrm{b},\mathrm{l})$ as functions of only $v$ at a given $T$ requires lengthy calculations. Thus, the straightforward application of the method proposed in Refs. [17–19] for constructing $W_i^{\mathrm{w}}$ as functions of both $v$ and $T$ and determining the threshold temperatures of hot and cold denaturations $T_{\mathrm{u}}^{+}$ and $T_{\mathrm{u}}^{-}$ and the corresponding unfolding times $t_{\mathrm{u}}^{+}$ and $t_{\mathrm{u}}^{-}$ would require time consuming computations. To avoid this problem, one can use the method proposed above where the temperature dependence of the protein folding time was determined by using Taylor series expansions.

Indeed, if $W_i^{\mathrm{w}}$ is known as a function of $v$ at some temperature $T_0$, one can evaluate $W_i^{\mathrm{w}}(v,T)$ for any $T$ in a close enough vicinity of $T_0$ using the series expansion

$$W_i^{\mathrm{w}}(v,T)=W_i^{\mathrm{w}}(v,T_0)\left[1-\alpha_i^{\mathrm{w}}(v,T_0)(T-T_0)-\beta_i^{\mathrm{w}}(v,T_0)(T-T_0)^2 -\gamma_i^{\mathrm{w}}(v,T_0)(T-T_0)^3+\cdots\right]\quad (w=+,-;\ i=\mathrm{b},\mathrm{l}), \tag{3.134}$$

where

$$\alpha_i^{\mathrm{w}}(v,T)=\frac{d\tau_i^{\mathrm{w}}(v,T)/dT}{\tau_i^{\mathrm{w}}(v,T)},\quad \beta_i^{\mathrm{w}}(v,T)=\frac{1}{2}\frac{d^2\tau_i^{\mathrm{w}}(v,T)/dT^2}{\tau_i^{\mathrm{w}}(v,T)},\quad \gamma_i^{\mathrm{w}}(v,T)=\frac{1}{6}\frac{d^3\tau_i^{\mathrm{w}}(v,T)/dT^3}{\tau_i^{w}(v,T)}. \tag{3.135}$$

Here, $\tau_i^{\mathrm{w}}(w=+,-)$ the mean first passage time necessary for the bead of type $i$ ($i$ = b, l) to escape from the well $w$. Equation 3.134 is derived from Equation 3.117 by assuming that $N_{\mathrm{b}}^{-}$, $N_{\mathrm{b}}^{+}$, $N_{\mathrm{l}}^{-}$, and $N_{\mathrm{l}}^{+}$, the numbers of beads of types b and l in the inner and outer potential wells, do not depend on $T$. Clearly, the range of validity of the expansions in Equation 3.134 becomes larger when higher-order (in $T-T_0$) terms are retained on its right-hand side.

## 3.5.5 Numerical Evaluations

In order to illustrate the above methods, we applied them to the folding and unfolding of two model proteins composed of 124 and 2500 amino acid residues. The first roughly mimics the bovine pancreatic ribonuclease (BPR) and was studied in Refs. [18,19,36]. In Refs. [18,19], the model BPR consisted of hydrophobic, hydrophilic, and neutral beads, whereas in Ref. [36] the neutral residues were treated as hydrophobic so that the whole heteropolymer was considered to consist of hydrophilic and hydrophobic residues with the mole fraction of the latter being $\chi_0 = 43/124 \simeq 0.347$. The second model protein is a representative of large proteins and was studied in Refs. [17,36,38]. It was a heteropolymer consisting of a total of 2500 hydrophobic and hydrophilic residues, with the mole fraction of hydrophobic residues $\chi_0 = 0.75$. We intentionally modeled such a large protein (relatively rare in nature) to demonstrate that our method provides a reasonable temperature dependence of the folding/unfolding times on a reasonable timescale even for the largest proteins. Straightforward MC or MD simulations of folding/unfolding of such large proteins are currently impossible within a sensible timeframe.

The presence of water molecules was not taken into account explicitly but was assumed to be included into the model via the potential parameters and diffusion coefficients. Clearly, hydrophilic residues are readily solvated by water molecules. The formation of hydrogen bonds between water molecules and amino acid residues (particularly hydrophilic ones) has various effects on the behavior of the polypeptide chain. The pairwise interactions of protein residues are affected by the water molecules hydrating them. Dihedral angle potential may be also affected due to changes in the properties of the hydrated residues involved. The direct interaction between water molecules hydrogen-bound to protein residues also contributes to residue–residue interactions. All these effects are included into the model via the choice of not only potential parameters of both pairwise and dihedral angle potentials, but also the temperature dependence of diffusion coefficients. Indeed, the destruction of hydrogen bonds between water molecules and protein residues with increasing temperature provides an additional physical mechanism for the increase of diffusion coefficients with increasing temperature which was implemented in the numerical calculations. Currently, we are developing a model that takes into account the effect of water–water hydrogen bonds on the total potential well around the cluster. The disruption of the hydrogen bond network in the first hydration shell around a cluster occurring when a protein residue (which can be itself surrounded by a network of hydrogen bounded water molecules) approaches the cluster from the unfolded part of the protein will lead to the decrease in the depth of the inner potential well which will likely lead to an increase in the protein folding time and a decrease in the unfolding time.

All numerical calculations were carried out for the following values of the interaction parameters: (1) $\beta = 105^0$, $\eta = 4.3 \times 10^{-8}$ cm, $\epsilon_{bb} = 4 \times 10^{-14}$ erg, $\epsilon_{ll} = (0.137/0.143)\,\epsilon_{bb}$, $\epsilon_{nn} = (0.151/0.143)\,\epsilon_{bb}$, $\epsilon_{ij} = \sqrt{\epsilon_{ii}\epsilon_{ij}}$ $(i, j = \text{b, l, n})$, $\epsilon'_\delta = \epsilon''_\delta = 1.2\epsilon_{bb}$ for the short protein model (124 residues); (2) $\beta = 105^0$, $\eta = 5.39 \times 10^{-8}$ cm, $\epsilon_{ll} = (2/700)\epsilon_{bb}$, $\epsilon_{bl} = \sqrt{\epsilon_{bb}\epsilon_{ll}}$, $\epsilon'_\delta = \epsilon''_\delta = 0.3\epsilon_{bb}$ for the large protein model (2500 residues). In both cases, $\epsilon_{bb} = 4 \times 10^{-14}$ erg (note that for the shorter protein, the parameters in the LJ potential were chosen in consistency with the optimized potentials for liquid simulations (OPLS) for proteins developed by Jorgensen and Tirado-Rives [44]). Here, $\beta$ is the bond angle, $\eta$ is the length parameter in the LJ potential, while $\epsilon_{bb}$, $\epsilon_{ll}$, and $\epsilon_{bl}$ are the energy parameters in the LJ potentials for the hydrophobic–hydrophobic, hydrophilic–hydrophilic, and hydrophobic–hydrophilic interactions, respectively (the mole fraction of hydrophobic beads was taken $\chi = 0.75$ both in the whole protein and in the cluster), and $\epsilon'_\delta = \epsilon''_\delta$ are the energy parameters in the dihedral angle potential (for more details, see Ref. [18]).

The typical density of the folded protein was evaluated using data from Refs. [45,46] and was set to $\rho_f\eta^3 = 1.05$, whereas the typical density of the unfolded protein in the compact configuration was set to be $\rho_u = 0.25\rho_f$. The temperature dependence of the diffusion coefficients $D^{iw}$ and $D^{ow}$ was modeled as a power law, $D_0^{iw}(1 - T_c/T)^\gamma$ and $D_0^{ow}(1 - T_c/T)^\gamma$ as predicted [47,48] by the mode coupling theory. Taking into account the results presented in Refs. [49,50], the diffusion coefficients in the ipw and the opw were assumed to be related by $D_0^{iw}\rho_f = D_0^{ow}\rho_u$. Because of the lack of reliable data on the diffusion coefficient of a residue in a native protein, $D_0 \equiv D_0^{iw}$ was assumed to vary between $10^{-7}$ and $10^{-9}$ $cm^2/s$ (much smaller than typical diffusion coefficients for gases but somewhat larger than typical values for solids). The parameters $T_c$ and $\gamma$ were chosen to be $T_c = 240$ K and $\gamma = 2.1$ (for the short protein) and $T_c = 250$ K and $\gamma = 2.3$ (for the large protein). The parameter $T_c$ determines the temperature at which the "freezing" of the protein residues occurs. Such a choice of $T_c$ is reasonable because, although water can be supercooled down to 235 K, the system "water molecules"–"protein residues" can be considered as a binary solution in which small particles are known [47,48] to remain mobile even when bigger particles have already stopped their motion. Note that the $T$ dependence of $D^{iw}$ arises not only from the direct temperature effect on the chaotic motion of protein residues but also from the indirect effect of either weakening (with increasing $T$) or strengthening (with decreasing $T$) of their hydrogen bonds with water molecules and between water molecules themselves.

#### 3.5.5.1 Results for Protein Folding

At temperature $T_0 = 293$ K, the model proteins are predicted [17–19,36] to fold into native states on a timescale of 1 to 100 s (depending on what numerical value is chosen for the factor $D_0$ in the diffusion coefficient $D^{iw}$). In fact, the above interaction parameters were chosen in order to obtain reasonable folding times of the order of 1 s for $D_0 = 10^{-6}$ $cm^2/s$ and of the order of 100 s for $D_0 = 10^{-8}$ $cm^2/s$ for purely two- and three-component heteropolymers (of both sizes considered) in which all the hydrophilic beads were assumed to be unable to carry any charge. However, these results are slightly altered when the ionizability of some hydrophilic residues is taken into account.

*3.5.5.1.1 Ionizability of Hydrophilic Residues* According to Elliott and Elliott [51], 5 out of 11 hydrophilic amino acids are ionizable even when they are incorporated into a protein. The acidic residues (aspartic and glutamic acids) have an extra –COOH on the side chains which can be negatively charged. On the other hand, the basic residues (lysine, arginine, and histidine) have extra $-NH_2$ groups on the side chains that can become positively charged.

The ionizability of these five hydrophilic beads was taken into account in Ref. [19], where some of the $N_l$ hydrophylic beads were negatively ionizable (the aspartic and glutamic acids), some were positively ionizable (lysine, arginine, and histidine), and all the others were nonionizable. More specifically, out of 81 hydrophilic residues in the BPR only 28 are ionizable, of which $\tilde{N}_l^- = 10$ and $\tilde{N}_l^+ = 18$ can be charged negatively and positively, respectively.

The actual numbers $N_-$ and $N_+$ of negatively and positively ionized residues in the protein can be smaller than $\tilde{N}_l^-$ and $\tilde{N}_l^+$, respectively, with the average dissociation coefficients $k_- = N_l^-/\tilde{N}_l^-$ and $k_+ = N_l^+/\tilde{N}_l^+$ depending on the pH of the surrounding medium and on the average dissociation coefficients $K_a$ and $K_b$ of the acidic and basic residues of the protein. In the BPR, there are 5 residues of aspartic acid (asp), 5 of glutamic acid (glu), 10 of lysine (lys), 4 of arginine (arg), and 4 of histidine (his). Thus, the average dissociation constant of acidic residues $K_a$ in BPR can be determined through the dissociation coefficients of the corresponding amino acids as $K_a = (5K_{asp} + 5K_{glu})/(5 + 5)$, whereas for the basic amino acids we have $K_b = (10K_{lys} + 4K_{arg} + 4K_{his})/(10 + 4 + 4)$. Data on $K_{asp}$, $K_{glu}$, $K_{lys}$, $K_{arg}$, and $K_{his}$ are available in the literature [51] (albeit they exhibit some scatter); hence, $K_a$ and $K_b$ can be estimated.

At any given pH, the dissociation coefficients $k_-$ and $k_+$ are completely determined by p$K_a$ and p$K_b$, respectively. We have carried out numerical calculations for pH = 7.3 (roughly that of living cells) at which $N_l^-/N_l^+ \simeq 0.69$, as well as for pH = 6.3 at which $N_l^-/N_l^+ \simeq 0.38$ and for pH = 8.3 at which $N_l^-/N_l^+ \simeq 1.59$.

The electrostatic interactions between nonadjacent beads were modeled by the Debye–Hückel approximation for the screened Coulomb potential where the inverse Debye length was assumed to be approximately constant and equal to 1/0.3 nm$^{-1}$, which roughly corresponds to an electrolyte concentration of 0.3 M (typical for living cells [51]) in the solution surrounding the protein.

The emission and absorption rates of a cluster, $W^-$ and $W^+$, respectively, are functions of three independent variables of its state. Figure 3.23 presents the dependence of $W^-$ and $W^+$ on the total number of beads in a cluster, $\nu = \nu_b + \nu_l + \nu_n$, at constant mole fractions of the hydrophobic and hydrophylic beads therein. Figure 3.23a presents the case of $x_b = 0.48$, $\chi_l = 0.49$, whereas Figure 3.23b is for $\chi_b = 0.31$, $\chi_l = 0.67$. The dashed (marked with b), dash-dotted (marked with l), and solid curves are for the hydrophobic, hydrophilic, and neutral beads, respectively.

Note that for the case shown in Figure 3.23a the emission rates (for all three components) are always greater than the absorption rates for all cluster sizes, i.e., the clusters with such a composition ($\chi_b = 0.48$,

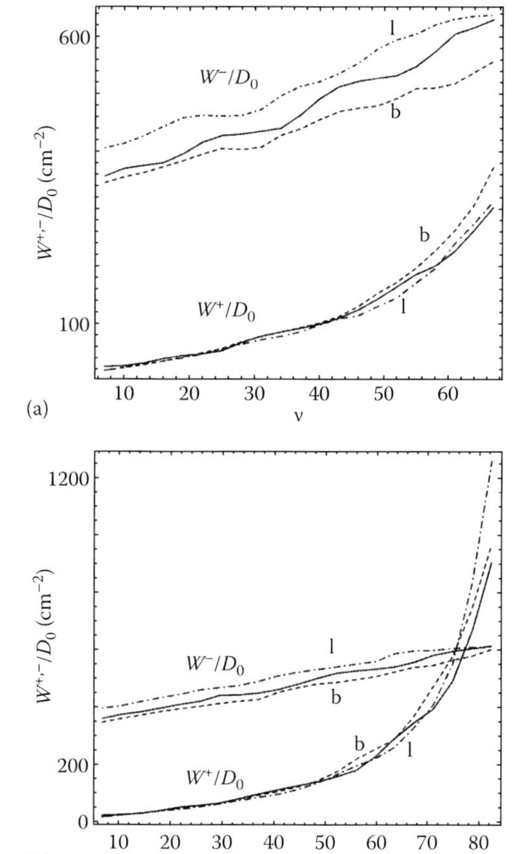


Figure 3.23

The dependence of $W_i^-$ and $W_i^+$ ($i$ = b, l, n) on the total number of beads in a cluster, $\nu = \nu_b + \nu_l + \nu_n$ for pH = 7.3 at constant mole fractions of the hydrophobic and hydrophylic beads therein: (a) $x_b = 0.48$, $\chi_l = 0.49$; (b) $\chi_b = 0.31$, $\chi_l = 0.67$. The dashed curves (marked with b) are for hydrophobic beads, the dash-dotted ones (marked with l) for hydrophilic, and the solid curves for neutral ones. The lower series is for $W_i^+$ and the upper one for $W_i^-$. The location of the simultaneous intersection of the three pairs of functions $W_i^-$ and $W_i^+$ ($i$ = b, l, n) determines the characteristics of the critical cluster, $\nu_{bc} \simeq 25$, $\nu_{lc} \simeq 55$, $\nu_{nc} \simeq 2.3$.

$\chi_l = 0.49$) have a strong tendency to disaggregate. However, for the case shown in Figure 3.23b ($\chi_b = 0.31$, $\chi_l = 0.67$), the emission rates are greater than the absorption rates, $W^- > W^+$, for smaller clusters, whereas for larger clusters absorption dominates emission, $W^- < W^+$. The location of the simultaneous intersection of the three pairs of functions $W_i^-$ and $W_i^+$ $(i = \text{b, l, n})$ determines the characteristics of the critical cluster, $\nu_{bc}$, $\nu_{lc}$, $\nu_{nc}$. Note that both $W^-$ and $W^+$ increase with increasing $\nu$, but $W^-$ increases roughly linearly with $\nu$ whereas $W^+$ shoots up by orders of magnitude after the intersection. This occurs because the width of the opw quickly decreases as the cluster grows, whereas the outer height of the barrier between the ipw and opw remains virtually unchanged. For this reason, it becomes increasingly easier for a bead located in the opw to cross the barrier and to fall into the ipw. The characteristics of the critical cluster, $\nu_{bc}$, $\nu_{lc}$, and $\nu_{nc}$ can be evaluated by numerically solving three simultaneous equations $W_i^-(\nu_b, \nu_l, \nu_n) = W_i^+(\nu_b, \nu_l, \nu_n)$ $(i = \text{b, l, n})$.

Alternatively, $\nu_{bc}$, $\nu_{lc}$, and $\nu_{nc}$ can be found as the solution of three simultaneous equations $\partial G(\nu_b, \nu_l, \nu_n)/\partial \nu_i = 0$ $(i = \text{b, l, n})$ satisfying the condition $\det(\mathbf{G}'') < 0$. We used the former method since it involves less computational effort. For the protein considered (bovine pancreatic ribonuclease with $N_0 = 124$) with the above choices of the system parameters our model predicts $\nu_{bc} \simeq 25$, $\nu_{lc} \simeq 55$, $\nu_{nc} \simeq 2.3$ (corresponding to $R_c \simeq 3.7\eta$) at pH = 7.3. Our estimates show that the time of protein folding is determined mainly by the time necessary for the first nucleation event to occur because $t_n \gg t_g$. For the diffusion coefficient $D_0^{iw}$ in the range from $10^{-6}$ to $10^{-8}$ cm$^2$/s, our model estimates the characteristic time of BPR folding to be in the range from about 3 to about 300 s at pH = 8.3, from about 2 to about 200 s at pH = 7.3, and from about 7 to about 700 s for pH = 6.3. Thus, when one takes into account the ionizability of some of the hydrophilic beads in BPR, an increase in the folding time occurs [19]. However, the effect is significant only for rather unusually low pH = 6.3, whereas for a physiological pH = 7.3 of living cells the effect is just slightly higher than the statistical uncertainty due to the Monte Carlo integration in calculating the average dihedral angle potential and to the numerical procedure of determining $\nu_{bc}$, $\nu_{lc}$, $\nu_{nc}$. Among all the pH values considered, the physiological pH provides the lowest folding time.

*3.5.5.1.2 Temperature Dependence of Folding Time* In order to evaluate the optimal temperature of folding as described in Section 3.5.3.2, calculations were carried out also at $T = 298$ K. Using the numerical results for $T_a = 293$ K and $T_b = 298$ K, we estimated the optimal folding temperatures for the above model protein to be $T^0 \simeq 291.5$ K (for the short protein) and $T^0 \simeq 290.5$ K (for the large protein). The accuracy of these estimates is rather unclear because of the uncertainty related to the use of the linear approximations for $W_c^+(T)$ and $g_e(\nu_c, T)$ (as functions of temperature) in the temperature interval from 293 K to 298 K.

The results of numerical calculations are presented in Figures 3.24 and 3.25. In Figure 3.24, the ratio $t_f(T)/t_f(T_0)$, given by Equation 3.130, is plotted as a function of temperature $T$ at fixed $T_0$. For the large protein (2500 residues) the calculations for $T_0 = 293$ K were carried out in Ref. [17], whereas for the short protein (124 residues) all the calculations were carried out in Ref. [36]. For the interaction potential parameters given above, the protein folding time $t_f(T)$ sharply increases (compared to $t_f(T_0)$) with relatively small deviations of the temperature $T$ from $T_0$: the ratio $t_f(T)/t_f(T_0)$ increases by several orders of magnitude as $T$ changes from $T_0$ to $T_0 \pm 5$ (for the larger protein) and from $T_0$ to $T_0 \pm 10$ (for the shorter protein). Thus, our model predicts

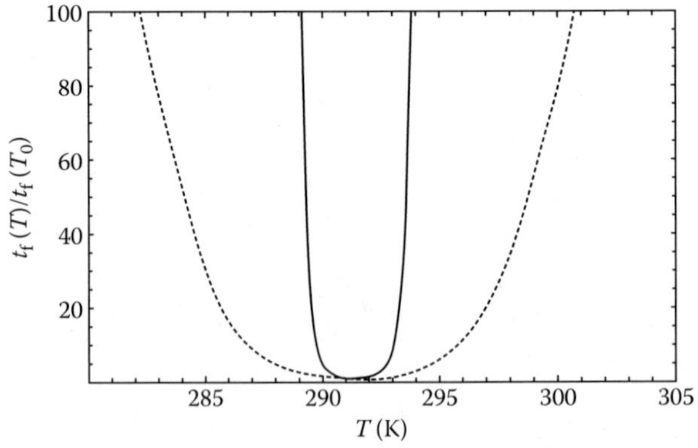


Figure 3.24

The ratio of the protein folding times at temperatures $T$ and $T_0$ as a function of temperature $T$ for $T_0 = 293$ K, i.e., $t_f(T)/t_f(T_0)$ versus $T$ as given by Equation 3.130. The solid curve is for the larger protein ($N = 2500$ residues) and the dashed curve is for the smaller one ($N = 124$ residues). The optimal temperature of protein folding was evaluated to be $T_0 \simeq 290.5$ K (for the larger protein) and $T_0 \simeq 291.5$ K (for the smaller protein).

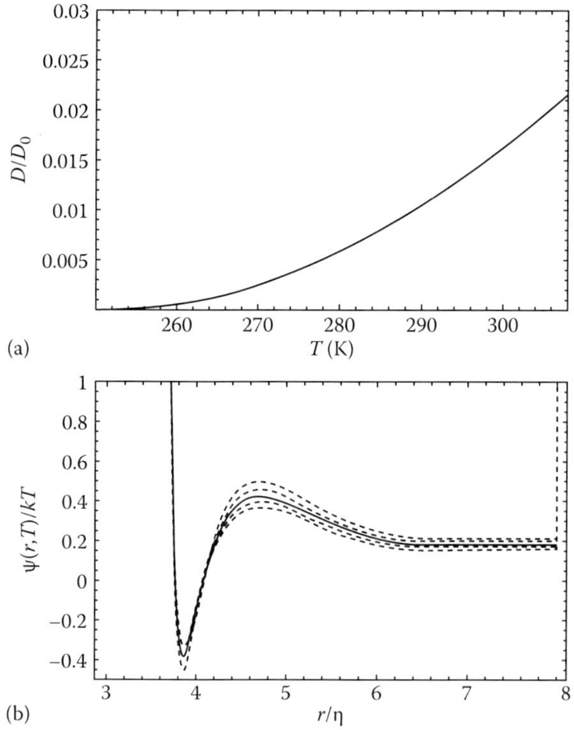


Figure 3.25

(a) The temperature dependence of the dimensionless diffusion coefficients $D^{iw}(T)/D_0^{iw}$ and $D^{ow}(T)/D_0^{ow}$ i.e., $(1 - T_c/T)^\gamma$ versus $T$ (for the protein of $N = 2500$ residues). (b) The temperature variations of the potential field around the cluster in $k_BT$ units presented as $\psi(r,T)/k_BT$ versus $r$ at various $T$ for the cluster of radius $R = 3\eta$. The curves are for $T = T_0 - 20$, $T_0 - 10$, $T_0$, $T_0 + 10$, and $T_0 + 20$ from top to bottom (the solid curve is for $T = T_0$, and the confining potential is shown as a vertical dashed line (for the protein of $N = 2500$ residues).

that while the considered proteins (initially unfolded) are very likely to fold on a reasonable timescale (of 1 to 100 s) at $T_0 = 293$ K, this becomes absolutely unlikely upon either relatively slight heating or relatively slight cooling of the system. Note that the folding time of the 2500-residue protein is much more sensitive to the temperature change than that of the 124-residue one.

A very similar, U (or V)-shaped dependence of the folding time on temperature was obtained by discontinuous molecular dynamics simulations of an all-atom model of the second α-hairpin fragment of protein G by Zhou et al. [31] They suggested that for a given protein there is an optimal (i.e., minimum) folding temperature because at high temperatures the native state is less stable, whereas at low temperatures there is an onset of low-energy nonnative intermediate states where the protein gets trapped during folding. Likewise, Faisca and Ball [32] reported a U (or V)-shaped dependence of the folding time obtained by Monte Carlo simulations of a simple lattice model of protein folding. Studying polypeptide chains of various lentgths, they pointed out that, as the length of the chain increases, the temperature range where folding can be observed becomes narrower. This corroborates the predictions of our model that for a protein of $N = 2500$ residues the range of temperatures (around the optimal temperature $T_0$), where the folding can occur on a biologically reasonable timescale, is significantly narrower than for a protein of $N = 124$ residues. A similar, U (or V)-shaped dependence of $t_f$ on $T$ was obtained by Li et al. [33], who studied the folding kinetics of three-dimensional lattice Go models with side chains by using Monte Carlo simulations. The results of all the aforementioned simulations [31–22] are consistent with the experimental data on the folding and unfolding kinetics of a 16-residue β-hairpin that undergoes cold denaturation at ambient temperatures and which were investigated by time-resolved infrared spectroscopy coupled with the laser-induced temperature jump initiation method [34].

The underlying physics that would explain the above behavior of $t_f(T)/t_f(T_0)$ as a function of $T$ is not immediately clear. The temperature dependence of the related folding rate is strongly non-Arrhenius, a known fact that has been discussed in the literature. Three possible explanations for this non-Arrhenius dependence were previously [52] suggested: (1) super-Arrhenius dependence associated to a rough energy landscape; (2) temperature dependence of the hydrophobic interaction, as reflected in a large activation heat capacity change; (3) temperature dependence of protein stability. In our model [17–19] for the nucleation mechanism of protein folding the temperature affects the folding time through two major factors. One is that of the diffusion coefficients in both inner and outer potential wells and the other is that of the depths of these wells in $k_BT$ units, i.e., $D^{iw}(T)$, $D^{ow}(T)$ and $\psi_{iw}(r,T)$, $\psi_{ow}(r,T)$, respectively. The temperature dependence of the diffusion coefficients is shown in Figure 3.25a as $(1 - T_c/T)^\gamma$ versus $T$ (note that $D^{iw}(T)/D_0^{iw} = D^{ow}(T)/D_0^{ow} = (1 - T_c/T)^\gamma$ and that $D_0^{iw} = D_0$). The temperature variations of the potential field around the cluster in $k_BT$ units is presented in Figure 3.25b as $\psi(r, T)/k_BT$ versus $r$ at various temperatures $T$ for a cluster of radius $R = 3\eta$. The curves are for $T = T_0 - 45$, $T_0 - 20$, $T_0$, $T_0 + 20$, and $T_0 + 45$ from top to bottom (the solid curve is for $T = T_0$, and the location of the confining potential is the same for all temperatures and given by the vertical dashed line at around $r/\eta \simeq 8$). Thus, with changing temperature, all these quantities behave monotonically. However, the protein folding time is a complex function of them and the interplay of their changes (with changing temperature) leads to predictions expected from experiments, namely, that a given protein acquires its native structure on a reasonable timescale under relatively narrow external conditions (temperature, pressure, etc.).

### 3.5.5.2 Results for Protein Unfolding

In the numerical calculations for the protein unfolding model, the parameters $T_c$ and $\gamma$ were chosen to be: (1) $T_c = 240$ K and $\gamma = 2.1$; (2) $T_c = 270$ K and $\gamma = 2.9$. The denaturation of the largest model protein (a two-component heteropolymer with $N = 2500$) was examined by using the Taylor series expansions [38] and by directly applying the MFPTA at several temperatures [36].

*3.5.5.2.1 Results by Taylor Series Expansions* Because the temperature dependence of $W^+(\nu, T)$ and $W^-(\nu, T)$ was obtained by using third-order Taylor series expansions (see Equation 3.134), the equation $W^-(N, T) = W^+(N, T)$ was cubic in $T - T_0$, i.e., had the form $q(T) \equiv a(T - T_0)^3 + b(T - T_0)^2 + c(T - T_0) + d = 0$, and hence can have up to three different real solutions. The discriminant of this equation is $\Delta = 4b^3d - b^2c^2 + 4ac^3 - 18abcd + 27a^2d^2$. If $\Delta = 0$, the equation $q(T) = 0$ has three real roots of which at least two coincide. All three roots may coincide yielding a triple real root or, alternatively, the equation has a double real root and another distinct single real root. If $\Delta > 0$, the equation has one real root and a pair of complex conjugate roots, the real root corresponding to either cold or hot denaturation depending on whether the root is smaller than or greater than $T_0$. The latter two situation were found in our calculations for two different sets of diffusion coefficient parameters. Finally, if $\Delta < 0$, the equation has three distinct real roots (as shown in Figure 3d of Ref. [38]): one at $T = T_u^+ > T_0$ corresponding to the threshold temperature of hot denaturation, another one at $T = T_u^- < T_0$ corresponding to the threshold temperature of cold denaturation, and the third one at some $T' > T_u^+$ or $T' < T_u^-$, which can be reasonably considered to be physically meaningless and hence disregarded as arising only due to the approximation of $W^+(\nu, T)$ and $W^-(\nu, T)$ by third-order Taylor series expansions.

Figure 3.22 presents $W^+(\nu, T)$ and $W^-(\nu, T)$, the emission and absorption rates of a cluster (of native residues), as functions of $\nu$ for the large protein model (2500 residues) at three temperatures: $T = T_0 = 293.15$ K (Figure 3.22a), $T = T_u^+ \simeq 358$ K (Figure 3.22b), and $T = T_u^+ + 5$ K (Figure 3.22c). At $T = T_0$ (Figure 3.22a), $W^+(\nu) < W^-(\nu)$ for $\nu < \nu_c$ and $W^+(\nu) > W^-(\nu)$ for $\nu_c < \nu < N$. Clearly, if the protein is initially unfolded at this temperature, it cannot fold by a regular growth of the folded cluster, but it can fold via nucleation. After the cluster of native residues attains the critical size $\nu_c$, the whole native structure forms immediately without passing through any detectable intermediate states, because the time of growth ($t_g$) of the cluster from size $\nu_c$ to size $N$ is much smaller than the time $t_n$ of the formation of the nucleus [17–19]. On the other hand, at temperature $T_u^+$ (Figure 3.22b), $W^+(\nu) < W^-(\nu)$ for $0 < \nu < N$. Therefore, if a protein is initially unfolded at $T_u^+$, it will never fold, whereas, if it is initially folded at this temperature, it will immediately start unfolding via the loss of residues from its native structure, i.e., in a barrierless way. If the temperature is increased beyond $T_u^+$ (Figure 3.22c), the time necessary for the protein to unfold via spinodal decomposition becomes smaller (compared to that time at $T = T_u^+$), because the higher is $T$ above $T_u^+$ the greater is the positive difference $W^-(\nu) - W^+(\nu)$. Similar behavior is observed when the temperature is decreased down from $T_u^-$.

Let us denote the times of barrierless protein denaturation upon heating and cooling at some temperature $T$ by $t_u^+(T)$ and $t_u'^- \equiv t_u'^-(T)$, respectively. By definition, $t_u^+ = t_u^+(T_u^+)$ and $t_u'^- = t_u'^-(T_u^-)$. The threshold temperatures of cold and hot denaturation were estimated to be $T_u^- \simeq 274$ K $= T_0 - 19$ K and $T_u^+ \simeq 358$ K $= T_0 + 65$ K, respectively (for the larger protein), and $T_u^- \simeq 265$ K $= T_0 - 28$ K and $T_u^+ \simeq 366$ K $= T_0 + 73$ K, respectively (for the smaller protein). The temperature dependence of $t_u'^-$ and $t_u'^+$ is presented in Figure 3.26a and b, respectively. In this figure, the large points are for the large protein model, whereas the small points are for the small protein model. The third-order Taylor series expansions (see Equation 3.134) are

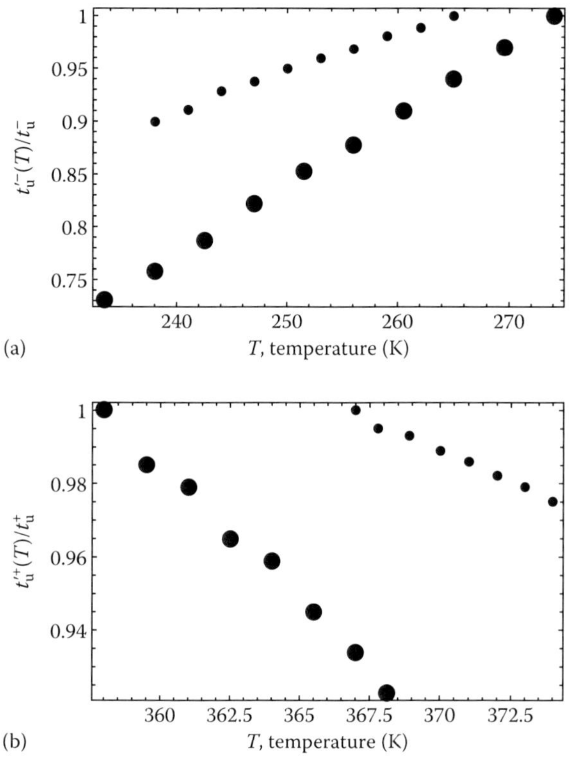


Figure 3.26

The temperature dependence of $t'^{-}_{u}/t^{-}_{u}$ (a) and $t'^{+}_{u}/t^{+}_{u}$ (b), the time of protein denaturation (via spinodal decomposition) upon cooling and heating, respectively, divided by their values at the corresponding threshold temperatures. In panels a and b, the large points (lower series) are for the large protein model ($N$ = 2500), whereas small points (upper series) are for the small protein model ($N$ = 124).

valid in the range $|T - T_0| < 80$ K, so that one can study the dependence of $t'^{-}_{u}$ on $T$ only in the range from $T = T^{+}_{u}$ to $T \simeq T_0 + 80$ K and the dependence of $t'^{-}_{u}$ on $T$ in the range from $T \simeq T_0 - 80$ K to $T = T'^{-}_{u} K$. The unfolding times are presented as the ratios $t'^{-}_{u}/t^{-}_{u}$ (Figure 3.26a) and $t'^{+}_{u}/t^{+}_{u}$ (Figure 3.26b). As made clear, the temperature sensitivity of the time of barrierless denaturation is predicted to be higher for the larger protein than for the smaller one.

The unfolding times $t^{-}_{u}$ and $t^{+}_{u}$ are inversely proportional to the coefficient $D_0$ in the temperature dependence of the diffusion coefficient $D^{iw}$. With $D_0 = 10^{-7}$ cm$^2$/s, the unfolding times of hot and cold denaturation are $t^{+}_{u} \simeq 0.05$ ms and $t^{-}_{u} \simeq 0.7$ ms, respectively, for the large protein model and $t^{+}_{u} \simeq 0.003$ ms and $t^{-}_{u} \simeq 0.06$ ms, respectively, for the small protein model. With $D_0 = 10^{-9}$ cm$^2$/s, the unfolding times of hot and cold denaturation are $t^{+}_{u} \simeq 5$ ms and $t^{-}_{u} \simeq 70$ ms, respectively, for the large protein model and $t^{+}_{u} \simeq 0.3$ ms and $t^{-}_{u} \simeq 6$ ms, respectively, for the small protein model. These estimates are consistent with experimental data on the unfolding times of hot and cold denaturation [53–57]. Many MD simulations report [58,59] much shorter timescales for hot denaturation, but those simulations are usually concerned with the sequence of structural changes occurring upon protein unfolding and hence use artificial means to speed up the process and make the unfolding occur on a timescale amenable to MD simulations, which is typically less than 100 ns.

*3.5.5.2.2 Results from Direct Application of the Method* The results of direct (without using Taylor series expansions) numerical calculations carried out for the thermal denaturation of a protein consisting of 2500 beads (only hydrophobic and hydrophilic) are presented in Figures 3.27 and 3.28. The temperature dependence of the difference between $W^{-}(N,T)$ and $W^{+}(N,T)$ is presented in Figure 3.27 as $\omega(T) \equiv [W^{+}(N,T) - W^{+}(N,T)]/W^{+}(N,T)$ versus $T$. The threshold temperatures $T^{-}_{u}$ and $T^{+}_{u}$ of cold and hot denaturation are determined as the roots of the equation $\omega(T) = 0$. For the model protein considered $T^{-}_{u} \simeq 267$ K $\simeq T_0 - 26$ K and $T^{+}_{u} \simeq 338$ K$\simeq T_0 + 45$ K. Although denaturation at both low and high temperatures is presumed to be a general property of proteins [4,5,53], the existence of two roots of the equation $\omega(T) = 0$ is not automatic but is possible due to the interplay between the temperature dependence of the diffusion coefficients in the inner and outer potential wells, $D^{iw}$ and $D^{ow}$. Therefore, such a temperature dependence is expected to

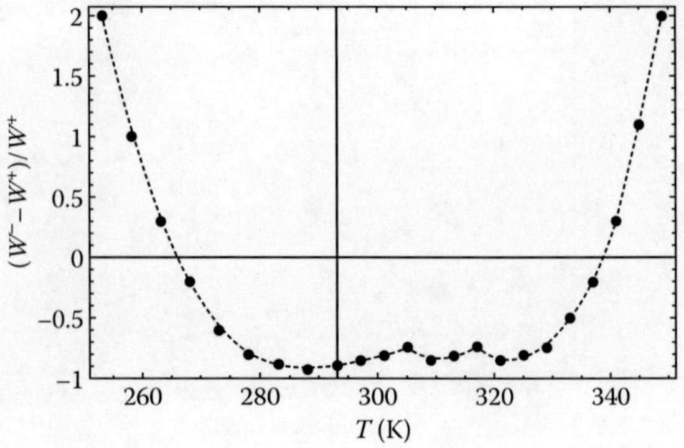


Figure 3.27

The function $\omega(T) = [W^-(N,T) - W^+(N,T)]/W^+(N,T)$ plotted versus $T$ for a protein exhibiting both cold and hot denaturation at $T_u^- \simeq 247$ K and $T_u^+ \simeq 338$ K, respectively (for more details, see the text). The dashed line through the points is just for guiding the eye.

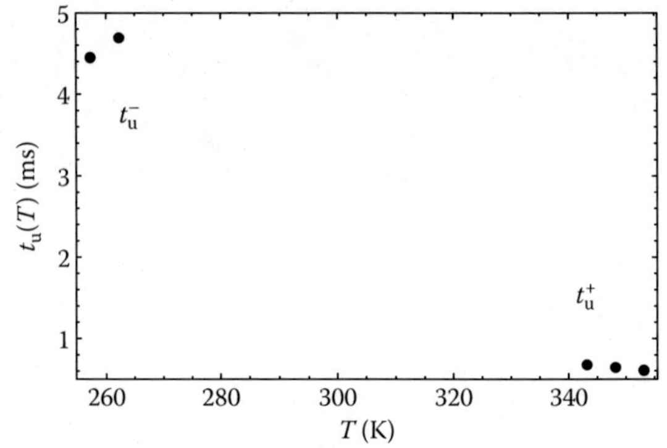


Figure 3.28

The temperature dependence of $t_u^-$ (two leftmost points) and $t_u^+$ (three rightmost points), the time of protein denaturation (via spinodal decomposition) upon cooling and heating, respectively.

be most adequate for real proteins exhibiting both cold and hot denaturation. The negative values of this function correspond to conditions when the protein unfolding cannot occur in a barrierless way (see also Figure 3.22a). Thus, the temperature interval from $T_u^-$ to $T_u^+$ roughly determines the protein stability range. Clearly, the real stability range may be somewhat narrower because the protein can unfold via nucleation under conditions when the barrierless process is not yet possible (i.e., when $T$ is slightly larger than $T_u^-$ or slightly smaller than $T_u^+$ The positive values of $\omega(T)$ correspond to conditions under which the protein unfolds in a barrierless way. The larger $\omega(T) > 0$, the faster denaturation is expected.

The temperature dependence of the cold denaturation time was calculated at two temperatures below $T_u^-$, namely, $T = T_u^- - 5$ K and $T = T_u^- - 10$ K. We have also calculated the time of hot denaturation at three temperatures above $T_u^+$, namely, $T = T_u^+ + 5$ K, $T = T_u^+ + 10$ K, and $T = T_u^+ + 15$ K. The temperature dependence of $t_u^-$ and $t_u^+$ is presented in Figure 3.28. As expected, the time of barrierless denaturation decreases as $T$ increases above $T_u^+$ or as $T$ decreases below $T_u^-$. The unfolding times $t_u^-$ and $t_u^+$ are inversely proportional to the coefficient $D_0$ in the temperature dependence of the diffusion coefficient $D^{iw}$. In Figure 3.28, the unfolding times $t_u^-$ and $t_u^+$ are shown for $D_0 = 10^{-8}$ cm²/s. Considering $D_0$ to be between $10^{-7}$ and $10^{-9}$ cm²/s, the unfolding time of hot denaturation $t_u^+$ varies from 0.07 to 7 ms, whereas the unfolding time of cold denaturation $t_u^-$ varies from 0.5 to 50 ms, consistent with the experimental data on the unfolding times of hot and cold denaturation [53–59].

### 3.5.6 Conclusions

We have presented a review of the kinetic models recently developed by the authors for the nucleation mechanism of protein folding and thermal denaturation of a native protein. Both models are based on the MFPTA. A protein is treated as a random heteropolymer with all the bonds having the same constant length and all the bond angles equal and fixed. The heteropolymer beads (modeling the protein residues) can be hydrophobic, hydrophilic, or neutral. The ionizability of some hydrophilic beads has been taken into account as well. As a crucial idea of the model, an overall potential around the cluster of native residues wherein a protein residue performs a chaotic motion is considered to be a combination of three terms representing the average dihedral, effective pairwise, and confining potentials. The overall potential as a function of the distance from the cluster center has a double well shape which allows one to determine both its emission and absorption rates by using a first passage time analysis. One can thus develop a self-consistent kinetic theory for the nucleation mechanism of protein folding and evaluate the size and composition of the nucleus and the protein folding time.

To avoid lengthy calculations, one can use Taylor series expansions in the formalism of the first passage time analysis. Once the folding time $t_f$ is evaluated at one temperature $T_0$, the proposed method allows one to evaluate $t_f$ at any other temperature $T$ in the vicinity of $T_0$. The width of the vicinity in which the method is valid can be increased by retaining incrementally higher order terms in the Taylor series expansions with respect to $T - T_0$.

An MFTPA-based kinetic model was also developed for the process of protein denaturation occurring in a barrierless way (i.e., as spinodal decomposition). The key elements of the proposed model are the temperature dependent rates with which a cluster of native residues within a compact (but not native) protein emits and absorbs residues. Considering a native protein, stable at temperature $T_0$, we have examined the temperature dependence of the emission and absorption rates of the cluster (which are also functions of the cluster size). To numerically obtain the emission and absorption rates as functions of the cluster size and temperature, the straightforward application of the method requires long computation. To reduce the computational effort, one can again use Taylor series expansions with respect to $T - T_0$ in the formalism of MFPTA. Once the emission and absorption rates are known as functions of cluster size at one temperature $T_0$, i.e., $W^-(\nu,T_0)$ and $W^+(\nu,T_0)$, one can employ the Taylor series representation of $W^-(\nu,T)$ and $W^+(\nu,T)$ to evaluate them at any other temperature $T$ in the vicinity of $T_0$. The width of the vicinity in which the method is valid depends on how many terms are retained in the Taylor series expansions.

To illustrate the proposed methods by numerical calculations, we have applied them to two model proteins previously studied in Refs. [16–18,34,36], namely, heteropolymers consisting of 2500 and 124 beads. The short protein model mimics the BPR (whereof 40 residues can be classified as hydrophobic, 81 as hydrophilic, and 3 as neutral). The larger heteropolymer (with the mole fraction of hydrophobic residues equal to 0.75 with no neutral ones) was considered to demonstrate that the method provides a reasonable temperature dependence of the folding time on a reasonable computational timescale even for the largest proteins, while straightforward MC/MD simulations of such proteins would require extremely lengthy computations.

With appropriate choices of the interaction parameters and densities and the diffusion coefficient of native residues in the range $10^{-6}$ to $10^{-8}$ cm$^2$/s, the characteristic time of protein folding was estimated to be in the range of 1 to 100 s. These values are consistent with the experimental data [5] on typical folding times of proteins as well as with estimates provided by other theoretical models [24,26] and simulations [7,14]. The predictive power of our model is mostly determined by the accuracy of Monte Carlo integration in calculating the average dihedral potential and the availability of reliable data on the diffusion coefficients and other parameters of the protein. If the ionizability of some of hydrophilic residues is taken into account, the protein folding time increases: [18] this more realistic model may predict protein folding times which may differ by as much as an order of magnitude from the predictions of the previous version of the model where this effect was neglected. However, these conclusions should not be generalized because they are specific to the particular protein considered.

For the temperature dependence of the protein folding time the model predicts that this time $t_f(T)$ sharply increases (compared to $t_f(T_0)$) with relatively small deviations of the temperature $T$ from $T_0$ and this increase is sharper for larger proteins. Thus, while the considered proteins (initially unfolded) reasonably quickly fold at $T_0 = 293$ K, their folding becomes highly unlikely upon either slight heating or slight cooling. These results are consistent with those from our model for barrierless protein denaturation that predicts both cold and hot denaturation, subject to an appropriate choice of diffusion coefficients (and their temperature dependence) in the folded and unfolded protein. The estimates for the unfolding times of hot and cold denaturation are essentially in the range of typical times of protein denaturation. One can notice that the unfolding time $t_u^-$ at the threshold temperature of cold denaturation is predicted to be an order-of-magnitude greater compared to that of hot denaturation, $t_u^+$.

## References


1. Anfinsen, C.B., *Science* 181, 223–320 (1973).
2. Ghelis, C., Yan, J., *Protein Folding*, Academic Press, New York, 1982.
3. Creighton, T.E., *Proteins: Structure and Molecular Properties*, W.H. Freeman, San Francisco, 1984.
4. Kauzmann, W., *Adv. Prot. Chem.* 14, 1 (1959).
5. Privalov, P.L., *Crit. Rev. Biochem. Mol. Biol.* 25, 281 (1990).
6. Stryer, L., *Biochemistry*, 3rd ed., W.H. Freeman, New York, 1988.
7. Nölting B., *Protein Folding Kinetics*, Springer–Verlag, Berlin, 2006.
8. Honeycutt, J.D., Thirumalai, D., *Proc. Natl. Acad. Sci. U.S.A.* 87, 3526 (1990).
9. Honeycutt, J.D., Thirumalai, D., *Biopolymers* 032, 695 (1992).
10. Nölting, B., *J. Theor. Biol.* 194, 419 (1998); 197, 113 (1999).
11. Weissman, J.S., Kim, P.S., *Science* 253, 1386–1393 (1991).
12. Creighton, T.E., *Nature (London)* 356, 194–195 (1992).
13. Pastore, A., Martin, S.R., Politou, A., Kondapalli, K.C., Stemmler, T., Temussi, P.A., *J. Am. Chem. Soc.* 129, 5374 (2007).
14. Marqués, M.I., *Phys Stat Sol A* 203, 1487 (2006).
15. Gursky, O., Atkinson, D., *Protein Sci.* 5, 1874 (1996).
16. Paci, E., Karplus, M., *Proc. Natl. Acad. Sci. U.S.A.* 97, 6521 (2000).
17. Djikaev, Y.S., Ruckenstein, E., *J. Phys. Chem. B* 111, 886 (2007). (Section 3.1 of this volume.)
18. Djikaev, Y.S., Ruckenstein, E., *J. Chem. Phys.* 126, 175103 (2007). (Section 3.2 of this volume.)
19. Djikaev, Y.S., Ruckenstein, E., *J. Chem. Phys.* 128, 025103 (2008). (Section 3.3 of this volume.)
20. Abkevich, V.I., Gutin, A.M., Shakhnovich, E.I., *Biochemistry* 33, 10026 (1994).
21. Guo, Z., Thirumalai, D., *Biopolymers* 36, 83–102 (1995).
22. Fersht, A.R., *Proc. Natl. Acad. Sci. U.S.A.* 92, 10869 (1995).
23. Fersht, A.R., *Curr. Opin. Struct. Biol.* 7, 3 (1997).
24. Chandrasekhar, S., *Rev. Mod. Phys.* 15, 1 (1943).
25. Gardiner, C.V., *Handbook of Stochastic Methods*, Springer, New York, 1983.
26. Szabo, A., Schulten, K., Schulten, Z., *J. Chem. Phys.* 139, 500 (1980).
27. Park, S., Sener, M.K., Lu, D., Schulten, K., *J. Chem. Phys.* 119, 1313 (2003).
28. Narsimhan, G., Ruckenstein, E., *J. Colloid Interface Sci.* 128, 549 (1989). (Section 1.2 of this volume.)
29. Ruckenstein, E., Nowakowski, B., *J. Colloid Interface Sci.* 137, 583 (1990). (Section 1.3 of this volume.)
30. Nowakowski, B., Ruckenstein, E., *J. Colloid Interface Sci.* 139, 500 (1990). (Section 1.4 of this volume.)
31. Zhou, Y., Zhang, C., Stell, G., Wang, J., *J. Am. Chem. Soc.* 125, 6300 (2003).
32. Faisca, P.F.N., Ball, R.C., *Phys. Rev. Lett.* 116, 7231 (2002).
33. Li, M.S., Klimov, D.K., Thirumalai, D., *Comput. Phys. Commun.* 147, 625–628 (2002).
34. Xu, Y., Wang, T., Gai, F., *Chem. Phys.* 323, 21–27 (2006).
35. Gutin, A.M., Abkevich, V.I., Shakhnovich, E.I., *Phys. Rev. Lett.* 77, 5433 (1996).
36. Djikaev, Y.S., Ruckenstein, E., *Phys. Chem. Chem. Phys.* 10, 6281 (2008). (Section 3.4 of this volume.)
37. Tsai, C.-J., Maizel, J.V., Nussinov, R., *Crit. Rev. Biochem. Mol. Biol.* 37, 55–69 (2002).
38. Djikaev, Y.S., Ruckenstein, E., *Phys. Rev. E* 78, 011909 (2008).
39. Koizumi, M., Hirai, H., Onai, T., Inoue, K., Hirai, M., *J. Appl. Crystallogr.* 40, 175–178 (2007).
40. Oleinikova, A., Smolin, N., Brovchenko, I., Geiger, A., Winter, R., *J. Phys. Chem. B* 109, 1988–1998 (2005).
41. Smolin, N., Oleinikova, A., Brovchenko, I., Geiger, A., Winter, R., *J. Phys. Chem. B* 109, 10995–11005 (2005).
42. Brovchenko, I., Krukau, A., Smolin, N., Oleinikova, A., Geiger, A., Winter, R., *J. Chem. Phys.* 123, 224905 (2005).
43. Trinkaus, H., *Phys. Rev. B* 27, 7372 (1983).
44. Jorgensen, W.L., Tirado-Rives, J., *J. Am. Chem. Soc.* 110, 1657 (1988).
45. Harpaz, Y., Gerstein, M., Chothia, C., *Structure* 2, 641–649 (1994).
46. Huang, D.M., Chandler, D., *Proc. Natl. Acad. Sci. U.S.A.* 15, 8324–8327 (2000).
47. Mukherjee, A., Bhattacharyya, S., Bagchi, B., *J. Chem. Phys.* 116, 4577–4586 (2002).
48. Bosse, J., Kaneko, Y., *Phys. Rev. Lett.* 74, 4023 (2002).
49. Hirshfelder, J.O., Curtiss, C.F., Bird, R.B., *Molecular Theory of Gases and Liquids*, Wiley, New York, 1964.

50. Wakeham, W.A., *J. Phys. B At. Mol. Phys.* 6, 372 (1964).
51. Elliott, W.H., Elliott, D.C., *Biochemistry and Molecular Biology,* Oxford University Press, New York, 2003.
52. Scalley, M.L., Barker, D., *Proc. Natl. Acad. Sci. U.S.A.* 94, 10636 (1997).
53. Georg, H., Wharton, C.W., Siebert, F., *Laser Chem.* 19, 233–235 (1999).
54. Mayor, U., Johnson, C.M., Daggett, V., Fersht, A.R., *Proc. Natl. Acad. Sci. U.S.A.* 97, 13518 (2000).
55. Jackson, S.E., Fersht, A.R., *Biochemistry* 30, 10428–10435 (1991).
56. Spector, S., Raleigh, D.P., *J. Mol. Biol.* 293, 763–768 (1999).
57. Tang, X., Pikal, M.J., *Pharm. Res.* 22, 1176 (2005).
58. Williams, M.A., Thornton, J.M., Goodfellow, J.M., *Protein Eng.* 10, 895 (1997).
59. Day, R., Daggett, V., *Proc. Natl. Acad. Sci. U.S.A.* 102, 13445 (2005).

## 3.6 Effect of Hydrogen Bond Networks on the Nucleation Mechanism of Protein Folding

*Yuri Djikaev and Eli Ruckenstein**

We have recently developed a kinetic model for the nucleation mechanism of protein folding (NMPF) in terms of ternary nucleation by using the first passage time analysis. A protein was considered as a random heteropolymer consisting of hydrophobic, hydrophilic (some of which are negatively or positively ionizable), and neutral beads. The main idea of the NMPF model consisted of averaging the dihedral potential in which a selected residue is involved over all possible configurations of all neighboring residues along the protein chain. The combination of the average dihedral, effective pairwise (due to Lennard–Jones-type and electrostatic interactions), and confining (due to the polymer connectivity constraint) potentials gives rise to an overall potential around the cluster that, as a function of the distance from the cluster center, has a double-well shape. This allows one to evaluate the protein folding time. In the original NMPF model hydrogen bonding was not taken into account explicitly. To improve the NMPF model and make it more realistic, in this paper we modify our (previously developed) probabilistic hydrogen bond model and combine it with the former. Thus, a contribution due to the disruption of hydrogen bond networks around the interacting particles (cluster of native residues and residue in the protein unfolded part) appears in the overall potential field around a cluster. The modified model is applied to the folding of the same model proteins that were examined in the original model: a short protein consisting of 124 residues (roughly mimicking bovine pancreatic ribonuclease) and a long one consisting of 2500 residues (as a representative of large proteins with superlong polypeptide chains), at pH = 8.3, 7.3, and 6.3. The hydrogen bond contribution now plays a dominant role in the total potential field around the cluster (except for very short distances thereto where the repulsive energy tends to infinity). It is by an order of magnitude stronger for hydrophobic residues than for hydrophilic ones. The range of "residue-cluster" distances, at which the hydrogen bond effect exists, is twice as long for hydrophobic residues as for hydrophilic ones.

### 3.6.1 Introduction

A well-defined three-dimensional structure [1,2] is necessary for a protein molecule to carry out a specific biological function. The formation of the native structure of a biologically active protein constitutes the core of the so-called "protein folding problem" [3,4]. Many thermodynamic and kinetic aspects of the protein folding process remain unexplained [5–8].

It is believed that initially an unfolded protein transforms quickly into a compact (but not native) configuration [9,10]. One of the pathways for the transition from a compact configuration to the native one is similar to nucleation, i.e., once a critical number of (native) tertiary contacts is established the native structure is formed rapidly without passing through any detectable intermediates [6,9–14].

We have recently developed [15–19] a model for the nucleation mechanism of protein folding (NMPF) in terms of ternary nucleation by using the first passage time analysis. A protein was considered as a random heteropolymer [20–22] consisting of hydrophobic, hydrophilic, and neutral beads with all the bonds in the heteropolymer having the same constant length and all the bond angles equal and fixed. The ionizability of some of protein residues (aspartic and glutamic acids, lysine, arginine, and histidine) was taken into account by considering the hydrophilic residues to be of three subtypes, namely, those which cannot be ionized at all and those which can (depending on the pH of the surrounding solution) be either positively or negatively ionized.

The main idea underlying the NMPF model consists of representing the overall potential field around a cluster of native protein residues (i.e., field in which a nonnative residue performs a chaotic motion) as a combination of the mean dihedral potential, effective pairwise potential, and confining potential. The latter is due to the polymer connectivity that confines the protein residues around its center. The effective pairwise potential (for pairwise interactions of a selected residue with those in the cluster) for an ionized residue contains an electrostatic contribution. The mean dihedral potential (in which a selected residue is involved) is calculated by averaging it over all possible configurations of all neighboring residues along the protein chain. As a function of the distance from the cluster center, the overall potential field has a double-well shape, which allows one to develop a self-consistent kinetic theory for the NMPF and evaluate its characteristic time (as well as the temperature dependence of the latter).

* *Phys. Rev. E* 80, 061918 (2009). Republished with permission.

In that model [15–19], hydrogen bonding (hb) was taken into account indirectly through its effect on the diffusion coefficients of protein residues. On the other hand, hydrogen bonding plays a crucial role in the formation, stability, and denaturation of the native structure of a biologically active protein, i.e., its folding, stability, and unfolding [1,3,4,23,24]. In addition to hydrogen bonds between the atoms of the polypeptide backbone (main-chain–main-chain N–H…O bonds), which are responsible for the formation of the secondary structure of proteins (both α helices and β sheets), many hydrogen-bonded interactions are provided by the polar groups of the side chains. Moreover, the biological activity of proteins appears to depend on the formation of a two-dimensional hydrogen-bonded network spanning most of the protein surface and connecting all the surface hydrogen-bonded water clusters [25–27].

In order to improve our kinetic model for protein folding, it appeared logical to find a way for explicitly taking into account the hydrogen bonding "water–water" and "water protein residue." As a first step, we developed a probabilistic model for the effect of hydrogen bond networks of water molecules around two solute particles (immersed in water) on their interaction [28,29]. The probabilistic hydrogen bond (PHB) approach was applied to the solvent-mediated interaction of (1) two spherical hydrophobic solutes [28] and (2) two infinite parallel plates whereof the surfaces facing each other have a composite hydrophobic–hydrophilic character [29] (i.e., they are covered with uniformly distributed hydrophobic and hydrophilic sites).

When two hydrophobic particles sufficiently approach each other, the disruption of boundary water–water hydrogen bonds in their first hydration layers can give rise to an additional contribution to their overall interaction potential. According to numerical evaluations, in the interplay between a decrease in the number of boundary bonds per water molecule (as a result of the proximity to the foreign hydrophobic particle) and the possible enhancement [30–32] of such a bond the former effect is predominant because the larger number of weaker bulk hydrogen bonds provide a more negative contribution to the free energy than the smaller number of stronger boundary hydrogen bonds (BHBs). Consequently, our model suggests that the disruption of the BHBs, which occurs when the first two hydration shells of two particles overlap, results in an attractive contribution between the particles. This attraction is naturally short range, appearing only when the separation between two particles becomes smaller than four lengths of a hydrogen bond.

To implement the PHB approach into the NMPF model, it is necessary to slightly modify the former to adapt it to the particular situations encountered on the nucleation pathway of protein folding. Indeed, the folded cluster of the protein consists of three kinds of residues; hence, its surface can be expected to have a composite (hydrophobic–hydrophilic) character. On the other hand, a single residue in the unfolded part of the protein is either hydrophobic or hydrophilic (neutral residues can be treated as hydrophobic as far as their hydrogen-bonding ability is concerned). Thus, the PHB model needs to be modified to examine the solvent-mediated interaction of a spherical particle of composite nature (modeling a folded cluster of a protein) with (1) a spherical hydrophobic particle (modeling hydrophobic and neutral protein residues) and (2) a spherical hydrophilic particle (modeling hydrophilic protein residues). The additional contributions to the interaction potentials, arising due to the disruption of hydrogen bond networks around the interacting particles, will then have to be added to the overall potential field around a cluster in the NMPF model. As a result, the water–water hydrogen bonding will be taken *explicitly* into account in a kinetic model for the NMPF.

The paper is structured as follows. In Section 3.6.2, we extend the PHB model to the cases where the spherical interacting particles have different nature (one composite the other either hydrophobic or hydrophilic) in addition to being of different radii. In Section 3.6.3, we present a modified version of our NMPF model based on the first passage time analysis in the framework of a ternary nucleation theory. The modified NMPF model will explicitly implement the effect of water–water hydrogen bond network by means of the PHB model. The numerical results of the application of the modified NMPF model to the folding of two model proteins are presented in Section 3.6.4. The results are discussed and conclusions are summarized in Section 3.6.5.

### 3.6.2 Probabilistic Model for the Effect of Water–Water Hydrogen Bonding on the Interaction of Solute Particles

Let us consider two solute particles (1 and 2) of spherical shape in water. The radii of the particles will be denoted by $R$ and $R'$ and the distance between them (i.e., between their centers) by $r$ (Figure 3.29). The radii of the particles are determined by the locus of the outermost molecules that constitute them. If the smaller solute consists of a single molecule, its radius $R'$ is equal to zero. (This case would represent a single residue in the unfolded part of the protein). The characteristic distance of pairwise interactions between molecules constituting particles $R$ and $R'$ will be denoted by $\eta$.

In order to make the model applicable to the problem of protein folding, we will consider the particle $R$ to have a composite hydrophobic–hydrophilic character so that its surface is covered with both hydrophobic

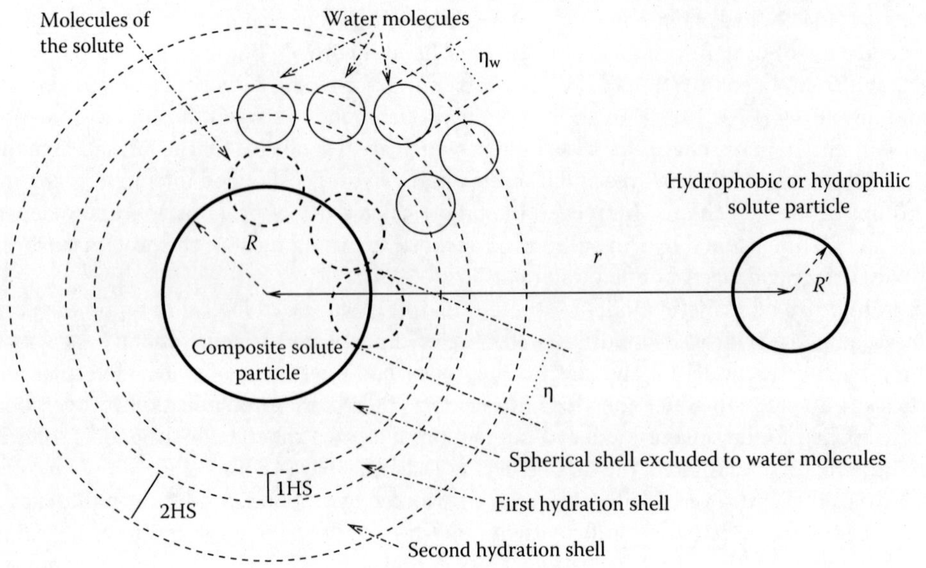


Figure 3.29

A schematic representation of the first two hydration layers two spherical solutes of radii $R$ and $R'$ at a distance $r$ between their centers. The surface of each particle is shown as a thick solid line. The circles of diameter $\eta_w$ represent water molecules. The dashed circle of radius $\eta$ represents a molecule of the solute particle.

and hydrophilic sites uniformly distributed over the whole surface. The smaller particle $R'$ will be either completely hydrophobic or completely hydrophilic. A water molecule is capable of forming a hydrogen bond with a hydrophilic site of the surface but not with a hydrophobic one. The probabilities $\omega_b$ and $\omega_l$ that a selected water molecule, adjacent to the particle $R$, will be in contact with hydrophobic or hydrophilic sites, respectively, are given by

$$\omega_b = \frac{A_b}{A},\ \omega_l = \frac{A_l}{A}, \tag{3.136}$$

where $A_b$ and $A_l$ are the total areas covered by hydrophobic and hydrophilic sites, respectively, and $A = A_b + A_l$ is the total surface area of the solute. Clearly, $\omega_b + \omega_l = 1$.

The location of a water molecule will be determined by the location of its center. The length of a water–water hydrogen bond (i.e., the distance between the centers of two bonded water molecules) will be denoted by $\eta_w$. Water molecules located in the layer of thickness $\frac{1}{2}\eta_w$ at distances to the particle surface from $\frac{1}{2}(\eta+\eta_w)$ to $\frac{1}{2}(\eta+\eta_w)+\frac{1}{2}\eta_w$ will be considered to belong to the first hydration shell (hereafter referred to as 1HS; see Figures 3.29 and 3.30). The second hydration shell (hereafter referred to as 2HS) of thickness $\eta_w$ is formed by water molecules at distances to the particle surface from $\frac{1}{2}(\eta+\eta_w)+\frac{1}{2}\eta_w$ to $\frac{1}{2}(\eta+\eta_w)+\frac{3}{2}\eta_w$. Hydrogen bonding being short ranged, the other layers are not affected by the presence of the particle surface.

Clearly, the properties of a hydrogen bond between a water molecule and a hydrophilic site on the solute may depend on the nature of the site. In a rough approximation, however, its length will be considered to be the same for all hydrophilic sites and equal to $\eta$. The characteristic length of pairwise interactions between a water molecule and that of the hydrophobic particle will be assumed to be equal to $\frac{1}{2}(\eta+\eta_w)$.

A water molecule itself is modeled to have four arms each capable of forming a single hydrogen bond. The configuration of the four hb arms is completely symmetric with the angle between any of them equal to $\alpha$ = 109.47°. This tetrahedral configuration is assumed to be rigid (independent of whether the water molecule is located in the bulk or in the 1HS). Each hb arm can adopt a continuum of orientations subject to the constraint of tetrahedral rigidity. A water molecule forms a hydrogen bond with another molecule when the tip of any hb arm of the first molecule exactly coincides with the second one. The length of an hb arm is thus equal to the length of a hydrogen bond, $\eta_w$.

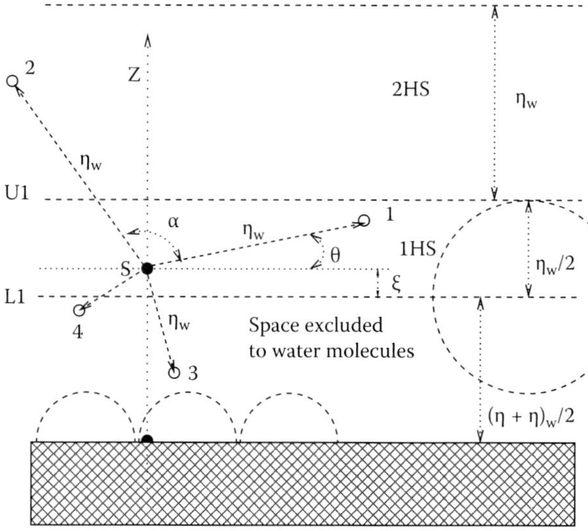


Figure 3.30

A schematic representation of a water molecule in the first hydration shell (1HS) of the planar substrate surface (shadowed area). The lower and upper boundaries of the 1HS are marked as the planes L1 and U1 and shown as long-dashed lines. The molecule, shown as a disk S, is at the distance $\xi$ from the lower boundary (closest to the hydrophobic surface) of 1HS. The four hb arms of the molecule are shown as short-dashed lines with the empty circles as the arm tips. Arms 1 and 2 are in the plane of the figure, whereas arms 3 and 4 are located out of the figure plane (one of them under it, the other above it). The angle between any two hb arms is $\alpha$. The angle $\Theta$ is the angle formed between the hb arm and its tangential projection (parallel to the lower boundary of the 1HS). In this figure, the origin of the Cartesian coordinate system lies at the hydrophobic surface with the $z$ axis being normal thereto. It is assumed that $-\pi/2 \le \Theta \le \pi/2$ with $\Theta < 0$ if the $z$ coordinate of the hb-arm tip is greater than $(\eta + \eta_w)/2 + \xi$ and $\Theta > 0$, otherwise. The large dashed circle represents a water molecule, while the smaller dashed semicircles represent the substrate molecules.

The water–water hydrogen bonds are treated along the lines of the Müller–Lee–Graziano (MLG) model [33,34]. Some of the hb arms of water molecules in the 1HS cannot form bonds because of the proximity to the hydrophobic regions of the solute. The bonds that such molecules do form (hereafter referred to as "boundary hydrogen bonds") can be somewhat stronger than the bulk ones [30,34], although such an enhancement is still a subject of discussion (see, e.g., Ref. [32] and references therein). However, some 1HS water molecules can form hydrogen bonds with the hydrophilic sites of the particle. The hydrogen bonds of such molecules are not altered compared to the bulk water bonds. Nevertheless, adopting a probabilistic approach one can consider the whole network of hydrogen bonds involving all 1HS molecules as a BHB network. A water–water hydrogen bond is enhanced (compared to its bulk value) if at least one of the two water molecules belongs to the 1HS and it does not form a hydrogen bond with the surface of the solute. The probability of the latter is $\omega_b$, whereas the probability that a 1HS water molecule forms a hydrogen bond with the solute is $\omega_l$. Therefore, one can assume that the properties of such a composite hydrogen bond network involving 1HS molecules can be obtained by averaging the properties of the corresponding networks for purely hydrophobic and purely hydrophilic particles with weights $\omega_b$ and $\omega_l$, respectively.

To some extent, our model for water–water hydrogen bonding is a selective combination of a three-dimensional analog of the Mercedes Benz (MB) water model [30] with the MLG model [33,34]. In the original MB model [30] (which is two-dimensional), the water molecules are modeled as Lennard–Jones (LJ) disks in a donor–acceptor approximation, with three hydrogen bonding arms (whereof the completely symmetric configuration resembles the MB logo, hence the name of the model). The interaction potential between two water molecules is the sum of two terms, with one representing the LJ interaction of disks and the other representing their hydrogen bonding ability. Although the latter is considered to be continuous, a hydrogen bond is modeled to be optimal at a specified distance and a relative orientation of the two molecules involved: if at this distance one hb-forming arm of one molecule aligns itself with an hb arm of the other molecule, then the hydrogen bond energy has its minimum value. Deviations from this lowest-energy hydrogen bond configuration (in distance and mutual orientation) are assumed to have a Gaussian distribution with a single width parameter for all degrees of freedom. Although the MB model allows continuous variations of the separation and orientation of the water molecules (disks), it is also consistent with the concept of bimodal character of the energetics of hydrogen bonds in the MLG model [33,34]. The discrete three-dimensional model presented in Refs. [28,29] can be regarded as a particular case of a three-dimensional version of the hydrogen bonding feature of the MB model where the Gaussian distribution would be infinitely narrow. However, the analytical treatment that we pursued in our model would be much more difficult to carry out if we considered the hydrogen bonding ability of water molecules to have a continuous (with respect to the separation and mutual orientation of molecules involved) character.

#### *3.6.2.1 Networks of Water–Water Hydrogen Bonds around Solutes*

Let us now examine how the BHB network affects the pairwise interaction potential $\Phi$ between two particles. Clearly, this effect can be neglected if the second hydration layers (2HS) around the two solutes do not overlap. However, if the two particles, $R$ and $R'$, are sufficiently close to each other, they have to share some parts of their BHB networks. The overlap of the 2HS of one particle with the 2HS of the other leads to a decrease in the total number of 2HS molecules but does not affect the total number of 1HS molecules. The latter decreases only as a result of the overlap of the 1HSs of the two particles. The overlap of the BHB networks of the two particles thus causes their mutual disruption.

If the distance $r$ between the two particles (i.e., between their centers) is greater than $\tilde{r} \equiv R + R' + \eta + 4\,\eta_w$, there is no overlapping of the first two hydration shells of the two particles so that they do not share their BHB networks and there is no contribution to the potential $\Phi = \Phi(r)$. However, when the particles are sufficiently close to each other, so that $r < \tilde{r}$, they share some parts of their first two hydration shells. Thus, with decreasing $r$ (at $r < \tilde{r}$), the total volume of the first two hydration shells decreases, which leads to the decrease in the total number of molecules in these shells. This, in turn, results in a decrease in the total number of BHBs in the first two hydration shells of the two particles, hereafter denoted by $N_s \equiv N_s(r)$. The total number of BHBs in the first two hydration shells of the two solutes at a distance $r$ between them will be denoted by $\nu_s \equiv \nu_s(r)$. A decrease in $N_s$ results in a decrease in $\nu_s$ (both occurring because of the overlapping of the first two hydration shells of the two particles): the corresponding quantities are given by $N_s^{r\infty} \equiv N_s^{r\infty}(r) = N_s(\infty) - N_s(r)$ and $\nu_s^{r\infty} \equiv \nu_s^{r\infty}(r) = \nu_s(\infty) - \nu_s(r)$, respectively. Clearly, $\nu_s^{r\infty}(r) = 0$ and $N_s^{r\infty}(r) = 0$ for $r \geq \tilde{r}$, whereas $\nu_s^{r\infty}(r) > 0$ and $N_s^{r\infty}(r) > 0$ for $r < \tilde{r}$.

The $N_s^{r\infty}$ molecules that left the 1HS and 2HS of the two solutes pass into the bulk water where they can form new hydrogen bonds with the energy $\epsilon_b < 0$ per bond. The average number of bonds per molecule of bulk water is denoted by $n_b$. The total energy $\Phi_b$ of the newly formed bonds is a function of $r$ (because so is $N_s^{r\infty}$): $\Phi_b \equiv \Phi_b(r) = \epsilon_b n_b N_s^{r\infty}(r)$. Denoting the number density of water molecules in the 1HS and 2HS by $\rho_w$, one can rewrite $\Phi_b$ as

$$\Phi_b(r) = \epsilon_b n_b \rho_w V_0(r), \qquad (3.137)$$

where $V_0(r)$ is the volume of the region resulting from the overlap of the first two hydration shells of the two solutes (note that $\rho_w$ may differ from the bulk water density $\rho_w^l$; see Section 3.6.4). The explicit expression for $V_0$ depends on whether the smaller particle $R'$ is hydrophobic or hydrophilic (see Appendix 3K).

On the other hand, the same $N_s^{r\infty}$ molecules were involved in $\nu_s^{r\infty}$ BHBs before leaving the first two hydration shells of particles $R$ and $R'$. Denoting the energy of a single BHB by $\epsilon_s < 0$, one can write the total energy of these $\nu_s^{r\infty}$ bonds as $\Phi_s \equiv \Phi_s(r) = \epsilon_s \nu_s^{r\infty}(r)$.

The contribution to the interaction potential between solutes $R$ and $R'$, arising from the disruption of the BHB networks in their vicinities because of their overlap, is given as

$$\phi^{hb} = \Phi_b - \Phi_s. \qquad (3.138)$$

This contribution is a function of $r$ (because so are $\Phi_s$ and $\Phi_b$ and $\omega_b$ (because so is $\Phi_s$), i.e., $\phi^{hb} \equiv \phi^{hb}(r, \omega_b)$.

The evaluation of $\Phi_s$ is complicated by the composite character of the solute particle $R$ because the water molecules belonging to its 1HS can form hydrogen bonds with the hydrophilic sites thereupon. According to the adopted model, one water molecule belonging to the 1HS can form only a single hydrogen bond with the particle surface; this happens when one of its hb arms is almost perpendicular to the surface and its tip pointing to a hydrophilic site of the latter (the probability of this event is $\omega_l$). In such a situation, the water molecule forms the same number of hydrogen bonds as in the bulk ($n_b$) with the same (bulk) energy per bond ($\epsilon_b$). Otherwise, the number $n_e$ of bonds per 1HS water molecule will be smaller, $n_e < n_b$, but the bonds may be energetically enhanced (with the energy per bond $\epsilon_e < \epsilon_b$) [30–32].

Explicit expressions for $\Phi_s$ can be obtained in various ways differing in their accuracy. For the solvent-mediated interaction of two infinitely large parallel plates of composite nature [29], we used a linear interpolation of $\Phi_s$ as a function of $\omega_b$. At $\omega_b = 0$, the function $\Phi_s(\omega_b)$ reduced to $\Phi_b$, whereas at $\omega_b = 1$ the function $\Phi_s(\omega_b)$ provided $\Phi_e$, the total energy of hydrogen bonds in which the $N_s^{r\infty}$ molecules would have been involved if the surface of both interacting plates had been completely hydrophobic. For the solvent-mediated interaction between two spherical solutes (either both composite or one composite the other hydrophobic or hydrophilic), in this (linear) approximation we would have $\Phi_s(\omega_b) = \omega_b\Phi_e + (1 - \omega_b)\Phi_b$, where $\Phi_b$ is given by Equation 3.137 and the energy $\Phi_e$ is given by the expression

$$\Phi_e = \epsilon_e n_e \left\{ f_{1,2}(1-\chi)[V_0(r) - V_m(r)] - 0.5\chi V_m(r) \right\}, \tag{3.139}$$

with $n_e$ being the average number of hydrogen bonds formed by a 1HS water molecule, where $\epsilon_e$ is the energy per such a bond; $\chi \approx 0.4$ being the probability that a hydrogen bond, formed by a selected 1HS molecule, is a bond of type 1 (when both water molecules involved belong to the 1HS; see Ref. [28]); $V_m(r)$ is the overlap volume of the 1HSs of the two solutes as a function of $r$ (see Appendix 3K); and coefficient $f_{1,2}$ relating the average density of BHBs of type 2 (between molecules of the 1HS and 2HS) in the overlap region of 2HS's of the two solutes to the number density of water molecules in the 1HS (see Ref. [28] for details).

Hereafter, we will adopt another more accurate approximation for $\Phi_s$. This energy as a function of $r$ can be found as

$$\Phi_s(r) = \epsilon_s v_s^{r\infty}(r), \tag{3.140}$$

where $\epsilon_s$ is the average energy per each hydrogen bond formed by a water molecule in the 1HSs of solutes $R$ and $R'$. Let us denote the average energy of such a bond in the 1HS of composite solute $R$ by $\epsilon_s^c$. On the other hand, the energy of a bond formed by a molecule in the 1HS of a hydrophobic solute is $\epsilon_e$, whereas the water molecules around the hydrophilic solute do not form BHBs at all (all hydrogen bonds there are the same as in bulk). Therefore, if a water molecule belongs to the overlap region of the 1HSs of the two particles, there is some uncertainty regarding the energy of water–water bonds that such a molecule forms in the case where the composite solute $R$ interacts with a hydrophobic solute $R'$: this energy is either $\epsilon_s^c$ or $\epsilon_e$, with equal probabilities. Adopting a most simple approach to get around this uncertainty, one can assume that $\epsilon_s$ is the arithmetic average of BHBs in the 1HSs of particles $R$ and $R'$ separately,

$$\epsilon_s = \begin{cases} \dfrac{1}{2}\left(\epsilon_s^c + \epsilon_e\right) & \text{(for a hydrophobic solute } R'\text{)} \\ \epsilon_s^c & \text{(for a hydropholic solute } R'\text{).} \end{cases} \tag{3.141}$$

We can then apply the linear interpolation with respect to $\omega_b$ (as described above) to $\epsilon_s^c$ as follows:

$$\epsilon_s^c = \omega_b \epsilon_e + (1 - \omega_b)\epsilon_b. \tag{3.142}$$

The quantity $v_s^{r\infty}$, a decrease in the total number of BHBs in the first two hydration shells at a distance $r$ between solutes, can be determined if one knows the average number $n_s$ of hydrogen bonds formed by one molecule of water in the overlap region of the 1HSs of particles $R$ and $R'$. If the latter is hydrophobic, then $n_s$ can be estimated to be the arithmetic mean of $n_e$ and $n_s^c$, with $n_s^c$ being the average number of such bonds in the 1HS of the composite particle $R$. If the solute $R'$ is hydrophilic, then $n_s$ simply equals $n_s^c$. Thus,

$$n_s = \begin{cases} \dfrac{1}{2}(n_s^c + n_c) & \text{(for a hydrophobic solute } R'\text{)} \\ n_s^c & \text{(for a hydropholic solute } R'\text{).} \end{cases} \tag{3.143}$$

Again, an explicit expression for $v_s^{r\infty}$ will depend whether the solute $R'$ is hydrophobic or hydrophilic (see Appendix 3K).

In order to determine $n_s^c$, consider a water molecule $m_1^c$ in the 1HS of the composite particle $R$. By analogy with Equation 11 of Ref. [28] and Equation 13 of Ref. [29], one can represent $n_s^c$ as the sum

$$n_s^c = p_1 + p_{2(1)} + p_{3(2,1)} + p_{4(3,2,1)}, \tag{3.144}$$

where $p_1$ is the probability that one of the hb arms of molecule $m_1^c$ forms a hydrogen bond, $p_{2(1)}$ is the probability that a second hb arm forms a hydrogen bond subject to the restriction that one of the hb arms has already formed the bond, $P_{3(2,1)}$ is the probability that a third hb arm forms a hydrogen bond subject to the restriction that two of the hb arms have already formed bonds, and $p_{4(3,2,1)}$ is the probability that the fourth hb arm forms a bond subject to the restriction that three of the hb arms have already formed bonds. If the solute surface were purely hydrophilic, then these probabilities would be $b_1, b_1^2, b_1^3$, and $b_1^4$, respectively, where $b_1$ is the probability that a bulk water molecule forms a hydrogen bond. If the solute surface were completely hydrophobic, then these probabilities would be equal to $s_1, s_{2(1)}, s_{3(2,1)}, s_{4(3,2,1)}$, respectively, related to $b_1$ by $s_1 = k_1 b_1$, $s_{2(1)} = k_2 b_1^2$, $s_{3(2,1)} = k_3 b_1^3$, $s_{4(3,2,1)} = k_4 b_1^4$ with the coefficients $k_1 \simeq 0.521694$, $k_2 \simeq 0.433148$, $k_3 \simeq 0.304122$, $k_4 \simeq 0.006433$ (see Refs. [28,29]). Taking into account the definition of the probabilities $s_1$, $s_{2(1)}$, $s_{3(2,1)}$, and $s_{4(3,2,1)}$, one can find their $\omega_b$ dependence via a linear interpolation between their values at $\omega_b = 0$ and $\omega_b = 1$, which provides

$$p_1 = K_1(\omega_b)b_1, \quad p_{2(1)} = K_2(\omega_b)b_1^2, \quad p_{3(2,1)} = K_3(\omega_b)b_1^3, \quad p_{4(3,2,1)} = K_4(\omega_b)b_1^4, \tag{3.145}$$

with the coefficients

$$K_i(\omega_b) = 1 - \omega_b + \omega_b k_i \quad (i = 1,2,3,4). \tag{3.146}$$

The probability $b_1$ is unambiguously determined by the thermodynamic state of the bulk water (temperature, pressure, etc.) as the solution of the equation $n_b = b_1 + b_1^2 + b_1^3 + b_1^4$ satisfying the constraint $0 < b_1 < 1$ (the bulk quantity $n_b$, whereof the dependence on thermodynamic conditions is well enough documented, is assumed to be given).

Substituting Equations 3.137 and 3.140 into Equation 3.138 and taking into account Equations 3.141 through 3.146 (as well as those in Appendix 3K), one can calculate the solvent-mediated contribution to the interaction between composite and hydrophobic or hydrophilic particles, contribution arising because of the overlap of the BHB networks around the two particles.

#### 3.6.2.2 Case of Protein Folding via Nucleation

It is clear that in order to apply the probabilistic approach to the NMPF, the role of the composite particle should be attributed to the cluster of native residues (i.e., with correct tertiary contacts as they exist in the native protein). The radius of this cluster evolves (increases) as folding advances. On the other hand, the role of the smaller particle is played by the residues of the protein unfolded part. Their radius is thus constant and equal to $R' = 0$. In terms of hydrogen-bonding ability, both hydrophobic and neutral protein residues can be considered as "hydrophobic" (i.e., unable to form hydrogen bonds with water molecules).

The water–water hydrogen bond contribution ($\phi^{hb}$ to the interactions of hydrophobic and hydrophilic residues with a protein folded cluster of radius $R = 5\ \eta_w$ is presented in Figure 3.31. The composition of the cluster was taken to be that of a bovine pancreatic ribonuclease (BPR), whereof the nucleation mechanism of folding we previously modeled [16–19]. The surface of the protein cluster was assumed to have the same composition as the whole protein so that the fraction of hydrophobic sites thereupon was taken to be $\omega_b = (N_b + N_n)/(N_b + N_l + N_n)$, where $N_b = 40$, $N_l = 81$, and $N_n = 3$ are the numbers of hydrophobic, hydrophilic, and neutral residues in BPR, respectively. The protein residues themselves are assumed to have the characteristic length of pairwise interaction $\eta = 1.3\eta_w$. Water was assumed to be under such thermodynamic conditions that $n_b = 3.65$. Quantitatively, the energetic enhancement of BHBs (in the 1HS of a hydrophobic surface) can be characterized by the $\epsilon_s/\epsilon_b$ ratio, where $\epsilon_e < 0$ is the energy of a hydrogen bond involving at least one 1HS molecule and $\epsilon_b < 0$ is the energy of a bond between two bulk molecules. In the PHB formalism [28,29], the $\epsilon_e/\epsilon_b$ ratio is allowed to take on any positive value, i.e., $0 < \epsilon_e/\epsilon_b < \infty$, although it is expected to be close to unity. As suggested in Ref. [34], the enhancement ratio $\epsilon_e/\epsilon_b$ was taken to be 1.1.

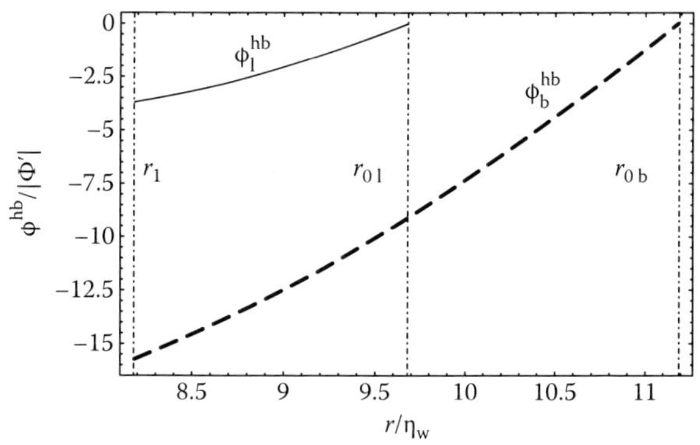


Figure 3.31

The boundary hydrogen bond networks contribution $\phi^{hb}$ to the total interaction potential between two spherical solutes, one of which is a composite particle of radius $R = 5\ \eta_w$ (and of hydrophobic surface fraction $\omega_b = 0.4$) and the other is either a hydrophobic (thick dashed curve) or hydrophilic (thin continuous curve) particle of radius $R' = 0$ (which corresponds to a single protein residue). The potential is plotted as $\phi^{hb}/|\Phi'|$ versus $r/\eta_w$, where $\Phi' = \epsilon_b n_b \rho_w \eta_w^3$ is twice the energy of water–water hydrogen bonds in the volume $\eta_w^3$ of bulk water. The leftmost dotted-dashed vertical line indicates the location of $r_1$, whereas the middle and rightmost lines indicate the location of $r_0$ for hydrophilic and hydrophobic beads, respectively. The solvent (water) is under such conditions that $n_b = 3.65$.

The contribution $\phi_l^{hb}$ to the "cluster–hydrophilic residue" interactions in Figure 3.31 is presented by the upper (thin solid curve), whereas the contribution $\phi_b^{hb}$ to the "cluster–hydrophobic residue" interactions is plotted as a thick dashed (lower) curve. The potentials are plotted as $\phi^{hb}/|\Phi'|$ versus $r/\eta_w$, where $\Phi' = \epsilon_b n_b \rho_w \eta_w^3$ is twice the energy of water–water hydrogen bonds in the volume $\eta_w^3$ of bulk water. The effect of BHB networks around the cluster and residues on "cluster–residue" interactions is clearly much (by an order of magnitude) stronger for the hydrophobic residues than for the hydrophilic ones. The range of $\phi^{hb}(r)$ for the former is twice as large as that for the latter.

### 3.6.3 Modified Model for NMPF

#### *3.6.3.1 Ternary Heteropolymer as a Protein Model*

The NMPF model [15–19] considered a protein polypeptide chain as a heteropolymer that consists of $N$ connected beads which can be thought of as representing the $\alpha$ carbons of various amino acids (see Figure 3.32). The heteropolymer consists of hydrophobic (b), hydrophilic (l), and neutral (n) beads. Besides, some (not all) hydrophilic beads are ionizable (some negatively and some positively), because 5 out of 11 hydrophilic amino acids in real proteins are ionizable. Two adjacent beads are connected by a covalent bond of a fixed length $\eta$. This model (and its variants) has been shown [6,10,12–14] to be capable of capturing the essential characteristics of protein folding even though it contains only some of the features of a real polypeptide chain.

The total energy of the heteropolymer (polypeptide chain) can contain three contributions of different types. First, the contribution from repulsive and attractive forces between pairs of nonadjacent beads (these can be, e.g., of LJ or other types) that are at least three links apart (the interaction between nearest neighbors is taken into account by the link of constant length between them, while the interactions between next-nearest neighbors are taken into account by the rigidity of the angle between neighboring links). Next is a contribution from the harmonic forces due to the oscillations of the bond angles, and finally, a contribution from the dihedral angle potential due to the rotation around the peptide bonds.

The bond angle forces are believed [6,10] to play a minor role in the protein folding and unfolding; hence, all bond angles can be set to be equal to $\beta_0$. It was shown [6,10] by molecular dynamics (MD) simulations, which used low-friction Langevin dynamics, that a proper balance between the remaining two contributions to the total energy of the heteropolymer ensures that the heteropolymer folds into a well-defined $\beta$-barrel structure. It was also found [6,10] that the balance between the dihedral angle potential, which tends to stretch the molecule into a state with all bonds in a *trans* configuration, and the attractive hydrophobic potential is crucial to induce folding into a $\beta$-barrel-like structure upon cooling. If attractive forces are excessively dominant, they make the heteropolymer fold into a globulelike structure, while an overwhelming dihedral angle potential forces the chain to remain in an elongated state (even at low temperatures) with bonds mainly in the *trans* configuration.

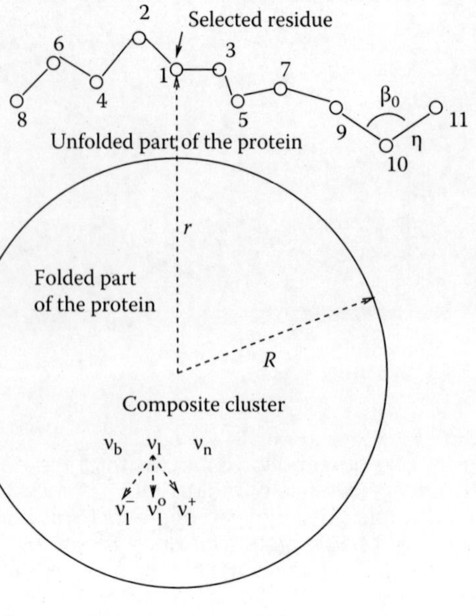


Figure 3.32

A piece of a heteropolymer chain around a spherical cluster consisting of $\nu_b$ hydrophobic, $\nu_l$ hydrophilic, and $\nu_n$ neutral beads. Among the hydrophilic beads themselves $\nu_l^0$ are uncharged, $\nu_l^-$ are negatively charged, and $\nu_l^+$ re positively charged. Bead 1 is in the plane of the figure, whereas other beads may all lie in different planes. All bond angles are equal to $\beta_0$ and their lengths are equal to $\eta$. The radius of the cluster is $R$ and the distance from the selected bead 1 to the cluster center is $r$.

Protein folding via nucleation is modeled to occur as the formation and evolution of a cluster of native residues within the protein. The cluster is assumed to have a spherical shape all the time. If the ionizability of (some) hydrophilic residues is taken into account, a cluster of native residues, involved in the NMPF, should be characterized by five independent variables and it would be necessary to use the formalism of a five-component nucleation. Such a model would require extremely lengthy numerical calculations when applied to real proteins. To avoid this difficulty, one can assume that, as the protein folds via nucleation, the "hydrophilic" mole fractions of positive and negative residues in the cluster remain equal to those in the whole protein. Under such assumptions, the cluster can be characterized by only three independent variables and a model for the NMPF can be again developed in terms of a ternary nucleation theory. In the framework of a three-component heteropolymer representation, the protein folding process can be regarded as a ternary phase transition with the ternary cluster (of native residues) within a ternary mother phase (unfolded part of the protein).

#### 3.6.3.2 Hydrogen Bond Contribution to the Potential Field around the Folded Part of a Protein

In our NMPF model, we considered the folded part of the protein to be a cluster of spherical shape immersed in a ternary fluid mixture (whereof the role is played by the unfolded part of the protein). Let us denote the number of molecules (beads) of component $i$ in such a ternary spherical cluster by $\nu_i$ ($i$ = b, l, n). The radius of the cluster will be denoted by $R$, implying that it plays the role of a composite solute particle discussed above. The total number of beads of component $i$ in the protein is denoted by $N_i$. In the original applications of the first passage time analysis to nucleation [35–37] a molecule of component $i$ (in our model $i$ = b, l, n) located in the surface layer of the cluster was considered to perform a thermal chaotic motion in a spherically symmetric potential well $\phi_i(r)$ resulting from the pair interactions (say, of LJ type) of this molecule with those in the cluster ($r$ is the distance between the center of the cluster and the molecule; here, it is the distance between the centers of the cluster and selected bead [see bead 1 in Figure 3.32]).

The total potential $\psi_i(r)$ for a residue of type $i$ around the cluster previously (i.e., in the original version of the NMPF model) had three different constituents, $\phi_i(r)$, $\bar{\phi}_i^{\delta}(r)$, and $\phi_{cp}$, which represented the effective LJ and electrostatic (for pairwise interactions of the selected residue with those in the cluster), average dihedral angle, and confining (representing the external boundary of the volume available to the unfolded residues) potentials, respectively,

$$\psi_i(r)=\phi_i(r)+\phi_{cp}(r)+\bar{\phi}_i^{\delta}(r)\ (i=\text{b, l, n}). \tag{3.147}$$

Complete details concerning the physical nature and calculation of $\phi_i(r)$, $\bar{\phi}_i^{\delta}(r)$, and $\phi_{cp}$ are given in Refs. [15–19].

In the original NMPF model, hydrogen bonding was taken into account just indirectly via the diffusion coefficients of amino-acid residues. We will hereafter improve that model by combining it with the above-presented BHB model for the solvent-mediated interactions of solute particles. The improvement consists of augmenting the overall potential field around the cluster, $\psi_i(r)$, by an additional term $\phi_i^{hb}(r)$ arising because of the disruption of the BHB networks around that residue and the cluster. As a result, instead of Equation 3.147, the overall potential $\psi_i(r)$ will now be

$$\psi_i(r) = \phi_i(r) + \phi_{cp}(r) + \bar{\phi}_i^{\delta}(r) + \phi_i^{hb}(r) \quad (i = \mathrm{b, l, n}), \tag{3.148}$$

where $\phi_i^{hb}(r)$ is given by Equation 3.138 and auxiliary equations in Section 3.6.2 with $R' = 0$ and $R = [3\upsilon(\nu_b + \nu_l + \nu_n)/4\pi]^{1/3}$.

As with Equation 3.147 (representing the original NMPF model [15–19]), the combination of potentials in Equation 3.148 gives rise to a double potential well around the cluster with a barrier between the two wells. Figure 3.33 presents typical shapes of the constituents $\phi_i(r)$ and $\bar{\phi}_i^{\delta}(r)$ as functions of the distance from the cluster center, as well as the overall potential well $\psi_i(r)$ itself (for details regarding the numerical calculations, see Refs. [15–19]). The contribution $\phi_i(r)$ arising from the pairwise interactions has a form reminiscent of the underlying LJ potential, whereas the contribution from the average dihedral potential has a rather remarkable behavior. Indeed, starting with a maximum value at the cluster surface, it monotonically decreases with increasing $r$ until it becomes constant for large enough values of $r$. Thus, except very short distances from the cluster surface where $\phi_i(r)$ sharply decreases from $\infty$ to its global minimum, the potential $\psi_i(r)$ is shaped by two competing terms, $\phi_i(r)$ and $\bar{\phi}_i^{\delta}(r)$, which increase and decrease, respectively, with increasing $r$, and by the confining potential $\phi_{cp}$. The double-well shape of the overall potential $\psi_i(r)$ is of crucial importance to the NMPF model (both original and modified) because it allows one to use the mean first passage time analysis [38,39] for the determination of the rates of both absorption and emission of beads by the cluster.

Once the rates of emission and absorption are found as functions of cluster independent variables, one can develop a self-consistent kinetic theory for the NMPF and evaluate the folding time and its temperature dependence. The time necessary for the protein to fold is evaluated as the sum of the times necessary for the appearance of the first nucleus and the time necessary for the nucleus to grow to the maximum size of the

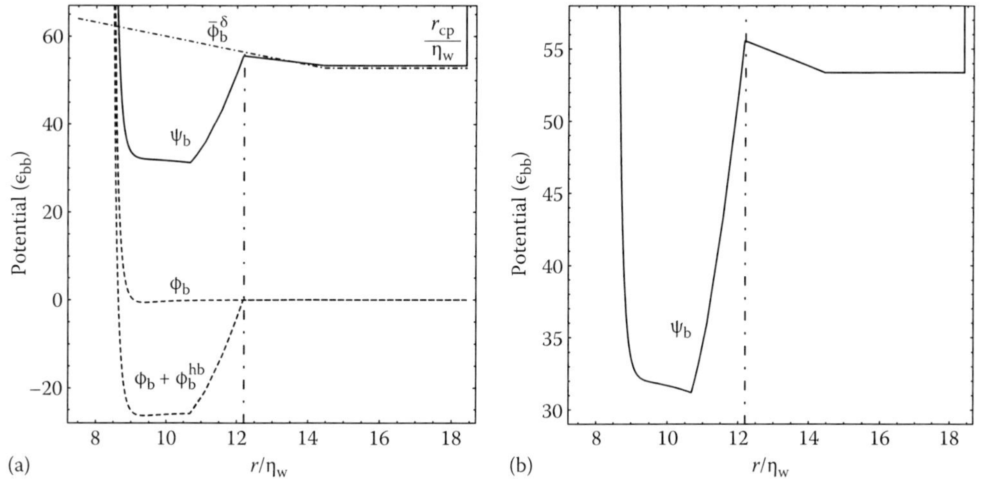


Figure 3.33

(a) The potentials $\phi_b(r)+\phi_b^{hb}$ (lower dashed curve), $\phi_b(r)$ (upper dashed curve), $\bar{\phi}_b^{\delta}(r)$ (dotted-dashed curve), and $\psi_b(r)$ (solid curve), for a hydrophobic bead around a cluster with $\nu_b = 23$, $\nu_l = 41$, $\nu_l^- = 4.5$, $\nu_l^+ = 6.5$, $\nu_n = 2$ at pH = 7.3 [$\phi_b(r)$, $\bar{\phi}_b^{\delta}(r)$, $\phi_{cp}$, $\phi^{hb}$, and $\psi_b(r)$ are the effective pairwise, average dihedral angle, confining, solvent-mediated (hydrogen bond), and total potentials for a hydrophobic bead around the cluster of folded protein]. The confining potential is shown as a solid vertical line at $r_{cp} \simeq 18.5\eta_w$, whereas the hydrogen bond contribution is plotted separately in Figure 3.31. All the potentials are given in units of $\epsilon_{bb}$, the energy parameter in the Lennard-Jones potential of pairwise interactions between two hydrophobic beads. (b) The curve $\psi_b(r)$ alone for the better visualization of its double-well shape.

folded protein in the native state (a brief description of the whole procedure is presented in Appendix 3L, while the reader is referred to Refs. [15–19] for details).

It is worth emphasizing that the double-well character of $\psi_i(r)$ is not related to the two-state (native-unfolded) nature of proteins. In the NMPF model, the latter is reflected in the function $G(\nu_b, \nu_l, \nu_n)$ (see Appendix 3L), which is the analog of the free energy of formation of a cluster $\nu_b$, $\nu_l$, $\nu_n$ and has the shape of a hyperbolic paraboloid (in a four-dimensional space of variables $\nu_b$, $\nu_l$, $\nu_n$, $G$). Its path of the steepest descent has at least one maximum at the point with the coordinates $\nu_{bc}$, $\nu_{lc}$, $\nu_{nc}$, $G(\nu_{bc}, \nu_{lc}, \nu_{nc})$ corresponding to the critical cluster. The initial point ($\nu_b = 0$, $\nu_l = 0$, $\nu_n = 0$) of that path corresponds to a completely unfolded protein and its final point ($\nu_b = N_b$, $\nu_l = N_l$, $\nu_n = N_n$) to a completely folded (native) structure. The height of the barrier between the unfolded and folded states, provided by $G_c \equiv G(\nu_{bc}, \nu_{lc}, \nu_{nc})$, plays a crucial role in the kinetics of protein folding and determines the time of this process.

The path of the steepest descent of $G(\nu_b, \nu_l, \nu_n)$ can have additional local maxima that can lie on either side of the global maximum. (Such a situation would arise if the rates of emission and absorption of residues by a cluster were nonmonotonic functions of cluster variables. This could be possible if the diffusion coefficients of protein residues in the inner potential well (ipw) and outer potential well (opw) were dependent on cluster variables.) The local minima would then correspond to intermediate (partially folded) metastable states of the protein [1,4,5]. The kinetics of the transition from the initial (unfolded) state into the first intermediate one, transitions between intermediate states, and transition from the last one into the final (completely folded native) state are governed mostly by the heights of the barriers between them. The latter [as well as the whole shape of the function $G(\nu_b, \nu_l, \nu_n)$] are determined by the emission and absorption rates as functions of the variables $\nu_b$, $\nu_l$, $\nu_n$ (see Appendix 3L).

Our models of a protein and its folding have so far always provided a single maximum on the path of the steepest descent of height from $26k_BT$ to $33k_BT$ ($k_B$ being the Boltzmann constant and $T$ being the temperature). This is a clear indication of a two-state nature of the model proteins considered, at least under external conditions at which they were studied. In principle, the character of the protein can change if under different external conditions the height of the barrier can decrease or increase. If it decreases down to a few $k_BT$s (or disappears at all), the protein would fold in a virtually (or purely) barrierless way, which is characteristic of fast-folding proteins [40,41]. On the other hand, this (i.e., virtually or purely barrierless folding) can be a result of some particular values for the set of parameters in the model interaction potentials. One can also expect that under favorable circumstances the existence of intermediate (metastable) states of a protein on a particular folding trajectory can significantly decrease protein folding time compared to the time on another folding trajectory (of the same protein) with just one global maximum (assuming that its height is comparable to that of the global maximum of the path with intermediate states). This can take place if all intermediate barriers between previous and next local minima are significantly smaller than the height of the global barrier.

### 3.6.4 Numerical Evaluations

As an illustration of the above theory, we have applied it to the folding of two model proteins, composed of 124 and 2500 amino acid residues. The first roughly mimics a BPR and was previously studied in Refs. [16–19] in the framework of the original NMPF model. According to the classification of Elliott and Elliott [42], BPR consists of $N_b = 40$ hydrophobic, $N_l = 81$ hydrophilic, and $N_n = 3$ neutral amino acids (with the total of $N = 124$). We have considered a model heteropolymer consisting of the same numbers of hydrophobic, hydrophilic, and neutral beads: $N_b = 40$ hydrophobic, $N_l = 81$ hydrophilic, and $N_n = 3$ neutral ones. Out of 81 hydrophilic beads, only 28 are ionizable, of which $\tilde{N}_l^- = 10$ can be charged negatively and $\tilde{N}_l^+ = 18$ positively. The presence of other solute molecules and ionized bead counterions was not taken into account explicitly. The second model protein was chosen as a representative of large proteins. As in Ref. [15], it was a heteropolymer consisting of a total of 2500 residues, with the same mole fractions of hydrophobic, hydrophilic, neutral, and positively and negatively ionizable residues as for a short protein, that is, $N_b = 807$, $N_l = 1633$, $N_n = 60$, and $\tilde{N}_l^+ = 363$ and $\tilde{N}_l^- = 202$ (with a total of 565 ionizable residues). We intentionally modeled such a large protein (relatively rare in nature) to demonstrate that our method provides reasonable folding times on a reasonable timescale even for the largest proteins. Straightforward Monte Carlo (MC) or MD simulations of folding and unfolding behavior of such large proteins are currently impossible.

The actual numbers $N_l^-$ and $N_l^+$ of negatively and positively ionized residues in the protein can be smaller than $\tilde{N}_l^-$ and $\tilde{N}_l^+$, respectively, with the average dissociation coefficients $k^- = N_l^-/\tilde{N}_l^-$ and $k^+ = N_l^+/\tilde{N}_l^+$ depending on the pH of the surrounding medium and on the average dissociation coefficients $K_a$ and $K_b$ of the acidic and basic residues in the protein. In the BPR, there are five residues of aspartic acid (asp), five of glutamic acid (glu), ten of lysine (lys), four of arginine (arg), and four of histidine (his) so that $K_a$ and $K_b$ in

BPR were taken to be $K_a = (5K_{asp} + 5K_{glu})/(5 + 5)$ and $K_b = (10K_{lys} + 4K_{arg} + 4K_{his})/(10 + 4 + 4)$, where the dissociation's constants $K_{asp}$, $K_{glu}$, $K_{lys}$, $K_{arg}$, and $K_{his}$ are available in the literature [42,43] (although they exhibit some scatters). At any given pH, the effective dissociation coefficients $k^-$ and $k^+$ are completely determined by $pK_a = -\log10\ K_a$, $pK_b = -\log_{10} K_b$, respectively, as $k^- = 1/(1+10^{-pH+pK_a})$, $k^+ = 1/\left(1+10^{-14+pH+pK_b}\right)$. We have carried out numerical calculations for pH = 7.3 (roughly that of living cells) at which $N_l^-/N_l^+ \simeq 0.69$, as well as for pH = 6.3 at which $N_l^-/N_l^+ \simeq 0.38$ and for pH = 8.3 at which $N_l^-/N_l^+ \simeq 1.59$. Note that the standard values [42,43] of the dissociation constants of ionizable residues are suitable for modeling unfolded states only. However, the corrections to $K_a$ and $K_b$ for the native state can be neglected because they affect only the electrostatic contributions to the total potential field around the cluster $\psi_l(r)$, which themselves constitute small corrections to the other terms in $\psi_l(r)$.

The contribution from the BHB networks to the overall potential was calculated as described in Section 3.6.2 with $R' = 0$ and $R = [3\upsilon(\nu_b + \nu_l + \nu_n)/4\pi]^{1/3}$. The electrostatic interactions between a selected charge $q_\pm$ at $\mathbf{r}$ and the elementary charge at point $\mathbf{r}'$ (within the cluster) is given in the Debye–Hückel approximation for the screened Coulomb potential by

$$w_\pm\left(|\mathbf{r}' - \mathbf{r}|\right) = \frac{k}{\varepsilon}\frac{eq_\pm}{|\mathbf{r}' - \mathbf{r}|}e^{-\kappa|r'-r|}, \tag{3.149}$$

where $k$ is the electrostatic constant (1 in cgs units and $1/4\pi\varepsilon_0$ in SI units, with $\varepsilon_0 = 8.854 \times 10^{-12}$ F/m being the dielectric permittivity of vacuum), $\varepsilon$ is the relative permittivity of the medium wherein the protein is present, and $\kappa$ is the inverse Debye length (in our numerical calculations, $1/\kappa \simeq 0.3$ nm, which roughly corresponds to an electrolyte concentration of 0.3 M, typical for living cells [42]). The total electrostatic potential $u_\pm(r)$ of a residue carrying a charge $q_\pm = \pm e$ around the cluster (at a distance $r$ from its center) was calculated as

$$u_\pm(r) = \int_V d\mathbf{r}'\rho_q(r')w_\pm\left(|\mathbf{r}' - \mathbf{r}|\right),$$

where $\mathbf{r}$ is the coordinate of the selected bead with a charge $q_\pm$ and the "number density" of charged residues at point $\mathbf{r}'$ within the cluster is assumed to be uniform, i.e., $\rho_q \equiv \rho_q(r') = \left(\nu_l^+ - \nu_l^-\right)/(4\pi R^3/3)$. The density $\rho_q$ can be negative, in which case the total charge of the cluster is negative. The integration in this equation has to be carried out over the whole volume of the system, but the contribution from the unfolded part is assumed to be small owing to the smaller density. The nonelectrostatic interactions between any two nonadjacent beads were modeled by LJ potentials whereas the potential due to the dihedral angle $\delta$ was represented by the expression $\phi_\delta = \epsilon'_\delta(1+\cos\delta) + \epsilon''_\delta(1+\cos 3\delta)$, where $\epsilon'_\delta$ and $\epsilon''_\delta$ are energy parameters which depend on the nature and sequence of the four beads involved in the dihedral angle $\delta$. The parameters of the LJ and electrostatic potentials were chosen as in Refs. [16–19], but the parameters of the dihedral angle potential needed a significant adjustment as explained below. The typical total densities $\rho_f$ and $\rho_u$ of protein residues in the folded and unfolded (but compact) states and the diffusion coefficients of residues $D^{iw}$ and $D^{ow}$ in the ipw and opw were taken the same as in Refs. [16–19]: $\rho_f\eta^3 = 0.57$, $\rho_u = 0.2\rho_f$, $D^{iw}\rho_f = D^{ow}\rho_u$, $D^{iw} = D_i^{iw}$ ($i$ = b, l, n), $D^{ow} = D_i^{ow}$ ($i$ = b, l, n), with $D^{iw}$ assumed to vary between $10^{-6}$ and $10^{-8}$ cm$^2$/s (because of the lack of reliable data on the diffusion coefficient of a residue in a protein chain).

For hydrophobic, neutral, and uncharged hydrophilic beads, the potential fields around the cluster are not affected by the charged residues in the cluster. For negatively and positively charged hydrophilic residues, the overall potentials around the cluster have the electrostatic contributions, $u_-(r)$ and $u_+(r)$, respectively. Included in the effective pairwise potential $\phi_j(r)$ ($j$ = +, −) in Equations 3.147 and 3.148, they are equal to each other in absolute values but have opposite signs, i.e., $u(r) \equiv u_+(r) = -u_-(r)$, so that $\phi_\pm(r) = \phi_0(r) \pm u(r)$, where $\phi_0(r)$ is the effective pairwise potential of uncharged hydrophilic beads (due exclusively to LJ interactions). As previously [16–19], the effective LJ and average dihedral angle potentials have been considered to be independent of whether a hydrophilic residue is positively or negatively charged or noncharged.

Figure 3.33 presents typical shapes of the overall potential well $\psi_b(r)$ and its different constituents as functions of distance from the cluster center. All the potentials are given in units of the energy parameter $\epsilon_{bb}$ of the LJ potential of pairwise "hydrophobic–bead–hydrophobic–bead" interactions. The curves shown are for a hydrophobic bead, but the curves for neutral and hydrophilic (both uncharged and positively and negatively charged) beads look very similar to those in Figure 3.33. The results shown are for pH = 7.3 and a cluster $\nu_b = 23$, $\nu_l^0 = 41$, $\nu_l^- = 4.5$, $\nu_l^+ = 6.5$, $\nu_n = 2$ carrying a total charge $q_\nu = +2e$.

Note that the pH of the medium (surrounding the protein) and the charge of the cluster do not affect the potentials (both overall and its constituents) for hydrophobic, neutral, and hydrophilic uncharged beads. The pairwise potential for hydrophilic charged residues, $\phi_j(r)$ ($j$ = +, –), is sensitive to the charge of the cluster because it contains the electrostatic contribution $\pm u(r)$ from the "charged bead–charged cluster" interactions. For example, for a positively charged cluster $\phi_0(r) + u_-(r) < \phi_0(r) < \phi_0(r) + u_+(r)$, i.e., the potential well (due to the well of the LJ potential) becomes shallower for positively charged hydrophilic beads and deeper for negatively charged ones as compared to uncharged ones [16–19]. However, the electrostatic contribution $u(r)$ to $\phi_j(r)$ ($j$ = +, –) is much weaker (by an order of magnitude) than the LJ contribution, i.e., $|u(r)/\phi_0(r)| \ll 1$. On the other hand, the entire effective pairwise contribution to $\psi_j$ is much weaker than the BHB networks contribution, i.e., $|\phi_j(r)/\phi_j^{hb}| \ll 1$. Therefore, the effective electrostatic potential $u(r)$ affects the overall potential field $\psi_j$ very weakly. Nevertheless, this weak effect results in the pH dependence of the protein folding time because it is magnified by the formalism of the first passage time analysis (involving the exponentials of the overall potential $\psi_j$).

As shown in Figure 3.31, the BHB network contribution $\phi^{hb}$ is a continuous negative function of $r$ (its first derivative has finite discontinuities at some $r = r_1$ and $r = r_0 > r_1$). However, the slope of $\phi^{hb}$ as a function of $r$ is clearly discontinuous at $r = r_1 \equiv R + (\eta + \eta_w)$ and $r = r_0$ with $r_0 \equiv R + \eta + 2\,\eta_w$, for hydrophilic beads and $r_0 \equiv R + \eta + 4\eta_w$, for hydrophobic ones. This is an artifact of the model and arises because of the sharp boundaries assumed for the 1HS and 2HS. When $r$ decreases from $\infty$ to $r_0$, the 1HS and 2HS of the particles are not affected; hence, $\phi^{hb}(r) = 0$ for $r \geq r_0$. When $r$ decreases from $r_0$ to $r_1$, 2HS and 1HS molecules are removed from the first two hydration shells of the particle(s), which leads to the decrease in $\phi^{hb}$ from zero to its minimum value (of strongest attraction) attained at $r_1$. When $r$ becomes smaller than $r_1$, virtually no water molecules can fit in between the particles and the change in $r$ practically does not lead to any change in the solvent-mediated interaction of the particles which hence remain constant (corresponding to the strongest interaction between particles). As a result, the attractive force between the particles is piecewise continuous with finite discontinuities at $r = r_1$ and $r = r_0$. (In Figure 3.31, the location of $r_1$ is shown by the leftmost dotted–dashed vertical line, whereas the middle and rightmost lines indicate the location of $r_0$ for hydrophilic and hydrophobic beads, respectively.) Thus, $\phi^{hb}$ represents an additional attraction between a cluster of native residues of the protein and a residue in the protein unfolded part at distances between their centers in the range $\eta < r < r_0$. Although this range is rather short, its strength is much higher than that of the effective pairwise potential at not too short distances between cluster and residue.

The effective pairwise potential $\phi_b(r)$ arising from elementary pairwise interactions (LJ for hydrophobic, uncharged hydrophilic, and neutral beads and LJ + electrostatic for positively and negatively charged hydrophilic ones) has a form reminiscent of the LJ potential. Although this contribution to $\psi_b(r)$ is much weaker than $\bar{\phi}_b^{\delta}(r)$ and $\phi_b^{hb}$ at $r \geq \eta$, it tends to $\infty$ as $r \to 0$, thus forming the inner (closer to the cluster) boundary of the inner potential well. Therefore, it cannot be neglected on the right-hand side of Equation 3.148 (likewise, one cannot neglect the confining potential that forms the outer boundary of the outer potential well, although it is zero everywhere at $r \leq r_{cp}$).

The average dihedral potential (assigned to a selected bead) $\bar{\phi}_b^{\delta}(r)$ has a maximum value at the cluster surface and decreases monotonically with increasing $r$ until it becomes constant for $r \geq R + \tilde{d}$, where $\tilde{d}$ is the maximum [15–19] distance between beads 1 and 6 (or beads 1 and 7) which depends on $\eta$ and $\beta_0$. Such a behavior of $\bar{\phi}_i^{\delta}(r)$ can be interpreted as a consequence of the decrease in entropy (hence an increase in the free energy) of the heteropolymer chain as the selected bead 1 approaches the cluster surface for $r < R + \tilde{d}$ which, in turn, is due to a decrease in the configurational space available to the neighboring beads (beads 2–7).

The overall potential $\psi_b(r)$ depends very much on the parameters for the potential associated with a single dihedral angle. Previously [15–19], their values were chosen so that the overall potential field around a cluster had a double-well shape. Using the same values for the dihedral angle potential parameters in Equation 3.148, corresponding to a modified (i.e., combined with the PHB model) NMPF model, would result in a potential field that has a single-well shape. This, in turn, would prevent us from applying the first passage time analysis to determining the absorption rate $W_i^+$ of the cluster (see Equation 3L.3 in Appendix 3L). Then, it remains rather unclear how one can determine $W_i^+$ and eventually evaluate the protein folding time.

In order to avoid this difficulty and conserve the double-well shape of the potential field around the cluster in the modified NMPF model, it is necessary to properly adjust the energy parameters in the dihedral angle potential. This adjustment must be carried out subject to a reasonable physical criterion, for instance, the requirement that the predicted folding times must be in the range of experimentally observed ones. To satisfy this requirement, we had to increase the two energy parameters (which, for simplicity, are taken to be equal to each other) in the dihedral angle potential by a factor of 12 compared to their values used in Refs. [16–19]. With such values, the potential field around the cluster again has a double-well shape (see Figure 3.33), although significantly different from the original NMPF model (because of the BHB networks contribution): the inner well (ipw) is separated by a potential barrier from the outer well (opw). The geometric characteristics of the wells (widths, depths, etc.) and the height and location of the barrier between them depend on the interaction parameters. The barrier has different heights for the ipw and opw beads. For a given protein the location $r_{cp}$ of the outer boundary of the opw is determined by the size of the cluster and densities $\rho_f$ and $\rho_u$. The existence of an opw allows one to consider the absorption of a bead by the cluster as an escape of the bead from the opw by crossing over the barrier into the ipw. One can therefore use the mean first passage time analysis to determine not only the emission rate but also the rate of absorption of beads by the cluster.

For the model proteins considered (a short BPR of $N = 124$ residues and a large protein of $N = 2500$ residues) with the above choices of the system parameters, our model estimates show that the time of protein folding is determined mainly by the time necessary for the first nucleation event to occur, as expected and as was the case in the original NMPF model. With the diffusion coefficient $D^{iw} = 10^{-6}$ cm$^2$/s, the modified NMPF model estimates the characteristic time of folding of the short protein to be about 6 s at pH = 8.3, 5 s at pH = 7.3, and 8 s at pH = 6.3, whereas the folding times for the long protein are about 165 s at pH = 8.3, 140 s at pH = 7.3, and 220 s at pH = 6.3. For $D^{iw} = 10^{-8}$ cm$^2$/s, the folding times of the short protein are predicted to be about 600, 500, and 800 s at pH = 8.3, 7.3, and 6.3, respectively, whereas for the long protein they are about 16,500, 14,000, and 22,000 s at pH = 8.3, 7.3, and 6.3, respectively. The folding times for the model BPR are in a good agreement with the experimentally observed folding times of real BPR (see Refs. [44,45], where the BPR folding time was reported to be on the order of 1000 s, and references therein). This suggests that the smaller value of the diffusion coefficient of protein residues in the unfolded state, $D^{iw} = 10^{-8}$ cm$^2$/s, is more appropriate to model the folding of proteins whereof the sequence and structure are similar to those of BPR. On the other hand, faster folding proteins are probably better characterized by faster diffusion of their residues in the unfolded state and hence are better modeled with $D^{iw} = 10^{-6}$ cm$^2$/s. (Note again that, besides linearly depending on $1/D^{iw}$, the folding times predicted by the above model are also sensitive to the energy parameter in the dihedral angle potential, the only interaction parameter of adjustable character.) For the short protein the effect of pH on the protein folding time is significantly less pronounced in the modified model than in the original one. This was also expected because the contribution of the electrostatic interactions to the overall potential field is less important in the modified model compared to the original one. As previously [16–19], among all three pH values considered, the physiological pH provides the lowest folding time. Clearly, one cannot necessarily conclude that the folding time as a function of pH has a minimum at 7.3, but one can suggest that this function does have a minimum at some $6.3 < \text{pH} < 8.3$. To more accurately determine the location of this minimum, it is necessary to calculate the folding times for more values of pH. The sensitivity of the folding time to pH is stronger for the large protein, which can be accounted for by larger net charges that the cluster of native residues can have during protein folding. Our results allow us to make some meaningful comparison of the folding times for two similar proteins differing (mainly) in the total number $N$ residues in the polypeptide chain (with all the mole fractions of different kind of residues being the same). The dependence of the folding time on the protein length (i.e., $N$) suggested by our model is in a qualitatively good agreement with the results previously reported in Refs. [46–48] and obtained by using extensive MC simulations of lattice model proteins. According to the latter (and several other theoretical studies), the folding time as a function of $N$ can be approximated by a power-law function with the exponent ranging from 3 to 6. Our results would suggest the value of 1.1 for this exponent, but more detailed studies of several proteins with different lengths (i.e., $N$s) are needed for more accurate conclusions.

### 3.6.5 Conclusions

We have recently developed [15–19] a kinetic model for the NMPF in terms of ternary nucleation by using the first passage time analysis. The main idea underlying the NMPF model consists of averaging the dihedral potential in which a selected residue is involved over all possible configurations of all neighboring residues along the protein chain. The combination of the average dihedral potential with the effective pairwise potential of a selected residue and with a confining potential caused by the bonds between the residues provides an overall potential around the cluster. As a function of the distance from the cluster center, the overall potential field has a double-well shape. This allows one to develop a self-consistent kinetic theory for the NMPF and evaluate its characteristic time.

In the original NMPF model hydrogen bonding was not taken into account explicitly. To improve the NMPF model, we have developed a PHB model for the effect of hydrogen bond networks of water molecules around two solute particles (immersed in water) on their interaction [28,29]. That model suggests that the disruption of the BHBs, which occurs when the first two hydration shells of two solute particles overlap, results in a short-ranged but strong attraction between the particles.

In this paper, we have combined the PHB [28,29] and NMPF [15–19] models by slightly modifying the former to adapt it to protein folding. The folded cluster of the protein consists of three kinds of residues; hence, its surface can be expected to have a composite (hydrophobic–hydrophilic) character. On the other hand, a single residue in the unfolded part of the protein is either hydrophobic or hydrophilic (neutral residues are treated as hydrophobic). The PHB model has been extended to the solvent-mediated interaction of a spherical particle of composite nature (modeling a cluster of folded protein) with (1) a spherical hydrophobic particle (modeling hydrophobic and neutral protein residues) and (2) a spherical hydrophilic particle (modeling hydrophilic protein residues). In such a way, the water–water hydrogen bonding is taken explicitly into account in a modified kinetic model for the NMPF. The additional contributions to the interaction potentials, arising due to the disruption of hydrogen bond networks around the interacting particles, have been thus added to the overall potential field around a cluster in the NMPF model.

For a numerical illustration we have again applied the model to the folding of two model proteins, one mimicking bovine pancreatic ribonuclease, protein consisting of $N$ = 124 residues whereof $N_b$ = 40 are hydrophobic, $N_l$ = 81 hydrophilic (of which ten are negatively and 18 positively ionizable), and $N_n$ = 3 neutral, and the other—representing large proteins—consisting of $N$ = 2500 residues with $N_b$ = 807, $N_l$ = 1633, $N_n$ = 60 (with 202 negatively and 363 positively ionizable residues). Numerical calculations, performed at pH = 8.3, 7.3, and 6.3, show that in the modified NMPF model the effect of pH on the protein folding time is less pronounced than in the original one (and the smaller the protein, the smaller this effect). This was expected because the contribution of the electrostatic interactions to the overall potential field is less important in the modified model compared to the original one. The hydrogen bond contribution now plays a dominant role in the total potential around the cluster (except for very short distances from the cluster surface where the LJ-type repulsion between the cluster and a residues increases infinitely thus giving rise to the inner wall of the double-well potential field). This contribution is by an order of magnitude stronger for hydrophobic residues than for hydrophilic ones. Besides, for the former the range of residue–cluster distances, at which the contribution exists at all, it is twice as large as for the latter.

In conclusion, one can note that, in principle, a similar approach can be used to develop a model for protein folding in a barrierless way, much like it was used in the model for barrierless protein denaturation [49]. The only difference would be that thermal denaturation is characteristic of most proteins, while fast-folding proteins (i.e., proteins folding in a barrierless way) are rather rare, so it is not straightforward to determine whether the model proposed above would capture this peculiar feature of folding of such rare proteins. A significant computational effort would be required to demonstrate that some specific proteins with specific sets of interaction parameters under specific conditions exhibit barrierless folding.

## Appendix 3K: Explicit Expressions for $V_0(r)$, $V_m(r)$, and $V_s^{r\infty}(r)$

In Equation 3.137, the volume of the region resulting from the overlap of the first two hydration shells of the solutes $R$ and $R'$ depends on whether the latter (i.e., the smaller solute) is hydrophobic or hydrophilic. If it is hydrophobic, then its 1HS is a spherical layer with the inner and outer radii $R'+\frac{1}{2}(\eta+\eta_w)$ and $R'+\frac{1}{2}(\eta+\eta_w)+\frac{1}{2}\eta_w$, respectively, whereas its 2HS is a spherical layer with the inner and outer radii $R'+\frac{1}{2}(\eta+\eta_w)+\frac{1}{2}\eta_w$ and $R'+\frac{1}{2}(\eta+\eta_w)+\frac{3}{2}\eta_w$, respectively. On the other hand, if the smaller solute is hydrophilic, then it does not have 1HS and 2HS, but it has an exclusion sphere of radius $R'+\frac{1}{2}(\eta+\eta_w)$, wherein no water molecules can be located. In both cases, the 1HS and 2HS of the composite solute are determined by three concentric spheres of radii $R+\frac{1}{2}(\eta+\eta_w)$, $R+\frac{1}{2}(\eta+\eta_w)+\frac{1}{2}\eta_w$, and $R+\frac{1}{2}(\eta+\eta_w)+\frac{3}{2}\eta_w$. Therefore, the overlap volume $V_0(r)$ can be found as

$$V_0(r)=\begin{cases} V_{bo}(r)\equiv V_{ex}\left(r,R+\frac{1}{2}[\eta+\eta_w]+\frac{3}{2}\eta_w,R'+\frac{1}{2}[\eta+\eta_w]+\frac{3}{2}\eta_w\right) & \text{(hydrophobic } R'\text{)} \\ V_{lo}(r)\equiv V_{ex}\left(r,R+\frac{1}{2}[\eta+\eta_w]+\frac{3}{2}\eta_w,R'+\frac{1}{2}[\eta+\eta_w]\right) & \text{(hydrophilic } R'\text{)} \end{cases} \tag{3K.1}$$

with

$$V_{ex}(r,a,b)=\frac{\pi}{12r}(a+b-r)(r^2+2ra-3a^2+2rb-3b^2+6ab), \tag{3K.2}$$

determining the volume of the region resulting from the overlap of two spheres of radii $a$ and $b$ as a function of the distance $r$ between them (i.e., between their centers). Likewise the overlap volume $V_m(r)$ (of the 1HSs of the solutes) in Equation 3.139 is determined as

$$V_m(r)=\begin{cases} V_{bm}(r)\equiv V_{ex}\left(r,R+\frac{1}{2}[\eta+\eta_w]+\frac{1}{2}\eta_w,R'+\frac{1}{2}[\eta+\eta_w]+\frac{1}{2}\eta_w\right) & \text{(hydrophobic } R'\text{)} \\ V_{lm}(r)\equiv V_{ex}\left(r,R+\frac{1}{2}[\eta+\eta_w]+\frac{1}{2}\eta_w,R'+\frac{1}{2}[\eta+\eta_w]\right) & \text{(hydrophilic } R'\text{).} \end{cases} \tag{3K.3}$$

In order to calculate the decrease in the total number of BHBs in the first two hydration shells at a distance $r$ between the solutes, it is convenient to classify them as follows. In the vicinity of a hydrophilic solute, all hydrogen bonds are assumed to be the same as in the bulk water. For a hydrophobic particle, a BHB bond can be of type 1 (when both water molecules involved belong to the 1HS) or of type 2 (when one of the water molecules involved belongs to the 2HS). For the composite solute, in addition to types 1 and 2, a BHB bond can be also of type 3 when it is formed with a hydrophilic site on the solute surface. The densities of types 1 and 2 BHBs are different for solutes $R$ and $R'$ not only because they are of different sizes but also because they are of different nature (the smaller solute $R'$ cannot give rise to type 3 bonds at all). They can be all calculated by following the same procedure used in Refs. [28,29]. The simplest way to estimate the average density of type 2 bonds in the 2HSs of the two solutes and the average density of type 1 bonds in the 1HSs of the particles is to take their arithmetic means. One can thus obtain expressions for $v_s^{r\infty}$. If the solute $R'$ is hydrophobic,

$$\begin{aligned} v_s^{r\infty}=\frac{1}{2}n_e\rho_w\Bigg\{&\left(\frac{n_s}{n_e}\chi^{(2)}\frac{V_1}{V_2}+(1-\chi)\frac{V_1'}{V_2'}\right)[V_{bo}(r)-V_{bm}(r)] \\ &+\left(\frac{n_s}{n_e}(0.5\chi^{(1)}+\chi^{(3)})+0.5\chi\right)V_{bm}(r)\Bigg\}, \end{aligned} \tag{3K.4}$$

whereas for the hydrophilic solute $R'$ we have

$$v_s^{r\infty}=n_e\rho_w\left\{\frac{n_s}{n_e}\chi^{(2)}\frac{V_1}{V_2}[V_{lo}(r)-V_{lm}(r)]+\frac{n_s}{n_e}\left(0.5\chi^{(1)}+\chi^{(3)}\right)V_{lm}(r)\right\}. \tag{3K.5}$$

In these expressions, $V_1$ and $V_2$ are the volumes of the 1HS and 2HS of the composite solute; $V_1'$ and $V_2'$ are the volumes of the 1HS and 2HS of the hydrophobic solute $R'$; $\chi$ is the probability that a hydrogen bond, formed by a 1HS molecule in the vicinity of a hydrophobic solute, is a bond of type 1 (previously [28,29], it was estimated to be $\chi \simeq 0.0831685$), whereas $\chi^{(1)}$, $\chi^{(2)}$, and $\chi^{(3)}$ are the probabilities that a hydrogen bond, formed by a 1HS molecule in the vicinity of a composite solute, is of type 1, 2, or 3, respectively. The latter three probabilities can be calculated as

$$\begin{aligned} \chi^{(1)}&=\frac{1}{C}\int_0^{0.5}d\kappa\int_{-\Theta_n(\kappa)}^{\Theta_x(\kappa)}d\Theta\sin\Theta,\ C \\ &=\int_0^{0.5}d\kappa\int_{-\Theta_0(\kappa,\omega_b)}^{\pi/2}d\Theta\sin\Theta, \end{aligned} \tag{3K.6}$$

$$\chi^{(2)} = \frac{1}{C}\int_0^{0.5} d\kappa \int_{\Theta_x(\kappa)}^{\pi/2} d\Theta \sin\Theta, \chi^{(3)}$$
$$= \frac{1}{C}\int_0^{0.5} d\kappa \int_{-\Theta_0(\kappa,\omega_b)}^{-\Theta_n(\kappa)} d\Theta \sin\Theta, \tag{3K.7}$$

with $\kappa = \xi/\eta_w$ ($\xi$ being the distance of the selected molecule from the inner boundary of the 1HS (see Figure 3.30), $\Theta_n(\xi) = \arcsin\kappa$, $\Theta_x(\xi) = \arcsin(0.5 - \kappa)$, and $\Theta_x(\xi) = \arcsin(0.5 - \kappa)$. Numerically, these expressions provide $\chi^{(1)} = 0.057898\,6$, $\chi^{(2)} = 0.638261$, and $\chi^{(3)} = 0.30384$.

It should be noted that in Equations 3K.4 and 3K.5, it is assumed that when the 2HSs of both particles overlap, a single molecule in the shared regions of their 2HSs can form type 2 BHBs only with the 1HS molecules of one solute. Although this assumption may lead to some inaccuracies in the expressions for $v_s^{r\infty}$, its effect on $\phi^{hb}$ can be expected to be significant only at large distances $r > 4\eta$, where the potential $\phi^{hb}$ itself is relatively small. With decreasing $r$ the effect of this assumption on $\phi^{hb}$ will also decrease. Indeed, because of strong orientational restriction on the 1HS molecules, the water molecules belonging to the 1HS of one plate can hardly form type 2 BHBs with the molecules belonging to the 1HS of the other plate. Besides, when one of the particles itself (with an excluded spherical shell of thickness $\eta$) overlap with the 2HS of the other particles, water molecules from the latter are removed without replacement hence type 2 BHBs that they were involved in are broken without the probability of being reformed. Therefore, Equations 3K.4 and 3K.5 can be considered to provide reasonably good estimates for $v_s^{r\infty}$ as a function of $r$.

## Appendix 3L: Evaluation of Protein Folding Time in the Framework of Ternary Nucleation Formalism

Let us use $W_i^- \equiv W_i^-(v_b, v_l, v_n)$ and $W_i^+ \equiv W_i^+(v_b, v_l, v_n)$ ($i$ = b, l, n) to denote the rates of emission and absorption, respectively, of beads of type $i$ by a cluster containing $v_b$ hydrophobic, $v_l$ hydrophilic, and $v_n$ neutral residues. These rates represent the fundamental kinetic characteristics of the protein folding and unfolding processes. At any given temperature, both functions $W^-$ and $W^+$ can be determined by using the first passage time analysis (the method was first [35–37] applied to calculating $W^- = W^-(v)$ in unary nucleation and later extended [15–19] to both $W^-$ and $W^+$ in protein folding and unfolding).

Consider a heteropolymer bead (i.e., a protein residue) of type $i$ ($i$ = b, l, n) performing a chaotic motion in a spherically symmetric potential well $\psi(r)$ with one boundary infinitely high (say, at $r = r_a$) and the other one of finite height (say, at $r = r_b$). The mean first passage time $\tau$ necessary for the molecule to escape from the well is

$$\tau = \frac{1}{Z}\frac{1}{D}\int_{r_a}^{r_b} dr r^2 e^{-\Psi(r)} \int_r^{r_b} dy y^{-2} e^{\Psi(y)} \int_{r_a}^{y} dx x^2 e^{-\Psi(x)}, \tag{3L.1}$$

where $D$ is the diffusion coefficient of a residue, $\Psi(r) = \psi(r)/k_BT$, and

$$Z = \int_{r_a}^{r_b} dr r^2 e^{-\Psi(r)}. \tag{3L.2}$$

The expression for $\tau$ was derived by solving a single-molecule master equation for the probability distribution function of a surface layer molecule (residue or bead) moving in a potential field $\psi(r)$ [15–19]. The diffusive motion of the bead is assumed to be governed by the Fokker–Planck equation [18,39]. The Fokker–Planck equation reduces to the Smoluchowski equation (which involves diffusion in an external field) if the relaxation time for the velocity distribution function of the molecule is very short and negligible compared to the characteristic timescale of the passage process.

The rates of emission and absorption of beads of type $i$ by the cluster (i.e., the numbers of residues of type $i$ escaping from the ipw into the opw and from the opw into the ipw, respectively, per unit time) are provided by

$$W_i^- = \frac{N_i^-}{\tau_i^-}, \quad W_i^+ = \frac{N_i^+}{\tau_i^+}, \tag{3L.3}$$

where $N_i^-$ and $N_i^+$ denote the numbers of molecules in the ipw and opw, respectively, and $\tau_i^-$ and $\tau_i^+$ are the mean first passage times for the transition of a bead of type $i$ from the opw into the ipw and from the ipw into the opw, respectively. Applying Equations 3L.1 through 3L.3 to calculate $W^-$, the locations of the boundaries of the ipw must be used, that is, $r_a = R$ and $r_b = R + \lambda^-$, with $R$ being the radius of the cluster and $\lambda^-$ being the width of the ipw; the diffusion coefficient must be taken to be $D^{iw}$. On the other hand, in calculating $W^+$, the locations of the boundaries of the opw must be used in Equations 3L.1 through 3L.3, that is, $r_a = R + \lambda^- + \lambda^+$ and $r_b = R + \lambda^-$, with $\lambda^+$ being the width of the opw; the diffusion coefficient must be taken to be $D^{ow}$. The quantities $N_i^-$ and $N_i^+$ can be calculated as the product of the "volume × number density,"

$$N_i^- = \frac{4\pi}{3}[(R+\lambda^-)^3 - R^3]\rho_f,$$

$$N_i^+ = \frac{4\pi}{3}[(R+\lambda^- + \lambda^+)^3 - (R+\lambda^-)^3]\rho_u,$$

where $\rho_f$ and $\rho_u$ are the number densities of residues in the folded and unfolded parts of the protein.

Knowing the emission and absorption rates as functions of $\nu_b$, $\nu_l$, $\nu_n$, one can find the nucleation rate $J_s$ and estimate the time $t_f$ necessary for the protein to fold via nucleation. Roughly speaking, the protein folding (via nucleation) consists of two stages. During the first stage, a critical cluster (nucleus) of native residues is formed (nucleation proper). Until the nucleus forms, the emission rate $W^-$ is larger than $W^+$, but the cluster still can attain the critical size and composition by means of fluctuations. At the second stage, the nucleus grows via regular absorption of native residues which dominates their emission, $W^- < W^+$. Thus, the folding time is given by

$$t_f \simeq t_n + t_g, \tag{3L.4}$$

where $t_n$ is the time necessary for one critical cluster to nucleate within a compact (but still unfolded) protein and $t_g$ is the time necessary for the nucleus to grow up to the maximum size, i.e., the size of the entirely folded protein. The time $t_n$ of the first nucleation event can be estimated as

$$t_n \simeq 1/[J_s V_0], \tag{3L.5}$$

where $J_s$ is the steady-state rate of ternary nucleation and $V_0$ is the volume of the unfolded protein in a compact configuration.

The nucleation rate $J_s$ is found by solving the steady-state version of the kinetic equation of ternary nucleation governing the temporal evolution of $g(\nu_b, \nu_l, \nu_n, t)$, the distribution of clusters with respect to their three independent variables of state at time $t$. This equation has to be solved in the vicinity of the saddle point of the function $G(\nu_b, \nu_l, \nu_n) = -k_B T \ln[g_e(\nu_b, \nu_l, \nu_n)/(\rho_{bu} + \rho_{lu} + \rho_{nu})]$, where $g_e(\nu_b, \nu_l, \nu_n)$ is the equilibrium distribution of clusters, which can be constructed once the emission and absorption rates $W_i^- = W_i^-(\nu_b, \nu_l, \nu_n)$ and $W_i^+ = W_i^+(\nu_b, \nu_l, \nu_n)$ are known as functions of $\nu_b$, $\nu_l$, $\nu_n$ (for more details, see Refs. [15–19]). Note that in the first passage time analysis-based approach to the kinetics of multicomponent nucleation the function $G(\nu_b, \nu_l, \nu_n)$ plays a role similar to the free energy of cluster formation in the classical nucleation theory (CNT). It determines a surface in a four-dimensional space that, under appropriate conditions (i.e., high enough metastability of the initial phase), is expected to have the shape of a hyperbolic paraboloid with at least one "saddle" point at the coordinates $\nu_{bc}$, $\nu_{lc}$, $\nu_{nc}$ (hereinafter the subscript "c" marks quantities at the saddle point). The steady-state kinetic equation of nucleation has to be solved subject to two boundary conditions, with one expressing the assumption that small clusters are in equilibrium and the other expressing the absence of too large clusters,

$$J_s = \frac{|\lambda_0|/2\pi k_B T}{\sqrt{-\det(\mathbf{G}''/2\pi k_B T)}} g_e(\nu_{bc}, \nu_{lc}, \nu_{nc}), \tag{3L.6}$$

where $\mathbf{G}''$ is the matrix of second derivatives of the function $G(\nu_b,\nu_l,\nu_n)$ with respect to $\nu_b,\nu_l,\nu_n$ and $\lambda_0$ is a negative eigenvalue of the matrix $\mathbf{A}{\cdot}\mathbf{G}''$, with the elements of the diagonal matrix $\mathbf{A}$ of the absorption rates given by $A_{ij} = \delta_{ij} W_i^+ \ (i, j = \mathrm{b}, \mathrm{l}, \mathrm{n})$, with $\delta_{ij}$ being the Kronecker delta. Note that although the form of the expression for $J_s$ in Equation 3L.6 is identical to that in Ref. [50], the latter was obtained (and applied to binary and ternary nucleation) in the framework of CNT. The crucial difference is hidden in the method for obtaining the equilibrium distribution $g_e(\nu_b, \nu_l, \nu_n)$. In the kinetic approach to a nucleation theory (originally proposed in Refs. [35–37]) it is obtained by using the mean first passage time analysis and the principle of detailed balance, while in CNT (whereupon its use had been previously based) the equilibrium distribution would have the form $(\rho_{bu} + \rho_{lu} + \rho_{nu})\exp[-F(\nu_b, \nu_l, \nu_n)]$, where $F(\nu_b, \nu_l, \nu_n)$ is the free energy of formation of a cluster, derived by using the concept of surface tension.

The growth time $t_g$ is provided by the integral

$$t_g \simeq \int_{\nu_{bc}}^{N_b} \frac{d\nu}{W_b^+(\nu_b, \nu_l(\nu_b), \nu_n(\nu_b)) - W_b^-(\nu_b, \nu_l(\nu_b), \nu_n(\nu_b))}. \tag{3L.7}$$

Here, the functions $\nu_l = \nu_l(\nu_b)$ and $\nu_n = \nu_n(\nu_b)$ determine the growth path in parametric form and can be found as the solution of a couple of simultaneous differential equations (see Refs. [15–19]):

$$\frac{d\nu_l}{d\nu_b} = \frac{W_l^+(\nu_b, \nu_l, \nu_n) - W_l^-(\nu_b, \nu_l, \nu_n)}{W_b^+(\nu_b, \nu_l, \nu_n) - W_b^-(\nu_b, \nu_l, \nu_n)},$$

$$\frac{d\nu_n}{d\nu_b} = \frac{W_n^+(\nu_b, \nu_l, \nu_n) - W_n^-(\nu_b, \nu_l, \nu_n)}{W_b^+(\nu_b, \nu_l, \nu_n) - W_b^-(\nu_b, \nu_l, \nu_n)}.$$

## References


1. Creighton, T.E., *Proteins: Structure and Molecular Properties*, W.H. Freeman, San Francisco, 1984.
2. Stryer, L., *Biochemistry*, 3rd ed., W.H. Freeman, San Francisco, 1988.
3. Anfinsen, C.B., *Science* 181, 223 (1973).
4. Ghelis, C., Yan, J., *Protein Folding*, Academic Press, New York, 1982.
5. Nölting, B., *Protein Folding Kinetics*, Springer-Verlag, Berlin, 2006.
6. Honeycutt, J.D., Thirumalai, D., *Biopolymers* 32, 695 (1992).
7. Weissman, J.S., Kim, P.S., *Science* 253, 1386 (1991).
8. Creighton, T.E., *Nature (London)* 356, 194 (1992).
9. Fersht, A.R., *Curr. Opin. Struct. Biol.* 7, 3 (1997).
10. Guo, Z., Thirumalai, D., *Biopolymers* 36, 83 (1995).
11. Abkevich, V.I., Gutin, A.M., Shakhnovich, E.I., *Biochemistry* 33, 10026 (1994).
12. Bryngelson, J.D., Wolynes, P.G., *Proc. Natl. Acad. Sci. U.S.A.* 84, 7524 (1987).
13. Bryngelson, J.D., Wolynes, P.G., *J. Phys. Chem.* 93, 6902 (1989).
14. Bryngelson, J.D., Wolynes, P.G., *Biopolymers* 30, 177 (1990).
15. Djikaev, Y.S., Ruckenstein, E., *J. Phys. Chem. B* 111, 886 (2007). (Section 3.1 of this volume.)
16. Djikaev, Y.S., Ruckenstein, E., *J. Chem. Phys.* 126, 175103 (2007). (Section 3.2 of this volume.)
17. Djikaev, Y.S., Ruckenstein, E., *J. Chem. Phys.* 128, 025103 (2008). (Section 3.3 of this volume.)

18. Djikaev, Y.S., Ruckenstein, E., *Phys. Chem. Chem. Phys.* 10, 6281 (2008). (Section 3.4 of this volume.)
19. Djikaev, Y.S., Ruckenstein, E., *Adv. Colloid Interface Sci.* 146, 18 (2009). (Section 3.5 of this volume.)
20. Shakhnovich, E.I., Gutin, A.M., *J. Phys. A* 22, 1647 (1989).
21. Bratko, D., Chakraborty, A.K., Shakhnovich, E.I., *J. Chem. Phys.* 106, 1264 (1997).
22. Konkoli, Z., Hertz, J., Franz, S., *Phys. Rev. E* 64, 051910 (2001).
23. Kauzmann, W., *Adv. Protein Chem.* 14, 1 (1959).
24. Privalov, P.L., *Crit. Rev. Biochem. Mol. Biol.* 25, 281 (1990).
25. Oleinikova, A., Smolin, N., Brovchenko, I., Geiger, A., Winter, R., *J. Phys. Chem. B* 109, 1988 (2005).
26. Koizumi, M., Hirai, H., Onai, T., Inoue, K., Hirai, M., *J. Appl. Crystallogr.* 40, s175 (2007).
27. Brovchenko, I., Krukau, A., Smolin, N., Oleinikova, A., Geiger, A., Winter, R., *J. Chem. Phys.* 123, 224905 (2005).
28. Djikaev, Y.S., Ruckenstein, E., *J. Chem. Phys.* 130, 124713 (2009).
29. Djikaev, Y.S., Ruckenstein, E., *J. Colloid Interface Sci.* 336, 575 (2009).
30. Silverstein, K.A.T., Haymet, A.D.J., Dill, K.A., *J. Chem. Phys.* 111, 8000 (1999).
31. Blokzijl, W., Engberts, J.B.F.N., *Angew. Chem., Int. Ed. Engl.* 32, 1545 (1993).
32. Meng, E.C., Kollman, P.A., *J. Phys. Chem.* 100, 11460 (1996).
33. Müller, N., *Acc. Chem. Res.* 23, 23 (1990).
34. Lee, B., Graziano, G., *J. Am. Chem. Soc.* 118, 5163 (1996).
35. Narsimhan, G., Ruckenstein, E., *J. Colloid Interface Sci.* 128, 549 (1989). (Section 1.2 of this volume.)
36. Ruckenstein, E., Nowakowski, B., *J. Colloid Interface Sci.* 137, 583 (1990). (Section 1.3 of this volume.)
37. Nowakowski, B., Ruckenstein, E., *J. Colloid Interface Sci.* 139, 500 (1990). (Section 1.4 of this volume.)
38. Chandrasekhar, S., *Rev. Mod. Phys.* 15, 1 (1943).
39. Gardiner, C.W., *Handbook of Stochastic Methods*, Springer, New York, 1983.
40. Kubelka, J., Hofrichter, J., Eaton, W.A., *Curr. Opin. Struct. Biol.* 14, 76 (2004).
41. Huang, F., Sato, S., Sharpe, T.D., Ying, L.M., Fersht, A.R., *Proc. Natl. Acad. Sci. U.S.A.* 104, 123 (2007).
42. Elliott, W.H., Elliott, D.C., *Biochemistry and Molecular Biology*, Oxford University Press, New York, 2003.
43. Schmidt, C.L.A., Kirk, P.L., Appleman, W.K., *J. Biol. Chem.* 88, 285 (1930).
44. Xu, G., Narayan, M., Scheraga, H.A., *Biochemistry* 44, 9817 (2005).
45. Pradeep, L., Shin, H.-C., Scheraga, H.A., *FEBS Lett.* 580, 5029 (2006).
46. Gutin, A.M., Abkevich, V.I., Shakhnovich, E.I., *Phys. Rev. Lett.* 77, 5433 (1996).
47. Li, M.S., Klimov, D.K., Thirumalai, D., *J. Phys. Chem. B* 106, 8302 (2002).
48. Cieplak, M., Hoang, T.X., Li, M.S., *Phys. Rev. Lett.* 83, 1684 (1999).
49. Djikaev, Y.S., Ruckenstein, E., *J. Chem. Phys.* 131, 045105 (2009).
50. Trinkaus, H., *Phys. Rev. B* 27, 7372 (1983).

# 4

# Kinetics of Cluster Growth on a Lattice

Until now, the understanding of the processes of phase transformations that involve nucleation and growth is far from being complete, even though much effort has been made in their investigation [1,2]. The complicated nature of these processes forces one to use a number of simplifications that allow us to obtain approximate solutions. The classical approach [1,2], based on the Becker–Döring equations, treated the dynamics of the number density of clusters of a new phase in the space of their size and used a thermodynamic approach to calculate the size of a critical cluster above which the clusters grow rapidly. In this way, the nucleation rate and time dependence of the cluster distribution function could be calculated by neglecting the splitting and coagulation of clusters and assuming spherical shapes. The kinetic approach to the problem, described in the preceding sections, is free of macroscopic thermodynamic arguments. However, the shape of the clusters was assumed spherical.

Another treatment of the kinetics of phase transformation is based on lattice models, which have been used to analyze the general properties of nucleation [3–9], the metastable states [10–12], and the cluster size distribution [13,14]. The intermolecular interactions in this treatment are limited to the molecules located only on the neighboring sites of the lattice and the state of the system is described by the Ising Hamiltonian,

$$H_s = -B\sum_j \sigma_j - J\sum_{\langle j,k\rangle} \sigma_j \sigma_{j+\Delta_k} \quad (J > 0),$$

where the variables $\sigma_j = \pm 1$ (spins) represent the two different phases (e.g., solid and liquid) to which a molecule at a site $j$ of a rigid lattice can belong, the parameter $B$ (the external field) represents the supersaturation $S$ ($B \propto \log S$) [8], and $J$ (the nearest-neighbor interaction energy, called below the energy) represents the intermolecular interactions. The subscript $\Delta_k$ refers to the $m$ neighboring lattice sites of site $j$, and $\langle j, k\rangle$ indicates that each pair of neighboring spins should be counted in the sum only once. Choosing appropriate rates of spin flips (liquid–solid transitions), which satisfy the detailed balance condition, and starting from a metastable initial state with almost all spins "down" (liquid state), the transition of the system to the final equilibrium state with all spin "up" (solid state) was examined with Monte Carlo simulations [3–7,12,13], and numerous results were obtained. For a theoretical interpretation of the simulation results, the classical theory was mostly used [5,6,13]. Attempts to supersede the last theory are demonstrated in Sections 4.1 through 4.3, where instead of the Becker–Döring equations, which imply a size distribution of clusters of a single shape (spherical), an Ising model on a two-dimensional lattice involving all the possible shapes of small clusters is considered

Another analytical approach to describe the phase transformation is based on the Glauber master equation [15], which treats the behavior of an Ising system in contact with a heat bath. This equation has the form

$$\frac{d}{dt}P(\sigma_1,\sigma_2,\ldots,\sigma_N;t) = -\sum_j W(\sigma_j)P(\sigma_1,\ldots,\sigma_j,\ldots,\sigma_N;t) + \sum_j W(-\sigma_j)P(\sigma_1,\ldots,-\sigma_j,\ldots,\sigma_N;t),$$

where $P(\sigma_1, \sigma_2,\ldots,\sigma_N; t)$ is the probability of finding an Ising spin system at time $t$ in the state $\{\sigma_1, \sigma_2,\ldots, \sigma_N\}$, $N$ is the total number of spins in the system, $W(\sigma_j)$ is the transition rate of spin $\sigma_j$ between its states $\sigma_j = 1$ and $\sigma_j = -1$, $W(\sigma_j)$ depends only on the states of the molecules located at site $j$ and on its nearest neighboring sites (i.e., on the values of the spin $\sigma_j$ and its neighboring spins), regardless of the kind of cluster (a group of bounded molecules in the same state or, in other words, a group of equally directed bounded spins) to which the neighboring molecules belong. For the latter reason, the Glauber master equation accounts implicitly for the splitting and coagulation of clusters of arbitrary shape, and this constitutes its main advantage compared with the classical Becker–Döring approach. The main disadvantage of the Glauber equation is that it leads to an infinite chain of coupled differential equations for the time-dependent spin correlation functions and exact solutions were obtained only for the case of zero external field [15,16].

In most cases, a mean-field or other similar approximations [17,18] were used to close the chain by representing the multiparticle correlation functions as products of lower order ones.

In Sections 4.4 through 4.7, this approach is applied to the Ising model on a one-dimensional lattice and on a Bethe lattice. In both cases, an exact analytical solution can be obtained at equilibrium and thus the method can be well illustrated. In addition to the rate of phase transformation, which coincides in the model with the relaxation rate of the polarization, the size distribution of the clusters of the "up" spins and time dependence of the cluster size are also calculated.

## References

1. Frenkel, J., *Kinetic Theory of Liquids*, Dover, New York, 1955.
2. Abraham, F.F., *Homogeneous Nucleation Theory*, Academic, New York, 1974.
3. Novotny, M.A., Brown, G., Rikvold, P.A., Large-scale computer investigations of finite-temperature nucleation and growth phenomena in magnetization reversal and hysteresis, *J. Appl. Phys.* 91, 6908–6913 (2002).
4. Tafa, K., Puri, S., Kumar, D., Kinetics of phase separation in ternary mixtures, *Phys. Rev. E* 64, 056139 (2001).
5. Schmelzer, J., Jr., Landau, D.P., Monte Carlo simulation of nucleation and growth in the 3D nearest-neighbor Ising model, *Int. J. Mod. Phys. C*, 12, 345–359 (2001).
6. Shneidman, V.A., Jackson, K.A., Beatty, K.M., Non-equilibrium interface of a two-dimensional low-temperature crystal, *J. Cryst. Growth* 212, 564–573 (2000).
7. Vehkamaki, H., Ford, I.J., Nucleation theorems applied to the Ising model, *Phys. Rev. E* 59, 6483–6488 (1999).
8. Wonczak, S., Strey, R., Stauffer, D., Confirmation of classical nucleation theory by Monte Carlo simulations in the 3-dimensional Ising model at low temperature, *J. Chem. Phys.* 113, 1976–1980 (2000).
9. Marchand, J.P., Martin, P.A., A microscopic derivation of the classical nucleation equation *Physica A* 127, 681–691 (1984).
10. Shneidman, V.A., Nita, G.M., Modulation of the nucleation rate preexponential in a low-temperature Ising system, *Phys. Rev. Lett.* 89, 025701 (2002).
11. Bovier, A., Manzo, F., Metastability in Glauber dynamics in the low-temperature limit: Beyond exponential asymptotics, *J. Stat. Phys.* 107, 757–779 (2002).
12. Park, K., Novotny, M.A., Rikvold, P.A., Scaling analysis of a divergent prefactor in the metastable lifetime of a square-lattice Ising ferromagnet at low temperatures, *Phys. Rev. E* 66, 056101 (2002).

13. Shneidman, V.A., Jackson, K.A., Beatty, K.M., On the applicability of the classical nucleation theory in an Ising system, *J. Chem. Phys.* 111, 6932–6941 (1999).
14. Ben-Naim, E., Krapivsky, P.L., Domain number distribution in the nonequilibrium Ising model, *J. Stat. Phys.* 93, 583–601 (1998).
15. Glauber, R., Time dependent statistics of the Ising model, *J. Math. Phys.* 4, 294–307 (1963).
16. Felderhof, B.U., Spin relaxation of the Ising chain, *Rep. Math. Phys.* 1, 215–234 (1971).
17. Suzuki, M., Kubo, R., Dynamics of Ising model near critical point: I. *J. Phys. Soc. Jpn.* 24, 51–60 (1968).
18. Binder, K., Müller-Krumbhaar, H., Investigation of metastable states and nucleation in kinetic Ising-model. *Phys. Rev. B* 9, 2328–2353 (1974).

## 4.1 Influence of Cluster Shape upon Its Growth in a Two-Dimensional Ising Model

*Gersh Berim and Eli Ruckenstein**

The formalism of the kinetic Ising model was used to investigate the initial growth of clusters with various number of spins and shapes in the two-dimensional Ising model on a square lattice. A general expression that provides the initial growth rate of a cluster was derived and applied to clusters of various shapes. For the simple shapes, such as rectangular and triangular, the problem was treated analytically for any cluster size and expressions for the shape-dependent critical size above which clusters of given shape initially grow derived. To analyze the case of arbitrary shapes, all possible configurations of clusters with up to 10 spins were generated by computer, and their initial growth rate was investigated numerically. It was shown that the initial growth rate of a cluster depends not only on its size but also on the cluster shape. For a given shape, there is a critical size above which the cluster initially grows. Because of this initial growth, the cluster can change its shape and will continue to grow only if its size is greater than the critical size for the new shape, and so on. (© *2002 American Institute of Physics.* doi: 10.1063/1.1497639.)

### 4.1.1 Introduction

It is a common assumption in the theory of nucleation that the nuclei have a spherical shape and hence that the nucleous growth rate depends only on the number of molecules it contains [1–4]. With such an assumption relatively simple equations, such as, for example, the Becker–Döring ones were derived [1], from which the critical size, i.e., the size above which a nucleous grows, and the rate of nucleation could be calculated. However, Monte Carlo simulations regarding the nucleation in an Ising model [5,6] as well as calculations involving the Wulff construction [7] showed that the shapes of clusters in nonequilibrium as well as in equilibrium states are not necessarily spherical. The consequence is that the coefficients in the Becker–Döring type of equations must depend not only on size but also on the cluster shape. In the present paper, we examine this issue for a two-dimensional Ising spin system interacting with a heat bath. The nonequilibrium behavior of such systems can be described by the Glauber kinetic equation [8], which for a long time was used to examine the cluster growth and nucleation processes [5,6,9–11]. We define a cluster as a group of bounded "up" spins in a sea of "down" spins and examine the initial evolution of a single cluster, neglecting the formation of other clusters inside the system and their coagulation. Such a restriction is reasonable when the concentration of clusters in the system is small.

### 4.1.2 The Model and the Basic Kinetic Equations

The Hamiltonian of the model under consideration has the following general form:

$$H = -h\sum_{j=1}^{N_{\text{tot}}}\sigma_j - \sum_{j=1}^{N_{\text{tot}}}\sum_{k} J_{\delta_k}\sigma_j\sigma_{j+\delta_k}\ (h>0), \tag{4.1}$$

where the spins $\sigma_j$ are located on the sites of a square lattice and have the values ±1, the subscript $\delta_k$ indicates the positions of the neighboring spins interacting with a spin $\sigma_j$ and $N_{\text{tot}}$ is the total number of spins. Each pair of interacting spins is counted only once. The Ising model was applied to the description of a wide range of systems, e.g., magnetic systems [12,13], polymers [14], and binary mixtures [15,16].

The physical meaning of the parameters $h$ and $J_{\delta_k}$, which have the dimension of energy, depends on the nature of the system under consideration. In the following, we will follow the tradition and will call $J_{\delta_k}$ "exchange integral" and $h$ "magnetic field." We will express $h$ and $J_{\delta_k}$ in units of the Boltzmann constant; consequently, they will have the dimension of temperature. The treatment can be easily extended to the two-dimensional liquid–solid nucleation.

To avoid unnecessary repetitions we will review only the main steps of the derivation of the Glauber kinetic equation [8,17,18]. The state of a spin system can be described by the probability $P(\sigma_1,\ldots,\sigma_j,\ldots,\sigma_{N_{\text{tot}}};t)$ to find an Ising system at time $t$ in the state $\{\sigma_1,\ldots,\sigma_j,\ldots,\sigma_{N_{\text{tot}}}\}$, and the time-dependent average of an arbitrary spin operator $A$ is given by

* *J. Chem. Phys.* 117, 4542 (2002). Republished with permission.

$$\langle A \rangle_t \equiv \sum_{\{\sigma\}} AP\left(\sigma_1,\ldots,\sigma_j,\ldots,\sigma_{N_{\text{tot}}};t\right). \tag{4.2}$$

The interaction of any spin $\sigma_j$ of the Ising system (Equation 4.1) with its neighbors and with an external field can be considered as its interaction with a local field $E_j$, which is given by

$$E_j = h + \sum_{k=1}^{z} J_k \sigma_{j+\delta_k}, \tag{4.3}$$

where $z$ is the number of spins interacting with $\sigma_j$. In the general case, this local field has $2^z$ different values $\varepsilon_\mu$ ($\mu = 1, 2, \ldots, 2^z$), corresponding to various combinations of signs in the expression $\varepsilon_\mu = \pm J_1 \pm J_2 \pm \cdots \pm J_z + h$.

Due to their interaction with a heat bath, all spins are subjected to flip-flop transitions, whose probability densities

$$W(\sigma_j) = \frac{1}{2} W_{\varepsilon_\mu}[1 - \sigma_j \tanh(\beta\varepsilon_\mu)] \tag{4.4}$$

satisfy the detailed balance condition [8]. Here, $\beta = 1/T$ and $T$ is the temperature in K. It is usually supposed [8,17,19] that $W_{\varepsilon_\mu}$ is a constant, that provides the timescale of the process. However, generally, $W_{\varepsilon_\mu}$ can depend on the details of the interactions between the system of spins and the heat bath and can be a function of $\varepsilon_\mu$ (i.e., a function of $\sigma_{j+\delta_k}$) [18,20]. In such cases, one can write $W_{\varepsilon_\mu} = W_\mu R_j(\mu)$, where $R_j(\mu)$ is equal to unity when the spins interacting with spin $\sigma_j$ create a local field equal to $\varepsilon_\mu$ and zero otherwise. Introducing these transition probabilities in the standard master equation [8,17] for $P\left(\sigma_1,\ldots,\sigma_j,\ldots,\sigma_{N_{\text{tot}}};t\right)$, one obtains the following kinetic equation for the time-dependent average $\langle\sigma_j\rangle_t$:

$$\frac{\mathrm{d}}{\mathrm{d}t}\langle\sigma_j\rangle_t = -\sum_{\mu} W_\mu[\langle\sigma_j R_j(\mu)\rangle_t - \langle R_j(\mu)\rangle_t \tanh(\beta\varepsilon_\mu)]. \tag{4.5}$$

In what follows, a kinetic equation for the total number of "up" spins in the lattice will be needed. The operator $n_{\text{total}}$ for such a quantity can be expressed in terms of the spin operators as

$$n_{\text{total}} = \sum_j \frac{1+\sigma_j}{2}. \tag{4.6}$$

Consequently

$$\frac{\mathrm{d}}{\mathrm{d}t}\langle n_{\text{total}}\rangle_t = -\frac{1}{2}\sum_j\sum_\mu W_\mu[\langle\sigma_j R_j(\mu)\rangle_t - \langle R_j(\mu)\rangle_t \tanh(\beta\varepsilon_\mu)]. \tag{4.7}$$

Since we consider a single cluster on the lattice, we suppose that in the initial state all spins outside the cluster have the value −1 ("down spins") and those in the cluster +1. To prohibit the creation of new clusters and flip-flop transitions inside the cluster, we exclude, as in Ref. [11], the flip-flop transitions of spins that interact with the equally directed nearest-neighbor spins (all these are "down" or "up"). For this reason, some of the probability densities $W_\mu$, which will be specified below, are equal to zero.

Under the above conditions, $\langle n_{\text{total}}\rangle_t$ provides the size of the selected cluster and $\mathrm{d}/\mathrm{d}t\langle n_{\text{total}}\rangle_t$ the rate of its size change. It should be emphasized that Equation 4.7 is not closed and therefore the time dependence of $\langle n_{\text{total}}\rangle_t$ cannot be obtained from it. However, at time $t = 0$, the right-hand side of Equation 4.7 can be calculated using the initial conditions. As a consequence, the derivative

$$n_1 = \frac{\mathrm{d}}{\mathrm{d}t}\langle n_{\text{total}}\rangle_t\Big|_{t=0} \tag{4.8}$$

can be obtained.

In this paper, the attention will be focused on the value of $n_1$, which provides the rate of growth of the cluster in the vicinity of time $t = 0$. The positive (negative) sign of $n_1$ means that the considered cluster tends to gain (lose) one or more spins, i.e., it grows (decays) at the initial moment. Evidently, the value of $n_1$ depends on the cluster shape and size (number of spins) and for each shape there is a size at which $n_1$ changes its sign. This size will be called shape-dependent critical size (SCS). The spectrum of SCSs can be considered as a characteristic of the ensemble of clusters present in the lattice at any nonzero temperature. Such an ensemble contains subensembles of clusters of different shapes, for instance, the subensemble of square clusters of different sizes. If one of the square clusters gains (loses) spins, it most likely leaves its subensemble and enters in another one, with clusters of another shape. The SCS of square clusters represents the size above (below) which a square cluster gains (loses) at the initial moment one or several spins and changes to a cluster, belonging to another shape, which has a larger (smaller) number of spins than the SCS of the initial square cluster and which can grow (decay) if its size is larger (smaller) than the SCS of the new shape, and so on. Thus, the shape-dependent critical cluster size differs from the "critical cluster size," which is used in the conventional nucleation theory, because it takes into account the shape of the cluster. However, if one assumes that all the clusters have the same shape, the two kinds of critical cluster sizes coincide.

Using Equation 4.7, Equation 4.8 becomes

$$n_1 = -\frac{1}{2}\sum_j\sum_\mu W_\mu[\langle\sigma_j R_j(\mu)\rangle_0 - \langle R_j(\mu)\rangle_0 \tanh(\beta\varepsilon_\mu)]. \tag{4.9}$$

To determine $n_1$, one must specify the characteristics of the initial cluster of "up" spins. Let us suppose that this cluster contains $N_C$ spins, $N_B$ of which are boundary spins (i.e., they have at least one neighboring "down" spin), and has $N_N$ neighbors (i.e., "down" spins that have at least one neighboring "up" spin). The manifolds of these spins will be denoted as $\{C\}$, $\{B\}$, and $\{N\}$, respectively. As an example, let us consider a simple cluster of eight "up" spins as shown in Figure 4.1. In this case, all the "up" spins constitute the manifold $\{C\}$ with $N_C = 8$, the squared ones the manifold $\{B\}$, with $N_B = 7$ and the circled "down" spins the manifold $\{N\}$, with $N_N = 11$. Due to the exclusion of some flip-flop transitions mentioned before and the properties

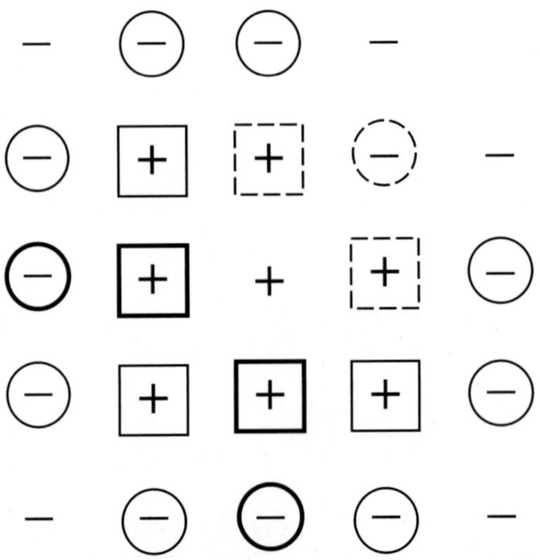

Figure 4.1

Example of eight spins cluster. The squared spins constitute the cluster boundary, the circled ones the cluster neighbors. The bold line squared spins, interacting with three cluster spins belong to the flat part of the boundary; the dashed line squared spins, interacting with two cluster spins belong to the rough part of the boundary.

of operators $R_j(\mu)$, only the boundary and the neighboring spins contribute to the initial growth rate of the cluster, and Equation 4.9 can be rewritten in the form

$$n_1 = -\frac{1}{2}\sum_{\mu} W_\mu \sum_{j\in\{B\}} \langle R_j(\mu)\rangle_0[1-\tanh(\beta\varepsilon_\mu)] + \frac{1}{2}\sum_{\mu} W_\mu \sum_{j\in\{N\}} \langle R_j(\mu)\rangle_0[1+\tanh(\beta\varepsilon_\mu)]. \quad (4.10)$$

Let us denote $n_{\mathrm{B}}^{\mu}\left(n_{\mathrm{N}}^{\mu}\right)$ the number of boundary (neighboring) cluster spins that are subjected to the local field $\varepsilon_\mu$. Then, the final expression of $n_1$ becomes

$$n_1 = \frac{1}{2}\sum_{\mu} W_\mu \left\{ n_{\mathrm{N}}^{\mu}[1+\tanh(\beta\varepsilon_\mu)] - n_{\mathrm{B}}^{\mu}[1-\tanh(\beta\varepsilon_\mu)]\right\}. \quad (4.11)$$

The first term in the curly brackets of Equation 4.11 represents the cluster growth at $t = 0$ due to the addition of neighboring spins to the cluster, and the second the cluster decay at $t = 0$ due to the removal of boundary spins. The right-hand side of this equation is a function of temperature, magnetic field and exchange integral.

For illustration purposes, all nonzero probability densities $W_\mu$ will be selected equal to 2. This selection does not affect the SCS values, which are provided by the equation $n_1 = 0$. In this particular case,

$$n_1 = N_{\mathrm{N}} - N_{\mathrm{B}} + \sum_{\mu}\left(n_{\mathrm{N}}^{\mu} + n_{\mathrm{B}}^{\mu}\right)\tanh(\beta\varepsilon_\mu). \quad (4.12)$$

Since $N_{\mathrm{N}} \geq N_{\mathrm{B}}$, one can immediately conclude that at infinite temperature ($\beta\varepsilon_\mu = 0$) as well as for those values of $h$ and $J_{\delta_k}$, for which $\varepsilon_\mu \geq 0$ for any of the values of $\mu$, the initial growth rate $n_1$ is positive. Hence, in these two ranges, all clusters of any shape and size will grow at the initial moment.

Below, we examine in more details two Ising models:

a. The model involving only the nearest-neighbor exchange interactions $J_{\mathrm{n}}$
b. The model involving interactions between nearest neighbors, next-nearest neighbors, and nearest neighbors in the diagonal directions, characterized by the exchange integrals $J_{\mathrm{n}}$, $J_{\mathrm{nn}}$, and $J_{\mathrm{d}}$, respectively

In model (a), the local field $E_j$ has the form (where the subscript $j$ was changed to the coordinates $m,l$ of the lattice sites)

$$E_{m,l} = h + J_{\mathrm{n}}\,(\sigma_{m+1,l} + \sigma_{m-1,l} + \sigma_{m,l+1} + \sigma_{m,l-1}) \quad (4.13)$$

and has the following five possible values:

$$\varepsilon_\mu = h + \mu J_{\mathrm{n}},\ \mu = 0, \pm 2, \pm 4. \quad (4.14)$$

For model (b), it is convenient to consider $\mu$ as a vector,

$$\mu \equiv \{\mu_1, \mu_2, \mu_3\},\ \mu_i = 0, \pm 2, \pm 4,\ i = 1, 2, 3, \quad (4.15)$$

and write the 125 possible values of the local field in the form

$$\varepsilon_\mu = h + \mu_1 J_{\mathrm{n}} + \mu_2 J_{\mathrm{d}} + \mu_3 J_{\mathrm{nn}},\ \mu_i = 0, \pm 2, \pm 4. \quad (4.16)$$

To prohibit the generation of new clusters and flip-flop transitions inside the cluster, one must take

$$W_4 = W_{-4} = 0 \quad (4.17)$$

for model (a) and

$$W_\mu = 0, \mu = \{\pm 4, \mu_2, \mu_3\}, \mu_i = 0, \pm 2, \pm 4\ (i = 2, 3) \tag{4.18}$$

for model (b).

### 4.1.3 Clusters with Simple Shape in the Ising Model with Nearest Neighbor Interactions

In this section, the initial growth rate of clusters with simple shapes in a two-dimensional Ising model with nearest-neighbor interactions is investigated. Only the clusters with rectangular, square, isosceles right angle triangular, zigzag, and rough boundary shapes (see Figure 4.2) are considered, and their excitation energy $E$ and initial growth rate $n_1$ as functions of the parameters of the Hamiltonian and temperature are calculated. The excitation energy is defined as the energy difference between the state of a system containing a cluster of "up" spins and all other "down" and one with all spins "down." It depends, generally, on the cluster shape.

The quantities in Equation 4.12 that are necessary to calculate the initial growth rate and the excitation energy are listed in Tables 4.1 and 4.2.

In the following, we provide expressions for the initial growth rate, the SCS and the excitation energy for clusters of different shapes.

#### 4.1.3.1 Rectangular Cluster

The excitation energy of this cluster is given by

$$E^{\text{rect}} = 4(M + L)J_n - 2MLh \tag{4.19}$$

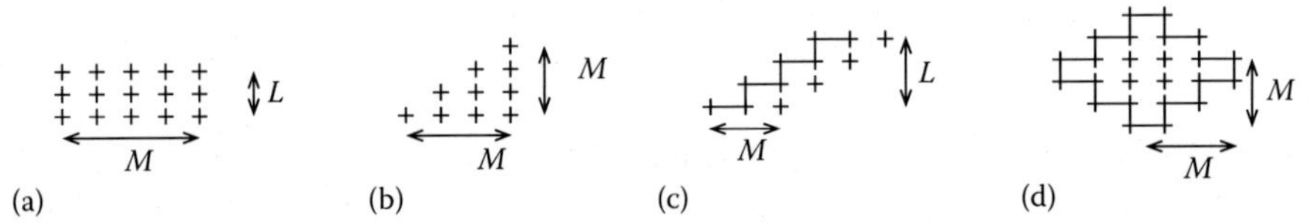


Figure 4.2

Clusters of different form (a) rectangular, (b) triangular, (c) zigzag, (d) rough boundary.

Table 4.1 Number $N_C$ of Spins in a Cluster and $N_B$ and $N_N$ (in Equation 4.12) for Clusters Presented in Figure 4.2

| Shape | $N_C$ | $N_B$ | $N_N$ |
|---|---|---|---|
| Rectangular | $ML$ | $2(M + L - 2)$ | $2(M + L)$ |
| Square ($M = L$) | $M^2$ | $4(M - 1)$ | $4M$ |
| Triangle | $M(M + 1)/2$ | $3(M - 1)$ | $3M + 1$ |
| Zigzag | $ML$ | $2(M + L - 2)$ | $2(M + L)$ |
| Zigzag ($M = L$) | $M^2$ | $4(M - 1)$ | $4M$ |
| Rough boundary | $2M(M + 1)$ | $4M$ | $4(M + 1)$ |

Table 4.2 The $n^\mu$ (in Equation 4.12) for Clusters Presented in Figure 4.2

| Shape | $n_B^{-2}$ | $n_B^0$ | $n_B^{+2}$ | $n_N^{-2}$ | $n_N^0$ | $n_N^{+2}$ |
|---|---|---|---|---|---|---|
| Rectangular | 0 | 4 | $2(M + L - 4)$ | $2(M + L)$ | 0 | 0 |
| Square ($M = L$) | 0 | 4 | $2(M - 2)$ | $4M$ | 0 | 0 |
| Triangle | 2 | $M - 1$ | $2(M - 2)$ | $2M + 2$ | $M - 1$ | 0 |
| Zigzag | 2 | $2(L - 1)$ | $2(M - 2)$ | $2(M + 1)$ | $2(L - 1)$ | 0 |
| Zigzag ($M = L$) | 2 | $2M - 2$ | $2M - 4$ | $2(M + 1)$ | $2(M - 1)$ | 0 |
| Rough boundary | 0 | $4M$ | 0 | 8 | $4(M - 1)$ | 0 |

and the initial growth rate $n_1$ by

$$n_1^{\text{rect}} = 4\left\{1+\tanh\beta\varepsilon_0 - 2\tanh\beta\varepsilon_{+2} + \frac{M+L}{2}[\tanh\beta\varepsilon_{+2} + \tanh+\beta\varepsilon_{-2}]\right\}, \tag{4.20}$$

where $\varepsilon_\mu$ is given by Equation 4.14. Because for $h > 0$, $\tanh\beta\varepsilon_{+2} + \tanh\beta\varepsilon_{-2} \geqq 0$, Equation 4.20 shows that a rectangular cluster of any size will grow at $t = 0$ if

$$1 + \tanh\beta\varepsilon_0 - 2\tanh\beta\varepsilon_{+2} > 0. \tag{4.21}$$

One can therefore conclude that, in addition to the cases of infinite temperature and $\varepsilon_\mu \geq 0$, there is a new range of initial growth at any size, provided by the inequality 4.21.

If condition 4.21 is not satisfied, then there is a SCS for which the geometrical characteristics can be found from equation $n_1 = 0$. They are related via the expression

$$\frac{1}{2}(M_{\text{crit}} + L_{\text{crit}}) = \frac{2\tanh\beta\varepsilon_{+2} - \tanh\beta\varepsilon_0 - 1}{\tanh\beta\varepsilon_{+2} + \tanh\beta\varepsilon_{-2}}. \tag{4.22}$$

Below we will consider the particular case $\beta J_n \gg 1$, $J_n \gg h$, and $\beta h \ll 1$, which implies low temperatures and low magnetic fields. These conditions will be denoted as the "LT–LMF" range. The geometrical characteristics of the SCS of the rectangular cluster are related in this case through the expression

$$M_{\text{crit}} + L_{\text{crit}} \simeq \frac{1}{4\beta h}\exp(4\beta J_n). \tag{4.23}$$

In the particular case of a square cluster ($M = L$), the excitation energy is given by

$$E^{\text{square}} = 8J_n\sqrt{N_C} - 2N_C h \tag{4.24}$$

and the initial growth rate in the LT–LMF range becomes

$$n_1^{\text{square}} \simeq 4[8M\beta h\exp(-4\beta J_n) - 1]. \tag{4.25}$$

Equation 4.25 provides the following expression for the number of spins, $N_{\text{crit}}$, of the SCS of a square cluster:

$$N_{\text{crit}}^{\text{square}} \simeq \frac{1}{(8\beta h)^2}\exp(8\beta J_n), \tag{4.26}$$

which decays exponentially with increasing temperature.

#### 4.1.3.2 Triangular Isosceles Right Angle Cluster

In this case, the excitation energy of the cluster is given by

$$E^{\text{triangle}} = 8MJ_n - M(M+1)h, \tag{4.27}$$

which for $M \geq 1$ becomes

$$E^{\text{triangle}} \simeq 8\sqrt{2}\sqrt{N_C}J_n - 2N_C h,\ N_C \gg 1. \tag{4.28}$$

For the initial growth rate,

$$\begin{aligned} n_1^{\text{triangle}} = {} & 4[1+\tanh\beta\varepsilon_{-2} - \tanh\beta\varepsilon_{+2}] - 2\tanh\beta\varepsilon_0 \\ & + 2M[\tanh\beta\varepsilon_0 + \tanh\beta\varepsilon_{+2} + \tanh\beta\varepsilon_{-2}], \end{aligned} \tag{4.29}$$

which in the LT–LMF range reduces to

$$n_1^{\text{triangle}} \simeq 4(M\beta h/2 - 1). \tag{4.30}$$

The SCS of this cluster in the LT–LMF range is given by the expression

$$N_{\text{crit}}^{\text{triangle}} \simeq \frac{2}{(\beta h)^2}, \tag{4.31}$$

which is independent of the exchange integral and increases with increasing temperature.

#### *4.1.3.3 Zigzag Cluster*

In this case, one obtains for the excitation energy

$$E^{\text{zigzag}} = 4(M + 2L - 1)J_n - 2MLh, \tag{4.32}$$

which for $M = L$ becomes

$$E^{\text{zigzag}} = 4(3\sqrt{N_C} - 1)J_n - 2N_C h. \tag{4.33}$$

The initial growth rate is given by the expression

$$\begin{aligned} n_1^{\text{zigzag}} &= 4[1 + \tanh\beta\varepsilon_{-2} - \tanh\beta\varepsilon_{+2} - \tanh\beta\varepsilon_0] \\ &\quad + 2M[\tanh\beta\varepsilon_{+2} + \tanh\beta\varepsilon_{-2}] + 4L\tanh\beta\varepsilon_0. \end{aligned} \tag{4.34}$$

In the LT–LMF range, the initial growth rate and the SCS of this cluster for $M = L$ are given by

$$n_1^{\text{zigzag}} \simeq 4(M\beta h - 1) \tag{4.35}$$

and

$$N_{\text{crit}}^{\text{zigzag}} \simeq \frac{1}{(\beta h)^2}. \tag{4.36}$$

#### *4.1.3.4 Rough Boundary Cluster*

In this case, the following expressions are obtained.

For the excitation energy,

$$E^{r\text{-bound}} = 16MJ_n - 4M(M + 1)h, \tag{4.37}$$

which can be rewritten as

$$E^{r\text{-bound}} \simeq 8\sqrt{2}\sqrt{N_C}J_n - 2N_C h,\ N_C \gg 1. \tag{4.38}$$

For the initial growth rate

$$n_1^{r\text{-bound}} = 8[1 - \tanh\beta\varepsilon_0 + 2\tanh\beta\varepsilon_{-2}] + 16M\tanh\beta\varepsilon_0,\ M \geq 2, \tag{4.39}$$

which in the LT–LMF range becomes

$$n_1^{r\text{-bound}} \simeq 8(2M\beta h - 1) \tag{4.40}$$

and leads to the following SCS:

$$N_{\text{crit}}^{r\text{-bound}} \simeq \frac{1}{2(\beta h)^2}. \quad (4.41)$$

The temperature dependencies of the SCSs of the zigzag and rough boundary clusters are similar to that of the triangular cluster, the latter having, however, the largest SCS.

For the LT–LMF range and cluster sizes larger than the SCS, the initial growth rates of the triangular, zigzag, and rough boundary clusters are much greater than that of the square cluster (for the same number of spins). The reason is that in the former cases, the boundaries are at least partially rough. One can provide the following simple arguments regarding the effect of a rough boundary.

According to Equation 4.4, the probability of an "up" → "down" transition of a boundary spin (which determines the decay of a cluster) and the probability of a "down" → "up" transition of a neighboring spin (which determines its growth) are proportional to $W^{(+)} = 1 - \tanh \beta H^{(+)}$ and $W^{(+)} = 1 + \tanh \beta H^{(-)}$, respectively, where $H^{(+)}$ and $H^{(-)}$ are the local fields acting on the corresponding spin. Each flat boundary spin (bold line squared in Figure 4.1) is subjected to the local field $H^{(+)} = h + 2J_n$ and interacts with its "down" directed neighbor which is subjected to the local field $H^{(-)} = h - 2J_n$. The contribution $n_1^{\text{flat}}$ of such spins to the total initial growth rate of the cluster is given by $n_1^{\text{flat}} = W^{(-)} - W^{(+)} = \tanh \beta H^{(-)} + \tanh \beta H^{(+)}$, which in the LT–LMF range leads to $n_1^{\text{flat}} \sim \beta h \exp(-4\beta J_n)$. However, for the spins located at the rough part of a boundary and their neighbors outside the cluster (dashed line framed in Figure 4.1), which are subjected to the local field $H^{(+)} = H^{(-)} = h$, one obtains $n_1^{\text{rough}} \sim \beta h$. Consequently, $n_1^{\text{rough}} \gg n_1^{\text{flat}}$.

It should also be emphasized that the rough boundary cluster has a smaller SCS than the clusters of other shapes.

### 4.1.4 Small Clusters of Arbitrary Shape in the Two-Dimensional Ising Model

In the preceding section, we considered clusters with simple shapes and with only nearest-neighbor interactions between spins. It was shown that their SCS is strongly affected by their shape. It is therefore appropriate to extend the analysis of the initial growth rate to clusters of arbitrary shape.

In this section, small clusters (up to 10 spins) of all possible shapes will be examined. All the configurations were generated and their excitation energies and initial growth rates calculated with the help of a computer. In the calculations, both model (a), which involves the nearest-neighbor interactions, and model (b), which involves, in addition, the first-order diagonal and next-nearest-neighbor interactions, were used.

Considering for $J_d$ and $J_{nn}$ in model (b) interactions similar to the van der Waals ones and taking the lattice constant as unit of length, one can write

$$J_{nn} = J_n / 2^6, \; J_d = J_n / (\sqrt{2})^6. \quad (4.42)$$

If not otherwise mentioned, in the numerical calculations the values of the exchange integral $J_n$ and temperature $T$ will be taken as $J_n = 5$ K and $T = 0.25$ K in order to have $\beta J_n \gg 1$. The temperature selected for calculation is much smaller than the temperature $T_c$ of the second-order phase transition at $h = 0$ between the paramagnetic and ferromagnetic states, which has the value $T_c = 11.3$ K. In model (b), the small additional interactions do not change $T_c$ appreciably and the inequality $T < T_c$ is still valid.

For both models, the excitation energy of an arbitrary cluster can be expressed in the following general form:

$$E = 2n_h h + n_n J_n + n_d J_d + n_{nn} J_{nn}, \quad (4.43)$$

where $n_h$, $n_n$, $n_d$, $n_{nn}$ are integers. In what follows, the simplified notation $E = \{n_h, n_n, n_d, n_{nn}\}$ will be used for the excitation energy.

#### 4.1.4.1 Number of Configurations

The number $N_{\text{conf}}$ of different configurations (the configurations are considered as "different" if they do not superpose by parallel shifting) for various numbers of spins are listed in Table 4.3. In the range considered, the dependence is well described by the "empirical" expression

$$\log_{10}(N_{\text{config}}) = 0.234563 + 0.534849(n - 2) \; (n \leq 10), \quad (4.44)$$

where $n$ is the number of spins of the cluster.

Table 4.3 Number $N_{conf}$ of Different Configurations for Clusters with Various Number $N_C$ of Spins

| $N_{conf}$ | $N_C$ |
|---|---|
| 2 | 2 |
| 6 | 3 |
| 19 | 4 |
| 63 | 5 |
| 216 | 6 |
| 760 | 7 |
| 2725 | 8 |
| 9910 | 9 |
| 36,446 | 10 |

In the following, we examine in detail the clusters that consist of five "up" spins. These clusters have a relatively small number of possible configurations and this simplifies the analysis.

### 4.1.4.2 Five-Spin Cluster

*4.1.4.2.1 Excitation Energy* Table 4.4 lists the possible excitation energies of the five-spin cluster, and shows that in the nearest-neighbor approximation the five-spin cluster has two values for the excitation energy, both highly degenerated. The additional interactions in model (b) do not affect the degeneracy of the lowest energy level (with $n_n = 20$), but split the upper energy into seven levels, among which six are degenerate.

*4.1.4.2.2 Initial Rate* In Figures 4.3 and 4.4, the $h$ dependence of the initial growth rate for model (a) is plotted for several particular situations. Figure 4.3 presents the $h$ dependence of the initial growth rate for three different configurations of the five-spin cluster that have the same excitation energy $E = \{-5, 24, 0, 0\}$. In all the cases, the curves have a similar qualitative behavior. For low magnetic fields, all the configurations have negative initial growth rates ($n_1 < 0$), whereas for high magnetic fields they grow initially ($n_1 \geq 0$). The transition point between these two states depends on the configuration. The most symmetric configuration, represented by the solid line, has a very small initial growth rate in a large range of magnetic fields ($0.5 \leq h \leq 9$), with small negative values for $h < h_c \simeq 4.8$ K and small positive values for $h > h_c$.

In Figure 4.4, the $h$ dependence of the initial growth rate for the configuration with the smallest excitation energy is plotted for three different temperatures. It shows that by increasing the temperature, the range in which the cluster does not initially decay is shifted towards smaller values of $h$, and that the $h$ dependence of the initial growth rate becomes smoother. The $h$ dependence of the initial growth rate exhibits a dramatic change at sufficiently low temperatures and has qualitatively the same behavior for all configurations.

To examine the effect of the longer range interactions upon the initial growth rate, its $h$ dependence for model (b) is plotted in Figures 4.5 and 4.6 for the configurations of the five-spin cluster for the smallest and largest excitation energies, together with the corresponding results for model (a). One can see that the

Table 4.4 Excitation Energies of the Five-Spin Cluster as Defined by Equation 4.43 and the Number $N_{conf}$ of Corresponding Configurations in Models (a) and (b)

| Model | $n_h$ | $n_n$ | $n_d$ | $n_{nn}$ | $N_{conf}$ |
|---|---|---|---|---|---|
| (a) | −5 | 20 | 0 | 0 | 8 |
| | −5 | 24 | 0 | 0 | 55 |
| (b) | −5 | 20 | 28 | 36 | 8 |
| | −5 | 24 | 24 | 32 | 1 |
| | −5 | 24 | 28 | 36 | 8 |
| | −5 | 24 | 28 | 40 | 4 |
| | −5 | 24 | 32 | 32 | 16 |
| | −5 | 24 | 32 | 36 | 12 |
| | −5 | 24 | 36 | 32 | 12 |
| | −5 | 24 | 40 | 28 | 2 |

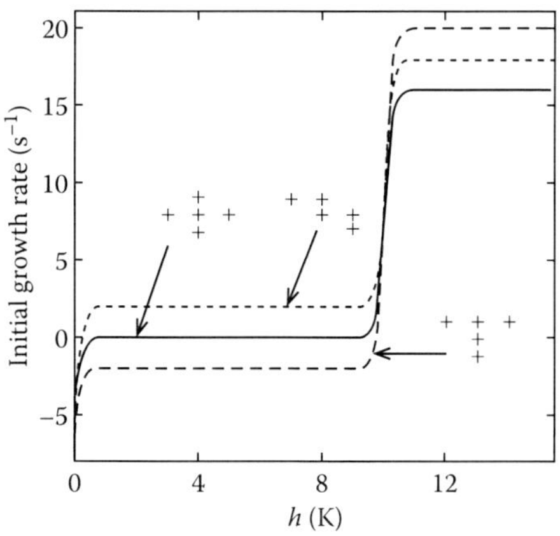


Figure 4.3

*h* dependence in model (a) of the initial growth rate for three different configurations of the five-spin cluster with the same excitation energy $E = \{-5, 24, 0, 0\}$. The results were obtained for $J_n = 5$ K, $T = 0.25$ K.

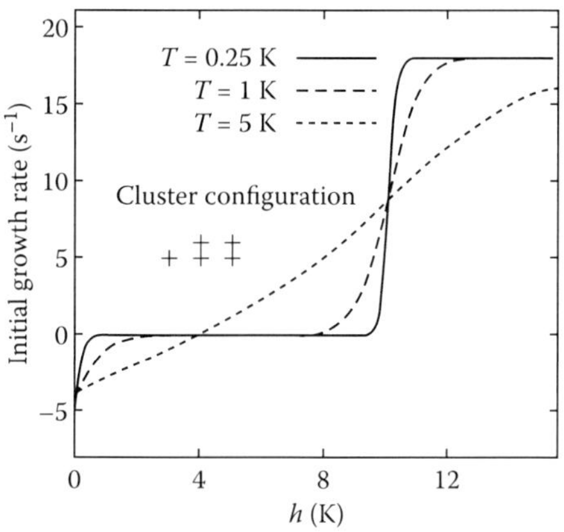


Figure 4.4

The *h* dependence in model (a) of the initial growth rate for the configuration of the five-spin cluster with minimum excitation energy at three different temperatures and $J_n = 5$ K.

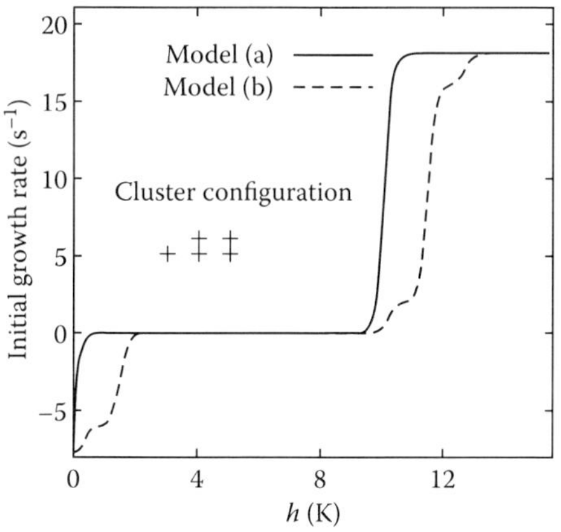


Figure 4.5

The *h* dependence of the initial growth rate for the configuration of five-spin cluster with the smallest excitation energy in models (a) ($E = \{-5, 20, 0, 0\}$) and (b) ($E = \{-5, 20, 28, 36\}$). The results were obtained for $J_n = 5$ K, $T = 0.25$ K.

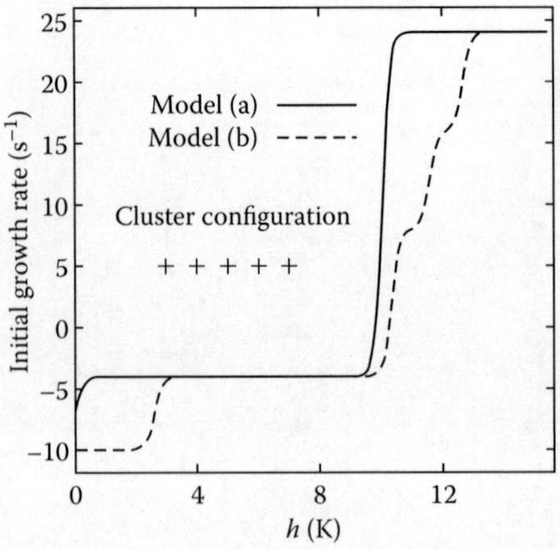


Figure 4.6

The $h$ dependence of the initial growth rate for the configuration of five-spin cluster with the largest excitation energy in model (a) ($E$ = {−5, 24, 0, 0}) and (b) ($E$ = {−5, 24, 40, 28}). The results were obtained for $J_n$ = 5 K, $T$ = 0.25 K.

additional interactions in model (b) lead in both cases to a somewhat more complicated $h$ dependence of the initial growth rate, but do not change the qualitative behavior.

In Table 4.5, the initial growth rates at a fixed value of the magnetic field $h = 4$ K are listed together with the corresponding number of configurations for the five- and nine-spin clusters. Because the values in Table 4.5 were calculated with limited accuracy, the small positive and negative values are collected together under the zero value. For the selected values of the parameters, the results for models (a) and (b) are the same (Table 4.5 is valid in both cases). The distribution of the fraction of configurations versus the initial growth rate is plotted in Figure 4.7, which shows that the fraction of cluster configurations with positive initial rates increases with increasing cluster size. Let us note the comparatively large fraction of configurations with zero initial growth rate, which is about 50% of the total number of configurations for the five-spin and about 30% for the nine-spin clusters.

The fraction of configurations which grow at $t = 0$ ($n_1 > 0$) also increases with increasing magnetic field. For the five-spin clusters, this dependence is presented in Figure 4.8, which shows that the fraction of configurations which grow initially increases from zero to unity as the magnetic field increases from zero to a value approximately equal to the largest local field due to the exchange interactions of a spin with its environment. At the latter magnetic field, all local fields $\varepsilon_\mu$ (Equation 4.14) involved in Equation 4.12 are positive, and hence the initial growth rate $n_1$ is positive at all temperatures.

Table 4.5 Initial Growth Rates $n_1$ and Number $N_{conf}$ of Corresponding Configurations for Clusters with Number of Spins $N_C = 5$ and $N_C = 9$

| $N_C$ | $n_1$ | $N_{conf}$ |
|---|---|---|
| | 2.0 | 4 |
| | 0.0 | 29 |
| 5 | −2.0 | 28 |
| | −4.0 | 2 |
| | 10.0 | 4 |
| | 8.0 | 44 |
| | 6.0 | 340 |
| 9 | 4.0 | 1692 |
| | 2.0 | 4196 |
| | 0.0 | 3100 |
| | −2.0 | 516 |
| | −4.0 | 18 |

*Note:* The results were obtained for $J_n$ = 5 K, $h$ = 4 K, and $T$ = 0.25 K.

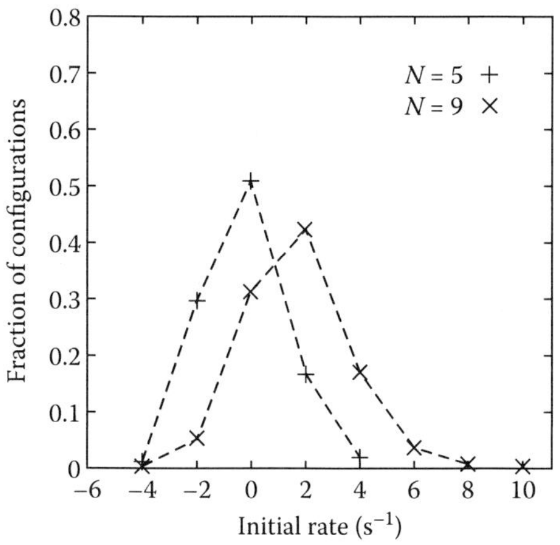


Figure 4.7

Distribution of the fraction of configurations versus the initial growth rate at a fixed value of the magnetic field for clusters with five and nine spins. The results were obtained for $J_n = 5$ K, $h = 4$ K, $T = 0.25$ K.

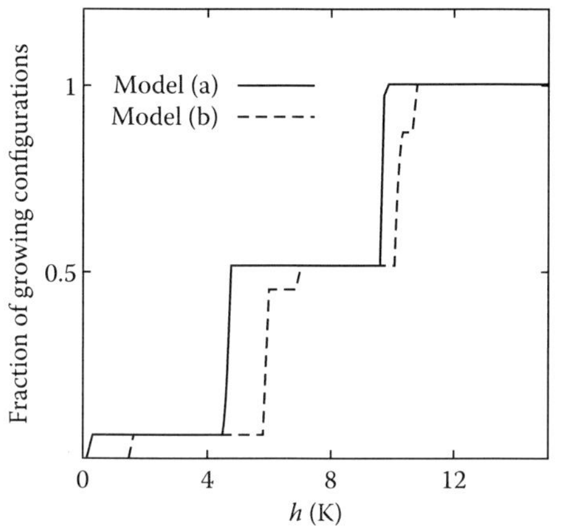


Figure 4.8

The fraction of growing configurations of five-spin clusters as a function of the magnetic field. The results were obtained for $J_n = 5$ K, $T = 0.25$ K.

## 4.1.5 Discussion and Conclusions

In this paper, the Glauber kinetic equation was applied to the two-dimensional Ising spin system interacting with a heat bath. On the basis of that equation, the initial growth rates of clusters of various sizes and shapes were calculated. One can conclude that the kinetic behavior of a cluster with a given number of spins depends not only on their number, but also on the shape of the cluster boundary. For example, in the low temperature–low magnetic field range and for cluster sizes larger than SCS, the initial growth rate of a square cluster, which is proportional to $\beta h e^{-4\beta J_n}$ (see Equation 4.25), is considerably smaller than that of a cluster with one or more rough boundaries (triangular, zigzag, rough boundary), whose initial growth rates are proportional to $\beta h$. The physical reason for such an inequality is the difference in the local fields acting upon the spins at or near the rough part of the boundary of a cluster and at or near the flat one.

For each shape, there is a SCS above which the cluster grows. As soon as a cluster starts to grow it will most likely change its shape and the SCS will change as well. Since the rougher clusters have smaller SCSs, a cluster that started to grow will continue to grow if the cluster acquires during its growth an increasingly rougher boundary.

The number of possible configurations on a square lattice, determined for clusters with up to 10 spins, grows almost exponentially with the cluster size with an exponent equal approximately to $0.535n$. The excitation energy of a cluster that constitutes one of its main physical characteristics is not determined uniquely by the number of spins in the cluster but depends also on the cluster shape. For the five-spin clusters, the initial growth rate of all possible shapes was investigated. For given values of the external parameters *h* and *T*, the initial growth rate $n_1$ depends strongly on the cluster shape, as one can see, for example, in Figure 4.3.

The initial growth rate is relatively independent on the excitation energy since configurations with different excitation energies can have almost equal *h* dependencies of the initial growth rate. Indeed, the initial growth rates of the configuration with excitation energy $E = \{-5,24,0,0\}$ (represented by the solid line in Figure 4.3) and the configuration with the smallest excitation energy $E = \{-5,20,0,0\}$ (represented by the solid line in Figure 4.4 for the same values of the parameters) nearly coincide. The initial cluster growth rate depends also on the external magnetic field; as the latter increases, the fraction of configurations which grow increases monotonically (see Figure 4.8). One of the features of the obtained results is the large number of configurations with zero or very small initial growth rates.

The next-nearest-neighbor interactions between spins lead to a decrease of the degeneracy of the excitation energies and to more complicated field and temperature dependencies of the initial growth rate.

## References


1. Frenkel, J., *Kinetic Theory of Liquids*, Dover, New York, 1955.
2. Abraham, F., *Homogeneous Nucleation Theory*, Academic, New York, 1974.
3. Nowakowski, B., Ruckenstein, E., *J. Chem. Phys.* 94, 1397 (1991). (Section 2.1 of this volume.)
4. Nowakowski, B., Ruckenstein, E., *J. Chem. Phys.* 94, 8487 (1991). (Section 2.3 of this volume.)
5. Schmelzer, J., Jr., Landau, D.P., *Int. J. Mod. Phys. C* 12, 345 (2001).
6. Wonczak, S., Strey, R., Stauffer, D., *J. Chem. Phys.* 113, 1976 (2000).
7. Rottman, C., Wortis, M., *Phys. Rev. B* 24, 6274 (1981).
8. Glauber, R., *J. Math. Phys.* 4, 294 (1963).
9. Shneidman, V.A., Jackson, K.A., Beauty, K.M., *J. Chem. Phys.* 111, 6932 (1999).
10. Shneidman, V.A., Jackson, K.A., Beauty, K.M., *Phys. Rev. B* 59, 3579 (1999).
11. Vehkamaki, H., Ford, I.J., *Phys. Rev. E* 59, 6483 (1999).
12. de Jongh, L.J., Miedema, A.R., *Adv. Phys.* 23, 1 (1974).
13. Steiner, M., Villain, J., Windsor, G.G., *Adv. Phys.* 25, 87 (1976).
14. Malakis, A., *J. Phys. A* 13, 631 (1980).
15. Kumar, A., Krishnamurty, H.R., Gopal, E.S.R., *Phys. Rep.* 98, 57 (1983).
16. Beysens, D., Bourgan, A., Calmettes, P., *Phys. Rev. A* 26, 3589 (1982).
17. Stoll, E., Binder, K., Schneider, T., *Phys. Rev. B* 8, 3266 (1973).
18. Berim, G.O., Kessel, A.R., *Physica A* 101, 112 (1980).
19. Suzuki, M., Kubo, R., *J. Phys. Soc. Jpn.* 24, 51 (1968).
20. Heims, S.P., *Phys. Rev. A* 138, 587 (1965).

## 4.2 Effect of Shape on the Critical Nucleus Size in a Three-Dimensional Ising Model: Energetic and Kinetic Approaches

*Gersh Berim and Eli Ruckenstein**

The initial growth rate and the excitation energy of spin clusters of different shapes (cubic, stair-like, pyramidal) on a three-dimensional cubic lattice were calculated, assuming Ising-type interactions between spins and the Glauber-type spin dynamics. Using energetic and kinetic approaches, the critical cluster size was obtained and compared with existing Monte Carlo "experiments." It was shown that, in most cases, the kinetic approach fits better the "experimental" results than the energetic one. A shape-independent kinetic criterium for a critical cluster size was also formulated on the basis of an analysis of all possible configurations of a cluster with a given number of spins. (© *2002 American Institute of Physics.* doi: 10.1063/1.1509051.)

### 4.2.1 Introduction

It was shown in a previous paper [1] that the growth of a cluster containing a given number of spins in the two-dimensional Ising model depends not only on their number, but also on the shape of the cluster boundary. A kinetic approach [2,3] based on the Glauber kinetic equation [4] was used to calculate the rate of growth of a cluster. For a given shape of the cluster boundary, there is a critical size above which the cluster initially grows. Because of this initial growth, the cluster can change its shape and will continue to grow only if its size is greater than the critical size for the new shape, and so on. Among the conclusions, it was noted that in the considered range of external parameters, the critical cluster size has decreased with increasing roughness of the cluster boundary.

In the present paper, we extend the treatment of the influence of the shape upon the growth of a cluster to the three-dimensional Ising model. This case is more interesting than the previous one, first because the three-dimensional Ising model is closer to real cases, and second because there are "experimental" Monte Carlo results [5–9] that can be compared with the present analytical ones. The critical size will be calculated both energetically (as in the classical nucleation theory) and kinetically and the results will be compared both between the two and with Monte Carlo experiments.

We will not rederive here the basic theoretical formulas obtained in Ref. [1] for the general case, but will only list them in Section 4.2.2.

### 4.2.2 Formulation of the Problem

The Ising model under consideration has the following Hamiltonian [8]:

$$H=-B\sum_{j}\sigma_j-J\sum_{\langle j,k\rangle}\sigma_j\sigma_k,(B>0), \tag{4.45}$$

where the spins $\sigma_j$ are located on the sites of a cubic lattice and have the values ±1, and $\langle j, k\rangle$ denotes summation over the nearest neighbors (each pair of interacting spins should be counted only once). The exchange integral $J$ and the magnetic field $B$ are expressed in units of the Boltzmann constant $k_B$; they have therefore the dimension of temperature.

As in our previous paper [1], we will consider here clusters of different sizes and shapes. We define a cluster as a group of bounded "up" spins in a sea of "down" spins and neglect the formation of other clusters in the lattice. The cluster growth is a result of the interaction of the spins with a thermal bath and is governed by the Glauber dynamics [7].

The number of spins in a single cluster coincides with the time-dependent average $\langle n_{\text{total}}\rangle_t$ of the operator $n_{\text{total}} = \Sigma_j(1 + \sigma_j)/2$ of the total number of spins in the lattice, where

$$\langle A\rangle_t \equiv \sum_{\{\sigma\}} AP\left(\sigma_1,\ldots,\sigma_j,\ldots,\sigma_{N_{\text{total}}};t\right). \tag{4.46}$$

* *J. Chem. Phys.* 117, 7732 (2002). Republished with permission.

In the above expression, $A$ is an arbitrary spin operator, $P\left(\sigma_1,\ldots,\sigma_j,\ldots,\sigma_{N_{\text{total}}};t\right)$ is the probability of finding an Ising system at time $t$ in the state $\left\{\sigma_1,\ldots,\sigma_j,\ldots,\sigma_{N_{\text{total}}}\right\}$, $N_{\text{total}}$ is the total number of spins in the system and the summation in Equation 4.46 is extended over all possible states of the system.

The differential equation for $\langle n_{\text{total}}\rangle_t$ has the form [1]

$$\frac{\mathrm{d}}{\mathrm{d}t}\left\langle n_{\text{total}}\right\rangle_t = -\frac{1}{2}\sum_j\sum_\mu W_\mu[\langle\sigma_j R_j(\mu)\rangle_t - \langle R_j(\mu)\rangle_t \tanh(\beta\varepsilon_\mu)], \tag{4.47}$$

where $\beta = 1/T$, $T$ is the temperature in K, $W_\mu$ are temperature-independent coefficients that provide the timescales of the process, $\varepsilon_\mu$ are the possible values of the local field that originates from the interactions of a spin with its nearest neighboring spins and $R_j(\mu)$ is equal to unity when the spins interacting with spin $\sigma_j$ create a local field equal to $\varepsilon_\mu$ and zero otherwise. Index $\mu$ has the following seven possible values: $\mu = 0, \pm2, \pm4, \pm6$. The corresponding values of the local field are

$$\varepsilon_\mu = B + \mu J. \tag{4.48}$$

The exclusion of formation of other clusters in the lattice and of flip-flop transitions of the spins of the cluster which are not located at the boundary leads to

$$W_6 = W_{-6} = 0. \tag{4.49}$$

This reduces the possible values of $\mu$ in Equation 4.47 to the following five ones: $\mu = 0, \pm2, \pm4$.

Equation 4.47 is not closed and, consequently, cannot be solved. As in Ref. [1], we calculate only the initial rate

$$n_1 = \frac{\mathrm{d}}{\mathrm{d}t}\langle n_{\text{total}}\rangle_t\bigg|_{t=0}, \tag{4.50}$$

which can be obtained exactly from Equation 4.47 using the initial conditions. The value of $n_1$ provides the rate of growth of the cluster in the vicinity of time $t = 0$. The positive (negative) sign of $n_1$ indicates that the considered cluster gains (loses) one or more spins, i.e., it grows (decays) at the initial moment.

The general expression for $n_1$ was derived in Ref. [1] and has the form

$$n_1 = \frac{1}{2}\sum_\mu W_\mu\left\{n_{\mathrm{N}}^\mu[1+\tanh(\beta\varepsilon_\mu)] - n_{\mathrm{B}}^\mu[1-\tanh(\beta\varepsilon_\mu)]\right\}, \tag{4.51}$$

where $n_{\mathrm{B}}^\mu$ and $n_{\mathrm{N}}^\mu$ are the numbers of boundary and nearest neighboring spins of the cluster subjected to the local field $\varepsilon_\mu$. (An "up" spin of a cluster is considered as a boundary one if it has at least one nearest neighbor "down" spin; a "down" spin out of a cluster is considered as neighboring a cluster if it has at least one cluster spin as its nearest neighbor.) For simplicity, all nonzero coefficients $W_\mu$ are selected equal to 2. In this case,

$$n_1 = N_{\mathrm{N}} - N_{\mathrm{B}} + \sum_\mu\left(n_{\mathrm{N}}^\mu + n_{\mathrm{B}}^\mu\right)\tanh(\beta\varepsilon_\mu), \tag{4.52}$$

where $N_{\mathrm{B}}$ and $N_{\mathrm{N}}$ are the numbers of boundary and neighboring spins of the cluster, respectively.

The excitation energy of a cluster is defined as the energy difference between that of a system containing a cluster of up spins in a sea of down spins and of a system with all spins down, and can be expressed in the following general form

$$E = -2BN_{\mathrm{C}} + 2N_{\mathrm{J}}J, \tag{4.53}$$

where $N_{\mathrm{C}}$ represents the total number of spins in the cluster and $N_{\mathrm{J}} = \Sigma_\mu[3-(\mu/2)]n_{\mathrm{B}}^\mu$.

Two kinds of critical clusters can be defined for an Ising system on a rigid lattice. One of them is similar to the critical nucleus of the traditional nucleation theory and its size $n^*$ is given by the maximum of the excitation energy of the cluster. The other one has a kinetic origin (see Refs. [2,3]) and its size $n^{\mathrm{kin}}$ is provided by the zero initial rate of growth of a cluster ($n_1 = 0$). At this cluster size, the probabilities of gaining an additional spin and of losing one become equal. The critical sizes $n^*$ and $n^{\mathrm{kin}}$ depend strongly on the shape of a cluster and, generally, do not coincide.

In the analysis of the results, we will denote $I = 4J/T$ and use instead of the magnetic field $B$ the dimensionless quantity $h = 2B/T$, which is considered as the supersaturation in the Ising model [8]. Two particular ranges of the supersaturation $h$ and temperature will be examined. The first one, called the low temperature–low supersaturation (LT–LS) range, is defined by the inequalities $I \gg 1$, $h \ll 1$, and the second one, low temperature–high supersaturation (LT–HS) range, is defined by the inequalities $I \gg 1$, $h \gg 1$.

In the numerical calculations, if not otherwise mentioned, the following values for the exchange integral $J$ and temperature $T$ will be used: $J = 5$ K, and $T = 5.64$ K; they provide $T/T_c = 0.25$, where [10] $T_c = J/0.221656$ is the second-order phase transition temperature in a three-dimensional Ising model without an external field. The same ratio was employed in the Monte Carlo simulations in the three-dimensional Ising model [8], with which the present results will be compared.

### 4.2.3 Clusters with Simple Topology

In this section, the excitation energy $E$, the initial growth rate $n_1$, and the critical sizes $n^*$ and $n^{\mathrm{kin}}$ of clusters with simple shapes (cubic, stairlike, pyramidal), sketched in Figure 4.9, will be calculated. The cubic cluster (Figure 4.9a) contains $l$ spins on each of its sides. A stairlike cluster (Figure 4.9b) can be generated by successive shifting of the layers of a cubic cluster in the direction of one of its sides. The pyramidal clusters, pyramid A (Figure 4.9c) and pyramid B (Figure 4.9d), consist of $2l + 1$ layers (where $l$ is an integer) whose projections onto the $XY$ plane are presented in Figure 4.9c and d by a succession of dashed and blank visible parts of the layers. The upper layer of both pyramids is a square with two spins on each side. The pyramidal clusters are symmetrical with respect to their largest layer.

The quantities in Equations 4.52 and 4.53, which are needed to calculate the excitation energy and the initial growth rate, are listed in Tables 4.6 and 4.7, respectively.

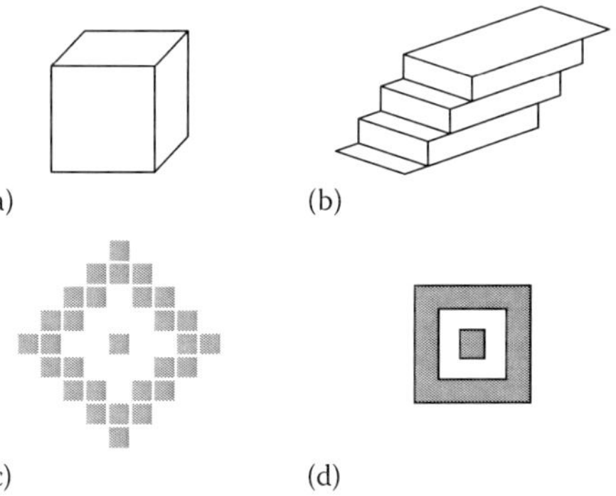

Figure 4.9

Some of the possible cluster shapes in the three-dimensional lattice: cubic (a), stairlike (b), pyramid A (c), and pyramid B (d). The last two clusters are presented as projections onto the $XY$ plane.

Table 4.6 Total Number $N_C$ of Spins and Numbers of Boundary $N_B$ and Neighboring $N_N$ Spins in Equation 4.52, and Value of Coefficient $N_J$ in Equation 4.53 for Clusters of Different Shapes

| Shape | $N_C$ | $N_B$ | $N_N$ | $N_J$ |
|---|---|---|---|---|
| Cubic | $l^3$ | $2(3l^3 - 6l + 4)$ | $6l^2$ | $6l^2$ |
| Stairlike | $l^3$ | $2(3l^3 - 6l + 4)$ | $6l^2$ | $2l(4l - 1)$ |
| Pyramid A | $\frac{16}{3}l(l^2+2)+12l^2+4$ | $4(4l^2 + 2l + 1)$ | $16(l + 1)^2$ | $8(4l^2 + 5l + 2)$ |
| Pyramid B | $\frac{4}{3}l(l+1)(2l^2+4l+3)$ | $4(2l^2 + 2l + 1)$ | $8(l^2 + 3l + 2)$ | $16(l + 1)^2$ |

Table 4.7 Nonzero Values of Coefficients $n_B^\mu$ and $n_N^\mu$ in Equations 4.51 and 4.52 for Clusters of Different Shapes

| Shape | $n_B^{-2}$ | $n_B^0$ | $n_B^2$ | $n_B^4$ | $n_N^{-4}$ | $n_N^{-2}$ | $n_N^0$ |
|---|---|---|---|---|---|---|---|
| Cubic | 0 | 8 | $12(l-2)$ | $6(l-2)^2$ | $6l^2$ | 0 | 0 |
| Stairlike | 4 | $6l-8$ | $2(l^2-l-2)$ | $4(l-2)^2$ | $2l(2l+1)$ | $2l(l-1)$ | 0 |
| Pyramid A | $8l+4$ | $8l^2$ | $8l$ | $8l(l-1)$ | $8(l^2+2l+2)$ | $24l$ | $8l(l-1)$ |
| Pyramid B | 4 | $16l$ | $8l(l-1)$ | 0 | $16(l+1)$ | $8l(l+1)$ | 0 |

One can see from Table 4.6 that the number $N_C$ of spins in a cluster contains a term proportional to $l^3$. Therefore, expression 4.53 for the excitation energy is a polynom of third degree in $l$. Because of the negative coefficient for $l^3$, $E$ decreases at large $l$ with increasing $l$ and this provides a maximum of $E$ for a critical size $n^*$. Considering $l$ as a continuous variable, one can find a critical $l^*$ from the equation

$$\left.\frac{dE}{dl}\right|_{l=l^*} = 0 \tag{4.54}$$

and a critical cluster size $n^*$ given by

$$n^* = N_C|_{l=l^*}. \tag{4.55}$$

Similarly, the expression for the initial growth rate $n_1$ of a cluster is a polynom of the second degree in $l$:

$$n_1 = al^2 + bl + c, \tag{4.56}$$

where the coefficients $a$, $b$, and $c$ depend on the exchange integral $J$, magnetic field $B$, and temperature. They are listed in Table 4.8. To find the kinetic critical size $n^{\text{kin}}$, one has to find the critical $l^{\text{kin}}$ of the cluster from the equation $n_1 = 0$ and then calculate $n^{\text{kin}}$ using the expression

$$n^{\text{kin}} = N_C\big|_{l=l^{\text{kin}}}. \tag{4.57}$$

The expression for $n^{\text{kin}}$ has, generally, a complicated form and will be provided below analytically for the LT–HS and LT–LS ranges, and numerically otherwise.

For comparison purposes, one may note that in the two-dimensional Ising model on a square lattice, the excitation energy of a cluster has the form of Equation 4.53, with $N_C \sim L^2$ and $N_J = \Sigma_\mu[2-(\mu/2)]n_B^\mu$, where $L$ is a geometric characteristic of the two-dimensional cluster and $n_B^\mu$ is proportional to $L$.

#### 4.2.3.1 Cubic Cluster

The critical cluster size $n^*_{\text{cube}}$, of this cluster calculated using Equations 4.55 and 4.54 is given by the expression

$$n^*_{\text{cube}} = \frac{8I^3}{h^3}, \tag{4.58}$$

where $I = 4J/T$ and $h = 2B/T$. The expression for the initial growth rate given by Equation 4.56 reduces in the LT–LS range to

$$n_1 \simeq 24he^{-2I}l^2 - 24e^{-I}l - 8, \ (h \ll 1) \tag{4.59}$$

Table 4.8 Coefficients $a$, $b$, and $c$ in Equation 4.56 for Clusters of Different Shapes

| Shape | a | b | c |
|---|---|---|---|
| Cubic | $6(z_{-4}+z_4)$ | $12(1+z_2-2z_4)$ | $-8(1-z_0)-24(z_2-z_4)$ |
| Stairlike | $2(z_{-2}+z_2)+4(z_{-4}+z_4)$ | $12+2z_{-4}-2z_{-2}+6z_0-2z_2-16z_4$ | $-8(1+z_0)+4z_{-2}+16z_4-4z_2$ |
| Pyramid A | $16z_0+8(z_{-4}+z_4)$ | $24+16z_{-4}+32z_{-2}-8z_0+8z_2-8z_4$ | $12+16z_{-4}+4z-2$ |
| Pyramid B | $8(z_{-2}+z_2)$ | $16(1+z_0)+16z_{-4}+8z_{-2}-8z_2$ | $12+16z_{-4}+4z_{-2}$ |

*Note:* Here $z_\mu = \tanh(\beta\varepsilon_\mu)$, $\varepsilon_\mu$ is defined by Equation 4.48.

and provides the following expression for the kinetic critical size:

$$n_{\text{cube}}^{\text{kin}} \simeq \frac{1}{h^3} e^{3I}. \tag{4.60}$$

Similarly, one obtains in the LT range

$$n_{\text{cube}}^{\text{kin}} \simeq \frac{8}{3\sqrt{3}} e^{-3h} e^{3I}, \ (h \gg 1). \tag{4.61}$$

One may observe that the energetic and kinetic critical sizes have in the LT–LS range the same $h$ dependence but with different coefficients $f(T)$:

$$n_{\text{cube}} \simeq \frac{f(T)}{h^3}. \tag{4.62}$$

In contrast, in the LT range, there is a major difference between $n_{\text{cube}}^{*}$ and $n_{\text{cube}}^{\text{kin}}$, because the latter depends exponentially on $h$.

In Figure 4.10, the $h$ dependencies of the kinetic and energetic critical sizes $n_{\text{cube}}^{\text{kin}}$ and $n_{\text{cube}}^{*}$ are plotted together with the Monte Carlo results [8]. It shows that, at the left and right ends of the considered range, the $n_{\text{cube}}^{\text{kin}}$ fits the simulation data better, whereas in the central part the best fit is provided by the $n_{\text{cube}}^{*}$. However, it should be noted that most of the simulation results involve numbers of spins that are not compatible with a cubic shape on a rigid lattice.

Let us compare qualitatively in the LT–LS range the expressions of the critical sizes in three- and two-dimensional models, using as examples the cubic cluster for the first model and the square cluster for the second one. The excitation energy of the square cluster with the side $L$ is given by

$$E_{\text{square}} = -2BL^2 + (12L - 8)J \tag{4.63}$$

and leads to the following energetical critical size:

$$n_{\text{square}}^{*} = \frac{9I^2}{4h^2}. \tag{4.64}$$

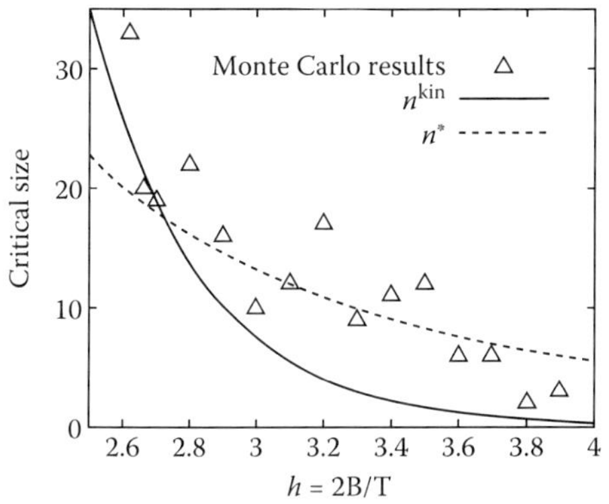


Figure 4.10

$h$ dependencies of the kinetic $n^{\text{kin}}$ and energetic $n^{*}$ critical sizes for a cubic cluster and the corresponding results of Monte Carlo simulations of Wonczak et al. [8].

The kinetic critical size was obtained in Ref. [1]. Using the notations of the present paper, it can be rewritten in the LT–LS range as

$$n_{\text{square}}^{\text{kin}} = \frac{1}{16h^2} e^{2I}. \tag{4.65}$$

Comparing Equations 4.64 and 4.65 with Equations 4.58 and 4.60, respectively, one can see that in both cases the corresponding expressions are similar, and the larger value is provided by the kinetic critical size.

#### 4.2.3.2 Stairlike Cluster

In this case, the energetic critical cluster is given by

$$l_{\text{stair}}^{\star} = \frac{4I}{3h}\left(1+\sqrt{1-\frac{3h}{16I}}\right) \tag{4.66}$$

and

$$n_{\text{stair}}^{\star} = \left(l_{\text{stair}}^{\star}\right)^3. \tag{4.67}$$

At low supersaturations ($h \ll I$),

$$n_{\text{stair}}^{\star} \simeq \frac{512I^3}{27h^3}. \tag{4.68}$$

For $h > 16I/3$, the excitation energy $E_{\text{stair}}$ decreases continuously with increasing $l$, and according to the classical (energetic) theory the stair cluster will grow at any size.

The kinetic critical size $n_{\text{stair}}^{\text{kin}}$ of a stairlike cluster is given in the LT–LS range by

$$n_{\text{stair}}^{\text{kin}} \simeq \frac{27}{64h^3} e^{3I}, \tag{4.69}$$

while in the LT range and when $I - 3h \gg 1$

$$n_{\text{stair}}^{\text{kin}} \simeq 27 e^{-6h} e^{3I}. \tag{4.70}$$

For intermediate values of $h$ ($h \sim 1$), the $h$ dependence of $n_{\text{stair}}^{\text{kin}}$ is plotted in Figure 4.11 together with the $n_{\text{stair}}^{\star}$ and the simulation results of Wonczak et al. [8]. One can see that in this case the kinetic critical cluster fits the simulations much better than the energetic one.

#### 4.2.3.3 Pyramidal Cluster

In this section, we list the results, obtained for two pyramidal clusters, that are sketched in Figure 4.9c and d.

The critical cluster size $l^{\star}$ is given by

$$l_{\text{pyramid A}}^{\star} = \frac{1}{12}\left(-9+\frac{12I}{h}+\sqrt{\frac{144I^2}{h^2}-\frac{36I}{h}-15}\right) \tag{4.71}$$

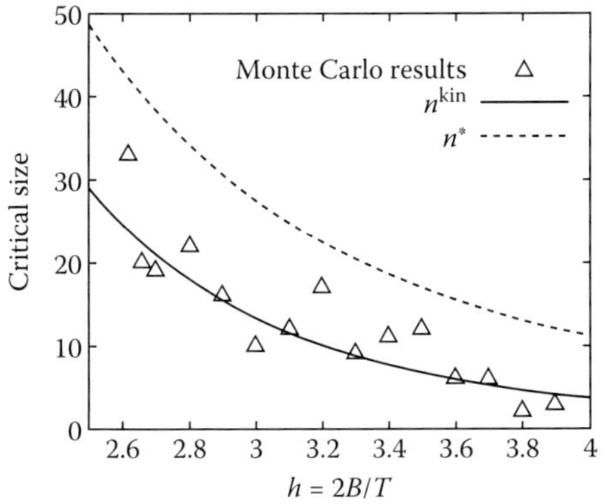


Figure 4.11

$h$ dependencies of the kinetic $n^{kin}$ and energetic $n^*$ critical sizes for a stairlike cluster and the corresponding results of Monte Carlo simulations of Wonczak et al. [8].

for pyramid A and by

$$l^*_{\text{pyramid B}} = \frac{I}{h} - 1 + \sqrt{\frac{I^2}{h^2} - \frac{1}{6}} \tag{4.72}$$

for pyramid B.

At low supersaturations ($h \ll I$), $l^*_{\text{pyramid A}} \simeq 2I/h$, $l^*_{\text{pyramid B}} \simeq I/h$, and the corresponding cluster sizes $n^*_{\text{pyramid A}}$ and $n^*_{\text{pyramid B}}$ are given by

$$n^*_{\text{pyramid A}} \simeq \frac{128 I^3}{3h^3}, \; n^*_{\text{pyramid B}} \simeq \frac{8I^3}{3h^3}. \tag{4.73}$$

For the kinetic critical sizes, one obtains

$$n^{\text{kin}}_{\text{pyramid A}} \simeq \frac{144}{h^3}, \; n^{\text{kin}}_{\text{pyramid B}} \simeq \frac{e^{3I}}{3h^3} \tag{4.74}$$

in the LT–LS range, and

$$\begin{aligned} n^{\text{kin}}_{\text{pyramid A}} &= 116, \; (I - 2h \gg 1), \\ n^{\text{kin}}_{\text{pyramid B}} &\simeq \frac{16}{3} e^{-6h} e^{3I}, \; (I - 3h \gg 1) \end{aligned} \tag{4.75}$$

in the LT–HS range.

In Figures 4.12 and 4.13, the $h$ dependencies of the kinetic critical sizes for the two types of pyramids are plotted together with the energetical critical size $n^*$ and simulation results [8]. The kinetic critical size fits the experimental points better than the energetic one.

### 4.2.4 Small Clusters of Arbitrary Shape

In this section, small clusters of up to eight spins of all possible shapes will be considered. All the configurations were generated and their excitation energies and initial growth rates calculated with the help of a computer. The excitation energy $E$ of an arbitrary cluster is given by Equation 4.53, and, in what follows, the simplified notation $E = \{N_C, N_J\}$ will be used.

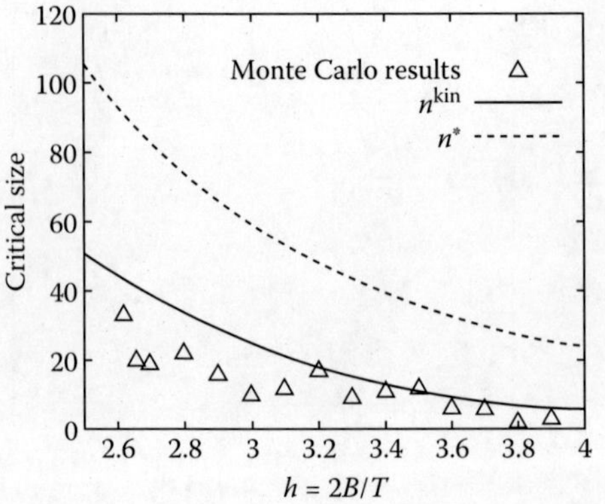


Figure 4.12

*h* dependencies of the kinetic $n^{kin}$ and energetic $n^*$ critical sizes for the pyramid A cluster and the corresponding results of Monte Carlo simulations of Wonczak et al. [8].

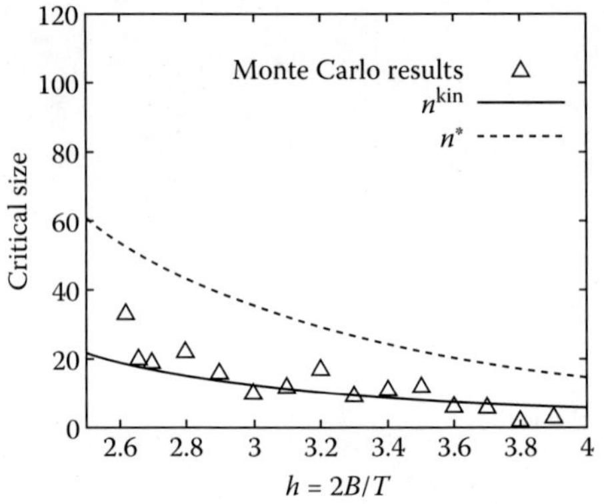


Figure 4.13

*h* dependencies of the kinetic $n^{kin}$ and energetic $n^*$ critical sizes for the pyramid B cluster and the corresponding results of Monte Carlo simulations of Wonczak et al. [8].

### 4.2.4.1 Number of Configurations

The number $N_{\text{config}}$ of different configurations (i.e., of configurations that cannot be superposed by parallel shifting) are listed in Table 4.9 for various numbers of spins in a cluster. As in the two-dimensional model, this number increases exponentially with the number of spins and is given by

$$\log_{10}(N_{\text{config}}) = 0.398388 + 0.792919(N_C - 2), (N_C \leq 8). \tag{4.76}$$

Consequently, $N_{\text{config}} \sim 10^{(0.792919 N_C - 1.187450)}$. In the two-dimensional case the analogous expression has the form: $N_{\text{config}} \sim 10^{(0.534849 N_C - 0.835135)}$. Comparing these two expressions, one can find that they can be represented by the single approximate formula

$$N_{\text{config}} \sim 10^{(\alpha_{\text{conf}} N_C - \beta_{\text{conf}})d}, (N_C \leq 8), \tag{4.77}$$

where $\alpha_{\text{conf}} \simeq 0.265866 \pm 0.0015595$, $\beta_{\text{conf}} \simeq 0.406693 \pm 0.0108755$, and $d$ is the dimensionality of the lattice.

Table 4.9 Number $N_{\text{conf}}$ of Configurations for Clusters with Various Number $N_C$ of Spins

| $N_C$ | 2 | 3 | 4 | 5 | 6 | 7 | 8 |
|---|---|---|---|---|---|---|---|
| $N_{\text{conf}}$ | 3 | 15 | 86 | 534 | 3481 | 23,502 | 162,913 |

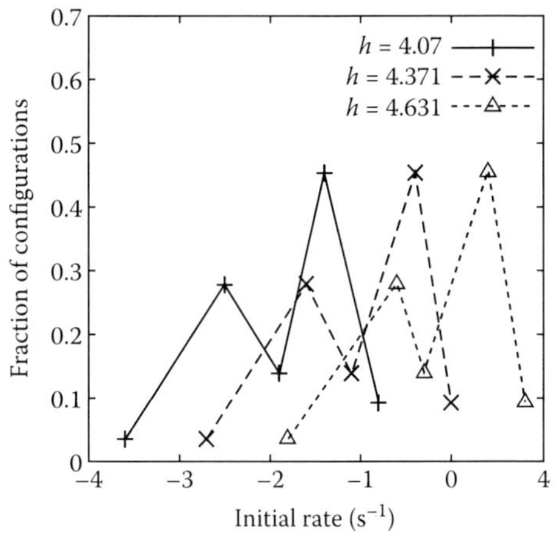


Figure 4.14

The dependence of the initial rate spectrum of a four-spin cluster as a function of supersaturation *h*.

#### 4.2.4.2 Excitation Energy and Initial Rate

It is obvious (see, e.g., Table 4.9) that the clusters with relatively large numbers of spins ($N_C \geq 4$) have a very large number of possible configurations and, consequently, a wide spectrum of initial rates. To simplify the analysis, we examine below the clusters that consist of four "up" spins because they have a relatively small number of configurations. The results obtained are still valid qualitatively for clusters of larger sizes.

The four-spin cluster has 86 possible configurations, three of them with the excitation energy $E = \{4, 32\}$ (square cluster, located in the $XY$, $XZ$, and $YZ$ planes) and the other 83 clusters with $E = \{4, 36\}$.

The $h$ dependence of the initial rate spectrum of this cluster at $T = 5.64$ K is presented in Figure 4.14. For $h < 4.371$, all the initial rates are negative, i.e., a four-spin cluster will decay at the initial moment regardless of its configuration. For $h > 4.371$, a fraction of configurations, which increases with increasing $h$, possess positive initial rates. The existence of such a fraction opens the channel for cluster growth and, hence, $h_{crit} \simeq 4.371$ can be considered as a critical supersaturation for a four-spin cluster at $T = 5.64$ K. In other words, one can consider the four-spin cluster as a critical cluster for $T = 5.64$ K and $h = 4.371$. This result is comparable with the extrapolation of those obtained by the Monte Carlo simulations, which have been carried out between $h = 2.5$ and $h = 4$. Note that this critical size is shape-independent.

Another kind of shape-independent kinetic critical size $n_{VF}^{kin}$ was defined in Ref. [9] through the condition that the *mean* value of the initial growth rate $n_1$ be equal to zero (the averaging was carried out over all possible configurations of clusters of a given size). Such a definition leads, obviously, to a larger critical size when compared to ours. Because of the large number of configurations, Vehkamaki and Ford [9] used the Monte Carlo method to perform the averaging. However, for a relatively small number of configurations such an average can be calculated exactly without difficulty. For example, for the case considered above ($T = 5.64$ K, $h = 4.371$) $n_{VF}^{kin} = 6$, whereas our definition leads to $n^{kin} = 4$. For comparison, one can note that the critical sizes for the cubic, stairlike, and pyramidal shapes of the clusters at these values of $T$ and $h$ are given by $n_{cube}^{kin} = 1$, $n_{stair}^{kin} = 8$, $n_{pyramid\,B}^{kin} = 24$, and $n_{pyramid\,A}^{kin} = 32$, respectively.

### 4.2.5 Conclusion

In this paper, expressions for the initial rate $n_1$, excitation energy $E$, and critical cluster sizes in the three-dimensional Ising model were derived. Exact analytical expressions for $n_1$ and $E$ for cubic, stairlike, and pyramidal clusters were obtained. Two kinds of critical clusters are defined: the energetic and the kinetic critical clusters. The first was obtained from the maximum of the excitation energy, while the second from the condition that the rate of growth be zero. In the LT–LS and LT–HS ranges, the kinetic critical size decreases with increasing roughness of cluster surface: $n_{cube}^{kin} > n_{stair}^{kin} > n_{pyramid\,B}^{kin} > n_{pyramid\,A}^{kin}$.

The results obtained have been compared with Monte Carlo simulations [8]. All kinetic critical sizes, except for the cubical one, fitted the simulations well, the stairlike critical cluster size providing the best fit. In most cases the energetic critical size was larger than the kinetic one. Among the energetic critical sizes only the cubic cluster fitted the simulation data well. It seems that the kinetic approach better describes the real situation than the energetic one.

To introduce a shape-independent critical cluster size, all possible configurations on a cubic lattice were generated by a computer for clusters with up to eight spins. The number of possible configurations increases exponentially with the cluster size with an exponent equal to approximately $0.793N_C$, $N_C$ being the number of spins in the cluster. A comparison with the corresponding result for the two-dimensional model, where this exponent is equal to $0.535N_C$ [1], leads to the formula: $N_{config} \simeq 10^{(\alpha_{conf} N_C - \beta_{conf})d}$ with $\alpha_{conf} \simeq 0.266$, and $\beta_{conf} \simeq 0.407$, with $d$ being the dimensionality of the lattice.

For the four-spin cluster, the initial growth rate of all possible configurations was calculated. At a given temperature, the spectrum of initial rates is shifted toward larger values with increasing $h$. For $h$ larger than a critical value $h_{crit}$ of the supersaturation, the initial rate becomes positive for a small fraction of configurations. Therefore, at a given temperature and for $h = h_{crit}$, the four-spin cluster can be considered as a critical one. At $T = 5.64$ K, the four-spin cluster becomes critical for $h = 4.371$.

## References


1. Berim, G.O., Ruckenstein, E., *J. Chem. Phys.* 117, 4542 (2002). (Section 4.1 of this volume.)
2. Nowakowski, B., Ruckenstein, E., *J. Chem. Phys.* 94, 1397 (1991). (Section 2.1 of this volume.)
3. Nowakowski, B., Ruckenstein, E., *J. Chem. Phys.* 94, 8487 (1991). (Section 2.3 of this volume.)
4. Glauber, R., *J. Math. Phys.* 4, 294 (1963).
5. Heermann, D.W., Coniglio, A., Klein, W., Stauffer, D., *J. Stat. Phys.* 36, 447 (1984).
6. Stauffer, D., Kertesz, J., *Physica A* 177, 381 (1991).
7. Schmelzer, J., Jr., Landau, D.P., *Int. J. Mod. Phys. C* 12, 345 (2001).
8. Wonczak, S., Strey, R., Stauffer, D., *J. Chem. Phys.* 113, 1976 (2000).
9. Vehkamaki, H., Ford, I.J., *Phys. Rev. E* 59, 6483 (1999).
10. Stauffer, D., *Int. J. Mod. Phys. C* 3, 287 (1992).

# 4.3 Shape-Dependent Small Cluster Kinetics in the Two-Dimensional Ising Model beyond the Classical Approximations

*Gersh Berim and Eli Ruckenstein**

The kinetics of small clusters of "up" spins in the two-dimensional Ising model on a square lattice is examined without the usual approximations of fixed cluster shape, constant number of "down" spins, and nonsplitting-noncoagulating dynamics. New kinetic equations for the number densities of clusters of various sizes and shapes are derived and solved numerically. It is shown that the kinetic behavior of small clusters depends on their shape and that the time dependence of the total number of down spins and the splitting and coagulation of clusters significantly affect various characteristics of the system, e.g., the range of validity of the steady-state approximation, the transient time to that state, and the values of mass fluxes. The influence of these factors grows with increasing temperature and supersaturation. (© *2003 American Institute of Physics.*)

## 4.3.1 Introduction

In the past few years, much attention has been paid to the investigation of kinetic phenomena in Ising spin systems on two- and three-dimensional rigid lattices [1–13]. In such systems, a cluster (nucleus) is defined as a group of bounded "up" spins in a sea of "down" ones. Two mechanisms of spin dynamics were considered. The first one, the most frequently used, is the Glauber [14], or Metropolis [15], single-spin dynamics, which does not conserve the total number of up spins in the system. In this case, the down spins constitute a reservoir from which additional up spins can be added to the existing clusters, or isolated one-spin clusters (single up spin with all its nearest neighbors down) can arise due to down–up transitions. Up spins cannot move along the lattice, but can arise and disappear at any of its sites with probabilities that satisfy the detailed balance condition. One can relate the up and down spins to two different phases.

The second mechanism, the Kawasaki two-spin dynamics [16], implies simultaneous flip-flop transitions of two neighboring oppositely oriented spins that conserve the total number of up spins. In this case the up and down spins can be considered as the solute and the solvent molecules of a two-component solution.

The Monte Carlo simulation technique was mainly employed [1–3,8–11] and the results (cluster size distribution, critical cluster size, nucleation rate, etc.) were compared with the predictions of the conventional nucleation theories [17,18] as well as with exact calculations based on the Ising model at the zero-temperature limit [19,20]. The conventional theories involve three important assumptions which considerably simplify the problem. First, it is assumed that the nucleus (cluster) of a new phase has a shape independent of the number of molecules it contains. In most cases this shape is considered to be spherical, but sometimes, when the classical theory is applied to nucleation on a rigid lattice, one assumes a shape, consistent with the lattice symmetry, e.g., the cubic one [5]. A fixed shape allows one to define uniquely a critical cluster size, $n^*$, using either energetic [17,18] or kinetic [21,22] approaches, and to obtain the nucleation rate $I$ as the total flux of clusters at steady state.

The second assumption is that the cluster can lose or gain only single molecules (monomers) with probabilities estimated usually for large spherical clusters. It is usually supposed that neither coagulation nor splitting of the clusters occur.

The third assumption of most of the traditional theories is that the number density of the monomers is constant; this assumption is applicable as long as the total number of monomers is much larger than the total number of molecules in the existing clusters.

The single-shape assumption can be applied to Ising systems only for the later stages of the kinetic process, when large clusters exist. At low temperatures, the shape of the clusters follows the symmetry of the lattice, while at high temperatures it is almost circular for a two-dimensional lattice (see, e.g., Refs. [23–25], where these issues were examined for two-dimensional Ising lattices).

The nonsplitting assumption is also valid only in the limit of large clusters, when the probability of cluster splitting due to the up–down transitions of a few spins is negligible. Obviously, the small clusters in Ising systems, which dominate during the early stages of the kinetic process, have a number of shapes and can split.

To describe this initial stage, it is necessary to develop an approach which takes into account the specific features of the Ising systems on rigid lattices and is free at least of some of the restrictions of the classical theories.

* *J. Chem. Phys.* 119, 806 (2003). Republished with permission.

In previous papers [12,13], we considered the effect of the shape on the critical cluster size for two- and three-dimensional Ising models on square and cubic lattices and suggested that the Becker–Döring conservation equation should be modified to account for the effect of the shape. Kinetic as well as energetic (classical) approaches were used in those papers. In the kinetic approach, the critical cluster was provided by the equality between the rates of growth and decay, while in the energetic one by the maximum of the cluster excess (excitation) energy. In both cases, the critical size was dependent on the shape of the cluster. Shape-dependent critical clusters have been calculated and the results were compared with existing Monte Carlo simulations. In most cases, the kinetic approach provided better agreement with the Monte Carlo simulations than the energetic one.

In the present paper, the Ising interactions between spins, and the Glauber single-spin-flip dynamics are used to derive kinetic equations for the densities of clusters of various sizes and shapes. In contrast to the classical Becker–Döring equation, which involves a single shape and a growth decay of clusters by single spins, the new equations involve all possible cluster shapes compatible with their size, as well as their coagulation and splitting. Two specific models will be employed. The first one (M1) is based on a single spin loss or gain assumption, but accounts for the depletion of the down spins. The existence of a quasi-steady state, state in which all the quantities change very slowly, will be thus proven. In the range of supersaturations in which a quasi-steady state does exist its characteristics as well as the transient behavior to that state will be examined. In the second model (M2), the possibility of splitting and coagulation of the small clusters because of the flip of a single spin of the cluster (or of the neighboring ones) is taken into account, by keeping, however, the assumption of constant density of down spins. The considerations are restricted to the case of a small density of up spins compared to a density of down ones. From our calculations, one can conclude that the shape should be considered as an important characteristic of the cluster along with its size and energy.

A similar but less general model, which accounts for various cluster shapes was developed independently by Shneidman and Nita [10]. Using a generalized Becker–Döring approach and the Metropolis spin dynamics, the steady-state total flux $I$ was calculated in that paper using the other two simplifying assumptions of the classical theory (constant number of down spins and the absence of splitting and coagulation).

### 4.3.2 Model 1

#### 4.3.2.1 Kinetic Equations

The Hamiltonian of the Ising model under consideration has the following form:

$$H = -B\sum_{j}\sigma_j - J_0\sum_{\langle j,k\rangle}\sigma_j\sigma_k, \tag{4.78}$$

where the spins $\sigma_j$ are located on the sites of a square lattice and have the values ±1, $\langle j, k\rangle$ denotes summation over the nearest neighbors, each pair of interacting spins being counted only once.* The magnetic field $B$ and the exchange integral $J_0 > 0$ are expressed in units of the Boltzmann constant $k_B$ and have therefore the dimension of temperature.

We define a cluster as a group of bounded up spins in a sea of down spins and suppose that at $t < 0$ the system is in equilibrium with most spins down. Such a state can be achieved in the presence of an external magnetic field $B_0$ in the down direction ($B_0 < 0$). We will call this state the *natural* initial state of the system. If at $t = 0$ the field changes direction to the up one and acquires a value $B > 0$, then the system will be in a metastable state far from equilibrium. In what follows, we consider a temperature range below the second-order phase transition temperature $T_c$ ($T_c = 2.269J_0$) [26], because in this range the lifetime of the metastable state is large.

One also can consider another (*artificial*) initial state, by supposing, for example, that at $t = 0$ all spins but one are down and the external field is positive. Such an initial state allows one to examine the independent growth of a single cluster [4,8,9], provided that the cluster is a critical or supercritical one. At very low temperatures, the field has to exceed $2J_0$ for a single spin cluster to become a nucleus. However, at intermediate and high temperatures much weaker fields are sufficient [12].

* In terms of the lattice-gas variables $p_j = (1 + \sigma_j)/2$, the Hamiltonian (Equation 4.78) has the form $H = -\mu\Sigma_j p_j - \varepsilon\Sigma(j,k)p_j p_k$, where $\mu = 2B - 4J_0$ and $\varepsilon = 4J_0$.

The further evolution from the initial state is provided by the Glauber single-spin-flip dynamics [14], which finally brings the system to a stable state with almost no down spins. Such a kind of kinetics was often used to describe the evolution of the total magnetization (see, e.g., Ref. [27] and references therein).

To examine the kinetics of small clusters we consider the time-dependent densities $f_{m,\alpha}$ of clusters with $m$ spins, where the index $\alpha$ labels the different shapes of the clusters. The number of shapes will be denoted $\alpha_m$. Two $m$-spin clusters are considered to have the same shape (and to belong to the same *class* $\{m,\alpha\}$ [10]) if they can be superposed by rotation and reflection. Several kinds of classes are sketched in Figure 4.15. The Glauber dynamics provides the transitions between classes by assuming that they lose or gain a single spin. The probabilities of these transitions are dependent on the energy difference between the initial and final clusters and satisfy the detailed balance condition. A cluster grows if one of its nearest neighbors makes a down–up transition (condensation). A cluster shrinks if one of its own boundary spins makes an up–down transition (evaporation). The probability densities (rates) of these transitions for the condensation, $C$, and evaporation, $E$, depend on the environment of the flipping spin and are given by

$$C(\mu)=\frac{1}{2}(1+z_\mu)\,W,\; E(\mu)=\frac{1}{2}(1-z_\mu)W, \tag{4.79}$$

where $W$ is a constant which provides the timescale of the process, $z_\mu = \tanh(h/2 + \mu J/4)$, with $\mu(= 0, \pm 2, \pm 4)$ the sum of the values of the nearest neighbors spins of a selected spin. Instead of the magnetic field $B$ and exchange integral $J_0$, we use the dimensionless quantities $J = 4J_0/T$ and $h = 2B/T$. The latter quantity ($h$) plays, in the context of nucleation, the role of supersaturation [5,10]. The coefficient $W$ in expressions (Equation 4.79) of the rates $C(\mu)$ and $E(\mu)$ will be selected equal to 2.

Each class has its own excitation energy $E_{m,\alpha}$ (which is defined as the energy needed to generate a cluster from the sea of down spins) and the rates $r^+_{m,\alpha}$ and $r^-_{m,\alpha}$ of growth and decay, respectively. To calculate $r^+_{m,\alpha}$, for the first model one should account for the down–up transitions of the nearest neighbor spins of a given cluster. To obtain $r^-_{m,\alpha}$, only those up–down transitions of the spins of the cluster will be considered which do not lead to splitting. The equality $r^+_{m,\alpha} = r^-_{m,\alpha}$ provides a temperature-dependent critical supersaturation $h^{(c)}_{m,\alpha}$ for the class $\{m,\alpha\}$ [12]. For $h > h^{(c)}_{m,\alpha}$, the clusters belonging to class $\{m, \alpha\}$ grow, and for $h < h^{(c)}_{m,\alpha}$ they shrink. Different classes, even with the same number of spins, have different critical supersaturations because their shapes are different. For example, at $T = 0.1T_c$ ($J = 17.63$), $h^{(c)}_{4,1} = 16.02$, $h^{(c)}_{4,2} = 15.43$, $h^{(c)}_{4,3} = 8.47$, $h^{(c)}_{4,4} = 8.12$, $h^{(c)}_{4,5} = 15.55$. Their excitation energy spectrum, calculated with the help of the Hamiltonian (1), is highly degenerate ($E_{4,1} = E_{4,2} = E_{4,4} = E_{4,5} = 20J_0 - 8B$, $E_{4,3} = 16J_0 - 8B$).

The following simple example clarifies the manner in which the kinetic equations for the densities $f_{m,\alpha}$ of clusters belonging to class $\{m,\alpha\}$ were derived. Let us consider the class {2, 1} (Figure 4.15). It can grow to two three-spin clusters; the first one, of class {3, 1}, has a density $f_{3,1}$, and can be obtained from class {2, 1} in two ways, each with the rate $C(-2)$. The second possible three-spin cluster belongs to class {3, 2}, and can be generated in four ways from the initial {2, 1} class, each with the rate $C(-2)$. The inverse transitions $\{3,1\to 2,1\}$ and $\{3,2\to 2,1\}$ can be achieved in two ways, each with the rate $E(-2)$.

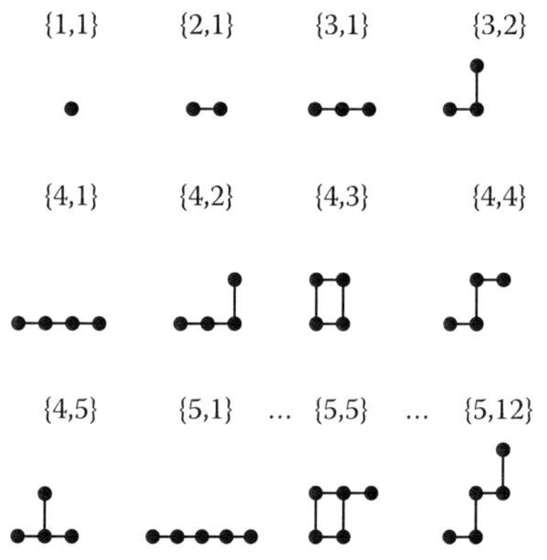


Figure 4.15

Some of the classes (clusters with different numbers of spins and different shapes) considered in M1. Filled circles represent the up spins.

The initial cluster can shrink in two ways, each with the rate $E(-2)$, to a one-spin cluster of class {1, 1}, which has the density $f_{1,1}$. The inverse transition {1,1→2,1} can occur in four ways, each with the rate $C(-2)$. The gain or loss of one spin by class {2, 1} decreases the density $f_{2,1}$, while the inverse transitions {3,1→2,1} and {3,2→2,1} increase the density $f_{2,1}$. The resulting kinetic equation for $f_{2,1}$ has, therefore, the form

$$\frac{df_{2,1}}{dt} = A_{1,1\to 2,1} f_{1,1} + A_{3,1\to 2,1} f_{3,1} + A_{3,2\to 2,1} f_{3,2} - A_{2,1} f_{2,1}, \tag{4.80}$$

where $A_{1.1\to 2.1} = 4C(-2)$, $A_{3,1\to 2,1} = A_{3,2\to 2,1} = 2E(-2)$, $A_{2,1} = 2E(-2) + 6C(-2)$. In the same way, one can write the kinetic equations for all cluster densities containing more than one spin.

To write a kinetic equation for the density $f_0$ of down spins, one has to take into account that $f_0$ increases due to shrinking of any of the existing clusters and decreases due to their growth and to the creation of new one spin clusters in the sea of down spins. The rate of the last process is proportional to the total number of those down spins whose nearest neighbors are all down spins. This number can be obtained by subtracting the number of down spins neighboring all the clusters of the system at time $t$ from the total number of down spins and is equal to $f_0 - \Sigma_{n,\beta} N_{n,\beta} f_{n,\beta}$, where $N_{n,\beta}$ is the number of nearest neighboring spins of a cluster belonging to class $\{n,\beta\}$. As a result, the equation for $f_0$ has the form

$$\frac{df_0}{dt} = -C(-4)\left(f_0 - \sum_{n,\beta} N_{n,\beta} f_{n,\beta}\right) + \sum_{n,\beta}\left(r_{n,\beta}^{-} - r_{n,\beta}^{+}\right) f_{n,\beta}. \tag{4.81}$$

In an analogous way, one can write a kinetic equation for $f_{1,1}$:

$$\frac{df1,1}{dt} = C(-4)\left(f_0 - \sum_{n,\beta} N_{n,\beta} f_{n,\beta}\right) + A_{2,1\to 1,1} f_{2,1} - A_{1,1} f_{1,1}. \tag{4.82}$$

The whole system of kinetic equations for the densities $f_{m,\alpha}$ has, therefore, the form

$$\begin{aligned}
&\frac{df_0}{dt} = -C(-4)\left(f_0 - \sum_{n,\beta} N_{n,\beta} f_{n,\beta}\right) + \sum_{n,\alpha}\left(r_{n,\beta}^{-} - r_{n,\beta}^{+}\right) f_{n,\beta},\\
&\frac{df_{1,1}}{dt} = C(-4)\left(f_0 - \sum_{n,\beta} N_{n,\beta} f_{n,\beta}\right) + A_{2,1\to 1,1} f_{2,1} - A_{1,1} f_{1,1},\\
&\vdots\\
&\frac{df_{m,\alpha}}{dt} = \sum_{m',\alpha'} A_{m',\alpha'\to m,\alpha} f_{m',\alpha'} - A_{m,\alpha} f_{m,\alpha},\\
&m = 1,2,\ldots;\ \alpha = 1,2,\ldots\alpha_m,
\end{aligned} \tag{4.83}$$

where the sum over $m'$ and $\alpha'$ is taken over all classes that provide transitions to class $\{m,\alpha\}$ with $m = 1, 2, \ldots$ and $\alpha = 1, 2, \ldots, \alpha_m$, and the coefficients $A_{i,j}$ are functions of $h$ and $J$, as already noted for the three-spin clusters. If not mentioned otherwise, we will use below the natural initial conditions by assuming for the initial densities $f_{m,\alpha}$ values corresponding to a magnetic field $B_0$ ($B_0 < 0$). They can be approximated by $f_{m,\alpha}(0) \approx w_{m,\alpha} \exp[-E_{m,\alpha}/T]$ [8,9], where $w_{m,\alpha}$ is a statistical weight and $E_{m,\alpha}$ is the excitation energy of a cluster $\{m,\alpha\}$.

If the densities $f_{m,\alpha}$ are known, one can calculate the time-dependent fluxes between different classes of clusters. For example, the flux $I_{2,1\to 3,1}$ between {2,1} and {3,1} classes can be written as

$$I_{2,1\to 3,1} = 2C(-2) f_{2,1} - 2E(-2) f_{3,1},$$

and so on. It should be noted that the right-hand side of Equation 4.83 can be rewritten in terms of fluxes, as in the traditional Becker–Döring theory. The latter form is convenient for the analysis of the steady state, when all the derivatives in the left-hand side of Equation 4.83 are equal to zero, but inconvenient for the analysis of time-dependent processes. The calculation of the total steady-state flux for an Ising lattice gas model was also performed by Shneidman and Nita [10] on the basis of kinetic equations for fluxes that account for the shapes of the clusters. The main difference between the present equations (Equation 4.83) and the equations in Ref. [10] is the presence in Equation 4.83 of an additional equation for the number density $f_0$ of the down spins as well as a term proportional to $C(-4)$ in the equation for the number density $f_{1,1}$ of the single spin clusters.

To close the system of equations (Equation 4.83), we will use the so-called Szillard boundary condition [18], according to which all clusters starting with a selected size $n_G$ are removed from the system. In the classical theory, which is based on the single-shape assumption [18], this size has to be larger than the critical size $n^*(h)$, which depends on the supersaturation $h$. If one fixes $n_G$, then such a choice restricts the possible values of $h$ for which $n^*(h) < n_G$. Because we examine the kinetics of small clusters we will select $n_G = 6$. There is no fixed value of $n^*(h)$ due to the additional dependence of the critical size on the shape of the cluster [12,13]. As the smallest value of $h$ we will choose that one at which at least one of the possible five-spin clusters is critical. For example, for $T = 0.1T_c$, this value of $h$ is 8, for $T = 0.4T_c$ it is 1.3, whereas for $T = 0.8T_c$ it is 0.5. At these values of $h$, the cluster of the class {5, 12} is the critical one. After this choice of $n_G$, the system of equation (Equation 4.83) reduces to 22 equations, which can be solved numerically.

Due to the linearity of Equation 4.83, the time behavior of all the quantities (densities $f_{m,\alpha}$, fluxes $I_{m',a'\to m,\alpha,\ldots}$) has the following general form:

$$Q(t) = \sum_k C_k \exp(-t/\tau_k), \tag{4.84}$$

where all the relaxation times $\tau_k$ are positive because of the physical nature of the problem, and $C_k$ can be constants, polynomials, or oscillatory functions of $t$.

At large times, all the quantities must be zero because of the depletion of the down spins. The characteristic time $\tau$ for the transition to the final state (lifetime of the metastable state) can be estimated by the largest $\tau_k$. If this time is much larger than the other $\tau_k$s, then the steady-state approximation can be used in the time interval $\tau \gg t \gg \tau_s$, where $\tau_s$ (the transient time to the quasi-steady state) is the second largest relaxation time among the $\tau_k$s. The quasi-stationary value of $Q(t)$ (i.e., the value it has for times $t$, $\tau \gg t \gg \tau_s$) is provided by the coefficient $C_k$ of the exponential $\exp(-t/\tau)$.

In Figure 4.16, the relaxation times $\tau$ (lifetime of the metastable state) and $\tau_s$ (the transient time) are plotted as functions of $h$ for three different temperatures and for the values of $h$ compatible with the Szillard

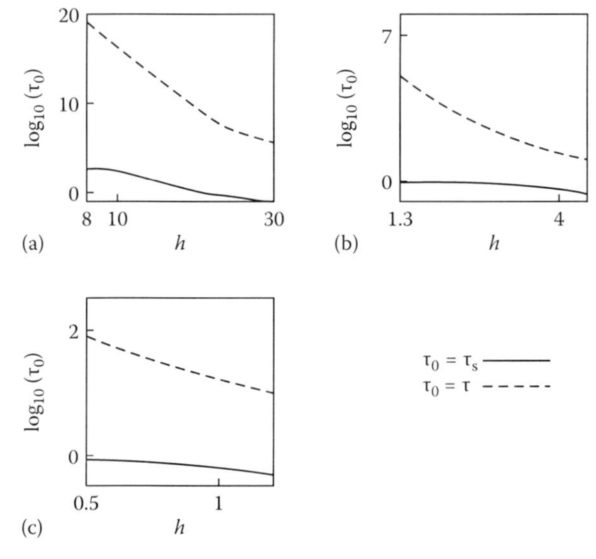


Figure 4.16

$h$ dependence of transient times to the final state ($\tau$) and to the quasi-steady state ($\tau_s$) at $T = 0.1T_c$ (a), $T = 0.4T_c$ (b), and $T = 0.8T_c$ (c) for M1.

boundary condition. One can see that the lifetime of the metastable state decreases with increasing temperature $T$ and supersaturation $h$. Qualitatively, the same dependence of the metastable lifetime was observed in Monte Carlo simulations for the two-dimensional Ising model [6]. Unfortunately, we cannot provide a quantitative comparison because the range of supersaturations considered here is outside that used in Ref. [6].

The transient time depends weakly on $T$ and $h$ at intermediate and high temperatures. In the entire ranges of $T$ and $h$ considered, $\tau \gg \tau_s$ and, consequently, the steady-state approximation can be used.

#### 4.3.2.2 Quasi-Steady State

In this section, the quasi-steady-state values of the fluxes between various classes (i.e., the values which they have for $\tau_s \ll t \ll \tau$) will be calculated for several temperatures and supersaturations $h$. For this time interval, the number of down spins can be considered constant $f_0(t) = N_0 = \text{const}$. For illustration purposes, only the fluxes from the four-spin classes (there are five such classes) to the five-spin classes (12 classes) will be calculated.

First a very low temperature $T/T_c \ll 1$ was selected, namely, $T = 0.1T_c$ (hence, $J = 17.63$). To compare the contributions of the critical and subcritical clusters with the same number of spins but different shapes to the total flux, a value of $h$ was selected for which a fraction of the four-spin clusters is critical or supercritical (i.e., has zero or a positive rate of growth) and the other fraction is subcritical. For example, for $h = 9$ the classes {4, 3; 4, 4} (Figure 4.15) are critical and the classes {4, 1; 4, 2; 4, 5} subcritical. It should be noted that from an energetic (classical) viewpoint all these clusters are subcritical.

The total flux $I_{\text{total}}$ is equal to the flux $I_{1,1\to 2,1}$, and for the selected values of the parameters $I_{\text{total}} = 4.60 \times 10^{-19} N_0$. The fluxes from each of the four-spin clusters to the five-spin ones, which can be obtained by adding one spin, are given by the following: $I_{4,1\to 5} = 1.41 \times 10^{-7} I_{\text{total}}$, $I_{4,2\to 5} = 3.80 \times 10^{-4} I_{\text{total}}$, $I_{4,3\to 5} = 0.999\, I_{\text{total}}$, $I_{4,4\to 5} = 2.40 \times 10^{-4}\, I$ total, $I_{4,5\to 5} = 3.24 \times 10^{-4}\, I_{\text{total}}$. The main contribution to the total flux is that of class {4, 3} (square cluster), for which the excitation energy $E_{4,3} = 16J_0 - 8B$ is the smallest; all the other ones have negligible contributions at the low temperature selected. Similar results were obtained for the fluxes from the five- to the six-spin clusters. In this case, the dominant contribution was provided by the flux $I_{5,5\to 6} = 0.999\,999 I_{\text{total}}$, which has its origin in the class {5, 5}. This class has the smallest excitation energy among the five-spin clusters. The above results are in agreement with those obtained by Shneidman and Nita [10]. The important role of rectangular clusters in the low-temperature kinetics of the Ising model was also noted in Refs. [19,28,29], where the "protuberance" mechanism of cluster growth was emphasized. In our case, this mechanism corresponds to the transition from the class {4, 3} to the class {5, 5}.

At any given temperature, the relative contribution of the lowest energy path to the total flux can be considerably affected by the supersaturation $h$ (see, e.g., Figure 4.17a). The class {4, 3} dominates the total flux up

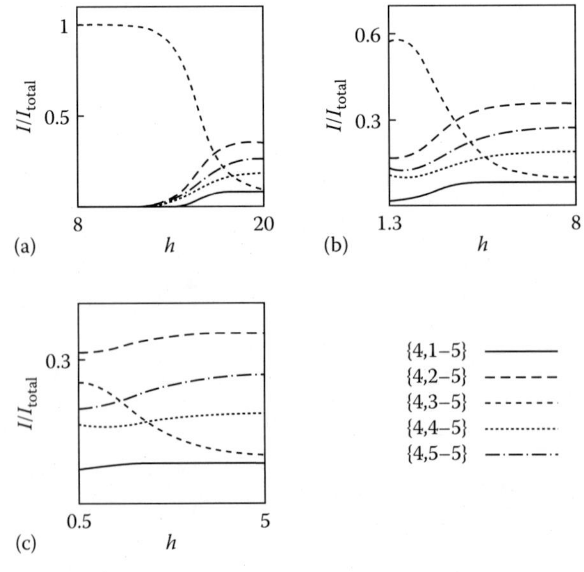


Figure 4.17

Relative steady state values of different fluxes $\{m, \alpha \to 5\}$ from four-spin to all five-spin classes as functions of supersaturation at $T = 0.1T_c$ (a), $T = 0.4T_c$ (b), and $T = 0.8T_c$ (c) for M1.

to $h \sim 14.5$. At this value of $h$, all the four-spin clusters become critical and the relative contribution of class {4, 3} represents about one-half of the total flux. For higher values of $h$, other classes dominate the total flux. In all the considered cases, the class {4, 3} has the minimum excitation energy among the four-spin clusters. This means that the lowest energy criterion does not provide always the main contribution to the total flux.

One can clearly see from Figure 4.17 that the contribution of the cluster with the smallest energy to the total flux decreases with increasing temperature. At a relatively high temperature, namely, $T \simeq 0.8T_c$, its contribution is no longer dominant regardless the value of $h$ (Figure 4.17c).

At all temperatures and supersaturations, the classes {4, 1}, {4, 2}, {4, 4}, {4, 5}, which have equal excitation energies, provide different contributions to the total flux. This fact demonstrates the importance of the shapes of the clusters in their kinetics.

#### 4.3.2.3 Transient Regime in Small Cluster Kinetics

In Figure 4.18, typical examples for the time dependence of the fluxes from four- to the five-spin clusters calculated numerically are plotted at the low temperature of $T = 0.1T_c$, for various supersaturations $h$. In Figure 4.18a and b, one can see that the transient time $\tau_s$ of a cluster belonging to class {4, $\alpha$} depends at low values of $h$ ($h = 9$) on the cluster shape and has a maximum for the square cluster {4, 3}, which has the smallest excitation energy. The values of the transient times are $\tau_s = 479.5$ (in units of $1/W$) for cluster {4, 3} and $\tau_s = 0.5$ for all the other four-spin clusters. It should be noted, that for the selected value of $h$ ($h = 9$) the latter classes provide negligible contributions to the total flux. For larger values of $h$ (Figure 4.18c and d), all the classes have approximately the same transient time $\tau_s$ and provide comparable contributions to the total flux.

In Figure 4.19, the $h$ dependence of the transient time $\tau_s$ for cluster {4, 3} is plotted for various temperatures. The common feature of these curves is the presence of a maximum. The position of this maximum

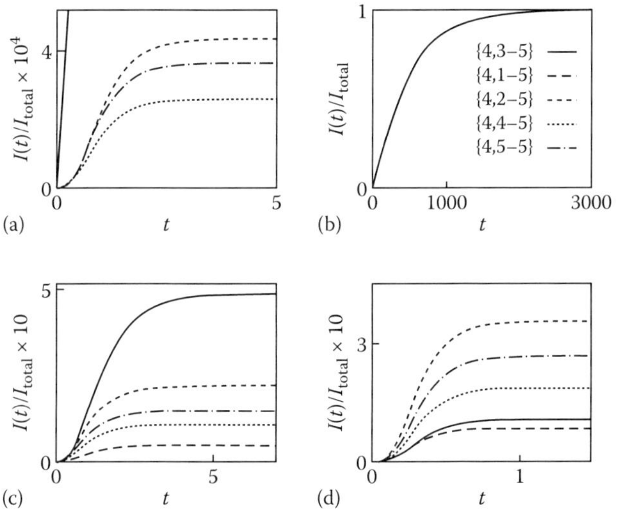


Figure 4.18

Time dependence of the relative values of the fluxes for M1 at (a) and (b) $T = 0.1T_c$, $h = 9$; (c) $T = 0.4T_c$, $h = 2$; and (d) $T = 0.8T_c$, $h = 0.7$. The flux $I_{4.1\to5}$ in part (a) has a very small value and it is not shown.

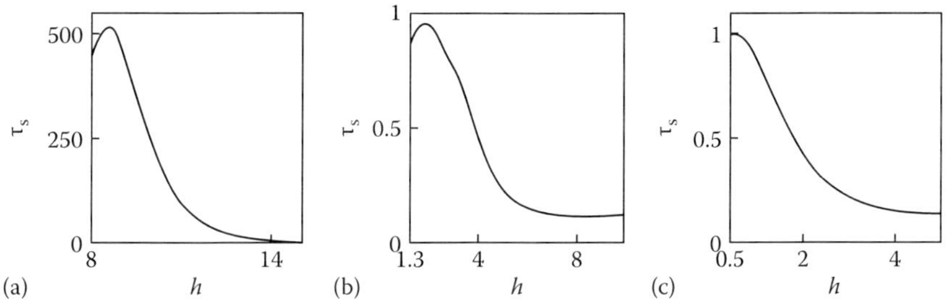


Figure 4.19

$h$ dependence of the transient time $\tau_s$ at (a) $T = 0.1T_c$, (b) $T = 0.4T_c$, and (c) $T = 0.8T_c$ for M1.

almost coincides with the value of $h$ for which the classes {4, 3} and {4, 4} are critical and the other four-spin classes are subcritical. It is not yet clear if this coincidence has some physical meaning.

As expected, at low temperatures the transient time is much larger than at high temperatures.

### 4.3.3 Model 2

In this section, we will examine a model which takes into account single-spin-flip transitions which lead to the splitting of the small clusters as well as to coagulation in some specific situations. Let us consider for illustration, the cluster of class {4, 1} (Figure 4.15). The flip of one of its internal up spins splits the cluster into two groups of spins becoming what we call a pseudo-cluster consisting of a one-spin cluster {1, 1} and a two-spin cluster {2, 1}, which are located in specific relative positions. Vice versa, if the clusters of classes {1, 1} and {2, 1} are in suitable positions (such as the ones they had after the splitting of the class {4, 1}), then the down–up transition of the intermediate spin leads to their coagulation and to the creation of one cluster of class {4, 1}.

Let us introduce now, in addition to the existing classes, a new class, {3, 3} (Figure 4.20), representing the pseudo-cluster described above as well as those, which can be obtained from this pseudo-cluster by rotation and reflection. In a similar way, other pseudo-clusters and the corresponding new classes can be generated by applying a similar procedure to all classes of M1. The main feature of the new classes is that their groups of spins can coagulate through a down–up transition of a single spin, generating one of the classes considered in M1. On the other hand, by allowing any of the spins of the clusters of M1 to make an up–down transition one can generate both new and old classes. By deriving kinetic equations for the densities $f_{m,\alpha}$ of all classes, one can account also for the splitting and coagulation of the small clusters.

Note that because of the Szillard boundary condition, which eliminates in our case all the clusters larger than the five-spin ones, the pseudo-clusters cannot contain more than four up spins. Examples of the new classes are sketched in Figure 4.20. The total number of classes (old and new) will be 1 for one- and two-spin classes, 6 for three-spin classes, 25 for four-spin classes, and 12 for five-spin classes. The smaller number of classes for the five- than for four-spin clusters is a consequence of the absence of the six-spin clusters from which five-spin pseudo-clusters can be generated.

To describe the evolution of a cluster or a pseudo-cluster, we will consider the up–down transitions of all their up spins, the down–up transitions of the nearest neighbors of the up spins of the cluster or pseudo-cluster, and, in addition, the down–up transitions of those of the next-nearest neighbors of the up spins which together with the spins of the cluster can form a pseudo-cluster (the nearest neighbors in the diagonal direction are also considered as next-nearest neighbors). If, as a result of these transitions, pseudo-clusters that contain five spins are formed, their constituent clusters will be treated independently. For example, the down–up transition of the spin marked with a cross in class {4, 10} (Figure 4.20) will generate clusters of classes {2, 1} and {3, 1} (Figure 4.15).

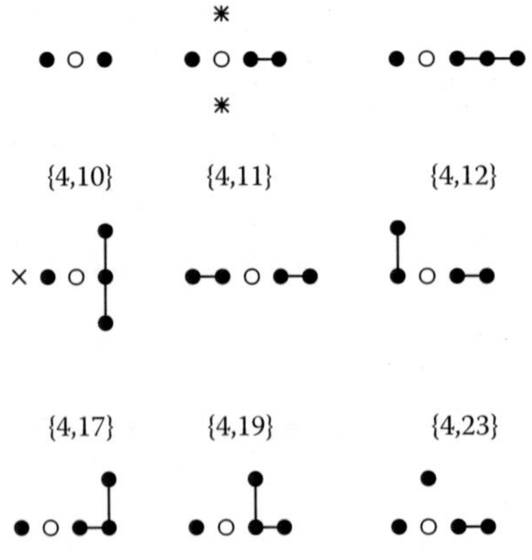


Figure 4.20

Some of the new classes considered in M2. The filled circles represent the up spins, the empty circles, the cross, and the stars represent the down spins.

To illustrate the above procedure, let us consider in some detail the derivation of the kinetic equation for the density $f_{3,3}(t)$ of a pseudo-cluster belonging to class {3, 3}. All classes, involved in this example are listed in Figures 4.15 and 4.20. The up–down transitions of both outermost up spins of cluster {3, 3} generate clusters of classes {2, 1} and {2, 2} with rates $E(-4)$ and $E(-2)$, respectively. The same up–down transition of the third up spin generates with the rate $E(-2)$ two one-spin clusters {1, 1}. The down–up transition of the spin marked with a circle in class {3, 3} (Figure 4.20) generates with rate $C(0)$ a cluster of class {4, 1}. The down–up transitions of the other nearest neighbors of the up spins of a given pseudo-cluster generate clusters of classes {4, 7}, {4, 11} (in one way each), {4, 12}, {4, 17}, and {4, 19} (in two ways each), each way with the rate $C(-2)$. The down–up transitions of the next-nearest neighbors, marked by stars (Figure 4.20), generate with the rate $C(-4)$ clusters of class {4, 23}. All the mentioned transitions decrease $f_{3,3}(t)$. The reverse transitions, i.e., up–down flips of the appropriate spins of the above listed four-spin clusters and pseudo-clusters increase $f_{3,3}(t)$.

As a result, the kinetic equation for $f_{3,3}(t)$ has the form

$$\frac{df_{3,3}}{dt} = -A_{3,3}f_{3,3} + \sum_{\{m,\alpha\}} A_{m,\alpha\to 3,3} f_{m,\alpha}, \tag{4.85}$$

where the index $\{m, \alpha\}$ runs over the values {2, 1}, {2, 2}, {4, 1}, {4, 7}, {4, 11}, {4, 12}, {4, 17}, {4, 19}, {4, 23} and

$$A_{3,3} = -2C(-4) - 8C(-2) - C(0) - E(-4) - 2E(-2),$$

$$A_{2,1\to 3,3} = 2C(-4), \quad A_{2,2\to 3,3} = 2C(-2),$$

$$A_{4,1\to 3,3} = 2E(0),$$

$$A_{4,7\to 3,3} = A_{4,11\to 3,3} = A_{4,12\to 3,3} = A_{4,17\to 3,3} = A_{4,19\to 3,3}$$

$$= A_{4,23\to 3,3} = E(-2)$$

The system of kinetic equations obtained in this manner contains 45 equations for 45 densities $f_{m,\alpha}$ and is too large to be written explicitly. The numerical solution of this system of equations showed that in the low-temperature range the difference between the same quantities ($f_{m,\alpha}$) in M1 and M2 does not exceed several percents, but grows with increasing temperature and supersaturation. Below, some results are provided that emphasize the new features that follow from the inclusion of splitting and coagulation in the kinetic equations.

The role of splitting and coagulation can be better demonstrated by using a special kind of Glauber dynamics and artificial initial conditions. Let us suppose, as in Refs. [8,9], that the flip-flop transition of the spins, whose all nearest neighbors are down, are prohibited, i.e., $C(-4) = E(-4) = 0$ (truncated Glauber dynamics). Such a kind of dynamics prevents the creation of new one-spin clusters in the sea of down spins and allows one to examine in a more simple way the evolution of single clusters. Suppose further, that in the initial state only the clusters of class {4, 1} (Figure 4.15) are present with $f_{4,1}(0) = 1$ and let us determine the density $f_{2,1}$ of clusters of class {2, 1} for $t > 0$. In Figure 4.21a, the time dependence of the density is presented for M1 and M2 at $T = 0.8T_c$ and $h = 2$. One can see that for M2 the value of $f_{21}$ is larger than that for M1. For M1, the cluster {2, 1} can be obtained from cluster {4, 1} only by successive up–down transitions of the outermost spins. For M2, in addition to such transitions, the splitting of the cluster {4, 1} into clusters {2, 1} and {1, 1} via the up–down transition of one of the internal spins can also occur. This explains the different time behaviors of $f_{2,1}$ shown in Figure 4.21a.

The effect of the coagulation of the small clusters is demonstrated in Figure 4.21b, where the time dependence of the density $f_{3,1}$ is plotted by assuming that at time $t = 0$ only $f_{1,1}(0)$ differs from zero for M1 ($f_{1,1}(0) = 2$), and only $f_{2,2}$ differs from zero ($f_{2,2}(0) = 1$) for M2. Thus, in both models, at $t = 0$ there are only two up spins per unit volume. For M2, these two one-spin clusters can coagulate via a down–up transition of an intermediate spin, thus generating one cluster of class {3, 1}; for M1, each of the two spins evolves independently and can grow via down–up transitions of neighboring spins, thus generating two clusters of class {3, 1}. For these reasons, at short times $f_{3,1}$ is larger for M2, and at intermediate times $f_{3,1}$ is larger for M1. At larger times, $f_{3,1}$ becomes again larger for M2 (Figure 4.21c), because of the splitting of the new clusters into independent ones, which serve as additional sources for the generation of {3, 1} clusters.

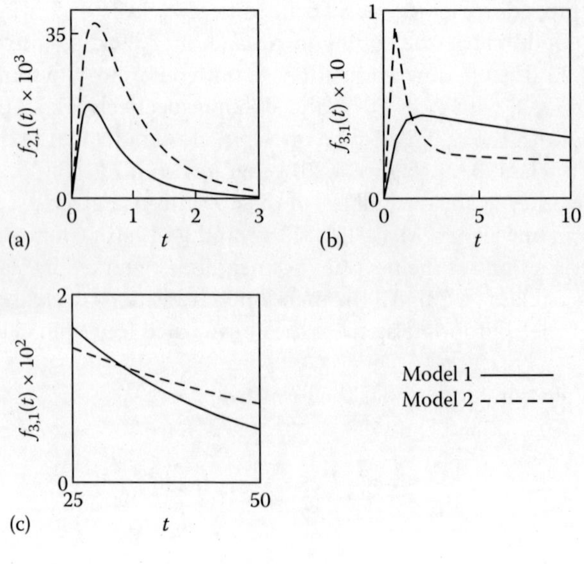


Figure 4.21

Time dependencies of densities $f_{2,1}$ and $f_{3,1}$ in M1 and M2 for artificial initial conditions at $T = 0.8T_c$: (a) $f_{4,1}(0) = 1$ both in M1 and M2, $h = 2$; (b and c) $f_{1,1}(0) = 2$ in M1, $f_{2,3}(0) = 1$ in M2. In both models, all other $f_{m,\alpha}(0)$ are equal to zero.

For the natural initial state described in Section 4.3.2.1, the effect of splitting and coagulation can be illustrated by examining the behaviors of the transient times $\tau_s$ and of the steady states of different fluxes.

The $h$ dependencies of the transient times $\tau_s$ for Models 1 and 2 are plotted in Figure 4.22 in the range of validity of our approximations. At low temperatures, there is no difference between the two, because the splitting transitions have negligible probabilities. The corresponding dependence was already shown in Figure 4.19a. At intermediate and high temperatures (Figure 4.22a and b), the difference between the two curves is greater, particularly at high supersaturations where $\tau_s$ for M2 is larger.

At intermediate and high temperatures, the relative distribution of fluxes between classes at steady state for M2 (Figure 4.23) differs from that for model M1 (Figure 4.17). However, all the qualitative conclusions concerning the energy and the shape dependence of fluxes that were valid for model M1 are equally valid for M2.

It should be noted that, in some cases, the solutions for the time dependence of the densities $f_{m,\alpha}(t)$ (and, consequently, for the fluxes) contain oscillatory terms of the form $e^{-t/\tau_k} \cos\omega_k t$. However, in all such cases the relaxation times $\tau_k$ are very short, and these oscillations are practically negligible.

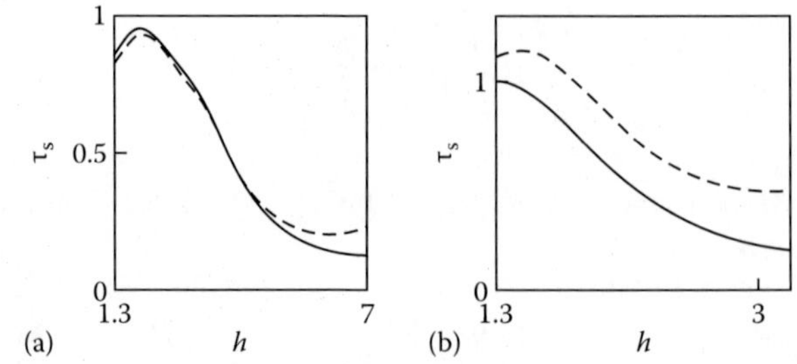


Figure 4.22

$h$ dependence of the transient time $\tau_s$ at (a) $T = 0.4T_c$ and (b) $T = 0.8T_c$ for M2 (dashed lines). For comparison, the corresponding curves for M1 are plotted as solid lines. At $T = 0.1T_c$ the results for M2 coincide with those for M1 (Figure 4.19a).

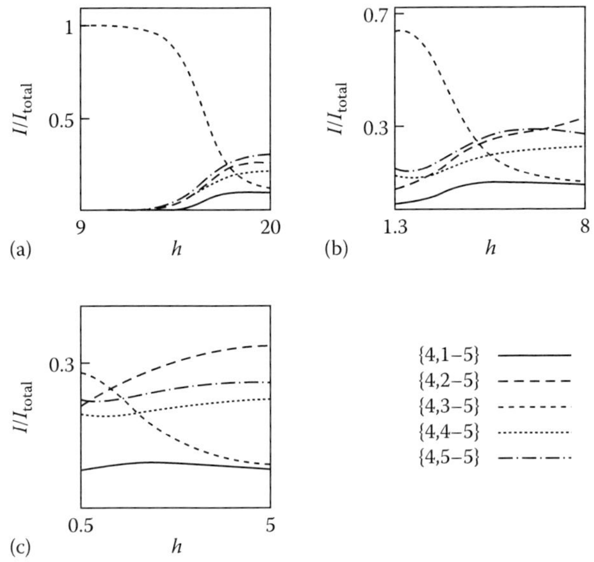


Figure 4.23

Relative steady-state values of different fluxes $\{m, \alpha \to 5\}$ from four-spin to all five-spin clusters as functions of supersaturation at (a) $T = 0.1T_c$, (b) $T = 0.4T_c$, and (c) $T = 0.8T_c$ in M2.

### 4.3.4 Conclusion

In this paper, the kinetics of small clusters in the two-dimensional Ising model on the square lattice was examined for a Glauber single-spin dynamics at temperatures lower than the critical one and beyond the classical approximations. New kinetic equations for time-dependent number densities of different clusters which take into account all possible cluster shapes as well as the time dependence of the number of down spins (Model 1) and the splitting and coagulation of clusters (Model 2) are derived and solved numerically. Some of the new results obtained from these equations are emphasized below.

Due to the linearity of the kinetic equations, the time dependence of all the densities is characterized by a set of relaxation times $\tau_k$, the largest of which, $\tau$, provides the lifetime of the metastable state, which is usually estimated as the reciprocal of the total flux at the steady state. If the second largest relaxation time, $\tau_s$, is much less than $\tau$ then in the time interval $\tau_s \ll t \ll \tau$, all quantities have a weak dependence on $t$ and the steady-state approximation can be employed. In this case, $\tau_s$ represents the transient time to the steady state that was not previously calculated in the investigations regarding the Ising model kinetics. In Model 2, the lifetime $\tau$ is infinite because of the supposition of constant number of down spins, and the steady state exists for all $t \gg \tau_s$. In Model 1, the depletion of the down spins leads to a finite lifetime $\tau$, which decreases with increasing temperature $T$ and supersaturation $h$, in qualitative agreement with the Monte Carlo simulations of Rikvold et al. [6]. However, for wide ranges of temperature $T$ and supersaturation $h$, the lifetime $\tau$ is much larger than $\tau_s$ and the steady-state approximation can be employed for $\tau_s \ll t \ll \tau$.

In both models, the kinetic behavior of the clusters depends on their shape and has some common features. At steady state, the clusters with the same energy and number of spins, but with different shapes, provide different contributions to the total mass transfer flux. The relative contributions of the fluxes between clusters to the total flux depend also on the supersaturation and temperature. It was shown that the contribution to the total flux of the clusters with the lowest excitation energy is dominant at low temperatures and low supersaturations, but no longer at high temperatures or high supersaturations where all clusters are involved in the kinetics. At very low temperatures, the transient times of the various fluxes to their steady-state values can be very different, but at high temperatures they are approximately the same.

The splitting and coagulation of small clusters, which are taken into account in Model 2, are important at intermediate and high temperatures, and their effect increases with increasing supersaturation. In particular, the transient time $\tau_s$ has the tendency to increase at high values of $h$. Such a behavior can decrease the range of applicability of the steady-state approximation.

One should note that the developed theory is applicable to the early stages of the kinetic process, when no large clusters are present in the system.

## References


1. Novotny, M.A., Brown, G., Rikvold, P.A., *J. Appl. Phys.* 91, 6908 (2002).
2. Cheon, M., Chang, I., *Phys. Rev. Lett.* 86, 4576 (2001).
3. Manzo, F., Olivieri, E., *J. Stat. Phys.* 104, 1029 (2001).
4. Vehkamaki, H., Ford, I.J., *Phys. Rev. E* 59, 6483 (1999).
5. Wonczak, S., Strey, R., Stauffer, D., *J. Chem. Phys.* 113, 1976 (2000).
6. Rikvold, P.A., Tomita, H., Miyashita, S., Sides, S.W., *Phys. Rev. E* 49, 5080 (1994).
7. Khorrami, M., Aghamohammadi, A., *Phys. Rev. E* 63, 042102 (2001).
8. Shneidman, V.A., Jackson, K.A., Beauty, K.M., *J. Chem. Phys.* 111, 6932 (1999).
9. Shneidman, V.A., Jackson, K.A., Beauty, K.M., *Phys. Rev. B* 59, 3579 (1999).
10. Shneidman, V.A., Nita, G.M., *Phys. Rev. Lett.* 89, 025701 (2002).
11. Schmelzer, J., Jr., Landau, D.P., *Int. J. Mod. Phys. C* 12, 345 (2001).
12. Berim, G.O., Ruckenstein, E., *J. Chem. Phys.* 117, 4542 (2002). (Section 4.1 of this volume.)
13. Berim, G.O., Ruckenstein, E., *J. Chem. Phys.* 117, 7732 (2002). (Section 4.2 of this volume.)
14. Glauber, R., *J. Math. Phys.* 4, 294 (1963).
15. Metropolis, N., Rosenbluth, A.W., Rosenbluth, M.N., Teller, A.H., Teller, E., *J. Chem. Phys.* 21, 1987 (1953).
16. Kawasaki, K., *Phys. Rev.* 145, 224 (1966).
17. Frenkel, J., *Kinetic Theory of Liquids*, Dover, New York, 1955.
18. Abraham, F., *Homogeneous Nucleation Theory*, Academic, New York, 1974.
19. Neves, E.J., Schonmann, R.H., *Commun. Math. Phys.* 137, 209 (1991).
20. Bovier, A., Manzo, F., *J. Stat. Phys.* 107, 757 (2002).
21. Nowakowski, B., Ruckenstein, E., *J. Chem. Phys.* 94, 1397 (1991). (Section 2.1 of this volume.)
22. Nowakowski, B., Ruckenstein, E., *J. Chem. Phys.* 94, 8487 (1991). (Section 2.3 of this volume.)
23. Shneidman, V.A., Zia, R.K.P., *Phys. Rev. E* 63, 085410 (2001).
24. Zia, R.K.P., *J. Stat. Phys.* 45, 801 (1986).
25. Avron, J.E., van Beijeren, H., Schulman, L.S., Zia, R.K.P., *J. Phys. A* 15, L81 (1982).
26. Honerkamp, J., *Statistical Physics*, Springer-Verlag, Berlin, 1998.
27. Binder, K., Müller-Krumbhaar, H., *Phys. Rev. B* 9, 2328 (1974).
28. Huiser, A.M.J., Marchand, J.P., Martin, Ph.A., *Helv. Phys. Acta* 55, 259 (1982).
29. Marchand, J.P., Martin, Ph.A., *Physica A* 127, 681 (1984).

# 4.4 A Closed Reduced Description of the Kinetics of Phase Transformation in a Lattice System Based on Glauber's Master Equation

*Gersh Berim and Eli Ruckenstein**

A generalized kinetic Ising model is applied to the description of phase transformations in lattice systems. A procedure, based on the conjecture that the probability distribution function of the states of the system is similar to the equilibrium one, is used for closing the infinite chain of kinetic equations. The method is illustrated by treating as an example the one-dimensional Ising model. The predicted rate of phase transformation (RPT) demonstrates various time behaviors dependent upon the details of the interactions between spins and a heat bath. If the parameters $W_0$ and $W$ the reciprocals of which characterize, respectively, the time scales of growth (decay) and splitting (coagulation) of clusters have the same order of magnitude, then the RPT is constant during almost the entire transformation process. For the case $W = 0$, which corresponds to the absence of splitting and coagulation of clusters, the phase transformation follows an exponential law in the final stage and is linear with respect to time during the initial one. It has a similar behavior for $W_0 \gg W \neq 0$; however, the RPT in the final stage is much smaller in the last case than for $W = 0$. In the absence of supersaturation, RPT decreases to zero as $T \to T_c$, where $T_c (= 0\text{ K})$ is the phase transition temperature for a one-dimensional model. The time-dependent size distribution of clusters is for all times exponential with respect to the cluster size. The average size of the cluster far from both equilibrium and initial state grows linearly in time. Both the above quantities behave in a manner similar to those obtained by Monte Carlo simulations for systems of higher dimension. (© *2003 American Institute of Physics*. doi: 10.1063/1.1615512.)

## 4.4.1 Introduction

The phase transformations involving nucleation and growth have been for a long time the focus of a large number of investigations [1,2], but until now the understanding of these processes is far from completion. Their complicated nature forces one to use a number of simplifications to obtain approximate solutions. The classical approach [1,2] considered the dynamics of the number density of clusters of a new phase in the space of their size (Becker–Döring equations) and employed thermodynamic arguments to calculate the size of a critical cluster above which the clusters grow rapidly in size. Neglecting the splitting and coagulation of clusters and assuming spherical shapes, the nucleation rate and time dependence of the cluster distribution function could be calculated. A more natural kinetic approach to the problem, free of macroscopic thermodynamic arguments, was suggested by Narsimhan and Ruckenstein [3] and Nowakowski and Ruckenstein [4–6]. However, the shape of the clusters was still assumed spherical.

Another line of study of the kinetics of phase transformation was based on lattice models, which have been used to analyze the general properties of nucleation [7–16], the metastable states [17–22], and the cluster size distribution [23–26]. In the treatment, it was assumed that the intermolecular interactions can be limited only to the molecules located on the neighboring sites of the lattice and the state of the system was mapped with the Ising Hamiltonian,

$$H_s = -B\sum_{j}\sigma_j - J\sum_{\langle j,k\rangle}\sigma_j\sigma_{j+\Delta_k} \quad (J > 0), \tag{4.86}$$

where the variables $\sigma_j = \pm 1$ (spins) represent the two different phases (e.g., solid and liquid) to which a molecule at a site $j$ of a rigid lattice can belong,† the parameter $B$ (the external field) models the supersaturation $S$ ($B \sim \log S$) [12], and $J$ (the nearest-neighbor interaction energy, referred below as the energy) models the intermolecular interactions. The subscript $\Delta_k$ refers to the $m$ neighboring lattice sites of site $j$, and $\langle j,k\rangle$ indicates that each pair of neighboring spins should be counted in the sum only once. Choosing appropriate rates of spin flips (liquid–solid transitions), which satisfy the detailed balance condition, and starting from a metastable initial state with almost all spins "down" (liquid state), the transition of the system to the final equilibrium state with all spin "up" (solid state· was examined by Monte Carlo simulations [7–11,13–15,19,20,22–24], and numerous results were obtained. For a theoretical interpretation of the

* *J. Chem. Phys.* 119, 9640 (2003). Republished with permission.

† The total number $M$ of the molecules in solid phase is given by the formula $M = \Sigma_j(1 + \sigma_j)/2$.

simulation results, the classical theory was mainly used [10–13,23,24]. Attempts to supersede the last theory were made in Refs. [17,27], where instead of the Becker–Döring equations, which imply a size distribution of clusters of a single shape (spherical), an Ising model on a two-dimensional lattice involving all the possible shapes of small clusters was considered. The coagulation and splitting were taken into account [27] or were disregarded [17], but only small clusters were considered.

In the present paper, we use another analytical approach to describe the phase transformation, which is based on the Glauber master equation [28] to treat the behavior of an Ising system in contact with a heat bath. This equation has the form

$$\begin{aligned}\frac{\mathrm{d}}{\mathrm{d}t}P(\sigma_1,\sigma_2,\ldots,\sigma_N;t)\\ =-\sum_j W(\sigma_j)P(\sigma_1,\ldots,\sigma_j,\ldots,\sigma_N;t)\\ +\sum_j W(-\sigma_j)P(\sigma_1,\ldots,-\sigma_j,\ldots,\sigma_N;t),\end{aligned} \tag{4.87}$$

where $P(\sigma_1, \sigma_2, \ldots, \sigma_N;t) \equiv P(\{\sigma\};t)$ is the probability to find an Ising spin system at time $t$ in the state $\{\sigma_1, \sigma_2, \ldots, \sigma_N\}$, $N$ is the total number of spins in the system, and $W(\sigma_j)$ is the transition rate of spin $\sigma_j$ between its states $\sigma_j = 1$ and $\sigma_j = -1$. $W(\sigma_j)$ depends only on the states of the molecules located at site $j$ and on its nearest neighboring sites (i.e., on the values of the spin $\sigma_j$ and its neighboring spins), regardless of the kind of cluster (a group of bounded molecules in the same state or, in other words, a group of equally directed bounded spins) to which the neighboring molecules belong. For the latter reason, the Glauber master equation implicitly accounts for the splitting and coagulation of clusters of arbitrary shape, and this constitutes its main advantage compared with the classical Becker–Döring approach. The main disadvantage of the Glauber equation is that it leads to an infinite chain of coupled differential equations for the time-dependent spin correlation functions (CFs) and exact solutions were obtained only for the case of zero external field [28,29]. In most cases, a mean field or other similar approximations [22,30–35] were used to close the chain by representing the multiparticle CFs as products of lower-order ones. In the present paper, a closing procedure is used, based on a conjecture regarding the probability distribution function (PDF) $P(\{\sigma\};t)$, which is chosen to have the form of the equilibrium density matrix, but with time-dependent external field and nearest-neighbor interaction energy. The approach is applied to the one-dimensional Ising model, because in this case an exact analytical solution can be obtained at equilibrium and thus the method can be well illustrated. In addition to the rate of phase transformation (RPT), which coincides in the model with the relaxation rate of the polarization, the size distribution of the clusters of the "up" spins will also be calculated.

## 4.4.2 Kinetic Equations and Closing Procedure

### *4.4.2.1 Kinetic Equations*

To write explicitly the kinetic equations for the time-dependent average,

$$\langle A\rangle_t \equiv \sum_{\{\sigma\}} AP(\{\sigma\};t) \tag{4.88}$$

of an arbitrary spin operator $A$ one has to select a specific form for the transition rate $W(\sigma_j)$ in Equation 4.87. We consider here the single-spin-flip relaxation mechanism according to which [22,28],

$$W(\sigma_j) = W_{E_j}[1-\sigma_j \tanh(\beta E_j)], \tag{4.89}$$

where $\beta = 1/T$, and $T$ denotes the temperature (the Boltzmann constant is considered equal to unity),

$$E_j = B + J\sum_{k=1}^{m} \sigma_{j+\Delta_k} \tag{4.90}$$

is the local field acting on spin $\sigma_j$ due to its interaction with the external field and with the neighboring spins, and the parameter $W_{E_j}$ depends, in general, on $E_j$.

Let us denote by $\mu$ the sum of the values of the spins that are neighbors of $\sigma_j$. Then, any value $\varepsilon_\mu$ of the local field (Equation 4.90) can be written as

$$\varepsilon_\mu = B + J\mu. \tag{4.91}$$

Introducing an operator $R_j(\mu)$ that is equal to unity when the local field that acts on spin $\sigma_j$ is equal to $\varepsilon_\mu$ and zero otherwise, the rate $W(\sigma_j)$ (Equation 4.89) can be rewritten in the following form:

$$W(\sigma_j) = \sum_\mu W_\mu R_j(\mu)[1 - \sigma_j \tanh(\beta\varepsilon_\mu)], \tag{4.92}$$

where $W_\mu$ is the value of $W_{E_j}$ for $E_j = \varepsilon_\mu$ and the sum is taken over all possible values of $\mu$ ($\mu = -m, -m+2, \ldots, m-2, m$). The coefficients $W_\mu$ are usually considered to be the same $\left(W_{E_j} \equiv 1/\tau\right)$, with $\tau$ playing the role of the timescale of the process. Here we consider a generalized version of the model and suppose that $W_\mu$ depends on $\mu$. The operator $R_j(\mu)$ can be constructed from the operators $\sigma_{j+\Delta k'}$ ($k = 1, \ldots, m$) of the neighboring spins of spin $\sigma_j$ as shown in Appendix 4A.

Using the master equation (Equation 4.87) and the transition rates (Equation 4.92), one can write the following kinetic equations for the averages $P(t) = \langle \Sigma_j \sigma_j \rangle_t$ and $E(t) = \langle \Sigma_{\langle j,k \rangle} \sigma_j \sigma_{j+\Delta_k} \rangle_t$, which represents the total polarization and the energy, respectively:

$$\begin{aligned} \frac{\mathrm{d}}{\mathrm{d}t} P(t) &= -\sum_\mu W_\mu \sum_j [\langle \sigma_j R_j(\mu) \rangle_t - z_\mu \langle R_j(\mu) \rangle_t] \\ \frac{\mathrm{d}}{\mathrm{d}t} E(t) &= -\sum_\mu \mu W_\mu [\langle \sigma_j R_j(\mu) \rangle_t - z_\mu \langle R_j(\mu) \rangle_t], \end{aligned} \tag{4.93}$$

where $z_\mu = \tanh(\beta\varepsilon_\mu)$. The time-dependent total number $M(t)$ of molecules in the solid phase and the total polarization $P(t)$ are connected through the relation $M(t) = [N + P(t)]/2$.

After substituting the operators $R_j(\mu)$, the right-hand side of Equation 4.93 involves the multispin CFs that contain up to $m + 1$ spins ($m$ being the number of neighboring spins of spin $\sigma_j$). One can write similar equations for higher-order CFs. However, each of them involves CFs of even higher order, and a closing equation is required.

#### 4.4.2.2 Closing Procedure

Let us suppose that the PDF $P(\{\sigma\};t)$ has the form

$$P(\{\sigma\}; t) = e^{-F(t)}/\mathrm{Tr}[e^{-F(t)}], \tag{4.94}$$

where

$$F(t) = -\beta_\sigma(t) \sum_j \sigma_j - \beta_e(t) \sum_{\langle j,k \rangle} \sigma_j \sigma_{j+\Delta_k} \tag{4.95}$$

with $\beta_\sigma(t)$, $\beta_e(t)$ unknown function of $t$. The plausibility of such a representation of the PDF follows from Bogolyubov's [36] and Zubarev's [37] representation of the nonequilibrium behavior of a macroscopic system as a hierarchy of relaxation times. The subsystems with short relaxation times reach quasi-equilibrium earlier than the entire system does. Consequently, the system can be characterized by a small number (reduced description) of quasi-integrals of motion which slowly change under the action of the weak interactions between the subsystems, establishing an equilibrium between them. Let us suppose that in addition to the interaction of the spins with the heat bath, there are direct or indirect non-Ising interactions between the spins (e.g., dipole–dipole interactions), which are much smaller than the main interactions (nearest-neighbor interaction $J$ and the interaction with the external field $B$), but much larger than the interactions of the spins with the heat bath. As shown by Provotorov [38], the main effect of such interactions upon the kinetics of the spin system is due to the pairwise flip-flop transitions of spins without changes in the polarization and energy of the system. In this case, the total polarization $\Sigma_j \sigma_j$ and the total energy $\Sigma_{(j,k)} \sigma_j \sigma_{j+\Delta k}$ are the most natural choices as quasi-integrals of motion, which slowly change due to the interactions of the spins with the heat bath. The choice of the PDF in the form of Equation 4.94 just emphasizes that these quantities determine the evolution of the system.

Due to the spatial homogeneity of the operator $F(t)$ (Equation 4.95), in the following we will use, instead of $P(t)$ and $E(t)$, the polarization per spin $\sigma = \langle \sigma_0 \rangle_t$ and the energy per bond $\epsilon = \langle \sigma_0, \sigma_1 \rangle_t$, respectively, where $\sigma_0$ is an arbitrarily selected spin and $\sigma_1$ represents one of its neighbors. Instead of the summation, the value $j = 0$ will be written in the right-hand side of Equation 4.93.

The choice of Equation 4.94 for the PDF allows one to close the kinetic equation (Equation 4.93), since all the CFs involved in Equation 4.93 can be expressed through the two unknown functions $\beta_e(t)$ and $\beta_\sigma(t)$ and thus, they can be rewritten in terms of these functions. However, in the present case it is more convenient to keep $\sigma$ and $\epsilon$ as variables and express through them all other CFs using $\beta_\sigma(t)$ and $\beta_e(t)$ as intermediate parameters.

Let us note that PDF (Equation 4.94) can be transformed into the equilibrium density matrix of the Ising model,

$$\rho_T = e^{-\beta H_s} / \mathrm{Tr}\left[ e^{-\beta H_s} \right], \tag{4.96}$$

where $H_s$ is given by Equation 4.86, by replacing $\beta_\sigma(t)$ and $\beta_e(t)$ by $\beta B$ and $\beta J$, respectively. Consequently, the time-dependent CFs in Equation 4.93 can be obtained from the equilibrium CFs of Model 1 by substituting $B$ with $\beta_\sigma(t)/\beta$ and $J$ with $\beta_e(t)/\beta$. For this reason, the above closing procedure can be most efficiently applied to those Ising models for which one can obtain exact analytical equilibrium solutions (such as the one-dimensional and some of the two-dimensional models). In the other cases, approximate methods must be employed to calculate the CFs involved in Equation 4.93.

In the following, it will be shown that, in spite of its simplicity, the suggested closing procedure can catch the main features of the kinetics of lattice models.

#### 4.4.2.3 Initial and Final Conditions

The initial values $\sigma(0)$ and $\epsilon(0)$ of the polarization and the energy can be calculated using Equations 4.88, 4.94, and 4.95, where the initial values $\beta_\sigma(0)$ and $\beta_e(0)$ of the functions $\beta_\sigma(t)$ and $\beta_e(t)$ can be selected from physical considerations. For example, to specify the initial conditions usually used in Monte Carlo simulations in the two- and three-dimensional Ising models (almost all spins have "down" direction, the liquid state) one can select a temperature $T < T_c$ ($T_c > 0$ is the phase transition temperature) and apply a small external field $B_{\text{init}}$ in the "down" direction ($B_{\text{init}} < 0$, supersaturation $S < 1$). In this case, $\sigma(0)$ and $\epsilon(0)$ can be obtained using $\beta_\sigma(0) = B_{\text{init}}/T$ and $\beta_e(0) = J/T$.

If at $t = 0$ the field $B$ changes direction ($B = B_{\text{fin}} > 0$ at $t > 0$, supersaturation $S > 1$), a transition process from the initial (liquid) state with most spins "down" to the final one with all spins "up" (solid state) is started. The final state is again an equilibrium one, corresponding to the same temperature $T$ and the new field $B_{\text{fin}} > 0$.

In the one-dimensional case the phase transition takes place at $T_c = 0$ K [39] and a state with $T < T_c$ is impossible. To model the initial and final states similar to those in models of higher dimension, one has to consider in this case a sufficiently low temperature and relatively strong fields $B_{\text{init}}$ and $B_{\text{fin}}$.

### 4.4.3 Phase Transformation on the One-Dimensional Lattice

In this case, the Ising Hamiltonian (Equation 4.86) has the form

$$H_s = -B \sum_j \sigma_j - J \sum_j \sigma_j \sigma_{j+1} \, (J > 0), \tag{4.97}$$

the local field $\varepsilon_\mu$ has three values $\varepsilon_\mu = B + \mu J$ with $\mu = 0, \pm 2$, and the operators $R_j(\mu)$ are given by Equation 4A.2 of Appendix 4A. The kinetic equations for the polarization per site $\sigma$ and the energy per bond $\epsilon$ are given by

$$\begin{aligned} \dot{\sigma} &= a_0 - a_1 \sigma - a_2 \epsilon - a_3 \epsilon_2 + a_4 \gamma_1, \\ \dot{\epsilon} &= b_0 - b_1 \sigma - b_2 \epsilon + b_0 \epsilon_2 - a_2 \gamma_1, \end{aligned} \tag{4.98}$$

where

$$
\begin{aligned}
a_0 &= z_0 W_0 + \frac{1}{2}(z_2 W_2 + z_{-2} W_{-2}),\\
a_1 &= W_0 + \frac{1}{2} W_2 (1 - 2z_2) + \frac{1}{2} W_{-2}(1 + 2z_{-2}),\\
a_2 &= W_2 - W_{-2},\\
a_3 &= z_0 W_0 - \frac{1}{2}(z_2 W_2 + z_{-2} W_{-2}),\\
a_4 &= W_0 - \frac{1}{2}(W_2 + W_{-2}),\\
b_0 &= z_2 W_2 - z_{-2} W_{-2},\\
b_1 &= W_2(1 - 2z_2) - W_{-2}(1 + 2z_{-2}),\\
b_2 &= 2(W_2 + W_{-2}).
\end{aligned}
\tag{4.99}
$$

They contain in the right-hand sides additional time-dependent CFs $\epsilon_k = \langle \sigma_0 \sigma_k \rangle_t$ and $\gamma_k = \langle \sigma_0 \sigma_k \sigma_{k+1} \rangle_t$, which can be expressed in terms of $\sigma$ and $\epsilon$ using the exact expressions of Appendix 4B for the corresponding equilibrium averages. This leads to the following closed system of nonlinear kinetic equations for $\sigma$ and $\epsilon$:

$$
\begin{aligned}
\dot{\sigma} &= a_0 - (a_1 - a_4)\sigma - a_2 \epsilon\\
&\quad - \frac{1}{1-\sigma^2}[a_3(\sigma^2 + \epsilon^2 - 2\epsilon\sigma^2) + a_4 \sigma (1-\epsilon)^2],\\
\dot{\epsilon} &= b_0 - (b_1 + a_2)\sigma\\
&\quad - b_2 \epsilon + \frac{1}{1-\sigma^2}[b_0(\sigma^2 + \epsilon^2 - 2\epsilon\sigma^2) + a_2 \sigma (1-\epsilon)^2].
\end{aligned}
\tag{4.100}
$$

The same equations were previously derived by Berim and Kessel [40,41] using a different more complex method. Below, some particular cases for which the solutions of these equations can be obtained in closed form will be considered. The RPT is defined as

$$
\mathrm{RPT} = \frac{\dot{\sigma}(t)}{\sigma(\infty) - \sigma(t)} = -\frac{d}{dt} \log \phi_\sigma(t), \tag{4.101}
$$

where

$$
\phi_\sigma(t) = \frac{\sigma(t) - \sigma(\infty)}{\sigma(0) - \sigma(\infty)} \tag{4.102}
$$

is the relaxation function of the polarization.

Before proceeding with the details, it is instructive to discuss with respect to the one-dimensional case the physical meaning of the terms in the kinetic equation (Equation 4.93), which are proportional to the parameters $W_\mu$. The latter parameters determine the rate of spin flips (liquid–solid transitions of molecules) with the energy changes $\Delta E_\mu = \pm(2B + 2\mu J)$, where $\mu = 0, \pm 2$ (see Equation 4.92). The flips with the energy changes $\Delta E_0$ and $\Delta E_{\pm 2}$ will be called *B* and *J* flips, respectively. A spin can make a *B* flip if its nearest neighbors have opposite directions to each other, and a *J* flip if they have the same directions. Consequently, the *B* flips provide the growth ($B^+$ flips) and decay ($B^-$ flips) of the cluster of "up" spins (see Figure 4.24a), and

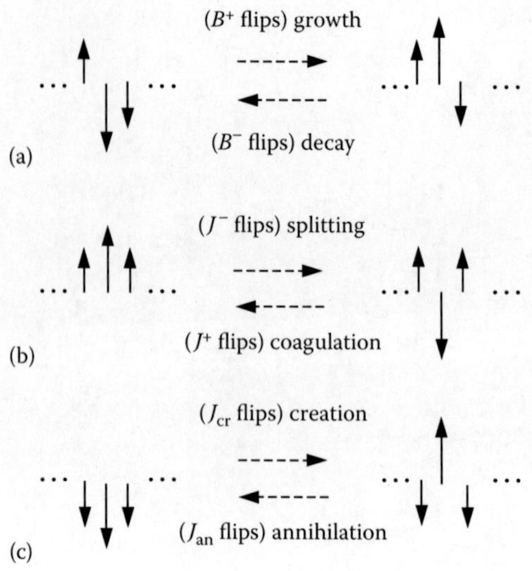


Figure 4.24

Examples of (a) $B^\pm$ flips leading to growth and decay of a cluster of "up" spins; (b) $J^\pm$ flips that generate splitting and coagulation of clusters; (c) $J_{cr}$ and $J_{an}$ flips, which create or annihilate the cluster with a single "up" spin. The flipping spin is depicted by a longer arrow.

the $J$ flips cause the splitting ($J^-$ flips) and coagulation ($J^+$ flips) of the clusters (Figure 4.24b) or creation $\left(J_{cr}^+ \text{ flips}\right)$ and annihilation $\left(J_{an}^- \text{ flips}\right)$ of the one-spin clusters (Figure 4.24c). Below, the spins that can make $B^\pm$ flips will be called $B^\pm$ spins. Both the $B$ and $J$ flips change the polarization of the system, whereas the energy is changed only by the $J$ flips. Thus, by choosing different values of the parameters $W_0$, $W_2$, $W_{-2}$, one can control the contribution of the growth decay, splitting–coagulation, and creation–annihilation processes, respectively, to the kinetics of the phase transformation. For instance, the case $W_0 \neq 0$, $W_2 = W_{-2} = 0$ corresponds to the case in which the clusters grow or decay by capturing or losing one spin, hence only $B^\pm$ flips take place, whereas coagulation (splitting) and creation (annihilation) are suppressed. If otherwise mentioned, in the following we will consider, for simplicity, the case when $W_2 = W_{-2} \equiv W$.

#### 4.4.3.1 Exact Solution for Growth Decay Regime

In this section, we consider the particular case of phase transformation when the splitting (coagulation) of clusters, as well as creation (annihilation) of the one-spin clusters are prohibited ($W = 0$). In this case, the total number of clusters is conserved and, as a consequence, at $t \to \infty$ the system does not reach equilibrium, with the exception of the case in which the initial number of clusters is equal to that at equilibrium. Such a situation can be considered as a limiting case of a more realistic one, in which $W$ *is* not equal to zero but is small compared to $W_0$. It should be noted that the exclusion of the above mentioned processes is a common assumption of most of the classical considerations of nucleation and phase transformation in lattice systems.

The condition $W = 0$ introduced in the second equation of the system (Equation 4.100) leads to $\dot{\epsilon} = 0$, with the trivial solution $\epsilon = \epsilon(0)$. In this case, the equation for $\sigma$ can be solved exactly. To examine the phase, transformation from a liquid to a solid due to a change of the supersaturation from $S < 1$ to $S > 1$, low temperatures [$e^{-4\beta J} \ll \sinh(2\beta B)$], and an external field that changes direction at $t = 0$, but not its magnitude ($B = -B_0$, $B_0 > 0$ at $t \le 0$, $B = B_0$ at $t > 0$), are selected. Then, one obtains

$$\sigma + \sigma_T - \frac{1-\epsilon_T}{4z_0}\log\left(\frac{\sigma_T - \sigma}{2\sigma_T}\right) = 2(1-\epsilon_T)z_0 W_0 t, \tag{4.103}$$

where $\sigma_T$ and $\epsilon_T$ are the equilibrium values of the polarization and energy corresponding to the external field $B_0$, both with magnitudes close to unity because of the conditions selected. The evolution of the polarization from the initial value $\sigma = -\sigma_T$ to the final one $\sigma = \sigma_T$ can be approximated by the linear expression $\sigma = -\sigma_T + 2(1-\epsilon_T)z_0 W_0 t$ for $0 < t < t_0 \simeq \sigma_T/[(1-\epsilon_T)z_0 W_0]$, and by an exponential one $\sigma - \sigma_T = 2\sigma_T e^{-t/\tau}$ with a relaxation rate $1/\tau = 8z_0 W_0$, for $t > t_0$. (See Figure 4.25, where the relaxation function $\phi_\sigma(t)$ is plotted.) Such a difference in relaxation regimes can be qualitatively understood by examining the dynamics of the number

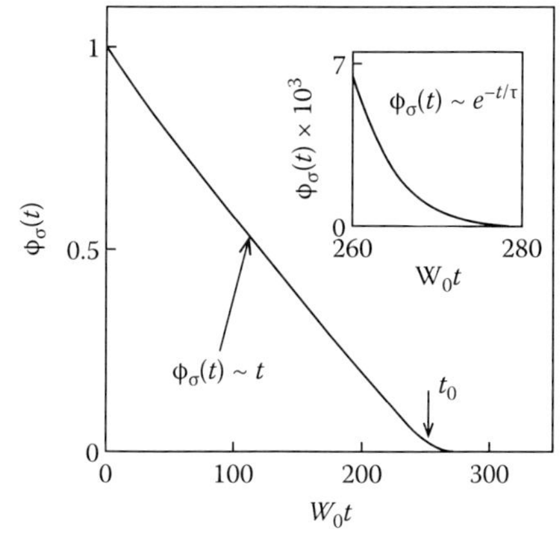


Figure 4.25

Time dependence of the relaxation function of the polarization for $W = 0$, $\beta J = 1.5$, $\beta B_0 = 0.2$. At time $t = t_0$, a crossover from the linear to the exponential regime take place. The insert shows the final exponential stage of the relaxation process.

densities $n(B^+)$ and $n(B^-)$ of $B^+$ and $B^-$ spins, which determine the time evolution of the polarization through the equation

$$\dot{\sigma} = n(B^+)W^+ - n(B^-)W^-, \tag{4.104}$$

where $W^\pm$ are the time-independent probabilities of the $B^\pm$ flips, with $W^+ > W^-$ for $B > 0$ due to the condition in Equation 4.92. In the absence of splitting and coagulation, the total number of clusters of "up" spins is the same at any time. The two boundary spins of each of such clusters with a size $n \geq 2$, where $n$ is the number of spins in the cluster, constitute $B^-$ spins, whereas each of the two "down" spins, which frame the cluster, constitute $B^+$ spins, provided that they have one "down" neighbor. An "up" spin framed by "down" spins constitutes a single-spin cluster, which has no $B^-$ spins. Consequently, the number density $n(B^-)$ increases with the decreasing number of single-spin clusters. The number of such clusters is the largest in the initial state, when most spins are "down," decreases quickly after the field reversal, due to the transformation of the single-spin clusters into larger ones, and remains further approximately constant until the end of the process. Analogously, the number density $n(B^+)$ of $B^+$ spins decreases when the clusters of "down" spins are transformed into single-spin clusters, which do not contain $B^+$ spins. The number density $n(B^+)$ remains approximately constant during the initial stage of relaxation, because the number of "down" single-spin clusters is small, and decreases during the final stage, because most spins are "up" and a large number of such clusters is formed. The time behaviors of $n(B^+)$ and $n(B^-)$ are presented in Figure 4.26. They have been

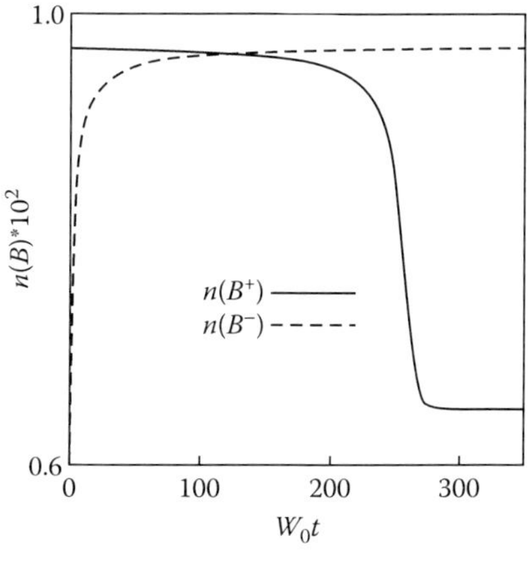


Figure 4.26

Time dependence of the number densities $n(B^\pm)$ of spins which can make a $B^\pm$ flip, for $\beta J = 1.5$, $\beta B_0 = 0.2$.

calculated using Equation 4B.4 (Appendix 4B). In the range in which $n(B^{\pm})$ are approximately constants, Equation 4.104 provides the following linear solution for the polarization $\sigma - \sigma(0) \sim t$. The rapid initial time dependence of $n(B^-)$ does not affect the slope of $\sigma$ because the probability $W^-$ is small.

#### *4.4.3.2 Small Deviation from Equilibrium*

When the deviations $\Delta\sigma \equiv \sigma - \sigma_T$, $\Delta\epsilon \equiv \epsilon - \epsilon_T$ of the nonequilibrium polarization and energy from their equilibrium values are small $(|\Delta\sigma| \ll 1, |\Delta\epsilon| \ll 1)$, then the kinetic equation (Equation 4.100) acquire the linear approximation form

$$\dot{\Delta\sigma} = -\frac{\Delta\sigma}{\tau_\sigma} - \frac{\Delta\epsilon}{\tau_{\sigma\epsilon}}, \quad \dot{\Delta\epsilon} = -\frac{\Delta\sigma}{\tau_{\epsilon\sigma}} - \frac{\Delta\epsilon}{\tau_\epsilon} \tag{4.105}$$

and have the following solution:

$$\begin{aligned} \Delta\sigma &= C^+ \exp(-t/\tau^+) + C^- \exp(-t/\tau^-), \\ \Delta\epsilon &= -\tau_{\sigma\epsilon}\left[\left(\frac{1}{\tau_\sigma} - \frac{1}{\tau^+}\right) C^+ \exp(-t/\tau^+) \right. \\ &\quad \left. + \left(\frac{1}{\tau_\sigma} - \frac{1}{\tau^-}\right) C^- \exp(-t/\tau^-)\right]. \end{aligned} \tag{4.106}$$

The expressions for the constants in Equations 4.105 and 4.106 are given in Appendix 4C.

The above solution shows that in the general case there are two relaxation times, $\tau^+$ and $\tau^-$ $(\tau^- \geq \tau^+)$, which characterize the approach of the system to equilibrium. When $\tau^- \gg \tau^+$, the relaxation for $t \gg \tau^+$ can be described by a single exponential function $e^{-t/\tau^-}$, where $1/\tau^-$ is the rate of the relaxation process (RPT). Its temperature dependence is shown in Figure 4.27 for various $B/J$ ratios and $W_0 = W$. For $B > 0$, the maximum of the curves is displaced towards lower temperatures with decreasing $B$ and its value increases. For low temperatures and $W_0 \gg W$, which corresponds to the slow splitting (coagulation) limit, one obtains for $\tau^-$ the simple expression

$$\tau^- = 2(1 - e^{-2\beta B})W. \tag{4.107}$$

In the absence of an external field, the polarization and the energy evolve independently

$$\Delta\sigma = \Delta\sigma(0)e^{-t/\tau_\sigma}, \quad \Delta\epsilon = \Delta\epsilon(0)e^{-t/\tau_\epsilon}, \tag{4.108}$$

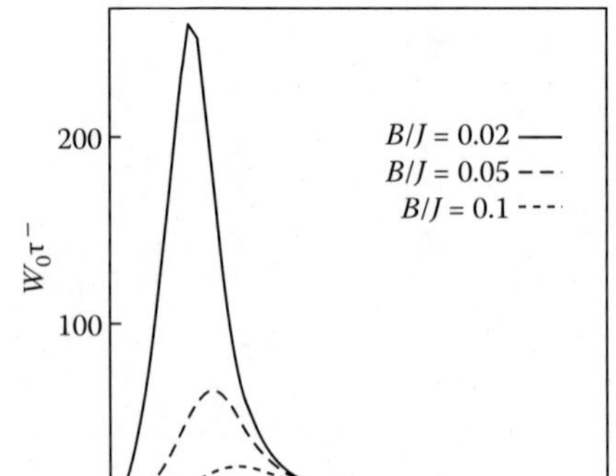


Figure 4.27

The temperature dependence of the relaxation time $\tau^-$ for various values of the ratio $B/J$ and $W_0 = W$.

where

$$\tau_\sigma^{-1} = \frac{1}{2}(W_0 - W)[1 - \tanh(\beta J]^2 + W[1 - \tanh(2\beta J)], \tag{4.109}$$

and

$$\tau_\epsilon^{-1} = 2W/\cosh(2\beta J). \tag{4.110}$$

In the particular case of $W_0 = W$, the expression of $\tau_\sigma^{-1}$ coincides with those obtained by Glauber [28] as an exact result and by Suzuki and Kubo [30] in the mean field approximation. Near the phase transition temperature ($T_c = 0$),

$$\begin{aligned}\tau_\sigma &= \xi^3/(W_0\xi + W),\\ \tau_\epsilon &= \xi/4W,\end{aligned} \tag{4.111}$$

where $\xi = \exp(2/\beta J)/2$ is the correlation length of the one-dimensional Ising model. Equation 4.111 show that the relaxation times $\tau_\sigma$ and $\tau_\epsilon$ grow without limit (the RPT, which is equal to $1/\tau_\sigma$, goes to zero) for $T \to T_c$. This phenomenon, the critical slowing down, is well known in the theory of dynamic critical phenomena [42] and is characterized by a dynamical critical exponent $z$ through the relation $\tau \sim \xi^z$. In the present case, the critical relaxation of the energy is characterized by an exponent $z_\epsilon = 1$, which is independent of $W_0$ and $W$. The dynamic critical exponent $z_\sigma$ is equal to $z_\sigma = 2$ for $W_0\xi \gg W$ and to $z_\sigma = 3$ for $W_0\xi \ll W$. The last case represents a restricted relaxation, which occurs when the $B$ flips have very low probabilities or are completely prohibited.

It should be noted that the results of this section can be also used to describe the final stage of the relaxation process.

#### 4.4.3.3 Reversal of External Field

Let us consider the relaxation after the reversal of the external field $B$ from its initial value $B = -B_0$, $B_0 > 0$ at $t = 0$ to the constant value $B = B_0$ at $t > 0$. This case is the most suitable for modeling the phase transformation, because it matches both the initial and final states of the system. Because the deviation of the system from equilibrium is large during this transition, the equations of the previous section are not applicable. Below, some results obtained via a numerical solution of the kinetic equations are presented. The intermediate temperatures ($\beta J = 1.5$) and fields ($\beta B_0 = 0.2$) were selected in the calculations. The results were qualitatively the same for lower temperatures. In the initial state, most of the spins are "down" [$\sigma(0) \simeq -0.97$, liquid state] and in the final, equilibrium, state most of the spins are "up" [$\sigma(\infty) = \sigma_T \simeq 0.97$, solid state]. The time dependence of the logarithm of the relaxation function $\phi_\sigma(t)$ is presented in Figure 4.28 for various values

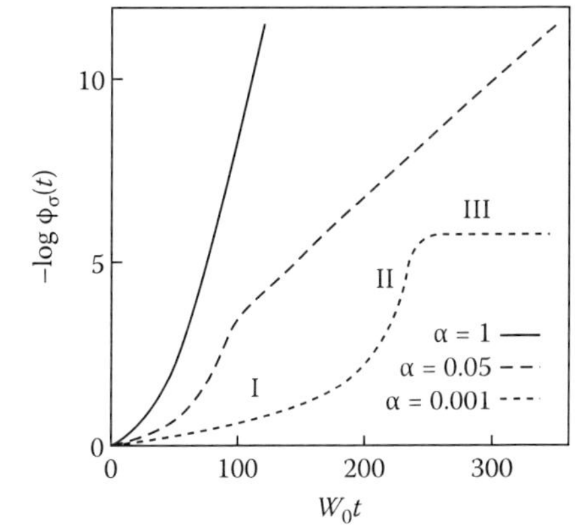


Figure 4.28

Semilog plot of the time dependence of the relaxation function of the polarization for various values of the ratio $\alpha = W/W_0$, $\beta J = 1.5$, and $\beta B_0 = 0.2$. For $\alpha = 0.001$, three distinct stages of relaxation are marked.

of the ratio $\alpha = W/W_0$. The linear parts of the curve correspond to the pure exponential relaxations and the slope of the line provides the instantaneous value of the relaxation rate. One can see that for $\alpha \simeq 1$ the relaxation outside of a short initial period can be described by the exponential function ($\Delta\sigma \sim e^{-1/\tau^-}$) with $1/\tau^-$ given by Equation 4C.3 of Appendix 4C. For $\alpha \ll 1$, which corresponds to small probabilities of splitting and coagulation of clusters, the relaxation process can be approximately divided into three stages, marked for the curve corresponding to $\alpha = 0.001$. In the first stage, the relaxation is nonexponential and the numerical analysis indicates that it follows a linear law similar to that for $W = 0$ (Section 4.4.3.1). In the second and third stages, the relaxation is exponential, with the relaxation rate the largest during the second stage, and the smallest during the third one, where it is equal $1/\tau^-$. Despite similar exponential forms, the relaxation rate of the polarization in the final stage differs considerably for $W = 0$ and $W_0 \gg W \neq 0$. While in the absence of splitting and coagulation ($W = 0$), the relaxation rate is equal to $1/\tau = 8z_0 W_0$, for $W \neq 0$ and $W \ll W_0$ it is given by $1/\tau^- \simeq 2(1-e^{-2\beta B_0})W \ll 1/\tau$. In the former case ($W = 0$), the relaxation is governed by $B^{\pm}$ flips, whose rate is proportional to $W_0$, and leads to a steady state, which does not coincide with the equilibrium one. In the latter case ($W \neq 0$, $W \ll W_0$), the approach of polarization in the final stage to equilibrium is governed by the slow $J^{\pm}$ flips whose rate is proportional to $W$. This latter stage is absent in the case $W = 0$.

### 4.4.4 Size Distribution of the Solid Phase Clusters

In this section, we examine the size distribution and the average size of the clusters of the solid phase (groups of bounded "up" spins framed by two "down" ones), using the results of Section 4.4.3.3. Let us introduce the time-dependent number density $f_l(t)$ of the solid phase clusters of size $l$, defined by the ratio $N_l(t)/N$, where $N_l(t)$ is the average number of clusters, containing $l$ "up" spins and $N$ is the number of spins in the system. (Since the lattice constant was selected equal to unity, the value of $l$ provides also the length of the cluster.) Then, the number density $f_{\text{up}}(t)$ of the "up" spins (molecules in the solid phase) and the number density $f_{\text{c}}(t)$ of the clusters of the solid phase are provided by the expressions

$$f_{\text{up}}(t)=\sum_{k=1}^{N} kf_k(t),\quad f_{\text{c}}(t)=\sum_{k=1}^{N} f_k(t), \tag{4.112}$$

and the average size $L_{\text{av}}(t)$ of the cluster is given by

$$L_{\text{av}}(t) = f_{\text{up}}(t)/f_{\text{c}}(t). \tag{4.113}$$

#### 4.4.4.1 Size Distribution

The density $f_l(t)$ can be written as follows:

$$f_l(t)=\left\langle n_0(-1)\left(\prod_{j=1}^{l} n_j(1)\right)n_{l+1}(-1)\right\rangle_t, \tag{4.114}$$

where the operators $n_j(\pm 1)$ provide the occupation probability of site $j$ by a spin $\sigma_j = \pm 1$ (see Appendix 4A). The operators $n_0(-1)$ and $n_{l+1}(-1)$ represent two "down" spins that frame the sequence of $l$ "up" spins represented by the product of operators $n_j(1)$. For small $l$, $f_l(t)$ can be calculated exactly. For example, for $l = 1$ and $l = 2$ (one- and two-spin clusters)

$$\begin{aligned} f_1(t)&=\frac{1}{8}(1-\sigma-2\epsilon+\epsilon_2+\gamma_1),\\ f_2(t)&=\frac{1}{16}(1-\epsilon-2\epsilon_2+\epsilon_3-2\gamma_1+2\gamma_2+\delta_2), \end{aligned} \tag{4.115}$$

where $\delta_k = \langle\sigma_0\sigma_{k-1}\sigma_k\sigma_{k+1}\rangle_t$. The time-dependent CFs $\epsilon_k$, $\gamma_k$, $\delta_k$ can be expressed through $\sigma$ and $\epsilon$ using expressions similar to those given by Equation 4B.3 for the equilibrium case.

For multiparticle clusters the corresponding exact expressions are complicated. To estimate the functions $f_l(t)$ for $l > 2$, one can employ a decoupling procedure which accounts for the correlations between the nearest neighboring spins,

$$
\begin{aligned}
&\left\langle n_0(-1)\left(\prod_{j=1}^{2k} n_j(1)\right)n_{2k+1}(-1)\right\rangle_t \\
&\quad\simeq\langle n_0(-1)n_1(1)\rangle_t\langle n_2(1)n_3(1)\rangle_t\ldots \\
&\quad\langle n_{2k}(1)n_{2k+1}(-1)\rangle_t \\
&\quad=\langle n_0(-1)n_1(1)\rangle_t^2\langle n_0(1)n_1(1)\rangle_t^{k-1}, \\
&\left\langle n_0(-1)\left(\prod_{j=1}^{2k+1} n_j(1)\right)n_{2k+2}(-1)\right\rangle_t \\
&\quad\simeq\langle n_0(-1)n_1(1)\rangle_t\langle n_2(1)n_3(1)\rangle_t\ldots \\
&\quad\langle n_{2k}(1)n_{2k+1}(1)n_{2k+2}(-1)\rangle_t \\
&\quad=\langle n_0(-1)n_1(1)\rangle_t\langle n_2(1)n_3(1)\rangle_t^{k-1} \\
&\quad\times\langle n_0(1)n_1(1)n_2(-1)\rangle_t
\end{aligned}
\tag{4.116}
$$

where $k \geq 1$. After simple transformations, one obtains for $l > 2$,

$$
f_l(t)=\begin{cases}\alpha_1\exp(-l/L(t)),\ l=2k+1,\\ \alpha_2\exp(-l/L(t)),\ l=2k+2,\end{cases}
\tag{4.117}
$$

where

$$
\begin{aligned}
\alpha_1&=\frac{(1-\epsilon)(1+\sigma-\epsilon_2-\gamma_1)}{4(1+2\sigma+\epsilon)^{3/2}},\\
\alpha_2&=\frac{(1-\epsilon)^2}{4(1+2\sigma+\epsilon)},\quad L(t)=-2/\log(1+2\sigma+\epsilon).
\end{aligned}
\tag{4.118}
$$

In Equation 4.117, $l$ is the size of the cluster, and $L(t)$ is its time-dependent characteristic size.

It is obvious that the decoupling procedure (Equation 4.116) is exact at an infinite temperature ($\beta J$, $\beta B \to 0$), when the spins are uncorrelated. To examine its applicability to a more interesting range of finite temperatures, one can take into account that the number density of "up" spins $f_{\text{up}}(t)$ provided by Equation 4.112 can be calculated using the expression

$$
f_{\text{up}}(t) = \langle n_0(1)\rangle_t = (1+\sigma)/2,
\tag{4.119}
$$

where $\sigma$ is provided by the solution of Equation 4.100, which can be considered an "exact" one. By comparing the right-hand sides of Equations 4.112 and 4.119, one can draw a conclusion about the applicability of the decoupling procedure (Equation 4.116). Figure 4.29 presents the ranges of parameters $\beta B$ and $\beta J$ in which the difference between the functions $f_{\text{up}}(t)$ provided by Equations 4.112 and 4.119 at any time does not exceed 10% for the relaxation process described in Section 4.4.3.3 involving $W_0 = W$.

The time dependencies of some of the functions $f_l(t)$ are presented in Figure 4.30. As expected, the density of the clusters containing $l$ particles grows with the time $t$, increasing from a very small value at $t = 0$ to a

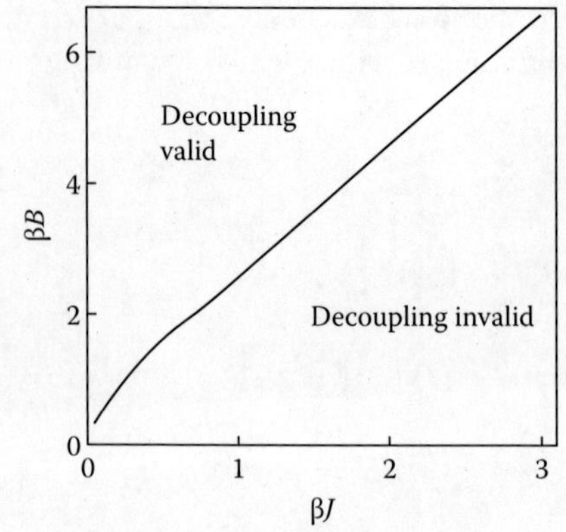


Figure 4.29

The range of validity of the decoupling procedure (Equation 4.116) for $W_0 = W$.

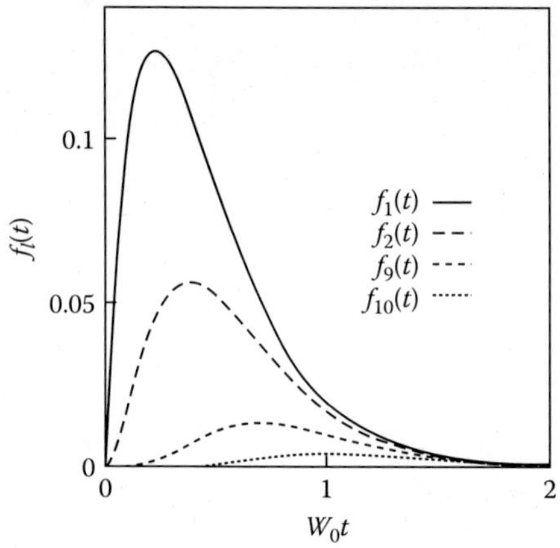


Figure 4.30

The time dependence of the size distribution function for various sizes in the range of validity of the decoupling procedure (Equation 4.116). $\beta J = 1.5$, $\beta B_0 = 3.6$, and $W_0 = W$.

maximum value at some time $t_l$, and then decreasing due to their transformation in large clusters. The time $t_l$ at which the density $f_l(t)$ passes through a maximum grows with increasing $l$.

The numerical analysis shows that, in the range of validity of approximation (Equation 4.116), the functions $f_1(t)$ and $f_2(t)$ can be represented by Equation 4.117, and that $\sigma_1 \simeq \sigma_2$ at all times except the small ones. Consequently, the size distribution $f_l(t)$ has the simple form

$$f_l(t) = \alpha_1 e^{-l/L(t)}. \tag{4.120}$$

A typical time dependence of the characteristic cluster size in the range of validity of approximation (Equation 4.116) is presented in Figure 4.31. Three different regimes can be recognized. In the initial stage ($0 < t < t_1$), $L(t)$ grows exponentially and reaches approximately one-fourth of the final, equilibrium, value. In the intermediate stage ($t_1 < t < t_2$), $L(t)$ grows linearly ($L \sim t^x$, with $x \simeq 1$) up to about three-fourths of the final value. At $t > t_2$, it approaches the equilibrium value in a more complicated way.

### *4.4.4.2 Average Cluster Size*

In the range of validity of approximation (Equation 4.116), the average cluster size can be calculated using Equations 4.112, 4.113, and 4.120, which lead to

$$L_{av}(t) = \frac{1}{1 - \exp(-1/L(t))}. \tag{4.121}$$

For $L(t) \gg 1$, one has $L_{av}(t) = L(t)$.

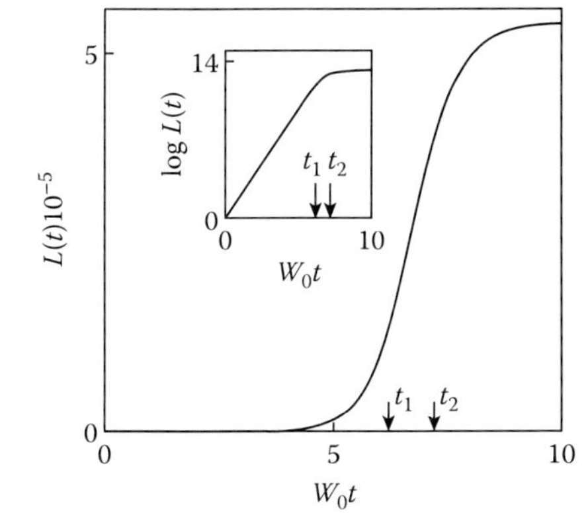


Figure 4.31

A typical time dependence of the characteristic cluster size $L(t)$ in the range of validity of the decoupling (Equation 4.116). $\beta J = 1.5$ and $\beta B_0 = 3.6$.

For the one-dimensional Ising model, $L_{av}(t)$ can be also calculated without using the distribution function $f_l(t)$. As mentioned earlier, the number density $f_{up}(t)$ of "up" spins is equal to $f_{up}(t) = (1 + \sigma)/2$. In addition, the number density $f_c(t)$ of clusters of "up" spins at large number of such clusters is equal to $f_c(t) = \frac{1}{2} n_{kink}(t)$, where $n_{kink}(t)$ is the number density of kinks (pairs of oppositely pointed spins) in a chain, which is equal to $n_{kink}(t) = \langle n_0(1) n_1(-1) \rangle_t \langle n_0(-1) n_1(1) \rangle_t = \frac{1}{2}(1 - \epsilon)$. Thus, one arrives at the simple expression

$$L_{av}(t) = \frac{2(1+\sigma)}{1-\epsilon}, \tag{4.122}$$

which allows us to obtain $L_{av}(t)$ without involving the distribution function $f_l(t)$.

In Figure 4.32, $L_{av}(t)$ is plotted for two limiting values of $W$, $W = 0$, and $W = W_0$. In the absence of coagulation and splitting ($W = 0$), almost the entire process of cluster growth follows a linear law [$L_{av}(t) \sim t$]. For $W = W_0$, the cluster growth is linear in the intermediate stage and exponential in the other ones.

One should note that the exponential form of $f_l(t)$ and the power law growth of $L_{av}(t)$ for large cluster sizes constitute general features of the phase transformation process in lattice systems. For example, in Refs. [8,12], $f_l(t)$ and $L_{av}(t)$ were determined via Monte Carlo simulations on lattices of higher dimensions (three and two dimensions, respectively). In both models, $f_l(t)$ decayed exponentially with respect to the cluster size

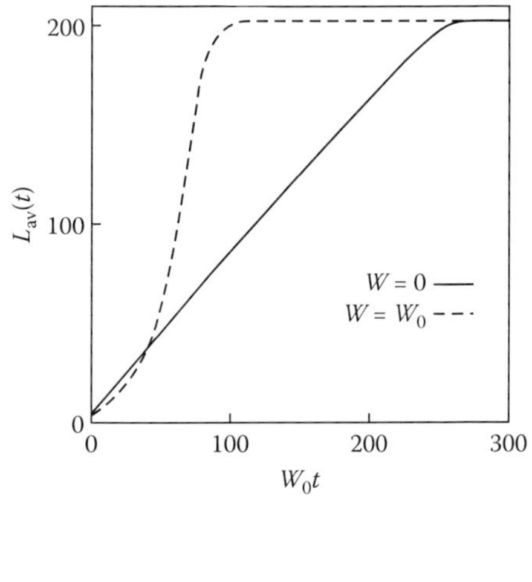


Figure 4.32

Time dependencies of the average cluster size for various values of $W$. $\beta J = 1.5$ and $\beta B_0 = 0.2$.

and $L_{av}(t)$ grew according to a power law, $L_{av}(t) \sim t^x$ with $x = 1$ in the initial stage and $x = 1/2$ in the later one. There are no analogous experiments concerning the one-dimensional system to compare with the results of the present paper.

### 4.4.5 Conclusion

In the present paper, a generalized kinetic Ising model was used to describe the phase transformations in lattice systems. A procedure was suggested to obtain a closed nonlinear system of kinetic equation for the polarization and energy of the system. These equations take into account implicitly the splitting and coagulation of clusters, and this constitutes an important advantage over the conventional classical theories. Using as an example a one-dimensional model, expressions for the RPT and the size distribution of clusters were obtained and analyzed in various situations. In the most representative case of phase transformation, namely, the case of field reversal (Section 4.4.3.3), which corresponds to the transition from an undersaturated to a supersaturated state, the relaxation of polarization exhibited an interesting time behavior in terms of the values of the parameters $W_0$ and $W$. The reciprocal of the first parameter, $1/W_0$, provides the timescale of the cluster growth or decay, and the reciprocal of the second one, $1/W$, the timescale for the splitting and coagulation of the clusters. For $W_0 \sim W$, the relaxation is pure exponential except during a short initial time and has a constant rate. For $W = 0$, which corresponds to the absence of splitting and coagulation of the clusters of "up" spins, the relaxation of polarization follows a linear law with respect to time, with a crossover to exponential one in the final stage. A similar behavior in the final stage occurred for $W_0 \gg W \neq 0$. However the RPT in that stage was much smaller than for $W = 0$. It was shown that the above behavior of the polarization could be qualitatively explained using the concept of $B$ and $J$ flips.

For small deviations from equilibrium and low temperatures ($T \to T_c = 0$), both the RPT and the rate of energy relaxation in the absence of an external field decreased exponentially with decreasing temperature. The critical exponent $z$ is equal to $z_e = 1$ for the energy relaxation rate and has two values: $z_\sigma = 2$ for $W\xi \ll W_0$ and $z_\sigma = 3$ for $W_0\xi \ll W$ ($\xi$ is the correlation length) for the RPT.

The time-dependent size distribution $f_l(t)$ of the clusters had at all times an exponential form with respect to the cluster size, the average cluster size $L_{av}(t)$ growing with increasing time. This growth was linear in $t$ far from equilibrium, and very slow near it. Such behaviors of $f_l(t)$ and $L_{av}(t)$ are similar to those found via Monte Carlo simulations for higher dimensions [8,12].

Along with obvious advantages, e.g., exact relations between the CFs, the existence of a parameter ($W$) that controls the splitting and coagulation of clusters, the one-dimensional Ising model does not allow to model the metastable state, because of the absence of an ordered state at a nonzero temperature. This can be obtained by using models in two or higher dimensions. Some results for a model on a Bethe lattice are in preparation.

## Acknowledgment

This work was supported by the National Science Foundation.

## Appendix 4A: Operator $R_j(\mu)$ and Equilibrium Averages

Let us consider an operator $R_j(\mu)$, which is equal to unity if the sum of the spins, neighboring to spin $\sigma_j$ is equal to $\mu$, and zero otherwise and introduce also the operators $n_j(1) = (1 + \sigma_j)/2$, $n_j(-1) = (1 - \sigma_j)/2$, which provide the occupation probabilities of a site $j$ by a spin in state $\sigma_j = 1$ and $\sigma_j = -1$, respectively. For example, $n_j(1) = 1$ when $\sigma_j = 1$ and $n_j(1) = 0$ when $\sigma_j = -1$. Obviously, the operator $n_{j+\Delta_1}(v_1)n_{j+\Delta_2}(v_2)\ldots n_{j+\Delta_m}(v_m)$ $(v_i = \pm 1)$ is equal to unity if the spins at the sites $j + \Delta_1, j + \Delta_2, \ldots, j + \Delta_m$ (neighboring sites of site $j$) are in the states $v_1, v_2, \ldots, v_m$, respectively. Consequently, the operator $R_j(\mu)$ can be written in the form

$$R_j(\mu) = \sum_{\left\{\sum_{n=1}^{m} \nu_n = \mu\right\}} n_{j+\Delta_1}(\nu_1)n_{j+\Delta_2}(\nu_2)\ldots n_{j+\Delta_m}(\nu_m). \tag{4A.1}$$

(Note the obvious property $\Sigma_\mu R_j(\mu) = 1$.)

For a one-dimensional Ising model, the sum of the two spins neighboring spin $\sigma_j$ acquires the values 0 and ± 2 and, correspondingly,

$$
\begin{aligned}
R_j(0) &= n_{j-1}(-1)n_{j+1}(1) + n_{j-1}(1)n_{j+1}(-1) \\
&= \frac{1}{2}(1-\sigma_{j-1}\sigma_{j+1}), \\
R_j(\pm 2) &= n_{j-1}(\pm 1)n_{j+1}(\pm 1) \\
&= \frac{1}{4}(1\pm\sigma_{j-1})(1\pm\sigma_{j+1}).
\end{aligned} \tag{4A.2}
$$

Starting with the definition of $R_j(\mu)$, the following properties of the equilibrium CFs containing $R_j(\mu)$ along with the arbitrary spin operator $A$ can be demonstrated for any dimension:

$$
\begin{aligned}
&\left\langle A\left(\sum_{k=1}^{m}\sigma_{j+\Delta_k}\right)R_j(\mu)\right\rangle_T = \mu\langle AR_j(\mu)\rangle_T, \\
&\langle \sigma_j A' R_j(\mu)\rangle_T = \tanh(\beta\varepsilon_\mu)\langle A' R_j(\mu)\rangle_T,
\end{aligned} \tag{4A.3}
$$

where the operator $A'$ does not include the spin $\sigma_j$, $\langle\ldots\rangle_T \equiv \mathrm{Tr}(\rho_T\ldots)$, and $\rho_T$ is given by Equation 4.96. The first property is obvious and follows from the definition of the operator $R_j(\mu)$. To prove the second one, let us write the Hamiltonian $H_s$ (Equation 4.86) in the form $H_s = H'_E - \sigma_j\left(B + J\sum_{k=1}^{m}\sigma_{j+\Delta k}\right)$, where the operator $H'_s$ does not contain the operator $\sigma_j$. Then,

$$
\begin{aligned}
\langle \sigma_j A' R_j(\mu)\rangle_T &= \frac{1}{Z}\mathrm{Tr}\left[\sigma_j A' R_j(\mu) e^{-\beta H'_s + \beta\sigma_j\left(B+J\sum_{k=1}^{m}\sigma_{j+\Delta_k}\right)}\right] \\
&= \frac{1}{Z}\mathrm{Tr}'\left\{A' R_j(\mu) e^{-\beta H'_s}\right. \\
&\quad \times\left.\left[e^{\beta(B+\mu J)} - e^{-\beta(B+\mu J)}\right]\right\} \\
&= \frac{2}{Z}\sinh(\beta\varepsilon_\mu)\mathrm{Tr}'\left[A' R_j(\mu) e^{-\beta H'_s}\right],
\end{aligned} \tag{4A.4}
$$

where $Z$ is the partition function of the system, Tr′ denotes the trace over all spin variables excluding $\sigma_j$, and $\varepsilon_\mu$ is given by Equation 4.91.

Similarly,

$$
\langle A' R_j(\mu)\rangle_T = \frac{2}{Z}\cosh(\beta\varepsilon_\mu)\mathrm{Tr}'\left[A' R_j(\mu) e^{-\beta H'_s}\right]. \tag{4A.5}
$$

Equations 4A.4 and 4A.5 lead to the second property in Equation 4A.3.

Note that the equilibrium average $\langle R_j(\mu)\rangle_T \equiv \mathrm{Tr}[R_j(\mu)\rho_T]$ provides the fraction of spins which are subjected to the local field $\varepsilon_\mu$.

## Appendix 4B: Relations between Equilibrium Averages

Exact expressions for the equilibrium averages of the one-dimensional Ising model, which are necessary in the present calculations, are provided. The CFs were calculated as suggested by Zhelifonov [43] and have the form

$$\begin{aligned}
\sigma_T &\equiv \langle \sigma_0 \rangle_T = \frac{\sinh(\beta B)}{\sqrt{\sinh^2(\beta B) + e^{-4\beta J}}}, \\
\epsilon_{kT} &\equiv \langle \sigma_0 \sigma_k \rangle_T = \sigma_T^2 + \left(1 - \sigma_T^2\right) f_T^k, \\
\gamma_{kT} &\equiv \langle \sigma_0 \sigma_k \sigma_{k+1} \rangle_T \\
&= \sigma_T \epsilon_T + f_T^k (1 + f_T)\left(1 - \sigma_T^2\right) \tanh(\beta B), \\
\delta_{kT} &\equiv \langle \sigma_0 \sigma_{k-1} \sigma_k \sigma_{k+1} \rangle_T \\
&= \sigma_T \langle \sigma_0 \sigma_1 \sigma_2 \rangle_T + \left(1 - \sigma_T^2\right) \phi_T f_T^k,
\end{aligned} \tag{4B.1}$$

where $\epsilon_T \equiv \langle \sigma_0 \sigma_1 \rangle_T$, and

$$\begin{aligned}
f_T &= \frac{\sigma_T - \tanh(\beta B)}{\sigma_T + \tanh(\beta B)} = \frac{\epsilon_T - \sigma_T^2}{1 - \sigma_T^2}, \\
\phi_T &= 2\sinh^2(\beta B)\coth(2\beta J) + \cosh(2\beta B).
\end{aligned} \tag{4B.2}$$

Using the identities $\tanh(\beta B) = \sigma_T\left[(1-\epsilon_T)/\left(1 - 2\sigma_T^2 + \epsilon_T\right)\right]$ and $\phi_T = \left(\epsilon_T - \sigma_T \langle \sigma_0 \sigma_1 \sigma_2 \rangle_T\right) f_T^{-1} / \left(1 - \sigma_T^2\right)$, one can write

$$\begin{aligned}
\epsilon_{kT} &= \sigma_T^2 + \left(\epsilon_T - \sigma_T^2\right)^k / \left(1 - \sigma_T^2\right)^{k-1}, \\
\gamma_{kT} &= \sigma_T\left[\epsilon_T + (1 - \epsilon_T)\left(\epsilon_T - \sigma_T^2\right)^k / \left(1 - \sigma_T^2\right)^k\right], \\
\delta_{kT} &= \sigma_T \langle \sigma_0 \sigma_1 \sigma_2 \rangle_T + \left(\epsilon_T - \sigma_T \langle \sigma_0 \sigma_1 \sigma_2 \rangle_T\right) \\
&\quad \times \left(\epsilon_T - \sigma_T^2\right)^{k-1} / \left(1 - \sigma_T^2\right)^{k-1}.
\end{aligned} \tag{4B.3}$$

The average number densities $n(B^+)$ and $n(B^-)$ of $B^\pm$ spins (see Section 4.4.3.1) are given at equilibrium by the expressions

$$\begin{aligned}
n_T(B^\pm) &= \langle n_{j-1}(1) n_j(\mp 1) n_{j+1}(-1) \\
&\quad + n_{j-1}(-1) n_j(\mp 1) n_{j+1}(1) \rangle_T \\
&= \frac{1}{4}(1 \mp \sigma_T - \epsilon_{2T} \pm \gamma_{1T}).
\end{aligned} \tag{4B.4}$$

Formally, the time-dependent correlation functions and densities $n(B^\pm)$ can be obtained from Equations 4B.3 and 4B.4 by eliminating from them the subscript $T$.

## Appendix 4C: Small Deviations from Equilibrium

In the linear approximation, the kinetic equations for $\Delta\sigma$ and $\Delta\epsilon$ have the forms (Equation 4.105)

$$\dot{\Delta\sigma} = -\frac{\Delta\sigma}{\tau_\sigma} - \frac{\Delta\epsilon}{\tau_{\sigma\epsilon}}, \quad \dot{\Delta\epsilon} = -\frac{\Delta\sigma}{\tau_{\epsilon\sigma}} - \frac{\Delta\epsilon}{\tau_\epsilon}, \tag{4C.1}$$

where

$$\tau_\sigma^{-1} = \sum_{\mu=0,\,\pm 2} W_\mu C_\mu, \quad \tau_\epsilon^{-1} = \sum_{\mu=0,\,\pm 2} \mu W_\mu d_\mu,$$
$$\tau_{\sigma\epsilon}^{-1} = \sum_{\mu=0,\,\pm 2} W_\mu d_\mu, \quad \tau_{\epsilon\sigma}^{-1} = \sum_{\mu=0,\,\pm 2} \mu W_\mu C_\mu,$$
$$C_\mu = \mu^2(2-\mu z_\mu)/8 + \frac{4}{4-3\mu^2}(1-f_T)^2\left(1+2z_\mu\sigma_T + \sigma_T^2\right), \tag{4C.2}$$
$$d_\mu = (\mu^2-4)z_0/2 + \mu[2+\mu\sigma_T(1-f_T) - \mu z_\mu f_T]/4.$$

The coefficients in Equation 4.106 are given by the following expressions:

$$\frac{1}{\tau^\pm} = \frac{1}{2}\left(\frac{1}{\tau_\sigma} + \frac{1}{\tau_\epsilon} \pm \frac{1}{\tau_D}\right), \quad \frac{1}{\tau_D^2} = \left(\frac{1}{\tau_\sigma} - \frac{1}{\tau_\epsilon}\right)^2 + \frac{4}{\tau_{\sigma\epsilon}\tau_{\epsilon\sigma}},$$
$$C^\pm = \pm\tau_D\left[\left(\frac{1}{\tau_\sigma} - \frac{1}{\tau^\mp}\right)\Delta\sigma(0) + \frac{\Delta\epsilon(0)}{\tau_{\sigma\epsilon}}\right]. \tag{4C.3}$$

## References


1. Frenkel, J., *Kinetic Theory of Liquids*, Dover, New York, 1955.
2. Abraham, F., *Homogeneous Nucleation Theory*, Academic, New York, 1974.
3. Narsimhan, G., Ruckenstein, E., *J. Colloid Interface Sci.* 128, 549 (1989). (Section 1.2 of this volume.)
4. Ruckenstein, E., Nowakowski, B., *J. Colloid Interface Sci.* 137, 583 (1990). (Section 1.3 of this volume.)
5. Nowakowski, B., Ruckenstein, E., *J. Chem. Phys.* 94, 1397 (1991). (Section 2.1 of this volume.)
6. Nowakowski, B., Ruckenstein, E., *J. Chem. Phys.* 94, 8487 (1991). (Section 2.3 of this volume.)
7. Novotny, M.A., Brown, G., Rikvold, P.A., *J. Appl. Phys.* 91, 6908 (2002).
8. Tafa, K., Puri, S., Kumar, D., *Phys. Rev. E* 63, 046115 (2001).
9. Tafa, K., Puri, S., Kumar, D., *Phys. Rev. E* 64, 056139 (2001).
10. Schmelzer, J., Jr., Landau, D.P., *Int. J. Mod. Phys. C* 12, 345 (2001).
11. Shneidman, V.A., Jackson, K.A., Beatty, K.M., *J. Cryst. Growth* 212, 564 (2000).
12. Wonczak, S., Strey, R., Stauffer, D., *J. Chem. Phys.* 113, 1976 (2000).
13. Vehkamaki, H., Ford, I.J., *Phys. Rev. E* 59, 6483 (1999).
14. Stauffer, D., *Physica A* 244, 344 (1997).
15. Stauffer, D., Kertesz, J., *Physica A* 177, 381 (1991).
16. Marchand, J.P., Martin, Ph.A., *Physica A* 127, 681 (1984).
17. Shneidman, V.A., Nita, G.M., *Phys. Rev. Lett.* 89, 025701 (2002).
18. Bovier, A., Manzo, F., *J. Stat. Phys.* 107, 757 (2002).
19. Park, K., Novotny, M.A., Rikvold, P.A., *Phys. Rev. E* 66, 056101 (2002).

20. Rikvold, P.A., Tomita, H., Miyashita, S., Sides, S.W., *Phys. Rev. E* 49, 5080 (1994).
21. Neves, E.J., Schonmann, R.H., *Commun. Math. Phys.* 137, 209 (1991).
22. Binder, K., Müller-Krumbhaar, H., *Phys. Rev. B* 9, 2328 (1974).
23. Shneidman, V.A., Jackson, K.A., Beauty, K.M., *J. Chem. Phys.* 111, 6932 (1999).
24. Shneidman, V.A., Jackson, K.A., Beauty, K.M., *Phys. Rev. B* 59, 3579 (1999).
25. Ben-Naim, E., Krapivsky, P.L., *J. Stat. Phys.* 93, 583 (1998).
26. Kuchanov, S.I., Aliev, M.A., *J. Phys. A* 30, 8479 (1997).
27. Berim, G.O., Ruckenstein, E., *J. Chem. Phys.* 119, 806 (2003). (Section 4.3 of this volume.)
28. Glauber, R., *J. Math. Phys.* 4, 294 (1963).
29. Felderhof, B.U., *Rep. Math. Phys.* 1, 215 (1971).
30. Suzuki, M., Kubo, R., *J. Phys. Soc. Jpn.* 24, 51 (1968).
31. Mazenko, G.F., Valls, O.T., Zhang, F.C., *Phys. Rev. B* 31, 4453 (1985).
32. Luque, J.J., Córdoba, A., *J. Chem. Phys.* 76, 6393 (1982).
33. Kumar, D., Stein, J., *J. Phys. C* 13, 3011 (1980).
34. Huang, H.W., *J. Chem. Phys.* 70, 2390 (1979).
35. Yahata, H., *J. Phys. Soc. Jpn.* 30, 657 (1971).
36. Bogolyubov, N.N., *The Problems of Dynamic Theory in Statistical Physics*, Gostekhizdat, Moscow, 1946 (in Russian).
37. Zubarev, D.N., *Nonequilibrium Statistical Thermodynamics*, Nauka, Moscow, 1971 (in Russian).
38. Provotorov, B.N., *Zh. Eksp. Teor. Fiz.* 41, 1582 (1961).
39. Baxter, R.J., *Exactly Solved Models in Statistical Mechanics*, Academic, New York, 1982.
40. Berim, G.O., Kessel, A.R., *Physica A* 101, 112 (1980).
41. Berim, G.O., Kessel, A.R., *Physica A* 101, 127 (1980).
42. Hohenberg, P.C., Halperin, B.I., *Rev. Mod. Phys.* 49, 435 (1977).
43. Zhelifonov, M.P., *Teor. Mater. Fiz.* 8, 401 (1971).

# 4.5 Kinetics of Phase Transformation on a Bethe Lattice

*Gersh Berim and Eli Ruckenstein**

A kinetic Ising model is applied to the description of phase transformations on a Bethe lattice. A closed set of kinetic equations for a model with the coordination number $q = 3$ is obtained using a procedure developed in Section 4.4. For $T$ close to $T_c$ ($T > T_c$), where $T_c$ is the phase transition temperature, and zero external field (absence of supersaturation), the rate of phase transformation (RPT) for small deviations from equilibrium is independent of time and tends to zero as $(T - T_c)$. At $T = T_c$, the RPT depends on time and for large times behaves as $t^{-1}$. For $T < T_c$, we examine the transformation from the initial state with almost all spins "down" to the state with almost all spin "up" after the external field jumped from $B_i < 0$ to $B_f > 0$. The role of different mechanisms responsible for growth (decay), splitting (coagulation), and creation (annihilation) of clusters are examined separately. In all cases, there is a critical value $B_c$ of the external field, such that the phase transformation takes place only for $B_f > B_c$. This result is also obtained from a more simple consideration involving spherical-like clusters on a Bethe lattice. The characteristic time $t_R$ at which the polarization becomes larger than zero diverges as $(B_f - B_c)^{-b}$ for $B_f \rightarrow B_c$ with $b = 0.47$. The RPT has a rapid growth near $t_R$ and remains constant for $t > t_R$. The average cluster size (number of spins in a cluster) exhibits a rapid unrestricted growth at a time $t_d \simeq t_R$, which indicates the creation of infinite clusters. The only exception to the latter behavior occurs when the kinetics is dominated by cluster growth and decay processes. In this case, the average cluster size remains finite during the transformation process. In contrast to the classical theory, the present approach does not separate the processes of creation of clusters of critical size (nucleation) and of their growth, both being accounted for by the kinetic equations used. (© *2004 American Institute of Physics.* doi: 10.1063/1.1629676.)

## 4.5.1 Introduction

In a previous paper [1], which will be referred to below as Paper I, the kinetic Ising model was applied to the description of phase transformations in lattice systems. The advantage of that approach over the classical nucleation theory is that the kinetic equations take into account implicitly the splitting and coagulation of clusters of equally pointed spins, independent of their size and shape. Previously, these processes were considered for an Ising model on a square lattice [2], but only small clusters were accounted for. A new closure procedure of the infinite chain of kinetic equations was suggested in Paper I, which provided a closed set of nonlinear kinetic equations for the spins polarization and their interaction energy. The one-dimensional Ising model was considered and the effectiveness of the closure procedure was demonstrated. Despite of numerous advantages, such as the existence of an analytical equilibrium solution, the existence of parameters which account for the splitting and coagulation, as well as for the growth and decay of clusters, that model has not allowed to examine the evolution of the system from a metastable state. Such states exist in the Ising spin systems for $T < T_c$, where $T_c$ is the phase transition temperature, which in the one-dimensional case is equal to zero. For this reason, the consideration of a model for which $T_c \neq 0$ is desirable.

The procedure of closing the kinetic equations used in Paper I is most effective when the model under consideration has an analytical equilibrium solution in the presence of an external field. Among the known models that have nonzero phase transition temperatures, the only one with such a solution is the Ising model on a Bethe lattice (IMBL) [3,4]. Despite its apparent artificiality, that model was successfully used to describe the ternary solutions [5,6], phase separation [7], polymers [8–10], spin glasses [11–13], magnetic bilayers [14], and superconductors [15,16]. In the present paper, we examine the phase transformation in IMBL using the approach developed in Paper I.

## 4.5.2 Kinetic Equations for the IMBL

### *4.5.2.1 General Case*

In this section, the kinetic equations for the present model will be derived using the methodology of Paper I.

* *J. Chem. Phys.* 120, 272 (2004). Republished with permission.

The Bethe lattice can be characterized by the coordination number $q$ that indicates the number of molecules interacting with any given one (see Figure 4.33a, in which $q = 3$). In the isotropic case, the general Hamiltonian for an IMBL can be written as

$$H_s = -B\sum_j \sigma_j - \frac{1}{2}J\sum_j\sum_{k=1}^{q}\sigma_j\sigma_{j+\delta_k} \quad (J>0), \tag{4.123}$$

where the variables $\sigma_j = \pm 1$ (spins) model two phases, e.g., a liquid ($\sigma_j = -1$) and a solid ($\sigma_j = 1$), to which a molecule at site $j$ can belong, $B$ is the external field, which models the supersaturation $S$ ($B \sim \log S$ [17]), $J$ is the interaction energy between two neighboring molecules (which will be called below as "energy"), and the index $\delta_k$ labels the neighboring spins, interacting with spin $\sigma_j$. This model has a phase transition at a critical temperature $T_c$ provided by the equation $\tanh \beta J = 1/(q-1)$ [4], where $\beta = 1/k_B T$, $T$ being the temperature in K and $k_B$ the Boltzmann constant. The other details concerning the equilibrium properties are provided in Appendix 4D.

For the single-spin-flip mechanism of spin dynamics [18], the kinetic equations for the dimensionless polarization $P = \Sigma_j \sigma_j$ and energy $E = \frac{1}{2}\Sigma_j \Sigma_{k=1}^{q} \sigma_j\sigma_{j+\delta_k}$ were provided in Ref. [1] in the following form, suitable for any Ising model of spin $s = 1/2$ ($\sigma = \pm 1$ in our case):

$$\begin{aligned}\frac{d}{dt}P(t) &= -\sum_{\mu=-q}^{q} W_\mu \sum_j [\langle \sigma_j R_j(\mu)\rangle_t - z_\mu \langle R_j(\mu)\rangle_t],\\ \frac{d}{dt}E(t) &= -\sum_{\mu=-q}^{q} \mu W_\mu \sum_j [\langle \sigma_j R_j(\mu)\rangle_t - z_\mu \langle R_j(\mu)\rangle_t],\end{aligned} \tag{4.124}$$

where $z_\mu = \tanh \beta\varepsilon_\mu$, $\varepsilon_\mu = B + J\mu$, $R_j(\mu)$ is an operator equal to unity if the sum of spins neighboring a spin $\sigma_j$ is equal to $\mu$ and zero otherwise. The terms in the right-hand side of Equation 4.124 that contain the operators $R_j(\mu)$ describe the dynamics of the spins whose nearest neighbor spins have a sum equal to $\mu$. The flip-flop transitions of such spins have rates proportional to $W_\mu$ and are accompanied by energy changes of $\Delta E_\mu = \pm 2\varepsilon_\mu$. In Equation 4.124, $P(t) = \langle P\rangle_t$, $E(t) = \langle E\rangle_t$, where the time-dependent averages $\langle \ldots\rangle_t$ are given by

$$\langle \ldots \rangle_t = \mathrm{Tr}[e^{-F(t)}\ldots]/\mathrm{Tr}[e^{-F(t)}] \tag{4.125}$$

with

$$F(t) = -\beta_\sigma(t)\sum_j \sigma_j - \frac{1}{2}\beta_\epsilon(t)\sum_j\sum_{k=1}^{q}\sigma_j\sigma_{j+\delta_k}, \tag{4.126}$$

where $\beta_\sigma(t)$ and $\beta_\epsilon(t)$ are functions of $t$.

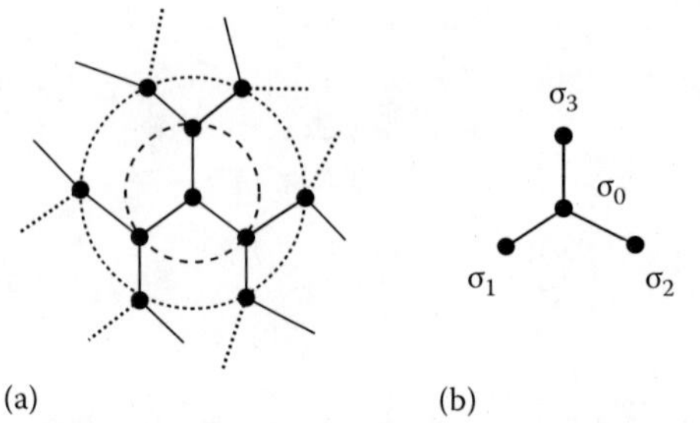


Figure 4.33

(a) A Bethe lattice for $q = 3$. The circles indicate the first and second shells of the central site and (b) central spin $\sigma_0$ and its three nearest neighbors.

One should note that the total number $M(t)$ of molecules in the solid phase is related to the total polarization $P(t)$ through the relation $M(t) = \frac{1}{2}[N + P(t)]$, where $N$ is the total number of molecules in the system.

### 4.5.2.2 The Case q = 3

In this paper, we select $q = 3$ that represents the simplest nontrivial version of IMBL [for $q = 2$ the Hamiltonian (Equation 4.123) reduces to the Hamiltonian of a linear Ising model]. The geometry of this lattice is presented in Figure 4.33a. In this case, $\mu = \pm 1, \pm 3$ and the operators $R_j(\mu)$ can be represented as follows (the coordinate $j$ being selected equal to zero, for convenience):

$$
\begin{aligned}
R_0(\pm 1) &= \frac{1}{8}(3 \pm \sigma_1 \pm \sigma_2 \pm \sigma_3 - \sigma_1\sigma_3 - \sigma_2\sigma_3 \\
&\quad - \sigma_1\sigma_2 \mp 3\sigma_1\sigma_2\sigma_3), \\
R_0(\pm 3) &= \frac{1}{8}(1 \pm \sigma_1 \pm \sigma_2 \pm \sigma_3 + \sigma_1\sigma_3 + \sigma_2\sigma_3 \\
&\quad + \sigma_1\sigma_2 \pm \sigma_1\sigma_2\sigma_3),
\end{aligned}
\tag{4.127}
$$

where the notations for the spins are indicated in Figure 4.33b.

Because of spatial homogeneity, the kinetic equations for $P(t)$ and $E(t)$ can be replaced by equations for $\sigma \equiv \langle\sigma_0\rangle_t$ (polarization per spin) and for $\epsilon \equiv \langle\sigma_0\sigma_1\rangle_t$ (interaction energy per bond):

$$
\begin{aligned}
\dot{\sigma} &= -\sum_{\mu=\pm 1,\pm 3} W_\mu\left[\langle\sigma_0 R_0(\mu)\rangle_t - z_\mu\langle R_0(\mu)\rangle_t\right], \\
\dot{\epsilon} &= -\frac{2}{3}\sum_{\mu=\pm 1,\pm 3} \mu W_\mu\left[\langle\sigma_0 R_0(\mu)\rangle_t - z_\mu\langle R_0(\mu)\rangle_t\right].
\end{aligned}
\tag{4.128}
$$

When it will be not confusing, the argument $t$ of the time-dependent functions $\sigma(t)$, $\epsilon(t)$, ... will be omitted. After substitution of Equation 4.127 in Equation 4.128, the kinetic equations become

$$
\dot{\sigma} = \sum_{i=0}^{6} \gamma_{\sigma i}\langle A_i\rangle_t, \quad \dot{\epsilon} = \sum_{i=0}^{6} \gamma_{\epsilon i}\langle A_i\rangle_t,
\tag{4.129}
$$

where

$$
\begin{aligned}
&A_0 = 1, \; A_1 = \sigma_0, \quad A_2 = \sigma_0\sigma_1, \quad A_3 = \sigma_1\sigma_3, \\
&A_4 = \sigma_1\sigma_2\sigma_3, \quad A_5 = \sigma_0\sigma_1\sigma_2, \quad A_6 = \sigma_0\sigma_1\sigma_2\sigma_3.
\end{aligned}
\tag{4.130}
$$

The coefficients $\gamma_{\sigma i}$ and $\gamma_{\epsilon i}$ depend on $W_\mu$ and $z_\mu$ and will be provided explicitly later for specific cases.

To transform Equation 4.129 to a closed form, all averages $\langle A_i\rangle_t$ should be expressed in terms of the two unknown functions $\beta_\sigma(t)$ and $\beta_e(t)$ using the similarity between the operator $F(t)$, Equation 4.126, and the Hamiltonian (Equation 4.1), and the exact analytical expressions for the equilibrium averages (see Appendices 4D and 4E for more details). In this way, a closed set of kinetic equations for $\beta_\sigma(t)$ and $\beta_e(t)$ can be obtained. However, in the present case, it is more convenient to use as variables the polarization $\sigma$ and the quantity $z \equiv \tanh \beta_e(t)$, using $\beta_\sigma(t)$ and $\beta_e(t)$ as intermediate parameters. As a result two nonlinear differential equations for $\sigma$ and $z$,

$$
\dot{\sigma} = f_\sigma(\sigma, z), \; \dot{z} = f_z(\sigma, z)
\tag{4.131}
$$

were derived. (The way to obtain the functions $f_\sigma(\sigma,z)$ and $f_z(\sigma,z)$ is sketched in Appendix 4E.)

Equation 4.131 provides the main equations for the considerations that follow. Before proceeding to their solution, it is instructive to emphasize the physical meaning of the parameters $W_\mu$ in connection with the dynamics of clusters of "up" spins on a Bethe lattice. The "up"–"down" flip of a spin whose neighbors are all "up" spins splits an existing cluster for $q = 3$ into three new clusters (Figure 4.34a). The reverse "down"–"up"

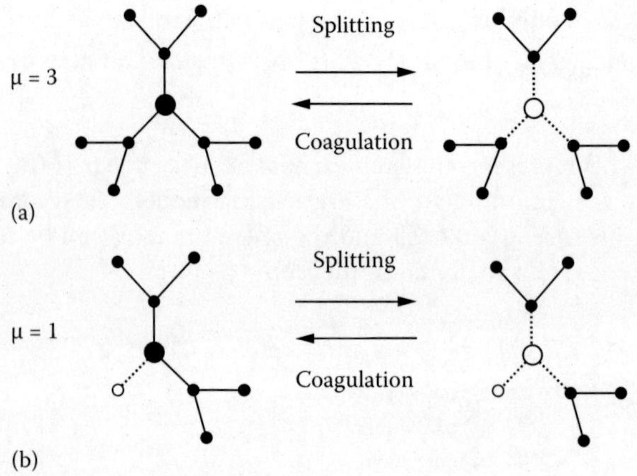


Figure 4.34

Splitting and coagulation of clusters of "up" spins (filled circles) due to the flip of connecting spin (shown by a larger circle). Nonfilled circles represent the "down" spins. (a) The case $\mu = 3$ and (b) the case $\mu = 1$.

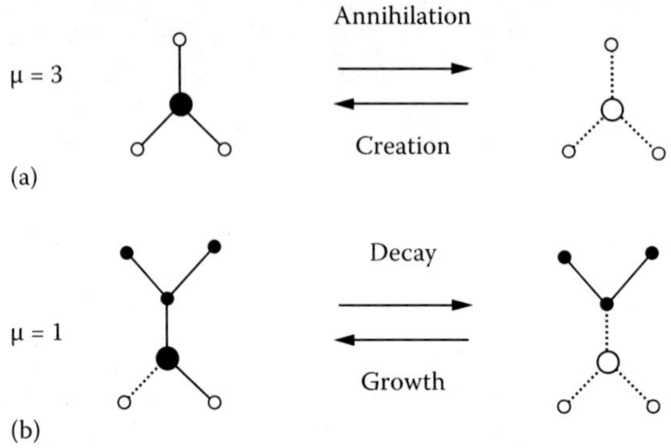


Figure 4.35

(a) Annihilation and creation of the single-spin cluster and (b) decay and growth of cluster.

flip in the same environment leads to the coagulation of the three clusters into a single one. The rates of the above two processes are proportional to $W_3$. A similar effect has the flips of the spins, which are associated with $W_1$. In this case, the existing cluster can be split into two clusters, and vice versa, two neighboring clusters can coagulate into a single one (Figure 4.34b). The spin flips that correspond to $\mu = -3$ provide the creation ("down"–"up" flips) or annihilation ("up"–"down" flips) of single-spin clusters on a Bethe lattice (Figure 4.35a). For $\mu = -1$, the corresponding flips provide growth ("down"–"up" flips) or decay ("up"–"down" flips) of the existing clusters (Figure 4.35b). By selecting some of the $W_\mu$ equal to zero, one can examine separately the different regimes of the lattice kinetics.

### 4.5.3 Phase Transformation in the Absence of Supersaturation

Let us suppose that the system of molecules on the Bethe lattice at $t < 0$ was in a supersaturated ($S > 1$) or an undersaturated $S < 1$ state, and that at $t > 0$ it is in the state with $S = 1$. Then the initial state of the related kinetic Ising model involves a nonzero external field ($B_i > 0$ for $S > 1$ and $B_i < 0$ for $S < 1$), which becomes zero in the final state ($B_f = 0$). In this case, $z_{-\mu} = -z_\mu$ and Equation 4.128 for $\sigma$ and $\epsilon$ acquire, when all $W_\mu$ are equal to $W$, the forms (Equation 4.129) with the coefficients $\gamma_{\sigma i}$, $\epsilon_{\sigma i}$ given by

$$\gamma_{\sigma 1} = W(-8+6z_1+6z_3),$$
$$\gamma_{\sigma 4} = W(-6z_1+2z_3),\ \gamma_{\epsilon 0} = 4W(z_1+z_3), \tag{4.132}$$
$$\gamma_{\epsilon 2} = -16W,\ \gamma_{\epsilon 3} = -4W(z_1-3z_3).$$

All the other coefficients ($\gamma_{\sigma i}$ and $\epsilon_{\sigma i}$) are zero.

Even in this simple case, the kinetic equations are very complicated after the present closure procedure is used (Appendix 4E). For this reason, we will restrict ourselves to some interesting particular cases, when the equations can be simplified.

#### 4.5.3.1 Small Deviations from Equilibrium at T > T$_c$

For $T > T_c$ and $B = 0$, Equations 4A.2 and 4A.5 provide for the equilibrium polarization $\sigma_T$ and equilibrium energy $\sigma_T$ the values $\sigma_T = 0$ and $\epsilon_T = z_T$. If the deviations $\Delta\sigma \equiv \sigma - \sigma_T$ and $\Delta\epsilon \equiv \epsilon - \epsilon_T$ of the nonequilibrium quantities $\sigma$ and $\epsilon$ from their equilibrium values are small, then one can obtain, after the closure and the linearization of Equation 4.129, the following simple solutions for $\Delta\sigma$ and $\Delta\epsilon$:

$$\Delta\sigma = \Delta\sigma(0)e^{-t/\tau_\sigma}, \tag{4.133}$$

$$\Delta\epsilon = \Delta\epsilon(0)e^{-t/\tau_\epsilon}, \tag{4.134}$$

where

$$1/\tau_\sigma = \frac{16W}{1+3z_T^2}(1-z_T)^2(1+z_T)\left(1+2z_T^2\right)(z_c - z_T), \tag{4.135}$$

with $z_c$, the value of $z_T$ at the critical temperature equal to 1/2, and

$$1/\tau_\epsilon = 16W\frac{1-z_T^2}{1+3z_T^2}. \tag{4.136}$$

By defining the rate of phase transformation (RPT) as

$$\text{RPT} = \frac{\dot{\sigma}(t)}{\sigma(\infty)-\sigma(t)}, \tag{4.137}$$

one can find that in the considered case the RPT has a constant value equal to $1/\tau_\sigma$ given by Equation 4.135.

Near the critical point $[(T - T_c)/T_c \ll 1]$, $\tau_\epsilon$ has a finite value ($\tau_\epsilon \simeq 7/48W$), and $\tau_\sigma$ diverges as $(T - T_c)^{-1}$:

$$\tau_\sigma \simeq \frac{7}{27W}\frac{T_c^2}{(T-T_c)J}. \tag{4.138}$$

This kind of divergence is known in the theory of phase transitions as the critical slowing down [19]. The divergence of $\tau_\sigma$ means that RPT tends to zero as $(T - T_c)$ when $T$ approaches $T_c$ from above.

#### 4.5.3.2 Small Deviations from Equilibrium at T = T$_c$

In this case, the linear approximation is no longer valid. Expanding all quantities up to the third order in $\Delta\sigma$ and first order in $\Delta z$, one obtains

$$\begin{aligned}\frac{1}{W}\frac{\dot{\Delta\sigma}}{\Delta\sigma} &= -\frac{12}{7}\Delta z + (\Delta\sigma)^2\left(-\frac{2}{7}+\frac{32}{21}\Delta z\right),\\ \frac{1}{W}\dot{\Delta z} &= -\left[\frac{48}{7}+\frac{16}{7}(\Delta\sigma)^2\right]\Delta z.\end{aligned} \tag{4.139}$$

The equation for $\Delta z$ has a simple particular solution $\Delta z(t) = 0$, which is compatible with the initial condition $\Delta z(0) = 0$. The corresponding solution for $\Delta\sigma$ is

$$\Delta\sigma = \frac{\Delta\sigma(0)}{\sqrt{1+4W[\Delta\sigma(0)]^2 t/7}}, \tag{4.140}$$

which, for large times, leads to the following power law relaxation of the polarization:

$$\Delta\sigma \simeq \sqrt{\frac{7}{4W}}\, t^{-1/2}. \tag{4.141}$$

The RPT for large times decreases as $t^{-1}$. The deviation $\Delta\epsilon$ of the energy from its equilibrium value $\epsilon_T = 1/2$ at $T = T_c$ is given by $\Delta\epsilon = (\Delta\sigma)^2/3 + (1 - 8(\Delta\sigma)^2/9)\Delta z$, which at large $t$ behaves as $t^{-1}$.

### 4.5.4 Kinetics for $T < T_c$ in the General Case

Let us consider the temperature range $T < T_c$ and suppose that at $t = 0$ the system is in equilibrium in an external field $B_i$ in "down" direction ($B_i < 0$, supersaturation $S < 1$). Regardless of how small such a field is, it breaks the "up"–"down" symmetry of the system and generates an initial state with almost all spin "down" (liquid state). If at $t = 0$ the external field changes direction and value to $B_f > 0$ (supersaturation $S > 1$), one expects that a transition process to the final state with all spins "up" (solid state) will be started. This final state has to be an equilibrium one, corresponding to the same temperature $T$ and new field $B_f$. Such a behavior was observed in numerous Monte Carlo experiments performed for $0 < T < T_c$ for an Ising model on a square lattice [20–24]. The experiments showed that the transition process takes place for any positive $B_f$, the relaxation time increasing with decreasing $B_f$. In most cases, all possible processes, such as creation–annihilation of single-spin clusters, splitting and coagulation as well as growth and decay of clusters, were taken into account in the simulations.

#### 4.5.4.1 Phase Transformation

In this section, the phase transformation on a Bethe lattice will be considered when all the coefficients $W_i$ ($i = \pm 1, \pm 3$) are equal to each other ($W_i \equiv W \neq 0$), thus accounting for all the above-mentioned processes. The initial field was chosen such as to have at any temperature $\beta B_i = -0.1$, and the range of temperatures was selected such as to have $\beta J > \text{arctanh}(0.5) \simeq 0.5493$, that corresponds to $T < T_c$.

The first specific feature of the phase transformation on a Bethe lattice is that the transition from a liquid ($\sigma \simeq -1$) to a solid $\sigma \simeq 1$ state at $T < T_c$ takes place only when the value $B_f$ of the external field at $t > 0$ is larger than a "critical" value $B_c$. For example, for $\beta J = 1.5$, this field is equal to $\beta B_c = 0.809\,34$. If $B_f < B_c$, then after an initial period of time the system remains in a steady state with a polarization $\sigma_{st} \neq \sigma_{eq}$, where $\sigma_{eq}$ is the equilibrium polarization in the field $B_f$. This metastable state has an infinite lifetime. The behavior of the polarization in this case for different values of $B_f$ is presented in Figure 4.36a. If $B_f > B_c$, the polarization behaves as shown in Figure 4.36b. If the difference between $B_f$ and $B_c$ is small [$(B_f - B_c)/B_c \ll 1$], the polarization changes its value from $\sigma = \sigma(0)$ to $\sigma = \sigma_{eq}$ in a very narrow time interval. The time $t_R$ at which the polarization becomes larger than zero can be considered as the relaxation time of the process, or the lifetime of the metastable state. It diverges as $B_f \to B_c$. The analysis of the values of $t_R$ obtained numerically for various values of $B_f$ shows that for $B_f \to B_c$, $t_R$ follows at any temperature the power law

$$Wt_R \sim \left(\frac{B_f - B_c}{B_c}\right)^b, \ (B_f \to B_c) \tag{4.142}$$

with $b = 0.47$. The critical field $B_c$ depends on temperature and decreases with increasing temperature (see Figure 4.37).

In Figure 4.38, a typical time dependence of RPT is presented. The RPT grows rapidly at $t \simeq t_R$, but remains constant for $t > t_R$.

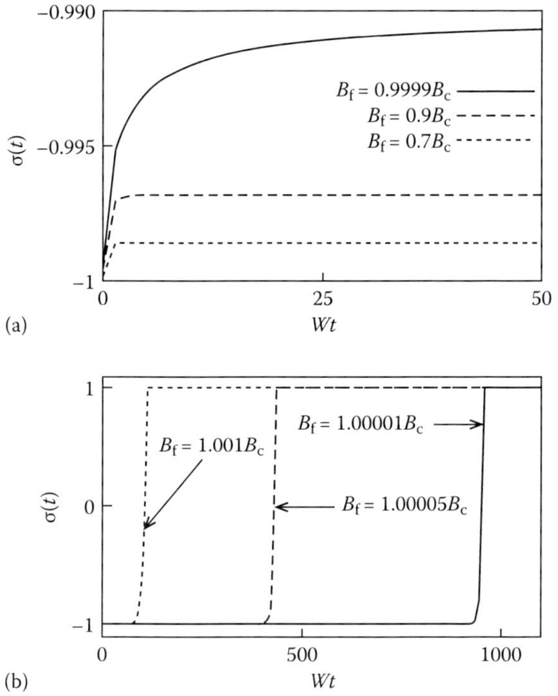


Figure 4.36

Time dependence of the polarization for $T < T_c$ ($\beta J = 1.5$) and various values of the final external field $B_f$. (a) $B_f < B_c$ and (b) $B_f > B_c$. The critical external field is given by $\beta B_c = 0.80934$.

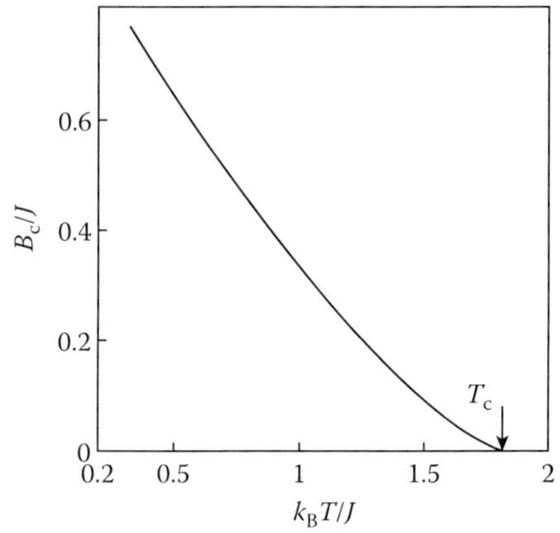


Figure 4.37

Temperature dependence of the critical field $B_c$. The critical temperature, $T_c$, is given by $k_BT_c/J = 1.8204$.

### *4.5.4.2 Cluster Dynamics*

Let us consider the dynamics of clusters of "up" spins on a Bethe lattice, focusing on the time-dependent average size (average number of spins in a cluster) $L_{av}(t)$, which is defined as the ratio of the number density $n_{up}(t)$ of "up" spins and of the number density $n_{cl}(t)$ of clusters (all number densities are defined with respect to the total number of spins):

$$L_{av}(t) = \frac{n_{up}(t)}{n_{cl}(t)}. \tag{4.143}$$

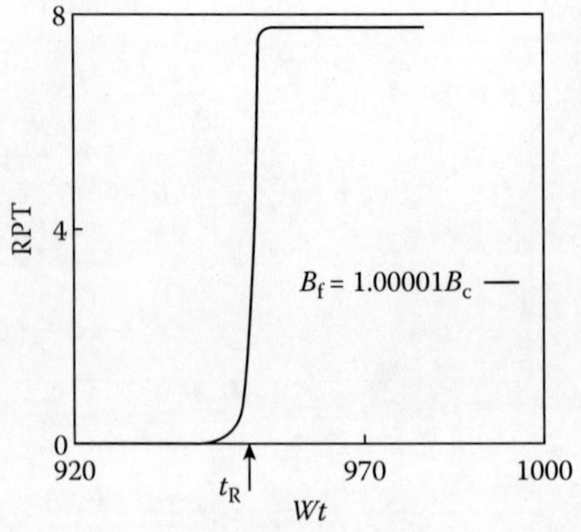


Figure 4.38

Time dependence of the rate of phase transformation for $\beta J = 1.5$, $B_f > B_c$. The time $t_R$ indicates the moment when a rapid growth of the RPT takes place ($Wt_R \simeq 951$).

The first quantity, $n_{up}(t)$, is changed by the flip of any of the spins, whereas the second, $n_{cl}(t)$, is increased (decreased) only by the flips related to the creation (annihilation) and splitting (coagulation) processes. In Appendix 4F, it is shown that

$$n_{up}(t) = \frac{1}{2}(1+\sigma) \tag{4.144}$$

and

$$n_{cl}(t) = \frac{1}{8}(1-2\sigma-3\epsilon). \tag{4.145}$$

Consequently,

$$L_{av}(t) = \frac{4(1+\sigma)}{1-2\sigma-3\epsilon}. \tag{4.146}$$

A typical time dependence of the number density $n_{cl}(t)$ given by Equation 4.145 is presented in Figure 4.39. In the first stage, the creation processes dominate, because of the small number of clusters in the initial state,

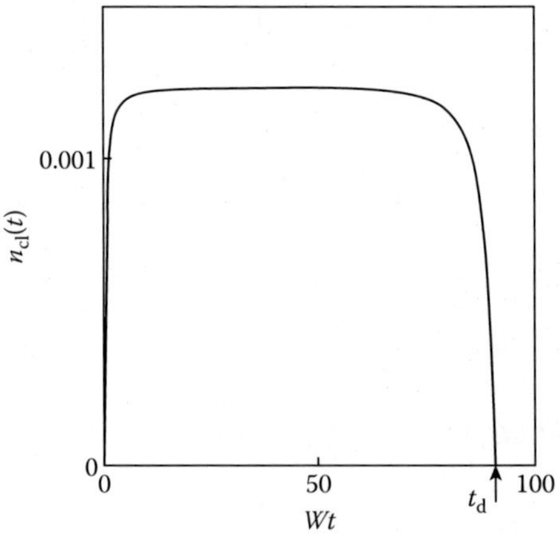


Figure 4.39

Time dependence of the number density of clusters for $\beta J = 1.5$, $B_f = 1.001B_c$. The time $t_d$ indicates the moment when infinite clusters appear ($Wt_d \simeq 91.2$).

and $n_{cl}(t)$ grows in time. In the second stage, the processes of creation and splitting compensate those of annihilation and coagulation and the number of clusters remains almost constant. In the third stage, the coagulation of the clusters leads to a decrease of $n_{cl}(t)$. This decrease takes place until most of the "up" spins coagulate into infinite clusters, whose number density is obviously small and tends to zero. This limit is reached at time $t = t_d$ for which $n_{cl}(t_d) = 0$. Equation 4.145 is no longer valid for $t > t_d$.

The time dependence of the average cluster size, calculated with Equation 4.146 is shown in Figure 4.40. One can see that $L_{av}(t)$ grows slowly when $t < t_d$ and rapidly near $t_d$. One can note some similarity between the average cluster size at which a rapid growth of $L_{av}(t)$ starts and the critical cluster size of the classical nucleation theory. The latter size is usually calculated by assuming a spherical (circular) shape of the critical cluster and using classical thermodynamics. It was shown, that the critical size for an Ising system depends considerably on the cluster shape. The $L_{av}(t)$ calculated here provides an average of the critical size over all possible shapes.

Let us calculate, using the kinetic approach of Refs. [25,26], the critical size of "spherical" clusters on a Bethe lattice, i.e., of a cluster that includes a central spin $\sigma_0$ and $m$ surrounding shells. The total number $N_{tot}$ of spins in such a cluster, the number $N_{nb}$ of neighboring "down" spins, and the number $N_b$ of spins on the outermost shell are given, respectively, by

$$N_{tot} = 3 \cdot 2^m - 2, \quad N_{nb} = 3 \cdot 2^m, \quad N_b = 3 \cdot 2^{m-1}.$$

The rate $n_l'$ of changing the size of the cluster due to the addition of neighboring spins and the removal of boundary ones is given by [26]

$$n_l' = \frac{1}{2} W_{-1}[N_{nb}(1+\tanh\beta\varepsilon_{-1}) - N_b(1-\tanh\beta\varepsilon_{-1})]. \qquad (4.147)$$

In addition, one must take into account the "up"–"down" transitions of internal spins whose number is equal to $N_{tot} - N_b = 3 \cdot 2^{m-1} - 2$. Their rate is

$$n_l'' = -\frac{1}{2} W_3(N_{tot} - N_b)(1-\tanh\beta\varepsilon_3). \qquad (4.148)$$

Consequently, the total rate $n_l = n_l' + n_l''$ of changing the size of a cluster is given by

$$\begin{aligned} 4n_l &= N_{tot}[W_{-1}(1+3\tanh\beta\varepsilon_{-1}) - W_3(1-\tanh\beta\varepsilon_3)] \\ &\quad + 2[W_{-1}(1+3\tanh\beta\varepsilon_{-1}) + W_3(1-\tanh\beta\varepsilon_3)]. \end{aligned} \qquad (4.149)$$

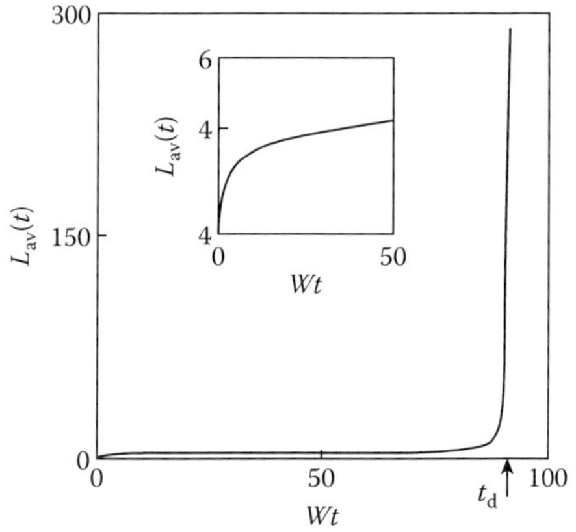


Figure 4.40

Time dependence of the average cluster size $L_{av}(t)$ for $\beta J = 1.5$, $B_f = 1.001B_c$. The initial increase of $L_{av}(t)$ is presented in an inset. The time $t_d$ has the same meaning as in Figure 4.39.

The analysis shows that the spherical cluster on a Bethe lattice always grows ($n_1 > 0$) when $B > B_c^{kin}$, where $B_c^{kin}$ is provided by the following equation

$$W_{-1}[1 + 3 \tanh \beta(B - J)] - W_3[1 - \tanh \beta(B + 3J)] = 0. \quad (4.150)$$

The spherical cluster always shrinks ($n_1 < 0$) when $B < B_{c1}^{kin}\left(B_{c1}^{kin} < B_c^{kin}\right)$, where $B_{c1}^{kin}$ is given by the equation

$$W_{-1}[1 + 3 \tanh \beta(B-J)] + W_3[1 - \tanh \beta(B + 3J)] = 0. \quad (4.151)$$

For $B_{c1}^{kin} < B < B_c^{kin}$, the cluster grows if

$$N_{tot} \leq N_1 = \frac{2[W_{-1}(1+3\tanh\beta\varepsilon_{-1})+W_3(1-\tanh\beta\varepsilon_3)]}{W_3(1-\tanh\beta\varepsilon_3)-W_{-1}(1+3\tanh\beta\varepsilon_{-1})}$$

and shrinks if

$$N_{tot} > N_1.$$

For $\beta J = 1.5$ and $W_{-1} = W_3 = 1$, one obtains $\beta B_c^{kin} = 1.153436$ and $\beta B_{c1}^{kin} = 1.153417$. Because of the very small difference between $\beta B_c^{kin}$ and $\beta B_{c1}^{kin}$ one can speak about a single critical field that one has to reach for initiating the phase transformation. This result is in agreement with that obtained on the basis of the kinetic equation 4.129. The value of $B_c^{kin}$ is much larger than the critical field $B_c$ calculated on the basis of Equation 4.129, $\beta B_c = 0.80934$. The obvious reason for this difference is that in the above calculations for spherical clusters the coagulation of clusters was not taken into account, whereas the kinetic Equation 4.129 takes into account clusters of all shapes and all the processes.

### 4.5.5 Kinetics in Growth-Decay Regime

Let us examine the kinetics of the IMBL when only the growth and decay of the clusters dominate ($W_{-1} \neq 0$), whereas the splitting and coagulation as well as the creation and annihilation of the single-spin clusters are negligible ($W_1 = W_3 = W_{-3} = 0$). In this case, the number density of clusters is constant.

#### *4.5.5.1 Phase Transformation*

Due to the absence of splitting and coagulation as well as creation and annihilation, the system does not reach equilibrium even for $B_f > B_c$. As a consequence, the steady state value $\sigma_{st}$ of the polarization never becomes equal to its equilibrium value (which is close to unity for $T < T_c$ in the field $B_f$). In Figure 4.41a, the time dependence of the polarization at $\beta J = 1.5$ is presented for various values of $\beta B_f$ larger than the critical value $\beta B_c = 0.807\,08$, which in this case is not equal (but is very close) to that for the general case ($\beta B_c = 0.809\,34$; Section 4.5.4.1). One can see that the polarization (and, consequently, the phase transformation dynamics) in the growth-decay regime is qualitatively the same as in the general case, with the exception that the steady state value $\sigma_{st}$ depends strongly on $B_f$ and becomes positive only for relatively high values of $B_f$. This means that in this regime large supersaturations are necessary to obtain a large volume of the new phase. The field dependence of the steady state polarization is presented in Figure 4.41b. This polarization remains almost constant for $B < B_c$, but changes rapidly when $B_f$ approaches $B_c$ and then grows smoothly with increasing $B_f$.

One should note that the time $t_R$ follows in this case the same behavior (Equation 4.142) as in the general case.

#### *4.5.5.2 Cluster Dynamics*

In the absence of coagulation and splitting, as well as creation and annihilation of single-spin clusters, the number density of clusters is constant during the entire transformation process. For ($\beta J = 1.5$ and $\beta B_i = -0.1$, $n_{cl}(0) = 1.08 \times 10^{-4}$ and $L_{av}(0) = 1.068$. The time dependence of $L_{av}(t)$ is presented in Figure 4.42a and the field dependence of its steady-state value $L_{av}(\infty)$ in Figure 4.42b. The latter quantity is finite for all values of the field $B_f$, the magnitude of $L_{av}(\infty)$ growing with decreasing temperature. This behavior of $L_{av}(\infty)$ is a

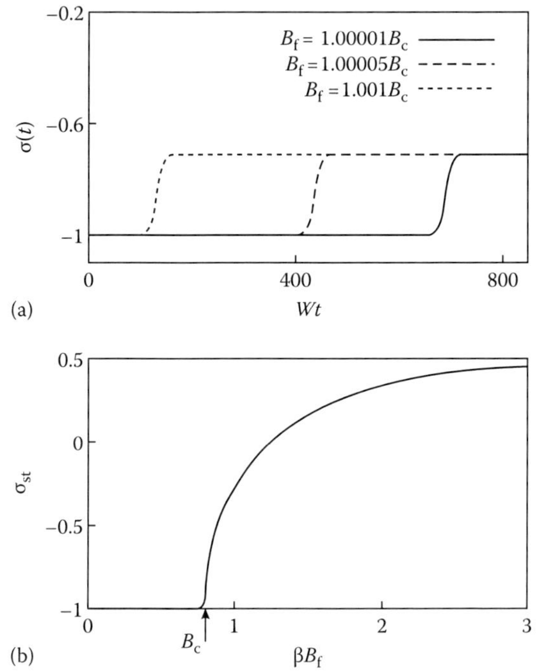


Figure 4.41

For $T < T_c$ ($\beta J = 1.5$) (a) time dependence of the polarization for various values of the final external field $B_f$ in the growth decay regime ($W_1 \neq 0$, $W_1 = W_3 = W_{-3} = 0$). The critical external field $B_c$ is given by $\beta B_c = 0.807\,09$; (b) the steady-state value of the polarization as function of $B_f$.

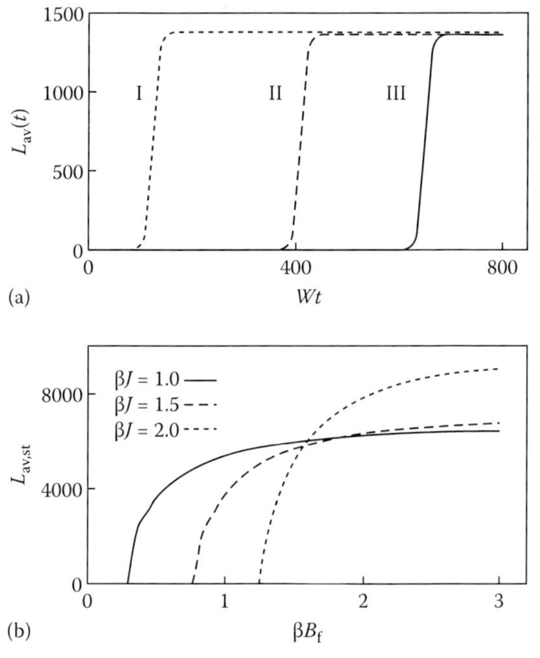


Figure 4.42

(a) Time dependence of the average cluster size $L_{av}(t)$ for various values of the final external field at $\beta J = 1.5$. Curve I corresponds to $B_f = 1.001B_c$, curve II to $B_f = 1.00005B_c$, and curve III to $B_f = 1.00001B_c$; (b) steady-state value $L_{av,st}$ of the time-dependent average cluster size as function of the final field for several temperatures.

consequence of the specific features of the transformation mechanism, which involves only growth and decay of clusters and no other mechanisms. This means that if a growing cluster encounters another one they cannot coagulate and the growth of one of them can continue only because of the shrinkage of the other. This process stops when the size of one of the clusters decreases down to a single spin which cannot be annihilated. For this reason the average cluster size is finite at steady state. The lower the temperature, the larger the average distance between clusters in the initial state and, as a consequence, the larger is the steady-state value of $L_{av}(t)$ because each cluster has a larger number of "down" spins that can contribute to their growth.

### 4.5.6 Kinetics in the Nonsplitting, Noncoagulation Regime

If the splitting and coagulation of the clusters is assumed negligible ($W_1 = W_3 = 0$) but the other processes are present ($W_{-1} = W_{-3} = \equiv W \neq 0$), then the kinetics of the system remains almost the same as in the general case. For $B_f > B_c$, the steady state coincides with the equilibrium one, with most of the phase transformation taking place in a narrow time interval near $t'_R$ at which a rapid growth of the RPT takes place (see Figures 4.36 and 4.38). The time $t'_R$ has the same field dependence, Equation 4.142), as in the general case, and is somewhat larger than $t_R$ due to the absence of coagulation. For $\beta J = 1.5$ and $\beta B_f = 0.80935$, one obtains $Wt_R$. 900, whereas $Wt'_R \simeq 950$.

The number density of clusters and the average cluster size behave in a similar manner as in Figures 4.39 and 4.40 for the general case. At the first sight, this result looks unreasonable, because the absence of coagulation seems to prevent the creation of infinite clusters as in the case of growth decay regime (Section 4.5.5.2). However, in the case under consideration, the annihilation process can remove the obstacles on the way of the growing cluster and thus create the possibility of infinite clusters. Obviously, the characteristic time for the generation of an infinite cluster ($t'_d$) has to be larger than in the general case ($t_d$) (Section 4.5.4.2). Indeed, for $\beta J = 1.5$ and $\beta B_f = 0.80935$, $Wt'_d \simeq 928$, whereas $Wt_d \simeq 884$.

### 4.5.7 Discussion

An approach developed in Paper I is applied in the present paper to the phase transformation on a Bethe lattice. The approach uses a formalism for the kinetic Ising model based on the Glauber master equation, with the additional assumption that the time-dependent averages can be represented in the form of Equation 4.125. The plausibility of such an assumption follows from Bogolyubov's [27] and Zubarev's [28] representation of the nonequilibrium behavior of a macroscopic system as a hierarchy of relaxation times (see Paper I for a more detailed discussion). As a consequence, a closed set of kinetic equation (Equation 4.129) was obtained, whose solution provided the time-dependent averages necessary for the analysis of the kinetics. It was shown that the parameters $W_\mu$ ($\mu = \pm 1, \pm 3$) can be associated with specific processes in cluster dynamics (growth and decay, coagulation and splitting, creation and annihilation). Using this fact, the different regimes involving all or only some of these processes were examined.

For temperatures $T$ larger than the phase transition temperature $T_c$ and a zero final magnetic field ($B_i \neq 0$, $B_f = 0$), the RPT calculated for small deviation of the initial state from equilibrium does not depend on time and approaches zero as $(T - T_c)$ when $T$ approaches $T_c$. At the critical point ($T = T_c$), the RPT is time-dependent and for large times it behaves as $t^{-1}$.

For lower temperatures ($T < T_c$), a phase transformation from a state with almost all spins "down" to a state with almost all spins "up" takes place only if the final external field is larger than a critical value $B_c$, which decreases with increasing temperature. The phase transformation occurs in a narrow time interval around a characteristic time $t_R$ near which the RPT grows rapidly. The latter time tends to infinity as $(B_f - B_c)^{-0.47}$, when $B_f \to B_c$ and such a behavior does not depend on the transformation mechanism. The RPT grows rapidly near $t_R$ but remains almost constant for $t > t_R$. Note that the critical temperature of the IMBL, which is given by the equation [3,4]

$$T_c = (2J/k)\log\frac{q}{q-2}, \tag{4.152}$$

increases with increasing $q$. Therefore, one can expect that the temperature range, in which the phase transformation can start after the external field becomes larger than a critical value, also increases with increasing $q$.

The steady state value $\sigma_{st}$ of the polarization when the kinetics is governed only by the growth and decay processes (growth-decay regime), strongly depends on the magnitude of the final external field $B_f$ and does not coincide with its equilibrium value $\sigma_{eq}$. For all other regimes, $\sigma_{st} = \sigma_{eq}$.

The average cluster size $L_{av}(t)$ in the growth-decay regime grows with increasing time and its steady state value increases with decreasing temperature. For other transformation mechanisms, $L_{av}(t)$ grows without limit, thus indicating the existence of infinite clusters on the lattice.

Finally, it is appropriate to compare the present approach and the classical theory describing the phase transformation. The classical theory considers the phase transformation as a two stage process. In the first stage, critical clusters (nucleus) are formed which can grow. In the second stage the growth of supercritical clusters occurs. The size of the critical cluster was calculated using thermodynamic [31,32] or kinetic [33,34] considerations. The rate of creation of such clusters (nucleation rate) was found from the Becker–Döring equations in the steady state approximation [31,32]. The present approach does not employ a division between nucleation and growth, but treats the phase transformation as a single whole. There is, however, some analogy with the classical nucleation theory. Indeed, in the classical theory there is a critical supersaturation below which the rate of nucleation is negligible and at which the nucleation rate becomes very large. Similarly, in the present case, there is a field $B_c$ (the field representing in our case the supersaturation) below which the rate of growth is zero, but above which the growth rate becomes large.

## Acknowledgment

This work was supported by the National Science Foundation.

## Appendix 4D: Some Equilibrium Averages for the IMBL

The equilibrium values $\langle\cdots\rangle_T \equiv \mathrm{Tr}[e^{-\beta H}\ldots]/\mathrm{Tr}e^{-\beta H}$ of one- and two-spin averages were calculated in Refs. [3,29,30]. For $q = 3$, they are

$$\langle\sigma_0\rangle_T \equiv \sigma_T = \tanh[3L/2 - \beta B/2], \tag{4D.1}$$

$$\langle\sigma_0\sigma_1\rangle_T \equiv \epsilon_T = \sigma_T^2 + \left(1-\sigma_T^2\right)\lambda_T,$$

$$\langle\sigma_1\sigma_2\rangle_T \equiv \epsilon_{2T} = \sigma_T^2 + \left(1-\sigma_T^2\right)\lambda_T^2, \tag{4D.2}$$

where

$$\lambda_T = z_T\frac{1-\tanh L}{1-z_T^2\tanh L}, \quad z_T = \tanh\beta J, \tag{4D.3}$$

$L$ being the root of equation [3]

$$L = \beta B + 2\,\mathrm{arctanh}(z_T \tanh L). \tag{4D.4}$$

Introducing the notation $\tanh L = y$, $z_{0T} = \tanh \beta B$, Equation 4D.4 becomes

$$z_T^2 y^3 + z_{0T}z_T(2-z_T)y^2 + (1-2z_T)y - z_{0T} = 0. \tag{4D.5}$$

The simplest way to calculate the necessary three- and four-spin averages $\langle A_{4-6}\rangle_T$ (see Equation 4.130 for the definition of $A_i$) is to use the following, easily demonstrable, relations (see Paper I, Appendix 4D, for details):

$$\langle\sigma_0 R_0(\mu)\rangle_T = z_{\mu T}\langle R_0(\mu)\rangle_T, \quad \mu = \pm 1, \pm 3, \tag{4D.6}$$

where $z_{\mu T} = \tanh(\beta\epsilon_\mu)$. Equation 4D.6 provide three linear independent equations from which one can obtain

$$\langle A_i \rangle_T = (a_i + b_i\sigma_T + c_i\epsilon_T + d_i\epsilon_{2T})/\Delta,\ i = 4,5,6, \tag{4D.7}$$

where

$$\begin{aligned}
&\Delta = -2z_{1T} + z_{-1T} + z_{3T},\\
&a_4 = -2z_{1T} - z_{-1T} - z_{3T},\quad a_5 = z_{1T}z_{-1T} - z_{1T}z_{3T},\\
&a_6 = -3z_{1T}z_{-1T} - z_{1T}z_{3T},\\
&b_4 = 4 - 2z_{1T} + z_{-1T} - 3z_{3T},\\
&b_5 = -z_{-1T} + z_{3T} - 2z_{1T}z_{3T} + 2z_{-1T}z_{3T},\\
&\frac{1}{2}b_6 = z_{1T} + z_{-1T} - z_{1T}z_{3T} - z_{-1T}z_{3T},\\
&c_4 = 4,\ c_5 = 2z_{1T} - 2z_{-1T},\quad c_6 = 3z_{-1T} + z_{3T},\\
&d_4 = 2z_{1T} + z_{-1T} - 3z_{3T},\\
&d_5 = -z_{-1T}z_{1T} - z_{1T}z_{3T} + 2z_{-1T}z_{3T},\\
&d_6 = 3z_{-1T}z_{1T} - z_{1T}z_{3T} - 2z_{-1T}z_{3T}.
\end{aligned} \tag{4D.8}$$

We sketch below the way to calculate the averages $\langle A_i \rangle_T$ as functions of $\sigma_T$ and $z_T$. First, using Equation 4D.1 together with Equation 4D.4, one can rewrite $\tanh L$ in the form

$$\begin{aligned}
\tanh L &= \frac{1}{2\sigma_T z_T}\left[1 + z_T - \sqrt{(1+z_T)^2 - 4\sigma_T^2 z_T}\right]\\
&\equiv f_L(\sigma_T, z_T).
\end{aligned} \tag{4D.9}$$

Second, using the relations

$$\begin{aligned}
&\beta B = 3\,\text{arctanh}[f_L(\sigma_T, z_T)] - 2\text{arctanh}(\sigma_T),\\
&\beta J = \text{arctanh}(z_T),
\end{aligned} \tag{4D.10}$$

which follow from Equation 4D.1 and from the definition of $z_T$, and Equations 4D.2, 4D.3, and 4D.7, one can arrive at expressions for $\langle A \rangle_T$ as functions of $\sigma_T$ and $z_T$, which are, however, too long to be written explicitly.

Let us note that, in the absence of a magnetic field ($B = 0$), the IMBL possesses a second-order phase transition at the critical temperature $T_c$, which for $q = 3$ is given by $\tanh(J/T_c) = 1/2$.

## Appendix 4E: Sketch of the Steps to Calculate Right-Hand Sides of the Kinetic Equations

In order to calculate the right-hand sides of Equation 4.131, one has first to express all time-dependent averages through $\sigma \equiv \langle \sigma_0 \rangle_t$ and $z = \tanh \beta_e(t)$. This can be performed by using the way sketched for the equilibrium averages in Appendix 4D. In the final expressions for these equilibrium averages, one has to replace $\sigma_T$ by $\sigma$, and $z_T$ by $z$. Instead of the time derivative $\epsilon$ of the average $\langle \sigma_0 \sigma_1 \rangle_t \equiv \epsilon$, which appears on the left-hand side of Equation 4.131, one has to use the sum

$$\frac{\partial \epsilon}{\partial \sigma}\frac{d\sigma}{dt} + \frac{\partial \epsilon}{\partial z}\frac{dz}{dt}.$$

Using the same method as in Appendix 4D and the following expression for the energy $\epsilon$:

$$\epsilon = \sigma^2 + z(1-\sigma^2)\frac{1-f_L(\sigma,z)}{1-z^2 f_L(\sigma,z)}, \tag{4E.1}$$

one obtains

$$\frac{\partial\epsilon}{\partial\sigma} = \frac{2\sigma(1-z)}{\sqrt{(1+z)^2-4\sigma^2 z}} \tag{4E.2}$$

and

$$\frac{\partial\epsilon}{\partial z} = (1+z)\left[\frac{z-1}{2z^2} + \frac{1+z^2-2\sigma^2 z}{\sqrt{(1+z)^2-4\sigma^2 z}}\right]. \tag{4E.3}$$

Equations 4E.2 and 4E.3 together with the expressions of $\langle A_i\rangle_t$ as functions of $\sigma$ and $z$ allows one to obtain the functions $f_\sigma(\sigma,z), f_z(\sigma,z)$ of Equation 4.131.

## Appendix 4F: Calculation of the Number of Clusters and of Average Cluster Size

Let us consider the operators $n_j(1) = (1 + \sigma_j)/2$, $n_j(-1) = (1 - \sigma_j)/2$, which provide the occupation probabilities of a site $j$ by a spin in the states $\sigma_j = 1$ and $\sigma_j = -1$, respectively. The above expressions show that $n_j(1) = 1$ when $\sigma_j = 1$ and $n_j(1) = 0$ when $\sigma_j = -1$. Obviously, the operator $n_j(v_1)n_k(v2)$ where $v_1,v_2 = \pm 1$ provide the occupation probability of sites $j$ and $k$ by spins in the states $v_1$ and $v_2$, respectively. As a consequence, the time-dependent average number density of "up" spins $n_{\text{up}}(t)$ is equal to $n_{\text{up}}(t) = (1/N)\sum_j\langle n_j(1)\rangle_t$, where $N$ is the total number of spins. Analogously, the number density $n_{\text{ud}}(t)$ of pairs of neighboring oppositely pointed spins is equal to $n_{\text{ud}}(t) = (1/2N)\sum_j\sum_{k=1}^{3}\langle n_j(1)n_{j+\delta_k}(-1)+n_j(-1)_{j+\delta_k}(1)\rangle$. In the homogeneous case, one has, after simple algebra, that

$$n_{\text{up}}(t) = \frac{1}{2}(1+\sigma), \tag{4F.1}$$

$$n_{\text{ud}}(t) = \frac{3}{4}(1-\epsilon). \tag{4F.2}$$

The number density of clusters, $n_{\text{cl}}(t)$, can be calculated in the following way. Let us denote by $f_l(t)$ the number density of finite clusters having $l$ spins. Then $n_{\text{up}}(t)$ and $n_{\text{cl}}(t)$ can be expressed in the forms

$$n_{\text{up}}(t) = \sum_l l f_l(t), \tag{4F.3}$$

$$n_{\text{cl}}(t) = \sum_l f_l(t). \tag{4F.4}$$

Let us now denote by $n_{\text{nb}}(t)$, the number density of spins neighboring the existing clusters. The $l$ spins cluster on a Bethe lattice has $l + 2$ neighbors and hence

$$n_{\text{nb}}(t) = \sum_l (l+2) f_l(t) = n_{\text{up}}(t) + 2n_{\text{cl}}(t). \tag{4F.5}$$

On the other hand, taking into account that $n_{nb}(t) = n_{ud}(t)$, and combining Equations 4F.1, 4F.2, and 4F.5, one obtains

$$n_{cl}(t) = \frac{1}{8}(1 - 2\sigma - 3\epsilon). \tag{4F.6}$$

Consequently, the time-dependent cluster average size is given by the expression

$$L_{av}(t) = \frac{4(1+\sigma)}{1 - 2\sigma - 3\epsilon}. \tag{4F.7}$$

## References


1. Berim, G.O., Ruckenstein, E., *J. Chem. Phys.* 119 9640 (2003). (Section 4.4 of this volume.)
2. Berim, G.O., Ruckenstein, E., *J. Chem. Phys.* 119, 806 (2003). (Section 4.3 of this volume.)
3. Katsura, S., Takizawa, M., *Prog. Theor. Phys.* 51, 82 (1974).
4. Baxter, R.J., *Exactly Solvable Models in Statistical Mechanics*, Academic, New York, 1982.
5. Buzatu, F.D., Huckaby, D.A., *Physica A* 299, 427 (2001).
6. Strout, D.L., Huckaby, D.A., *Physica A* 173, 60 (1991).
7. Majumdar, S.N., Sire, C., *Phys. Rev. Lett.* 70, 4022 (1993).
8. Tobita, H., *J. Polym. Sci., Part B: Polym.* 32, 901 (1994).
9. Tobita, H., *Makromol. Chem.* 2, 761 (1993).
10. Gujrati, P.D., *J. Chem. Phys.* 98, 1613 (1992).
11. Akagi, E., Seino, M., Katsura, S., *Physica A* 178, 406 (1991).
12. Doty, C.A., Fisher, D.S., *Phys. Rev. B* 39, 12098 (1989).
13. Mélin, R., Anglès d'Auriac, J.C., Chandra, P., Douçot, D., *J. Phys. A* 29, 5773 (1996).
14. Tucker, J.W., Balcerzak, T., Gzik, M., Sukiennicki, A., *J. Magn. Magn. Mater.* 187, 381 (1998).
15. Lyra, M.L., Cavalcanti, S.B., *Phys. Rev. B* 45, 8021 (1992).
16. Paraskevaidis, C.E., Papatriantafillou, C., *J. Stat. Phys.* 64, 385 (1991).
17. Wonczak, S., Strey, R., Stauffer, D., *J. Chem. Phys.* 113, 1976 (2000).
18. Glauber, R., *J. Math. Phys.* 4, 294 (1963).
19. Hohenberg, P.C., Halperin, B.I., *Rev. Mod. Phys.* 49, 435 (1977).
20. Novotny, M.A., Brown, G., Rikvold, P.A., *J. Appl. Phys.* 91, 6908 (2002).
21. Wonczak, S., Strey, R., Stauffer, D., *J. Chem. Phys.* 113, 1976 (2000).
22. Vehkamaki, H., Ford, I.J., *Phys. Rev. E* 59, 6483 (1999).
23. Shneidman, V.A., Jackson, K.A., Beauty, K.M., *J. Chem. Phys.* 111, 6932 (1999).
24. Shneidman, V.A., Jackson, K.A., Beauty, K.M., *Phys. Rev. B* 59, 3579 (1999).
25. Berim, G.O., Ruckenstein, E., *J. Chem. Phys.* 117, 4542 (2002). (Section 4.1 of this volume.)
26. Berim, G.O., Ruckenstein, E., *J. Chem. Phys.* 117, 7732 (2002). (Section 4.2 of this volume.)
27. Bogolyubov, N.N., *The Problems of Dynamic Theory in Statistical Physics*, Gostekhizdat, Moscow, 1946 (in Russian).
28. Zubarev, D.N., *Nonequilibrium Statistical Thermodynamics*, Nauka, Moscow, 1971 (in Russian).
29. Hu, C.-K., Izmailian, N.Sh., *Phys. Rev. B* 58, 1644 (1998).
30. Mukamel, D., *Phys. Lett. A* 50, 339 (1974).
31. Frenkel, J., *Kinetic Theory of Liquids*, Dover, New York, 1955.
32. Abraham, F., *Homogeneous Nucleation Theory*, Academic, New York, 1974.
33. Nowakowski, B., Ruckenstein, E., *J. Chem. Phys.* 94, 1397 (1991). (Section 2.1 of this volume.)
34. Nowakowski, B., Ruckenstein, E., *J. Chem. Phys.* 94, 8487 (1991). (Section 2.3 of this volume.)

# 4.6 Phase Transformation in a Lattice System in the Presence of Spin-Exchange Dynamics

*Gersh Berim and Eli Ruckenstein**

A joint action of the Glauber single spin-flip and the Kawasaki spin-exchange mechanisms upon the processes of phase transformation is examined in the framework of the one-dimensional kinetic Ising model. It is shown that the addition of the Kawasaki dynamics to that of Glauber accelerates the process of phase transformation in the initial stage, but slows it down in later stages. For the truncated form of Glauber dynamics, which excludes the processes of splitting and coagulation of clusters, the addition of the Kawasaki dynamics always accelerates the phase transformation process. Acting alone, the Kawasaki mechanism provides a cluster growth proportional to $t^{1/2}$ (where $t$ is the time) in the initial stage and proportional to $t^{1/3}$ (Lifshitz–Slyozov–Wagner law) in the intermediate stage. In the final stage, a cluster size approaches exponentially its equilibrium value. (© *2004 American Institute of Physics*. doi: 10.1063/1.1638376.)

## 4.6.1 Introduction

In previous papers [1,2], the kinetics of phase transformation (the rate of phase transformation, time-dependent size distribution of clusters, etc.) in various lattice systems was investigated in different regimes using the kinetic Ising model with the Glauber one-spin-flip dynamics [3]. In that model the transitions of the molecules situated on the sites of a rigid lattice from one phase to another (e.g., from the liquid to the solid phase) were modeled by the flips of the spins $\sigma_j = \pm 1$ ($j = 1, \ldots, N$) assigned to each of the $N$ molecules, and the motion of molecules along the lattice was ignored. It is obvious that such a motion can affect the cluster growth and therefore has to be included into the model. As an appropriate mechanism, the Kawasaki pair-spin exchange dynamics [4], which implies the simultaneous flips of pair of oppositely directed nearest neighbor spins, providing the motion of "up" and "down" spins in opposite directions, can be employed. Applied alone, the Kawasaki dynamics was used to examine the spin diffusion [4,5], as well as for Monte Carlo simulations of phase separation and cluster growth in binary alloys [6–8], in a lattice gas [9], and in ternary mixtures [10,11]. In those papers, only the case of zero external field was considered. In that case, the nonequilibrium process started after the temperature was quenched from an initial value $T_i$ to a final value $T_f$, and the total spin polarization was taken equal to zero during the entire process. Under those circumstances the final state coincides with the equilibrium one. Note that, in the presence of an external field, the Kawasaki dynamics alone cannot provide the final equilibrium state, because of the conservation of the total spin, or, in other words, because of the conservation of the number of molecules belonging to each of the phases.

The combination of the spin-flip mechanisms of the Glauber and the Kawasaki dynamics was employed to investigate the nonequilibrium steady state of a lattice system in the presence of an energy flux from an external source [12–22]. Only those pair-spin transitions were considered which increased the energy of the system. The steady state in that case does not coincide with the equilibrium one and its characteristics were investigated in the absence of an external field.

In the present paper, the Kawasaki dynamics in its original form [4] is included into the kinetic Ising model, i.e., the pairs of neighboring oppositely directed spins flip due to the interactions with a thermostat, and the rates of these flips satisfy the detailed balance condition. The one-dimensional lattice is chosen to illustrate the method in order to avoid too complicated calculations. It is reasonable to consider that, in systems of higher dimensions, the influence of the Kawasaki dynamics upon the phase transformation will have similar features. In Section 4.6.2, the main equations are presented. In Section 4.6.3, the influence of the Kawasaki dynamics on the transformation of a state with almost all spins "down" (liquid state) to a state with almost all spins "up" (solid state) after the reversal of the external field is considered. The separation of different phases from a completely disordered state after quenching the temperature of the thermostat is examined in Section 4.6.4. The peculiarities of the phase transformation after the simultaneous changes of the temperature and the external field are considered in Section 4.6.5.

* *J. Chem. Phys.* 120, 2851 (2004). Republished with permission.

## 4.6.2 Background

Let us consider a system of $N$ spins, $\sigma_j = \pm 1$, representing the molecules located on the sites of a one-dimensional lattice, the energy of which is provided by the Hamiltonian

$$H_s = -B\sum_j \sigma_j - J\sum_j \sigma_j\sigma_{j+1} \quad (J>0), \tag{4.153}$$

where the parameter $B$, the external field, models the supersaturation $S$ ($B \sim \log S$ [23]), and $J$ (the nearest-neighbor interaction energy, referred below as energy) models the intermolecular interactions. The system is described by a time-dependent probability distribution function (PDF) $P(\{\sigma\};t)$, where $\{\sigma\}$ denotes the state $\{\sigma_1, \sigma_2, \ldots, \sigma_N\}$ of the system. The time evolution of PDF is provided by the master equation

$$\begin{aligned}\frac{\mathrm{d}}{\mathrm{d}t}P(\{\sigma\};t) = \sum_{\{\sigma'\}}\sum_{\alpha=G,K}\big[&-W^{\alpha}(\{\sigma\}\to\{\sigma'\})P(\{\sigma\};t)\\ &+W^{\alpha}(\{\sigma'\}\to\{\sigma\})P(\{\sigma'\};t)\big],\end{aligned} \tag{4.154}$$

where the superscript $\alpha$ labels the transition rates in the Glauber ($\alpha = G$) and Kawasaki ($\alpha = K$) dynamics, respectively.

The transition rates $W^{\alpha}(\{\sigma\} \to \{\sigma'\})$ are different from zero only if the state $\{\sigma\}$ differs from the state $\{\sigma'\}$ by the sign of only one spin, say $\sigma_j$, for the Glauber dynamics, or by the orientation of a pair of neighboring oppositely directed spins, say $\sigma_j$ and $\sigma_{j+1}$ for the Kawasaki dynamics. To satisfy the detailed balance condition the rate $W^{G}(\{\sigma\} \to \{\sigma'\})$ can be represented as follows [1]:

$$W^{G}(\{\sigma\}\to\{\sigma'\}) = \sum_{\mu} W_{\mu}^{G} R_j(\mu)(1-\sigma_j \tanh\beta\varepsilon_{\mu}), \tag{4.155}$$

where $\beta = 1/kT$, $T$ is the temperature in $K$, $k$ is the Boltzmann constant, $\mu$ is the sum of the neighboring spins, which takes the values $0, \pm 2$, $\varepsilon_{\mu} = B + J\mu$, and the operator $R_j(\mu)$ is equal to unity when the sum of the neighboring spins of spin $\sigma_j$ is equal to $\mu$, and zero otherwise, $W_{\mu}^{G}$ are constants, defining the specific features of the relaxation mechanisms [1]. In the one-dimensional case, the operators $R_j(\mu)$ are given by [1]

$$\begin{aligned}R_j(0) &= \frac{1}{2}(1-\sigma_{j-1}\sigma_{j+1}),\\ R_j(\pm 2) &= \frac{1}{4}(1\pm\sigma_{j-1})(1\pm\sigma_{j+1}).\end{aligned} \tag{4.156}$$

When $W_{\mu}^{G} \equiv W^{G}$ ($\mu = 0, \pm 2$) and $B = 0$, the rates (Equation 4.155) acquire the form [4]

$$W^{G}(\{\sigma\}\to\{\sigma'\}) = \frac{1}{2}W^{G}\left[1-\frac{z}{2}\sigma_j(\sigma_{j-1}+\sigma_{j+1})\right], \tag{4.157}$$

where $z = \tanh 2J$.

The form of $W^{K}(\{\sigma\} \to \{\sigma'\})$ is much more complicated than Equation 4.155, being for the one-dimensional lattice given by

$$\begin{aligned}W^{K}(\{\sigma\}\to\{\sigma'\}) = \frac{1}{4}(1-\sigma_j\sigma_{j+1})\big[&W_0^{K}(1+\sigma_{j-1}\sigma_{j+2})\\ &-W_2^{K}\sigma_j(1-\sigma_{j-1}\sigma_{j+2})(\sigma_{j+1}+z\sigma_{j-1})\big].\end{aligned} \tag{4.158}$$

In the above expression, the first factor shows that only oppositely directed spins provide nonzero contributions to the rate of transition; the second factor (in square brackets) takes into account the state of the neighboring spins of the flipping spins $\sigma_j$ and $\sigma_{j+1}$. When $W_0^{\mathrm{K}} = W_2^{\mathrm{K}} \equiv W^{\mathrm{K}}$,

$$W^{\mathrm{K}}(\{\sigma\} \rightarrow \{\sigma'\}) = \frac{W^{\mathrm{K}}}{2}(1-\sigma_j\sigma_{j+1})\left[1-\frac{z}{2}(\sigma_{j-1}\sigma_j+\sigma_{j+1}\sigma_{j+2})\right]. \tag{4.159}$$

In what follows, an approximate form of the PDF $P(\{\sigma\};t)$ will be employed that was used previously [1,2] for decoupling the infinite chain of kinetic equations that usually arise in the kinetic Ising models. In this approximation, the form of PDF is selected to be similar to the equilibrium density matrix

$$P(\{\sigma\};t) = e^{-F(t)}/\mathrm{Tr}[e^{-F(t)}], \tag{4.160}$$

where

$$F(t) = -\beta_\sigma(t)\sum_j \sigma_j - \beta_e(t)\sum_j \sigma_j\sigma_{j+1} \tag{4.161}$$

with $\beta_\sigma(t)$ and $\beta_e(t)$ unknown function of $t$. The PDF (Equation 4.160) can be obtained from the equilibrium density matrix

$$\rho_T = e^{-\beta H_s}\Big/\mathrm{Tr}\left[e^{-\beta H_s}\right] \tag{4.162}$$

of the Ising model (Equation 4.153), using the substitutions

$$\beta B \rightarrow \beta_\sigma(t), \quad \beta J \rightarrow \beta_\epsilon(t). \tag{4.163}$$

Because of such a choice, all time-dependent averages

$$\langle A\rangle_t \equiv \sum_{\{\sigma\}} AP(\{\sigma\};t) \tag{4.164}$$

are spatially homogeneous and can be calculated as functions of $\beta_\sigma(t)$ and $\beta_\epsilon(t)$, using the methods developed for the equilibrium Ising model.

Using the master equation (Equation 4.154) and the transition rates Equations 4.155 and 4.159, and taking into account the spatial homogeneity of the averages $\langle\cdots\rangle_t$, the following kinetic equations for the time-dependent averages $\sigma \equiv \langle\sigma_0\rangle_t$ and $\epsilon \equiv \langle\sigma_0\sigma_1\rangle_t$, representing the polarization of spin $\sigma_0$ and the interaction energy of spins $\sigma_0$ and $\sigma_1$, are obtained for the one-dimensional Ising model,

$$\begin{aligned}\dot{\sigma} &= R_\sigma^{\mathrm{G}},\\ \dot{\epsilon} &= R_\epsilon^{\mathrm{G}} + R_\epsilon^{\mathrm{K}}\end{aligned} \tag{4.165}$$

with the relaxation terms $R_\sigma^{\mathrm{G}}$, $R_\epsilon^{\alpha}(\alpha = \mathrm{G},\mathrm{K})$ given by

$$\begin{aligned}R_\sigma^{\mathrm{G}} &= a_0 - a_1\sigma - a_2\epsilon - a_3\epsilon_2 + a_4\gamma,\\ R_\epsilon^{\mathrm{G}} &= b_0 - b_1\sigma - b_2\epsilon + b_0\epsilon_2 - a_2\gamma,\\ R_\epsilon^{\mathrm{K}} &= W_2^{\mathrm{K}}[-2(\epsilon_1-\epsilon_2) + z(1-\epsilon_1-\epsilon_3+\delta],\end{aligned} \tag{4.166}$$

where $\epsilon_k = \langle\sigma_0\sigma_k\rangle_t$, $\gamma = \langle\sigma_0\sigma_1\sigma_2\rangle_t$, $\delta = \langle\sigma_0\sigma_1\sigma_2\sigma_3\rangle_t$. The coefficients $a_i$, $b_i$ depend on the temperature and the parameters $W_\mu^G$. Their explicit expressions are provided in Appendix 4G. The argument $t$ of the time-dependent functions $\sigma(t)$, $\epsilon(t)$, ... will be omitted.

To close Equation 4.189, the same procedure used in Ref. [1] is employed. Since the details are provided in that paper, only the main steps of the closure procedure will be mentioned. First, the time-dependent averages $\epsilon_k$, $\gamma$, and $\delta$ are calculated using the exact equilibrium solution for the one-dimensional Ising model and the substitutions (Equation 4.163). Second, all averages are represented as functions of only two of them, $\sigma$ and $\epsilon$, as follows:

$$\epsilon_k = \sigma^2 + (\epsilon - \sigma^2)^k/(1-\sigma^2)^{k-1} \quad (\epsilon_1 \equiv \epsilon),$$
$$\gamma = \sigma[\epsilon + (1-\epsilon)(\epsilon-\sigma^2)/(1-\sigma^2)], \tag{4.167}$$
$$\delta = \sigma\gamma + (\epsilon - \sigma\gamma)(\epsilon-\sigma^2)/(1-\sigma^2).$$

Introducing expression 4.167 in Equation 4.165, one obtains a closed system of nonlinear differential equations, which are, however, too cumbersome to be written explicitely.

The goal of the present paper is to consider the phase transformation from initial states far from equilibrium. In that range, the usual approximation, namely the linearization of the equations for small deviations from equilibrium, cannot be made, and most of the results had to be obtained by numerical integration of Equation 4.165.

In the calculations provided below the case $W_\mu^G \equiv W^G$ ($\mu$ = 0, ± 2) will be considered (if not otherwise mentioned) and a low temperature range, which is the most interesting for modeling of the phase transformation, will be selected.

### 4.6.3 Kinetics after Field Reversal (Liquid–Solid Transformation)

In this section, the initial conditions usually employed to model the phase transformation in a lattice system [1] are considered. Let us assume that at $t = 0$ an external field $B$ is changed from an initial value $B = B_i < 0$ to a final value $B = B_f > 0$, that corresponds to a jump of the supersaturation $S$ from $S_i < 1$ to $S_f > 1$. For low thermostat temperatures, almost all spins have "down" directions in the initial, and "up" directions in the final state. At any moment, the average size $L(t)$ of the clusters of "up" spins (clusters of the solid phase) can be calculated according to the expression

$$L = \frac{2(1+\sigma)}{1-\epsilon} \tag{4.168}$$

(see Appendix 4H for details). In Figure 4.43, the time dependence of the average cluster size for $\beta J = 3$, $\beta B_f = -\beta B_i = 0.1$, and for two extreme values of the parameter $p = W^K/W^G$, $p = 0$ (absence of Kawasaki dynamics),

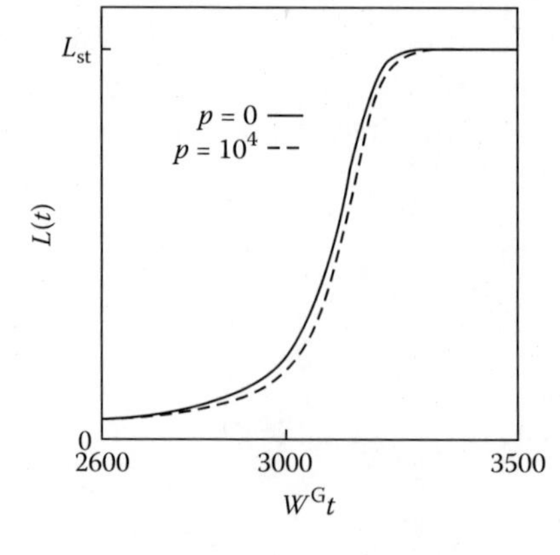


Figure 4.43

Time dependence of average cluster size for the field reversal regime for $\beta B_f = -\beta B_i = 0.1$ and $\beta J = 3$ and for various values of the parameter $p = W^K/W^G$. The steady-state value $L_{st} \simeq 36{,}000$ of $L(t)$ is equal to the equilibrium average cluster size.

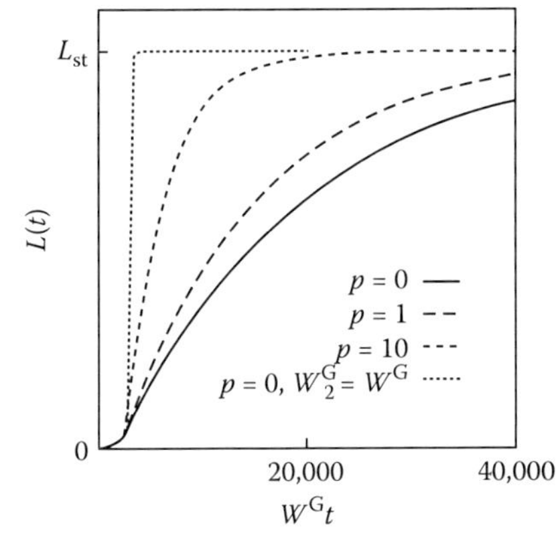


Figure 4.44

Time dependence of average cluster size for the field reversal regime for nonsplitting, noncoagulation Glauber dynamics $\left(W_2^G = 0\right)$ for various values of the parameter $p = W^K/W^G$. Here, $\beta B_f = -\beta B_i = 0.1$, $\beta J = 3$, and $L_{st} \simeq 36{,}000$. For comparison purposes, the curve corresponding to the general case of Glauber dynamics $\left(W_2^G \neq 0\right)$ is also presented.

and $p = 10{,}000$ ($W^K \gg W^G$) is presented. One can see that the difference between the two cases is very small, i.e., the Glauber dynamics dominates in the case considered above.

It should be noted that the rapid growth of the average cluster size, which one can see in Figure 4.43, is due to the process of coagulation of clusters during their growth. The rate of this process is proportional to the parameter $W_2^G$, that controls the "down"–"up" flips of a spin, whose nearest neighboring spins are "up" ones. The cluster growth in the noncoagulating, nonsplitting regime $\left(W_2^G = 0,\ W_0^G = W_{-2}^G = W^G\right)$ is shown in Figure 4.44. Because of the absence of coagulation, the relaxation process runs much slower than for $W_2^G \neq 0$, and the Kawasaki dynamics accelerates this process even for $p = 1$. To explain the physical reasons of such a behavior, let us consider a fragment of the chain consisting of two two-"up"-spin clusters separated by a "down" spin, ↓↑↑↓↑↑↓. In the presence of coagulation $\left(W_2^G \neq 0\right)$ a "down" spin makes a "down"–"up" transition with a rate $\frac{1}{2}W_2^G(1+\tanh\beta\varepsilon_2)$ and for this reason the mentioned two-spin clusters coagulate in one step. In the absence of coagulation $\left(W_2^G = 0\right)$, such a flip is forbidden and the same transformation can be achieved by the Glauber dynamics as a sequence of five flips, e.g., ↓↑↑↓↑↑↓→↓↑↑↓↓↑↓→↓↑↑↓↓↓↓→↓↑↑↑↓↓↓→↓↑↑↑↑↓↓→↓↑↑↑↑↑↓, the first two having the rates $\frac{1}{2}W^G(1-\tanh\beta B_f)$, and the next three the rates $\frac{1}{2}W^G(1+\tanh\beta B_f)$, (the rates are estimated by accounting for the detailed balance condition). Addition of the Kawasaki mechanism shortens the "coagulation" to three steps ↓↑↑↓↑↑↓→↓↑↑↑↓↑↓→↓↑↑↑↑↓↓→↓↑↑↑↑↑↓ with the rates $W^K/2$, $\frac{1}{2}W^K(1+\tanh\beta\varepsilon_2)$ and $\frac{1}{2}W^G(1+\tanh\beta B_f)$, respectively. If $W^K \geq W^G$, the latter rates are larger than those for the Glauber dynamics with $W_2^G = 0$. For this reason and due to the smaller number of steps involved in the latter sequence, the Kawasaki dynamics accelerates the cluster growth.

### 4.6.4 Temperature Quench in the Absence of an External Field (Phase Separation)

In the absence of an external field, the kinetic equation for polarization has a simple form

$$\dot{\sigma} = -2W^G(1-z)\sigma \tag{4.169}$$

and provides an exponential decay of the polarization, $\sigma = \sigma(0)e^{-t/\tau_\sigma}$, with a relaxation time $\tau_\sigma = (1 - z)W^G/2$. If in the initial state the system is completely disordered, because, for example, of its contact with a thermostat of infinite temperature, then $\sigma(0) = 0$ and the polarization is equal to zero during the entire transformation process. In this case, the kinetic equation for the average energy is

$$\dot{\varepsilon} = W^K z(\epsilon - \epsilon_-)(\epsilon_+ - \epsilon)(a - \epsilon), \tag{4.170}$$

where $\epsilon_- = \tanh J/kT$, $\epsilon_+ = \coth J/kT$, $a = 1 + 2/p$, $p = W^K/W^G$, and $T$ is the temperature of the thermostat at $t > 0$. Equation 4.170 has the following solution:

$$\left[\frac{\epsilon-\epsilon_-}{\epsilon(0)-\epsilon_-}\right]^{\epsilon_+-a}\left[\frac{\epsilon-\epsilon_+}{\epsilon(0)-\epsilon_+}\right]^{a-\epsilon_-}\left[\frac{\epsilon-a}{\epsilon(0)-a}\right]^{\epsilon_--\epsilon_+} = \exp[-zW^K(\epsilon_+-\epsilon_-)(a-\epsilon_-)(\epsilon_+-a)t]. \tag{4.171}$$

The average energy $\epsilon$ has three steady states: $\epsilon_-$, $\epsilon_+$, and $a$. Since the latter two are larger than unity, they are unphysical. The only physically reasonable steady state value $\epsilon = \epsilon_-$ coincides with the equilibrium value. In the final stage of the transformation process, when $\epsilon$ is close to $\epsilon_-$, one can substitute $\epsilon$ by $\epsilon_-$ in the second and third factors of the left-hand side of Equation 4.171. Then, the following simple equation for the time-dependent average energy is obtained:

$$\epsilon - \epsilon_- \sim Ae^{-t/\tau_\epsilon}(t > \tau_\epsilon), \tag{4.172}$$

where

$$\tau_\epsilon^{-1} = zW^K(\epsilon_+ - \epsilon_-)(a - \epsilon_-) \tag{4.173}$$

and

$$A = [\epsilon(0)-\epsilon_-]\left[\frac{\epsilon(0)-\epsilon_+}{\epsilon_--\epsilon_+}\right]^{(a-\epsilon_-)/(\epsilon_+-a)} \times\left[\frac{\epsilon(0)-a}{\epsilon_--a}\right]^{(\epsilon_--\epsilon_+)/(\epsilon_+-a)}. \tag{4.174}$$

At low temperatures, the above rate becomes $\tau_\epsilon^{-1} \simeq 8W^G e^{-2\beta J}$, which shows that in this case the Kawasaki dynamics does not participate to the final stage of the relaxation process.

Note that the condition $\sigma \equiv 0$ means that during the entire relaxation process the average numbers of "up" and "down" spins are equal to each other, or, in other words, the equality of the amounts of solid and liquid phases which exist in clusters of equal average size.

In Figure 4.45, the time-dependent cluster size, calculated according to Equations 4.168 and 4.171 is plotted for several values of the parameter $p$. One can see that, whereas an increase of $p = W^K/W^G$ accelerates the phase separation, it does not affect the qualitative behavior of $L(t)$. Large values of $W^K$ are needed to reach a visible influence of the Kawasaki mechanism. It follows from Equations 4.168 and 4.172 that for $t \gg \tau_\epsilon$ the average cluster size $L(t)$ approaches exponentially a steady state value $L_{st} = 2/(1-\epsilon_-)$:

$$L_{st} - L(t) \sim e^{-t/\tau_\epsilon}, \tag{4.175}$$

where the relaxation time $\tau_\epsilon$ varies from $\tau_\epsilon = 50.42/W^G$ for $p = 0$ to $\tau_\epsilon = 40.43/W^G$ for $p = 100$. Numerical calculations have shown that the same exponential law is applicable with high accuracy for times smaller than $\tau_\epsilon$, i.e., the relaxation has an exponential character at any time.

If the process is governed by Kawasaki dynamics only ($W^K \neq 0$, $W^G = 0$), the solution of Equation 4.170 corresponding to the initial condition $\sigma(0) = 0$ is

$$\left[\frac{\epsilon-\epsilon_-}{\epsilon(0)-\epsilon_-}\right]^{\epsilon_+-1}\left[\frac{\epsilon-\epsilon_+}{\epsilon(0)-\epsilon_+}\right]^{1-\epsilon_-}\left[\frac{\epsilon-1}{\epsilon(0)-1}\right]^{\epsilon_--\epsilon_+} = \exp[-zW^K(\epsilon_+-\epsilon_-)(a-\epsilon_-)(\epsilon_+-a)t]. \tag{4.176}$$

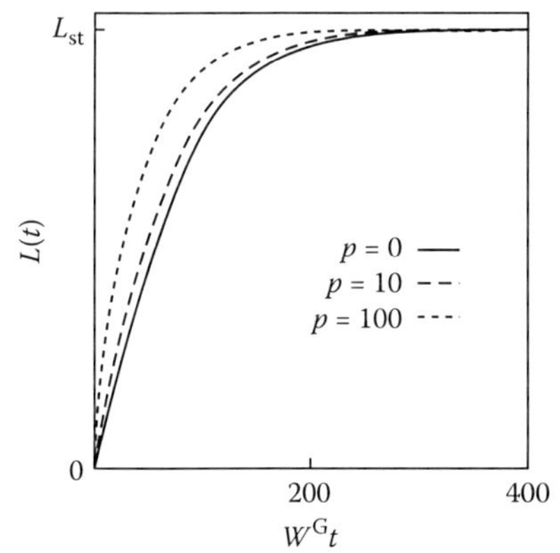


Figure 4.45

Time dependence of average cluster size in the absence of an external field ($\beta B_f = \beta B_i = 0$) for various values of parameter $p = W^K/W^G$, $\beta J = 3$, and for a completely disordered initial state [$\sigma(0) = 0$, $\epsilon(0) = 0$]. The steady-state value $L_{st}$ of $L(t)$ is equal in this case to $L_{st} \simeq 400$.

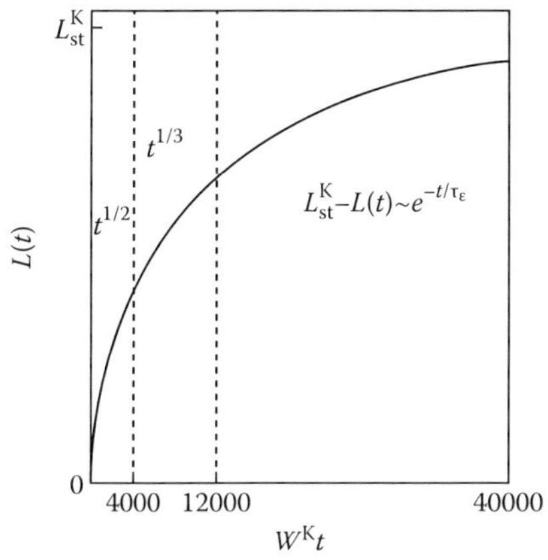


Figure 4.46

Time dependence of average cluster size for $\beta J = 3$ in the absence of an external field ($\beta B_f = \beta B_i = 0$) and in the absence of Glauber dynamics. The steady-state value $L_{st}^K$ of $L(t)$ is equal in this case to $L_{st}^K \simeq 400$.

In the final stage ($t > \tau_\epsilon$), the average cluster size also follows the exponential law, Equation 4.175, with $\tau_\epsilon^{-1} = zW^K(\epsilon_+ - \epsilon_-)(1-\epsilon_-)$. However, in the initial stage, the time behavior of $L(t)$ differs from that for $W^G \neq 0$. For small values of $t$, the average cluster size grows as $t^{1/2}$. Then there is a crossover to the well-known phenomenological Lifshitz–Slyozov–Wagner [24,25] growth law $L(t) \sim t^{1/3}$. In Figure 4.46, the ranges of the three above-mentioned regimes are presented for $\beta J = 3$. It is not yet clear whether $t^{1/2}$ law is a common feature of the phase separation in a system of any dimensions, or it is specific to the one-dimensional case. Note that $t^{1/3}$ law for two- and three-dimensional kinetic Ising model was observed in Refs. [6,7] via Monte Carlo simulations. On the basis of general qualitative considerations [7], it was concluded that the $t^{1/3}$ law has to be valid in systems of any dimension. Our results confirm this conclusion and provide the range in which it is valid in the one-dimensional case.

### 4.6.5 Kinetics after Temperature Quenching and External Field Change

Let us consider a disordered initial state, as described in Section 4.6.4, with the only difference that at $t > 0$ the external field $B_f$ is not equal to zero ($B_f > 0$), and estimate the effect of the Kawasaki mechanism upon the relaxation process. Due to the presence of an external field and of the Glauber component of the dynamics, the final polarization will be close to unity for low temperatures of the thermostat, i.e., the average size of the clusters of "up" spins (clusters of the solid phase) will be much larger than for $B_f = 0$. In this case, the system of kinetic equation (Equation 4.165) can be solved only numerically and the results for the case $\beta J = 3$, $\beta B_f = 0.1$ are presented in Figure 4.47. For the parameters selected, $L_{st} = 36{,}041.0$. In the initial stage, the

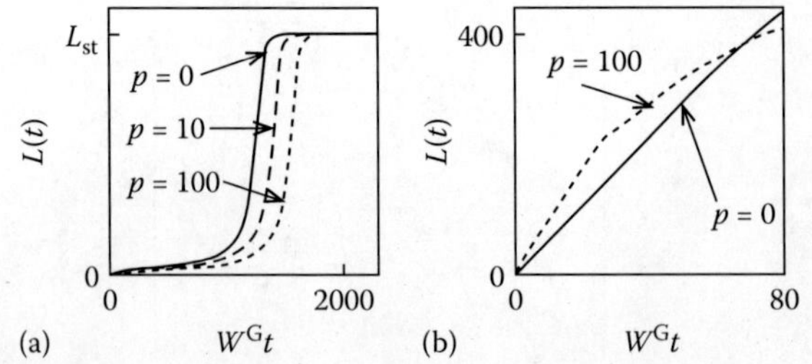


Figure 4.47

(a) Time dependence of average cluster size for $\beta B_i = 0$, $\beta B_f = 0.1$, various values of the parameter $p = W^K/W^G$, and $\beta J = 3$. Their initial state is completely disordered [$\sigma(0) = 0$, $\epsilon(0) = 0$]. The steady state value $L_{st}$ of $L(t)$ is equal to $L_{st} \simeq 36{,}000$. (b) The same for the initial stage of the relaxation process.

Kawasaki contribution to the dynamics accelerates the cluster growth, but in the later stage, it slows it down. Note that the average cluster size at which such a change of behavior takes place is approximately equal to the steady-state value $L^K_{st} \simeq 400$ of $L(t)$ in the absence of the Glauber dynamics ($W^G = 0$) (see Section 4.6.4). This fact clarifies the physical reason for which the effect of Kawasaki dynamics is changed during the transformation process. Indeed, for $W^G = 0$, the Kawasaki mechanism provides a growth of the clusters if their size is smaller than $L^K_{st}$ and a shrinking in the opposite case. In the presence of the Glauber mechanism, the steady-state value of $L(t)$, $L_{st}$ (~36,000), is much larger than $L^K_{st}$ (~ 400). $L_{st}$ and the average cluster size grows from the small value $L(0) \sim 2$, surpasses the size $L^K_{st}$, and grows further up to $L(t) = L_{st}$. In the initial stage, when $L(t) < L^K_{st}$, the Kawasaki mechanism accelerates the process. However, when $L(t)$ becomes larger than $L^K_{st}$, its action results in the slowing down of the cluster growth.

### 4.6.6 Conclusion

In the present paper, the joint action of two mechanisms (Glauber and Kawasaki) of phase transformation on the one-dimensional lattice is examined on the basis of a quasi-equilibrium approximation of the solution of the kinetic equations. It is shown that the influence of the Kawasaki spin-exchange mechanism depends on the details of the initial change of the external parameters (field and temperature) and on the specific features of the Glauber dynamics. In the case of the field reversal and constant (low) temperature of a thermostat, the influence of the Kawasaki mechanism is negligible up to very large values of the parameter $p = W^K/W^G$ ($p \sim 10{,}000$). The dominant role in the cluster growth is played in this case by the coagulation of clusters, which is provided by the Glauber dynamics. However, in the noncoagulation, nonsplitting regime of the Glauber dynamics, the Kawasaki dynamics accelerates the phase transformation at much smaller values of $p$ ($p \sim 1$) due to the additional mechanism of cluster growth which it provides.

For temperature quenchings from high to low values and in the absence of an external field, the relaxation process manifests as a phase separation. In this case, the Kawasaki mechanism added to the Glauber mechanism accelerates the cluster growth. Alone, the Kawasaki mechanism provides for the same value of the relaxation constant ($W^G = W^K$) much slower phase separations than the Glauber mechanism. In the initial stage, the average cluster size for the Kawasaki dynamics grows as $t^{1/2}$, with a crossover to the Lifshitz–Slyozow–Wagner, $t^{1/3}$, law for the intermediate stage. In the final stage, the approach of the average cluster size to the steady state value follows an exponential law with a relaxation rate $\tau_\epsilon^{-1} \sim e^{-2\beta J}$ for the Glauber mechanism alone or together with the Kawasaki mechanism, and $\tau_\epsilon^{-1} \sim e^{-4\beta J}$ for the Kawasaki dynamics alone.

If the temperature quench is accompanied by a field change, then in the initial stage of the phase transformation, the Kawasaki mechanism accelerates the cluster growth. However, as soon as the cluster size exceeds the steady state value for the Kawasaki dynamics, the latter mechanism slows down the process of phase transformation.

## Acknowledgment

This work was supported by the National Science Foundation.

## Appendix 4G: Coefficients of the Kinetic Equations

The coefficients $a_i$ and $b_i$ in Equation 4.166 are given by

$$
\begin{aligned}
a_0 &= z_0 W_0^{G} + \frac{1}{2}\left(z_2 W_2^{G} + z_{-2} W_{-2}^{G}\right),\\
a_1 &= W_0^{G} + \frac{1}{2} W_2^{G}(1-2z_2) + \frac{1}{2} W_{-2}^{G}(1+2z_{-2}),\\
a_2 &= W_2^{G} - W_{-2}^{G},\\
a_3 &= z_0 W_0^{G} - \frac{1}{2}\left(z_2 W_2^{G} + z_{-2} W_{-2}^{G}\right),\\
a_4 &= W_0^{G} - \frac{1}{2}\left(W_2^{G} + W_{-2}^{G}\right),\\
b_0 &= z_2 W_2^{G} - z_{-2} W_{-2}^{G},\\
b_1 &= W_2^{G}(1-2z_2) - W_{-2}^{G}(1+2z_{-2}),\\
b_2 &= 2\left(W_2^{G} + W_{-2}^{G}\right),
\end{aligned}
\tag{4G.1}
$$

where $z_\mu = \tanh \beta(h + \mu J)$, $\mu = 0, \pm 2$.

## Appendix 4H: Average Cluster Size

Let us define an average size $L(t)$ of clusters of "up" spins as the ratio of the average number of "up" spins $f_{\mathrm{up}}(t)$ and the average number of such clusters $f_c(t)$, $L(t) = f_{\mathrm{up}}(t)/f_c(t)$. Introducing the occupation number operator $n_j(\nu) = (1 + \nu\sigma_j)/2$, $\nu = \pm 1$, which is equal to unity if the spin $\sigma_j$ at site $j$ has the value $\sigma_j = \nu$, and zero, if $\sigma_j = -\nu$, one obtains

$$
f_{\mathrm{up}}(t) = \sum_j \langle n_j(1)\rangle_t = \frac{N}{2}(1+\sigma). \tag{4H.1}
$$

The number of clusters of "up" spins $f_c(t)$ is given by

$$
f_c(t) = \frac{1}{2} n_{\mathrm{kink}}(t), \tag{4H.2}
$$

where $n_{\mathrm{kink}}(t)$ is the average number of kinks, pair of oppositely directed nearest neighboring spins. This number can be calculated using the formula

$$
n_{\mathrm{kink}}(t) = \sum_j \left(\left\langle n_j(1) n_{j+1}(-1)\right\rangle + \left\langle n_j(-1) n_{j+1}(1)\right\rangle_t\right) = \frac{N}{2}(1-\epsilon). \tag{4H.3}
$$

Therefore,

$$
L(t) = \frac{2(1+\sigma)}{1-\epsilon}. \tag{4H.4}
$$

## References


1. Berim, G.O., Ruckenstein, E., *J. Chem. Phys.* 119, 9640 (2003). (Section 4.4 of this volume.)
2. Berim, G.O., Ruckenstein, E., *J. Chem. Phys.* 120, 272 (2004). (Section 4.5 of this volume.)
3. Glauber, R., *J. Math. Phys.* 4, 294 (1963).
4. Kawasaki, K., *Phys. Rev.* 145, 224 (1966).
5. Zhu, H., Zhu, J.-Y., *Phys. Rev. E* 66, 017102 (2002).
6. Schmelzer, J., Jr., Landau, D.P., *Int. J. Mod. Phys. C* 12, 345 (2001).
7. Marko, J.F., Barkema, G.T., *Phys. Rev. E* 52, 2522 (1995).
8. Fratzl, P., Penrose, O., *Phys. Rev. B* 50, 3477 (1994).
9. Szolnoki, A., Szabó, G., Mouritsen, O.G., *Phys. Rev. E* 55, 2255 (1997).
10. Tafa, K., Puri, S., Kumar, D., *Phys. Rev. E* 64, 056139 (2001).
11. Puri, S., *Phys. Rev. E* 55, 1752 (1997).
12. Tomé, T., de Oliveira, M.J., *Phys. Rev. A* 40, 6643 (1989).
13. Grandi, B.C.S., Figueiredo, W., *Phys. Rev. B* 50, 12595 (1994).
14. Grandi, B.C.S., Figueiredo, W., *Phys. Rev. E* 54, 4722 (1996).
15. Grandi, B.C.S., Figueiredo, W., *Phys. Rev. E* 56, 5240 (1997).
16. Grandi, B.C.S., Figueiredo, W., *Phys. Rev. E* 56, 4922 (1999).
17. Grandi, B.C.S., Figueiredo, W., *Physica A* 234, 764 (1997).
18. Ma, Y., Liu, J., *Phys. Lett. A* 238, 159 (1998).
19. Leão, J.R.S., Grandi, B.C.S., Figueiredo, W., *Phys. Rev. E* 60, 5367 (1999).
20. Szolnoki, A., *Phys. Rev. E* 62, 7466 (2000).
21. Godoy, M., Figueiredo, W., *Phys. Rev. E* 66, 036131 (2002).
22. Zhu, H., Zhu, J., Zhou, Y., *Phys. Rev. E* 66, 036106 (2002).
23. Wonczak, S., Strey, R., Stauffer, D., *J. Chem. Phys.* 113, 1976 (2000).
24. Lifshitz, I.M., Slyozov, V.V., *J. Phys. Chem. Solids* 19, 35 (1961).
25. Wagner, C., *Z. Elektrochem.* 65, 581 (1961).

# 4.7 Kinetics of Phase Transformation on a Bethe Lattice in the Presence of Spin Exchange

*Gersh Berim and Eli Ruckenstein**

Kinetics of phase transformation on a Bethe lattice governed by single-spin-flip Glauber and spin-exchange Kawasaki dynamics is examined. For a general Glauber dynamics for which all processes (splitting and coagulation, growth and decay of clusters, as well as creation and annihilation of single-spin clusters) take place, the addition of the Kawasaki dynamics accelerates the transformation process without changing the qualitative behavior. In the growth-decay regime of the Glauber dynamics, regime in which the splitting and coagulation, and creation and annihilation processes due to single-spin flips are negligible, the Kawasaki dynamics strongly increases the fraction of transformed phase because of the splitting and coagulation of clusters induced by the spin-exchange processes. Acting alone, the Kawasaki dynamics leads to the growth of the clusters of each of the phases after the quenching of the temperature to a lower value. When the final temperature $T_f$ is smaller than a certain temperature $T_{f0}$, the average cluster radius grows linearly with time during both the initial and intermediate stages of the kinetic process, and diverges as $\log_2(t_d - t)^{-1}$ when the time $t$ approaches the value $t_d$ at which infinite clusters arise. It is shown that, among the various spin-exchange processes involved in Kawasaki dynamics, the main contribution is provided by those which decrease or increase the number of clusters by unity. (© *2004 American Institute of Physics.* doi: 10.1063/1.1710855.)

## 4.7.1 Introduction

In the study of the kinetics of multiphase systems on a rigid lattice two different mechanisms of particles dynamics—the Glauber dynamics (GD) and Kawasaki dynamics (KD)—are usually employed [1,2]. The first mechanism describes the interactions of the particles with a thermostat and their transition from one phase, say phase A, to another one, phase B. The transition rate of a given particle between possible states depends on the state of the neighboring particles and satisfies the detailed balance condition. The number of particles belonging to different phases is not conserved in this case, and one of the phases can be transformed entirely to the other one. In contrast, the Kawasaki dynamics describes the pairwise exchange of states of neighboring particles belonging to different phases and, therefore, it conserves the number of particles of both phases. Similar to the GD, the rate of such an exchange depends on the states of the particles neighboring those involved into the exchange process. If KD is the only mechanism of particle dynamics, then the latter can be considered as a diffusion of particles belonging to phase A among those of phase B and vice versa [2]. The properties of such a diffusion on a square lattice were examined, for example, in Refs. [3–8] using the Monte Carlo method.

In the case of joint action of GD and KD, the Kawasaki mechanism was employed to model the energy flux into the system from external sources, by taking into account only those exchange processes which increase the energy of the system [9–18]. The Glauber mechanism describes the contact of the system with a thermostat and changes the fractions of the phases involved. The competition between the two mechanisms results at steady state, in various phases, which, obviously, do not coincide with the equilibrium state, which occurs in the absence of an energy flux.

On the other hand, the Kawasaki mechanism can also be considered as providing an exchange of the energy between the system and a thermostat. In this case, the transitions generated by KD satisfy, as in the case of GD, the detailed balance condition and can lead to phases in equilibrium. The importance of investigating the phase transformation during the transition of the system to equilibrium via both the Glauber and Kawasaki mechanisms was emphasized by Dattagupta [19] using as an example the structural transition in materials with two sites per unit cell that are occupied by molecules of types A and B. In such a system the interchange of the atoms in one cell is described by the Glauber mechanism, and the exchange of the configurations in two neighboring cells by the Kawasaki mechanism.

In a previous paper [20], we considered the joint action of GD and KD for the case of phase transformation on a one-dimensional lattice using a closure procedure of the kinetic equations introduced earlier [21]. The main feature of the addition of KD to GD was an acceleration of the transformation process in the initial stages and a slowing down in later stages. In some particular cases of the Glauber relaxation

* *J. Chem. Phys.* 120, 9800 (2004). Republished with permission.

mechanism, the presence of KD always accelerated the phase transformation. In the absence of GD, the Kawasaki mechanism provided a cluster growth proportional to $t^{1/2}$ in the initial stages and to $t^{1/3}$ in the intermediate stage. However, the one-dimensional lattice implies a phase transition at zero critical temperature ($T_c = 0$) [22], and it will be of interest to consider models in which the phase transition occurs at a finite temperature $T_c > 0$. The kinetics of one of such model, an Ising model on a Bethe lattice, was investigated previously in Ref. [23] in the framework of the Glauber dynamics. In the present paper, we consider the kinetics of such a model in the presence of both the Glauber and the Kawasaki dynamics. In Section 4.7.2, the kinetic equations are derived and some results obtained in a previous paper [23] are summarized. In Section 4.7.3, the consequences of the addition of KD to the GD for the phase transformation are discussed. In Section 4.7.4, the phase separation involving the KD mechanism acting alone is considered.

### 4.7.2 Kinetic Equations

A Bethe lattice is an ordered structure which can be generated by adding $q$ neighbors (first shell) to a core site, then adding to each of the new sites $q - 1$ new neighbors (second shell), and so on [22]. Such a lattice for $q = 3$ is presented in Figure 4.48a. If the number of shells is infinite, then the system is spatially homogeneous, i.e., any site can be considered as a core site.

If each of the sites is occupied by a molecule belonging to one of the two phases (e.g., liquid or solid) then the state of the system can be described by the Ising model with the Hamiltonian

$$H_s = -B\sum_j \sigma_j - \frac{1}{2}J\sum_j \sum_{k=1}^{q} \sigma_j \sigma_{j+\delta_k} \,(J > 0), \tag{4.177}$$

where $\sigma_j = \pm 1$ (spins) model the two phases, e.g., a liquid ($\sigma_j = -1$) and a solid ($\sigma_j = 1$), $B$ is the external field, which models the supersaturation $S$ ($B \sim \log S$) [24], $J$ is the interaction energy between two neighboring molecules, and the index $\delta_k$ labels the neighboring spins, that interact with spin $\sigma_j$. The thermodynamics of this model was examined [22,25] and the equilibrium averages calculated exactly [26,27]. The model has a phase transition at a critical temperature $T_c$ provided by the equation [22] $\tanh \beta J = 1/(q - 1)$, where $\beta = 1/k_B T$, $T$ being the temperature in K and $k_B$ the Boltzmann constant. In the present paper, the model with $q = 3$ is examined. For such a model, the critical temperature is provided by the expression $\tanh(J/k_B T_c) = 1/2$ ($J/k_B T_c \simeq 0.549$).

The nonequilibrium behavior of the system is described by a time-dependent probability distribution function (PDF) $P(\{\sigma\};t)$, where $\{\sigma\}$ denotes the state $\{\sigma_1, \sigma_2, \ldots, \sigma_N\}$ of the particles of the system. The time evolution of PDF is provided by the master equation [1,2,9–18]

$$\begin{aligned}\frac{d}{dt}P(\{\sigma\};t) = \sum_{\{\sigma'\}} \sum_{\alpha=G,K} [&-W^{\alpha}(\{\sigma\} \to \{\sigma'\})P(\{\sigma\};t) \\ &+ W^{\alpha}(\{\sigma'\}) \to \{\sigma\})P(\{\sigma'\};t)],\end{aligned} \tag{4.178}$$

where the superscript $\alpha$ labels the transition rates in the Glauber ($\alpha$ = G) and Kawasaki ($\alpha$ = K) dynamics, respectively.

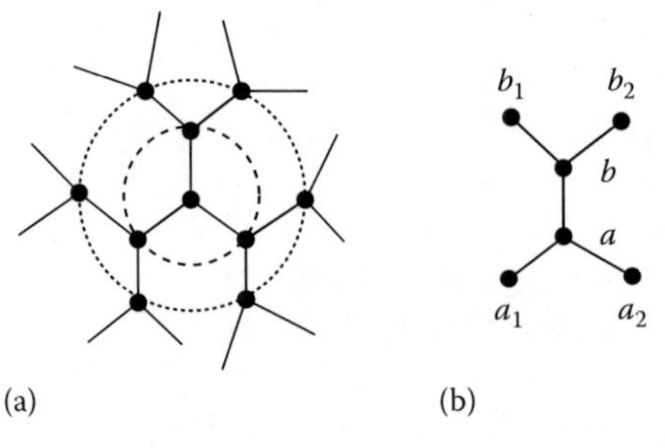


Figure 4.48

(a) A Bethe lattice for $q = 3$. The dashed circles indicate the first and second shells of the central site; (b) neighboring sites $a$ and $b$ and their four nearest neighbors, where the spins $\sigma_a, \sigma_b, \sigma_{a_1}, \sigma_{a_2}, \sigma_{b_1}, \sigma_{b_2}$ are located.

The Glauber transition rates $W^G(\{\sigma\}\to\{\sigma'\})$ differ from zero only if the state $\{\sigma\}$ differs from the state $\{\sigma'\}$ by the sign of only one arbitrary chosen spin, say $\sigma_a$ (Figure 4.48b). To satisfy the detailed balance condition, the flipping rate $W^G(\{\sigma\} \to \{\sigma'\})$ can be represented by [21,23]

$$W^{G}(\{\sigma\}\to\{\sigma'\})=\sum_{a}\sum_{\mu}W_{\mu}^{G}R_{a}(\mu)(1-\sigma_{a}\tanh\beta\varepsilon_{\mu}), \tag{4.179}$$

where $\mu$ takes the values $-q, -q+2, \ldots, q$ ($\mu = \pm1, \pm 3$ for $q = 3$), $\varepsilon_\mu = B + J\mu$, the operator $R_a(\mu)$ is unity when the sum of the neighboring spins of spin $\sigma_a$ is equal to $\mu$, and zero otherwise, and $W_\mu$ are constants, defining the specific features of the relaxation mechanisms [21,23]. ($W_{-1}^G$ characterizes the growth and decay, $W_1^G$ and $W_3^G$ the splitting and coagulation, and $W_{-3}^G$ the creation and annihilation of clusters, all due to single-spin flips.) For example, the case $W_{-1}^G \neq 0$, $W_1^G = W_{-3}^G = W_3^G = 0$ corresponds to a process during which there is no splitting (coagulation) and no creation (annihilation) of clusters on the lattice [23].

The transition probability $W^K(\{\sigma\}\to\{\sigma'\})$ has a more complicated form than $W^G(\{\sigma\}\to\{\sigma'\})$. It differs from zero only if the state $\{\sigma\}$ differs from the state $\{\sigma'\}$ by the orientation of a pair of neighboring oppositely directed spins, say $\sigma_a$ and $\sigma_b$ (Figure 4.48b) and can be written in the form

$$W^{K}(\{\sigma\}\to\{\sigma'\})=\frac{1}{2}\sum_{\langle a,b\rangle}\sum_{v=0,\pm2,\pm4}W_{v}^{K}R_{a,b}(v)[1-\sigma_{a}\sigma_{b}-(\sigma_{a}-\sigma_{b})\tanh\beta vJ], \tag{4.180}$$

where the operator $R_{a,b}(v)$ is unity if the configuration (the set of values) of spins $\sigma_{a_1}, \sigma_{a_2}, \sigma_{b_1}$, and $\sigma_{b_2}$ provides a change of the energy of the system equal to $\Delta E = 2vJ$ due to the transition of spins $\sigma_a$, $\sigma_b$ from the state $\sigma_a = -\sigma_b = 1$ to the state $\sigma_a = -\sigma_b = -1$ and $\langle a, b\rangle$ indicates that each pair of neighboring spins should be counted in the sum only ones. The expression in square brackets accounts for the detailed balance condition for all possible transitions. In Table 4.10, only one-half of the possible configurations and the corresponding energy changes $\Delta E$ are listed together with the changes $\Delta N_{cl}$ in the number of clusters on the lattice. The second half of possible configurations can be obtained by changing the signs of all spins as well as of $\Delta E$ and $\Delta N_{cl}$. For illustration, some of the transitions are presented in Figure 4.49. Note that the processes with $\Delta N_{cl} = -1$ and $\Delta N_{cl} = -2$ correspond, respectively, to the coagulation of two or three clusters into a single one, whereas the processes with $\Delta N_{cl} = +1$ and $\Delta N_{cl} = +2$ correspond to the splitting of a cluster into two or three clusters, respectively. The processes with $\Delta N_{cl} = \pm1$ are characterized by the coefficients $w_{\pm2}^K$ and those with $\Delta N_{cl} = \pm 2$ by $w_{\pm4}^K$.

Employing the probabilities (Equations 4.179 and 4.180), and the master equation (Equation 4.178), one can obtain an infinite chain of kinetic equations for the time-dependent averages

$$\langle A\rangle_t \equiv \sum_{\{\sigma\}}AP(\{\sigma\};t) \tag{4.181}$$

of various spin operators. To close this chain, we will use the previously employed approximation [20,21,23] for the PDF $P(\{\sigma\};t)$ according to which it can be represented in a form similar to the equilibrium density matrix

$$P(\{\sigma\};t) = e^{-F(t)}/\mathrm{Tr}[e^{-F(t)}], \tag{4.182}$$

Table 4.10 Possible Spin Configurations of the Neighbors of the $\sigma_a$, $\sigma_b$ Spins and the Change in Energy ($\Delta E$) and Number of Clusters $\Delta N_{cl}$ Due to the Spin-Exchange Transition from the State with $\sigma_a = -\sigma_b = 1$ to the State with $\sigma_a = -\sigma_b = -1$

| $\sigma_{a_1}$ | $\sigma_{a_2}$ | $\sigma_{b_1}$ | $\sigma_{b_2}$ | $\Delta E$ | $\Delta N_{cl}$ |
|---|---|---|---|---|---|
| 1 | 1 | −1 | −1 | $8J$ | +2 |
| 1 | 1 | 1 | −1 | $4J$ | +1 |
| 1 | 1 | −1 | 1 | $4J$ | +1 |
| 1 | 1 | 1 | 1 | 0 | 0 |
| 1 | −1 | 1 | −1 | 0 | 0 |
| −1 | 1 | 1 | −1 | 0 | 0 |
| 1 | −1 | 1 | 1 | $-4J$ | −1 |
| −1 | 1 | 1 | 1 | $-4J$ | −1 |

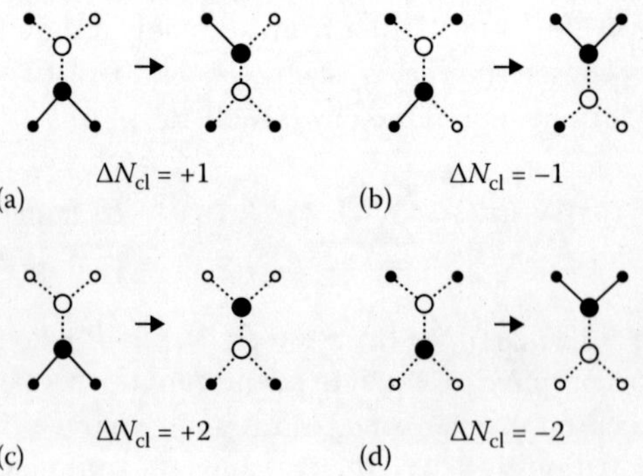


Figure 4.49

Schematic representation of the possible transitions generated by the Kawasaki spin-exchange dynamics. The empty circles represent the "down" and the filled ones represent the "up" spins. The flipping spins are represented by the larger circles. $\Delta N_{cl}$ is the change of the number of clusters due to the corresponding transition.

where

$$F(t) = -\beta_\sigma(t)\sum_j \sigma_j - \beta_e(t)\sum_j \sum_{k=1}^{q} \sigma_j\sigma_{j+\delta_k} \tag{4.183}$$

with $\beta_\sigma(t)$ and $\beta_e(t)$ unknown function of $t$. The PDF (Equation 4.182) can be obtained from the equilibrium density matrix

$$\rho_T = e^{-\beta H_s}/\mathrm{Tr}\left[e^{-\beta H_s}\right] \tag{4.184}$$

of the Ising model (Equation 4.177), using the substitutions

$$\beta B \to \beta_\sigma(t),\ \beta J \to \beta_\epsilon(t). \tag{4.185}$$

Because of this choice of the PDF, all time-dependent averages $\langle A\rangle_t$ are spatially homogeneous and can be calculated as functions of $\beta_\sigma(t)$ and $\beta_\epsilon(t)$, using the methods developed for the equilibrium Ising model.

In the framework of approximation (Equation 4.182), one has to consider only two equations from the infinite chain, namely, the kinetic equations for the time-dependent averages $\sigma(t) \equiv \langle\sigma_a\rangle_t$ and $\epsilon(t) = \langle\sigma_a\sigma_b\rangle_t$, representing the spin polarization and the interaction energy of neighboring spins, respectively (below, the argument $t$ of these and similar functions will be omitted). These equations can be derived in the standard way [1,2], using Equations 4.178 through 4.180, and have the form

$$\begin{aligned} \dot{\sigma} &= R^{\mathrm{G}}_\sigma, \\ \dot{\epsilon} &= R^{\mathrm{G}}_\epsilon, + R^{\mathrm{K}}_\epsilon, \end{aligned} \tag{4.186}$$

where the relaxation terms are given by

$$\begin{aligned} R^{\mathrm{G}}_\sigma &= -\sum_{\mu=\pm1,\pm3} W^{\mathrm{G}}_\mu\left[\langle\sigma_a R_a(\mu)\rangle_t - z_\mu\langle R_a(\mu)\rangle_t\right], \\ R^{\mathrm{G}}_\epsilon &= -\frac{2}{3}\sum_{\mu=\pm1,\pm3} \mu W^{\mathrm{G}}_\mu\left[\langle\sigma_a R_a(\mu)\rangle_t - z_\mu\langle R_a(\mu)_t\rangle\right], \\ R^{\mathrm{K}}_\epsilon &= -\frac{8}{3}\sum_{\nu=\pm2,\pm4} W^{\mathrm{K}}_\nu\Big[\langle\sigma_{a_1}(\sigma_a - \sigma_b)R_{a,b}(\nu)\rangle_t \\ &\quad - z^0_\nu\langle\sigma_{a_1}(1-\sigma_a\sigma_b)R_{a,b}(\nu)\rangle_t\Big], \end{aligned} \tag{4.187}$$

with $z_\mu = \tanh \beta(B+\mu J)$ and $z_\nu^0 = \tanh(\nu\beta j)$. The explicit forms of the operators $R_a(\mu)$ are given in Ref. [23] and of the operators $R_{a,b}(\nu)$ in Appendix 4I. At equilibrium, one has

$$\langle \sigma_a R_a(\mu)\rangle_T = z_\mu \langle R_a(\mu)\rangle_T \ (\mu = 0, \pm 1, \pm 2), \tag{4.188}$$

$$\left\langle \sigma_{a_1}(\sigma_a - \sigma_b)R_{a,b}(\nu)\right\rangle_T = z_\nu^0 \left\langle \sigma_{a_1}(1-\sigma_a\sigma_b)R_{a,b}(\nu)\right\rangle_T \ (\nu = \pm 2, \pm 4), \tag{4.189}$$

where $\langle\ldots\rangle_T = \mathrm{Tr}(\rho_T \ldots)$. Equation 4.188 was proven in Appendix A of Ref. [21], and Equation 4.189 is demonstrated in Appendix 4I. Using substitutions (Equation 4.185), Equations 4.188 and 4.189 can be rewritten for time-dependent averages, and the relaxation terms acquire the forms

$$\begin{aligned}
R_\sigma^{\mathrm{G}} &= -\sum_{\mu=\pm1,\pm3} W_\mu^{\mathrm{G}}[(z_{\mu q} - z_\mu)\langle R_a(\mu)\rangle_t], \\
R_\epsilon^{\mathrm{G}} &= -\frac{2}{3}\sum_{\mu=\pm1,\pm3} \mu W_\mu^{\mathrm{G}}(z_{\mu q} - z_\mu)\langle R_a(\mu)\rangle_t, \\
R_\epsilon^{\mathrm{K}} &= -\frac{8}{3}\sum_{\mu=\pm2,\pm4} W_\nu^{\mathrm{K}}\left(z_{\nu q}^0 - z_\nu^0\right)\left\langle \sigma_{a_1}(1-\sigma_a\sigma_b)R_{a,b}(\nu)\right\rangle_t,
\end{aligned} \tag{4.190}$$

where

$$z_{\mu q} = \tanh[\beta_\sigma(t) + \mu\beta_\epsilon(t)] \text{ and } z_{\nu q}^0 = \tanh \nu\beta_\epsilon(t). \tag{4.191}$$

Using explicit forms for the operators $R_a(\mu)$ and $R_{a,b}(\nu)$, the time-dependent averages $\langle R_a(\mu)\rangle_t$, $\left\langle \sigma_{a_1}(1-\sigma_a\sigma_b)R_{a,b}(\nu)\right\rangle_t$ can be represented as linear combinations of the time-dependent averages of various products of the spin operators $\sigma_a, \sigma_{a_1}, \sigma_{a_2}, \sigma_b, \sigma_{b_1}, \sigma_{b_2}$. After simple algebra, one obtains for the case in which $W_\nu^{\mathrm{K}} = W_{-\nu}^{\mathrm{K}}$,

$$\begin{aligned}
R_\sigma^{\mathrm{G}} &= -\sum_{\mu=\pm1,\pm3} W_\mu^{\mathrm{G}} A_\mu (z_{\mu q} - z_\mu), \\
R_\epsilon^{\mathrm{G}} &= -\frac{2}{3}\sum_{\mu=\pm1,\pm3} \mu W_\mu^{\mathrm{G}} A_\mu (z_{\mu q} - z_\mu), \\
R_\epsilon^{\mathrm{K}} &= -\sum_{\nu=2,4} W_\nu^{\mathrm{K}} B_\nu \left(z_{\nu q}^0 - z_\nu^0\right),
\end{aligned} \tag{4.192}$$

where

$$\begin{aligned}
A_{\pm1} &= \frac{3}{8}(1 \pm \sigma - \epsilon_2 \mp \gamma_1), \\
A_{\pm3} &= \frac{1}{8}(1 \pm 3\sigma + 3\epsilon_2 \mp \gamma_1), \\
B_2 &= \frac{2}{3}(1 - \epsilon - \delta_3 + \kappa), \\
B_4 &= \frac{1}{3}(1 - \epsilon + 2\epsilon_2 - 4\epsilon_3 - 2\delta_1 + 4\delta_2 + \delta_3 - \kappa)
\end{aligned} \tag{4.193}$$

and

$$\sigma=\langle\sigma_a\rangle_t,\ \epsilon=\langle\sigma_a\sigma_b\rangle_t,\ \epsilon_2=\langle\sigma_a\sigma_{b_1}\rangle_t,\ \epsilon_3=\langle\sigma_{a_1}\sigma_{b_1}\rangle_t,$$

$$\gamma_1=\langle\sigma_b\sigma_{a_1}\sigma_{a_2}\rangle_t,\ \delta_1=\langle\sigma_{a_1}\sigma_{a_2}\sigma_a\sigma_b\rangle_t,$$

$$\delta_2=\langle\sigma_{a_1}\sigma_a\sigma_b\sigma_{b_1}\rangle_t, \tag{4.194}$$

$$\delta_3=\langle\sigma_{a_1}\sigma_{a_2}\sigma_{b_1}\sigma_{b_2}\rangle_t,\ \kappa=\langle\sigma_a\sigma_{a_1}\sigma_{a_2}\sigma_b\sigma_{b_1}\sigma_{b_2}\rangle_t.$$

One should note that the term with $v = 0$ is absent in the expression of $R_\epsilon^K$. That term describes the spin transitions between states with the same energy and for this reason it does not contribute to the corresponding relaxation term.

The procedure of transforming Equation 4.186 to a closed system of equations is similar to that discussed by Berim and Ruckenstein [23] for the case $R_\epsilon^K = 0$. According to that procedure all averages (Equation 4.194) are expressed in terms of the two functions, $\beta_\sigma(t)$ and $\beta_\epsilon(t)$, by employing the known exact equilibrium solution for the Ising model on a Bethe lattice and the substitutions (Equation 4.185). This procedure provides a closed set of kinetic equations for $\beta_\sigma(t)$ and $\beta_\epsilon(t)$. Furthermore, using as independent variables the quantities $z = \tanh\beta_e(t) \equiv z_{1q}^0$ and $\sigma \equiv \langle\sigma_a\rangle_t$ the latter equations can be transformed into

$$\dot{\sigma}=f_\sigma(\sigma,z),\ \dot{z}=f_z(\sigma,z), \tag{4.195}$$

which can be solved numerically. Some specific details of this procedure for the case under consideration are given in Appendix 4J.

### 4.7.3 Phase Transformation after Field Reversal for $T < T_c$

Let us suppose that at $t = 0$ the system is at equilibrium in an external field $B = B_i$ in the "down" direction ($B_i < 0$, supersaturation $S < 1$). For $T < T_c$, almost all spins are "down" (liquid phase state). If at $t = 0$ the external field changes direction and acquires a value $B_f > 0$ (supersaturation $S > 1$), the transformation to a new, solid state, starts. In Ref. [23], such a transformation was examined when the Glauber mechanism was acting alone. It was shown that the final state coincides with the equilibrium one corresponding to the new field $B_f$, only when $B_f$ is larger than a certain critical value $B_c$. When $B_f$ is smaller, the final state is a steady state one. The main transformation occurred near a time $t_R$ which, for $B_f \to B_c$ ($B_f > B_c$) depends on $B_f$ as $t_R \simeq (B_f - B_c)^{-b}$ with $b \simeq 0.47$. The average cluster size (the number of spins in a cluster) grew unrestrictedly at time $t_d \simeq t_R$, indicating the creation of infinite clusters on the lattice.

In this section, we consider the kinetics of a model in which KD is taken into account along with GD. For the considerations that follow a temperature smaller than the critical one ($J/k_BT = 1.5$) and an initial field $\beta B_i = -0.1$ were selected. In this case, the critical field $B_c$ in the absence of KD is equal to $\beta B_c = 0.80934$.

Regarding the effect of KD upon the phase transformation, three main types of Kawasaki mechanisms, which account for different processes in the cluster dynamics (Figure 4.49), will be compared:

1. Type 1: $W_2^K = W_4^K = W^K$ ($\Delta N_{cl} = \pm 1$ and $\pm 2$ processes).
2. Type 2: $W_2^K = W^K, W_4^K = 0$ ($\Delta N_{cl} = \pm 1$ process, Figure 4.49a and b).
3. Type 3: $W_2^K = 0,\ W_4^K = W^K$ ($\Delta N_{cl} = \pm 2$ processes, Figure 4.49c and d).

Because none of these types of Kawasaki mechanisms affect the value of the critical field $B_c$. the selected value of the final field $\beta B_f = 0.80935$ will be larger than $B_c$ regardless the values of the parameters $W_v^K$.

In Figure 4.50a, the time dependence of the polarization for the first type of KD, which involves all the possible spin-exchange transitions, is presented for several values of the ratio $W^K/W^G = p$ and $W_v^G \equiv W^G$. It shows that an increase of $p$ does not affect the qualitative behavior of the polarization, but decreases the time $t_R$, i.e., KD accelerates the process of phase transformation. A similar behavior of the polarization occurs for the second and third type of KD. In Figure 4.50b, the $p$ dependencies of the time $t_R$ are plotted for all three types of KD. One can see that both $\Delta N_{cl} = \pm 1$ (the second type) and $\Delta N_{cl} = \pm 2$ (the third type) processes decrease for $p \neq 0$ the value of $t_R$ compared to the case with $p = 0$ ($W^K = 0$), the $\Delta N_{cl} = \pm 1$ processes providing the largest contribution. The time $t_d$ near which a rapid growth of the average cluster radius takes place (see Figure 4.51a) depends on the type of relaxation process in a way similar to $t_R$ (see Figure 4.51b). To

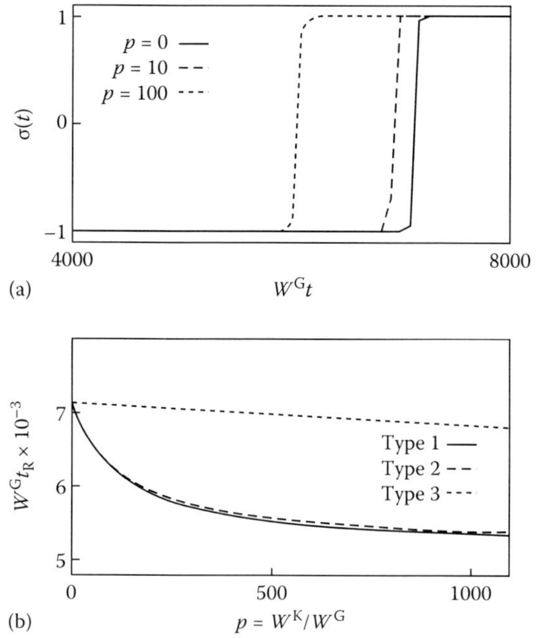


Figure 4.50

(a) Time dependence of the polarization for various values of the parameter $p = W^K/W^G$ for the case $W^G_{\pm1} = W^G_{\pm2} = W^G$, $W^K_{\pm2} = W^K_{\pm4} = W^K$ after field reversal for $T < T_c$. (b) $p$ dependence of the time $t_R$ for the three types of Kawasaki dynamics. (1) $W^K_2 = W^K_4 = W^K$; (2) $W^K_2 = W^K$, $W^K_4 = 0$; (3) $W^K_2 = 0$, $W^K_4 = W^K$.

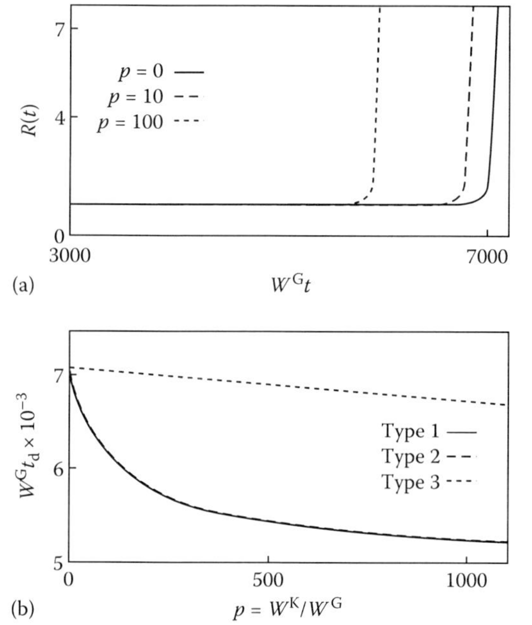


Figure 4.51

(a) Time dependence of the average cluster radius for the case $W^G_{\pm1} = W^G_{\pm2} = W^G$, $W^K_{\pm2} = W^K_{\pm4} = W^K$ for several values of $p = W^K/W^G$ after field reversal for $T < T_c$. (b) $p$ dependence of the time $t_d$ for the three types of Kawasaki dynamics. (1) $W^K_2 = W^K_4 = W^K$; (2) $W^K_2 = W^K$, $W^K_4 = 0$; (3) $W^K_2 = 0$, $W^K_4 = W^K$.

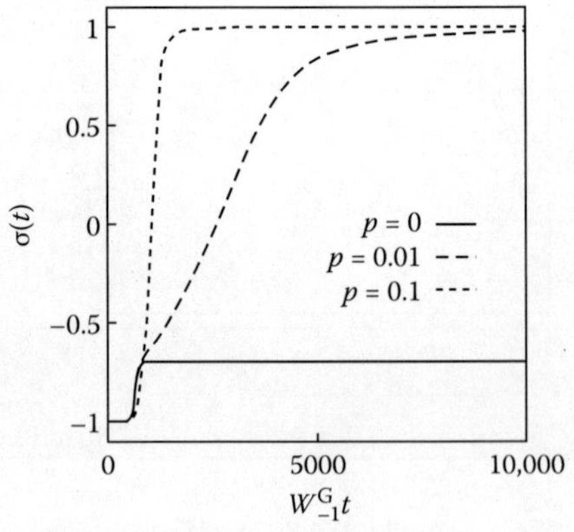


Figure 4.52

Typical time dependence of the polarization for the growth decay regime of GD in the presence and absence of KD.

understand the physical reasons for the effects of the different Kawasaki mechanisms one should note that the main role in the above phase transformation is played by the Glauber single-spin-flip mechanism. The "down"–"up" single spin flips generated by that mechanism have the highest rate when all the spins neighboring the flipping one have "up" direction, because for such a transition the energy of the system decreases by the largest amount. As one can see from Figure 4.49a and c, the Kawasaki dynamics with both $\Delta N_{cl} = +1$ ($\Delta E = 4J$) and $\Delta N_{cl} = +2$ ($\Delta E = 8J$) contribute to the increase of the number of favorable configurations for the Glauber flips to occur. At low temperatures ($\beta J > 1$), the rates of processes $\Delta N_{cl} = +1$ and $\Delta N_{cl} = +2$ are proportional to $e^{-4\beta J}$ and $e^{-8\beta J}$, respectively, and therefore the effectiveness of the $\Delta N_{cl} = +1$ process is higher than that of the $\Delta N_{cl} = +2$ process.

In addition to the general GD case considered earlier, it is also instructive to examine the role of KD in a truncated GD, in which the processes of splitting and coagulation, as well as creation and annihilation of clusters are negligible (growth-decay regime). In this growth-decay regime, $W^G_{-1} \neq 0$ and $W^G_1 = W^G_{\pm 3} = 0$. Furthermore, because of constraints, this regime does not lead to thermodynamic equilibrium, but only to a final steady state for which the polarization $a$ is much smaller than that for the equilibrium, which is reached when all the relaxation processes take place. In Figure 4.52, the time dependence of the polarization in that regime is presented in the presence of the Kawasaki dynamics. For comparison purposes, the curve in which KD is absent is also included. One can see that even for small values of $p = W^K/W^G_{-1}$, the KD qualitatively changes the relaxation process and provides an almost complete phase transformation [$\sigma(\infty) \simeq 1$]. It is clear that this kind of transformation is a result of the coagulation (splitting) processes generated by KD.

One should note again that, in the regimes considered above, the Glauber dynamics plays the key role in the phase transformation, independently of the value of $p$, because only this dynamics can change the number of "up" spins. KD provides only an acceleration of the transformation process.

### 4.7.4 Temperature Quench in the Absence of an External Field

Let us examine the kinetics for a model in which the single-spin-flip processes are negligible $\left(W^G_\mu = 0\right)$. In this case, the kinetics is governed by the spin-exchange Kawasaki dynamics alone and the average polarization $\sigma$ is conserved ($\sigma$ = const). The conservation of $\sigma$ implies the conservation of the average (with respect to the total number of particles) number densities $n_{up} = (1 + \sigma)/2$ and $n_{down} = (1 - \sigma)/2$ of the particles belonging to the phases associated with the "up" and "down" spins, respectively. In this case, the state of the system can be characterized by the average size $L(t)$ of the clusters with "up" spins which is given by the expression [23]

$$L(t) = \frac{4(1+\sigma)}{1 - 2\sigma - 3\epsilon}, \tag{4.196}$$

where the time dependence of the right-hand side is provided by the time dependence of the average energy $\epsilon$. The average size of the clusters of "down" spins can be obtained from Equation 4.196 by changing the sign of the polarization $\sigma$. Along with $L(t)$, one can also use other characteristic quantities. Let us define a "spherical" cluster on a Bethe lattice as a cluster consisting of a central spin $\sigma_a$ and a number of surrounding shells, and name the number $R$ of such shells as the radius of a spherical cluster. The total number $N_{total}$ of spins in such a cluster is given by $N_{total} = 3 \times 2^R - 2$ and hence $R = \log_2[(N_{tot} + 2)/3]$. Then the cluster of average size $L(t)$ can be characterized by the average radius $R(t)$ of a spherical cluster with the same number of spins through the relation

$$R(t) = \log_2 \frac{2 + L(t)}{3}. \tag{4.197}$$

Below, we consider the case in which the initial state of the system is an equilibrium one at a temperature $T_i > T_c$, and no external field is present ($B = 0$). In that case, the initial polarization $\sigma(0)$ and, consequently, $\sigma(t)$ are equal to zero and the number densities of "up" and "down" spins are equal to each other ($n_{up} = n_{down} = 1/2$). If at $t = 0$, the temperature is changed to $T_f < T_i$, the system starts to approach the steady state with an average size of clusters of "up" spins different from the initial one. If $T_f$ is larger than the critical temperature $T_c$, then the steady state coincides with the equilibrium one, provided by the Ising model on a Bethe lattice. When $T_f < T_c$, the steady state is obviously not the equilibrium one because the polarization $\sigma(t) = 0$ does not correspond to the equilibrium state for $T < T_c$, for which it should have a nonzero value [22].

#### 4.7.4.1 The Case $T_f < T_c$, $B = 0$

Let us select an initial temperature $T_i$ much larger than $T_c$ ($T_i = 20T_c$) and a final temperature smaller than $T_c$ ($T_f = 0.9T_c$). In Figure 4.53a, the time dependence of $R(t)$ is presented for the first type of KD $\left(W_2^K = W_4^K = W^K\right)$. As for the Glauber dynamics, which was considered by Berim and Ruckenstein [23], there is a characteristic time $t_d$ near which $R(t)$ grows unrestrictedly. In the initial stage, $R(t)$ grows linearly with time.

A better estimation of the time dependence of $R(t)$ can be obtained on the basis of the analysis of the average number density $n_{up}(t) = (1 - 2\sigma - 3\epsilon)/8$ of finite clusters of "up" spins which was derived in Ref. [23]. In the absence of an external field, this number density for $\sigma(t) = 0$ is given by

$$n_{up}^0(t) = \frac{1}{8}(1 - 3\epsilon), \tag{4.198}$$

which becomes smaller than zero for $t > t_d$ (Figure 4.53b), when the average energy $\epsilon$ exceeds the value $\epsilon_c = 1/3$. For $\epsilon_c = 1/3$, infinite clusters arise and expression 4.198 is no longer valid. A numerical analysis of

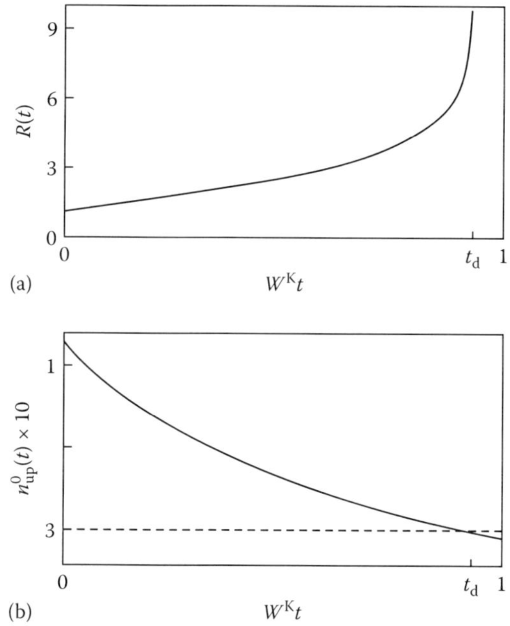


Figure 4.53

(a) Typical time dependence of the average cluster radius $R(t)$ for $T_f < T_c$ in the absence of GD. (b) The same for the number density of finite clusters of "up" spins.

$n^0_{\text{up}}(t)$ has shown that in the initial and intermediate stages it can be approximated by a simple exponential function

$$n^0_{\text{up}}(t) \simeq n^0_{\text{up}}(0)e^{-t/\tau}. \tag{4.199}$$

The time $\tau$ can be obtained by comparing the two first terms of the Taylor expansions of $n^0_{\text{up}}(t)$ given by Equations 4.198 and 4.199. This comparison yields

$$1/\tau = \frac{3}{8n^0_{\text{up}}(0)}\dot{\epsilon}(0) = \frac{3}{8n^0_{\text{up}}(0)} R^{\text{K}}_{\epsilon}\bigg|_{t=0}, \tag{4.200}$$

where the initial value $R^{\text{K}}_{\epsilon}\big|_{t=0}$ of the relaxation term $R^{\text{K}}_{\epsilon}$ can be obtained from expressions 4.192 and 4.193, and from the expressions for the equilibrium values of the spin averages involved in $R^{\text{K}}_{\epsilon}$ for $T = T_{\text{f}}$. The equilibrium averages for $T_{\text{i}} > T_{\text{c}}$ are: $\epsilon_T = z_T$, $\epsilon_{2T} = z_T^2, \epsilon_{3T} = z_T^3$, $\delta_{1T} = z_T^3$, $\delta_{2T} = z_T^2$, $\delta_{3T} = z_T^4$, $\kappa_T = z_T^5$, where $z_T = \tanh \beta J$.

The average cluster size, which for $B = 0$ is equal to $L^0(t) = 1/\left[2n^0_{\text{up}}(t)\right]$, grows exponentially $\left[L^0(t) = e^{t/\tau}/2n^0_{\text{up}}(0)\right]$ in the initial and intermediate stages, and the average radius for $L^0(t) \gg 1$ grows linearly $\left\{R(t) \simeq t/\tau - \log_2\left[6n^0_{\text{up}}(0)\right]\right\}$ as one can see in Figure 4.53a.

In the vicinity of the time $t_{\text{d}}$, the average number of clusters can be represented by $n^0_{\text{up}}(t) \simeq (t - t_{\text{d}})/\tau_1$, where $1/\tau_1 = -\frac{3}{8} R^{\text{K}}_{\epsilon}\big|_{t=t_{\text{d}}}$. Therefore, $L^0(t)$ grows near $t_{\text{d}}$ as $(t - t_{\text{d}})^{-1}$ when $t \to t_{\text{d}}$ $[L_0(t) \simeq (\tau_1/2)(t - t_{\text{d}})^{-1}]$ and the average radius $R(t)$ has the following time dependence:

$$R(t) \simeq \log_2[\tau_1/(t_{\text{d}} - t)],\ t \to t_{\text{d}}. \tag{4.201}$$

The linear growth of the average radius can be qualitatively understood from the simple example of the growth of a spherical cluster in the absence of coagulation and splitting. The spherical cluster of radius $R$ contains $N_{\text{tot}} = 3 \times 2^R - 2$ spins and has $N_{\text{bound}} = 3 \times 2^{R-1}$ boundary spins and $N_{\text{nb}} = 3 \times 2^R$ neighboring spins [22]. If the rate of "up"–"down" transitions of the boundary spins is equal to $W_-$ and the rate of "down"–"up" transitions of neighboring spins is equal to $W_+$, then one can write the following growth (decay) equation for a given cluster:

$$\frac{\text{d}}{\text{d}t} N_{\text{tot}} = -W_- N_{\text{bound}} + W_+ N_{\text{nb}}, \tag{4.202}$$

which can be transformed into the following equation for $R$:

$$\frac{\text{d}}{\text{d}t} 2^R = \alpha 2^R, \tag{4.203}$$

where $\alpha = (3W_+ - 3\,W_-/2)$. The latter equation has the following solution:

$$R(t) = R(0) + \frac{\alpha}{\log 2} t, \tag{4.204}$$

which provides a linear growth for the average cluster radius. One should note that this result is obtained because on a Bethe lattice the number of spins in a spherical cluster as well as the number of boundary and neighboring spins grow exponentially with increasing radius.

#### 4.7.4.2 The Case $T_f > T_c$, $B = 0$

Let us define a characteristic temperature $T_{\text{f0}}$ as the temperature at which the equilibrium number density $n^0_{\text{up}T}$ of finite clusters becomes zero, i.e., $\epsilon_T = 1/3$. Because for $T > T_{\text{c}}$ and $B = 0$, $\epsilon_T = \tanh \beta J$ [26,27], one obtains $J/(k_{\text{B}}T_{\text{f0}}) = \text{arctanh}(1/3)$. Hence, $T_{\text{f0}}$ is larger than $T_{\text{c}}$, which is given by $J/(k_{\text{B}}T_{\text{c}}) = \text{arctanh}(1/2)$. If the final temperature $T_{\text{f}}$ is larger than $T_{\text{f0}}$, then the average radius increases with time and reaches a finite value.

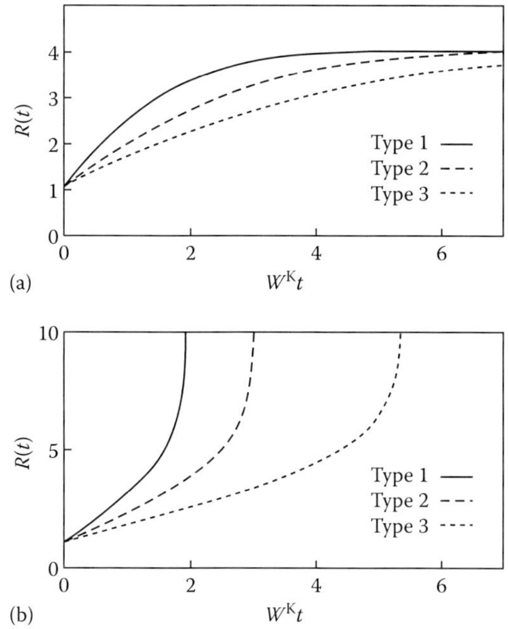


Figure 4.54

(a) Time dependence of the average cluster radius $R(t)$ for $T_f > T_{f0} > T_c$ ($T_f = 1.1T_{f0}$) in the absence of GD for the three types of the Kawasaki dynamics. (1) $W_2^K = W_4^K = W^K$; (2) $W_2^K = W^K$, $W_4^K = 0$; (3) $W_2^K = 0$, $W_4^K = W^K$. (b) The same for $T_{f0} > T_f > T_c$ ($T_f = 0.9T_{f0}$).

As an example, in Figure 4.54a, the time dependence of $R(t)$ is plotted for all types of Kawasaki mechanisms and for $T_f = 1.1T_{f0}$. As for $T_f < T_c$, the main contribution to the growth of the average radius is provided by the processes involved in the second type of Kawasaki dynamics.

For $T_c < T_f < T_{f0}$, the average cluster radius grows unrestrictedly near $t = t_d$ (Figure 4.54b), which is larger than the analogous time for $T_f < T_c$ (compare $t_d \simeq 2.08\ W^K$ for the present case with $t_d \simeq 0.95W^K$ for $T_f < T_c$ for the first type of KD). The lower the temperature, the larger are the rates of $\Delta N_{cl} = -1$ and $\Delta N_{cl} = -2$ spin-exchange transitions, which lead to the coagulation and growth of the clusters, and therefore the shorter is the time $t_d$ for the formation of infinite clusters.

### 4.7.5 Conclusion

In the present paper, the kinetics of phase transformation on a Bethe lattice is considered in the presence of both single-spin-flip Glauber and spin-exchange Kawasaki relaxation mechanisms, as well as for the Kawasaki mechanism acting alone. The physical meaning of different spin-exchange processes of the Kawasaki dynamics was clarified and their contributions to the entire relaxation process estimated.

For the general Glauber dynamics, when all processes (splitting and coagulation, growth and decay of clusters, and creation and annihilation of single spin clusters) are in effect, the Kawasaki dynamics does not change the qualitative features of the phase transformation presented in Ref. [23], but provides an acceleration of the transformation process. If the splitting (coagulation) and creation (annihilation) processes are forbidden in the Glauber dynamics, then the addition of the Kawasaki dynamics considerably increases the fraction of the transformed phase (the average polarization tends to unity for large times) compared with the case of pure Glauber dynamics, for which the average polarization grows very little.

The Kawasaki mechanism acting alone provides the growth of clusters of "up" spins after a quench from an initial temperature $T_i > T_c$ to a lower temperature $T_f$. If $T_f$ is larger than $T_{f0} = J/[k_B \operatorname{arctanh}(1/3)] > T_c$, the average radius of the clusters grows with time and reaches a finite value (a steady state) as $t \to \infty$. For $T_f < T_{f0}$, this radius grows linearly in the initial and intermediate stages and diverges as $\log_2(t_d - t)^{-1}$ when $t$ approaches the time $t_d$ at which infinite clusters arise. The specific value of the time $t_d$ could be determined from the numerical analysis of the kinetic equations. It was shown using as an example a spherical cluster that the linear dependence of the cluster radius on time is a consequence of the exponential dependence of the number of spins in the cluster and of the number of boundary and neighboring spins on the radius. This constitutes a specific feature of a Bethe lattice which has infinite dimensions [22]. For this reason the linear law cannot be compared with the "classical" Lifshits–Slyozov–Wagner $t^{1/3}$ law for the average cluster radius which is typical for systems with restricted dimensionality.

One should note that, in all considered cases (mixed Glauber–Kawasaki dynamics, and KD acting alone), the main contribution provided by the spin-exchange processes involved in KD are those that decrease or increase the number of clusters by one.

## Acknowledgment

This work was supported by the National Science Foundation.

## Appendix 4I: Operators $R_{a,b}(v)$ and Their Properties

The operator $R_{a,b}(v)$ is associated with those combinations of various values of the four neighboring spins to the $\sigma_a$, $\sigma_b$ spins which provide transition energies $\Delta E = 2vJ$ ($v = 0, \pm 2, \pm 4$) when $\sigma_a$ and $\sigma_b$ exchange their values from $\sigma_a = -\sigma_b = 1$ to $\sigma_a = -\sigma_b = -1$. There are 16 such possible combinations and each of them can be associated with the operators

$$r(\mu_{a_1}, \mu_{a_2}, \mu_{b_1}, \mu_{b_2}) = \frac{1}{16}(1+\mu_{a_1}\sigma_{a_1})(1+\mu_{a_2}\sigma_{a_2})(1+\mu_{b_1}\sigma_{b_1})(1+\mu_{b_2}\sigma_{b_2}), \tag{4I.1}$$

where all $\mu$ can take the values +1 or −1 and the sites $a_1$, $a_2$, $b_1$, and $b_2$ are those defined in Figure 4.48b. The operator $r(\mu_{a_1}, \mu_{a_2}, \mu_{b_1}, \mu_{b_2})$ is equal to unity when $\sigma_k = \mu_k$ ($k = a_1, a_2, b_1, b_2$) and to zero in all the other possible cases. Considering the spin-exchange transition from the state $\sigma_a = -\sigma_b = 1$ to the state $\sigma_a = -\sigma_b = -1$ for all possible combinations of the neighboring spins $\sigma_{a_1}, \sigma_{a_2}, \sigma_{b_1}, \sigma_{b_2}$, and collecting together those compatible with $v = 0, \pm 2, \pm 4$, one obtains for $R_{a,b}(v)$ the following expressions:

$$\begin{aligned}
R_{a,b}(0) &= r(1,1,1,1) + r(1,-1,1,-1) + r(-1,1,1,-1) \\
&\quad + r(1,-1,-1,1) + r(-1,1,-1,1) + r(-1,-1,-1,-1), \\
R_{a,b}(2) &= r(1,1,1,-1) + r(1,1,-1,1) + r(1,-1,-1,-1) + r(-1,1,-1,-1), \\
R_{a,b}(-2) &= r(1,-1,1,1) + r(-1,1,1,1) + r(-1,-1,1,-1) + r(-1,-1,-1,1), \\
R_{a,b}(4) &= r(1,1,-1,1), \quad R_{a,b}(-4) = r(-1,-1,1,1).
\end{aligned} \tag{4I.2}$$

These operators satisfy the equality

$$\langle A(\sigma_a - \sigma_b)R_{a,b}(v)\rangle_T = z_v^0 \langle A(1-\sigma_a\sigma_b)R_{a,b}(v)\rangle_T \; (v = 0, \pm 2, \pm 4), \tag{4I.3}$$

where $z_v^0 = \tanh \beta v J$, and $A$ is an operator which does not involve the spins $\sigma_a$ and $\sigma_b$. Let us proof this equality for the operator $R_{a,b}(2)$. For all other operators $R_{a,b}(v)$ the proof is analogous. First we consider the left-hand side of Equation 4I.3 for $v = 2$ and replace the operator $R_{a,b}(2)$ with the more simple operator $r(1,1,1,-1)$, which is a part of $R_{a,b}(2)$. After calculations, one obtains

$$\begin{aligned}
&\langle A(\sigma_a - \sigma_b) r(1,1,1,-1)\rangle_T \\
&\equiv \frac{1}{Z}\mathrm{Tr}\{A(\sigma_a - \sigma_b) r(1,1,1,-1) e^{-\beta H_s'}\} \\
&= \frac{2}{Z}\mathrm{Tr}'\{A r(1,1,1,-1) e^{-\beta H_s'}\} \\
&\quad \times \left[e^{\beta J(\sigma_{a_1}+\sigma_{a_2}-\sigma_{b_1}-\sigma_{b_2}-1)} - e^{\beta J(-\sigma_{a_1}-\sigma_{a_2}-\sigma_{b_1}+\sigma_{b_2}-1}\right] \\
&= \frac{2}{Z} e^{\beta J}(1 - e^{-4\beta J})\mathrm{Tr}'\left[A r(1,1,1,-1) e^{-\beta H_s'}\right],
\end{aligned} \tag{4I.4}$$

where $Z = \mathrm{Tr}[e^{-\beta H_s}]$ is the partition function of the system, Tr′ denotes the trace over all spins with the exclusion of $\sigma_a$ and $\sigma_b$, and $H_s'$ is the part of the Hamiltonian $H_s$ which does not contain $\sigma_a$ and $\sigma_b$. The last right-hand side of expression 4I.4 is obtained by taking into account the definition of the operator $r(1, 1, 1, -1)$, which differs from zero only for $\sigma_{a_1} = 1$, $\sigma_{a2} = 1$, $\sigma_{b_1} = 1$, $\sigma_{b_2} = -1$. In a similar way, one can calculate the right-hand side of 4I.3. Finally, one obtains

$$z_2^0 \langle A(1-\sigma_a\sigma_b)r(1,1,1,-1)\rangle T$$
$$= \frac{2}{Z} e^{\beta J}(1-e^{-4\beta J})\mathrm{Tr}'\left[Ar(1,1,1,-1)e^{-\beta H_s'}\right]. \quad (4I.5)$$

Comparing the right-hand sides of Equations 4I.4 and 4I.5, one can arrive at Equation 4I.3 in which $R_{a,b}(v)$ is replaced by $r(1,1,1,-1)$. The same equality can be proven for all other operators $r(\mu_{a_1}, \mu_{a_2}, \mu_{b_1}, \mu_{b_2})$ involved in the expression of $R_{a,b}(2)$ and due to the linearity of the latter operator with respect to the operators $r(\mu_{a_1}, \mu_{a_2}, \mu_{b_1}, \mu_{b_2})$ one arrives at the equality 4I.3 for $R_{a,b}(2)$.

## Appendix 4J: Some Details of the Closure Procedure of Kinetic Equations

To transform Equation 4.186, which contains relaxation terms given by Equations 4.192 and 4.193 to a closed form, expressions for all averages (Equation 4.194) as functions of $\beta_\sigma(t)$ and $\beta_\epsilon(t)$ involved in PDF (Equation 4.182) are necessary. To obtain such expressions, the equilibrium values of the averages (Equation 4.194), and the substitutions (Equation 4.185) have to be employed. The equilibrium values $\sigma_T$, $\epsilon_T$, $\epsilon_{2T}$, $\gamma_{1T}$, $\delta_{1T}$ of the averages $\sigma$, $\epsilon$, $\epsilon_2$, $\gamma_1$, $\delta_1$ were obtained in Appendix A of Ref. [23], using the exact expressions of Refs. [26,27] and relation 4.188. The equilibrium value $\epsilon_{3T}$ of the average $\epsilon_3$ can be obtained in a similar way as

$$\epsilon_{3T} = \sigma_T^2 + \left(1-\sigma_T^2\right)\lambda_T^3, \quad (4J.1)$$

where

$$\lambda_T = z_T \frac{1-y}{1-z_T^2 y}, \; z_T = \tanh\beta J, \quad (4J.2)$$

and $y$ is the root of the equation (with $z_{0T} = \tanh \beta B$)

$$z_T^2 y^3 + z_{0T} z_T (2-z_T) y^2 + (1-2z_T) y - z_{0T} = 0. \quad (4J.3)$$

To calculate the four- and six-spin averages $\delta_{2T}$, $\delta_{3T}$, $\kappa_T$, we have used Equation 4.189, which contains the additional unknown averages $\langle\sigma_{a_1}\sigma_{a_2}\sigma_{b_1}\rangle_T$, $\langle\sigma_{a_1}\sigma_a\sigma_{b_1}\rangle_T$, $\langle\sigma_{a_1}\sigma_{a_2}\sigma_b\sigma_{b_1}\sigma_{b_2}\rangle_T$, $\langle\sigma_{a_1}\sigma_{a_2}\sigma_a\sigma_b\sigma_{b_2}\rangle_T$. The total number of equations (four) is smaller than the number of unknowns (seven). To find the averages $\delta_{2T}$, $\delta_{3T}$ and $\kappa_T$, the first three of the new averages are approximated as follows:

$$\langle\sigma_{a_1}\sigma_{a_2}\sigma_{b_1}\rangle_T = \langle\sigma_{a_1}\sigma_{a_2}\rangle_T\langle\sigma_{b_1}\rangle_T \equiv \sigma_T\epsilon_{2T},$$

$$\langle\sigma_{a_1}\sigma_a\sigma_{b_1}\rangle_T = \langle\sigma_{a_1}\sigma_a\rangle_T\langle\sigma_{b_1}\rangle_T \equiv \sigma_T\epsilon_T,$$

$$\langle\sigma_{a_1}\sigma_{a_2}\sigma_b\sigma_{b_1}\sigma_{b_2}\rangle_T = \langle\sigma_{a_1}\sigma_{a_2}\sigma_b\rangle_T\langle\sigma_{b_1}\sigma_{b_2}\rangle_T \equiv \gamma_{1T}\epsilon_{2T}.$$

Then, using the substitutions (Equation 4.185), all the obtained averages can be represented as functions of $\beta_\sigma(t)$ and $\beta_\epsilon(t)$. A further transformation to equations of the type of Equation 4.195 was described in Ref. [23], Appendices 4I and 4J.

## References


1. Glauber, R., *J. Math. Phys.* 4, 294 (1963).
2. Kawasaki, K., *Phys. Rev.* 145, 224 (1966).
3. Fratzl, P., Penrose, O., *Phys. Rev. B* 50, 3477 (1994).
4. Schmelzer, J., Jr., Landau, D.P., *Int. J. Mod. Phys. C* 12, 345 (2001).
5. Marko, J.F., Barkema, G.T., *Phys. Rev. E* 52, 2522 (1995).
6. Szolnoki, A., Szabó, G., Mouritsen, O.G., *Phys. Rev. E* 55, 2255 (1997).
7. Tafa, K., Puri, S., Kumar, D., *Phys. Rev. E* 64, 056139 (2001).
8. Puri, S., *Phys. Rev. E* 55, 1752 (1997).
9. Grandi, B.C.S., Figueiredo, W., *Phys. Rev. B* 50, 12595 (1994).
10. Grandi, B.C.S., Figueiredo, W., *Phys. Rev. E* 54, 4722 (1996).
11. Grandi, B.C.S., Figueiredo, W., *Phys. Rev. E* 56, 5240 (1997).
12. Grandi, B.C.S., Figueiredo, W., *Phys. Rev. E* 56, 4922 (1999).
13. Grandi, B.C.S., Figueiredo, W., *Physica A* 234, 764 (1997).
14. Ma, Y., Liu, J., *Phys. Lett. A* 238, 159 (1998).
15. Leao, J.R.S., Grandi, B.C.S., Figueiredo, W., *Phys. Rev. E* 60, 5367 (1999).
16. Szolnoki, A., *Phys. Rev. E* 62, 7466 (2000).
17. Godoy, M., Figueiredo, W., *Phys. Rev. E* 66, 036131 (2002).
18. Zhu, H., Zhu, J., Zhou, Y., *Phys. Rev. E* 66, 036106 (2002).
19. Dattagupta, S., *Physica A* 194, 137 (1993).
20. Berim, G.O., Ruckenstein, E., *J. Chem. Phys.* 120, 2851 (2004). (Section 4.6 of this volume.)
21. Berim, G.O., Ruckenstein, E., *J. Chem. Phys.* 119, 9640 (2003). (Section 4.4 of this volume.)
22. Baxter, R.J., *Exactly Solvable Models in Statistical Mechanics*, Academic, New York, 1982.
23. Berim, G.O., Ruckenstein, E., *J. Chem. Phys.* 120, 272 (2004). (Section 4.5 of this volume.)
24. Wonczak, S., Strey, R., Stauffer, D., *J. Chem. Phys.* 113, 1976 (2000).
25. Katsura, S., Takizawa, M., *Prog. Theor. Phys.* 51, 82 (1974).
26. Hu, C.-K., Izmailian, N.Sh., *Phys. Rev. E* 58, 1644 (1998).
27. Mukamel, D., *Phys. Lett. A* 50, 339 (1974).

# Index

Page numbers followed by f and t indicate figures and tables, respectively.


## A

Absorption rates, determination of, 238, 258–259, 283–285
  of residues by cluster, 328–329
Aerosol growth, alternate mechanisms of, 9–10
Amino acids, 232, 277
Amorphous spherical cluster, 13
Average cluster size, 416–418, 437–438, 447
Average dihedral angle potential, 311–316
Avogadro number, 95

## B

Backward Smoluchowski equation, 32, 247, 248
Barrierless deliquescence, 64–65; *see also* Deliquescence, kinetics of
Barrierless denaturation; *see also* Kinetic nucleation theory to protein folding, application of
  Taylor series expansions for emission/absorption rates, 335
  threshold temperatures of cold/hot denaturation/unfolding times, 334
  time of barrierless protein denaturation, 333–334
Barrierless thermal denaturation of native protein, temperature on, *see* Temperature on nucleation mechanism of protein folding
Bead–cluster interactions, 281
Becker–Döring approach, 368, 394
Binary classical nucleation theory (BCNT), *see* Binary nucleation, first passage time analysis-based
Binary clusters, equilibrium distribution of, 181–182
Binary nucleation, first passage time analysis-based
  determination of function $\omega_i(v_1,v_2)$ by DFT methods, 174
  equilibrium distribution of binary clusters, 181–182
  first passage time analysis, 170–174
  kinetic equation, 168–170
  kinetics of, 166–174
  model binary system with surface active component, 174–177
  numerical evaluations, 177–179
  overview, 165–166
  thermodynamics of binary nucleation, on capillarity approximation, 167–168
Boltzmann constant, 117, 137, 147, 253
Bond angle forces, 279
Born repulsion, 4
Boundary hydrogen bonds (BHB), 347, 350, 351
Bovine pancreatic ribonuclease (BPR), 262, 286, 335, 352
Brownian coagulation of small particles, 3
Brownian motion, 118, 128, 188
  of surface monomer, 15, 17

## C

Capillarity approximation (CA), 165, 167, 194
Capillarity approximation-based binary nucleation, 167–168
Cartesian system of coordinates, 245, 268
Classical nucleation theory (CNT)
  characteristics, 127, 132, 150, 156, 198
  for nucleation mechanism of protein folding, 232
  prediction with, 24–28; *see also* Nucleation rate in liquids, prediction of
  shortcoming of, 208
  using capillarity approximation, 87
Classical theory of heterogeneous nucleation, 186
Classical theory of nucleation, 190
Cluster
  binary, equilibrium distribution of, 181–182
  cubic, 385–386
  decay of, 132
  dynamics, 429–432, 432–434
  evaporation rate from
    heterogeneous nucleation, 186–188
    nucleation in gases, 116–118
  growth of, 119, 188–189
  macroscopic thermodynamics for, 132
  of native residues, potential well around, 234–238

number, calculation of, 437–438
one-dimensional flux density of, 184
population of, equations for, 146–149
potential field for protein residue around, 279–281
for selected bead, potential field around
average dihedral angle potential, 311–316
confining potential, 311
effective pairwise potential, 310–311
size, 19, 158
size distribution function of, 146
spherical cap, 192–193
Cluster–hydrophilic residue, 353
Cluster(s)
with simple shape in Ising model; *see also* Kinetics of cluster growth on lattice
rectangular cluster, 374–375
rough boundary cluster, 376–377
triangular isosceles right angle cluster, 375–376
zigzag cluster, 376
with simple topology; *see also* Kinetics of cluster growth on lattice
about, 385–386
cubic cluster, 386–388
pyramidal clusters, 388–389
stairlike cluster, 388
Cluster-vapor system, 160
Composite droplets, equilibrium distribution of, 70–71
Computer simulations, 251
Condensation rate, 34, 81; *see also* Solute-solute/solute-solvent interactions on nucleation in liquids
Condensation rate *vs.* supersaturation ratio, 39f
Confining potential, 311
Contact angle on nucleation rate, 214
Coulomb potential, 283
Critical cluster, 168, 189
of native residues, 241
radius of, 40
Critical cluster radius, asymptotic expansion of equation for, 40–41
Critical cluster size, 12, 141, 372
Critical nucleus, 35
Critical particle size, 4
Crystallization in liquid, 88
Crystal nucleation, theory of, 98
Crystal nuclei
free energy of heterogeneous formation of, *see* Free energy of heterogeneous formation of crystal nuclei
free energy of homogeneous formation of
in surface-stimulated/volume-based modes, 105–109
Cubic cluster, 385–386
Curve of zero flux of condensate molecules, 67
Cylindrical drop size, defined, 220

## D

Debye–Huckel approximation, 283
Deliquescence, kinetics of
barrierless deliquescence, 64–65
fluctuational deliquescence
equilibrium distribution/steady-state nucleation rate, 63–64
first passage time analysis, 60–62
profile of solute concentration within film, 64
Deliquescence point, 57
Deliquescence theory, background of; *see also* Liquid-to-solid nucleation, kinetic theory of
kinetic equation of deliquescence, 58–59
overview, 56–57
thermodynamics of deliquescence, 57–58
Denaturation
threshold temperatures of cold/hot, 334
Denaturation (unfolding) of protein, 301
Density functional formalism, 175–177
Density-functional theory (DFT)
in BCNT, 166
determination of function $\omega_i(v_1,v_2)$ by, 174
and fluid density distribution (FDD) in nuclei, 218
improvement of RNN approach by, 160–161
in nucleation theory, 159–160
Density uniformity assumption, 161
Diffusion coefficient, 118, 130
Dihedral angle forces, 235
Dihedral angle potential, 233, 305, 311–316
Dimers, 177–178, 178f
Dirac delta function, 177
Dissociation
of doublet, 4–8
kinetics of, 32–34
rate calculation, 79–81; *see also* Solute-solute/solute-solvent interactions on nucleation in liquids
Doublet consisting of equal-size aerosol particles
aerosol growth, alternate mechanisms of, 9–10
dissociation rate of doublet, 5–8
formation rate of doublet, 4–5
Droplets, heterogeneously formed, 206
Droplet surface, deformation, 87

## E

Eigenvalue equation, 150, 152t
Emission of residues by cluster, rates of, 328–329
Emission rates, determination of, 238, 258–259, 283–285
Energy, dimensionless, 5
Energy space, diffusion coefficient in, 211
Enthalpy, 96
Ethanol, 124t
Euler-Lagrange equation, 159, 175, 219, 224
Evaporation rate from cluster
heterogeneous nucleation, 186–188
theory of nucleation in gases, 116–118
Excitation energy, 374, 378, 391

## F

Face-centered cubic (fcc) crystalline clusters, 30
Face-centered cubic (fcc) packing, 13
FDD, *see* Fluid density distribution
Field reversal, kinetics after, 442–443
First hydration shell (1HS), 348, 349f
First passage time analysis; *see also* Vapor-to-liquid nucleation, kinetic theory of
binary nucleation based on
determination of function $\omega_i(v_1,v_2)$ by DFT methods, 174

first passage time analysis, 170–174
kinetic equation, 168–170
model binary system with surface active component, 175–177
overview, 165–166
thermodynamics of binary nucleation based on capillarity approximation, 167–168
emission rate via, 204–206
nucleation based on
DFT methods in nucleation theory, 159–160
improvement of RNN approach by DFT methods, 160–161
numerical results, 161–163
RNN approach to kinetics of nucleation, 156–159
for nucleation mechanism of protein folding, 295–296
nucleation mechanism of protein folding based on, 254–255
First passage time analysis, 60–62; *see also* Fluctuational deliquescence
First passage time analysis of protein folding via nucleation
about, 324–325
barrierless denaturation
Taylor series expansions for emission/absorption rates, 335
threshold temperatures of cold/hot denaturation/unfolding times, 334
time of barrierless protein denaturation, 333–334
MFPTA in protein folding/unfolding problems
cluster of native residues, potential well around, 326–328
heteropolymer as protein model, 326
rates of emission/absorption of residues by cluster, 328–329
protein folding
protein folding time in ternary nucleation formalism, evaluation, 329–330
temperature dependence of protein folding time, 330–332
Five-spin cluster
excitation energy, 378
initial rate, 378–381
Fixed bond length/bond angle, 237
Fluctuational deliquescence; *see also* Deliquescence, kinetics of
equilibrium distribution/steady-state nucleation rate, 63–64
first passage time analysis, 60–62
profile of solute concentration within film, 64
Fluid density distribution (FDD)
about, 218
Euler-Lagrange equation for, 224
Fluid–solid interaction, 222
Fokker-Planck equation (FPE)
about, 116, 127
three-dimensional, for evaporation from clusters
kinetics of growth (evaporation) of a single cluster, 140–142
nucleation rate, 142–143
numerical results/discussion, 143–145
overview, 136
probability density distribution function of energy, equation for, 136–140
unidimensional, in treatment of nucleation in gases
evaporation rate of molecules from cluster, 131
growth rate of cluster, 131–132
kinetic equation in energy space, 127–131
nucleation rate, 132–133
overview, 127
results/discussion, 133–135
Fokker–Planck operator, 149
Folded protein, 253
Foreign particle
completely immersed in liquid, 88–91
in contact with liquid-vapor interface, 91–94
Foreign particle–crystal–vapor, three-phase contact, 96
Free energy
droplet formation, 58
of formation of droplet, 200
of heterogeneous formation of crystal nuclei; *see also* Liquid-to-solid nucleation, kinetic theory of
foreign particle completely immersed in liquid, 88–91
foreign particle in contact with liquid-vapor interface, 91–94
immersion and contact modes, comparison of, 94–95
overview, 87–88
of homogeneous formation of crystal nuclei
in surface-stimulated/volume-based modes, 105–109
of nucleus formation, 167, 266
Friction coefficient, 118, 137

G

Gaussian distribution, 349
Gibbs adsorption equation, 168
Gibbs–Curie theorem, 92, 98
Gibbs–Duhem relation, 168
Gibbs energy of hydration, 294
Gibbs free energy, 90
Gibbsian model, 165
Gibbsian thermodynamics, 156
Glauber dynamics (GD), 439–446, 449, 456
Glauber master equation, 368
Glauber single spin-flip mechanism, 395, 439
Goodrich approach, 133
Goodrich expansions, 51, 52–53
Green's function, 32
Growth-decay regime, kinetics in
cluster dynamics, 432–434
phase transformation, 432

H

Hamaker constant, 6, 7f, 8f
Hamiltonian equation, 131, 137
Hard-sphere system, 177
Helmholtz free energy, 88, 159, 175, 176, 219
Heterogeneous condensation, 194
Heterogeneous crystal nucleation, thermodynamics of; *see also* Liquid-to-solid nucleation, kinetic theory of
about, 85–87
correction to, 103

crystal nucleus, 98–102
free energy of heterogeneous formation of crystal nuclei
foreign particle completely immersed in liquid, 88–91
foreign particle in contact with liquid-vapor interface, 91–94
immersion and contact modes, comparison of, 94–95
overview, 87–88
numerical evaluations and discussions, 95–97
Heterogeneous droplet formation, 196
Heterogeneous nucleation; *see also* Vapor-to-liquid nucleation, kinetic theory of
calculations and discussion, 191–192
contributions to free energy of system/Euler-Lagrange equation for FDD, 224
evaporation rate from cluster, 186–188
growth of cluster, 188–189
interaction potential between molecule/spherical cap cluster, 192–193
kinetic equation of, 200–203
KNT basic equations of, 224–228
nucleation rate, 189–190
overview, 186
results and discussion, 221–224
on rough surface
basic equations of kinetic theory, 210–213
overview, 208
results, 214
system and interaction potentials, 209–210
Heterogeneous nucleation theory on soluble particles; *see also* Liquid-to-solid nucleation, kinetic theory of
about, 55–56
deliquescence theory, background of
kinetic equation of deliquescence, 58–59
overview, 56–57
thermodynamics of deliquescence, 57–58
kinetics of deliquescence, 59–65
numerical evaluations, 65–69
Heterogeneous unary nucleation on liquid aerosols
emission rate via first passage time analysis, 204–206
equilibrium distribution of heterogeneously formed droplets, 206
heterogeneous nucleation, kinetic equation of, 200–203
heterogeneous nucleation, kinetics of, 198–200
kinetics of nucleation, 196–198
numerical evaluations, 200–203
overview, 194–196
Heteropolymer as protein model, 232–234, 278–279, 326
Homogeneous crystal nucleation in droplet, thermodynamics/kinetics of
free energy of homogeneous formation of crystal nuclei in surface-stimulated/volume-based modes, 105–109
surface-stimulated mode on crystal nucleation kinetics, effect of, 109–110
Homogeneous nucleation in gases, 136–145; *see also* Fokker-Planck equation
Hydrogen bond networks on NMPF; *see also* Kinetic nucleation theory to protein folding, application of
about, 346–347
NMPF model, modified
contribution to potential field around folded part of protein, 354–356
ternary heteropolymer as protein model, 353–354
numerical evaluations, 356–359
water-water hydrogen bonding on interaction of solute particles
networks of water-water hydrogen bonds around solutes, 350–352
overview, 347–350
protein folding via nucleation, case of, 352–353
Hydrophilic beads, 232, 251, 282
Hydrophilic residues, ionizability of, 336–338
Hydrophobic beads, 232, 251

## I

Ice–(liquid) water–(water) vapor contact, line tension of, 111
Icosahedral layers, 45
IMBL, *see* Ising model on Bethe lattice
Immersion and contact modes, comparison of, 94–95
Inner potential well (ipw), 238, 242, 283, 356
Interfacial tension, 27, 59
of macroscopic clusters, 22–23
Intermolecular potential, 13
Inverted micelle, 196
Ionizability of hydrophilic residues, 336–338
Ionized protein residues on nucleation pathway of protein folding
emission and absorption rates, determination of, 283–285
ionizability of protein residues, 281–283
numerical evaluations, 286–289
protein folding time evaluation, 285
ternary nucleation model, based on first passage time analysis, 278–281
Ising model (two-dimensional)
cluster shape and, *see under* Kinetics of cluster growth on lattice
shape-dependent small cluster kinetics in
about, 393–394
model 1, 394–400
model 2, 400–403
Ising model (three-dimensional), shape on critical nucleus size, *see under* Kinetics of cluster growth on lattice
Ising model on Bethe lattice (IMBL)
equilibrium averages for, 435–436
kinetic equations for, 423–426
Ising spin system, 368

## J

Jacobian of transformation, 183

## K

Kawasaki dynamics (KD), 439, 449, 454
Kawasaki spin-exchange mechanisms, 439

Kelvin equation, 119, 120, 135, 189
Kinetic equation of deliquescence, 58–59
Kinetic equation of ternary nucleation, 274–275
Kinetic nucleation theory (KNT)
about, 218
basic equations for, 224–228
objectives of, 1
Kinetic nucleation theory to protein folding, application of
first passage time analysis of, via nucleation/of barrierless protein denaturation
about, 324–325
barrierless denaturation, 333–335
MFPTA in protein folding/unfolding problems, 326–329
numerical evaluations, 335–342
protein folding, 329–332
hydrogen bond networks on NMPF
about, 346–347
NMPF model, modified, 353–356
numerical evaluations, 356–359
water-water hydrogen bonding on interaction of solute particles, 347–353
ionized protein residues on nucleation pathway of protein folding
about, 277–278
emission and absorption rates, determination of, 283–285
ionizability of protein residues, 281–283
numerical evaluations, 286–289
protein folding time evaluation, 285
ternary nucleation model, based on first passage time analysis, 278–281
overview, 231
protein folding
about, 231–232
CNT-based model for nucleation mechanism of, 232–234
heteropolymer as protein model, 232–234
mean first passage time analysis for emission/absorption rates calculation, 247–249
nucleation mechanism of, 234–241
numerical evaluations, 241–243
temperature effects on
about, 292–294
numerical evaluations, 304–309
temperature expansions for protein folding time, 295–301
time of protein denaturation via spinodal decomposition, 301–304
ternary nucleation model for nucleation pathway of protein folding
about, 251–252
CNT-based model for nucleation mechanism of protein folding, 253
equilibrium distribution of ternary clusters, 272–274
heteropolymer as protein model, 252–253
kinetic equation of ternary nucleation, 274–275
mean first passage time analysis for emission/absorption rate, 271–272
nucleation mechanism of protein folding based on first passage time analysis, 254–255
numerical evaluations, 262–267
ternary nucleation mechanism of protein folding, 255–262
Kinetics for temperature range
cluster dynamics, 429–432
phase transformation, 428–429
Kinetics of cluster growth on lattice
cluster shape upon growth in 2D Ising model
and basic kinetic equations, 370–374
clusters with simple shape in Ising model with nearest neighbor interactions, 374–377
discussion, 381–382
small clusters of arbitrary shape in, 377–381
overview, 367–368
phase transformation in lattice system, Gauber's master equation-based
about, 405–406
closing procedure, 407–408
equilibrium averages, relations between, 420
initial/final conditions, 408
kinetic equations, 406–407
one-dimensional lattice, phase transformation on, 408–414
size distribution of solid phase clusters, 414–418
small deviations from equilibrium, 421
phase transformation in lattice system in presence of spin-exchange dynamics
about, 439
average cluster size, 447
background, 440–442
coefficients of kinetic equations, 447
kinetics after field reversal (liquid-solid transformation), 442–443
kinetics after temperature quenching/external field change, 445–446
temperature quench in absence of external field, 443–445
phase transformation on Bethe lattice
calculation of kinetic equations, 436–437
discussion, 434–435
equilibrium averages for IMBL, 435–436
growth-decay regime, kinetics in, 432–434
kinetic equations for IMBL, 423–426
kinetics for temperature range, 428–432
nonsplitting, noncoagulation regime, 434
number of clusters/average cluster size, calculation of, 437–438
phase transformation in absence of supersaturation, 426–428
phase transformation on Bethe lattice in presence of spin exchange
closure procedure of kinetic equations, 461
kinetic equations, 450–454
operators and properties, 460–461
overview, 449–450
phase transformation after field reversal, 454–456
temperature quench in absence of external field, 456–459

shape-dependent small cluster kinetics in 2D Ising model
about, 393–394
model 1, 394–400
model 2, 400–403
shape on critical nucleus size in 3D Ising model
clusters with simple topology, 385–389
formulation of problem, 383–385
small clusters of arbitrary shape, 389–391
Kinetic theory of liquid-to-solid nucleation, *see* Liquid-to-solid nucleation, kinetic theory of
Köhler activation, 195, 197
Kramers equation, 127

## L

Lagrange multiplier, 90
Langevin dynamics, 233, 252, 326
Langevin equation, 128
Laplace equation, 159
Lennard–Jones (LJ) monomers, 165
Lennard–Jones (LJ) potentials, 200
Lennard–Jones (LJ) type potentials, 241, 304
Lifshitz–Slyozov–Wagner law, 439, 445
Limit of large clusters/thermodynamic relations, 35–36
Line tension of crystal-liquid-vapor contact, determination of
about, 104–105
homogeneous crystal nucleation in droplet, thermodynamics/kinetics of
free energy of homogeneous formation of crystal nuclei in surface-stimulated/volume-based modes, 105–109
surface-stimulated mode on crystal nucleation kinetics, effect of, 109–110
Line tension of ice-water-vapor contact, determination of
numerical evaluations, 111–112
overview, 110–111
Line tension of three-phase contact, 85
Liquid aerosols, 194
Liquid–crystal–foreign particle contact line, 96
Liquid–liquid–fluid system, line tension in, 104
Liquid-solid transformation, 442–443
Liquid-to-solid nucleation, kinetic theory of
asymptotic expansion of equation for critical cluster radius, 40–41
coefficients of polynomial, explicit expressions for, 74
equilibrium distribution of composite droplets, 70–71
heterogeneous crystal nucleation, thermodynamics of
about, 85–87
correction to, 103
crystal nucleus, 98–102
free energy of heterogeneous formation of crystal nuclei, 87–95
numerical evaluations and discussions, 95–97
heterogeneous nucleation theory on soluble particles
about, 55–56
deliquescence theory, background of, 56–59
kinetics of deliquescence, 59–65
numerical evaluations, 65–69
line tension of crystal-liquid-vapor contact, determination of
about, 104–105
homogeneous crystal nucleation in droplet, thermodynamics/kinetics of, 105–110
line tension of ice-water-vapor contact, determination of
numerical evaluations, 111–112
overview, 110–111
nucleation in liquids, kinetic theory of
about, 30
condensation rate, 34
critical nucleus, 35
kinetics of dissociation, 32–34
limit of large clusters/thermodynamic relations, 35–36
molecule and cluster, interaction between, 30–31
numerical results/comparison with classical theory, 37–39
rate of nucleation, 36–37
nucleation rate calculation, comparison
discussion, 54
Goodrich expansions, 52–53
rate of nucleation, 53–54
Zeldovich Expansion, 51–52
nucleation rate in liquids, prediction of
about, 12–13
classical nucleation theory, prediction with, 24–28
dissociation rate of surface monomer, 15–19
interfacial tension of macroscopic clusters, 22–23
rate of nucleation, 20–22
solubility, 19–20
structure of molecular clusters, 13–14
surface monomer, interaction potential of, 14–19
nucleation rate in liquids for face-centered cubic/icosahedral clusters
about, 42
mean passage time approach, 42–44
results/discussion, 46–49
structure of clusters, 44–45
surface atom with cluster, interaction of, 45–46
for small aerosol particles growth
about, 3–4
aerosol growth, alternate mechanisms of, 9–10
formation rates/dissociation of doublet, comparison, 4–8
solute-solute/solute-solvent interactions
condensation rate, 81
dissociation rate, 79–81
interaction potentials, 76–77
nucleation rate, 78
numerical estimation of, 81–84
Local-density approximation (LDA), 159
London–van der Waals attraction, 186, 210
Lorentz transformation, 72, 173, 183

## M

Macroscopic clusters, interfacial tension of, 22–23
Macroscopic thermodynamics, 157, 186, 192
Marine aerosols, 196
Maxwellian distribution, 255

Mean first passage time, 205
Mean first passage time analysis (MFPTA)
for emission/absorption rate, 247–249, 271–272, 316–317
in protein folding/unfolding problems
cluster of native residues, potential well around, 326–328
heteropolymer as protein model, 326
rates of emission/absorption of residues by cluster, 328–329
Mean first passage times, 259
Mean passage time approach, 42–44
Mercedes Benz (MB) water model, 349
Methanol, 124t
Metropolis spin dynamics, 394
MFPTA, *see* Mean first passage time analysis
Model binary system with surface active component, 175–177
Molecular clusters, structure of; *see also* Nucleation rate in liquids, prediction of
amorphous spherical cluster, 13
spherical face-centered cubic (fcc) crystalline cluster, 14
Molecular dynamics (MD) simulations, 231, 251
Molecule and cluster, interaction between, 30–31
Monomer–monomer interaction energy, 24f, 26
Monomers, 177–178, 178f
Monomer–solvent interaction, 15
Monte Carlo (MC) simulations, 231, 251, 370, 393
Muller–Lee–Graziano (MLG) model, 349

## N

Neutral beads, 232, 234–235
NMPF, *see* Nucleation mechanism of protein folding
Nonadjacent beads, 252
Non-Arrhenius temperature dependence, 307
Nonsplitting, noncoagulation regime, kinetics in, 434
Nonsplitting assumption, 393
Nonuniform density in nuclei, 218–228
Nucleating centers, 194
Nucleation, kinetics of, 196–198
Nucleation in gases, theory of; *see also* Vapor-to-liquid nucleation, kinetic theory of
evaporation rate from cluster, 116–118
growth rate of cluster, 119
nucleation rate, 119–120
overview, 116
results/comparison with classical theory/experiment, 121–125
Nucleation in liquids; *see also* Liquid-to-solid nucleation, kinetic theory of
about, 30
condensation rate, 34
critical nucleus, 35
kinetics of dissociation, 32–34
limit of large clusters/thermodynamic relations, 35–36
molecule and cluster, interaction between, 30–31
numerical results/comparison with classical theory, 37–39
rate of nucleation, 36–37
Nucleation mechanism of protein folding (NMPF)
based on first passage time analysis, 254–255
CNT-based model for, 253
emission and absorption rates, determination of, 238
hydrogen bond networks on, *see* Hydrogen bond networks on NMPF
nucleation rate, equilibrium distribution/steady-state, 238–240
potential well around cluster of native residues, 234–238
protein folding time evaluation, 241
Nucleation mechanism of protein folding (NMPF) model, modified
contribution to potential field around folded part of protein, 354–356
ternary heteropolymer as protein model, 353–354
Nucleation of liquid phase in nanocavities, 208–217; *see also* Heterogeneous nucleation on rough surface
Nucleation pathway of protein folding
ionized protein residues on, *see* Ionized protein residues on nucleation pathway of protein folding
ternary nucleation mechanism of, *see* Ternary nucleation mechanism of nucleation pathway of protein folding
Nucleation rate
calculation, comparison; *see also* Liquid-to-solid nucleation, kinetic theory of
discussion, 54
Goodrich expansions, 52–53
rate of nucleation, 53–54
Zeldovich Expansion, 51–52
heterogeneous nucleation, 189–190
in solute–solute and solute–solvent interactions, 78
steady-state, 63–64, 238–240, 259–261
structure of molecular clusters, 20–22
theory of nucleation in gases, 119–120
*vs.* supersaturation, 134f
Nucleation rate in liquids, prediction of
about, 12–13
classical nucleation theory, prediction with, 24–28
dissociation rate of surface monomer, 15–19
interfacial tension of macroscopic clusters, 22–23
molecular clusters, structure of
amorphous spherical cluster, 13
spherical face-centered cubic (fcc) crystalline cluster, 14
rate of nucleation, 20–22
solubility, 19–20
surface monomer, interaction potential of, 14–19
Nucleation rate in liquids for face-centered cubic/icosahedral clusters; *see also* Liquid-to-solid nucleation, kinetic theory of
about, 42
mean passage time approach, 42–44
results/discussion, 46–49
structure of clusters, 44–45
surface atom with cluster, interaction of, 45–46

## O

One-dimensional lattice, phase transformation on, *see* Phase transformation on one-dimensional lattice
Optimal temperature of protein folding, 331–332
Optimized potentials for liquid simulations (OPLS), 336
Outer potential well (opw), 238, 242, 328, 356

## P

Pairwise potential, 310–311
Peptide bonds, 252, 255
Phase separation, 443–445
Phase transformation
  in absence of supersaturation, 426–428
    about, 426–427
    small deviations from equilibrium, 427–428
  on Bethe lattice
    calculation of kinetic equations, 436–437
    discussion, 434–435
    equilibrium averages for IMBL, 435–436
    growth-decay regime, kinetics in, 432–434
    kinetic equations for IMBL, 423–426
    kinetics for temperature range, 428–432
    nonsplitting, noncoagulation regime, 434
    number of clusters/average cluster size, calculation of, 437–438
    phase transformation in absence of supersaturation, 426–428
  on Bethe lattice in presence of spin exchange
    closure procedure of kinetic equations, 461
    kinetic equations, 450–454
    operators and properties, 460–461
    overview, 449–450
    phase transformation after field reversal, 454–456
    temperature quench in absence of external field, 456–459
  growth-decay regime, kinetics in, 432
  kinetics for temperature range, 428–429
  in lattice system, Gauber's master equation-based, *see under* Kinetics of cluster growth on lattice
  in lattice system in presence of spin-exchange dynamics
    about, 439
    average cluster size, 447
    background, 440–442
    coefficients of kinetic equations, 447
    kinetics after field reversal (liquid-solid transformation), 442–443
    kinetics after temperature quenching/external field change, 445–446
    temperature quench in absence of external field, 443–445
  on one-dimensional lattice
    about, 408–410
    growth decay regime, exact solution for, 410–412
    reversal of external field, 413–414
    small deviation from equilibrium, 412–413
Polarization, 432
Polynomial, coefficients of, 74
Polypeptide chain, 232
Population of clusters, equations for, 146–149
Primary line tension effects, 95
Probabilistic hydrogen bond (PHB), 347
Probability density function, 127, 129, 137
Probability distribution function (PDF), 440, 450
Probability distribution of energy, 117
Propanol, 124t
Protein denaturation
  defined, 294
  via spinodal decomposition, 301–304
Protein folding
  about, 231–232
  CNT-based model for NMPF, 232–234
  heteropolymer as protein model, 232–234
  nucleation mechanism of
    emission and absorption rates, determination of, 238
    nucleation rate, equilibrium distribution/steady-state, 238–240
    potential well around cluster of native residues, 234–238
    protein folding time evaluation, 241
  optimal temperature of, 298–300
  results for, 306–307
    ionizability of hydrophilic residues, 336–338
    temperature dependence of folding time, 338–340
  ternary nucleation mechanism of, *see* Ternary nucleation mechanism of protein folding
  via nucleation, 354
  via nucleation, case of, 352–353
Protein folding time
  evaluation, 241, 261–262, 285
  Taylor series expansions for, 300–301
  temperature dependence of
    optimal temperature of protein folding, 331–332
    simplified model, 330–331
    Taylor series expansions for, 332
  temperature expansions for, 295–301
  in ternary nucleation formalism, evaluation, 329–330
Protein residues, 305
  ionizability of, 281–283
Proteins, importance of, 231
Protein unfolding
  about, 325
  results for, 308–309
    direct application of method, 341–342
    by Taylor series expansions, 340–341
Pyramidal clusters, 388–389

## Q

Quasi-steady-state values of fluxes, 398

## R

Random-phase approximation (RPA), 159
Rate of phase transformation (RPT), 423, 434
Rectangular cluster, 374–375
RNN approach, *see* Ruckenstein-Nowakowski-Narsimhan (RNN) approach
Rough boundary cluster, 376–377
Ruckenstein–Narsimhan–Nowakowski theory (RNNT), 115

Ruckenstein-Nowakowski-Narsimhan (RNN) approach
improvement of, by DFT methods, 160–161
to kinetics of nucleation, 156–159

## S

Schrodinger-type equation, 152
Second hydration layers (2HS), 350
Shape-dependent critical size (SCS), 372, 374
Small aerosol particles growth; *see also* Liquid-to-solid nucleation, kinetic theory of
about, 3–4
alternate mechanisms of, 9–10
doublet consisting of equal-size aerosol particles
dissociation rate of doublet, 5–8
formation rate of doublet, 4–5
formation rates/dissociation of doublet, comparison, 4–8
Small clusters of arbitrary shape
about, 389
excitation energy and initial rate, 391
number of configurations, 390
Small clusters of arbitrary shape in 2D Ising model
five-spin cluster
excitation energy, 378
initial rate, 378–381
number of configurations, 377–378
Smoluchowski approximation, 16, 127, 170, 198
Solid–liquid–vapor systems, line tension in, 104
Solid phase clusters, size distribution of
average cluster size, 416–418
size distribution, 414–416
Solid substrate–ice–liquid water–water vapor systems, 97
Solubility, 19–20; *see also* Nucleation rate in liquids, prediction of
Soluble nuclei, 194
Soluble particle, 57
Solute concentration within film, profile of, 64
Solutes, water-water hydrogen bonds around, 350–352
Solute-solute/solute-solvent interactions on nucleation in liquids
condensation rate, 81
dissociation rate, 79–81
interaction potentials, 76–77
nucleation rate, 78
numerical estimation of, 81–84
Spherical amorphous cluster, 19
Spherical cap cluster
interaction potential between molecule and, 192–193
Spherical face-centered cubic (fcc) crystalline cluster, 14
Spherical fcc clusters, 44
Spherical fcc crystalline cluster, 19, 20
Spin-flip mechanisms, 439
Spinodal decomposition, 301–304
Stairlike cluster, 388
Steady-state cluster size distribution, 21
Steady-state nucleation rate, 21, 23, 63–64, 261
Subsurface nucleation (SSN) layer, 109–110
Super-Arrhenius temperature dependence, 307
Supersaturation, 24
anout, 119, 120, 123f, 216
defined, 212
Surface active component, model binary system with, 174–177, 175–177
Surface atom with cluster, interaction of, 45–46
Surface molecule, 55
Surface monomer, 13, 13f, 14f
dissociation rate of, 15–19
interaction potential of, 14–19
Surface-stimulated mode
on crystal nucleation kinetics, effect of, 109–110
free energy of homogeneous formation of crystal nuclei in, 105–109
Surface-stimulation term, 96
Surface tension, 159, 166
calculation, 144
Survival probability, 61, 171, 271, 316
Szillard boundary condition, 397, 400

## T

Taylor series expansion
coefficients in, 319–322
for emission/absorption rates, 335; *see also* Barrierless denaturation
for protein folding time, 300–301, 332
Temperature dependence of folding time, 330–332, 338–340
Temperature expansions for protein folding time
first passage time analysis-based model for, 295–296
optimal temperature of protein folding, 298–300
Taylor series expansions, 300–301
$W^+ = W^+(v)$ and $W^- = W^-(v)$, determination of, 296–298
Temperature on nucleation mechanism of protein folding
about, 292–294
numerical evaluations, 304–309
temperature expansions for protein folding time, 295–301
time of protein denaturation via spinodal decomposition, 301–304
Temperature quenching
in absence of external field, 443–445
and external field change, 445–446
Ternary clusters, equilibrium distribution of, 272–274
Ternary heteropolymer as protein model, 353–354
Ternary nucleation, kinetic equation of, 274–275
Ternary nucleation formalism, 277
Ternary nucleation mechanism of nucleation pathway of protein folding; *see also under* Kinetic nucleation theory to protein folding, application of
emission and absorption rates, determination of, 258–259
equilibrium distribution and steady-state nucleation rate, 259–261
potential well around cluster of native residues, 255–258
protein folding time evaluation, 261–262
Ternary nucleation model of protein folding, first passage time analysis-based
heteropolymer as protein model, 278–279
potential field for protein residue around cluster, 279–281

Thermal denaturation, 333
Thermodynamics of deliquescence, 57–58
Threshold temperature of hot denaturation, 304
Threshold temperatures of cold/hot denaturation, 334
Time-dependent cluster distribution in nucleation; *see also* Vapor-to-liquid nucleation, kinetic theory of
  numerical calculations, results of, 150–154
  overview, 146
  population of clusters, equations for, 146–149
  time-dependent equations, solution of, 149–150
Time-dependent equations, solution of, 149–150
Time lag
  calculation of, 154
  defined, 153
Time of barrierless protein denaturation, 333–334
Time of protein denaturation, 301–304
Transition probability, 247, 451
Transition state mechanism, 251
Triangular isosceles right angle cluster, 375–376
Trinkaus' expression, 261

## U

Unary nucleation, 59
Unfolded protein, 241, 251

## V

Van der Waals interactions, 138
Vapor–liquid–ice contact line, 97
Vapor supersaturation, 159
Vapor-to-liquid nucleation, kinetic theory of
  binary nucleation based on first passage time analysis
    determination of function $\omega_i(v_1,v_2)$ by DFT methods, 174
    equilibrium distribution of binary clusters, 181–182
    first passage time analysis, 170–174
    kinetic equation, 168–170
    kinetics of, 166–174
    model binary system with surface active component, 174–177
    numerical evaluations, 177–179
    overview, 165–166
    thermodynamics of binary nucleation, on capillarity approximation, 167–168
  first passage time analysis, nucleation based on
    DFT methods in nucleation theory, 159–160
    improvement of RNN approach by DFT methods, 160–161
    numerical results, 161–163
    RNN approach to kinetics of nucleation, 156–159
  heterogeneous nucleation
    calculations and discussion, 191–192
    evaporation rate from cluster, 186–188
    growth of cluster, 188–189
    interaction potential between molecule/spherical cap cluster, 192–193
    nucleation rate, 189–190
    overview, 186
  heterogeneous nucleation, kinetic theory of
    contributions to free energy of system/Euler–Lagrange equation for FDD, 224
    kinetic nucleation theory (KNT), basic equations of, 220–221
    KNT basic equations of, 224–228
    results and discussion, 221–224
    system/interaction potentials/definition of nucleus, 218–220
  heterogeneous nucleation on rough surface
    basic equations of kinetic theory, 210–213
    discussion, 216–217
    overview, 208
    results, 214
    system and interaction potentials, 209–210
  heterogeneous unary nucleation on liquid aerosols of binary solution
    emission rate via first passage time analysis, 204–206
    equilibrium distribution of heterogeneously formed droplets, 206
    heterogeneous nucleation, kinetic equation of, 203–204
    heterogeneous nucleation, kinetics of, 198–200
    kinetics of nucleation, 196–198
    numerical evaluations, 200–203
    overview, 194–196
  nucleation in gases, theory of
    evaporation rate from cluster, 116–118
    growth rate of cluster, 119
    nucleation rate, 119–120
    overview, 116
    results/comparison with classical theory/experiment, 121–125
  three-dimensional Fokker-Planck equation for evaporation from clusters
    kinetics of growth (evaporation) of a single cluster, 140–142
    nucleation rate, 142–143
    numerical results/discussion, 143–145
    overview, 136
    probability density distribution function of energy, equation for, 136–140
  time-dependent cluster distribution in nucleation
    numerical calculations, results of, 150–154
    overview, 146
    population of clusters, equations for, 146–149
    time-dependent equations, solution of, 149–150
  unidimensional Fokker-Planck equation for nucleation in gases
    evaporation rate of molecules from cluster, 131
    growth rate of cluster, 131–132
    kinetic equation in energy space, 127–131
    nucleation rate, 132–133
    overview, 127
    results/discussion, 133–135

## W

Water–methanol vapor mixture, 168
Water protein residue, 347
Water vapor, heterogeneous nucleation of, 202
Water–water hydrogen bond, 349, 352
Water-water hydrogen bonding on interaction of solute particles

networks of water-water hydrogen bonds around solutes, 350–352
overview, 347–350
protein folding via nucleation, case of, 352–353
Weeks–Chandler–Andersen (WCA) separation, 177
Wulff relations
about, 90
illustration, 106f, 107
Wulff (equilibrium) shape, 85
$W^+ = W^+(v)$ and $W^- = W^-(v)$, determination of, 296–298; *see also* Temperature expansions for protein folding time

## Z

Zeldovich approach, 142, 174
Zeldovich Expansion, 51–52
Zigzag cluster, 376
Zwitterionic structures, 277